CW00548387

TOPOLOGICAL VECTOR SPACES AND DISTRIBUTIONS

John Horváth

Professor Emeritus of Mathematics
University of Maryland

Dover Publications, Inc.
Mineola, New York

Bibliographical Note

This Dover edition, first published in 2012, is a corrected, unabridged republication of the work originally published in 1966 by the Addison-Wesley Publishing Company, Reading, Massachusetts.

Library of Congress Cataloging-in-Publication Data

Horváth, John, 1924–
Topological vector spaces and distributions / John Horváth. — Dover ed.
p. cm.
Originally published: Reading, Mass. : Addison-Wesley, 1966.
Includes bibliographical references and index.
ISBN-13: 978-0-486-48850-9
ISBN-10: 0-486-48850-0
1. Linear topological spaces. 2. Theory of distributions (Functional analysis). I. Title.

QA322.H6 2012
515'.73—dc23

2012002508

Manufactured in the United States by Courier Corporation
48850001
www.doverpublications.com

To

Jean Dieudonné,

Alexandre Grothendieck,

and

Laurent Schwartz

Preface

In spite of the several excellent books published recently, there still seems to be room for an elementary introduction to topological vector spaces and their most important application: the theory of distributions of Laurent Schwartz.

The present text grew out of courses taught at the University of Maryland, the Université de Nancy, and the Universidad de los Andes, Bogotá. I tried to make it reasonably self-contained by including in small print all the necessary definitions and results from algebra and topology, giving complete proofs for all results which are not immediate consequences of the relevant definitions. This arrangement should enable the undergraduate mathematics major to get into contact with functional analysis early, and I hope to have served also the applied mathematician, the physicist, and the engineer whose training does not usually include abstract algebra and topology. A student who has had a course of advanced calculus and a minimum of abstract algebra and topology (if only that of metric spaces) should have no difficulties reading the book, though a few examples also use the theory of analytic functions of one complex variable. I even hope that the book might serve as an introduction to general topology and, by exhibiting an important application of the theory, stimulate and motivate its study. The reader who is already familiar with algebra and topology should skip the small print, consulting it for the terminology with help of the index, if necessary. In preference to other convergence theories I use exclusively filters. Not only do I think that their theory is the most beautiful, simple, and (because of the theorem of ultra-filters) powerful (cf. the miraculous proof of Tihonov's theorem), but the main object of our study, namely the system of neighborhoods of the origin in a topological vector space, *is* a filter. Of course a complete exposition of the theory of filters is given in the text (Chapter 2, §§2 and 8). On the other hand, I do not need or use the theory of uniform spaces.

The first chapter is of an introductory character. It deals with the theory of Banach spaces and serves as a motivation and guide for the generalizations which are to follow in the next two chapters. With the exception of a few properties of Hilbert spaces and some quantitative theorems involving norms, the results of the first chapter are proved again in a more general setting, and therefore the more advanced or impatient reader is advised to start with Chapter 2. The treatment is not encyclopaedic, and many important results, such as the Krein-Milman, Eberlein, Šmulian theorems, and Choquet's beautiful theory of the integral representation of points of a compact, convex set [5, 108], are completely omitted. The interested reader is referred in the first place to the superb monograph of Köthe [52] for further study. I postponed some parts of the theory of topological vector spaces to the place where they are first needed, to ensure a better motivation and livelier pace. Thus the theory of bilinear maps is treated in Chapter 4, §7 in connection with the multiplication of distributions.

The theory of distributions is presented, as originally by Schwartz, as a theory of duality of topological vector spaces. Many so-called elementary approaches to distributions have been devised in the past few years, but, to quote Dieudonné, "none offers, in my opinion, the flexibility and power of the original description of Schwartz" [*Amer. Math. Monthly*, **71** (1964), p. 241]. The treatment in Chapter 4 is again not exhaustive, and no serious student can forgo to read Schwartz's original masterwork [81, 82], where, in the introduction, he will find an excellent account of the precursors and origins of the theory. I also want to call the reader's attention to the monumental work in five volumes by I. M. Gelfand and his collaborators on "generalized functions" (cf. Example 4.1.1), written in a leisurely and informal style and containing an enormous wealth of material [30 through 34].

Chapter 4 contains only those results on measures and distributions which do not use Lebesgue's theory of integration. Volume II, which I hope will appear some day, should start with an exposition of the theory of integration and introduce some important Banach spaces of distributions: the L^p spaces discovered by Frederick Riesz in 1908. Following that, a chapter on miscellaneous subjects is planned, such as convolutions and Fourier transforms in L^p spaces, $\mathcal{H}^s$ spaces, distributions defined by analytic continuation (distributions of Marcel Riesz), Sobolev's inequalities, the Paley-Wiener theorem, Laplace transforms, Lions' theorem of supports, Schwartz's kernel theorem, etc. Finally, a few selected applications to partial differential equations will be given. These chapters will be very short, since the reader has the splendid monographs of Hörmander [46] and Treves [100] at his disposal.

Most exercises are simple verifications and serve the reader to check whether he has understood the theory. Some are results which are too simple to waste space for them in the text; these results are used freely later on. The reader who wants more challenging exercises will find his fill in the books by Bourbaki [9] and Grothendieck [43].

The three great mathematicians to whom this book is dedicated are not only the authors of most of what is contained in it but also the persons from whom I learned almost everything I know about the subject. It was Schwartz who pointed out to me more than ten years ago the merits of writing an introductory textbook, and both he and Dieudonné gave me constant help and encouragement while I was writing this book. I want to express my gratitude to Jacques Deny who first initiated me in 1947 into the theory of distributions. The book owes a lot to many other friends and colleagues, in the first place to Robert Freeman, who read carefully the first version of the first three chapters and the galleys of Chapter 4, to my audiences in College Park, Nancy, and Bogotá, to Mohamed Salah Baouendi, John Brace, Ákos Császár, Arthur Du Pré, H. G. Garnir, Pierre Grisvard, Adam Kleppner, Germán Lemoine, Jacques-Louis Lions, Robert Nielsen, Jean-Louis Ovaert, Mrs. Jacques Rebibo, Hugo Sun, Henri Yerly, and others who pointed out errors, taught me proofs, made suggestions, and encouraged me in every possible way. I want to thank Professor Lynn H. Loomis for accepting this book in the prestigious series of which he is the editor and the Addison-Wesley Publishing Co. for producing it according to their usual high standards.

J. H.

College Park, Maryland
June 1966

Contents

Chapter 3

Duality

Chapter 4

Distributions

Terminology and Notations

The empty set is denoted by $\emptyset$. The notation $A \subset B$ means that A is a subset of B, where the case $A = B$ is not excluded. If A and B are both subsets of the same set X, then $A \cup B$ is their union, $A \cap B$ their intersection, and $\complement_X A$, or simply $\complement A$, the complement of A with respect to X, i.e., the set of those points of X which are not in A. We shall often denote a set by writing between $\{\ \}$ first its generic element and then, separated by $\mid$, the relations which define the set, e.g.,

$$A = \{x \mid x \in A\}, \qquad \complement_X A = \{x \mid x \in X, x \notin A\}.$$

The *cartesian product* $X \times Y$ of two sets X and Y is the set of ordered pairs (x, y) with $x \in X$, $y \in Y$. A *map* $f: X \to Y$ from X into Y (or a *function* defined on X with values in Y) is determined by a subset $G = G(f)$ of $X \times Y$, called the *graph* of f, such that: (a) for every $x \in X$ there exists $y \in Y$ such that $(x, y) \in G$; (b) if $(x, y) \in G$ and $(x, y') \in G$, then $y = y'$. If $(x, y) \in G(f)$, we write $y = f(x)$ or $f: x \mapsto y$. If $A \subset X$, then

$$f(A) = \{f(x) \mid x \in A\} \subset Y.$$

In particular, $f(X) = \operatorname{Im}(f)$ is called the *image* of f. If $B \subset Y$, then

$$f^{-1}(B) = \{x \mid f(x) \in B\} \subset X.$$

If $f(x) = f(x')$ implies $x = x'$, we say that f is an *injection* or an *injective* map. In particular, if $A \subset B$, then the map $x \mapsto x$ is injective; it is called the *canonical injection* and denoted by $A \hookrightarrow B$. If for every $y \in Y$ there exists $x \in X$ such that $f(x) = y$, i.e., if $\operatorname{Im}(f) = Y$, then we say that f is a *surjection* or a *surjective* map or that f maps X *onto* Y. A map which is both injective and surjective is said to be a *bijection* or a *bijective* map.

The *diagonal* Δ of the set $X \times X$ is formed by all the pairs (x, x). If $A \subset X \times Y$, then

$$A^{-1} = \{(y, x) \mid (x, y) \in A\} \subset Y \times X.$$

If A is the graph of the map $f: X \to Y$ and A^{-1} is the graph of a map (i.e., satisfies conditions (a) and (b) above), then we denote by f^{-1} the map from Y into X determined by A^{-1}, and call it the map *inverse* to f. Clearly $f: X \to Y$ has an inverse if and only if it is bijective.

If $A \subset X \times Y$ and $B \subset Y \times Z$, then $B \circ A$ is the subset of $X \times Z$ formed by all the pairs (x, z) for which there exists $y \in Y$ such that $(x, y) \in A$ and $(y, z) \in B$. If A is the graph of a function f and B the graph of a function g, then $B \circ A$ is the graph of the *composite function* $g \circ f$.

An *equivalence relation* defined on a set X is determined by a subset R of $X \times X$ such that: (a) $\Delta \subset R$, (b) $R^{-1} = R$, (c) $R \circ R \subset R$. If $(x, y) \in R$, we shall say that x is equivalent to y modulo R and write $x \sim y$. The three conditions can be restated in the following form: (a′) for every $x \in X$ we have $x \sim x$ (reflexivity); (b′) if $x \sim y$, then $y \sim x$ (symmetry); (c′) if $x \sim y$ and $y \sim z$, then $x \sim z$ (transitivity). An equivalence relation R on X defines a partitioning of X into disjoint subsets called *equivalence classes* (modulo R); two elements of X belong to the same equivalence class if they are equivalent modulo R. The set X/R whose elements are the equivalence classes modulo R is called the *quotient set* of X modulo R. The map $X \to X/R$ which associates with an element of X its equivalence class (i.e., the equivalence class to which it belongs) is called the *canonical surjection.*

An *order* on X is determined by a subset C of $X \times X$ such that: (a) $C \circ C \subset C$, (b) $C \cap C^{-1} = \Delta$. If $(x, y) \in C$, we usually write $x \leqq y$ or something similar. Then the two conditions can be restated in the following form: (a′) if $x \leqq y$ and $y \leqq z$, then $x \leqq z$; (b′) we have $x \leqq y$ and $y \leqq x$ if and only if $x = y$. An order is *total* if for any pair x, y of elements we have either $x \leqq y$ or $y \leqq x$. An ordered set is *directed* (to the right) if for any pair x, y of elements there exists an element z such that $x \leqq z$ and $y \leqq z$. Given a subset A of an ordered set X, we say that the element $x \in X$ is the *least upper bound* (sup) of A if: (α) $a \leqq x$ for all $a \in A$; (β) if $y \in X$ is such that $a \leqq y$ for all $a \in A$, then $x \leqq y$. The definition of the *greatest lower bound* (inf) of A is similar. An element x of an ordered set is *maximal* if $x \leqq y$ implies $x = y$. An ordered set is *inductive* if every totally ordered subset has a least upper bound.

Zorn's lemma: An inductive ordered set possesses at least one maximal element.

We shall employ the following notations: $\mathbf{N}$ is the set of natural numbers, $\mathbf{Z}$ the set of integers, $\mathbf{Q}$ the set of rational numbers, $\mathbf{R}$ the set of real numbers, and $\mathbf{C}$ the set of complex numbers. We shall use the same letters if we consider any of these sets together with any of the structures (additive group, ring, field, order, metric space) defined on it. An asterisk, as in $\mathbf{N}^*$, indicates that the set is to be considered without the element 0. A real number x such that $x \geqq 0$ will be called *positive,* while a number $x > 0$ will be called *strictly positive.* The symbols $\mathbf{Z}_+$, $\mathbf{Q}_+$, $\mathbf{R}_+$ will stand for the subsets formed by the positive elements of the respective sets. If $z = x + iy \in \mathbf{C}$, we denote by $\bar{z} = x - iy$ its conjugate, by $\mathcal{R}e z = x$ its real part, and by $\mathcal{I}m z = y$ its imaginary part.

A surjection $f: I \to A$ defines a *family* $(a_\iota)_{\iota \in I}$, where $f(\iota) = a_\iota$ ($\iota \in I$, $a_\iota \in A$) and I is the *index set*. If I is a subset of $\mathbf{N}$, then $(a_n)_{n \in I}$ is a *sequence*. If $(A_\iota)_{\iota \in I}$ is a family of subsets of a given set, then $\bigcup_{\iota \in I} A_\iota$ is the union and $\bigcap_{\iota \in I} A_\iota$ the intersection of the family. Given a set A, we can always define a family whose elements are those of the set A with the help of the identity map of A onto itself.

Definitions, propositions, theorems, examples, remarks, and exercises are numbered within each section. Within the same section they will be referred to by their numbers (e.g., Definition 2); within the same chapter by the section number and their numbers (e.g., Proposition 3.4); and in another chapter by the chapter number, the section number, and their own numbers (e.g., Theorem 2.12.2). The end of a proof is marked with the symbol ▌. References to the bibliography are in brackets.

CHAPTER 1

Banach Spaces

§1. *The definition of Banach spaces*

In the whole book the letter **K** will always stand either for the field **R** of real numbers or for the field **C** of complex numbers. The elements of **K** are called *scalars* and will be denoted mostly by small Greek letters.

A structure of *vector space* on a set E is defined by two maps:

(1) a map $(x, y) \mapsto x + y$ from $E \times E$ into E, called *addition*,

(2) a map $(\lambda, x) \mapsto \lambda x$ from $\mathbf{K} \times E$ into E, called *multiplication by a scalar*.

These maps (or algebraic operations) must satisfy the following axioms:

(VS 1) $x + y = y + x$ (commutativity).

(VS 2) $(x + y) + z = x + (y + z)$ (associativity).

(VS 3) There exists an element 0 in E such that $x + 0 = x$ for all $x \in E$. This element 0 is called the *zero vector* or the *origin* of E.

(VS 4) For every element $x \in E$, there exists an element $-x \in E$ such that $x + (-x) = 0$. The element $-x$ is called the *opposite* of x.

(VS 5) $\lambda(x + y) = \lambda x + \lambda y$.

(VS 6) $(\lambda + \mu)x = \lambda x + \mu x$.

(VS 7) $(\lambda\mu)x = \lambda(\mu x)$.

(VS 8) $1 \cdot x = x$ for all $x \in E$.

If these axioms are satisfied, we say that E is a vector space (or linear space) over the field **K**; the elements of a vector space are called *vectors*. The first four axioms express the fact that E is an abelian group under addition. Note that the symbol 0 is used to denote both the zero scalar and the zero vector; this ambiguity does not in general lead to any confusion. Axioms 5 and 6 express the distributivity of multiplication by a scalar with respect to the two kinds of addition. Axiom 7 expresses a kind of associativity.

A vector space over **R** will also be called a real vector space and a vector space over **C** a complex vector space.

Here are some easy consequences of the above definitions.

(a) $0 \cdot x = 0$ for all $x \in E$.

(Here on the left-hand side we have the scalar zero and on the right-hand side the vector zero.)

Proof. We have $x = 1 \cdot x = (1 + 0) \cdot x = 1 \cdot x + 0 \cdot x = x + 0 \cdot x$. Adding $-x$ to both sides and using (VS 2), we obtain $0 = 0 \cdot x$.

(b) $\lambda \cdot 0 = 0$ for all $\lambda \in \mathbf{K}$.

Proof. We have $\lambda \cdot 0 + \lambda \cdot 0 = \lambda(0 + 0) = \lambda \cdot 0$. Adding $-(\lambda \cdot 0)$ to both sides, we obtain $\lambda \cdot 0 = 0$.

(c) The vector 0 is unique.

Indeed, suppose we have two vectors 0 and $0'$ such that $x + 0 = x$ and $x + 0' = x$ for all x. Then $0 = 0 + 0' = 0'$.

(d) The opposite vector is unique.

Indeed, if $x + x' = 0$ and $x + x'' = 0$, then

$$x' = x' + (x + x'') = (x' + x) + x'' = x''.$$

(e) $(-1) \cdot x = -x$ for all $x \in E$.

Indeed, $x + (-1) \cdot x = 1 \cdot x + (-1) \cdot x = [1 + (-1)] \cdot x = 0 \cdot x = 0$ by (a) and the assertion follows from (d).

(f) Given two vectors a and b, there exists a unique vector x such that $a + x = b$.

Proof. $x = b + (-a)$ satisfies the equation and its uniqueness follows from the fact that x must be the opposite of $a + (-b)$.

The element $a + (-b)$ is usually written as $a - b$ and called the *difference* of the vectors a and b.

DEFINITION 1. *Given a vector space E, a norm on E is a map $x \mapsto \|x\|$ from E into the set $\mathbf{R}_+$ of positive real numbers which satisfies the following axioms:*

(N 1) $\|x\| = 0$ *if and only if* $x = 0$.

(N 2) $\|\lambda x\| = |\lambda| \cdot \|x\|$ *for all* $\lambda \in \mathbf{K}$ *and* $x \in E$.

(N 3) $\|x + y\| \leqq \|x\| + \|y\|$ *(the triangle inequality).*

A vector space on which a norm is defined is called a normed vector space or simply a normed space.

EXAMPLE 1. The *n-dimensional real Euclidean space* $\mathbf{R}^n$ is the set of all n-tuples $x = (x_1, \ldots, x_n)$ of real numbers, where addition and multiplication by a scalar $\lambda \in \mathbf{R}$ are defined by

$$(x_1, \ldots, x_n) + (y_1, \ldots, y_n) = (x_1 + y_1, \ldots, x_n + y_n),$$
$$\lambda(x_1, \ldots, x_n) = (\lambda x_1, \ldots, \lambda x_n).$$

The zero vector is $(0, \ldots, 0)$. The eight axioms of a real vector space can be easily verified. The norm on $\mathbf{R}^n$ will be denoted by $|x|$ rather than

$\|x\|$, and is defined by

$$|x| = \left(\sum_{i=1}^{n} x_i^2\right)^{1/2}.$$

The axioms (N 1) and (N 2) can be readily verified. To prove the triangle inequality, let us first prove the *Cauchy-Schwarz inequality:*

$$\left|\sum_{i=1}^{n} x_i y_i\right| \leqq |x| \cdot |y|. \tag{1}$$

The quadratic polynomial in λ,

$$\sum_{i=1}^{n} (x_i + \lambda y_i)^2 = \sum_{i=1}^{n} x_i^2 + 2\lambda \sum_{i=1}^{n} x_i y_i + \lambda^2 \sum_{i=1}^{n} y_i^2, \tag{2}$$

is positive for every real λ. Now we know from elementary algebra that if $a\lambda^2 + b\lambda + c \geqq 0$ for every real λ, then necessarily $b^2 - 4ac \leqq 0$. Applying this to the polynomial (2), we obtain

$$\left(\sum_{i=1}^{n} x_i y_i\right)^2 - \left(\sum_{i=1}^{n} x_i^2\right)\left(\sum_{i=1}^{n} y_i^2\right) \leqq 0,$$

which is just another form of (1).

We have, using the Cauchy-Schwarz inequality,

$$\begin{aligned} |x + y|^2 &= \sum_{i=1}^{n} (x_i + y_i)^2 = \sum_{i=1}^{n} x_i^2 + 2 \sum_{i=1}^{n} x_i y_i + \sum_{i=1}^{n} y_i^2 \\ &\leqq |x|^2 + 2|x| \cdot |y| + |y|^2 = (|x| + |y|)^2. \end{aligned}$$

If we take the positive square roots on both sides, we have also verified (N 3).

The norm $|x|$ can be thought of as the length of the "line segment" going from the origin 0 to the point x. The corresponding expression for $n = 2$ and $n = 3$ is well known from analytic geometry. The triangle inequality expresses the fact that the length of one side of a triangle is less than or equal to the sum of the lengths of the other two sides. This accounts for its name.

In a similar fashion the set $\mathbf{C}^n$ of all n-tuples $z = (z_1, \ldots, z_n)$ of complex numbers is a complex normed vector space if we define

$$|z| = \left(\sum_{i=1}^{n} |z_i|^2\right)^{1/2}.$$

See Example 10.

EXAMPLE 2. Let $I = [a, b]$ be a finite closed interval of the real line and let $\mathcal{C}(I)$ or $\mathcal{C}_{\mathbf{R}}(I)$ be the set of all continuous functions $x \mapsto f(x)$ defined on I and whose values are real numbers. The function $f + g$ is defined by $x \mapsto f(x) + g(x)$ and the function λf ($\lambda \in \mathbf{R}$) is defined by $x \mapsto \lambda f(x)$. It can be immediately verified that with these operations $\mathcal{C}(I)$ becomes a vector space over $\mathbf{R}$, where the zero vector is the function which is identically zero for every $x \in I$. The norm of $f \in \mathcal{C}(I)$ is defined by

$$\|f\| = \max_{x \in I} |f(x)|.$$

It is well known that this maximum exists (Weierstrass' theorem [2], Theorem 4–20, p. 73). The first two properties of a norm are quite obviously satisfied. To see the third one we observe that for any $x \in I$ we have

$$|f(x) + g(x)| \leqq |f(x)| + |g(x)| \leqq \|f\| + \|g\|,$$

whence

$$\|f + g\| \leqq \|f\| + \|g\|.$$

If we consider the set $\mathcal{C}_{\mathbf{C}}(I)$ of complex-valued functions, we obtain in a similar way a normed vector space over $\mathbf{C}$.

The norm of a vector space defines a metric in a natural way. Let us recall that a *metric* on a set X is a map $(x, y) \mapsto \delta(x, y)$ from $X \times X$ into the set $\mathbf{R}_+$ of positive real numbers which satisfies the following axioms:

(M 1) $\delta(x, y) = 0$ if and only if $x = y$.

(M 2) $\delta(x, y) = \delta(y, x)$.

(M 3) $\delta(x, y) \leqq \delta(x, z) + \delta(z, y)$ (the triangle inequality).

The value $\delta(x, y)$ is called the *distance* between the points x and y. A set X on which a metric δ is defined is called a *metric space,* and we say that X is equipped with the metric δ.

If E is a normed vector space, we set $\delta(x, y) = \|x - y\|$. The three axioms for a metric are verified:

(1) $\delta(x, y) = 0 \Leftrightarrow \|x - y\| = 0 \Leftrightarrow x - y = 0 \Leftrightarrow x = y$, where we use (N 1).

(2) $y - x = (-1)(x - y)$ and thus by Axiom (N 2)

$$\|y - x\| = |-1| \cdot \|x - y\| = \|x - y\|,$$

i.e., $\delta(y, x) = \delta(x, y)$.

(3) $\|x - y\| = \|(x - z) + (z - y)\| \leqq \|x - z\| + \|z - y\|$ by (N 3).

In the sequel we shall always consider a normed vector space as a metric space equipped with the metric just defined.

In a metric space X (and thus in a normed vector space) we have the usual notions of topology: closed and open sets, neighborhoods, convergence, etc. (see §2). Let us recall in particular that a sequence $(x_n)_{n \in \mathbf{N}}$ of points of X is said to *converge* to a point x, if for every $\epsilon > 0$ there exists an $N = N(\epsilon) \in \mathbf{N}$ such that $\delta(x, x_n) < \epsilon$ for every $n > N$. A sequence is said to be *convergent* if it converges to some point.

Let $(x_n)_{n \in \mathbf{N}}$ be a convergent sequence. Then it satisfies the following so-called *Cauchy condition:* For every $\epsilon > 0$ there exists $N = N(\epsilon) \in \mathbf{N}$ such that $\delta(x_n, x_m) < \epsilon$ for $n > N$ and $m > N$. Indeed we have $\delta(x, x_n) < \epsilon/2$ for $n > N$, $\delta(x, x_m) < \epsilon/2$ for $m > N$, and hence by the triangle inequality

$$\delta(x_n, x_m) \leqq \delta(x, x_n) + \delta(x, x_m) < \tfrac{1}{2}\epsilon + \tfrac{1}{2}\epsilon = \epsilon \qquad \text{for} \qquad n, m > N.$$

Let us call a sequence which satisfies the Cauchy condition a *Cauchy sequence.* We have just proved that every convergent sequence is a Cauchy sequence. If, conversely, every Cauchy sequence is convergent, then X is said to be a *complete* metric space. It is a basic property of the real line $\mathbf{R}$ (and also of the complex plane $\mathbf{C}$) that it is a complete metric space. The rational line $\mathbf{Q}$ is not a complete metric space.

Definition 2. *A normed vector space E is called a Banach space if it is complete as a metric space.*

Example 3. The space $\mathbf{R}^n$ is complete. Indeed, let

$$(x^{(m)}) = ((x_1{}^{(m)}, \ldots, x_n{}^{(m)}))$$

be a Cauchy sequence. It follows from the obvious inequality

$$|x_i| \leqq \left(\sum_{i=1}^{n} x_i^2\right)^{1/2} = |x|$$

that for every i with $1 \leqq i \leqq n$, the sequence of real numbers $(x_i^{(m)})$ is a Cauchy sequence. Thus, by the completeness of $\mathbf{R}$, there exist real numbers x_i such that $|x_i - x_i^{(m)}| < \epsilon/\sqrt{n}$ for $m > M(i, \epsilon)$. Setting $x = (x_1, \ldots, x_n)$, we have

$$|x - x^{(m)}| = \left(\sum_{i=1}^{n} |x_i - x_i^{(m)}|^2\right)^{1/2} < \left(\sum_{i=1}^{n} \frac{\epsilon^2}{n}\right)^{1/2} = \epsilon$$

for $m > \max_{1 \leqq i \leqq n} M(i, \epsilon)$; in other words, $(x^{(m)})$ converges to x.

It can be shown in an entirely analogous fashion that $\mathbf{C}^n$ is a complex Banach space.

Example 4. The space $\mathcal{C}(I)$ is a Banach space. Indeed, let (f_n) be a Cauchy sequence in $\mathcal{C}(I)$. For every $x \in I$ we have

$$|f_n(x) - f_m(x)| \leqq \|f_n - f_m\|.$$

Thus for every $x \in I$ the sequence of numbers $(f_n(x))$ converges to some number which we shall denote by $f(x)$. The function $f: x \mapsto f(x)$ defined on I is continuous, and (f_n) converges to f in the space $\mathcal{C}(I)$. Indeed, for every $\epsilon > 0$ there exists an integer M such that $\|f_n - f_m\| < \epsilon$ for $n, m > M$, and thus $|f_n(x) - f_m(x)| < \epsilon$ for every $x \in I$ and $n, m > M$. It follows that

$$|f(x) - f_n(x)| \leqq \epsilon \qquad \text{for every} \qquad x \in I \quad \text{and} \quad n > M, \qquad (3)$$

i.e., the sequence (f_n) of continuous functions converges uniformly to the function f; hence f is continuous, i.e., $f \in \mathcal{C}(I)$. From (3) we have $\|f - f_n\| \leqq \epsilon$ for $n > M$, and the proof is complete.

For the convenience of the reader let us prove the theorem we have just used, according to which *the limit f of a uniformly convergent sequence (f_n) of continuous functions is itself continuous.* Given $\epsilon > 0$ there exists an integer N such that

$$|f(x) - f_n(x)| < \frac{\epsilon}{3}$$

if $n > N$. Let n be a fixed index such that $n > N$. For every $x_0 \in I$ there exists an $\alpha = \alpha(x_0, \epsilon) > 0$ such that

$$|f_n(x_0) - f_n(x)| < \frac{\epsilon}{3}$$

if $|x_0 - x| < \alpha$. Thus

$$|f(x_0) - f(x)| \leqq |f(x_0) - f_n(x_0)| + |f_n(x_0) - f_n(x)| + |f_n(x) - f(x)|$$
$$< \frac{\epsilon}{3} + \frac{\epsilon}{3} + \frac{\epsilon}{3} = \epsilon$$

if $|x_0 - x| < \alpha$, i.e., f is continuous at x_0.

Let us also observe that a sequence (f_n) converges to f in the space $\mathcal{C}(I)$ if and only if it converges to f uniformly on I. Indeed, the relation $\|f - f_n\| < \epsilon$ is equivalent to "$|f(x) - f_n(x)| < \epsilon$ for every $x \in I$."

EXAMPLE 5. Let $\mathcal{C}_0(\mathbf{R}^n)$ be the set of all continuous functions defined on $\mathbf{R}^n$ and which "vanish at infinity." By this last condition the following is meant: for every $f \in \mathcal{C}_0(\mathbf{R}^n)$ and every $\epsilon > 0$ there exists a $\rho = \rho(\epsilon, f) > 0$ such that $|f(x)| < \epsilon$ if $|x| > \rho$. It is clear that $\mathcal{C}_0(\mathbf{R}^n)$ is a vector space (real or complex according as we consider real- or complex-valued functions), and if we define again $\|f\| = \max_{x \in \mathbf{R}^n} |f(x)|$, then the requirements of a norm are satisfied. The maximum exists again; indeed, suppose that $f(x)$ is not identically zero. Then for some $x \in \mathbf{R}^n$ we have $|f(x)| = \eta > 0$. Now $|f(x)| < \eta/2$ for $|x| > \rho$. Thus

$$\max_{x \in \mathbf{R}^n} |f(x)| = \max_{|x| \leqq \rho} |f(x)|,$$

and the second maximum exists by Weierstrass' theorem.

Finally, $\mathcal{C}_0(\mathbf{R}^n)$ is complete. Indeed, it can be shown, exactly as in the previous example, that a Cauchy sequence (f_n) tends uniformly to a continuous function f. Thus we have only to prove that $f \in \mathcal{C}_0(\mathbf{R}^n)$, i.e., that f vanishes at infinity. Let $\epsilon > 0$. Then $|f(x) - f_n(x)| < \epsilon/2$ for some n, and $|f_n(x)| < \epsilon/2$ for $|x| > \rho$. Thus $|f(x)| < \epsilon$ for $|x| > \rho$.

EXAMPLE 6. Denote by c_0 the set of all infinite sequences $x = (\xi_n)_{n \in \mathbf{N}}$ which tend to zero. Let $y = (\eta_n)$, and define

$$x + y = (\xi_n + \eta_n), \qquad \lambda x = (\lambda \xi_n).$$

Then c_0 is clearly a vector space (real or complex according as we consider sequences of real or complex numbers). Define

$$\|x\| = \max_{n \in \mathbf{N}} |\xi_n|.$$

Since (ξ_n) tends to zero, this maximum is attained, and the properties of the norm are trivially verified. We leave it as an exercise to prove that c_0 is complete.

EXAMPLE 7. Let p be a real number, $1 \leqq p < \infty$. Let l^p be the set of all sequences $x = (\xi_n)_{n \in \mathbf{N}}$ of elements of $\mathbf{K}$ for which the infinite series

$$\sum_{n=0}^{\infty} |\xi_n|^p$$

converges. In contrast to the previous examples, it is not at all trivial that l^p is a vector space over $\mathbf{K}$. We shall prove this and prove at the same time that

$$\|x\| = \left(\sum_{n=0}^{\infty} |\xi_n|^p \right)^{1/p}$$

is a norm on l^p (the first two axioms of a norm are obviously satisfied, so that we must only prove the triangle inequality).

We start with the *inequality between the weighted arithmetic and geometric mean:* Let $\alpha > 0$, $\beta > 0$, and $\alpha + \beta = 1$. Then for every $u > 0$, $v > 0$ we have

$$u^\alpha v^\beta \leqq \alpha u + \beta v. \tag{4}$$

To prove this, let us observe that the function t^α is concave downward for $t > 0$, since its second derivative $\alpha(\alpha - 1)t^{\alpha-2}$ is negative because $0 < \alpha < 1$. It follows that the curve is below its tangent line at the point $t = 1$, i.e.,

$$t^\alpha \leqq \alpha t + \beta.$$

Setting $t = u/v$ and multiplying both sides by v, we obtain (4).

Next we prove the *Hölder inequality*. Let $1 < p < \infty$ and $1/p + 1/q = 1$. Then we have

$$\sum_{i=1}^{n} |\xi_i \eta_i| \leqq \left(\sum_{i=1}^{n} |\xi_i|^p\right)^{1/p} \left(\sum_{i=1}^{n} |\eta_i|^q\right)^{1/q}, \tag{5}$$

where (ξ_i) and (η_i) are two arbitrary finite sequences of real or complex numbers. The case $p = q = \frac{1}{2}$ yields the Cauchy-Schwarz inequality (since obviously $|\sum \xi_i \eta_i| \leqq \sum |\xi_i \eta_i|$). For the proof let us set

$$u_i = \frac{|\xi_i|^p}{\sum_{j=1}^{n} |\xi_j|^p}, \qquad v_i = \frac{|\eta_i|^q}{\sum_{j=1}^{n} |\eta_j|^q}, \qquad \alpha = \frac{1}{p}, \qquad \beta = \frac{1}{q}.$$

Then it follows from (4) that

$$\begin{aligned}\sum_{i=1}^{n} \frac{|\xi_i| \cdot |\eta_i|}{(\sum |\xi_j|^p)^{1/p} (\sum |\eta_j|^q)^{1/q}} &= \sum u_i^\alpha v_i^\beta \leqq \sum (\alpha u_i + \beta v_i) \\ &= \alpha \sum u_i + \beta \sum v_i \\ &= \alpha + \beta = 1,\end{aligned}$$

since $\sum u_i = \sum v_i = 1$. Multiplying by $(\sum |\xi_j|^p)^{1/p} (\sum |\eta_j|^q)^{1/q}$, we obtain (5).

Finally we prove the *Minkowski inequality*,

$$\left(\sum_{i=1}^{n} |\xi_i + \eta_i|^p\right)^{1/p} \leqq \left(\sum_{i=1}^{n} |\xi_i|^p\right)^{1/p} + \left(\sum_{i=1}^{n} |\eta_i|^p\right)^{1/p}. \tag{6}$$

This is clear for $p = 1$. For $1 < p < \infty$ we have

$$\begin{aligned}\sum |\xi_i + \eta_i|^p &= \sum |\xi_i + \eta_i|^{p-1} \cdot |\xi_i + \eta_i| \\ &\leqq \sum |\xi_i + \eta_i|^{p-1} (|\xi_i| + |\eta_i|).\end{aligned}$$

Using the Hölder inequality, we have

$$\begin{aligned}\sum |\xi_i + \eta_i|^{p-1} \cdot |\xi_i| &\leqq \left(\sum |\xi_i + \eta_i|^{(p-1)q}\right)^{1/q} \cdot \left(\sum |\xi_i|^p\right)^{1/p} \\ &= \left(\sum |\xi_i + \eta_i|^p\right)^{1/q} \cdot \left(\sum |\xi_i|^p\right)^{1/p},\end{aligned}$$

since $(p - 1)q = p$. Adding the similar estimate for $\sum |\xi_i + \eta_i|^{p-1} \cdot |\eta_i|$, we obtain

$$\sum |\xi_i + \eta_i|^p \leqq \left(\sum |\xi_i + \eta_i|^p\right)^{1/q} \left[\left(\sum |\xi_i|^p\right)^{1/p} + \left(\sum |\eta_i|^p\right)^{1/p}\right].$$

Dividing both sides by $(\sum |\xi_i + \eta_i|^p)^{1/q}$, we obtain (6).

Let us now return to the space l^p. If $(\xi_i) \in l^p$ and $(\eta_i) \in l^p$, then it follows from (6) that

$$\left(\sum_{i=0}^{n} |\xi_i + \eta_i|^p\right)^{1/p} \leqq \left(\sum_{i=0}^{\infty} |\xi_i|^p\right)^{1/p} + \left(\sum_{i=0}^{\infty} |\eta_i|^p\right)^{1/p}$$

for any $n \in \mathbf{N}$, and thus

$$\left(\sum_{i=0}^{\infty} |\xi_i + \eta_i|^p\right)^{1/p} \leqq \left(\sum_{i=0}^{\infty} |\xi_i|^p\right)^{1/p} + \left(\sum_{i=0}^{\infty} |\eta_i|^p\right)^{1/p}. \tag{7}$$

Consequently, $(\xi_i + \eta_i) \in l^p$, and the axioms of a vector space can now be immediately verified. The inequality (7) is just the triangle inequality.

To conclude we prove that l^p is complete. Let $(x^{(m)}) = ((\xi_i{}^{(m)}))$ be a Cauchy sequence in l^p. Then for each $i \in \mathbf{N}$ the numerical sequence $(\xi_i{}^{(m)})_{m \in \mathbf{N}}$ is a Cauchy sequence and therefore $(\xi_i{}^{(m)})$ converges to some number ξ_i. There exists a constant $\mu > 0$ such that

$$\left(\sum_{i=0}^{N} |\xi_i^{(n)} - \xi_i^{(m)}|^p\right)^{1/p} \leqq \|x^{(m)} - x^{(n)}\| \leqq \mu$$

for all $m, n, N \in \mathbf{N}$. Letting $n \to \infty$, it follows that

$$\left(\sum_{i=0}^{N} |\xi_i - \xi_i^{(m)}|^p\right)^{1/p} \leqq \mu,$$

and hence by the Minkowski inequality

$$\left(\sum_{i=0}^{N} |\xi_i|^p\right)^{1/p} \leqq \|x^{(m)}\| + \mu$$

for all $N \in \mathbf{N}$; i.e., the sequence $x = (\xi_i)$ belongs to l^p. Furthermore, for any $\epsilon > 0$ there exists $n_0 \in \mathbf{N}$ such that

$$\left(\sum_{i=0}^{N} |\xi_i^{(n)} - \xi_i^{(m)}|^p\right)^{1/p} \leqq \|x^{(m)} - x^{(n)}\| < \epsilon$$

for $m, n > n_0$ and for all $N \in \mathbf{N}$. Letting $n \to \infty$, it follows that

$$\left(\sum_{i=0}^{N} |\xi_i - \xi_i^{(m)}|^p\right)^{1/p} \leqq \epsilon$$

for $m > n_0$; and letting $N \to \infty$, we see that $\|x - x^{(m)}\| \leq \epsilon$ for $m > n_0$. Thus $(x^{(m)})$ tends to x, and we have proved completely that l^p is a Banach space.

EXAMPLE 8. Our last example is the space m of all bounded sequences $x = (\xi_n)$. It is clear that if $x = (\xi_n) \in m$ and $y = (\eta_n) \in m$, then

$$x + y = (\xi_n + \eta_n) \in m$$

and

$$\lambda x = (\lambda \xi_n) \in m.$$

The norm is now defined as

$$\|x\| = \sup_{n \in \mathbf{N}} |\xi_n|,$$

since a maximum of the values $|\xi_n|$ does not necessarily exist. The first two properties of a norm are again trivial to verify. To prove the third one, let us observe that $|\xi_n + \eta_n| \leqq |\xi_n| + |\eta_n| \leqq \|x\| + \|y\|$. Thus $\|x + y\| \leqq \|x\| + \|y\|$. The proof of the completeness of m is left to the reader.

A very important class of Banach spaces is constituted by the Hilbert spaces, to the definition of which we now turn.

DEFINITION 3. *A vector space E over $\mathbf{K}$ is called an inner product space if there is defined a map $(x, y) \mapsto (x \mid y)$ from $E \times E$ into $\mathbf{K}$ which has the following properties:*

(SP 1) $(x \mid x) \geqq 0$ *for every* $x \in E$.

(SP 2) $(x \mid x) = 0$ *if and only if* $x = 0$.

(SP 3) $(x \mid y) = \overline{(y \mid x)}$ *for every* $x \in E$, $y \in E$.

(SP 4) $(\lambda x + \mu y \mid z) = \lambda(x \mid z) + \mu(y \mid z)$ *for every* $\lambda, \mu \in \mathbf{K}$, x, y, *and* $z \in E$.

The value $(x \mid y)$ is called the inner or scalar product of the vectors x and y.

Let us observe that in the case when $\mathbf{K} = \mathbf{R}$, Axiom (SP 3) simply means that $(x \mid y) = (y \mid x)$ for all $x \in E$, $y \in E$. It follows from (SP 3) and (SP 4) that

$$(x \mid \lambda y + \mu z) = \bar{\lambda}(x \mid y) + \bar{\mu}(x \mid z)$$

for $\lambda, \mu \in \mathbf{K}$, $x, y, z \in E$, where of course the right-hand side is simply $\lambda(x \mid y) + \mu(x \mid z)$ if $\mathbf{K} = \mathbf{R}$.

The scalar product defines a norm on E in a natural way. Indeed, setting

$$\|x\| = +\sqrt{(x \mid x)} \tag{8}$$

we have $\|x\| \in \mathbf{R}_+$ by (SP 1) and $\|x\| = 0$ if and only if $x = 0$ by (SP 2). Furthermore, $(\lambda x \mid \lambda x) = \lambda\bar{\lambda}(x \mid x) = |\lambda|^2(x \mid x)$. Thus $\|\lambda x\| = |\lambda| \cdot \|x\|$.

To prove the triangle inequality, let us first prove the inequality

$$|\mathcal{R}e(x \mid y)| \leqq \|x\| \cdot \|y\|. \tag{9}$$

By (SP 1) the expression

$$(x + \lambda y \mid x + \lambda y) = (x \mid x) + \lambda\overline{(x \mid y)} + \bar{\lambda}(x \mid y) + |\lambda|^2(y \mid y)$$

is always positive, and in particular for real λ the quadratic polynomial in λ with real coefficients,

$$(x \mid x) + 2\mathcal{R}e(x \mid y) \cdot \lambda + (y \mid y)\lambda^2,$$

is always positive. As in the proof of (1), we have

$$(\mathcal{R}e(x \mid y))^2 - (x \mid x) \cdot (y \mid y) \leqq 0,$$

which is just another form of (9). We now have

$$\begin{aligned} \|x + y\|^2 &= (x + y \mid x + y) = (x \mid x) + (x \mid y) + (y \mid x) + (y \mid y) \\ &= \|x\|^2 + 2\mathcal{R}e(x \mid y) + \|y\|^2 \leqq \|x\|^2 + 2\|x\| \cdot \|y\| + \|y\|^2 \\ &= (\|x\| + \|y\|)^2. \end{aligned}$$

Taking the positive square roots on both sides, we obtain the triangle inequality.

DEFINITION 4. *Let E be an inner product space and $\|x\|$ the norm defined by* (8). *If E is complete for this norm* (i.e., E is a Banach space), *then E is said to be a Hilbert space.*

EXAMPLE 9. In the vector space $\mathbf{R}^n$ we define a scalar product by setting

$$(x \mid y) = \sum_{i=1}^{n} x_i y_i.$$

Axioms (SP 1) through (SP 4) are trivial to check and the norm obtained from the scalar product is the same as the norm introduced in Example 1. Thus $\mathbf{R}^n$ is a real Hilbert space.

EXAMPLE 10. In the vector space $\mathbf{C}^n$ we define the scalar product of two vectors $z = (z_1, \ldots, z_n)$ and $w = (w_1, \ldots, w_n)$ by

$$(z \mid w) = \sum_{i=1}^{n} z_i \bar{w}_i.$$

It is again trivial to verify that Axioms (SP 1) through (SP 4) hold, and so $\mathbf{C}^n$ is an inner product space over $\mathbf{C}$. But we can prove exactly as in Example 3 that $\mathbf{C}^n$ is complete; hence $\mathbf{C}^n$ is a complex Hilbert space.

EXAMPLE 11. For two vectors $x = (\xi_i)$ and $y = (\eta_i)$ of the space l^2 we define the scalar product by

$$(x \mid y) = \sum_{i=0}^{\infty} \xi_i \eta_i \qquad \text{resp.} \qquad (x \mid y) = \sum_{i=0}^{\infty} \xi_i \bar{\eta}_i$$

according as we consider real or complex sequences. The right-hand side series converge by virtue of inequality (5). The norms obtained from these scalar products are the same as the norms defined in Example 7, and thus l^2 is a real or a complex Hilbert space.

EXAMPLE 12. The set of all continuous functions on an interval $I = [a, b]$ with the inner product

$$(f \mid g) = \int_a^b f(x)\overline{g(x)}\, dx$$

is an inner product space. We shall see later (Chapter 5) that it is not complete.

We conclude this section with the proof of the general *Cauchy-Schwarz inequality:*

PROPOSITION 1. *Let E be an inner product space. Then*

$$|(x \mid y)| \leqq \|x\| \cdot \|y\|$$

for all $x \in E$, $y \in E$.

Proof. For a real E the statement is equivalent to inequality (9). Let E be a complex inner product space. For $x \in E$, $y \in E$ there exists an $e^{i\theta}$ such that $e^{i\theta}(x \mid y) = (e^{i\theta}x \mid y) \geqq 0$. But then $|(x \mid y)| = e^{i\theta}(x \mid y)$, and by (9)

$$(e^{i\theta}x \mid y) = |\mathcal{R}e(e^{i\theta}x \mid y)| \leqq \|e^{i\theta}x\| \cdot \|y\| = \|x\| \cdot \|y\|. \quad \blacksquare$$

EXERCISES

1. Prove that the spaces c_0 and m are complete.

2. Let c be the set of all sequences $x = (\xi_n)$ which converge to some finite limit. Define on c a structure of Banach space.

3. Let $I = [a, b]$ be a finite closed interval of the real line and let m be a strictly positive integer. Let $\mathcal{C}^m(I)$ be the set of all functions $x \mapsto f(x)$ defined on I which have continuous m-th derivatives. Prove that with the norm

$$\|f\| = \max_{x \in I} |f(x)| + \max_{x \in I} |f^{(m)}(x)|$$

the set $\mathcal{C}^m(I)$ becomes a Banach space.

§2. *Some notions from algebra and topology*

Let E be a vector space over a field $\mathbf{K}$. A nonempty subset F of E is said to be a *linear subspace* (or simply subspace) of E if the following two conditions are satisfied:

(SS 1) If $x \in F$ and $y \in F$, then $x + y \in F$.

(SS 2) If $x \in F$ and $\lambda \in \mathbf{K}$, then $\lambda x \in F$.

If these conditions are satisfied, then for $x \in F$ we have also $-x = (-1) \cdot x \in F$ and $x + (-x) = 0 \in F$. The operations of addition and multiplication by a scalar are defined on F, because F is a subset of E, and by our conditions they always yield elements of F; we say that the operations on E *induce* the operations on F. It is trivial to check that under the induced operations F is a vector space over $\mathbf{K}$. There are two trivial subspaces of E (except when $E = \{0\}$): the space E itself and the subspace $\{0\}$. All other subspaces are called *proper*.

A linear subspace of a normed space or of an inner product space is in a natural way a normed space or an inner product space itself.

If $(F_\iota)_{\iota \in I}$ is an arbitrary family of linear subspaces of E, then the intersection $\bigcap_{\iota \in I} F_\iota$ is also a subspace of E. Given any set A of elements of E, we can therefore speak of the smallest linear subspace containing A, i.e., the intersection of all linear subspaces containing A. This smallest linear subspace F will be called the linear subspace *generated* by A (or the linear hull of A), and A is a set of *generators* of F. The subspace F is the set of all linear combinations of elements of A, i.e., the set of all expressions $\lambda_1 x_1 + \lambda_2 x_2 + \cdots + \lambda_n x_n$, where $\lambda_i \in \mathbf{K}$ and $x_i \in A$.

A family $(x_\iota)_{\iota \in I}$ of elements of E is said to be *algebraically free* if a relation $\sum_{\iota \in I} \lambda_\iota x_\iota = 0$ can hold only if all $\lambda_\iota = 0$. Equivalently, $(x_\iota)_{\iota \in I}$ is algebraically free if no x_ι belongs to the subspace generated by the x_κ with $\kappa \neq \iota$. A family consisting of a single nonzero element is algebraically free, and if (x_ι) is an algebraically free family, then $x_\iota \neq x_\kappa$ for $\iota \neq \kappa$. A subset L of E is said to be algebraically free if the family defined by the identical bijection of L onto itself is algebraically free. The elements of an algebraically free set are also said to be *linearly independent*. A family $(x_\iota)_{\iota \in I}$ which is not algebraically free is said to be *linearly dependent;* this means that there exists a family $(\lambda_\iota)_{\iota \in I}$ of scalars, a finite number of which are different from zero, such that $\sum_{\iota \in I} \lambda_\iota x_\iota = 0$.

An algebraically free family of generators of a subspace F is called an *algebraic basis* (or Hamel basis) of F. It follows from Zorn's lemma that every subspace F has an algebraic basis. More precisely, given a set of generators S of F and an algebraically free subset L in F such that $S \supset L$, there exists an algebraic basis B of F such that $S \supset B \supset L$. In particular, an algebraic basis of a subspace F of E can always be completed to become an algebraic basis of E. An algebraic basis is a maximal algebraically free subset and also a minimal set of generators. A family $(y_\iota)_{\iota \in I}$ of elements of F is an algebraic basis of F if and only if every $x \in F$ has a unique representation of the form $x = \sum_{\iota \in I} \lambda_\iota y_\iota$ (where all except a finite number of λ_ι are equal to zero).

If a vector space E has an algebraic basis with a finite number of elements, then every other algebraic basis of E is finite and has the same number of elements. It is clearly enough to prove that if an algebraic basis B of E has n elements, then any other algebraic basis C has *at most* n elements; this is a consequence of the following lemma:

Let $y_1, \ldots, y_{n+1}$ be $n + 1$ elements of E which are all linear combinations of n elements $x_1, \ldots, x_n$ of E. Then the elements $y_1, \ldots, y_{n+1}$ are linearly dependent.

The proof goes by induction. For $n = 1$ the lemma is true, since if $y_1 = \lambda x$ and $y_2 = \mu x$, then $\mu y_1 - \lambda y_2 = 0$. Assume therefore that the lemma is true for $n - 1$. We can write

$$y_j = \sum_{i=1}^{n} \lambda_{ji} x_i \qquad (j = 1, \ldots, n + 1)$$

and it is no restriction to suppose that $\lambda_{11} \neq 0$. The n elements

$$y_j' = y_j - \frac{\lambda_{j1}}{\lambda_{11}} y_1 \qquad (j = 2, \ldots, n + 1) \tag{1}$$

are linear combinations of $x_2, \ldots, x_n$, and therefore by our induction hypothesis we have a linear relation

$$\sum_{j=2}^{n+1} \mu_j y_j' = 0,$$

where not all μ_j are zero. From (1) we obtain the relation

$$\sum_{j=2}^{n+1} \mu_j y_j - \sum_{j=2}^{n+1} \mu_j \frac{\lambda_{j1}}{\lambda_{11}} y_1 = 0,$$

which proves that the y_j $(j = 1, 2, \ldots, n + 1)$ are linearly dependent. Q.E.D.

Coming back to our original statement, suppose that C has more than n elements. Then we can pick $n + 1$ elements from C, and these $n + 1$ elements are linear combinations of the n elements of B. Hence, by the lemma, the $n + 1$ elements are linearly dependent, which contradicts the fact that C is an algebraic basis.

If E has a finite algebraic basis, then the number n of elements in any algebraic basis of E is called the *algebraic dimension* of E and denoted by $\dim_{\mathbf{K}} E$ or $\dim E$. If E does not have a finite algebraic basis, then we say that E is infinite-dimensional. It can be proved that even in the infinite-dimensional case two algebraic bases of E have the same cardinal ([7], §1, no. 12, Corollary 12 of Proposition 23 or [52], §7, 4.(2)).

EXAMPLE 1. An algebraic basis of the vector space $\mathbf{R}^n$ is clearly given by the n elements $e_1 = (1, 0, \ldots, 0)$, $e_2 = (0, 1, \ldots, 0), \ldots, e_n = (0, 0, \ldots, 1)$. Thus the dimension of $\mathbf{R}^n$ over $\mathbf{R}$ is n, which justifies its name. The basis $(e_i)_{1 \leq i \leq n}$ is called the *canonical basis* of $\mathbf{R}^n$.

Similarly, it can be seen that $\dim_{\mathbf{C}} \mathbf{C}^n = n$.

If E is a vector space $x \in E$ and $A \subset E$, then $x + A$ is the set of all vectors $x + y$ with $y \in A$ and is called the *translate* of A by x. Similarly, if $A \subset E$, $B \subset E$, then $A + B$ is the set of all vectors $x + y$ with $x \in A$, $y \in B$, and $A - B$ is the set of all vectors $x - y$ with $x \in A$, $y \in B$. Finally, if $\lambda \in \mathbf{K}$ and $A \subset E$, then λA is the set of all vectors λx with $x \in A$.

Before we link the algebraic concepts just recalled with the norm on the vector space, we must also recall some notions and results from the topology of metric spaces. This will be an excellent occasion to review some topological properties of the space $\mathbf{R}^n$, with which the reader is anyhow expected to be familiar.

Let X be a metric space with distance $\delta(x, y)$ defined on it (cf. §1). A subset A of X is in a natural way a metric space under the metric *induced* by that of X. Let a be a point of X and ρ a strictly positive real number. The set $B_\rho(a)$ of all points $x \in X$ such that $\delta(a, x) \leqq \rho$ is called the *closed ball* with center a and radius ρ. A set $V \subset X$ is a *neighborhood* of a if it contains some closed ball with center a. A sequence $(x_n)_{n \in \mathbf{N}}$ of points of X converges to a if and only if for every neighborhood V of a there exists an integer N such that $x_n \in V$ for every $n > N$.

Let A be a subset of X. A point $x \in X$ is said to *adhere* to A if every neighborhood of x contains points of A. The point x adheres to A if and only if there exists a sequence of points of A which converges to x. The set $\overline{A}$ of all points adherent to A is called the *adherence* or *closure* of A. Clearly, $A \subset \overline{A}$. If $A = \overline{A}$, then the set A is said to be *closed*. The intersection of any family of closed sets is closed; the union of a *finite* family of closed sets is closed.

If A and B are two subsets of X, we denote by $\delta(A, B)$ the greatest lower bound of all numbers $\delta(x, y)$, where x varies in A, y varies in B, and we call it the distance between the sets A and B. If $A = \{a\}$ is reduced to one element, we write $\delta(a, B)$ instead of $\delta(\{a\}, B)$. We have $\delta(x, A) = 0$ if and only if $x \in \overline{A}$.

The *interior* $\mathring{A}$ of a subset A of X is the set of those points $x \in X$ which possess a neighborhood contained in A. Clearly $\mathring{A} \subset A$. If $\mathring{A} = A$, then the set A is said to be *open*. A is open if and only if its complementary $\complement A$ is closed. The union of any family of open sets is open; the intersection of a *finite* family of open sets is open. The total space X and the empty set $\emptyset$ are both open and closed. For any set $A \subset X$ the set $\overline{A}$ is closed and $\mathring{A}$ is open. If $A \subset X$, $B \subset X$, and $\overline{A} \supset B$, then A is *dense* in B. If $\overline{A} = X$ then A is *everywhere dense*.

Let X and Y be two metric spaces with their respective metrics δ and η. A map f from X into Y is said to be *continuous* at the point $a \in X$, if for every neighborhood W of the point $f(a) \in Y$ there exists a neighborhood V of a such that $f(x) \in W$ for every $x \in V$ (i.e., $f(V) \subset W$). This is equivalent to the following: to every $\epsilon > 0$, there exists $\alpha > 0$ such that $\delta(a, x) < \alpha$ implies $\eta(f(a), f(x)) < \epsilon$, and also to the following: for every sequence (x_n) tending to a the sequence $(f(x_n))$ tends to $f(a)$. A map $f\colon X \to Y$ is said to be continuous if it is continuous at every point of X. The map f is continuous if and only if for every closed (resp. open) set A in Y the set $f^{-1}(A)$ is closed (resp. open) in X. More generally, let X_i $(1 \leqq i \leqq n)$ be a finite family of metric spaces and let f be a map of the cartesian product $\prod_{i=1}^n X_i$ into Y (i.e., a function $(x_1, \ldots, x_n) \mapsto f(x_1, \ldots, x_n)$ of n variables, where x_i varies in X_i). The map f is said to be continuous at the point $(a_1, \ldots, a_n)$ if for every neighborhood W of the point $f(a_1, \ldots, a_n)$ there exist neighborhoods V_i of the points a_i $(1 \leqq i \leqq n)$ such that $f(x_1, \ldots, x_n) \in W$ for $x_i \in V_i$ $(1 \leqq i \leqq n)$ (i.e., $f(\prod_{i=1}^n V_i) \subset W$). A continuous map from $\prod_{i=1}^n X_i$ into Y is a map which is continuous at every point of $\prod_{i=1}^n X_i$.

Two metric spaces X and Y are *isometric* if there exists a bijection $f: X \to Y$ (called *isometry*) such that $\eta(f(x), f(y)) = \delta(x, y)$ for every $x, y \in X$. Two metric spaces X and Y are *homeomorphic* if there exists a bijection $f: X \to Y$ (called *homeomorphism*) which is continuous together with its inverse f^{-1}. Clearly, an isometry is a homeomorphism, but the converse is not necessarily true. A homeomorphism transforms closed sets into closed sets and open sets into open sets.

Let E be a normed vector space and $a \in E$. The bijection $x \mapsto x + a$, called *translation* by a, is an isometry of E onto itself, since

$$\|x + a - (y + a)\| = \|x - y\|.$$

If $B_\rho = B_\rho(0)$ then $B_\rho(a) = B_\rho + a$; i.e., the ball with center a and radius ρ is obtained through translation by a from the ball with center 0 and radius ρ. If λ is a nonzero scalar, the bijection $x \mapsto \lambda x$, called *homothecy* by λ, is a homeomorphism of E onto itself. Indeed,

$$\|\lambda x - \lambda y\| = |\lambda| \cdot \|x - y\| < \epsilon$$

if $\|x - y\| < \epsilon/|\lambda|$, and the inverse map, given by $x \mapsto \lambda^{-1}x$, is also continuous.

PROPOSITION 1. *Let E be a normed vector space. Then the map $(x, y) \mapsto x + y$ from $E \times E$ into E, the map $(\lambda, x) \mapsto \lambda x$ from $\mathbf{K} \times E$ into E, and the map $x \mapsto \|x\|$ from E into $\mathbf{R}_+$ are continuous. If E is an inner product space, then the map $(x, y) \mapsto (x \mid y)$ from $E \times E$ into $\mathbf{K}$ is also continuous.*

Proof. We have $\|a + b - (x + y)\| \leqq \|a - x\| + \|b - y\|$. Thus

$$\|a + b - (x + y)\| < \epsilon$$

if

$$\|a - x\| < \tfrac{1}{2}\epsilon \quad \text{and} \quad \|b - y\| < \tfrac{1}{2}\epsilon.$$

Next we have the identity

$$\xi x - \lambda a = (\xi - \lambda)(x - a) + (\xi - \lambda)a + \lambda(x - a);$$

hence $\|\xi x - \lambda a\| < \epsilon$, provided:

$$|\xi| < \sqrt{\epsilon}, \quad \|x\| < \sqrt{\epsilon} \quad \text{if} \quad \lambda = 0, \quad a = 0,$$

$$|\xi| < \min\left(\sqrt{\frac{\epsilon}{2}}, \frac{\epsilon}{2\|a\|}\right), \quad \|x - a\| < \sqrt{\frac{\epsilon}{2}} \quad \text{if} \quad \lambda = 0, \quad a \neq 0,$$

$$|\xi - \lambda| < \sqrt{\frac{\epsilon}{2}}, \quad \|x\| < \min\left(\sqrt{\frac{\epsilon}{2}}, \frac{\epsilon}{2|\lambda|}\right) \quad \text{if} \quad \lambda \neq 0, \quad a = 0,$$

$$|\xi - \lambda| < \min\left(\sqrt{\frac{\epsilon}{3}}, \frac{\epsilon}{3\|a\|}\right), \quad \|x - a\| < \min\left(\sqrt{\frac{\epsilon}{3}}, \frac{\epsilon}{3|\lambda|}\right)$$
$$\text{if} \quad \lambda \neq 0, \quad a \neq 0.$$

The continuity of the norm follows from the inequality

$$|\,\|x\| - \|y\|\,| \leqq \|x - y\|. \tag{2}$$

To prove this inequality let us observe that

$$\|x\| = \|x - y + y\| \leqq \|x - y\| + \|y\|,$$

i.e., $\|x\| - \|y\| \leqq \|x - y\|$. For reasons of symmetry we also have $\|y\| - \|x\| \leqq \|x - y\|$. The last two inequalities imply (2).

Finally, the continuity of the scalar product follows from the identity

$$(x \mid y) - (a \mid b) = (x - a \mid y - b) + (x - a \mid b) + (a \mid y - b)$$

similarly as in the proof of the continuity of multiplication by a scalar. ∎

An *open cover* of a set A in a metric space X is a family $(B_\iota)_{\iota \in I}$ of open sets such that $A \subset \bigcup_{\iota \in I} B_\iota$. If $J \subset I$ and $(B_\iota)_{\iota \in J}$ is a cover of A, we say that $(B_\iota)_{\iota \in J}$ is a *subcover* of $(B_\iota)_{\iota \in I}$. A set $A \subset X$ is *compact* if any of the following equivalent conditions is satisfied:

(1) Every sequence $(x_n)_{n \in \mathbf{N}}$ of points of A has a subsequence which converges to a point of A.

(2) Every open cover of A has a finite subcover (Heine-Borel-Lebesgue theorem).

(3) If $(F_\iota)_{\iota \in I}$ is any family of closed sets such that $\bigcap_{\iota \in I} F_\iota$ is disjoint from A, then there exists a finite subfamily of $(F_\iota)_{\iota \in I}$ whose intersection is already disjoint from A (Cantor's theorem).

Let us make the important remark that these conditions are equivalent only in metric spaces and not in general topological spaces (see Chapter 2, §10). A compact set is always closed. If f is a continuous map of a metric space X into a metric space Y and A is compact in X, then $f(A)$ is compact in Y. A set A is *relatively compact* if its closure is compact or, equivalently, if every sequence of points of A has a subsequence which converges to a point of X (but not necessarily of A).

In a metric space X a set A is *bounded* if it is contained in some ball $B_\rho(a)$.

In a normed vector space a bounded set is always contained in a ball $B_\sigma(0) = B_\sigma$ with center at the origin. Indeed, if $\|x - a\| \leqq \rho$ for every x in A, then

$$\|x\| = \|x - a + a\| \leqq \|x - a\| + \|a\| \leqq \rho + \|a\|,$$

i.e., A is contained in the ball B_σ with center 0 and radius $\sigma = \rho + \|a\|$.

A compact set is always bounded. It is a basic property of the spaces $\mathbf{R}^n$ and $\mathbf{C}^n$ that in them the converse is also true:

Weierstrass-Bolzano theorem: A closed bounded subset of $\mathbf{R}^n$ *or* $\mathbf{C}^n$ *is compact. A bounded set of* $\mathbf{R}^n$ *or* $\mathbf{C}^n$ *is relatively compact.*

This theorem is not true for infinite-dimensional Banach spaces; we shall give an example below. In fact we shall see in Chapter 3, §9 that if in a Banach space E every bounded set is relatively compact, then E is necessarily finite dimensional.

EXAMPLE 2. Consider in l^2 (Example 1.7) the set U of all unit vectors $e_0 = (1, 0, 0, \ldots)$, $e_1 = (0, 1, 0, \ldots), \ldots$. The set U is bounded, since $\|e_n\| = 1$ for all $n \in \mathbf{N}$. Since $\|e_n - e_m\| = \sqrt{2} > 1$ for $n \neq m$, any ball with radius $< \frac{1}{2}$ contains at most one element of U. Therefore U is also closed. But U is not compact, since the sets $V_n = \{x \mid \|x - e_n\| < 1\}$ form an open cover of U, and clearly no proper subfamily of (V_n) covers U.

A continuous function defined on a compact set is always bounded, and attains its maximum and its minimum (Weierstrass' theorem; we have used this property in Examples 2 and 5 of §1).

We have already introduced in the previous section the concept of a complete metric space. A compact metric space is always complete. A complete subset of a metric space is always closed. If a metric space X is not complete, then we can construct a complete metric space $\hat{X}$, called the *completion* of X, such that X is isometric with a subset X_0 of $\hat{X}$ and X_0 is dense in $\hat{X}$. Most often X will be identified with X_0, and hence X is considered to be a subset of $\hat{X}$. The construction of $\hat{X}$ generalizes the familiar construction of the set $\mathbf{R}$ of real numbers from the set $\mathbf{Q}$ of rational numbers; we shall recall it briefly. Let S be the set of all Cauchy sequences of X. In S we introduce an equivalence relation R: two Cauchy sequences (x_n) and (y_n) are equivalent if the sequence $(\delta(x_n, y_n))$ tends to zero. It is trivial to check that R is indeed an equivalence relation. $\hat{X}$ is now the set of all equivalence classes of S modulo R (i.e., $\hat{X}$ is the quotient set S/R). Let us define a metric on $\hat{X}$. If $\dot{x}$ and $\dot{y}$ are two elements of $\hat{X}$, let (x_n) and (y_n) be elements of S representing these equivalence classes. Then we set

$$\hat{\delta}(\dot{x}, \dot{y}) = \lim_{n \to \infty} \delta(x_n, y_n).$$

The following points are now easy to check: (1) the limit on the right-hand side really exists, (2) it is independent of the choice of the representatives (x_n) and (y_n) in the equivalence classes $\dot{x}$ and $\dot{y}$, (3) $\hat{\delta}(\dot{x}, \dot{y})$ is a metric on $\hat{X}$. Now let

$x \in X$, let (x_n) be the Cauchy sequence in which $x_n = x$ for every $n \in \mathbf{N}$, and let $\dot{x}$ be the class of (x_n) modulo R. The map $x \mapsto \dot{x}$ is clearly an isometry of X onto a subset X_0 of $\hat{X}$. Let $\dot{x} \in \hat{X}$, let (x_n) be a representative of $\dot{x}$ in S, and let $\dot{x}_n$ be the image in X_0 of $x_n \in X$ under the isometry just defined. Then $(\dot{x}_n)$ converges to $\dot{x}$, since

$$\hat{\delta}(\dot{x}, \dot{x}_n) = \lim_{m \to \infty} \delta(x_m, x_n).$$

This proves that X_0 is dense in $\hat{X}$. Finally, we have to prove that $\hat{X}$ is complete. Let $(\dot{x}_n)_{n \in \mathbf{N}}$ be a Cauchy sequence in $\hat{X}$. For every $n \in \mathbf{N}$ there exists a $\dot{y}_n \in X_0$ such that $\hat{\delta}(\dot{x}_n, \dot{y}_n) < 2^{-n}$. Let $\dot{y}_n$ be the image of $y_n \in X$. Then (y_n) is a Cauchy sequence in X, since

$$\delta(y_n, y_m) = \hat{\delta}(\dot{y}_n, \dot{y}_m) \leqq \hat{\delta}(\dot{y}_n, \dot{x}_n) + \hat{\delta}(\dot{x}_n, \dot{x}_m) + \hat{\delta}(\dot{x}_m, \dot{y}_m);$$

and thus (y_n) defines an element $\dot{x} \in \hat{X}$. By what we have said above, $(\dot{y}_n)$ converges to $\dot{x}$; but then $(\dot{x}_n)$ also converges to $\dot{x}$.

Let X and Y be two metric spaces with their respective metrics δ and η. A map f from X into Y is said to be *uniformly continuous* if to every $\epsilon > 0$ there exists an $\alpha > 0$ such that $\eta(f(x), f(y)) < \epsilon$ for any two points $x, y \in X$ satisfying $\delta(x, y) < \alpha$. We have a similar definition for a map from a product space $\prod_{i=1}^{n} X_i$ into Y. A uniformly continuous map is clearly continuous; the converse is in general false. On a compact set, however, every continuous map is uniformly continuous. If $f: X \to Y$ is uniformly continuous and (x_n) is a Cauchy sequence in X, then $(f(x_n))$ is a Cauchy sequence in Y.

PROPOSITION 2. *Let f be a map from a normed vector space E into a normed vector space F. The map f is uniformly continuous if for every neighborhood W of the origin in F there exists a neighborhood V of the origin in E such that $x - y \in V$ implies $f(x) - f(y) \in W$ for $x, y \in E$.*

Proof. Both conditions express the fact that to every $\epsilon > 0$ there exists an $\alpha > 0$ such that $\|f(x) - f(y)\| < \epsilon$ whenever $\|x - y\| < \alpha$. ▮

REMARK 1. Though we shall not need nor use the notion of uniform structure in this book, it may be worth while to note for the interested reader that the uniform structure deduced from the metric space structure of a normed vector space coincides with the uniform structure deduced from the abelian group structure. Proposition 2 is an immediate consequence of this rather obvious fact (see also Chapter 2, §9).

PROPOSITION 3. *Let E be a normed vector space. The map $(x, y) \mapsto x + y$ from $E \times E$ into E and the map $x \mapsto \|x\|$ from E into $\mathbf{R}_+$ are uniformly continuous. For every $\lambda \in \mathbf{K}$ the map $x \mapsto \lambda x$ from E into E is uniformly continuous, and for any $a \in E$ the map $\xi \mapsto \xi a$ from $\mathbf{K}$ into E is uniformly continuous. If E is an inner product space and $a \in E$, then the map $x \mapsto (x \mid a)$ from E into $\mathbf{K}$ is uniformly continuous.*

This follows immediately from the inequalities

$$\begin{aligned} \|(a+b)-(x+y)\| &\leqq \|a-x\|+\|b-y\|, \\ |\,\|x\|-\|y\|\,| &\leqq \|x-y\|, \\ \|\lambda x-\lambda y\| &\leqq |\lambda|\cdot\|x-y\|, \\ \|\xi a-\eta a\| &\leqq |\xi-\eta|\cdot\|a\|, \\ |(x\mid a)-(y\mid a)| &\leqq \|x-y\|\cdot\|a\|. \ \blacksquare \end{aligned}$$

Let X be a metric space, Y a complete metric space and A a subset of X. Then to every uniformly continuous map f from A into Y there corresponds a unique uniformly continuous map $\bar{f}$ from $\bar{A}$ into Y such that $\bar{f}(x) = f(x)$ for every $x \in A$ (i.e., such that the *restriction* of $\bar{f}$ to A coincides with f). Let us recall the construction of $\bar{f}$. Let x be a point which adheres to A but does not belong to A, and let (x_n) be a sequence of points of A which converges to x. Then (x_n) is a Cauchy sequence in X and, since f is uniformly continuous, $(f(x_n))$ is a Cauchy sequence in Y which converges to some point $y \in Y$ because Y is complete. Define $y = \bar{f}(x)$. It is easy to check that y is independent of the particular choice of the sequence (x_n) tending to x. For $x \in A$, we define, of course, $\bar{f}(x) = f(x)$. Let us prove that $\bar{f}$ is uniformly continuous on $\bar{A}$. Let $\epsilon > 0$ be arbitrary and let $\alpha > 0$ be such that $\delta(x, y) < 3\alpha$ implies $\eta(f(x), f(y)) < \epsilon/3$ for $x, y \in A$. We claim that $\delta(x, y) < \alpha$ implies

$$\eta(\bar{f}(x), \bar{f}(y)) < \epsilon \qquad \text{for} \qquad x, y \in \bar{A}.$$

Indeed, let $x, y \in \bar{A}$ be such that $\delta(x, y) < \alpha$. There exists a sequence (x_n) of points of A tending to x and a sequence (y_n) of points of A tending to y. It follows that there exists an $m \in \mathbf{N}$ such that $\delta(x_n, x_m) \leqq \alpha/2$ and $\delta(y_n, y_m) \leqq \alpha/2$ for $n \geqq m$. Hence $\eta(f(x_n), f(x_m)) < \epsilon/3$, $\eta(f(y_n), f(y_m)) < \epsilon/3$ for $n \geqq m$, and thus $\eta(\bar{f}(x), f(x_m)) \leqq \epsilon/3$, $\eta(\bar{f}(y), f(y_m)) \leqq \epsilon/3$. We also have $\delta(x, x_m) < \alpha$, $\delta(y, y_m) < \alpha$, and thus

$$\delta(x_m, y_m) \leqq \delta(x_m, x) + \delta(x, y) + \delta(y, y_m) < \alpha + \alpha + \alpha = 3\alpha,$$

which implies that $\eta(f(x_m), f(y_m)) < \epsilon/3$. Finally, we obtain

$$\begin{aligned} \eta(\bar{f}(x), \bar{f}(y)) &\leqq \eta(\bar{f}(x), f(x_m)) + \eta(f(x_m), f(y_m)) + \eta(f(y_m), \bar{f}(y)) \\ &< \frac{\epsilon}{3} + \frac{\epsilon}{3} + \frac{\epsilon}{3} = \epsilon. \end{aligned}$$

It is quite obvious that $\bar{f}$ is unique.

Of course, we have an entirely analogous theorem for a uniformly continuous map from a subset of a product of metric spaces into a complete space.

If f is an isometry from A onto $f(A)$, then $\bar{f}$ is an isometry from $\bar{A}$ onto $\bar{f}(\bar{A})$. This shows that the completion $\hat{X}$ of a metric space X is unique up to an isometry.

A subset A of a metric space X is said to be *precompact* (or totally bounded) if its closure in $\hat{X}$ is compact. A is precompact if and only if for every $\epsilon > 0$ there exists a finite family $a_1, a_2, \ldots, a_n$ of elements of X such that given $x \in A$ we can find an a_j which verifies $\delta(x, a_j) < \epsilon$. In other words, A is precompact if and only if for every $\epsilon > 0$ it can be covered by a finite number of open balls (see Exercise 2) $O_\epsilon(a_1), \ldots, O_\epsilon(a_n)$ with radius ϵ. Indeed, the balls $O_\epsilon(a)$, where a runs through A, cover the closure $\overline{A}$ of A in $\hat{X}$; and since $\overline{A}$ is compact, it can be covered by a finite number of these balls, which cover A *a fortiori*. To prove that the condition is also sufficient, it is enough to show that if it is satisfied, then every sequence (x_n) of elements of A has a subsequence which is a Cauchy sequence. First, cover A by balls of radius 1. One of these balls contains infinitely many terms of the sequence (x_n); pick one of these terms and call it x_{n_1}. Next, cover A by balls of radius $\frac{1}{2}$. The intersection of one of these balls with the first ball contains infinitely many terms of (x_n). Pick one whose index is larger than n_1 and call it x_{n_2}. Continuing this way we obtain a sequence $(x_{n_k})_{k \in \mathbf{N}}$, where $\delta(x_{n_k}, x_{n_l}) \leqq 1/2^{m-1}$ if $k, l > m$.

If X is complete, then of course precompact sets are the same as relatively compact sets.

DEFINITION 1. *Two normed vector spaces E and F over the same field $\mathbf{K}$ are isometrically isomorphic if there exists a bijection $f: E \to F$ such that*

$$f(x + y) = f(x) + f(y), \qquad f(\lambda x) = \lambda f(x), \qquad \textit{and} \qquad \|f(x)\| = \|x\|$$

for all $x, y \in E$, $\lambda \in \mathbf{K}$. Two inner product spaces E and F over the same field $\mathbf{K}$ are isometrically isomorphic if there exists a bijection $f: E \to F$ such that

$$f(x + y) = f(x) + f(y), \quad f(\lambda x) = \lambda f(x), \quad \textit{and} \quad (f(x) \mid f(y)) = (x \mid y)$$

for all $x, y \in E$, $\lambda \in \mathbf{K}$.

PROPOSITION 4. *Let E be a normed vector space. Then there exists a Banach space $\hat{E}$ such that E is isometrically isomorphic to a linear subspace E_0 of $\hat{E}$ and that E_0 is dense in $\hat{E}$. The space $\hat{E}$ is unique up to an isometric isomorphism.*

Let E be an inner product space. Then there exists a Hilbert space $\hat{E}$ such that E is isometrically isomorphic to a linear subspace E_0 of $\hat{E}$ and that E_0 is dense in $\hat{E}$. The space $\hat{E}$ is unique up to an isometric isomorphism.

Proof. Let $\hat{E}$ be the completion of the underlying metric space of E, equipped with the metric $\hat{\delta}$, and let E_0 be the dense subspace of $\hat{E}$ isometric with E. The functions

$$(x, y) \mapsto x + y, \qquad x \mapsto \lambda x \quad (\lambda \in \mathbf{K}), \qquad x \mapsto \|x\|$$

can be transported to E_0, and since they are uniformly continuous (Proposition 3), they can be extended uniquely to $\hat{E}$. Similarly, if E is an inner product space, for each $a \in E_0$ the map $x \mapsto (x \mid a)$ is uniformly continuous on E_0 and can be extended uniquely to $\hat{E}$, and then for each $b \in \hat{E}$ the map $y \mapsto (b \mid y)$ is uniformly continuous on E_0 and can be extended uniquely to $\hat{E}$. The maps so defined satisfy Axioms (VS 1) through (VS 8), (N 2), (N 3) (resp. (SP 1), (SP 3), (SP 4)) trivially because E_0 is dense in $\hat{E}$. Let us show this for instance for (VS 1). Let $x, y \in \hat{E}$. Then there exist two sequences (x_n), (y_n) in E_0 which tend to x and y respectively. By the continuity of addition $(x_n + y_n)$ tends to $x + y$ and $(y_n + x_n)$ tends to $y + x$. But $x_n + y_n = y_n + x_n$, and since a sequence has at most one limit, $x + y = y + x$. We can see similarly that the distance $\hat{\delta}(x, y)$ already defined on $\hat{E}$ coincides with the value $\|x - y\|$ of the extended norm (cf. Exercise 3 below). This implies that Axioms (N 1), resp. (SP 2), are also satisfied on $\hat{E}$. ∎

REMARK 2. $\hat{E}$ is called the *completion* of E. We usually identify E with E_0 and thus imbed E into $\hat{E}$.

EXERCISES

1. Supply complete proofs of all the statements made without proof.

2. Prove that in a metric space X a closed ball is a closed set. The set $O_\rho(a)$ of all points $x \in X$ such that $\delta(a, x) < \rho$ is called the *open ball* with center a and radius ρ. Prove that an open ball is an open set. Prove that V is a neighborhood of a point a if and only if it contains some open ball with center a.

3. Prove that if X is a metric space, the mapping $(x, y) \mapsto \delta(x, y)$ from $E \times E$ into $\mathbf{R}_+$ is uniformly continuous.

4. Prove that if A is a bounded subset of $\mathbf{K}$ and B a bounded subset of the normed vector space E, then the map $(\lambda, x) \mapsto \lambda x$ from $A \times B$ into E is uniformly continuous. Similarly prove that if A and B are bounded subsets of the inner product space E, then the map $(x, y) \mapsto (x \mid y)$ from $A \times B$ into $\mathbf{K}$ is uniformly continuous.

5. (a) Prove that the identity

$$(x \mid y) = \left\| \frac{x + y}{2} \right\|^2 - \left\| \frac{x - y}{2} \right\|^2$$

holds for any real inner product space.

(b) Prove that the identity

$$(x \mid y) = \left\| \frac{x + y}{2} \right\|^2 - \left\| \frac{x - y}{2} \right\|^2 + i \left\| \frac{x + iy}{2} \right\|^2 - i \left\| \frac{x - iy}{2} \right\|^2$$

holds for any complex inner product space.

(c) Prove that if the underlying normed vector spaces of two inner product spaces are isometrically isomorphic, then the inner product spaces are also isometrically isomorphic.

6. Let K be a compact subset of $\mathbf{R}^n$ and let $\mathcal{C}(K)$ be the set of all real- (or complex-) valued continuous functions defined on K. Generalizing Examples 2 and 4 of §1, define a structure of Banach space on $\mathcal{C}(K)$.

7. Let Ω be an open subset of $\mathbf{R}^n$ and let $\mathcal{C}_0(\Omega)$ be the set of all real- (or complex-) valued continuous functions defined on Ω which "vanish at the boundary of Ω" in the following sense: for every $f \in \mathcal{C}_0(\Omega)$ and every $\epsilon > 0$ there exists a compact subset K of Ω such that $|f(x)| < \epsilon$ for $x \in \Omega \cap \complement K$. Define a structure of Banach space on $\mathcal{C}_0(\Omega)$ (cf. Example 1.5).

§3. *Subspaces*

A *closed subspace* F of a normed space E is a linear subspace of the vector space E which is a closed subset of the metric space E.

PROPOSITION 1. *The closure $\overline{F}$ of a linear subspace F of a normed space E is a closed subspace of E.*

Proof. We know that $\overline{F}$ is a closed set; we have to prove that $\overline{F}$ is a subspace of E. Let $x \in \overline{F}$, $y \in \overline{F}$. Then there exist sequences (x_n), (y_n) of points of F which tend to x and y respectively. Since F is a linear subspace, $x_n + y_n \in F$ for every $n \in \mathbf{N}$, and by the continuity of addition (Proposition 2.1) the sequence $(x_n + y_n)$ tends to $x + y$. Thus $x + y \in \overline{F}$. Similarly, (λx_n) tends to λx for any $\lambda \in \mathbf{K}$; i.e., $\lambda x \in \overline{F}$. ▮

The intersection of an arbitrary family of closed subspaces is a closed subspace, since it is both a linear subspace and a closed set. Therefore, given a set A of elements of E, we can speak of the smallest closed subspace containing A, i.e., the intersection of all the closed subspaces containing A.

DEFINITION 1. *A set A is total in a closed subspace F of a normed space E if F is the smallest closed subspace containing A. Then F is said to be the closed subspace generated by A* (or the closed linear hull of A).

The closed subspace generated by a set A is the closure F of the subspace M formed by all linear combinations of elements of A. Indeed, F is a closed subspace (Proposition 1) containing A. Conversely, any closed subspace containing A necessarily contains every linear combination of elements of A as well as the limit of every sequence of such linear combinations.

EXAMPLE 1. The set of the n unit vectors $e_1, e_2, \ldots, e_n$ is total in $\mathbf{R}^n$ (resp. in $\mathbf{C}^n$).

EXAMPLE 2. Let I be a finite closed interval of the real line. A well-known theorem of Weierstrass asserts that every continuous function on I can be approximated uniformly by polynomials; i.e., given a continuous function f on I, for every $\epsilon > 0$ there exists a polynomial

$$P(x) = a_0 + a_1 x + \cdots + a_n x^n$$

such that

$$|f(x) - P(x)| < \epsilon \qquad \text{for every} \qquad x \in I. \tag{1}$$

Now this theorem can be restated in the following way: *In the Banach space $\mathcal{C}(I)$ the set $(x^n)_{n \in \mathbf{N}}$ is total.* Indeed, (1) is equivalent to $\|f - P\| < \epsilon$, i.e., f is adherent to the space of linear combinations of the functions x^n. Later (Example 6.1) we shall prove Weierstrass' theorem in this form.

Another approximation theorem of Weierstrass states that every continuous function f on the closed interval $[0, 2\pi]$ such that $f(0) = f(2\pi)$ can be approximated uniformly by trigonometric polynomials, i.e., expressions of the form

$$a_0 + \sum_{k=1}^{n} a_k \cos kx + b_k \sin kx.$$

This can be expressed by saying that the sequence $1, \cos x, \sin x, \ldots, \cos nx, \sin nx, \ldots$ is total in the closed subspace F of $\mathcal{C}([0, 2\pi])$ formed by the functions f which satisfy $f(0) = f(2\pi)$.

EXAMPLE 3. It can be shown that the set $(x^n e^{-|x|^\alpha})_{n \in \mathbf{N}}$ of functions is total in $\mathcal{C}_0(\mathbf{R})$ if and only if $\alpha \geqq 1$. More generally, let F be a strictly positive, even function defined on $\mathbf{R}$. Assume that

$$\lim_{x \to \infty} (x^n / F(x)) = 0$$

for all $n \in \mathbf{N}$ and that $\log F(x)$ is a convex function of $\log |x|$. Then it can be shown that the set $(x^n / F(x))_{n \in \mathbf{N}}$ is total in $\mathcal{C}_0(\mathbf{R})$ if and only if the integral $\int_1^\infty x^{-2} \log F(x)\, dx$ diverges [65]. Even more general results have been obtained recently by Nachbin [69 through 73].

DEFINITION 2. *A family $(x_\iota)_{\iota \in I}$ of elements of a normed space is topologically free if no x_ι belongs to the closed subspace generated by the x_κ with $\kappa \neq \iota$.*

Clearly a topologically free family is also algebraically free, but the converse is not necessarily true.

EXAMPLE 4. We shall see later (Example 6.1) that the family $(x^n)_{n \in \mathbf{N}}$ is not topologically free in $\mathcal{C}([0, 1])$, although it is algebraically free. We shall also see (Example 6.2) that the sequence $1, \cos x, \sin x, \ldots, \cos nx, \sin nx, \ldots$ is topologically free in $\mathcal{C}([0, 2\pi])$.

The set of all topologically free families, when ordered by inclusion, is not inductive and it is not true that every topologically free family is contained in a maximal topologically free family (see Exercise 3(b)). We shall now see that for certain topologically free families in a Hilbert space the situation is more favorable.

DEFINITION 3. *A set A in a vector space is convex if for $x \in A$, $y \in A$, and $0 \leqq \alpha \leqq 1$ we have $\alpha x + (1 - \alpha)y \in A$.*

EXAMPLE 5. A linear subspace of a vector space is clearly a convex set.

EXAMPLE 6. A closed ball $B_\rho(a)$ in a normed vector space is a convex set. Indeed, $x \in B_\rho(a)$ and $y \in B_\rho(a)$ mean $\|a - x\| \leqq \rho$, $\|a - y\| \leqq \rho$; hence setting $\beta = 1 - \alpha$, we have

$$\begin{aligned}\|a - \alpha x - \beta y\| = \|\alpha(a - x) + \beta(a - y)\| &\leqq \alpha\|a - x\| + \beta\|a - y\| \\ &\leqq \alpha\rho + \beta\rho = \rho,\end{aligned}$$

i.e., $\alpha x + \beta y \in B_\rho(a)$.

PROPOSITION 2 (Parallelogram law). *For any two vectors x, y in an inner product space, we have the identity*

$$\|x + y\|^2 + \|x - y\|^2 = 2\|x\|^2 + 2\|y\|^2.$$

Proof.

$$\begin{aligned}\|x + y\|^2 + \|x - y\|^2 &= (x + y \mid x + y) + (x - y \mid x - y) \\ &= (x \mid x) + (x \mid y) + (y \mid x) + (y \mid y) \\ &\quad + (x \mid x) - (x \mid y) - (y \mid x) + (y \mid y) \\ &= 2(x \mid x) + 2(y \mid y) = 2\|x\|^2 + 2\|y\|^2. \quad \blacksquare\end{aligned}$$

PROPOSITION 3 (Lemma of F. Riesz). *Let A be a closed convex set in a Hilbert space. Then there exists a unique $x \in A$ such that $\|x\| \leqq \|y\|$ for every $y \in A$.*

Proof. Let $d = \inf_{y \in A} \|y\|$. There exists a sequence of elements $y_n \in A$ such that $\|y_n\| \to d$. By the parallelogram law (Proposition 2) we have

$$\left\|\frac{y_n - y_m}{2}\right\|^2 = \frac{1}{2}\|y_n\|^2 + \frac{1}{2}\|y_m\|^2 - \left\|\frac{y_n + y_m}{2}\right\|^2. \tag{2}$$

Given $\epsilon > 0$, there exists $N \in \mathbf{N}$ such that $\|y_n\| < d + \epsilon$ for $n > N$. Furthermore, since A is convex, $\frac{1}{2}(y_n + y_m) \in A$, and so $\|\frac{1}{2}(y_n + y_m)\| \geqq d$. Thus it follows from (2) that

$$\left\|\frac{y_n - y_m}{2}\right\|^2 \leqq 2\epsilon d + \epsilon^2 \qquad \text{for} \qquad n, m > N,$$

i.e., (y_n) is a Cauchy sequence. Since the Hilbert space is complete, (y_n) converges to an element x which belongs to the closed set A. By the continuity of the norm (Proposition 2.1) $\|x\| = \lim \|y_n\| = d$.

To prove the uniqueness of x let us suppose that we have two vectors $x_1 \neq x_2$ with $\|x_1\| = \|x_2\| = d$. Then

$$\left\|\frac{x_1 + x_2}{2}\right\|^2 = \frac{1}{2}\|x_1\|^2 + \frac{1}{2}\|x_2\|^2 - \left\|\frac{x_1 - x_2}{2}\right\|^2 < d^2,$$

since $\|x_1 - x_2\| > 0$. But $\frac{1}{2}(x_1 + x_2) \in A$ and the norm of every element in A is $\geqq d$. ▮

REMARK 1. It would be clearly sufficient to assume that A is a complete convex subset in an inner product space.

COROLLARY. *Let A be a closed convex set in a Hilbert space E and a a vector in E. Then there exists a unique $x \in A$ such that $\|x - a\| \leqq \|y - a\|$ for every $y \in A$.*

Proof. Let $B = A - a$ be the set of all vectors $y - a$, where y runs through A. The set B is obviously closed and convex; hence it has a point x_0 such that $\|x_0\| \leqq \|y_0\|$ for all $y_0 \in B$. Setting $x = x_0 + a$, we have $\|x - a\| \leqq \|y - a\|$ for all $y \in A$. ▮

DEFINITION 4. *Two vectors x, y in an inner product space are orthogonal if $(x \mid y) = 0$. A vector x is orthogonal to a subset M if $(x \mid y) = 0$ for all $y \in M$. Two subsets M and N are orthogonal if $(x \mid y) = 0$ for every $x \in M$ and $y \in N$.*

The only vector which is orthogonal to itself is 0 by Axiom (SP 2).

PROPOSITION 4. *Let F be a closed subspace of a Hilbert space E. Then every vector $x \in E$ can be uniquely written in the form $x = y + z$, where $y \in F$ and z is orthogonal to F.*

Proof. Since F is a closed convex set, by the above corollary there exists a vector $y \in F$ such that

$$\|x - y\| \leqq \|x - v\| \quad \text{for all} \quad v \in F. \tag{3}$$

Set $z = x - y$. We must show that z is orthogonal to F. Let w be an arbitrary nonzero vector in F. Then for all $\lambda \in \mathbf{K}$ the vector $v = y + \lambda w$ belongs to F; hence by (3) we have $\|z\| \leqq \|z - \lambda w\|$, i.e.,

$$0 \leqq -2\mathcal{R}e\lambda(w \mid z) + |\lambda|^2\|w\|^2 \quad \text{for all} \quad \lambda \in \mathbf{K}.$$

This is possible only if $(w \mid z) = 0$. Otherwise, setting

$$\lambda = \overline{(w \mid z)}/\|w\|^2$$

would give a contradiction.

To prove uniqueness, let us suppose that $x = y + z = y' + z'$, where $y \in F$, $y' \in F$, z and z' are orthogonal to F. Then $z - z'$ is also orthogonal to F; and since $z - z' = y' - y \in F$, we have $z - z' = 0$. Thus $z = z'$ and $y = y'$. ∎

COROLLARY. *Let F be a closed subspace of a Hilbert space E. If $F \neq E$, then there exists a nonzero vector in E which is orthogonal to F.*

DEFINITION 5. *A family $(a_\iota)_{\iota \in I}$ of elements of an inner product space E is orthogonal if $(a_\iota \mid a_\kappa) = 0$ for $\iota \neq \kappa$. It is orthonormal if furthermore $\|a_\iota\| = 1$ for every $\iota \in I$.*

PROPOSITION 5. *An orthonormal family in an inner product space is topologically free.*

Proof. Let $(\lambda_\kappa)_{\kappa \neq \iota}$ be any family of scalars of which only a finite number are different from 0. Then

$$\Big\| a_\iota - \sum_{\kappa \neq \iota} \lambda_\kappa a_\kappa \Big\|^2 = \Big(a_\iota - \sum_{\kappa \neq \iota} \lambda_\kappa a_\kappa \mid a_\iota - \sum_{\kappa \neq \iota} \lambda_\kappa a_\kappa \Big)$$
$$= \|a_\iota\|^2 + \sum_{\kappa \neq \iota} |\lambda_\kappa|^2 \cdot \|a_\kappa\|^2 \geqq 1. \blacksquare$$

PROPOSITION 6. *Any orthonormal family in an inner product space is included in a maximal orthonormal family.*

Proof. Consider the set $\mathfrak{S}$ of all orthonormal families containing the given orthonormal family. If we order $\mathfrak{S}$ by inclusion, then this set is inductive. Indeed, let (O_α) be a totally ordered family in $\mathfrak{S}$ and let $O = \bigcup_\alpha O_\alpha$. Then every $x \in O$ belongs to some O_α; hence $\|x\| = 1$. Furthermore, if $x, y \in O$, $x \neq y$, then $x \in O_\alpha$, $y \in O_\beta$; and if for instance $O_\alpha \subset O_\beta$, then $x \in O_\beta$; thus $(x \mid y) = 0$. Consequently O belongs to $\mathfrak{S}$ and is the least upper bound of the family (O_α). The existence of a maximal element in $\mathfrak{S}$ now follows from Zorn's lemma. ∎

PROPOSITION 7. *In a Hilbert space E a maximal orthonormal family is total.*

Proof. Let F be the closed subspace generated by the family. If $F \neq E$, then by the corollary of Proposition 4 there exists a vector $y \neq 0$ which is orthogonal to F and in particular to every element of the family. Thus enlarging the original family by $y/\|y\|$, we would still obtain an orthonormal family, which contradicts the hypothesis of maximality. ∎

REMARK 2. A maximal orthonormal family in a Hilbert space is also called an *orthonormal basis*. It should not be confused with the notion of algebraic basis (§2). It can be shown that two orthogonal bases of a Hilbert space have the same cardinal. This common cardinal is the *Hilbert dimension* of the space and should not be confused with the algebraic dimension (§2).

DEFINITION 6. *Let* $(a_\iota)_{\iota\in I}$ *be an orthonormal family in a Hilbert space* E. *For* $x \in E$ *the numbers* $(x \mid a_\iota)$ $(\iota \in I)$ *are called the Fourier coefficients of* x *with respect to* (a_ι). *The expression* $\sum_{\iota\in I} (x \mid a_\iota)a_\iota$ *is called the Fourier expansion of* x.

DEFINITION 7. *A family* $(x_\iota)_{\iota\in I}$ *of elements in a normed vector space is summable to an element* x *if for any* $\epsilon > 0$ *there exists a finite subset* J *of* I *such that* $\|x - \sum_{\iota\in H} x_\iota\| < \epsilon$ *for any finite subset* H *of* I *satisfying* $H \supset J$. *In this case, we write* $x = \sum_{\iota\in I} x_\iota$.

PROPOSITION 8. (a) *A family* $(x_\iota)_{\iota\in I}$ *of elements in a Banach space is summable if and only if for every* $\epsilon > 0$ *there exists a finite subset* J *of* I *such that* $\|\sum_{\iota\in H} x_\iota\| < \epsilon$ *for any finite subset* H *of* I, *verifying* $H \cap J = \emptyset$.

(b) *A family* $(\alpha_\iota)_{\iota\in I}$ *of positive real numbers is summable if and only if there exists a number* β *such that* $\sum_{\iota\in H} \alpha_\iota \leqq \beta$ *for all finite subsets* H *of* I. *In this case, we have also* $\sum_{\iota\in I} \alpha_\iota \leqq \beta$.

Proof. (a) If $(x_\iota)_{\iota\in I}$ is summable, then for $\epsilon > 0$ we have a finite subset J of I such that

$$\left\|x - \sum_{\iota\in J} x_\iota\right\| < \tfrac{1}{2}\epsilon \quad \text{and} \quad \left\|x - \sum_{\iota\in J\cup H} x_\iota\right\| < \tfrac{1}{2}\epsilon$$

whenever $H \cap J = \emptyset$. Thus $\|\sum_{\iota\in H} x_\iota\| < \epsilon$.

Conversely, let the condition of the proposition be satisfied, and let $(\epsilon_n)_{n\in\mathbf{N}}$ be a sequence of positive numbers tending to zero. To every ϵ_n there corresponds a finite subset J_n of I such that $\|\sum_{\iota\in H} x_\iota\| < \epsilon_n$ if $H \cap J_n = \emptyset$, and we may suppose that $J_n \subset J_{n+1}$ $(n \in \mathbf{N})$. Set $y_n = \sum_{\iota\in J_n} x_\iota$. Then $(y_n)_{n\in\mathbf{N}}$ is a Cauchy sequence. Indeed,

$$y_m - y_n = \sum_{\iota\in J_m\cap\complement J_n} x_\iota$$

if $m \geqq n$, and thus $\|y_m - y_n\| < \epsilon_n$ for $m \geqq n$. Hence $(y_n)_{n\in\mathbf{N}}$ converges to a vector x, and we have $\|x - y_n\| \leqq \epsilon_n$. If H is a finite subset of I such that $H \supset J_n$, then

$$\left\|x - \sum_{\iota\in H} x_\iota\right\| \leqq \left\|x - \sum_{\iota\in J_n} x_\iota\right\| + \left\|\sum_{\iota\in H\cap\complement J_n} x_\iota\right\| < \epsilon_n + \epsilon_n = 2\epsilon_n.$$

Thus (x_ι) is summable to x.

(b) Let λ be the least upper bound of all expressions $\sum_{\iota\in H} \alpha_\iota$, where H is any finite subset of I. For any $\epsilon > 0$ there exists a finite subset J of I such that $\lambda - \epsilon < \sum_{\iota\in J} \alpha_\iota$; and thus, since the α_ι are positive,

$$\lambda - \epsilon < \sum_{\iota\in H} \alpha_\iota \leqq \lambda$$

for all finite subsets H of I containing J. Hence (α_ι) is summable to λ.

If (α_ι) is summable to a number λ, then there exists a finite subset J of I such that $|\lambda - \sum_{\iota \in H} \alpha_\iota| < 1$ for any finite subset H of I such that $H \supset J$. It follows clearly that

$$\sum_{\iota \in H} \alpha_\iota < 1 + \lambda$$

for any finite subset H of I. ▮

PROPOSITION 9 (Bessel's inequality). *Let $(a_\iota)_{\iota \in I}$ be an orthonormal family in an inner product space E. Then for every $x \in E$ we have*

$$\sum_{\iota \in I} |(x \mid a_\iota)|^2 \leqq \|x\|^2.$$

Proof. For any finite subset H of I we have

$$\begin{aligned} 0 &\leqq \Big\| x - \sum_{\iota \in H} (x \mid a_\iota) a_\iota \Big\|^2 \\ &= \Big(x - \sum_{\iota \in H} (x \mid a_\iota) a_\iota \mid x - \sum_{\iota \in H} (x \mid a_\iota) a_\iota \Big) \\ &= \|x\|^2 - \sum_{\iota \in H} (x \mid a_\iota)(a_\iota \mid x) - \sum_{\iota \in H} (x \mid a_\iota)^2 + \sum_{\iota \in H} (x \mid a_\iota)^2 \\ &= \|x\|^2 - \sum_{\iota \in H} |(x \mid a_\iota)|^2. \end{aligned}$$

The result now follows from Proposition 8(b). ▮

COROLLARY. *In a Hilbert space E, for any $x \in E$ the family $((x \mid a_\iota)a_\iota)_{\iota \in I}$ is summable.*

Indeed, for any finite subset H of I we have

$$\Big\| \sum_{\iota \in H} (x \mid a_\iota) a_\iota \Big\|^2 = \Big(\sum_{\iota \in H} (x \mid a_\iota) a_\iota \mid \sum_{\iota \in H} (x \mid a_\iota) a_\iota \Big) = \sum_{\iota \in H} |(x \mid a_\iota)|^2.$$

The corollary now follows by using Proposition 8(a). ▮

PROPOSITION 10. *Let $(x_\iota)_{\iota \in I}$ be a summable family in an inner product space E, and let $y \in E$. Then the family $((x_\iota \mid y))_{\iota \in I}$ is summable and*

$$\Big(\sum_{\iota \in I} x_\iota \mid y \Big) = \sum_{\iota \in I} (x_\iota \mid y).$$

Proof. Set $x = \sum_{\iota \in I} x_\iota$. Given $\epsilon > 0$, there exists a finite subset J of I such that

$$\Big\| x - \sum_{\iota \in H} x_\iota \Big\| < \frac{\epsilon}{\|y\|}$$

for every finite subset H of I verifying $H \supset J$. But then

$$\left|(x \mid y) - \sum_{\iota \in H} (x_\iota \mid y)\right| = \left|\left(x - \sum_{\iota \in H} x_\iota \mid y\right)\right| \leqq \left\|x - \sum_{\iota \in H} x_\iota\right\| \cdot \|y\| < \epsilon$$

whenever H contains J. ▮

PROPOSITION 11. *Let $(a_\iota)_{\iota \in I}$ be an orthonormal family in a Hilbert space E. The following conditions are equivalent:*

(a) *The family $(a_\iota)_{\iota \in I}$ is maximal.*

(b) *For every $x \in E$ we have $x = \sum_{\iota \in I} (x \mid a_\iota) a_\iota$.*

(c) *For every $x \in E$ we have $\|x\|^2 = \sum_{\iota \in I} |(x \mid a_\iota)|^2$ (Parseval's relation).*

Proof. (a) $\Rightarrow$ (b). Let $y = \sum_{\iota \in I} (x \mid a_\iota) a_\iota$ (this vector exists by the corollary of Proposition 9). Using Proposition 10, we have $(y \mid a_\iota) = (x \mid a_\iota)$ for every $\iota \in I$, i.e., $x - y$ is orthogonal to every a_ι. Since $(a_\iota)_{\iota \in I}$ is maximal, we have $x = y$.

(b) $\Rightarrow$ (c). If $x = \sum_{\iota \in I} (x \mid a_\iota) a_\iota$, then by Proposition 10

$$\|x\|^2 = \left(\sum_{\iota \in I} (x \mid a_\iota) a_\iota \mid x\right) = \sum_{\iota \in I} |(x \mid a_\iota)|^2.$$

(c) $\Rightarrow$ (a). If $(a_\iota)_{\iota \in I}$ is not maximal, then there exists an $x \neq 0$ which is orthogonal to every a_ι $(\iota \in I)$. But then $\|x\|^2 \neq \sum_{\iota \in I} |(x \mid a_\iota)|^2$, since the last expression is 0. ▮

EXERCISES

1. Why is the subspace F of Example 2 closed in $\mathcal{C}([0, 2\pi])$?

2. Prove that the family $(x^n)_{n \in \mathbf{N}}$ is algebraically free in $\mathcal{C}(I)$ (cf. Example 4).

3. (a) Prove that the unit vectors e_n (see Example 2.2) form a maximal orthonormal family in l^2.

(b) Set

$$a_0 = e_0, \qquad a_k = e_0 + \frac{1}{k} e_k \qquad \text{for} \qquad k \in \mathbf{N}^*.$$

Prove that for any $n \in \mathbf{N}$ the family $(a_k)_{0 \leqq k \leqq n}$ is topologically free, but the family $(a_k)_{k \in \mathbf{N}}$ is not topologically free.

4. A Banach space is said to be *uniformly convex* if for every ϵ such that $0 < \epsilon < 2$ there exists a $\delta > 0$ such that the relations $\|x\| \leqq 1$, $\|y\| \leqq 1$, $\|x - y\| \geqq \epsilon$ imply

$$\left\|\frac{x + y}{2}\right\| \leqq 1 - \delta.$$

(a) Show that a Hilbert space is uniformly convex. (*Hint:* Use the parallelogram law.)

(b) Show that Proposition 3 also holds in a uniformly convex Banach space.

(c) Show that if $2 \leqq p < \infty$ and α, β are any two complex numbers, then

$$|\alpha + \beta|^p + |\alpha - \beta|^p \leqq 2^{p-1}(|\alpha|^p + |\beta|^p).$$

(d) Prove that for $2 \leqq p < \infty$ the Banach space l^p is uniformly convex.

REMARK 3. It can be proved that l^p is uniformly convex for $1 < p < \infty$ ([52], §26, 7.(12)).

5. Let $(x_\iota)_{\iota \in I}$ and $(y_\iota)_{\iota \in I}$ be two summable families in a normed space.

(a) Show that at most countably many x_ι are different from zero.

(b) Show that the family $(x_\iota + y_\iota)_{\iota \in I}$ is summable and that

$$\sum_{\iota \in I} (x_\iota + y_\iota) = \sum_{\iota \in I} x_\iota + \sum_{\iota \in I} y_\iota.$$

(c) Show that for any $\lambda \in \mathbf{K}$ the family $(\lambda x_\iota)_{\iota \in I}$ is summable and that

$$\lambda \sum_{\iota \in I} x_\iota = \sum_{\iota \in I} \lambda x_\iota.$$

6. Let $(a_\iota)_{\iota \in I}$ be a maximal orthonormal family in a Hilbert space E. Let x and y be two elements of E, and $\sum_{\iota \in I} \xi_\iota a_\iota$ and $\sum_{\iota \in I} \eta_\iota a_\iota$ their Fourier expansions. Show that

$$(x \mid y) = \sum_{\iota \in I} \xi_\iota \bar{\eta}_\iota.$$

7. Let I be a set of indices. Let $L^2(I)$ be the set of all families $(\xi_\iota)_{\iota \in I}$ of (real or complex) numbers such that the family $(|\xi_\iota|^2)_{\iota \in I}$ is summable. Show that $L^2(I)$ is a Hilbert space, where the scalar product is given by

$$((\xi_\iota) \mid (\eta_\iota)) = \sum_{\iota \in I} \xi_\iota \bar{\eta}_\iota.$$

Show that every Hilbert space is isometrically isomorphic with some space $L^2(I)$. [It follows from Remark 2 that two spaces $L^2(I)$ and $L^2(K)$ are isometrically isomorphic if and only if I and K are equipotent.]

§4. *Linear maps*

Let E and F be two vector spaces over the same field of scalars $\mathbf{K}$. We recall that a map f from E into F is said to be *linear* if it satisfies the identity

$$f(\lambda a + \mu b) = \lambda f(a) + \mu f(b)$$

for all $a, b \in E$ and $\lambda, \mu \in \mathbf{K}$.

If $f: E \to F$ is a linear map from a vector space E into a vector space F, then the set of all those vectors $x \in E$ for which $f(x) = 0$ is a linear subspace of E, called the *kernel* of f and denoted by $f^{-1}(0)$ or $\operatorname{Ker}(f)$. The map f is injective if and only if $\operatorname{Ker}(f)$ is reduced to 0.

If f is a continuous linear map from a normed space E into the normed space F, then $\operatorname{Ker}(f)$ is a closed subspace of E, since it is the inverse image of the closed subset $\{0\}$ of F.

PROPOSITION 1. *If a linear map f from a normed vector space E into a normed vector space F is continuous at the origin, then it is continuous everywhere and even uniformly continuous.*

Proof. Clearly, f maps the origin of E into the origin of F. Let us assume that for every $\epsilon > 0$ there exists an $\alpha > 0$ such that $\|x\| < \alpha$ implies $\|f(x)\| < \epsilon$. But then $\|x - y\| < \alpha$ implies $\|f(x) - f(y)\| = \|f(x - y)\| < \epsilon$. ∎

PROPOSITION 2. *A linear map f from a normed space E into a normed space F is continuous if and only if there exists a positive number M such that $\|f(x)\| \leqq M\|x\|$ for every $x \in E$.*

Proof. Suppose that such an M exists. Then for $\epsilon > 0$ we set $\alpha = \epsilon/M$; thus $\|x\| < \alpha$ implies $\|f(x)\| < M \cdot \alpha = \epsilon$. By Proposition 1 the map f is then continuous. For the converse it will be instructive to give two proofs, each using a different definition of continuity.

(1) If f is continuous, then there exists a $\gamma > 0$ such that $\|x\| \leqq \gamma$ implies $\|f(x)\| \leqq 1$. We can take $M = 1/\gamma$, for if $x \in E$, then

$$\left\|\frac{\gamma x}{\|x\|}\right\| \leqq \gamma,$$

and thus

$$\left\|f\left(\frac{\gamma x}{\|x\|}\right)\right\| = \frac{\gamma}{\|x\|}\|f(x)\| \leqq 1,$$

i.e., $\|f(x)\| \leqq (1/\gamma)\|x\|$.

(2) Suppose that f is continuous, but that there exists no M satisfying the condition of the theorem. Then for every $n \in \mathbf{N}^*$ there exists $x_n \in E$ such that $\|f(x_n)\| \geqq n\|x_n\|$. Set $y_n = x_n/n\|x_n\|$. Then $\|y_n\| = 1/n$ and (y_n) tends to zero, but $\|f(y_n)\| \geqq 1$, i.e., $(f(y_n))$ does not tend to zero, which contradicts the continuity of f. ∎

DEFINITION 1. *A bijective continuous linear map f from a normed space E onto a normed space F is an isomorphism if the inverse map f^{-1} is continuous. An injective continuous linear map f from a normed space E into a normed space F is a strict morphism* (or topological homomorphism) *if it is an isomorphism from E onto $f(E) = \operatorname{Im}(f)$.*

To make this definition meaningful, it should be observed that if $f\colon E \to F$ is a linear map from a vector space E into a vector space F, then $f(E) = \operatorname{Im}(f)$ is a linear subspace of F. If, furthermore, f is bijective, then f^{-1} is a linear map from F onto E.

A continuous linear map f from a normed space E into a normed space F is an injective strict morphism if and only if there exists a strictly positive number m such that $\|f(x)\| \geqq m\|x\|$ for every $x \in E$. Indeed, if this relation is satisfied, then $f(x) = 0$ implies $x = 0$; i.e., f is injective. Furthermore, if we denote by g the inverse of the bijection $f: E \rightarrow f(E)$, then the relation is equivalent to $\|g(y)\| \leqq (1/m)\|y\|$ for every $y \in f(E)$, i.e., to the continuity of g (Proposition 2). In particular, a *surjective* linear map $f: E \rightarrow F$ is an isomorphism if and only if there exist two constants $0 < m \leqq M$ such that

$$m\|x\| \leqq \|f(x)\| \leqq M\|x\|$$

for every $x \in E$. If, furthermore, $m = M = 1$, then f is an isometric isomorphism (Definition 2.1).

We want to extend the notion of strict morphism also to continuous linear maps which are not injective, and for this we need the concept of the quotient space of a normed space modulo a closed subspace. Let E be a vector space over $\mathbf{K}$ and M a linear subspace of E. We say that two elements $x, y \in E$ are congruent modulo M, and write $x \equiv y \bmod M$, if $x - y \in M$. By the definition of a linear subspace, the congruence is an equivalence relation on E. We denote by E/M the quotient set of E modulo the equivalence relation defined by M. We can define a structure of vector space over $\mathbf{K}$ on E/M, by setting $\dot{x} + \dot{y} = (x + y)^{\bullet}$ and $\lambda\dot{x} = (\lambda x)^{\bullet}$, where for instance x is a representative in E of the equivalence class $\dot{x} \in E/M$ and $\lambda \in \mathbf{K}$. It is trivial to check that this definition of the algebraic operations in E/M is independent of the choice of the representatives x and y, and also that all axioms of a vector space are satisfied. In particular, M is an equivalence class in itself and is the zero vector of E/M. The quotient set E/M equipped with the vector space structure just defined is the *quotient vector space* (or simply quotient space) of E modulo M.

Now let E be a normed vector space and M a closed subspace of E. We define a norm on the quotient vector space E/M by setting

$$\|\dot{x}\| = \inf_{x \in \dot{x}} \|x\|, \tag{1}$$

and call it the *quotient norm*. Let us prove that Axioms (N 1), (N 2), (N 3) are satisfied. We first observe that every equivalence class $\dot{x}$ in E is a set of the form $\dot{x} = x + M$, where x is an arbitrary element of $\dot{x}$, and therefore $\dot{x}$ is a closed set in E. If now $\|\dot{x}\| = 0$, then 0 adheres to $\dot{x}$ in E and consequently belongs to it. Thus $\dot{x} = M$ or $\dot{x} = \dot{0}$, where $\dot{0}$ denotes the zero vector of E/M. We also have

$$\|\lambda\dot{x}\| = \inf_{x \in \dot{x}} \|\lambda x\| = |\lambda| \inf_{x \in \dot{x}} \|x\| = |\lambda| \cdot \|\dot{x}\|.$$

Next let $\epsilon > 0$ and $\dot{x}, \dot{y} \in E/M$. Then there exist $x \in \dot{x}$ and $y \in \dot{y}$ such that $\|x\| \leqq \|\dot{x}\| + \epsilon$ and $\|y\| \leqq \|\dot{y}\| + \epsilon$, and thus

$$\|x + y\| \leqq \|x\| + \|y\| \leqq \|\dot{x}\| + \|\dot{y}\| + 2\epsilon;$$

and since ϵ is arbitrary,

$$\|\dot{x} + \dot{y}\| = \inf \|x + y\| \leqq \|\dot{x}\| + \|\dot{y}\|.$$

PROPOSITION 3. *Let E be a Banach space and M a closed subspace of E. Then the quotient vector space E/M equipped with the quotient norm* (1) *is complete, i.e., a Banach space.*

Proof. Let $(\dot{x}_n)_{n \in \mathbf{N}}$ be a Cauchy sequence in E/M. Then there exists a subsequence $(\dot{x}_{n_k})$ of $(\dot{x}_n)$ such that $\|\dot{x}_{n_k} - \dot{x}_m\| < 2^{-k}$ for $m > n_k$, in particular that $\|\dot{x}_{n_{k+1}} - \dot{x}_{n_k}\| < 2^{-k}$ for every $k \in \mathbf{N}$, and we can find elements $x_{n_k} \in \dot{x}_{n_k}$ such that $\|x_{n_{k+1}} - x_{n_k}\| < 2^{-k}$. The sequence (x_{n_k}) is a Cauchy sequence, for if $l > k$, then

$$\begin{aligned}\|x_{n_l} - x_{n_k}\| &\leqq \|x_{n_l} - x_{n_{l-1}}\| + \cdots + \|x_{n_{k+1}} - x_{n_k}\| \\ &< 2^{-l+1} + \cdots + 2^{-k} < 2^{-k+1}.\end{aligned}$$

Thus (x_{n_k}) converges to some element x of E, and if $\dot{x}$ is the class of x modulo M, then the relation $\|\dot{x}_{n_k} - \dot{x}\| \leqq \|x_{n_k} - x\|$ shows that $(\dot{x}_{n_k})$ converges to $\dot{x}$ in E/M. For every $\epsilon > 0$ there exists an $N \in \mathbf{N}$ such that $\|\dot{x}_n - \dot{x}_m\| \leqq \epsilon/2$ if $n, m > N$. Thus if we choose $n > N$ and $n_k > N$, we have

$$\|\dot{x} - \dot{x}_n\| \leqq \|\dot{x} - \dot{x}_{n_k}\| + \|\dot{x}_{n_k} - \dot{x}_n\| \leqq \tfrac{1}{2}\epsilon + \tfrac{1}{2}\epsilon = \epsilon;$$

i.e., the sequence $(\dot{x}_n)$ converges to $\dot{x}$ and the space E/M is complete. ∎

Let E be a vector space and M a linear subspace of E. The canonical surjection φ of E onto E/M which associates with every element $x \in E$ its equivalence class $\varphi(x) = \dot{x}$ modulo M is a linear map by the very definition of the algebraic operations in E/M. If f is a linear map from E into another vector space F such that $f(x) = 0$ for all $x \in M$ [i.e., such that $\mathrm{Ker}(f) \supset M$], then we can define a map $\bar{f}$ from E/M into F by setting $\bar{f}(\dot{x}) = f(x)$ for $\dot{x} \in E/M$, where $x \in E$ is an element representing $\dot{x}$. If y is another element representing $\dot{x}$, then $x - y \in M$, $f(x - y) = 0$, and $f(x) = f(y)$, which proves that $\bar{f}$ is well-defined. It is clear that $\bar{f}$ is linear and that the definition of $\bar{f}$ is equivalent to the relation $f = \bar{f} \circ \varphi$. If we choose in particular $M = \mathrm{Ker}(f)$, then $\bar{f}$ is injective and will be called the *injection associated with f.*

Now let E be a normed vector space and M a closed subspace of E. The canonical surjection $\varphi: E \to E/M$ is continuous, since by (1) we have $\|\varphi(x)\| \leqq \|x\|$. If f is a continuous linear map from E into another normed space F such that $f(x) = 0$ for all x belonging to the closed subspace M of E, then the map $\bar{f}: E/M \to F$ defined by $f = \bar{f} \circ \varphi$ is also continuous. Indeed, we have $\|f(x)\| \leqq \mu\|x\|$ for some $\mu \geqq 0$ and all $x \in E$ (Proposition 2), and therefore $\|\bar{f}(\dot{x})\| \leqq \mu\|x\|$ for all $x \in \dot{x}$. Hence by (1) we obtain $\|\bar{f}(\dot{x})\| \leqq \mu\|\dot{x}\|$, $\dot{x} \in E/M$, which proves the continuity of $\bar{f}$. In particular, the injection $\bar{f}$: $E/\mathrm{Ker}(f) \to F$ associated with f is continuous.

Definition 2. *A continuous linear map f from a normed space E into a normed space F is a strict morphism* (or topological homomorphism) *if the associated injection $\bar{f}$: $E/\mathrm{Ker}(f) \to F$ is a strict morphism* (Definition 1).

If E is a normed space and M a closed subspace of E, then the canonical surjection φ: $E \to E/M$ is a strict morphism since the injection associated with φ is precisely the identical isomorphism of E/M onto itself. If V is a neighborhood of 0 in E, then $\varphi(V)$ is a neighborhood of 0 in E/M. Indeed, there exists an $\epsilon > 0$ such that V contains all x which satisfy $\|x\| < \epsilon$. If $\dot{x} \in E/M$ is such that $\|\dot{x}\| < \epsilon$, then there exists an $x \in \dot{x}$ for which $\|x\| < \epsilon$, i.e., this x belongs to V and $\dot{x} = \varphi(x) \in \varphi(V)$. Consequently $\varphi(V)$ contains all $\dot{x}$ such that $\|\dot{x}\| < \epsilon$; i.e., $\varphi(V)$ is a neighborhood of 0 in E/M. If A is an open set in E, then $\varphi(A)$ is an open set in E/M. Indeed, if $\dot{x} \in \varphi(A)$, then there exists a point $x \in A$ such that $\dot{x} = \varphi(x)$. Since A is open, there exists a neighborhood V of x satisfying $V \subset A$. But then $\varphi(V)$ is a neighborhood of $\dot{x}$ satisfying $\varphi(V) \subset \varphi(A)$; i.e., $\varphi(A)$ is open.

Proposition 4. *If f is a continuous linear map from a normed space E into a normed space F, then the following conditions are equivalent:*

(a) *f is a strict morphism,*
(b) *f maps every neighborhood of 0 in E onto a neighborhood of 0 in $f(E)$,*
(c) *f maps every open set of E onto an open set of $f(E)$.*

Proof. Let N be the kernel of f, φ the canonical surjection of E onto E/N and $\bar{f}$: $E/N \to F$ the injection associated with f. We have $f = \bar{f} \circ \varphi$. Let g be the inverse map of the bijection $\bar{f}$: $E/N \to f(E)$.

First assume that f is a strict morphism. If V is a neighborhood of 0 in E, then $W = \varphi(V)$ is a neighborhood of 0 in E/N. Since g is continuous, $g^{-1}(W) = \bar{f}(W) = f(V)$ is a neighborhood of 0 in $f(E)$. Thus (a) $\Rightarrow$ (b). If A is an open set in E, then $B = \varphi(A)$ is an open set in E/N, and $g^{-1}(B) = \bar{f}(B) = f(A)$ is an open set in $f(E)$. Thus (a) $\Rightarrow$ (c).

Conversely, suppose that (b) is satisfied. Let W be a neighborhood of 0 in E/N. Then $V = \varphi^{-1}(W)$ is a neighborhood of 0 in E, and by (b) $f(V) = \bar{f}(W)$ is a neighborhood of 0 in $f(E)$. Thus g is continuous; i.e., f is a strict morphism and (b) $\Rightarrow$ (a). Similarly, let (c) be satisfied. If B is an open set in E/N, then $\varphi^{-1}(B) = A$ is an open set in E, and by (c) $\bar{f}(B) = f(A)$ is an open set in $f(E)$. Thus g is continuous; i.e., (c) $\Rightarrow$ (a). ∎

Remark 1. Proposition 4 shows why a strict morphism is sometimes called a *relatively open* (continuous linear) *map* and an *open map* if it is surjective.

Exercises

1. Supply complete proofs of all the algebraic statements made without proof.

2. Let c_0 and l^1 be the Banach spaces introduced in Examples 6 and 7 of §1. Clearly l^1 is a subset of c_0. Show that the canonical injection of l^1 into c_0 is a continuous linear map but not a strict morphism.

3. Prove that if f is a linear map from a normed vector space E into a normed vector space F, which maps every bounded set of E into a bounded set of F, then there exists a positive number M such that $\|f(x)\| \leqq M\|x\|$, and in particular f is continuous. (*Hint:* Use a device similar to that employed in part (2) of the proof of Proposition 2.)

4. Let E and F be two normed vector spaces over the same field $\mathbf{K}$. Show that the operations

$$(x, y) + (x_1, y_1) = (x + x_1, y + y_1), \qquad \lambda(x, y) = (\lambda x, \lambda y),$$

where x, $x_1 \in E$, y, $y_1 \in F$, $\lambda \in \mathbf{K}$, define a structure of vector space on the cartesian product $E \times F$ of the two spaces. Show that the map

$$(x, y) \mapsto (\|x\|^2 + \|y\|^2)^{1/2}$$

is a norm on $E \times F$. Show that if E and F are Banach spaces, then $E \times F$ is a Banach space. Show that if E and F are Hilbert spaces, then the scalar product corresponding to this norm defines a structure of Hilbert space on $E \times F$. Show, furthermore, that the projection $(x, y) \mapsto x$ from $E \times F$ onto E is continuous.

5. Let $(H_n)_{n\in\mathbf{N}}$ be a sequence of Hilbert spaces. Let H be the set of all sequences $(x_n)_{n\in\mathbf{N}}$, where $x_n \in H_n$ and $\sum_{n=0}^{\infty} \|x_n\|^2 < \infty$. Clearly, H is a vector space in a natural way. Show that the scalar product

$$(x \mid y) = \sum_{n=0}^{\infty} (x_n \mid y_n)$$

defines a structure of Hilbert space on H.

§5. *Linear forms*

Let E be a vector space over the field $\mathbf{K}$. A *linear form* (or linear functional) on E is a map f from E into $\mathbf{K}$ which satisfies the identity

$$f(\lambda a + \mu b) = \lambda f(a) + \mu f(b)$$

for all a, $b \in E$ and λ, $\mu \in \mathbf{K}$. We can consider $\mathbf{K}$ as a vector space over itself. Indeed, the operations $(x, y) \mapsto x + y$ and $(\lambda, x) \mapsto \lambda x$ are defined if $x, y, \lambda \in \mathbf{K}$ and satisfy Axioms (VS 1) through (VS 8). Furthermore, $\mathbf{K}$ has dimension 1 over itself, since any nonzero element constitutes a basis. A linear form on E is then a linear map from E into the vector space $\mathbf{K}$ and thus we can apply our results concerning linear maps to obtain results on linear forms.

Equipped with the absolute value as norm, the vector space $\mathbf{K}$ is a Banach space over $\mathbf{K}$; it is simply the Banach space $\mathbf{R}^1$ or $\mathbf{C}^1$ introduced in Examples 1 and 3 of §1. Therefore, if we want to consider $\mathbf{K}$ as a Banach space over itself, we shall denote it by $\mathbf{K}^1$. It follows from Proposition 4.2 that a linear form f on a normed space E is continuous if and only if there exists a positive number M such that $|f(x)| \leqq M\|x\|$ for every $x \in E$.

A maximal proper linear subspace H of a vector space E is called a *hyperplane*. In other words, a linear subspace H of E is a hyperplane if $H \neq E$ and if for any linear subspace M of E such that $H \subset M \subset E$ we have either $M = H$ or $M = E$.

A linear subspace H of a vector space E is a hyperplane if and only if the quotient space E/H is one-dimensional. In fact, if H is a hyperplane and $a \notin H$, then the set $\{x \mid x = h + \lambda a, h \in H, \lambda \in \mathbf{K}\}$ is a linear subspace of E, containing H but different from it, and thus necessarily equal to E. Thus the image $\mathring{a}$ of a in E/H generates E/H; i.e., E/H is one-dimensional. Conversely, assume that E/H is generated by the element $\mathring{a}$, and let $a \in E$ be a representative of $\mathring{a}$. Then every element $x \in E$ can be written in the form $x = h + \lambda a$ with $h \in H$ and $\lambda \in \mathbf{K}$. If M is a linear subspace of E containing H, then either $a \notin M$, in which case $M = H$, or $a \in M$, in which case $M = E$. This proves that H is a hyperplane.

It follows from the previous discussion that if H is a hyperplane in E, and a a vector in E which does not belong to H, then every $x \in E$ can be represented in the form $x = h + \lambda a$, where $h \in H$ and $\lambda \in \mathbf{K}$. This representation is unique, since if $h + \lambda a = h' + \lambda' a$, then $(\lambda - \lambda')a = h' - h \in H$. Therefore, $\lambda - \lambda' = 0$ and consequently $\lambda = \lambda'$, $h = h'$.

If f is a not identically zero linear form on E, then $\operatorname{Ker}(f)$ is a hyperplane. Indeed, f is clearly surjective, and therefore the associated injection $\bar{f}: E/\operatorname{Ker}(f) \to \mathbf{K}$ is a bijective linear map, i.e., an isomorphism of the two vector spaces in the sense of algebra. Since $\mathbf{K}$ is one-dimensional, so is $E/\operatorname{Ker}(f)$.

Conversely, if H is a hyperplane, there exists a not identically zero linear form f on E such that $H = \operatorname{Ker}(f)$; we say that H has *equation* $f(x) = 0$. In fact, let a be a fixed vector in E not belonging to H, and for $x \in E$ let us write $x = h + \lambda a$, where $h \in H$, $\lambda \in \mathbf{K}$. Then we can define f by $f(x) = \lambda$.

Finally, let us observe that if f and g are two not identically zero linear forms on E such that $f(x) = 0$ and $g(x) = 0$ are the equations of the same hyperplane H, then there exists a nonzero scalar α such that $f = \alpha g$; i.e., $f(x) = \alpha \cdot g(x)$ for all $x \in E$. Indeed, if $a \notin H$, we can choose $\alpha = f(a)/g(a)$.

If f is a continuous linear form on a normed vector space E, then $\operatorname{Ker}(f)$ is a *closed hyperplane*. In fact, we have just seen that it is a hyperplane, and we observed in the previous section that the kernel of a continuous linear map is a closed subspace.

We define the *norm* $\|f\|$ of a continuous linear form f as the greatest lower bound of all numbers M for which the relation $|f(x)| \leqq M\|x\|$ is satisfied. In other words,

$$\|f\| = \sup_{\substack{x\in E\\ x\neq 0}} \frac{|f(x)|}{\|x\|} = \sup_{\substack{\|x\|=1\\ x\in E}} |f(x)|.$$

The equality of the last two expressions follows immediately from the fact that for any $x \in E$ the norm of $x/\|x\|$ is 1. We have, furthermore,

$$\|f\| = \sup_{\|x\|\leqq 1} |f(x)|, \tag{1}$$

since

$$\|f\| = \sup_{\|x\|=1} |f(x)| \leqq \sup_{\|x\|\leqq 1} |f(x)| = \sup_{\substack{\|x\|\leqq 1\\ x\neq 0}} \|x\| \cdot \left| f\left(\frac{x}{\|x\|}\right)\right| \leqq \sup_{\|x\|=1} |f(x)|.$$

Now let E be an inner product space and a a fixed vector in E. The map $f\colon x \mapsto (x \mid a)$ from E into $\mathbf{K}$ is linear by Axiom (SP 4) and continuous, since by Proposition 1.1 we have $|(x \mid a)| \leqq \|a\| \cdot \|x\|$. We also see that $\|f\| \leqq \|a\|$, but for $x = a$ we get $f(a) = \|a\|^2$, and thus $\|f\| = \|a\|$. For a Hilbert space we shall now prove the converse of this result:

PROPOSITION 1 (Fréchet-Riesz). *Let E be a Hilbert space and f a continuous linear form on E. Then there exists a unique vector $a \in E$ such that $f(x) = (x \mid a)$ for all $x \in E$.*

Proof. If f is identically zero, we choose $a = 0$. If f is not identically zero, then its kernel H is distinct from E, and by the corollary to Proposition 3.4 there exists a nonzero vector $b \in E$ which is orthogonal to H. Set $a = b\overline{f(b)}/\|b\|^2$. Then

$$f(a) = \|a\|^2 = \frac{|f(b)|^2}{\|b\|^2}.$$

Since H is a hyperplane, every vector $x \in E$ can be written in the form $y + \lambda a$, where $y \in H$ and $\lambda \in \mathbf{K}$. Thus

$$f(x) = f(y + \lambda a) = \lambda f(a) = \lambda\|a\|^2 = (\lambda a \mid a) = (y + \lambda a \mid a) = (x \mid a).$$

The uniqueness of a is clear. ∎

REMARK 1. We do not need to use explicitly the fact that H is a hyperplane. Indeed, if z is a vector orthogonal to H, then $f(a)z - f(z)a$ belongs to H and is also orthogonal to it; i.e., $z = f(z)a/f(a)$. Using Proposition 3.4, we get the decomposition $x = y + \lambda a$, where $\lambda = f(z)/f(a)$.

If E is a vector space over a field $\mathbf{K}$, we say that a map f from E into $\mathbf{K}$ is a *semi-linear form* if

$$f(\lambda a + \mu b) = \bar{\lambda} f(a) + \bar{\mu} f(b)$$

for all $a, b \in E$ and $\lambda, \mu \in \mathbf{K}$. This is, of course, a new concept only in the case $\mathbf{K} = \mathbf{C}$; for $\mathbf{K} = \mathbf{R}$ a semi-linear form is simply linear. Now let E and F be two vector spaces over the same field $\mathbf{K}$. A map $(x, y) \mapsto B(x, y)$ from $E \times F$ into $\mathbf{K}$ is called a *sesquilinear form* if for every $y \in F$ the map $x \mapsto B(x, y)$ is a linear form on E, and for every $x \in E$ the map $y \mapsto B(x, y)$ is a semi-linear form on F. If, instead, $y \mapsto B(x, y)$ is a linear form, then we say that $(x, y) \mapsto B(x, y)$ is a *bilinear form*. Of course, if $\mathbf{K} = \mathbf{R}$, then sesquilinear and bilinear forms are the same.

PROPOSITION 2. *Let E and F be two normed spaces over the same field. A sesquilinear form B on $E \times F$ is continuous if and only if there exists a positive number M such that $|B(x, y)| \leqq M\|x\| \cdot \|y\|$ for every $x \in E$, $y \in F$.*

Proof (cf. the proof of Proposition 2.1). Suppose that such an $M > 0$ exists. Let (a, b) be a point of $E \times F$ and $\epsilon > 0$. We have the identity

$$B(x, y) - B(a, b) = B(x - a, y - b) + B(x - a, b) + B(a, y - b);$$

hence $|B(x, y) - B(a, b)| < \epsilon$, provided

$$\|x\| < \sqrt{\epsilon/M}, \qquad \|y\| < \sqrt{\epsilon/M}, \qquad \text{if } a = b = 0,$$

$$\|x\| < \min\left(\sqrt{\frac{\epsilon}{2M}}, \frac{\epsilon}{2M\|b\|}\right), \quad \|y - b\| < \sqrt{\frac{\epsilon}{2M}}, \quad \text{if } a = 0, b \neq 0,$$

$$\|x - a\| < \sqrt{\frac{\epsilon}{2M}}, \quad \|y\| < \min\left(\sqrt{\frac{\epsilon}{2M}}, \frac{\epsilon}{2M\|a\|}\right), \quad \text{if } a \neq 0, b = 0,$$

$$\|x - a\| < \min\left(\sqrt{\frac{\epsilon}{3M}}, \frac{\epsilon}{3M\|b\|}\right), \quad \|y - b\| < \min\left(\sqrt{\frac{\epsilon}{3M}}, \frac{\epsilon}{3M\|a\|}\right),$$
$$\text{if } a \neq 0, b \neq 0.$$

Conversely, if B is continuous, then there exist $\alpha > 0$, $\beta > 0$ such that $\|x\| \leqq \alpha$, $\|y\| \leqq \beta$ imply $|B(x, y)| \leqq 1$. Then we can take $M = 1/\alpha\beta$, for if $x \in E$, $y \in F$, then

$$\left\|\frac{\alpha x}{\|x\|}\right\| \leqq \alpha, \qquad \left\|\frac{\beta y}{\|y\|}\right\| \leqq \beta$$

and thus

$$\left|B\left(\frac{\alpha x}{\|x\|}, \frac{\beta y}{\|y\|}\right)\right| = \frac{\alpha\beta}{\|x\| \cdot \|y\|} |B(x, y)| \leqq 1;$$

that is,

$$|B(x, y)| \leqq \frac{1}{\alpha\beta} \|x\| \cdot \|y\|. \blacksquare$$

If E is an inner product space, then the scalar product is a continuous sesquilinear form on $E \times E$. More generally, if f is a continuous linear map from E into E satisfying $\|f(x)\| \leqq M\|x\|$ for every $x \in E$, then $(x, y) \mapsto (x \mid f(y))$ is a sesquilinear form on $E \times E$ satisfying

$$|(x \mid f(y))| \leqq M \cdot \|x\| \cdot \|y\|.$$

For a Hilbert space E we shall now prove the converse of this statement as an easy corollary to Proposition 1.

PROPOSITION 3. *Let E be a Hilbert space and B a continuous sesquilinear form on $E \times E$. Then there exists a continuous linear map f from E into E such that $B(x, y) = (x \mid f(y))$ for all $x, y \in E$.*

Proof. For a fixed $y \in E$ the map $x \mapsto B(x, y)$ is a linear form on E. Thus by Proposition 1 there exists a unique vector $f(y)$ such that

$$B(x, y) = (x \mid f(y)).$$

The map $y \mapsto f(y)$ is clearly linear. Finally, we have

$$|B(x, y)| \leqq M\|x\| \cdot \|y\|,$$

or setting $x = f(y)$ and using the identity just proved,

$$\|f(y)\|^2 \leqq M\|f(y)\| \cdot \|y\|,$$

i.e., $\|f(y)\| \leqq M\|y\|$. ∎

PROPOSITION 4 (Lax-Milgram). *Let B be a continuous sesquilinear form on a Hilbert space E. Suppose that there exists an $m > 0$ such that $|B(x, x)| \geqq m\|x\|^2$ for every $x \in E$. Let f be a continuous linear form on E. Then there exists a unique vector $a \in E$ such that $f(x) = B(x, a)$ for all $x \in E$.*

Proof. By Proposition 3 we have $B(x, y) = (x \mid g(y))$ for all $x, y \in E$, where g is a continuous linear map from E into E. Let us show that g is an isomorphism. In the first place, g is an injective strict morphism, since

$$m\|x\|^2 \leqq |B(x, x)| = |(x \mid g(x))| \leqq \|x\| \cdot \|g(x)\|$$

implies

$$m\|x\| \leqq \|g(x)\|.$$

Next, $g(E)$ is a closed subspace of E. Indeed, if the sequence $y_n = g(x_n)$ converges to $y \in E$, then x_n is a Cauchy sequence in E and therefore converges to some element $x \in E$. But then, by the continuity of g, the sequence $g(x_n)$ converges to $g(x)$, and we have $y = g(x) \in g(E)$. Now let x be a vector orthogonal to $g(E)$. Then in particular we have $(x \mid g(x)) = 0$; hence $|(x \mid g(x))| \geqq m\|x\|^2$ implies $x = 0$. Thus by the corollary to Proposition 3.4 we have $g(E) = E$.

By Proposition 1 there exists $b \in E$ such that $f(x) = (x \mid b)$ for every $x \in E$. By what we have just proved, there exists $a \in E$ such that $b = g(a)$. But then $f(x) = (x \mid b) = (x \mid g(a)) = B(x, a)$ for all $x \in E$.

Finally, let $B(x, a) = B(x, a')$ for all $x \in E$. Then $B(x, a - a') = 0$ for all x, and in particular $0 = |B(a - a', a - a')| \geqq m\|a - a'\|$; i.e., $a = a'$. ∎

Exercises

1. Prove that a continuous linear form which is not identically zero is a surjective strict morphism onto $\mathbf{K}^1$.

2. Prove that if f is a map from a real normed space E into $\mathbf{R}$ which satisfies $f(a + b) = f(a) + f(b)$ for every $a \in E$, $b \in E$, and if f is bounded in some neighborhood of the zero vector in E, then f is a continuous linear form on E.

3. Prove that a sesquilinear or a bilinear form B defined on the product $E \times F$ of two normed spaces is continuous if it is continuous at the point $(0, 0)$.

§6. *The Hahn-Banach theorem*

We now arrive at the first of the three main theorems of the theory of normed vector spaces.

THEOREM 1. *Let E be a normed space and M a subspace of E. Let f be a continuous linear form defined on M. Then there exists a continuous linear form g on E which coincides with f on M and for which $\|g\| = \|f\|$.*

We first observe that we can suppose the subspace M to be closed. Indeed, by Proposition 4.1 if f is continuous on M, then it is uniformly continuous and thus can be extended to $\overline{M}$ (see §2), which according to Proposition 3.1 is a closed subspace of E. We know that the extended map $\bar{f}$ is continuous on $\overline{M}$. Let us show that it is linear. Given $x \in \overline{M}$ and $y \in \overline{M}$ for every $\epsilon > 0$, there exist $x' \in M$, $y' \in M$ such that $|\bar{f}(x) - f(x')| < \epsilon$, $|\bar{f}(y) - f(y')| < \epsilon$, $|\bar{f}(x + y) - f(x' + y')| < \epsilon$. Thus

$$\begin{aligned}
|\bar{f}(x) &+ \bar{f}(y) - \bar{f}(x + y)| \\
&= |\bar{f}(x) + \bar{f}(y) - \bar{f}(x + y) - f(x') - f(y') + f(x' + y')| \\
&\leqq |\bar{f}(x) - f(x')| + |\bar{f}(y) - f(y')| + |\bar{f}(x + y) - f(x' + y')| < 3\epsilon,
\end{aligned}$$

because $f(x' + y') = f(x') + f(y')$. Since ϵ was arbitrary, we have $\bar{f}(x + y) = \bar{f}(x) + \bar{f}(y)$. Similarly, if $x \in \overline{M}$, $\lambda \in \mathbf{K}$ and $\epsilon > 0$, there exists $x' \in M$ such that $|\bar{f}(x) - f(x')| < \epsilon$, and since $\lambda x'$ tends to λx as x' tends to x, we can choose x' such that also $|\bar{f}(\lambda x) - f(\lambda x')| \leqq \epsilon$. Then

$$\begin{aligned}
|\bar{f}(\lambda x) - \lambda \bar{f}(x)| &= |\bar{f}(\lambda x) - \lambda \bar{f}(x) - f(\lambda x') + \lambda f(x')| \\
&\leqq |\bar{f}(\lambda x) - f(\lambda x')| + |\lambda| \cdot |\bar{f}(x) - f(x')| \leqq (1 + |\lambda|)\epsilon.
\end{aligned}$$

Thus we have $\tilde{f}(\lambda x) = \lambda \tilde{f}(x)$. Finally, for any $x \in \overline{M}$ and $\epsilon > 0$ there exists an $x' \in M$ such that

$$|\tilde{f}(x)| < |f(x')| + \epsilon \leqq \|f\| \cdot \|x'\| + \epsilon \leqq \|f\| \cdot \|x\| + 2\epsilon;$$

thus $\|\tilde{f}\| \leqq \|f\|$. Since obviously $\|f\| \leqq \|\tilde{f}\|$, we have $\|\tilde{f}\| = \|f\|$.

LEMMA 1. *Let E be a normed vector space, z a nonzero element of E, and $L = \{\lambda z \mid \lambda \in \mathbf{K}\}$ the linear subspace generated by z* (the straight line through the origin and z). *Then L is a closed set.*

Proof. If the point x adheres to L, there exists a sequence $(\lambda_n z)$ which converges to x. From $\|\lambda_n z - \lambda_m z\| = |\lambda_n - \lambda_m| \cdot \|z\|$ it follows that (λ_n) is a Cauchy sequence. Since $\mathbf{K}$ is complete (λ_n) converges to some $\lambda \in \mathbf{K}$. But then by Proposition 2.1 the sequence $(\lambda_n z)$ tends to λz. Thus $x = \lambda z \in L$. ∎

LEMMA 2. *Let M be a closed subspace of the normed space E, and let z be a vector in E which does not belong to M. Let N be the closed subspace of E generated by M and by z* (Definition 3.1). *Then every element $x \in N$ can be written in a unique way in the form $x = y + \lambda z$, where $y \in M$, $\lambda \in \mathbf{K}$.*

Proof. The set $P = \{y + \lambda z \mid y \in M, \lambda \in \mathbf{K}\}$ is clearly the smallest linear subspace of E containing M and z. Thus to show that $P = N$, we have only to show that P is closed. Let φ be the canonical surjection from E onto E/M, and set $\dot{z} = \varphi(z)$. By Lemma 1 the straight line $L = \{\lambda \dot{z} \mid \lambda \in \mathbf{K}\}$ is a closed set in E/M, and therefore $P = \varphi^{-1}(L)$ is closed in E. ∎

The next lemma is the heart of the proof of the Hahn-Banach theorem.

LEMMA 3. *Let E be a real normed space and M, z and N as in Lemma 2. Let f be a continuous linear form on M. Then there exists a continuous linear form g on N which coincides with f on M and for which $\|g\| = \|f\|$.*

Proof. For $y_1, y_2 \in M$ we have

$$\begin{aligned} f(y_1) - f(y_2) &= f(y_1 - y_2) \leqq \|f\| \cdot \|y_1 - y_2\| \\ &= \|f\| \cdot \|y_1 + z - (y_2 + z)\| \\ &\leqq \|f\| \cdot \|y_1 + z\| + \|f\| \cdot \|y_2 + z\|; \end{aligned}$$

hence

$$-\|f\| \cdot \|y_2 + z\| - f(y_2) \leqq \|f\| \cdot \|y_1 + z\| - f(y_1). \tag{1}$$

Set

$$\xi = \sup_{y \in M} (-\|f\| \cdot \|y + z\| - f(y)),$$

$$\Xi = \inf_{y \in M} (\|f\| \cdot \|y + z\| - f(y)).$$

It follows from (1) that $\xi \leqq \Xi$. Let μ be any real number such that $\xi \leqq \mu \leqq \Xi$. We define

$$g(y + \lambda z) = f(y) + \lambda\mu;$$

g is obviously a linear form on N. We have furthermore, if $\lambda > 0$,

$$g(y + \lambda z) = f(y) + \lambda\mu = \lambda\{f(\lambda^{-1}y) + \mu\}$$
$$\begin{cases} \leqq \lambda\{f(\lambda^{-1}y) + \Xi\} \leqq |\lambda| \cdot \|f\| \cdot \|\lambda^{-1}y + z\| = \|f\| \cdot \|y + \lambda z\|, \\ \geqq \lambda\{f(\lambda^{-1}y) + \xi\} \geqq -|\lambda| \cdot \|f\| \cdot \|\lambda^{-1}y + z\| = -\|f\| \cdot \|y + \lambda z\|, \end{cases}$$

and if $\lambda < 0$,

$$g(y + \lambda z) = f(y) + \lambda\mu = \lambda\{f(\lambda^{-1}y) + \mu\}$$
$$\begin{cases} \geqq \lambda\{f(\lambda^{-1}y) + \Xi\} \geqq -|\lambda| \cdot \|f\| \cdot \|\lambda^{-1}y + z\| = -\|f\| \cdot \|y + \lambda z\|, \\ \leqq \lambda\{f(\lambda^{-1}y) + \xi\} \leqq |\lambda| \cdot \|f\| \cdot \|\lambda^{-1}y + z\| = \|f\| \cdot \|y + \lambda z\|. \end{cases}$$

It follows from these inequalities that

$$|g(y + \lambda z)| \leqq \|f\| \cdot \|y + \lambda z\|,$$

which proves the continuity of g and the relation $\|g\| \leqq \|f\|$. Since the relation $\|g\| \geqq \|f\|$ is trivial, we have the asserted equality. ∎

We are now in the position to prove Theorem 1 for the case of *real scalars*. Let $\mathfrak{F}$ be the set of all continuous linear forms h, defined on subspaces N of E containing M, which coincide with f on M and are such that $\|h\| = \|f\|$. We order $\mathfrak{F}$ by setting $h \leqq h'$ if the corresponding subspace N and N' satisfy $N \subset N'$ and if h' coincides with h on N. The set $\mathfrak{F}$ is inductive. Indeed, let (h_α) be a totally ordered family in $\mathfrak{F}$ and let N_α be the subspace on which h_α is defined. Set $N = \bigcup_\alpha N_\alpha$. If $x \in N$, then $x \in N_\alpha$ for some α and we define $h(x) = h_\alpha(x)$. Clearly, the definition of h is independent of the choice of α; and h is linear, since if $x \in N_\alpha$, $y \in N_\beta$, and for instance $N_\alpha \subset N_\beta$, then

$$h(x) + h(y) = h_\alpha(x) + h_\beta(y) = h_\beta(x) + h_\beta(y) = h_\beta(x + y) = h(x + y).$$

Furthermore, h coincides with f on M and verifies $\|h\| = \|f\|$. Thus h is the least upper bound of (h_α) in $\mathfrak{F}$.

Zorn's lemma proves the existence of a maximal element g in $\mathfrak{F}$. The subspace N on which g is defined is closed, since otherwise we could extend g to $\overline{N}$ by continuity. Furthermore, $N = E$, since otherwise we could, by Lemma 3, extend g to a subspace which contains N properly, in contradiction to the maximality of g. Hence g has all the properties required in Theorem 1.

Next we turn to the proof of Theorem 1 in the case of *complex scalars*. In the first place, if E is a vector space over $\mathbf{C}$, we can, by restricting ourselves to real scalars only, consider the set E as a vector space E_0 over $\mathbf{R}$. If E is a complex normed space, then E_0 is a real normed space. We say that E_0 is the real normed space *underlying* the complex normed space E. Similarly if M is a subspace of E, we can consider the set M as a subspace M_0 of E_0.

With the notations of Theorem 1, set $f_1(x) = \mathcal{R}e\, f(x)$ and $f_2(x) = \mathcal{I}m\, f(x)$. We have $f(x) = f_1(x) + if_2(x)$, and clearly $x \mapsto f_1(x)$ is linear on M_0; i.e.,

$$f_1(\lambda x + \mu y) = \lambda f_1(x) + \mu f_1(y)$$

for real λ and μ. Since $|f_1(x)| \leqq |f(x)| \leqq \|f\| \cdot \|x\|$, we have $\|f_1\| \leqq \|f\|$, and f_1 is a continuous linear form on M_0. We also have

$$f(ix) = f_1(ix) + if_2(ix)$$

and

$$f(ix) = if(x) = if_1(x) - f_2(x).$$

Comparing the last two identities, we obtain $f_2(x) = -f_1(ix)$ and

$$f(x) = f_1(x) - if_1(ix). \tag{2}$$

By the part of Theorem 1 we have already proved, there exists an extension g_1 of f_1 to E_0 which satisfies $\|g_1\| = \|f_1\|$. Define

$$g(x) = g_1(x) - ig_1(ix)$$

for $x \in E$. It follows from (2) that g coincides with f on M. Since g_1 is linear on E_0, it follows that g is real linear on E; and to prove that g is complex linear we have only to show that $g(ix) = ig(x)$. But, indeed,

$$g(ix) = g_1(ix) - ig_1(-x) = i(g_1(x) - ig_1(ix)) = ig(x).$$

Finally, let $x \in E$ and suppose that $g(x) = \rho e^{i\theta}$, where ρ is a positive real number. Then

$$\begin{aligned} |g(x)| = \rho = e^{-i\theta}g(x) &= g(e^{-i\theta}x) = g_1(e^{-i\theta}x) \\ &\leqq \|g_1\| \cdot \|x\| = \|f_1\| \cdot \|x\| \leqq \|f\| \cdot \|x\|. \end{aligned}$$

Consequently, $\|g\| \leqq \|f\|$ and since obviously $\|g\| \geqq \|f\|$, we have $\|g\| = \|f\|$. ∎

REMARK 1. Let E and F be two Banach spaces, M a closed subspace of E and f a continuous linear map from M into F. For the problem whether it is possible to extend f into a continuous linear map $g: E \to F$, see the recent report of Nachbin [68].

Let us prove a few immediate corollaries of the Hahn-Banach theorem.

Proposition 1. *Let M be a closed subspace of the normed space E and let z be a vector in E which does not belong to M. Then there exists a continuous linear form f defined on E such that $f(z) = 1$ and $f(x) = 0$ for $x \in M$.*

Proof. We know from Lemma 2 that every element x of the closed subspace N generated by M and z can be written uniquely in the form $x = y + \lambda z$, $y \in M$, $\lambda \in \mathbf{K}$. For $x \in N$ define $f(x) = f(y + \lambda z) = \lambda$. Then clearly f is linear, $f(z) = 1$, and $f(x) = 0$ for $x \in M$. If we set

$$\alpha = \inf_{y \in M} \|z + y\|,$$

then $\alpha > 0$ (since M is closed and $z \notin M$), and we have for $\lambda \neq 0$

$$\alpha|\lambda| \leqq |\lambda| \cdot \|\lambda^{-1} y + z\| = \|y + \lambda z\|,$$

i.e.,

$$|f(y + \lambda z)| \leqq \frac{1}{\alpha} \|y + \lambda z\|.$$

This shows that f is continuous on N. Extending f to E by Theorem 1, we obtain the required continuous linear form. ∎

Proposition 2. *A set G is total* (Definition 3.1) *in a normed space E if and only if every continuous linear form f on E which satisfies $f(x) = 0$ for all $x \in G$ is identically zero.*

This is an immediate consequence of the following more general result.

Proposition 3. *A set G is total* (Definition 3.1) *in a closed subspace M of a normed space E if and only if every continuous linear form f on E which satisfies $f(x) = 0$ for all $x \in G$ is zero for every element x of M.*

Proof. Let us first prove the trivial part of the proposition, and let us suppose that G is total in M. Let $x \in M$, and let f be a continuous linear form on E which vanishes on G. For every $\epsilon > 0$ there exists a linear combination $\sum \lambda_i x_i$ with $x_i \in G$ such that $\|x - \sum \lambda_i x_i\| < \epsilon$. Then

$$|f(x)| = |f(x - \sum \lambda_i x_i)| \leqq \|f\| \cdot \epsilon,$$

and consequently $f(x) = 0$.

Conversely, suppose that G is not total in M. Then the closed subspace P generated by G is different from M. Let z be a vector in M but not in P. By Proposition 1 there exists a continuous linear form f on E such that $f(z) = 1$ and $f(x) = 0$ for all $x \in P$, and in particular for all $x \in G$. ∎

Example 1. Anticipating the Riesz representation theorem which will be proved in Chapter 5, and using a few results from function theory, we prove here the Weierstrass approximation theorem (Example 3.2). Let us observe first that it is sufficient to consider the case $I = [0, 1]$. Indeed, the function $x \mapsto t = (x - a)/(b - a)$ maps the interval $a \leqq x \leqq b$ onto the interval $0 \leqq t \leqq 1$. If f is a continuous function on $a \leqq x \leqq b$, then g, defined by $g(t) = f(a + (b - a)t)$, is a continuous function on $0 \leqq t \leqq 1$. If the polynomial $Q(t)$ satisfies $|g(t) - Q(t)| < \epsilon$ on $0 \leqq t \leqq 1$, then the polynomial

$$P(x) = Q\left(\frac{x - a}{b - a}\right)$$

satisfies $|f(x) - P(x)| < \epsilon$ on $a \leqq x \leqq b$.

We shall now prove the more general result that if $(\lambda_n)_{n \in \mathbf{N}}$ is a strictly increasing sequence of positive numbers with $\lambda_0 = 0$ such that $\sum_{n=1}^{\infty} 1/\lambda_n = \infty$, then $(x^{\lambda_n})_{n \in \mathbf{N}}$ is total in $\mathfrak{C}([0, 1])$ (theorem of Müntz). By Proposition 2 we have to show that if F is a continuous linear form on $\mathfrak{C}([0, 1])$ such that $F(x^{\lambda_n}) = 0$ for $n \in \mathbf{N}$, then F is identically zero. According to Riesz' representation theorem, there corresponds to F a function of bounded variation $\mu(x)$ on $0 \leqq x \leqq 1$ such that

$$F(f) = \int_0^1 f(x)\, d\mu(x)$$

for all $f \in \mathfrak{C}([0, 1])$. The function

$$\Phi(\zeta) = \int_0^1 x^{\zeta}\, d\mu(x) \tag{3}$$

of the complex variable $\zeta = \xi + i\eta$ is holomorphic for $\xi > 0$ and bounded for $\xi \geqq 0$, since

$$|\Phi(\zeta)| \leqq \int_0^1 x^{\xi}\, |d\mu(x)| \leqq \int_0^1 |d\mu(x)|.$$

The condition $F(x^{\lambda_n}) = 0$ $(n \geqq 1)$ then means that $\Phi(\lambda_n) = 0$ and it follows from Carlson's theorem ([6], Theorem 9.3.8, p. 156) that $\Phi = 0$. Now, the substitution $x = e^{-y}$ transforms (3) into the Laplace-Stieltjes transform

$$\Phi(\zeta) = \int_0^{\infty} e^{-y\zeta}\, d\mu(e^{-y}),$$

and it follows from the uniqueness theorem ([103], Chapter II, Theorem 6.3, p. 63 and Remark on p. 70) that $\mu(e^{-y})$ is equivalent to a constant for $0 \leqq y < \infty$; i.e., μ is equivalent to a constant for $0 < x \leqq 1$. But μ cannot have a jump at $x = 0$ either, since $F(x^{\lambda_n}) = 0$ also for $n = 0$; i.e., $\int_0^1 d\mu(x) = 0$. Hence μ is equivalent to a constant and thus $F = 0$. Q.E.D.

It follows from the result just proved that (x^n) is not topologically free, as asserted in Example 3.4. Indeed, if we omit one term of this sequence, the remaining set is still total in $\mathfrak{C}([0, 1])$, and thus every element of the sequence (x^n) belongs to the closed subspace generated by the remaining elements.

It can be proved that if $\sum_{n=1}^{\infty} 1/\lambda_n < \infty$ then $(x^{\lambda_n})_{n \in \mathbf{N}}$ is not total in $\mathcal{C}([0, 1])$ (Müntz) and that it is topologically free (Laurent Schwartz [80]).

PROPOSITION 4. *A family $(a_\iota)_{\iota \in I}$ of elements in a normed space E is topologically free if and only if there exists a family $(f_\iota)_{\iota \in I}$ of continuous linear forms on E such that $f_\iota(a_\iota) = 1$ and $f_\iota(a_\kappa) = 0$ if $\iota \neq \kappa$.*

Proof. If $(a_\iota)_{\iota \in I}$ is topologically free, let M_ι be the closed subspace generated by the a_κ with $\kappa \neq \iota$. Then $a_\iota \notin M_\iota$, and by Proposition 1 there exists a continuous linear form f_ι on E such that $f_\iota(a_\iota) = 1$ and $f_\iota(x) = 0$ for $x \in M_\iota$, and in particular $f_\iota(a_\kappa) = 0$ for $\kappa \neq \iota$.

Conversely, if $(a_\iota)_{\iota \in I}$ is not topologically free, then there exists an element a_κ which is contained in the closed subspace generated by the a_ι with $\iota \neq \kappa$. But then for every continuous linear form f such that $f(a_\iota) = 0$ for $\iota \neq \kappa$, we also have $f(a_\kappa) = 0$ (cf. first half of the proof of Proposition 3). ▌

EXAMPLE 2. Consider the continuous linear form

$$f \mapsto F_n(f) = \frac{1}{\pi} \int_0^{2\pi} f(x) \cos nx \, dx \qquad (n \in \mathbf{N}^*)$$

on $\mathcal{C}([0, 2\pi])$. A simple calculation shows that

$$F_n(\cos nx) = 1, \quad F_n(\cos mx) = 0 \quad \text{if} \quad m \neq n,\ m \in \mathbf{N},$$
$$F_n(\sin mx) = 0 \quad \text{if} \quad m \in \mathbf{N}^*.$$

Similarly, for the continuous linear form

$$f \mapsto G_n(f) = \frac{1}{\pi} \int_0^{2\pi} f(x) \sin nx \, dx \qquad (n \in \mathbf{N}^*)$$

we have

$$G_n(\cos mx) = 0 \quad \text{if} \quad m \in \mathbf{N},$$
$$G_n(\sin nx) = 1, \quad G_n(\sin mx) = 0 \quad \text{if} \quad m \neq n,\ m \in \mathbf{N}^*.$$

Finally, for

$$f \mapsto F_0(f) = \frac{1}{2\pi} \int_0^{2\pi} f(x) \, dx$$

we have

$$F_0(1) = 1, \quad F_0(\cos nx) = 0 \quad \text{if} \quad n \in \mathbf{N}^*,$$
$$F_0(\sin nx) = 0 \quad \text{if} \quad n \in \mathbf{N}^*.$$

It follows that the sequence 1, $\cos x$, $\sin x, \ldots, \cos nx$, $\sin nx, \ldots$ is topologically free in $\mathcal{C}([0, 2\pi])$, as asserted in Example 3.4.

PROPOSITION 5 (F. Riesz). *Let $(x_\iota)_{\iota \in I}$ be a family of elements in a normed space E and let $(\alpha_\iota)_{\iota \in I}$ be a family of scalars. A necessary and sufficient condition for the existence of a continuous linear form f on E which verifies $f(x_\iota) = \alpha_\iota$ is the existence of a positive constant M such that*

$$\left|\sum_{\iota \in I} \lambda_\iota \alpha_\iota\right| \leqq M \left\|\sum_{\iota \in I} \lambda_\iota x_\iota\right\| \tag{4}$$

for any family $(\lambda_\iota)_{\iota \in I}$ of scalars, of which only a finite number of elements are different from zero.

Proof. If such an f exists, then

$$\left|\sum_{\iota \in I} \lambda_\iota \alpha_\iota\right| = \left|\sum_{\iota \in I} \lambda_\iota f(x_\iota)\right| = \left|f\left(\sum_{\iota \in I} \lambda_\iota x_\iota\right)\right| \leqq \|f\| \cdot \left\|\sum_{\iota \in I} \lambda_\iota x_\iota\right\|,$$

and we can take $M = \|f\|$.

Conversely, define

$$f\left(\sum_{\iota \in I} \lambda_\iota x_\iota\right) = \sum_{\iota \in I} \lambda_\iota \alpha_\iota.$$

If $\sum_{\iota \in I} \lambda_\iota x_\iota = 0$, then it follows from condition (4) that $f(\sum_{\iota \in I} \lambda_\iota x_\iota) = 0$; thus f is a well-defined, continuous linear form on the subspace generated by the family $(x_\iota)_{\iota \in I}$. The extension of f to E, which exists by virtue of Theorem 1, satisfies the requirements of the proposition. ∎

EXERCISE

1. Show that for a Hilbert space Proposition 3 can be proved independently of the Hahn-Banach theorem. (*Hint:* Use the corollary to Proposition 3.4.)

§7. *The dual space*

Let E be a normed vector space over a field $\mathbf{K}$, and let E' be the set of all continuous linear forms on E. We shall now define in a very natural way a structure of Banach space over $\mathbf{K}$ on the set E'. If f and g are two continuous linear forms on E, then $f + g$ is defined by

$$(f + g)(x) = f(x) + g(x) \qquad \text{for all} \qquad x \in E.$$

It is quite clear that $f + g$ is a linear form; its continuity follows from the inequality

$$|(f + g)(x)| \leqq |f(x)| + |g(x)| \leqq (\|f\| + \|g\|)\|x\|. \tag{1}$$

If λ is a scalar, then λf is defined by

$$(\lambda f)(x) = \lambda \cdot f(x) \qquad \text{for all} \qquad x \in E.$$

It is again clear that λf is a linear form; its continuity follows from

$$|(\lambda f)(x)| \leqq |\lambda| \cdot \|f\| \cdot \|x\|.$$

Axioms (VS 1) through (VS 8) are readily verified; in particular the zero element of E' is the map which to every $x \in E$ assigns the zero of $\mathbf{K}$. The norm of a continuous linear form has already been defined in §5. It clearly satisfies Axiom (N 1), and it follows from (1) that it also satisfies (N 3). Axiom (N 2) is also verified since

$$\|\lambda f\| = \sup \frac{|\lambda f(x)|}{\|x\|} = \sup \frac{|\lambda| \cdot \|f(x)\|}{\|x\|} = |\lambda| \sup \frac{\|f(x)\|}{\|x\|} = |\lambda| \cdot \|f\|.$$

Finally, we prove the remarkable fact that the normed space E' is complete. Let $(f_n)_{n \in \mathbf{N}}$ be a Cauchy sequence in E'. Then for $x \in E$ we have

$$|f_n(x) - f_m(x)| \leqq \|f_n - f_m\| \cdot \|x\| < \epsilon\|x\| \tag{2}$$

whenever $n, m > N(\epsilon)$. This means that the sequence $(f_n(x))_{n \in \mathbf{N}}$ of scalars is convergent. Let us denote its limit by $f(x)$. The map $f: x \mapsto f(x)$ from E into $\mathbf{K}$ is linear, since

$$\begin{aligned} f(\lambda x + \mu y) &= \lim_{n \to \infty} f_n(\lambda x + \mu y) = \lambda \lim_{n \to \infty} f_n(x) + \mu \lim_{n \to \infty} f_n(y) \\ &= \lambda f(x) + \mu f(y). \end{aligned}$$

It follows from (2) that $|f(x) - f_m(x)| \leqq \epsilon\|x\|$ if $m > N$. Thus

$$|f(x)| \leqq |f(x) - f_m(x)| + |f_m(x)| \leqq (\epsilon + \|f_m\|)\|x\|;$$

i.e., f is continuous. Furthermore,

$$\|f - f_m\| = \sup_x \frac{|f(x) - f_m(x)|}{\|x\|} \leqq \epsilon \qquad \text{if} \qquad m > N,$$

which signifies that (f_n) converges to f in the sense of the norm of E'. We have thus completely proved that E' is a Banach space. This space is called the *dual* (conjugate or adjoint) space of E.

EXAMPLE 1. Let l^p $(1 < p < \infty)$ be the space introduced in Example 1.7, and let the conjugate exponent q be defined by

$$\frac{1}{p} + \frac{1}{q} = 1.$$

Every element $f = (\varphi_n)_{n \in \mathbf{N}}$ of the space l^q defines a continuous linear form $\tilde{f}$ on l^p according to the formula

$$\tilde{f}(x) = \sum_{n=0}^{\infty} \varphi_n \xi_n, \qquad x = (\xi_n) \in l^p.$$

Indeed, the right-hand side converges absolutely, since by Hölder's inequality we have

$$\sum_{n=0}^{m} |\varphi_n \xi_n| \leqq \left(\sum_{n=0}^{m} |\varphi_n|^q\right)^{1/q} \left(\sum_{n=0}^{m} |\xi_n|^p\right)^{1/p} \leqq \|f\|_q \|x\|_p,$$

where, of course, $\|f\|_q$ denotes the norm of f in l^q and $\|x\|_p$ the norm of x in l^p. It also follows that $|\tilde{f}(x)| \leqq \|f\|_q \|x\|_p$, i.e., $\|\tilde{f}\| \leqq \|f\|_q$. We shall prove that *the map $f \mapsto \tilde{f}$ is an isometric isomorphism of l^q onto the dual space $(l^p)'$ of l^p* (F. Riesz). It is evident that this map is injective. Let $\tilde{f}$ be an element of $(l^p)'$. Denoting the n-th unit vector (i.e., the vector with components $\xi_n = 1$, $\xi_m = 0$ if $m \neq n$) of l^p by e_n, we define $\varphi_n = \tilde{f}(e_n)$. If $x = (\xi_n)$ is an element of l^p such that $\xi_n = 0$ for all n larger than some integer m, then

$$\tilde{f}(x) = \tilde{f}\left(\sum_{n=0}^{m} \xi_n e_n\right) = \sum_{n=0}^{m} \xi_n \tilde{f}(e_n) = \sum_{n=0}^{m} \varphi_n \xi_n.$$

Set $\varphi_n = e^{i\theta_n}|\varphi_n|$ and define $y_m = (\eta_{mn})_{n \in \mathbf{N}}$, where

$$\eta_{mn} = \begin{cases} e^{-i\theta_n}|\varphi_n|^{q-1} & \text{if } 0 \leqq n \leqq m, \\ 0 & \text{if } n > m. \end{cases}$$

Then we have

$$|\tilde{f}(y_m)| = \sum_{n=0}^{m} |\varphi_n|^q \leqq \|\tilde{f}\| \cdot \|y_m\|_p = \|\tilde{f}\| \cdot \left(\sum_{n=0}^{m} |\varphi_n|^{p(q-1)}\right)^{1/p}.$$

Using the fact that $p(q-1) = q$ and $1 - 1/p = 1/q$, we obtain

$$\left(\sum_{n=0}^{m} |\varphi_n|^q\right)^{1/q} \leqq \|\tilde{f}\|.$$

This proves that the sequence $f = (\varphi_n)_{n \in \mathbf{N}}$ belongs to l^q and that $\|f\|_q \leqq \|\tilde{f}\|$. If $x = (\xi_n) \in l^p$, let $x_m = (\xi_{mn})_{n \in \mathbf{N}}$ be defined by

$$\xi_{mn} = \begin{cases} \xi_n & \text{if } 0 \leqq n \leqq m, \\ 0 & \text{if } n > m. \end{cases}$$

For every $\epsilon > 0$ we have

$$\|x - x_m\|_p = \left(\sum_{n=m+1}^{\infty} |\xi_n|^p\right)^{1/p} < \epsilon$$

if $m > M$. Furthermore,

$$\tilde{f}(x_m) = \sum_{n=0}^{m} \varphi_n \xi_n \tag{3}$$

and

$$|\tilde{f}(x) - \tilde{f}(x_m)| \leqq \|\tilde{f}\| \cdot \|x - x_m\|_p \leqq \epsilon \cdot \|\tilde{f}\|. \tag{4}$$

By Hölder's inequality we have

$$\left| \sum_{n=0}^{\infty} \varphi_n \xi_n - \sum_{n=0}^{m} \varphi_n \xi_n \right| = \left| \sum_{n=m+1}^{\infty} \varphi_n \xi_n \right| \leqq \|f\|_q \cdot \|x - x_m\|_p \leqq \|\tilde{f}\| \cdot \epsilon. \tag{5}$$

It follows from (3), (4), and (5) that

$$\left| \tilde{f}(x) - \sum_{n=0}^{\infty} \varphi_n \xi_n \right| \leqq 2\epsilon \cdot \|\tilde{f}\|,$$

i.e.,

$$\tilde{f}(x) = \sum_{n=0}^{\infty} \varphi_n \xi_n,$$

which concludes the proof.

By virtue of this result we usually identify the spaces $(l^p)'$ and l^q and say that *the dual of l^p is l^q*.

EXAMPLE 2. Consider now the space c_0 defined in Example 1.6. If $f = (\varphi_n)$ is an element of the space l^1, then the relation

$$\tilde{f}(x) = \sum_{n=0}^{\infty} \varphi_n \xi_n, \qquad x = (\xi_n) \in c_0,$$

defines a continuous linear form $\tilde{f}$ on c_0, since evidently

$$\sum_{n=0}^{\infty} |\varphi_n \xi_n| \leqq \max |\xi_n| \cdot \sum_{n=0}^{\infty} |\varphi_n| = \|f\|_1 \cdot \|x\|.$$

It also follows that $\|\tilde{f}\| \leqq \|f\|_1$. Let us prove that the map $f \mapsto \tilde{f}$ establishes an isometric isomorphism between l^1 and $(c_0)'$. Let $\tilde{f}$ be an element of $(c_0)'$ and define, similarly as in the previous example, $\varphi_n = \tilde{f}(e_n)$, where e_n is the n-th unit vector of c_0. If $x = (\xi_n)$ is such that $\xi_n = 0$ for $n > m$, then we again have

$$\tilde{f}(x) = \sum_{n=0}^{m} \varphi_n \xi_n.$$

Set $\varphi_n = e^{i\theta_n}|\varphi_n|$ and define $y_m = (\eta_{mn})_{n\in\mathbf{N}}$, where

$$\eta_{mn} = \begin{cases} e^{-i\theta_n} & \text{if } 0 \leqq n \leqq m, \\ 0 & \text{if } n > m. \end{cases}$$

Then we have

$$|\tilde{f}(y_m)| = \sum_{n=0}^{m} |\varphi_n| \leqq \|\tilde{f}\| \cdot \|y_m\| = \|\tilde{f}\|.$$

It follows that $f = (\varphi_n) \in l^1$ and $\|f\|_1 \leqq \|\tilde{f}\|$. If $x = (\xi_n) \in c_0$, then we define again $x_m = (\xi_{mn})_{n \in \mathbf{N}}$ by

$$\xi_{mn} = \begin{cases} \xi_n & \text{if } 0 \leqq n \leqq m, \\ 0 & \text{if } n > m. \end{cases}$$

For any $\epsilon > 0$ we have $\|x - x_m\| = \max_{n>m} |\xi_n| < \epsilon$ for $m > M$, since by the definition of c_0 the sequence $(\xi_n)_{n \in \mathbf{N}}$ tends to zero. Furthermore,

$$\tilde{f}(x_m) = \sum_{n=0}^{m} \varphi_n \xi_n, \tag{6}$$

$$|\tilde{f}(x) - \tilde{f}(x_m)| \leqq \|\tilde{f}\| \cdot \|x - x_m\| \leqq \epsilon \cdot \|\tilde{f}\| \tag{7}$$

and

$$\begin{aligned} \left| \sum_{n=0}^{\infty} \varphi_n \xi_n - \sum_{n=0}^{m} \varphi_n \xi_n \right| &= \left| \sum_{n=m+1}^{\infty} \varphi_n \xi_n \right| \\ &\leqq \max_{n>m} |\xi_n| \cdot \sum_{n=m+1}^{\infty} |\varphi_n| \leqq \epsilon \cdot \|f\|_1 \leqq \epsilon \cdot \|\tilde{f}\|. \end{aligned} \tag{8}$$

It follows from (6), (7), and (8) that

$$\left| \tilde{f}(x) - \sum_{n=0}^{\infty} \varphi_n \xi_n \right| \leqq 2\epsilon \cdot \|\tilde{f}\|,$$

i.e.,

$$\tilde{f}(x) = \sum_{n=0}^{\infty} \varphi_n \xi_n,$$

which concludes the proof.

We usually identify the spaces $(c_0)'$ and l^1.

EXAMPLE 3. As our final example, let us find the dual of l^1. If $f = (\varphi_n)$ is an element of the space m (Example 1.8), then the formula

$$\tilde{f}(x) = \sum_{n=0}^{\infty} \varphi_n \xi_n, \qquad x = (\xi_n) \in l^1,$$

defines a continuous linear form $\tilde{f}$ on l^1, since

$$\sum_{n=0}^{\infty} |\varphi_n \xi_n| \leqq \sup_n |\varphi_n| \cdot \sum_{n=0}^{\infty} |\xi_n| = \|f\|_\infty \cdot \|x\|_1,$$

where $\|f\|_\infty$ is the norm of f in m. It also follows that $\|\tilde{f}\| \leqq \|f\|_\infty$. The map $f \mapsto \tilde{f}$ is an isometric isomorphism of m onto $(l^1)'$. Indeed, given an $\tilde{f} \in (l^1)'$, we define, as in the previous two examples, $\varphi_n = \tilde{f}(e_n)$ and obtain

$$\tilde{f}(x) = \sum_{n=0}^{m} \varphi_n \xi_n$$

if $\xi_n = 0$ for $n > m$. Set as before $\varphi_n = e^{i\theta_n}|\varphi_n|$ and let $y_m = (\eta_{mn})_{n\in\mathbf{N}}$ be defined by

$$\eta_{mn} = \begin{cases} e^{-i\theta_n} & \text{if } n = m, \\ 0 & \text{if } n \neq m. \end{cases}$$

Then we have

$$|\tilde{f}(y_m)| = |\varphi_m| \leqq \|\tilde{f}\| \cdot \|y_m\|_1 = \|\tilde{f}\|.$$

This implies that $f = (\varphi_n) \in m$ and that $\|f\|_\infty \leqq \|\tilde{f}\|$. The proof that for an arbitrary $x = (\xi_n) \in l^1$ we have

$$\tilde{f}(x) = \sum_{n=0}^{\infty} \varphi_n \xi_n$$

is similar to that of the two previous examples and can be left to the reader.

We usually identify the spaces $(l^1)'$ and m.

Let E be a normed space and E' its dual space. With a very useful change of notation we shall denote the elements of E' by primed letters $x', y', \ldots$ and define the symbol $\langle x, x' \rangle$ by

$$\langle x, x' \rangle = x'(x),$$

where the right-hand side is the value of the linear form x' at the point x. The map $(x, x') \mapsto \langle x, x' \rangle$ is a bilinear form on $E \times E'$, called the *canonical bilinear form*. Linearity in x follows from the fact that x' is a linear form; linearity in x' follows from the definition of the operations in E'. We have, by the definition of the norm in E',

$$|\langle x, x' \rangle| \leqq \|x\| \cdot \|x'\|,$$

which shows in particular that the canonical bilinear form is continuous. Two other properties are the following:

(a) if for an $x' \in E'$ we have $\langle x, x' \rangle = 0$ for every $x \in E$, then $x' = 0$;

(b) if for an $x \in E$ we have $\langle x, x' \rangle = 0$ for every $x' \in E'$, then $x = 0$.

Property (a) is just the definition of the zero vector in E'; property (b), however, lies much deeper and is an immediate consequence of Proposition 6.1 for $M = \{0\}$.

Definition 1. *Two vectors $x \in E$ and $x' \in E'$ are said to be orthogonal if $\langle x, x' \rangle = 0$. The vector $x' \in E'$ is said to be orthogonal to the set $M \subset E$ if x' is orthogonal to every $x \in M$. The vector $x \in E$ is said to be orthogonal to the set $M' \subset E'$ if x is orthogonal to every $x' \in M'$.*

Given a set $M \subset E$, we denote by $M^\perp$ the set of all vectors of E' which are orthogonal to M. Given a set $M' \subset E'$, we denote by $M'^\perp$ the set of all vectors of E which are orthogonal to M'.

With this terminology we can restate Proposition 6.3 in the following form: A set G is total in a closed subspace M of a normed space E if and only if every element $x' \in E'$ which is orthogonal to G is also orthogonal to M.

Proposition 1. *Let G be a subset of a normed space E. Then $G^\perp$ is a closed subspace of the dual space E' and $G^{\perp\perp} = (G^\perp)^\perp$ is the closed subspace of E generated by G.*

Proof. It is clear that $G^\perp$ is a subspace of E'. Let x' adhere to $G^\perp$. Then for every $\epsilon > 0$ there exists $y' \in G^\perp$ such that $\|x' - y'\| < \epsilon$. For $x \in G$ we have

$$|\langle x, x' \rangle| = |\langle x, x' \rangle - \langle x, y' \rangle| \leqq \|x\| \cdot \|x' - y'\| \leqq \|x\| \cdot \epsilon,$$

i.e., $\langle x, x' \rangle = 0$, $x' \in G^\perp$. Thus $G^\perp$ is closed.

A similar reasoning shows that $G^{\perp\perp}$ is a closed subspace of E. Clearly, $G \subset G^{\perp\perp}$. If x' is orthogonal to G, then $x' \in G^\perp$ and thus x' is orthogonal to $G^{\perp\perp}$. By the new version of Proposition 6.3 this implies that G generates $G^{\perp\perp}$. ▮

Before stating the next result let us introduce a notation. If M and N are two closed subspaces of a normed space E, then $M \vee N$ will denote the closed subspace of E generated by $M \cup N$. We recall that $M + N$ denotes the subspace of E formed by all vectors of the form $y + z$, where $y \in M$, $z \in N$. It then follows from the remark made after Definition 3.1 that $M \vee N = \overline{M + N}$.

Proposition 2. *Let M and N be two closed subspaces of a normed space E. Then*

$$(M \vee N)^\perp = M^\perp \cap N^\perp.$$

Proof. Let $x' \in (M \vee N)^\perp$. Then in particular x' is orthogonal to M and N; i.e., $x' \in M^\perp \cap N^\perp$. Conversely, let $x' \in M^\perp \cap N^\perp$. Then x' is orthogonal to both M and N, i.e., to $M \cup N$. It follows from the trivial half of Proposition 6.3 that x' is orthogonal to the subspace $M \vee N$ generated by $M \cup N$; i.e., $x' \in (M \vee N)^\perp$. ▮

Let E be a normed space. Every element $x \in E$ defines a linear form $\tilde{x}$ on the dual space E' according to the formula

$$\tilde{x}(x') = \langle x, x' \rangle \qquad \text{for all} \qquad x' \in E'.$$

It follows from the inequality

$$|\tilde{x}(x')| = |\langle x, x' \rangle| \leqq \|x\| \cdot \|x'\|$$

that $\tilde{x}$ is continuous and that $\|\tilde{x}\| \leqq \|x\|$. Thus $x \mapsto \tilde{x}$ is a continuous linear map from E into the space $E'' = (E')'$, called the *bidual* (or second conjugate) of E. We have $\|x\| = \|\tilde{x}\|$. Indeed, let the linear form y' be given on the subspace $\{\lambda x\}$ of E by $y'(\lambda x) = \lambda \|x\|$. Clearly, $\|y'\| = 1$, and extending to E by the Hahn-Banach theorem (Theorem 6.1), we obtain an element $y' \in E'$ which verifies $\langle x, y' \rangle = \|x\|$ and $\|y'\| = 1$. But then

$$\|\tilde{x}\| = \sup_{x' \in E'} \frac{|\langle x, x' \rangle|}{\|x'\|} \geqq \frac{\langle x, y' \rangle}{\|y'\|} = \|x\|.$$

It follows that the map $x \mapsto \tilde{x}$ establishes an isometric isomorphism of E onto a subspace $\widetilde{E}$ of E'' (we usually identify E with $\widetilde{E}$). It may happen that $\widetilde{E} \neq E''$. If $\widetilde{E} = E''$, then we say that the normed space E is *reflexive*. Clearly, a reflexive normed space is complete, i.e., a Banach space.

EXAMPLE 4. For $1 < p < \infty$ the space l^p is reflexive. Indeed, observe that as p increases from 1 to ∞, the conjugate exponent q decreases from ∞ to 1, and therefore the roles of p and q are symmetric. Now, in Example 1 we have precisely proved that the map $f \mapsto \tilde{f}$ from l^q into $(l^p)' = (l^q)''$ is surjective.

EXAMPLE 5. The space c_0 is not reflexive. Indeed, we know from Examples 2 and 3 that $(c_0)'' = (l^1)' = m$, and the map $x \mapsto \tilde{x}$ is here clearly the canonical injection $c_0 \hookrightarrow m$, which is not surjective.

Now let E be a Hilbert space. If to an element $a \in E$ we assign the linear form f_a defined by

$$f_a(x) = (x \mid a), \qquad x \in E,$$

then $a \mapsto f_a$ is a semi-linear map from E into E', since we have

$$(\bar{\lambda} f_a + \bar{\mu} f_b)(x) = (x \mid \lambda a + \mu b).$$

If $\mathbf{K} = \mathbf{R}$, then, of course, the map $a \mapsto f_a$ is just linear. It follows from the Fréchet-Riesz theorem (Proposition 5.1) and the remarks preceding it that $a \mapsto f_a$ maps E onto E' and conserves the norms, i.e., $\|a\| = \|f_a\|$.

We shall sometimes identify the spaces E and E' by this "isometric semi-isomorphism." If we define $(f_a \mid f_b) = (b \mid a)$, then the Banach space E' even becomes a Hilbert space.

It follows that *every Hilbert space is reflexive.* To show this, let us denote by Φ the map $a \mapsto f_a$ from E onto E' and by Ψ the analogous map $b' \mapsto f_{b'}$ from E' onto E''. We have by definition

$$\langle x, \Phi(a)\rangle = (x \mid a) \qquad \text{for all} \qquad x \in E,$$
$$\langle x', \Psi(b')\rangle = (x' \mid b') \qquad \text{for all} \qquad x' \in E'.$$

Let $a \in E$. Then we have for any $x' \in E'$:

$$\tilde{a}(x') = \langle a, x'\rangle.$$

Now, $x' = \Phi(x)$ for some $x \in E$, and hence

$$\begin{aligned}\tilde{a}(x') &= \langle a, \Phi(x)\rangle = (a \mid x)\\ &= (\Phi(x) \mid \Phi(a)) = (x' \mid \Phi(a)) = \langle x', \Psi(\Phi(a))\rangle;\end{aligned}$$

i.e., $\tilde{a} = \Psi(\Phi(a))$. Thus the map $a \mapsto \tilde{a}$ is surjective as the composite of the two surjective maps Φ and Ψ.

If E is a Hilbert space, M a closed subspace of E, and if we identify E with E', then $M^\perp$ will be the subspace of E consisting of all vectors z which verify $(y \mid z) = 0$ for every $y \in M$. We know (Proposition 1) that $M^\perp$ is a closed subspace of E; it is called the *orthogonal complement* of M. We have clearly $M \cap M^\perp = \{0\}$ and it follows from Proposition 3.4 that $M + M^\perp = E$.

Let us go back to a Banach space E. With the help of the dual space E' we can define on E a topology, distinct from the one defined by the norm, and which is called the *weak topology*. We can say roughly that a sequence $(x_n)_{n\in\mathbf{N}}$ of elements in E tends weakly to zero in E if for every $x' \in E'$ the sequence $(\langle x_n, x'\rangle)$ of scalars tends to zero.

EXAMPLE 6. The sequence (e_n) of unit vectors in l^2 does not tend to zero in l^2 since $\|e_n\| = 1$ for every n. It tends, however, to zero weakly, since if $y = (\eta_n)$ is an element of l^2, then $(e_n \mid y) = \bar{\eta}_n$ and $\bar{\eta}_n$ tends to zero because $\sum|\eta_n|^2$ is convergent.

Sequences are, however, not adequate to describe the weak topology and an exact definition must be given in terms of neighborhoods. Let $a \in E$. A neighborhood V of a in the weak topology is determined by a finite sequence $(x'_i)_{1\leqq i\leqq n}$ of elements of E'; V is the set of all those vectors $x \in E$ which verify $|\langle(x - a), x'_i\rangle| \leqq 1$ for $1 \leqq i \leqq n$. We shall prove in the next chapter that the sets so defined satisfy the axioms of neighborhoods and also that the two operations on E are continuous for the weak topology.

The weak topology cannot be defined in terms of a norm (except in trivial cases) and thus we have here the first example of a topological vector space which is not a normed space. The next two chapters of this book will be devoted to the theory of general topological vector spaces.

The weak topology has many useful properties which the strong topology does not have. Thus the closed unit ball B_1 of a Hilbert space (and more generally of any reflexive space) is weakly compact, which we know (Example 2.2) is not true for the topology defined by the norm.

Exercises

1. Complete the proof in Example 3.

2. Let E be a normed space and $(M_\iota)_{\iota \in I}$ a family of closed subspaces of E. We denote by $\bigvee_{\iota \in I} M_\iota$ the closed subspace of E generated by $\bigcup_{\iota \in I} M_\iota$. Prove the relation

$$(\bigvee_{\iota \in I} M_\iota)^\perp = \bigcap_{\iota \in I} M_\iota^\perp.$$

3. Let E be a Hilbert space (which we identify with E'). Let M and N be closed subspaces of E.

(a) Give a proof of $M^{\perp\perp} = M$ independent of the Hahn-Banach theorem. (*Hint:* Use the corollary to Proposition 3.4.)

(b) Prove that if the subspaces M and N are orthogonal (Definition 3.4), then $M + N$ is closed, i.e., $M \vee N = M + N$. (*Hint:* If $x = y + z$ with $y \in M$, $z \in N$, then

$$\|x\|^2 = \|y\|^2 + \|z\|^2.)$$

REMARK 1. If M and N are not orthogonal, then $M + N$ is not necessarily closed ([44], §15).

(c) Prove the relation $(M \cap N)^\perp = M^\perp \vee N^\perp$. (*Hint:* Use Proposition 2 and (a).)

(d) Deduce $M + M^\perp = E$ from (a), (b), and (c). (*Hint:* $M + M^\perp = M^{\perp\perp} + M^\perp = M^{\perp\perp} \vee M^\perp = (M^\perp \cap M)^\perp = \{0\}^\perp = E$.)

4. Let $\mathcal{L}(E, F)$ be the set of all continuous linear maps from the normed space E into the Banach space F. Defining the algebraic operations on $\mathcal{L}(E, F)$ in a natural way and setting

$$\|f\| = \sup_{\substack{x \in E \\ x \neq 0}} \frac{\|f(x)\|}{\|x\|},$$

show that $\mathcal{L}(E, F)$ is a Banach space.

5. Prove that if the dual E' of a Banach space E is reflexive, then E itself is reflexive. (*Hint:* First show that the image of E in E'' under the map $x \mapsto \tilde{x}$ is closed (cf. the proof of Proposition 5.4); then apply Proposition 6.1.)

§8. The Banach-Steinhaus theorem

The second fundamental result concerning normed spaces is the following:

THEOREM 1 (Banach-Steinhaus theorem or the principle of uniform boundedness). *Let $(u_\iota)_{\iota \in I}$ be a family of continuous linear forms defined on a Banach space E. Suppose that for each $x \in E$ the family of scalars $(u_\iota(x))_{\iota \in I}$ is bounded. Then there exists a constant $M > 0$ such that $|u_\iota(x)| \leqq M\|x\|$ for all $x \in E$ and $\iota \in I$.*

In this theorem the completeness of E is essential and we shall first present the general result from the theory of metric spaces on which the proof is based.

We say that a subset A of a metric space is *rare* (or nowhere dense) if the closure of A has a void interior, i.e., if $\mathring{\bar{A}} = \emptyset$.

BAIRE'S THEOREM. *If X is a complete metric space and $(F_n)_{n \in \mathbf{N}}$ a sequence of closed rare subsets of X, then $\bigcup_{n \in \mathbf{N}} F_n \neq X$.*

Proof. Write $A_n = \complement F_n$ for $n \in \mathbf{N}$. The set A_1 is open and therefore contains a closed ball $B_{\rho_1}(x_1)$ with $\rho_1 < 1$. The intersection $A_2 \cap B_{\rho_1}(x_1)$ is not empty because F_2 is rare, and therefore there exists a closed ball $B_{\rho_2}(x_2)$ with $\rho_2 < \frac{1}{2}$ such that $B_{\rho_2}(x_2) \subset A_2 \cap B_{\rho_1}(x_1)$. Continuing this construction, we obtain a nested sequence

$$B_{\rho_1}(x_1) \supset B_{\rho_2}(x_2) \supset \cdots \supset B_{\rho_n}(x_n) \supset \cdots$$

of closed balls such that $B_{\rho_n}(x_n) \subset A_n$ and $\rho_n < 1/n$. Then $(x_n)_{n \in \mathbf{N}}$ is a Cauchy sequence, since for $p, q \geqq n$ we have $x_p \in B_{\rho_n}(x_n)$, $x_q \in B_{\rho_n}(x_n)$ and thus

$$\delta(x_p, x_q) \leqq \delta(x_p, x_n) + \delta(x_q, x_n) \leqq 2\rho_n < 2/n \cdot$$

Since X is complete, the sequence (x_n) converges to some point $x \in X$. Now x adheres to every ball $B_{\rho_n}(x_n)$, because $x_p \in B_{\rho_n}(x_n)$ for $p \geqq n$. Thus $x \in B_{\rho_n}(x_n)$ and *a fortiori* $x \in A_n$ for all $n \in \mathbf{N}$. Hence the point $x \in X$ does not belong to any set F_n. ▮

COROLLARY. *If the complete metric space X is the union of the sequence $(F_n)_{n \in \mathbf{N}}$ of closed sets, then at least one set F_n contains some ball $B_\rho(z)$.*

OSGOOD'S THEOREM. *Let $(u_\iota)_{\iota \in I}$ be a family of continuous functions defined on a complete metric space X. Suppose that for each $x \in X$ the family $(u_\iota(x))_{\iota \in I}$ of numbers is bounded. Then there exists a ball $B_\rho(z)$ in X and a constant $M > 0$ such that $|u_\iota(x)| \leqq M$ for all $x \in B_\rho(z)$ and $\iota \in I$.*

Proof. For $\iota \in I$ and $n \in \mathbf{N}$ let $G_{\iota n}$ be the set of those points $x \in X$ in which $|u_\iota(x)| \leqq n$. Since u_ι is continuous, the set $G_{\iota n}$ is closed. The intersection $F_n = \bigcap_{\iota \in I} G_{\iota n}$ is also closed, and it consists of those points $x \in X$ in which $|u_\iota(x)| \leqq n$ for every $\iota \in I$. The hypothesis that $(u_\iota(x))_{\iota \in I}$ is bounded at every point means that $X = \bigcup_{n \in \mathbf{N}} F_n$. By the previous corollary there exists a ball $B_\rho(z)$ which is contained in some F_n, i.e., $|u_\iota(x)| \leqq n$ for all $x \in B_\rho(z)$ and $\iota \in I$. ∎

Proof of Theorem 1. By Osgood's theorem there exists a ball $B_\rho(z)$ in E and $N > 0$ such that $|u_\iota(x)| \leqq N$ for all $x \in B_\rho(z)$ and $\iota \in I$. If x is arbitrary in E, then we have

$$\frac{x}{\|x\|}\,\rho + z \in B_\rho(z)$$

and therefore

$$\left|u_\iota\left(\frac{x}{\|x\|}\,\rho + z\right)\right| \leqq N$$

for every $\iota \in I$. It follows that

$$\left|u_\iota\left(\frac{x}{\|x\|}\,\rho\right)\right| = \left|u_\iota\left(\frac{x}{\|x\|}\,\rho + z - z\right)\right|$$
$$\leqq \left|u_\iota\left(\frac{x}{\|x\|}\,\rho + z\right)\right| + |u_\iota(z)| \leqq 2N$$

and

$$|u_\iota(x)| \leqq \frac{2N}{\rho}\,\|x\|$$

for every $x \in E$ and $\iota \in I$. Taking $M = 2N/\rho$, we obtain the desired conclusion. ∎

REMARK 1. The method just used, based on Baire's theorem, will serve us in Chapter 3 to generalize the Banach-Steinhaus theorem. There is, however, a different proof, using the method of the "gliding hump," which is also of interest (see [52], §20, 11.(3), (b), p. 254 and [9], Chapter IV, §5, Exercise 4(a)).

As a corollary of Theorem 1 we get:

PROPOSITION 1. *Let $(u_n)_{n \in \mathbf{N}}$ be a sequence of continuous linear forms on a Banach space E. Suppose that for each $x \in E$ the sequence $(u_n(x))$ converges to some value $u(x)$. Then $u\colon x \mapsto u(x)$ is a continuous linear form on E.*

Proof. The map u is a linear form since

$$u(x + y) = \lim_{n \to \infty} u_n(x + y) = \lim_{n \to \infty} u_n(x) + \lim_{n \to \infty} u_n(y) = u(x) + u(y)$$

and

$$u(\lambda x) = \lim_{n \to \infty} u_n(\lambda x) = \lambda \lim_{n \to \infty} u_n(x) = \lambda u(x).$$

For every $x \in E$ the sequence $(u_n(x))$ is bounded; hence by Theorem 1 there exists $M > 0$ such that $|u_n(x)| \leqq M\|x\|$ for all $x \in E$ and $n \in N$. But then $|u(x)| \leqq M\|x\|$ for all $x \in E$; i.e., u is continuous. ▮

In general, to ensure continuity of the limit of a sequence of continuous functions, we need uniform convergence (cf. Example 1.4). Proposition 1 shows that in the case of linear functions on a Banach space, pointwise convergence already implies the continuity of the limit function.

EXAMPLE 1. As a striking application of the Banach-Steinhaus theorem, we shall prove that *there exist continuous functions whose Fourier series diverges at one point.*

Let us recall a few facts concerning Fourier series. A series of the form

$$\frac{a_0}{2} + \sum_{k=1}^{\infty} a_k \cos kx + b_k \sin kx \tag{1}$$

is called a trigonometric series. If the series (1) is uniformly convergent, then its sum $f(x)$ is a continuous periodic function with period 2π. In this case if we multiply by $\cos lx$ or by $\sin lx$, the resulting series will still be uniformly convergent. Integrating termwise and using the "orthogonality formulas" (cf. Example 6.2)

$$\int_0^{2\pi} \cos kx \cos lx \, dx = \begin{cases} \pi, & k = l \neq 0, \\ 0, & k \neq l, \end{cases}$$

$$\int_0^{2\pi} \sin kx \sin lx \, dx = \begin{cases} \pi, & k = l \neq 0, \\ 0, & k \neq l, \end{cases}$$

$$\int_0^{2\pi} \cos kx \sin lx \, dx = 0, \qquad \int_0^{2\pi} dx = 2\pi,$$

we obtain the expressions

$$\begin{aligned} a_k &= \frac{1}{\pi} \int_0^{2\pi} f(x) \cos kx \, dx \qquad (k \in \mathbf{N}), \\ b_k &= \frac{1}{\pi} \int_0^{2\pi} f(x) \sin kx \, dx \qquad (k \in \mathbf{N}^*). \end{aligned} \tag{2}$$

Now let $f(x)$ be a continuous periodic function with period 2π. Then we can *define* the so-called *Fourier coefficients* of $f(x)$ by the formulas (2), and the series (1) formed with these coefficients is called the *Fourier series* of f. The first question we ask is, of course, whether the Fourier series of f converges. To answer this question we represent the partial sum

$$s_n(x) = \frac{a_0}{2} + \sum_{k=1}^{n} a_k \cos kx + b_k \sin kx$$

in a closed form.

To do this we start out with the geometric progression

$$\sum_{k=0}^{n} e^{ik\theta} = \frac{e^{i(n+1)\theta} - 1}{e^{i\theta} - 1}.$$

It follows that

$$\frac{1}{2} + \sum_{k=1}^{n} e^{ik\theta} = \frac{e^{i(n+1)\theta} - 1}{e^{i\theta} - 1} - \frac{1}{2} = \frac{2e^{i(n+1/2)\theta} - e^{(1/2)i\theta} - e^{-(1/2)i\theta}}{2(e^{(1/2)i\theta}) - e^{-(1/2)i\theta})}$$

$$= \frac{e^{i(n+1/2)\theta} - \cos \frac{1}{2}\theta}{2i \sin \frac{1}{2}\theta},$$

and taking the real parts,

$$\frac{1}{2} + \sum_{k=1}^{n} \cos k\theta = \frac{\sin \frac{1}{2}(2n+1)\theta}{2 \sin \frac{1}{2}\theta}.$$

From this formula we obtain

$$s_n(x) = s_n(x; f)$$

$$= \frac{1}{2\pi} \int_0^{2\pi} f(t)\, dt + \sum_{k=1}^{n} \frac{1}{\pi} \int_0^{2\pi} f(t)(\cos kt \cos kx + \sin kt \sin kx)\, dt$$

$$= \frac{1}{\pi} \int_0^{2\pi} f(t) \left(\frac{1}{2} + \sum_{k=1}^{n} \cos k(t - x) \right) dt$$

$$= \frac{1}{2\pi} \int_0^{2\pi} f(t) \frac{\sin \frac{1}{2}(2n+1)(t-x)}{\sin \frac{1}{2}(t-x)}\, dt.$$

Let E be the Banach space formed by the continuous periodic functions with period 2π, with the norm $\|f\| = \sup_{0 \leq x \leq 2\pi} |f(x)|$. Of course E can be identified with the closed subspace of $\mathcal{C}([0, 2\pi])$ formed by the functions f which satisfy $f(0) = f(2\pi)$ (cf. Example 3.2). The continuous linear form

$$u_n: f \mapsto s_n(0; f) = \frac{1}{2\pi} \int_0^{2\pi} f(t) \frac{\sin \frac{1}{2}(2n+1)t}{\sin \frac{1}{2}t}\, dt$$

has the norm

$$\|u_n\| = \frac{1}{2\pi} \int_0^{2\pi} \left| \frac{\sin \frac{1}{2}(2n+1)t}{\sin \frac{1}{2}t} \right| dt = l_n, \tag{3}$$

and the number l_n ($n \in \mathbf{N}$) is called the n-th *Lebesgue constant*. Let us prove (3). It is clear that if $|f(x)| \leq 1$ for all x, then $|u_n(f)| \leq l_n$. Therefore we have only to show that for any $\epsilon > 0$ there exists an $f \in E$ such that $\|f\| \leqq 1$ and $|u_n(f)| \geqq l_n - \epsilon$. The function $\sin \frac{1}{2}(2n+1)x$ is zero at the points $x = 2l\pi/(2n+1)$ $(l \in \mathbf{Z})$. Suppose that $\delta < \pi/(2n+1)$ and define the continuous function f_δ by

$$f_\delta(x) = \begin{cases} \operatorname{sgn} \dfrac{\sin \frac{1}{2}(2n+1)t}{\sin \frac{1}{2}t} & \text{if} \quad \dfrac{2l\pi}{2n+1} + \delta \leqq x \leqq \dfrac{2(l+1)\pi}{2n+1} - \delta, \\ \text{linear} \quad \text{if} \quad \dfrac{2l\pi}{2n+1} - \delta \leqq x \leqq \dfrac{2l\pi}{2n+1} + \delta, \end{cases}$$

where $\operatorname{sgn} \xi$ stands for $\xi/|\xi|$. Clearly, $\|f_\delta\| = 1$. Furthermore, we have

$$\begin{aligned} |u_n(f_\delta)| &\geqq l_n - \sum_{l=0}^{2n+1} \frac{1}{2\pi} \int_{2l\pi/(2n+1)-\delta}^{2l\pi/(2n+1)+\delta} \left| \frac{\sin \frac{1}{2}(2n+1)t}{\sin \frac{1}{2}t} \right| dt \\ &\geqq l_n - \sum_{l=0}^{2n+1} \frac{\delta}{\pi} (2n+1) = l_n - \frac{\delta}{\pi} (2n+1)(2n+2), \end{aligned}$$

since

$$\left| \frac{\sin \frac{1}{2}(2n+1)t}{\sin \frac{1}{2}t} \right| = \left| 1 + 2 \sum_{k=1}^{n} \cos kt \right| \leqq 2n+1.$$

If now the sequence $\big(s_n(0; f)\big)_{n \in \mathbf{N}}$ converged for every $f \in E$, then in particular it would be bounded for every $f \in E$. It would then follow from the Banach-Steinhaus theorem that the sequence (l_n) is bounded. Thus, to prove the existence of a continuous function whose Fourier series diverges at the origin, it is sufficient to show that l_n tends to $+\infty$ as $n \to \infty$. In the intervals

$$\frac{(4l+1)\pi}{4n+2} \leqq x \leqq \frac{(4l+3)\pi}{4n+2}$$

we have

$$\left| \sin \frac{1}{2}(2n+1)x \right| \geqq \frac{1}{\sqrt{2}}$$

and consequently

$$l_n > \frac{\sqrt{2}}{4\pi} \sum_{l=0}^{2n} \int_{(4l+1)\pi/(4n+2)}^{(4l+3)\pi/(4n+2)} \frac{dt}{|\sin \frac{1}{2}t|}.$$

Furthermore, since $\sin \frac{1}{2}t < \frac{1}{2}t$ for $t > 0$, we have

$$l_n > \frac{\sqrt{2}}{4\pi} \sum_{l=0}^{2n} \int_{(4l+1)\pi/(4n+2)}^{(4l+3)\pi/(4n+2)} \frac{2}{t}\, dt > \frac{\sqrt{2}}{\pi} \sum_{l=0}^{2n} \frac{1}{4l+3},$$

and the last sum tends to $+\infty$ as $n \to \infty$.

REMARK 2. In Definition 3.6 we introduced the name Fourier coefficient in a sense which appears to be completely different from the present one. We shall see in Chapter 5 that the terminology is consistent.

EXERCISES

1. Let E be a Banach space, F a normed space, and let $(u_\iota)_{\iota \in I}$ be a family of continuous linear maps from E into F. Suppose that for each $x \in E$ the family $(u_\iota(x))_{\iota \in I}$ is bounded in F. Show that there exists a constant $M > 0$ such that $\|u_\iota(x)\| \leqq M\|x\|$ for all $x \in E$ and $\iota \in I$.

2. Show that Baire's theorem does not hold in the space $\mathbf{Q}$ of rational numbers.

3. Give an example of a sequence (u_n) of continuous functions defined on the interval $[0, 1]$ such that the sequence $(u_n(x))$ is bounded for every $x \in [0, 1]$ but $\max_{0 \leqq x \leqq 1} |u_n(x)|$ is unbounded.

4. (a) Let $(u_{mn})_{(m,n) \in \mathbf{N} \times \mathbf{N}}$ be a double sequence of continuous linear forms defined on a Banach space E. Suppose that for every $m \in \mathbf{N}$ there exists an element x_m such that the sequence $(u_{mn}(x_m))_{n \in \mathbf{N}}$ is unbounded. Prove that there exists a common element $x \in E$ such that for every $m \in \mathbf{N}$ the sequence $(u_{mn}(x))_{n \in \mathbf{N}}$ is unbounded (principle of condensation of singularities).

(b) Prove that if A is any countable set in $[0, 2\pi]$, then there exists a continuous function whose Fourier series diverges at every point of A.

5. Let $(\alpha_n)_{n \in \mathbf{N}}$ be a sequence of numbers such that for every $(\xi_n) \in c_0$ the series $\sum_{n=0}^{\infty} \alpha_n \xi_n$ converges. Prove that $(\alpha_n) \in l^1$. (*Hint:* Use Example 7.2 and Proposition 1.)

6. (a) Let E be a Banach space and $(x, y) \mapsto B(x, y)$ a bilinear form defined on $E \times E$. Suppose that for each $x \in E$ there exists $\mu(x) > 0$ such that $|B(x, y)| \leqq \mu(x)\|y\|$ for all $y \in E$ and that for all $y \in E$ there exists a $\nu(y) > 0$ such that $|B(x, y)| \leqq \nu(y)\|x\|$ for all $x \in E$. Show that there exists a constant $M > 0$ such that $|B(x, y)| \leqq M\|x\| \cdot \|y\|$ for all $x \in E$ and $y \in E$. (*Hint:* Consider the linear forms $x \mapsto B(x, y)$ as y varies in the unit ball $\{y \mid \|y\| \leqq 1\}$ of E.)

(b) Let $(\alpha_{ij})_{(i,j) \in \mathbf{N} \times \mathbf{N}}$ be an infinite square matrix such that for any two elements $x = (\xi_i)$ and $y = (\eta_i)$ of l^2 the double sum

$$\sum_{i=0}^{\infty} \sum_{j=0}^{\infty} \alpha_{ij} \xi_i \eta_j$$

converges. Show that there exists a constant $M > 0$ such that

$$\left| \sum_{i=0}^{\infty} \sum_{j=0}^{\infty} \alpha_{ij} \xi_i \eta_j \right| \leqq M\|x\| \cdot \|y\|$$

for all $x, y \in l^2$ (theorem of Hellinger-Toeplitz). (*Hint:* First prove with the method of Exercise 5 that if $(\zeta_i)_{i \in \mathbf{N}}$ is a sequence of numbers such that for every $(\xi_i) \in l^2$ the series $\sum_{i=0}^{\infty} \xi_i \zeta_i$ converges, then $(\zeta_i) \in l^2$.)

§9. *Banach's homomorphism theorem and the closed-graph theorem*

Let us start with a result from topology. Let X be a compact metric space, Y an arbitrary metric space, and f a continuous bijection from X onto Y. Then the inverse map is also continuous; i.e., f is a homeomorphism. Indeed, if A is a closed set in X, then it is compact; thus $f(A)$ is compact and hence also closed in Y.

We now prove a somewhat analogous result for linear maps of Banach spaces.

THEOREM 1. *Let E and F be two Banach spaces and f a continuous bijective linear map from E onto F. Then f is an isomorphism.*

We shall need two lemmas.

LEMMA 1. *If the image $f(B_1)$ of the unit ball B_1 of E is dense in some ball B'_ρ of F, then $f(B_1)$ contains the interior $\mathring{B}'_\rho$ of B'_ρ.*

Proof. The set $A = f(B_1) \cap B'_\rho$ is dense in B'_ρ. Indeed, let $y \in B'_\rho$ and $\epsilon > 0$. Take $z \in F$ such that $\|z\| < \rho$ and $\|y - z\| < \frac{1}{2}\epsilon$. There exists $x \in B_1$ such that $\|f(x) - z\| \leqq \min(\frac{1}{2}\epsilon, \rho - \|z\|)$. But then

$$\|y - f(x)\| < \epsilon \quad \text{and} \quad \|f(x)\| < \rho.$$

Let $y \in \mathring{B}'_\rho$. Take $0 < \epsilon < 1$ and let y_0 be the zero vector in F. We shall define by induction a sequence (y_n) of elements of F such that

$$y_{n+1} - y_n \in \epsilon^n A, \qquad \|y_{n+1} - y\| < \epsilon^{n+1}\rho \qquad (n \in \mathbf{N}).$$

Suppose that the elements y_i for $0 \leqq i \leqq n$ have already been selected and satisfy the conditions indicated. The set $y_n + \epsilon^n A$ is dense in the ball

$$y_n + \epsilon^n B'_\rho = B'_{\epsilon^n\rho}(y_n),$$

and $y \in y_n + \epsilon^n \mathring{B}'_\rho$. Hence there exists an element $y_{n+1} \in y_n + \epsilon^n A$ such that $\|y_{n+1} - y\| < \epsilon^{n+1}\rho$.

There exists a sequence $(x_n)_{n\in\mathbf{N}}$ of elements of E such that $x_0 = 0$, $f(x_{n+1}) = y_{n+1} - y_n$ and $\|x_{n+1}\| \leqq \epsilon^n$. Since the series with positive terms $\sum_{n=0}^{\infty} \epsilon^n$ is convergent, it follows from Proposition 3.8 that (x_n) is summable. Let $x = \sum_{n\in\mathbf{N}} x_n$. Then we have

$$\|x\| \leqq \sum_{n=0}^{\infty} \|x_n\| \leqq \sum_{n=0}^{\infty} \epsilon^n = \frac{1}{1-\epsilon}$$

and

$$f(x) = \lim_{n\to\infty} \sum_{k=1}^{n} f(x_k) = \lim_{n\to\infty} \sum_{k=1}^{n} (y_k - y_{k-1}) = \lim_{n\to\infty} y_n = y.$$

We thus have $f(B_{1/(1-\epsilon)}) \supset B'_\rho$; that is, $f(B_1) \supset B'_{\rho(1-\epsilon)}$ for every $\epsilon > 0$. Hence $f(B_1)$ contains the interior of B'_ρ. ∎

LEMMA 2. *For every $B_\rho \subset E$ the set $\overline{f(B_\rho)}$ is a neighborhood of 0 in F.*

Proof. Since f is surjective, we have $\bigcup_{n \in \mathbf{N}} \overline{f(nB_\rho)} = F$. By the corollary to Baire's theorem (§8) one of the closed sets $\overline{f(nB_\rho)}$ contains some ball and consequently $\overline{f(B_\rho)}$ contains some ball $B'_\sigma(y_0)$. Clearly also $-B'_\sigma(y_0) \subset \overline{f(B_\rho)}$. We show that $\overline{f(B_\rho)}$ contains the ball B'_σ, from which the conclusion of the lemma follows. Let $\|y\| \leqq \sigma$; then

$$y = \tfrac{1}{2}(y_0 + y) + \tfrac{1}{2}(y - y_0),$$

and we have

$$y_0 + y \in B'_\sigma(y_0), \qquad y - y_0 \in - B'_\sigma(y_0).$$

Since $\overline{f(B_\rho)}$ is clearly convex (Definition 3.3), it follows that $y \in \overline{f(B_\rho)}$. ∎

Proof of Theorem 1. If B_ρ is a ball in E, it follows from Lemma 2 that $\overline{f(B_\rho)}$ is a neighborhood of 0 in F; i.e., $f(B_\rho)$ is dense in some ball B'_σ of F. But then, by Lemma 1, the set $f(B_\rho)$ contains the interior of B'_σ; i.e., $f(B_\rho)$ is a neighborhood of 0 in F. Thus f^{-1} is continuous at the origin. The theorem now follows from Proposition 4.1. ∎

Let us now deduce a few simple consequences of Theorem 1.

PROPOSITION 1 (Banach's homomorphism theorem or the open-mapping theorem). *Let E and F be two Banach spaces and f a continuous linear map from E onto F. Then f is a strict morphism* (Definition 4.2).

Proof. Let N be the kernel of f. The associated injection $\bar{f}$ is a continuous linear bijection from E/N onto F. Hence by Theorem 1 it is an isomorphism. ∎

PROPOSITION 2. *Let E and F be two Banach spaces and f a continuous linear map from E into F. Then f is a strict morphism if and only if $f(E)$ is closed in F.*

Proof. If $f(E)$ is closed in F, then it is a Banach space. Hence by Proposition 1 the map f is a strict morphism of E onto $f(E)$.

Conversely, if f is a strict morphism and N its kernel, then the associated injection $\bar{f}$ is an isomorphism of E/N onto $f(E)$. Since E/N is complete (Proposition 4.3), so is $f(E)$ and hence it is also closed in F. ∎

Observe that a particular case of the second (trivial) half of the proposition has already been used in the proof of Proposition 5.4.

Let X and Y be two metric spaces and f a map from X into Y. The graph of f is the subset of the cartesian product $X \times Y$ formed by all pairs of the form $(x, f(x))$. If f is continuous, then its graph G is a closed set in $X \times Y$. Indeed, let $(x, y) \in \overline{G}$. Then there exists a sequence (x_n)

of elements of X such that $\lim_{n\to\infty} x_n = x$ and $\lim_{n\to\infty} f(x_n) = y$. But the continuity of f implies that $\lim_{n\to\infty} f(x_n) = f(x)$. Thus $y = f(x)$ and $(x, y) = (x, f(x)) \in G$. We shall now see that for linear maps of Banach spaces the converse is also true.

PROPOSITION 3 (Closed-graph theorem). *Let E and F be two Banach spaces and f a linear map from E into F. If the graph of f in $E \times F$ is closed, then f is continuous.*

Proof. The graph G of f is a closed subspace of the Banach space $E \times F$ provided with the norm $\|(x, y)\| = (\|x\|^2 + \|y\|^2)^{1/2}$ (cf. Exercise 4.4). Thus G is also a Banach space, and it follows from the inequality $\|x\| \leqq \|(x, y)\|$ that the projection $(x, f(x)) \mapsto x$ from G onto E is continuous. By Theorem 1 the map $x \mapsto (x, f(x))$ is then also continuous; and since the projection $(x, f(x)) \mapsto f(x)$ is clearly continuous, it follows that the composite map $x \mapsto f(x)$ is continuous. ▮

COROLLARY. *Let f be a linear map from a Banach space E into a Banach space F. If for every sequence (x_n) of points in E which tends to zero and for which $f(x_n)$ tends to a vector y we have $y = 0$, then f is continuous.*

Proof. Let (x, z) adhere to the graph G of f in $E \times F$. Then there exists a sequence (x_n) of points of E such that (x_n) tends to x and $f(x_n)$ tends to z. But then $(x_n - x)$ tends to 0 and $f(x_n - x)$ tends to $z - f(x)$. By hypothesis $z = f(x)$; thus $(x, z) \in G$ and G is closed. Proposition 3 shows that f is continuous. ▮

EXERCISES

1. Show that Lemmas 1 and 2 are true if f is a continuous linear map from E onto F which is not necessarily injective. Then deduce Proposition 1 directly from the lemmas.

2. Deduce Theorem 1 from the closed-graph theorem. (*Hint:* Observe that the map $(x, y) \mapsto (y, x)$ from $E \times F$ onto $F \times E$ establishes an isomorphism between the graph of f and that of f^{-1}.)

CHAPTER 2

Locally Convex Spaces

§1. Some notions from topology

A *topology* is defined on a set X by assigning to every point $x \in X$ a collection $\mathfrak{B}(x)$ of subsets of X so that the following axioms are satisfied:

(NB 1) If $W \subset X$ and W contains a set $V \in \mathfrak{B}(x)$, then $W \in \mathfrak{B}(x)$.

(NB 2) The intersection of a finite collection of sets in $\mathfrak{B}(x)$ belongs to $\mathfrak{B}(x)$.

(NB 3) If $V \in \mathfrak{B}(x)$, then $x \in V$.

(NB 4) For every $V \in \mathfrak{B}(x)$ there exists $W \in \mathfrak{B}(x)$ such that for every $y \in W$ we have $V \in \mathfrak{B}(y)$.

A set X with a topology $\mathcal{T}$ defined on it is called a *topological space* and we say that X is equipped with the topology $\mathcal{T}$. The sets of $\mathfrak{B}(x)$ are called *neighborhoods* of x.

If X is a metric space, then it is trivial to check that the neighborhoods of the points $x \in X$, as defined in §2 of Chapter 1, satisfy the axioms (NB 1) through (NB 4). Thus every metric space is equipped in a natural way with a topology.

Let X be a topological space and $A \subset X$. The *interior* $\mathring{A}$ of A is the set of all those points x for which there exists a neighborhood $V \in \mathfrak{B}(x)$ such that $V \subset A$. Clearly, $\mathring{A} \subset A$. If $\mathring{A} = A$, then the set A is *open*. The set X and the empty subset of X are always open. Let $\mathfrak{O}$ be the collection of all open sets of X, then it is easy to verify that we have the following two properties:

(O 1) The union of any collection of sets in $\mathfrak{O}$ belongs to $\mathfrak{O}$.

(O 2) The intersection of a finite collection of sets in $\mathfrak{O}$ belongs to $\mathfrak{O}$.

Conversely, suppose that we have a set X and a collection $\mathfrak{O}$ of subsets of X which satisfies the conditions (O 1) and (O 2). Observe that $X \in \mathfrak{O}$ as the intersection of the empty collection of sets in $\mathfrak{O}$ and that $\emptyset \in \mathfrak{O}$ as the union of the empty collection of sets in $\mathfrak{O}$. We say that a set V is a neighborhood of $x \in X$ if there exists a set $U \in \mathfrak{O}$ such that $V \supset U$ and $x \in U$. The neighborhoods defined in this fashion satisfy Axioms (NB 1) through (NB 4). Indeed, (NB 1) and (NB 3) are obvious; (NB 2) follows from (O 2). As for (NB 4), we can choose $W = U$. Furthermore, for the topology $\mathcal{T}$ defined by these neighborhoods, $\mathfrak{O}$ is exactly the collection of open sets. To show this, let us denote by $\mathfrak{O}'$ the collection of all open sets for $\mathcal{T}$. If $A \in \mathfrak{O}'$, then for each $x \in A$ there exists a set $U_x \in \mathfrak{O}$ such that $x \in U_x \subset A$. Thus $A = \bigcup_{x \in A} U_x$, and by (O 1) we have $A \in \mathfrak{O}$. Conversely, if $A \in \mathfrak{O}$, then for every $x \in A$ the set A itself is a neighborhood of x contained in A; hence $A \in \mathfrak{O}'$. We see therefore that a topology can also be defined in terms of its open sets, and the system of Axioms (O 1) and (O 2) is equivalent to (NB 1) through (NB 4). A set is open if and only if it is the neighborhood of each of its points. We also define a neighborhood of an arbitrary subset A of a topological space X as a set V which contains an open set containing A. Thus a subset A of X is open if and only if it is a neighborhood of itself.

If A is a subset of a topological space X, then a point $x \in X$ is said to *adhere* to A if for any neighborhood V of x we have $V \cap A \neq \emptyset$. The set $\overline{A}$ of all points adherent to A is called the *adherence* or *closure* of A. Clearly, $A \subset \overline{A}$. If $A = \overline{A}$, then A is *closed*. A set A is closed if and only if $\complement A$ is open. The intersection of any collection of closed sets is a closed set. The union of a finite collection of closed sets is a closed set. If A is any subset of X, the set $\mathring{A}$ is open and the set $\overline{A}$ is closed. If $A \subset X$, $B \subset X$, and $\overline{A} \supset B$, then A is *dense* in B. If $\overline{A} = X$, then A is *everywhere dense*.

Let X and Y be two topological spaces. A bijection from X onto Y is called a *homeomorphism* if it transforms the collection of all open sets of X into the collection of all open sets of Y. Two topological spaces are *homeomorphic* if there exists a homeomorphism between them.

Let X and Y be two topological spaces and let f be a map from X into Y. We say that f is *continuous* at the point $x \in X$ if for every neighborhood W of $f(x)$ the set $f^{-1}(W)$ is a neighborhood of x. We say that f is continuous on X (or simply continuous) if it is continuous at every point of X. For this the following equivalent conditions are necessary and sufficient:

(a) For every open set A in Y the set $f^{-1}(A)$ is open in X.
(b) For every closed set A in Y the set $f^{-1}(A)$ is closed in X.

A bijection $f: X \to Y$ is a homeomorphism if and only if f and f^{-1} are both continuous. If $f: X \to Y$ and $g: Y \to Z$ are continuous maps, then $g \circ f$ is a continuous map from X into Z. Similarly, if f is continuous at $x \in X$ and g is continuous at $f(x)$, then $g \circ f$ is continuous at x.

Let $\mathcal{T}_1$ and $\mathcal{T}_2$ be two topologies on the same set X and let $\mathfrak{O}_i$ $(i = 1, 2)$ be the collection of all open sets for the topology $\mathcal{T}_i$. We say that $\mathcal{T}_1$ is *finer* (or larger) than $\mathcal{T}_2$ and that $\mathcal{T}_2$ is *coarser* (or smaller) than $\mathcal{T}_1$ if $\mathfrak{O}_1 \supset \mathfrak{O}_2$. The following conditions are necessary and sufficient for $\mathcal{T}_1$ to be finer than $\mathcal{T}_2$:

(a) For every $x \in X$ we have $\mathfrak{B}_1(x) \supset \mathfrak{B}_2(x)$, where $\mathfrak{B}_i(x)$ is the collection of all neighborhoods of x for $\mathcal{T}_i$ $(i = 1, 2)$.

(b) The identical bijection $x \mapsto x$ from X_1 onto X_2 is continuous, where X_i denotes X equipped with the topology $\mathcal{T}_i$ $(i = 1, 2)$.

Clearly, the relation "$\mathcal{T}_1$ finer than $\mathcal{T}_2$" is an order on the set of all topologies on the set X. The finest topology on X (i.e., the largest element of the set of all topologies) is the *discrete* topology for which every subset of X is open. The coarsest topology on X (i.e., the smallest topology) is the one for which $\emptyset$ and X are the only open sets (*chaotic* or indiscrete topology).

Let $(\mathcal{T}_\iota)_{\iota \in I}$ be a nonempty family of topologies on a set X and denote by $\mathfrak{O}_\iota$ the collection of all open sets for the topology $\mathcal{T}_\iota$ $(\iota \in I)$. There exists a topology $\mathcal{T}$ on X which is the *greatest lower bound* of the topologies $\mathcal{T}_\iota$, i.e., a topology $\mathcal{T}$ which has the following two properties:

(a) $\mathcal{T}$ is coarser than any $\mathcal{T}_\iota$ $(\iota \in I)$.

(b) If $\mathcal{T}'$ is coarser than every $\mathcal{T}_\iota$, then $\mathcal{T}'$ is coarser than $\mathcal{T}$.

Indeed, the collection $\mathfrak{O}$ of all open sets for $\mathcal{T}$ is simply given by $\mathfrak{O} = \bigcap_{\iota \in I} \mathfrak{O}_\iota$.

Similarly, there exists a topology $\overline{\mathcal{T}}$ on X which is the *least upper bound* of the topologies $\mathcal{T}_\iota$, i.e., a topology $\overline{\mathcal{T}}$ which has the following two properties:

(a) $\overline{\mathcal{T}}$ is finer than any $\overline{\mathcal{T}}_\iota$ $(\iota \in I)$.

(b) If $\overline{\mathcal{T}}'$ is finer than every $\mathcal{T}_\iota$, then $\overline{\mathcal{T}}'$ is finer than $\overline{\mathcal{T}}$.

Let Φ be the set of all topologies finer than any $\mathcal{T}_\iota$ $(\iota \in I)$. The set Φ is nonempty since the discrete topology on X belongs to it. Then $\overline{\mathcal{T}}$ is the greatest lower bound of Φ. In other words, $\overline{\mathcal{T}}$ is the smallest element of the set of all topologies which are finer than all the $\mathcal{T}_\iota$ $(\iota \in I)$. Similarly, of course, the greatest lower bound $\mathcal{T}$ of the topologies $\mathcal{T}_\iota$ $(\iota \in I)$ is the largest element of the set of all topologies which are coarser than all the $\mathcal{T}_\iota$ $(\iota \in I)$.

Let $\mathfrak{S}$ be a collection of subsets of a set X. There exists a coarsest topology $\mathcal{T}$ for which all the sets of $\mathfrak{S}$ are open, namely, the greatest lower bound of the set Φ of all topologies for which the sets of $\mathfrak{S}$ are open. The set Φ is not empty, since it contains the discrete topology. We say that $\mathfrak{S}$ is a *subbasis* (or a system of generators) of $\mathcal{T}$ and that $\mathcal{T}$ is the topology generated by $\mathfrak{S}$. Let $\mathfrak{S}'$ be the collection of all finite intersections of sets in $\mathfrak{S}$, then the collection of all open sets for the topology $\mathcal{T}$ is formed by all the unions of sets belonging to $\mathfrak{S}'$.

In particular, if $(\mathcal{T}_\iota)_{\iota \in I}$ is a nonempty family of topologies on a set X and $\mathfrak{O}_\iota$ is the collection of open sets for $\mathcal{T}_\iota$, then $\bigcup_{\iota \in I} \mathfrak{O}_\iota$ is a subbasis of the least upper bound of the topologies $\mathcal{T}_\iota$.

If $\mathfrak{B}$ is a collection of subsets of a set X with the property that every open set in the topology $\mathcal{T}$ generated by $\mathfrak{B}$ is the union of sets belonging to $\mathfrak{B}$, then we say that $\mathfrak{B}$ is a *basis* for $\mathcal{T}$. Thus if $\mathfrak{S}$ is a subbasis of the topology $\mathcal{T}$, then the collection $\mathfrak{S}'$ of all finite intersections of sets in $\mathfrak{S}$ is a basis of $\mathcal{T}$.

Let X be a set, $(Y_\iota)_{\iota \in I}$ a family of topological spaces, and for each index $\iota \in I$ let f_ι be a map from X into Y_ι. There exists a *coarsest topology* $\mathcal{T}$ on X *for which all the maps f_ι are continuous*, also called the *initial topology* on X for the family $(f_\iota)_{\iota \in I}$. If $\mathfrak{O}_\iota$ denotes the collection of all open subsets of Y_ι, then a subbasis of $\mathcal{T}$ is formed by all the sets $f_\iota^{-1}(A)$, where $A \in \mathfrak{O}_\iota$, $\iota \in I$. A map g from a topological space Z into X is continuous for $\mathcal{T}$ if and only if $f_\iota \circ g$ is a continuous map from Z into Y_ι for every $\iota \in I$.

There are two important examples for this method of defining a topology. First let X be a topological space and A a subset of X. We call the *induced topology* on A the coarsest topology on A for which the canonical injection $A \hookrightarrow X$ is continuous. A subset $U \subset A$ is open for the induced topology if and only if $U = A \cap G$, where G is an open subset of X. The topological space A equipped with the induced topology is called a (topological) *subspace* of X.

Next let X_1 and X_2 be two topological spaces and let π_i $(i = 1, 2)$ denote the projection of the cartesian product $X_1 \times X_2$ onto the factor space X_i, i.e., $\pi_i((x_1, x_2)) = x_i$. The *product topology* on $X_1 \times X_2$ is defined as the coarsest topology for which the projections π_i $(i = 1, 2)$ are continuous. A basis for the product topology on $X_1 \times X_2$ is formed by all the sets

$$U_1 \times U_2 = \overset{-1}{\pi_1}(U_1) \cap \overset{-1}{\pi_2}(U_2),$$

where U_i $(i = 1, 2)$ is an open set in X_i. Let $f\colon (x, y) \mapsto f(x, y)$ be a map from $X_1 \times X_2$ into a topological space Z. If $y \in X_2$, we define the partial map f_y from X_1 into Z by setting $f_y(x) = f(x, y)$ for all $x \in X_1$. If f is continuous, then f_y is also continuous.

Now let $(X_\iota)_{\iota \in I}$ be a family of topological spaces, Y a set, and for each index $\iota \in I$ let f_ι be a map from X_ι into Y. There exists a *finest topology* $\mathcal{T}$ on Y *for which all the maps* f_ι *are continuous;* we also call this the *final topology* on Y for the family (f_ι). A set $U \subset Y$ is open for $\mathcal{T}$ if and only if $f_\iota^{-1}(U)$ is open in X_ι for every $\iota \in I$. A map g from Y into a topological space Z is continuous for $\mathcal{T}$ if and only if $g \circ f_\iota$ is a continuous map from X_ι into Z for every $\iota \in I$.

An important example for this method of defining a topology is the *quotient topology*, which we shall consider in §5.

A topological space X is said to be a *Hausdorff space* if for any two distinct points x and y of X there exists a neighborhood V of x and a neighborhood W of y such that $V \cap W = \emptyset$. Any finite subset of a Hausdorff space is a closed set. A subspace of a Hausdorff space is a Hausdorff space. A product space $X_1 \times X_2$ is a Hausdorff space if and only if each factor space X_i $(i = 1, 2)$ is a Hausdorff space.

Exercises

1. Supply complete proofs for all the statements made without proof in this section.

2. Let x be a point of a topological space and V a neighborhood of x. Prove that $\mathring{V}$ is a neighborhood of x. (*Hint:* V contains the set W whose existence is postulated in Axiom (NB 4).)

3. (a) Prove that a collection $\mathfrak{S}$ of subsets of a set X is a subbasis for the topology $\mathcal{T}$ if and only if for every point $x \in X$ and every neighborhood V of x there exists a finite collection of sets in $\mathfrak{S}$ whose intersection is contained in V and contains x.

(b) State and prove a condition for a collection of subsets to be a basis of $\mathcal{T}$.

4. Prove that a collection $\mathfrak{S}$ of subsets of X is a basis of the topology it generates if and only if for any two sets U and V in $\mathfrak{S}$ and any point $x \in U \cap V$ there exists a set $W \in \mathfrak{S}$ such that $x \in W \subset U \cap V$.

5. (a) Let X be a set, Y a topological space equipped with the topology $\mathcal{T}$, and f a map from X into Y. The coarsest topology on X for which f is continuous is called the inverse image of $\mathcal{T}$ and denoted by $f^{-1}(\mathcal{T})$. Show that a subset A of X is open (resp. closed) in X for $f^{-1}(\mathcal{T})$ if and only if $A = f^{-1}(B)$,

where B is an open (resp. closed) set in Y. Show that a subset V of X is a neighborhood of the point $x \in X$ for $f^{-1}(\mathcal{T})$ if and only if $V = f^{-1}(W)$, where W is a neighborhood of $f(x)$ in Y. Finally, show that if $X \subset Y$ and f is the canonical injection $X \hookrightarrow Y$, then $f^{-1}(\mathcal{T})$ is the induced topology on X.

(b) Let X be a set, $(Y_\iota)_{\iota \in I}$ a family of topological spaces, and for each $\iota \in I$ let f_ι be a map from X into Y_ι. Call $\mathcal{T}_\iota$ the topology of Y_ι. Show that the coarsest topology on X for which all the maps f_ι are continuous is the least upper bound of all the inverse-image topologies $f_\iota^{-1}(\mathcal{T}_\iota)$ on X ($\iota \in I$).

§2. *Filters*

A collection $\mathfrak{F}$ of subsets of a set X is called a *filter* on X if it satisfies the following axioms:

(F 1) If $A \subset X$ and A contains a set $B \in \mathfrak{F}$, then $A \in \mathfrak{F}$.

(F 2) The intersection of a finite collection of sets in $\mathfrak{F}$ belongs to $\mathfrak{F}$.

(F 3) The empty subset of X does not belong to $\mathfrak{F}$.

It follows from the Axioms (F 2) and (F 3) that the intersection of a finite collection of sets in a filter is never empty. Since the intersection of an empty family of subsets of X is X itself, it follows from (F 2) that X belongs to every filter.

EXAMPLE 1. Let x be a point of a set X. The collection of all the subsets of X containing x is a filter on X.

EXAMPLE 2. Let X be a topological space and $x \in X$. The collection $\mathfrak{B}(x)$ of all neighborhoods of x forms a filter on X according to Axioms (NB 1) through (NB 3). Observe that Example 1 is a special case: the filter considered there is the filter of all neighborhoods of x for the discrete topology on X.

EXAMPLE 3. Let A be a nonempty subset of a set X. The collection of all the subsets of X containing A is a filter on X. Example 1 is of course again a special case.

EXAMPLE 4. Let X be a set having infinitely many elements. The collection of all subsets of X whose complement is finite is a filter on X. In particular, if $X = \mathbf{N}$, we obtain in this manner a filter called the *Fréchet filter*.

If $\mathfrak{F}_1$ and $\mathfrak{F}_2$ are two filters on the same set X, we say that $\mathfrak{F}_1$ is *finer* than $\mathfrak{F}_2$ (or that $\mathfrak{F}_2$ is *coarser* than $\mathfrak{F}_1$) if $\mathfrak{F}_1 \supset \mathfrak{F}_2$.

The relation "$\mathfrak{F}_1$ is finer than $\mathfrak{F}_2$" is an order on the set of all filters on X. The filter which consists of the set X alone is clearly the "smallest" element of the set of all filters. If X has two or more elements, then there is no "largest" element in the set of all filters on X.

Let $(\mathfrak{F}_\iota)_{\iota \in I}$ be a nonempty family of filters on a set X. Then it can be immediately verified that the set $\mathfrak{F} = \bigcap_{\iota \in I} \mathfrak{F}_\iota$ is a filter on X which has the following two properties:

(a) $\mathfrak{F}$ is coarser than any $\mathfrak{F}_\iota$ $(\iota \in I)$.

(b) If $\mathfrak{F}'$ is a filter coarser than every $\mathfrak{F}_\iota$ $(\iota \in I)$, then $\mathfrak{F}$ is finer than $\mathfrak{F}'$.

In other words, the filter $\mathfrak{F}$ is the *greatest lower bound* of the family $(\mathfrak{F}_\iota)_{\iota \in I}$ with respect to the order defined above on the set of all filters on X.

Again let $(\mathfrak{F}_\iota)_{\iota \in I}$ be a family of filters on a set X. This family has a *least upper bound* in the set of all filters on X if and only if there exists a filter which is finer than every $\mathfrak{F}_\iota$ $(\iota \in I)$. Indeed, if such a least upper bound exists, it is finer than every $\mathfrak{F}_\iota$ $(\iota \in I)$. Conversely, suppose that the set Φ of filters finer than any $\mathfrak{F}_\iota$ $(\iota \in I)$ is not empty. Then the greatest lower bound $\overline{\mathfrak{F}}$ of Φ is the least upper bound of $(\mathfrak{F}_\iota)_{\iota \in I}$. This filter $\overline{\mathfrak{F}}$ is characterized by the following two properties:

(a) $\overline{\mathfrak{F}}$ is finer than any $\mathfrak{F}_\iota$ $(\iota \in I)$.

(b) If $\overline{\mathfrak{F}}'$ is a filter finer than every $\mathfrak{F}_\iota$ $(\iota \in I)$, then $\overline{\mathfrak{F}}$ is coarser than $\overline{\mathfrak{F}}'$.

In particular, if $(\mathfrak{F}_\iota)_{\iota \in I}$ is a totally ordered family of filters on X, then $\bigcup_{\iota \in I} \mathfrak{F}_\iota$ is a filter, and clearly, the least upper bound of the family $(\mathfrak{F}_\iota)_{\iota \in I}$. Thus the set of all filters on a set X is inductive. A maximal element of the set of all filters is called an *ultrafilter*. In other words, a filter $\mathfrak{U}$ is an ultrafilter if there exists no filter finer than $\mathfrak{U}$ and distinct from $\mathfrak{U}$. It follows from Zorn's lemma that *every filter is contained in an ultrafilter.*

A filter $\mathfrak{F}$ *is an ultrafilter if and only if* $A \cup B \in \mathfrak{F}$ *implies that either* $A \in \mathfrak{F}$ *or* $B \in \mathfrak{F}$. Indeed, suppose that $\mathfrak{F}$ is an ultrafilter, $A \notin \mathfrak{F}$, $B \notin \mathfrak{F}$, and $A \cup B \in \mathfrak{F}$. Let $\mathfrak{G}$ be the collection of subsets Z of X such that $A \cup Z \in \mathfrak{F}$. It is trivial to verify that $\mathfrak{G}$ is a filter finer than $\mathfrak{F}$. In particular $\emptyset \notin \mathfrak{G}$ follows from the hypothesis that $A \notin \mathfrak{F}$. But $\mathfrak{G}$ is distinct from $\mathfrak{F}$ since $B \in \mathfrak{G}$, $B \notin \mathfrak{F}$. This contradicts the hypothesis that $\mathfrak{F}$ is an ultrafilter. Conversely, suppose that $A \cup B \in \mathfrak{F}$ implies either $A \in \mathfrak{F}$ or $B \in \mathfrak{F}$. Let $\mathfrak{G}$ be a filter finer than $\mathfrak{F}$ and let $Z \in \mathfrak{G}$. Then $\complement Z \notin \mathfrak{G}$ and thus $\complement Z \notin \mathfrak{F}$. Since $X = Z \cup \complement Z \in \mathfrak{F}$, we must have $Z \in \mathfrak{F}$; i.e., $\mathfrak{G} = \mathfrak{F}$. ∎

Let $\mathfrak{G}$ be a collection of subsets of a set X and let $\mathfrak{G}'$ be the collection of all finite intersections of sets in $\mathfrak{G}$. A necessary and sufficient condition for the existence of a filter $\mathfrak{F}$ such that $\mathfrak{F} \supset \mathfrak{G}$ is that none of the sets in $\mathfrak{G}'$ be empty. Indeed, if such a filter exists, then by Axiom (F 2) it must contain $\mathfrak{G}'$, and thus by Axiom (F 3) the empty set cannot belong to $\mathfrak{G}'$. Conversely, suppose that no set of $\mathfrak{G}'$ is empty and let $\mathfrak{F}$ be the collection of all subsets of X which contain a set belonging to $\mathfrak{G}'$. It can be immediately verified that $\mathfrak{F}$ is a filter containing $\mathfrak{G}$. Actually the filter $\mathfrak{F}$ just defined is clearly the coarsest filter containing $\mathfrak{G}$. We say that $\mathfrak{G}$ is a *subbasis* (or a system of generators) of $\mathfrak{F}$ and that $\mathfrak{F}$ is the filter *generated* by $\mathfrak{G}$.

In particular, if $(\mathfrak{F}_\iota)_{\iota \in I}$ is a family of filters, then $\bigcup_{\iota \in I} \mathfrak{F}_\iota$ is a subbasis of the least upper bound $\overline{\mathfrak{F}}$ of the filters $\mathfrak{F}_\iota$, provided this least upper bound exists. For $\overline{\mathfrak{F}}$ to exist it is necessary and sufficient that the intersection of a finite number of sets selected from various filters $\mathfrak{F}_\iota$ be never empty.

Let $\mathfrak{B}$ be a collection of subsets of a set X and let $\mathfrak{F}$ be the collection of all subsets of X which contain a set of $\mathfrak{B}$. Clearly, $\mathfrak{F}$ is a filter if and only if $\mathfrak{B}$ satisfies the following two conditions:

(FB 1) The intersection of two sets of $\mathfrak{B}$ contains a set of $\mathfrak{B}$.

(FB 2) $\mathfrak{B}$ is not empty and the empty subset of X does not belong to $\mathfrak{B}$.

A collection $\mathfrak{B}$ of subsets of X is called a *filter basis* if it satisfies Axioms (FB 1) and (FB 2). We also say that $\mathfrak{B}$ is a *basis* of the filter $\mathfrak{F}$ it generates. Two filter bases are said to be *equivalent* if they generate the same filter. Let $\mathfrak{B}$ and $\mathfrak{B}'$ be two filter bases generating the filters $\mathfrak{F}$ and $\mathfrak{F}'$ respectively. Then $\mathfrak{F}$ is finer than $\mathfrak{F}'$ if and only if every set in $\mathfrak{B}'$ contains a set in $\mathfrak{B}$. In particular, $\mathfrak{B}$ and $\mathfrak{B}'$ are equivalent if and only if every set in $\mathfrak{B}$ contains a set in $\mathfrak{B}'$ and every set in $\mathfrak{B}'$ contains a set in $\mathfrak{B}$.

A subcollection $\mathfrak{B}$ of sets belonging to a filter $\mathfrak{F}$ is a basis of $\mathfrak{F}$ if and only if every set of $\mathfrak{F}$ contains a set of $\mathfrak{B}$. Indeed, if each set in $\mathfrak{F}$ contains a set in $\mathfrak{B}$, then clearly $\mathfrak{B}$ is a filter basis, and by Axiom (F 1), $\mathfrak{F}$ is the collection of all sets containing a set of $\mathfrak{B}$. Conversely, if $\mathfrak{B}$ is a basis of $\mathfrak{F}$, then by definition every set in $\mathfrak{F}$ must contain a set in $\mathfrak{B}$.

EXAMPLE 5. Let X be a directed set and for each $x \in X$ denote by $S(x)$ the set $\{y \mid y \geqq x\}$. The collection $\mathfrak{S}$ of all the sets $S(x)$ is a filter basis on X. Indeed, Axiom (FB 2) is evidently satisfied. Furthermore, if x and y are elements of X, then there exists an element $z \in X$ such that $x \leqq z$ and $y \leqq z$, and thus $S(x) \cap S(y) \supset S(z)$. The filter generated by $\mathfrak{S}$ is called the *filter of sections* of X. In particular, if $X = \mathbf{N}$, the filter of sections is the Fréchet filter.

Let X be a topological space and x a point of X. A basis of the filter $\mathfrak{B}(x)$ of all neighborhoods of x is called a *fundamental system of neighborhoods* of x. A collection $\mathfrak{S}$ of neighborhoods of x is a fundamental system of neighborhoods of x if and only if every neighborhood of x contains a set $W \in \mathfrak{S}$.

The open neighborhoods of a point x in a topological space form a fundamental system of neighborhoods of x (see Exercise 1.2). If X is a metric space and x a point of X, the set of all closed balls $B_\rho(x)$, where ρ runs through the set of all strictly positive real numbers (or only through the set of all numbers of the form $1/n$, $n \in \mathbf{N}^*$), forms a fundamental system of neighborhoods of x (see p. 19).

Let X_1 and X_2 be two topological spaces and $X_1 \times X_2$ their product equipped with the product topology. A fundamental system of neighborhoods of a point $(x_1, x_2) \in X_1 \times X_2$ is given by all sets of the form $V_1 \times V_2$, where V_i is a neighborhood of x_i in X_i $(i = 1, 2)$. More generally, if $\mathfrak{S}_i$ $(i = 1, 2)$ is a fundamental system of neighborhoods of the point x_i in X_i, then the sets $W_1 \times W_2$, where W_i runs through $\mathfrak{S}_i$, form a fundamental system of neighborhoods of (x_1, x_2).

A topological space is said to be *regular* if it is a Hausdorff space and if the closed neighborhoods of each of its points form a fundamental system of neighborhoods of that point. In other words, a Hausdorff space is regular if every neighborhood contains a closed neighborhood.

Let $\mathfrak{B}$ be a filter basis on a set X and let f be a mapping from X into a set Y. We denote by $f(\mathfrak{B})$ the collection of all sets $f(A)$, where $A \in \mathfrak{B}$. Then $f(\mathfrak{B})$ is a filter basis on Y since $f(A) \cap f(B) \supset f(A \cap B)$ and $A \neq \emptyset$ implies $f(A) \neq \emptyset$.

Let $\mathfrak{F}$ be the filter on X generated by $\mathfrak{B}$ and $\mathfrak{G}$ the filter on Y generated by $f(\mathfrak{B})$. *If $\mathfrak{F}$ is an ultrafilter, then so is $\mathfrak{G}$.* Indeed, let $A \cup B \in \mathfrak{G}$; we have to show that either $A \in \mathfrak{G}$ or $B \in \mathfrak{G}$. There exists $C \in \mathfrak{B}$ such that $A \cup B \supset f(C)$, that is, $f^{-1}(A) \cup f^{-1}(B) \supset f^{-1}(f(C)) \supset C$. Thus $f^{-1}(A) \cup f^{-1}(B) \in \mathfrak{F}$; and since $\mathfrak{F}$ is an ultrafilter, either $f^{-1}(A) \in \mathfrak{F}$ or $f^{-1}(B) \in \mathfrak{F}$. Suppose, for instance, that $f^{-1}(A) \in \mathfrak{F}$. Then

$$f^{-1}(A) \supset D, \quad D \in \mathfrak{B},$$

and thus $A \supset f(D)$, that is, $A \in \mathfrak{G}$. Q.E.D.

Now let $\mathfrak{B}$ be a filter basis on a set Y and let f be a mapping from a set X into Y. We denote by $f^{-1}(\mathfrak{B})$ the collection of all sets $f^{-1}(A)$, where $A \in \mathfrak{B}$. Then $f^{-1}(\mathfrak{B})$ is a filter basis on X if and only if $f^{-1}(A) \neq \emptyset$ for every $A \in \mathfrak{B}$, i.e., if $f(X)$ meets every $A \in \mathfrak{B}$, as can be seen from the relation $f^{-1}(A \cap B) = f^{-1}(A) \cap f^{-1}(B)$. In particular, $f^{-1}(\mathfrak{B})$ is a filter basis if f is a surjection.

If X is a set, A a nonempty subset of X, and $\mathfrak{F}$ a filter on X, we denote by $\mathfrak{F}_A$ the *trace* of $\mathfrak{F}$ on A, i.e., the collection of all sets $A \cap B$, where $B \in \mathfrak{F}$. Then $\mathfrak{F}_A$ is a filter on A if and only if $A \cap B \neq \emptyset$ for every $B \in \mathfrak{F}$. Indeed, this last condition is necessary and sufficient for $\mathfrak{F}_A$ to satisfy Axiom (F 3). The other two axioms are always satisfied since

$$\left(\bigcap_{i=1}^{n} B_i\right) \cap A = \bigcap_{i=1}^{n} (B_i \cap A)$$

and

$$B \cap A \subset C \subset A \quad \text{implies} \quad C = (B \cup C) \cap A.$$

If the condition is satisfied, we say that $\mathfrak{F}_A$ is the filter *induced* by $\mathfrak{F}$ on A.

Similarly, if $\mathfrak{B}$ is a filter basis on X, then the trace

$$\mathfrak{B}_A = \{A \cap B \mid B \in \mathfrak{B}\}$$

of $\mathfrak{B}$ on A is a filter basis on A if and only if A meets every $B \in \mathfrak{B}$. This follows from our considerations above, since $\mathfrak{B}_A = f^{-1}(\mathfrak{B})$, where f is the canonical injection $A \hookrightarrow X$.

Exercises

1. Prove that a Hausdorff space X is regular if and only if given a point $x \in X$ and a closed subset A of X such that $x \notin A$, there exist disjoint open sets U and V such that $x \in U$ and $A \subset V$.

2. Let $\mathfrak{U}$ be an ultrafilter on a set X and A a subset of X. Prove that the trace of $\mathfrak{U}$ on A is a filter on A if and only if $A \in \mathfrak{U}$.

3. Let $\mathfrak{F}$ be a filter on a set X and A a subset of X. There exists a filter $\mathfrak{F}'$ finer than $\mathfrak{F}$ and containing A if and only if $A \cap B \neq \emptyset$ for every $B \in \mathfrak{F}$.

4. Show that if the map $f: X \to Y$ is surjective and $\mathfrak{F}$ is a filter on X, then $f(\mathfrak{F})$ is a filter on Y. (*Hint:* A map $f: X \to Y$ is surjective if and only if $f(f^{-1}(A)) = A$ for every $A \subset Y$.)

5. Let X and Y be two sets and f a map from X into Y.

(a) Prove that if $\mathfrak{B}$ is a filter basis on X, then $f^{-1}(f(\mathfrak{B}))$ is a filter basis on X and that the filter generated by $\mathfrak{B}$ is finer than the filter generated by $f^{-1}(f(\mathfrak{B}))$.

(b) Prove that $f^{-1}(f(\mathfrak{B}))$ is equivalent to $\mathfrak{B}$ for every filter basis $\mathfrak{B}$ on X if and only if f is injective.

§3. *Topological vector spaces*

Definition 1. *A set E on which a structure of vector space over* $\mathbf{K}$ *and a topology are defined is a topological vector space if*

(TVS 1) *the map $(x, y) \mapsto x + y$ from $E \times E$ into E is continuous;*

(TVS 2) *the map $(\lambda, x) \mapsto \lambda x$ from $\mathbf{K} \times E$ into E is continuous.*

It is understood, of course, that in Axioms (TVS 1) and (TVS 2) we consider the product topology on the product spaces. A vector space structure and a topology on a set are said to be *compatible* if Axioms (TVS 1) and (TVS 2) are satisfied.

Example 1. According to Proposition 1.2.1 a normed vector space, equipped with the topology defined by its norm, is a topological vector space.

It follows from (TVS 1) that for any $a \in E$ the translation by a (p. 20) is a continuous map from E into E. Since the inverse of the bijection $x \mapsto x + a$ is the map $x \mapsto x - a$, we find that the translation by a is a homeomorphism of E onto itself. In particular, the neighborhoods of a are the sets of the form $V + a$, where V is a neighborhood of 0. Thus we know the topology of a topological vector space if we know the neighborhood of 0. We shall now investigate the latter in greater detail.

In the first place, it follows from (TVS 1) that for any neighborhood V of 0 there exists a neighborhood U of 0 such that $U + U \subset V$.

DEFINITION 2. *A set A in a vector space E over* **K** *is absorbing* (or radial at 0) *if for every $x \in E$ there exists an $\alpha > 0$ such that $x \in \lambda A$ for all $\lambda \in$* **K**, *such that $|\lambda| \geqq \alpha$.*

Since this means that $\lambda x \in A$ for all λ such that $|\lambda| \leqq \alpha^{-1}$, it follows from (TVS 2) that every neighborhood of 0 in E is absorbing.

DEFINITION 3. *A set A in a vector space E over* **K** *is balanced* (or circled) *if $\lambda A \subset A$ for every $\lambda \in$* **K** *such that $|\lambda| \leqq 1$.*

If A is balanced and for every $x \in E$ there exists a $\mu \in \mathbf{K}$ such that $x \in \mu A$, then A is absorbing. Indeed, if $|\lambda| \geqq |\mu|$, then $|\lambda^{-1}\mu| \leqq 1$ and thus $x \in \mu A = \lambda(\lambda^{-1}\mu)A \subset \lambda A$.

The intersection of an arbitrary family of balanced sets is balanced. Given an arbitrary set B in E, there exists a smallest balanced set A containing B. This set A is called the *balanced hull* of B and is the intersection of all balanced sets containing B.

The union of an arbitrary family of balanced sets is balanced. Therefore given any subset A of E, there exists a largest balanced set B contained in A, namely, the union of all balanced sets contained in A. The set B is called the *balanced core* of A. It is nonempty if and only if A contains the origin.

A point $x \in E$ belongs to the balanced core B of A if and only if $\lambda x \in A$ for all $\lambda \in \mathbf{K}$ such that $|\lambda| \leqq 1$. Indeed, the set $C(x) = \{\lambda x \mid |\lambda| \leqq 1\}$ is clearly balanced; hence if $C(x) \subset A$, then $C(x) \subset B$, and in particular, $x \in B$. Conversely, if $x \in B$, then $\lambda x \in B$ and *a fortiori* $\lambda x \in A$ for all $\lambda \in \mathbf{K}$ with $|\lambda| \leqq 1$.

It follows that if $B \neq \emptyset$, then it is given by

$$B = \bigcap_{|\lambda| \geqq 1} \lambda A, \tag{1}$$

since in this case "$x \in \lambda A$ for every $\lambda \in \mathbf{K}$ with $|\lambda| \geqq 1$" is equivalent to "$\mu x \in A$ for every $\mu \in \mathbf{K}$ with $|\mu| \leqq 1$."

Let E and F be two vector spaces over **K** and $f: E \to F$ a linear map. If A is a balanced set in E, then $f(A)$ is a balanced set in F. If B is a balanced set in F, then $f^{-1}(B)$ is a balanced set in E.

Let us now suppose that E is a topological vector space. For $\alpha \neq 0$ the bijection $x \mapsto \alpha x$ from E onto E is continuous by (TVS 2). Since its inverse is the bijection $x \mapsto \alpha^{-1}x$, it follows that $x \mapsto \alpha x$ is a homeomorphism from E onto itself. Hence formula (1) shows that if A is closed, then its balanced core is also closed.

In a topological vector space the balanced core of a neighborhood V of 0 is a neighborhood of 0. Indeed, by (TVS 2) there exists an $\alpha > 0$ and a neighborhood U of 0 such that $|\lambda| \leqq \alpha$ and $x \in U$ imply $\lambda x \in V$. Since the map $x \mapsto \alpha x$ is a homeomorphism of E onto itself, αU is a neighborhood of 0. Furthermore, αU is contained in the balanced core of V. For let $\alpha x \in \alpha U$ and $|\mu| \leqq 1$; then $|\mu\alpha| \leqq \alpha$ and thus

$$\mu(\alpha x) = (\mu\alpha)x \in V.$$

We have now proved the first part of the following:

THEOREM 1. *In a topological vector space E there exists a fundamental system $\mathfrak{N}$ of neighborhoods of 0 such that:*

(NS 1) *Every $V \in \mathfrak{N}$ is absorbing.*

(NS 2) *Every $V \in \mathfrak{N}$ is balanced.*

(NS 3) *For every $V \in \mathfrak{N}$ there exists a $U \in \mathfrak{N}$ such that $U + U \subset V$.*

Conversely, let E be a vector space over **K** *and let $\mathfrak{N}$ be a filter basis on E which satisfies conditions* (NS 1) *through* (NS 3). *Then there exists a unique topology on E for which E is a topological vector space and for which $\mathfrak{N}$ is a fundamental system of neighborhoods of* 0.

Proof. Let us first observe that if there exists a topology on E for which E is a topological vector space and $\mathfrak{N}$ a fundamental system of neighborhoods of 0, then in that topology a set W is a neighborhood of a point $a \in E$ if and only if W contains a set of the form $V + a$, where $V \in \mathfrak{N}$. This proves that the topology, if it exists, must be unique. Let us now prove that defining the neighborhoods in E in the way just indicated, we obtain a structure of topological vector space on E.

Clearly, if $W_1 \supset W$ and $W \supset V + a$, where $V \in \mathfrak{N}$, then $W_1 \supset V + a$; that is, Axiom (NB 1) is satisfied. Next let W_i $(1 \leqq i \leqq n)$ be a finite number of neighborhoods of a. Then each W_i contains a set $V_i + a$, where $V_i \in \mathfrak{N}$. Since $\mathfrak{N}$ is a filter basis, there exists a set $V \in \mathfrak{N}$ contained in $\bigcap_{i=1}^n V_i$. Then

$$\bigcap_{i=1}^{n} W_i \supset V + a;$$

i.e., $\bigcap_{i=1}^n W_i$ is a neighborhood of a. This proves (NB 2). For every $V \in \mathfrak{N}$ we have $0 \in V$ since, by the definition of a filter basis, no V is empty and if $x \in V$, then by (NS 2) we have $0 \cdot x \in V$. It follows that if $W \supset V + a$, $V \in \mathfrak{N}$, then $a \in W$; i.e., (NB 3) is satisfied. Finally, let

$W \supset V + a$, where $V \in \mathfrak{N}$. By (NS 3) there exists $U \in \mathfrak{N}$ such that $U + U \subset V$. Then $U + a$ is a neighborhood of a, and if $b \in U + a$, then $U + b \subset U + U + a \subset V + a \subset W$; i.e., for every $b \in U + a$ the set W is a neighborhood of b. Thus we have also verified Axiom (NB 4).

Let $a + b = c$ and W be a neighborhood of c. Then W contains a set $V + c$, where $V \in \mathfrak{N}$ and by (NS 3) there exists $U \in \mathfrak{N}$ such that $U + U \subset V$. Then $U + a$ is a neighborhood of a, $U + b$ is a neighborhood of b, and we have

$$(U + a) + (U + b) \subset V + a + b = V + c \subset W;$$

i.e., Axiom (TVS 1) is satisfied.

Next let us prove that given $V \in \mathfrak{N}$ and $\lambda \in \mathbf{K}$, there exists $U \in \mathfrak{N}$ such that $\lambda U \subset V$. In the first place, it follows from (NS 3) by mathematical induction that given $V \in \mathfrak{N}$, for any $n \in \mathbf{N}$ there exists $U \in \mathfrak{N}$ such that $2^n U \subset V$. Now let n be so large that $|\lambda| \leqq 2^n$. If $U \in \mathfrak{N}$ is such that $2^n U \subset V$, then by (NS 2) we have $\lambda 2^{-n} U \subset U$, i.e. $\lambda U \subset 2^n U \subset V$.

Let $a \in E$, $\lambda \in \mathbf{K}$, and let W be a neighborhood of λa. There exists $V \in \mathfrak{N}$ such that $W \supset V + \lambda a$ and furthermore, applying (NS 3) twice, one sees that there exists $U \in \mathfrak{N}$ such that $U + U + U \subset V$. Because of (NS 1) there exists $\epsilon > 0$ such that $|\eta| < \epsilon$ implies $\eta a \in U$. By the remark just proved, there exists $T \in \mathfrak{N}$ such that $\lambda T \subset U$. Furthermore, if $|\eta| \leqq 1$ and $x - a \in U$, then by (NS 2) we have $\eta(x - a) \in U$. Let $S \in \mathfrak{N}$ be such that $S \subset T \cap U$. It follows from the identity

$$\xi x - \lambda a = (\xi - \lambda)a + \lambda(x - a) + (\xi - \lambda)(x - a)$$

that if $|\xi - \lambda| \leqq \min(1, \epsilon)$ and $x \in S + a$, then

$$\xi x - \lambda a \in U + U + U \subset V,$$

i.e., $\xi x \in W$. This proves (TVS 2). ∎

PROPOSITION 1. *Let E be a vector space over $\mathbf{K}$ and let $\mathfrak{S}$ be a collection of absorbing, balanced subsets of E such that for every $V \in \mathfrak{S}$ there exists $U \in \mathfrak{S}$ such that $U + U \subset V$. Then there exists a unique topology on E for which E is a topological vector space and for which the finite intersections of elements of $\mathfrak{S}$ form a fundamental system of neighborhoods of* 0.

Proof. An absorbing set is nonempty; hence every $V \in \mathfrak{S}$ contains 0 because it is balanced. Thus the finite intersections of elements of $\mathfrak{S}$ form a filter basis $\mathfrak{N}$ on E. It is completely clear that $\mathfrak{N}$ satisfies the conditions (NS 1) through (NS 3) of Theorem 1. ∎

EXAMPLE 2. In a normed vector space the balls $B_\rho = \{x \mid \|x\| \leqq \rho\}$ form a filter basis satisfying conditions (NS 1) through (NS 3).

EXAMPLE 3. Let Ω be an open subset of $\mathbf{R}^n$ and let $\mathcal{C}(\Omega)$ be the vector space of all continuous functions f defined on Ω. For every compact subset K of Ω and strictly positive number ϵ let $V_{K,\epsilon}$ be the set of all $f \in \mathcal{C}(\Omega)$ such that $|f(x)| \leqq \epsilon$ for $x \in K$. Since the union of two compact sets is compact, the sets $V_{K,\epsilon}$ form a filter basis on $\mathcal{C}(\Omega)$ which clearly satisfies conditions (NS 1) through (NS 3). Thus the $V_{K,\epsilon}$ form a fundamental system of neighborhoods of 0 for a topology on $\mathcal{C}(\Omega)$ which is called the topology of uniform convergence on compact sets.

EXAMPLE 4. Let K be a compact subset of $\mathbf{R}^n$. We denote by $\mathfrak{D}(K)$ the vector space of all functions which are defined in some open set containing K, which vanish outside K and whose partial derivatives of all orders exist and are continuous. We can, of course, extend the definition of the functions of $\mathfrak{D}(K)$ to the whole space $\mathbf{R}^n$ simply by assigning the value zero to each point outside the original domain of definition and consider the elements of $\mathfrak{D}(K)$ as infinitely differentiable functions defined on $\mathbf{R}^n$.

Let f be a continuous function defined in some open subset Ω of $\mathbf{R}^n$. The *support* (or carrier) of f is the closure of the set on which f is different from zero; i.e., if $A = \{x \mid f(x) \neq 0\}$, then $\operatorname{Supp} f = \bar{A}$. With this terminology we can say that $\mathfrak{D}(K)$ is the vector space of all functions defined on $\mathbf{R}^n$ whose partial derivatives of all orders exist and are continuous and whose support is contained in K.

Before defining the topology of $\mathfrak{D}(K)$, let us introduce some important notations which will be of constant use in this book. We denote by $\mathbf{N}^n$ the set of all n-tuples $p = (p_1, \ldots, p_n)$, where each $p_j \in \mathbf{N}$. In other words, an element of $\mathbf{N}^n$ is a vector whose components are natural numbers and we shall also call such an element a *multi-index*. The *order* $|p|$ of a multi-index $p = (p_1, \ldots, p_n) \in \mathbf{N}^n$ is defined by $|p| = p_1 + \cdots + p_n$. We shall use the abbreviated symbol ∂_j for the symbol of derivation $\partial/\partial x_j$ $(j = 1, \ldots, n)$, and for any $p \in \mathbf{N}^n$ we set

$$\partial^p = \partial_1^{p_1} \ldots \partial_n^{p_n} = \frac{\partial^{|p|}}{\partial x_1^{p_1} \partial x_2^{p_2} \cdots \partial x_n^{p_n}},$$

where the order $|p|$ of p is also the order of the derivation.

For any $p \in \mathbf{N}^n$ and $\epsilon > 0$ let $V_{p,\epsilon}$ be the set of all $f \in \mathfrak{D}(K)$ such that $|\partial^p f(x)| \leqq \epsilon$ for all $x \in K$. Clearly, the sets $V_{p,\epsilon}$ satisfy conditions (NS 1) through (NS 3) and thus by Proposition 1 their finite intersections form a fundamental system of neighborhoods of 0 for a topology for which $\mathfrak{D}(K)$ is a topological vector space.

We shall see many more examples of topological vector spaces in the next section; now we prove some additional results.

PROPOSITION 2. *A topological vector space is a Hausdorff space if for every element $a \neq 0$ there exists a neighborhood V of 0 which does not contain a.*

Proof. It is sufficient to prove that if $a \neq 0$, then there exists a neighborhood U of 0 and a neighborhood W of a such that $U \cap W \neq \emptyset$. Indeed, if $a \neq b$, then $a - b \neq 0$. Let U be a neighborhood of 0 and W a neighborhood of $a - b$ such that $U \cap W = \emptyset$. Then $W + b$ is a neighborhood of a, $U + b$ a neighborhood of b, and $(W + b) \cap (U + b) = \emptyset$.

Let $a \neq 0$ and let V be a neighborhood of 0 which does not contain a. There exists a balanced neighborhood U of 0 such that $U + U \subset V$. Then U is a neighborhood of 0 and $U + a$ is a neighborhood of a. Furthermore, $U \cap (U + a) = \emptyset$ since $x = y + a \in U$, $y \in U$ would imply $a = x - y \in U + U \subset V$. ∎

The spaces considered in Examples 2, 3, and 4 are Hausdorff spaces.

PROPOSITION 3. *In a topological vector space every neighborhood of* 0 *contains a closed neighborhood of* 0.

Proof. Let V be a neighborhood of 0. There exists a balanced neighborhood U of 0 such that $U + U \subset V$. Let us show that $\overline{U} \subset V$. If $x \in \overline{U}$, then $(x + U) \cap U \neq \emptyset$; i.e., there exists $y \in U$ such that $x + y \in U$. But then

$$x \in -y + U \subset U + U \subset V. \blacksquare$$

COROLLARY. *If a topological vector space is a Hausdorff space, then it is regular.*

EXERCISES

1. Prove that if in a topological vector space E the set $\{0\}$ is closed, then E is a Hausdorff space (cf. Proposition 5.5).

2. Prove that the intersection of a finite collection of absorbing sets in a vector space is absorbing.

3. (a) Prove that the balanced hull of a set B in a vector space is given by $\bigcup_{|\lambda| \leqq 1} \lambda B$.

(b) Show that if for each $\iota \in I$ the set A_ι is the balanced hull of B_ι, then $\bigcup_{\iota \in I} A_\iota$ is the balanced hull of $\bigcup_{\iota \in I} B_\iota$.

4. Show by an example that if $0 \notin A$, then $\bigcap_{|\lambda| \geqq 1} \lambda A$ can be nonempty (i.e., different from the balanced core of A).

5. Prove that the closure of a balanced set in a topological vector space is also balanced. (*Hint:* See the proof of Proposition 4.3.)

§4. *Locally convex spaces*

We have already defined convex sets in a vector space E (Definition 1.3.3). We can say that a set A is convex if for $\alpha \geqq 0$, $\beta \geqq 0$, $\alpha + \beta = 1$ we have $\alpha A + \beta A \subset A$. If A is a convex set in E, then $A + a$ is convex for every $a \in E$ and λA is convex for every $\lambda \in \mathbf{K}$. Let E and F be two vector spaces and f a linear map from E into F. If A is convex in E, then $f(A)$ is convex in F. If B is convex in F, then $f^{-1}(B)$ is convex in E.

The intersection of an arbitrary family of convex sets is a convex set. Given an arbitrary set B in E, there exists a smallest convex set A containing B, namely, the intersection of all convex sets containing B. This set A is called the *convex hull* of B.

PROPOSITION 1. *Let A be a convex subset of a vector space, $(x_i)_{1 \leqq i \leqq n}$ a finite sequence of elements of A, and $(\lambda_i)_{1 \leqq i \leqq n}$ a sequence of scalars such that $\lambda_i \geqq 0$ $(1 \leqq i \leqq n)$ and $\sum_{i=1}^n \lambda_i = 1$. Then the linear combination $\sum_{i=1}^n \lambda_i x_i$ belongs to A.*

Proof. For $n = 1$ the proposition is trivial, and for $n = 2$ it is the definition of a convex set. Assume therefore that the proposition is true for $n - 1$, and we may clearly suppose that $\lambda_i > 0$ for $1 \leqq i \leqq n$. Set

$$\alpha = \sum_{i=1}^{n-1} \lambda_i, \qquad \beta = \lambda_n, \qquad \mu_i = \frac{\lambda_i}{\alpha}$$

for $1 \leqq i \leqq n - 1$. By the induction hypothesis we have

$$\sum_{i=1}^{n-1} \mu_i x_i \in A,$$

and therefore by the definition of a convex set

$$\sum_{i=1}^{n} \lambda_i x_i = \alpha \left(\sum_{i=1}^{n-1} \mu_i x_i \right) + \beta x_n \in A. \blacksquare$$

PROPOSITION 2. *Let $(A_\iota)_{\iota \in I}$ be a family of convex subsets of a vector space. Then the convex hull of $\bigcup_{\iota \in I} A_\iota$ is the set C of all linear combinations $\sum_{\iota \in I} \lambda_\iota x_\iota$, where $x_\iota \in A_\iota$, $\lambda_\iota \geqq 0$, $\sum_{\iota \in I} \lambda_\iota = 1$, and, of course, only finitely many λ_ι are different from zero.*

Proof. By Proposition 1 the set C is contained in every convex set containing the sets A_ι. On the other hand, taking $\lambda_\iota = 1$, we see that $A_\iota \subset C$ for every $\iota \in I$.

Thus we have only to prove that C is convex. Let $x = \sum_{\iota \in I} \lambda_\iota x_\iota$ and $y = \sum_{\iota \in I} \mu_\iota y_\iota$ be two points of C, $\alpha > 0$, $\beta > 0$, and $\alpha + \beta = 1$. For each $\iota \in I$ we set $\nu_\iota = \alpha\lambda_\iota + \beta\mu_\iota$. Let J be the finite subset of I for which $\nu_\iota > 0$. Then

$$z_\iota = \frac{\alpha\lambda_\iota x_\iota + \beta\mu_\iota y_\iota}{\alpha\lambda_\iota + \beta\mu_\iota} \in A_\iota \qquad \text{for} \qquad \iota \in J$$

and

$$\alpha x + \beta y = \sum_{\iota \in J} \nu_\iota z_\iota \in C$$

since $\sum \nu_\iota = \alpha \sum \lambda_\iota + \beta \sum \mu_\iota = 1$. $\blacksquare$

COROLLARY. *The convex hull of a set A in a vector space is the set of all linear combinations $\sum_i \lambda_i x_i$, where (x_i) is a finite family of elements of A, $\lambda_i \geqq 0$ and $\sum_i \lambda_i = 1$.*

Indeed, each point of A is a convex set and A is their union.

It follows from §3, formula (1) that the balanced core of a convex set is convex. A set A is balanced and convex ("absolutkonvex" in [52]) if and only if for all $x, y \in A$ and $\lambda, \mu \in \mathbf{K}$ such that $|\lambda| + |\mu| \leqq 1$ we have $\lambda x + \mu y \in A$. Indeed, if this condition is satisfied, we see that A is balanced, choosing $\mu = 0$; and that A is convex, choosing $\lambda > 0$, $\mu > 0$, $\lambda + \mu = 1$. Conversely, if A is both balanced and convex, we see that $\lambda x + \mu y \in A$ from the identity

$$\lambda x + \mu y = (|\lambda| + |\mu|)\left\{\frac{|\lambda|}{|\lambda| + |\mu|}\left(\frac{\lambda}{|\lambda|}x\right) + \frac{|\mu|}{|\lambda| + |\mu|}\left(\frac{\mu}{|\mu|}y\right)\right\}.$$

PROPOSITION 3. *In a topological vector space the closure of a convex set is convex.*

Proof. Let A be a convex set, $x \in \overline{A}$, $y \in \overline{A}$, $\alpha > 0$, $\beta > 0$, $\alpha + \beta = 1$. Let W be a neighborhood of $\alpha x + \beta y$. Since the map $(u, v) \mapsto \alpha u + \beta v$ is continuous from $E \times E$ into E, there exists a neighborhood U of x and a neighborhood V of y such that $\alpha U + \beta V \subset W$. Now, $U \cap A \neq \emptyset$ and $V \cap A \neq \emptyset$. Let $z \in U \cap A$, $w \in V \cap A$. Then $\alpha z + \beta w \in W \cap A$, which proves that $\alpha x + \beta y \in \overline{A}$. ∎

DEFINITION 1. *A topological vector space is locally convex if each point has a fundamental system of convex neighborhoods.*

A locally convex topological vector space will be called simply a locally convex space. We shall deal almost exclusively with locally convex spaces, which are the only ones occurring in functional analysis. Clearly, a topological vector space is locally convex if 0 has a fundamental system of convex neighborhoods. If $\mathfrak{S}$ is a system of subsets of a vector space E which satisfies the conditions of Proposition 3.1, and furthermore, every $V \in \mathfrak{S}$ is convex, then the topology defined by $\mathfrak{S}$ on E is locally convex.

EXAMPLE 1. The spaces considered in Examples 2, 3, and 4 of §3 are locally convex. Indeed, we have seen in Example 1.3.6 that the balls B_ρ are convex. Furthermore, it is easy to verify directly that the sets $V_{K,\epsilon}$ in $\mathcal{C}(\Omega)$ and the sets $V_{p,\epsilon}$ in $\mathfrak{D}(K)$ are convex.

EXAMPLE 2. To give an example of a not locally convex topological vector space, let E be the vector space of all real-valued functions continuous on the interval $[0, 1]$. For any pair ϵ, δ of numbers such that $\epsilon > 0$, $0 < \delta < 1$, we denote by $V_{\epsilon,\delta}$ the set of all $f \in E$ such that $|f(t)| \leqq \epsilon$ for all t not belonging to some open subset $A = \bigcup_{i=1}^{\infty}]\alpha_i, \beta_i[$ of $[0, 1]$

such that $\sum_{i=1}^{\infty} (\beta_i - \alpha_i) \leqq \delta$ (the subset A depends on f). Let us recall that every open subset of [0, 1] is the union of a countable family of disjoint open intervals.

The sets $V_{\epsilon,\delta}$ form a filter basis on E since $V_{\epsilon',\delta'} \cap V_{\epsilon'',\delta''}$ contains $V_{\epsilon,\delta}$ with $\epsilon = \min(\epsilon', \epsilon'')$, $\delta = \min(\delta', \delta'')$. It is easy to check that the $V_{\epsilon,\delta}$ satisfy conditions (NS 1) and (NS 2) of Theorem 3.1. Furthermore, (NS 3) is also satisfied since $V_{\epsilon/2,\delta} + V_{\epsilon/2,\delta} \subset V_{\epsilon,\delta}$. Thus the sets $V_{\epsilon,\delta}$ form a fundamental system of neighborhoods of 0 in E, and E becomes a topological vector space.

To prove that E is not locally convex it is enough to show that if a set $U \subset E$ satisfies $V_{\epsilon,\delta} \supset U \supset V_{\epsilon',\delta'}$, then U cannot be convex. We may suppose that $\delta < \frac{1}{2}$. Let I_i $(1 \leqq i \leqq n)$ be n disjoint open intervals in [0, 1] each of length δ' and such that the sum of their lengths exceeds 2δ. Denote by J_i the interval of length $\delta'/2$ lying in the middle of I_i and let the function $f_i \in E$ be equal to zero outside I_i and such that $f_i(t) > n\epsilon$ in J_i. Then $f_i \in V_{\epsilon',\delta'} \subset U$ for $1 \leqq i \leqq n$, but

$$\sum_{i=1}^{n} \frac{1}{n} f_i \notin V_{\epsilon,\delta},$$

and thus $\sum_{i=1}^{n} (1/n) f_i \notin U$ (cf. Proposition 1).

PROPOSITION 4. *In a locally convex space the balanced, closed, convex neighborhoods of* 0 *form a fundamental system of neighborhoods of* 0.

Proof. Let W be a neighborhood of 0. Then W contains a closed neighborhood V of 0 by Proposition 3.3. By hypothesis V contains a convex neighborhood U of 0 and $\overline{U} \subset V$. Finally, the balanced core of $\overline{U}$ is a balanced, closed, convex neighborhood of 0 contained in W. ▮

Let us prove a result in the opposite direction.

PROPOSITION 5. *Let E be a vector space and let $\mathfrak{B}$ be a filter basis on E formed by absorbing, balanced, convex sets. Let $\mathfrak{N}$ be the collection of all sets λV, where $\lambda > 0$ and $V \in \mathfrak{B}$. Then there exists a unique topology on E for which E is a locally convex space and for which $\mathfrak{N}$ is a fundamental system of neighborhoods of* 0.

Proof. Clearly, $\mathfrak{N}$ is a filter basis on E which satisfies conditions (NS 1) and (NS 2) of Theorem 3.1, and furthermore, every $V \in \mathfrak{N}$ is convex. But $\mathfrak{N}$ also satisfies condition (NS 3) since if $V \in \mathfrak{N}$, then $\frac{1}{2}V \in \mathfrak{N}$ and $\frac{1}{2}V + \frac{1}{2}V \subset V$. ▮

EXAMPLE 3. Let E be a vector space and $\mathfrak{N}$ the collection of *all* absorbing, balanced, convex sets. Clearly, $\mathfrak{N}$ forms a filter basis; and if $V \in \mathfrak{N}$, then $\lambda V \in \mathfrak{N}$ for all $\lambda > 0$. Thus $\mathfrak{N}$ defines a topology on E called the *finest locally convex topology*.

EXAMPLE 4. In the space $\mathcal{C}(\Omega)$ of Example 3.3 the sets $V_K = V_{K,1}$ satisfy the conditions of Proposition 5. Since $V_{K,\epsilon} = \epsilon V_K$, we see again that the $V_{K,\epsilon}$ form a fundamental system of neighborhoods of 0 for a locally convex topology on $\mathcal{C}(\Omega)$.

PROPOSITION 6. *Let E be a vector space and let $\mathfrak{S}$ be a collection of absorbing, balanced, convex subsets of E. Let $\mathfrak{N}$ be the collection of all finite intersections of sets of the form λV, where $\lambda > 0$ and $V \in \mathfrak{S}$. Then there exists a unique topology on E for which E is a locally convex space and for which $\mathfrak{N}$ is a fundamental system of neighborhoods of* 0.

Proof. We see, exactly as in the proof of Proposition 3.1, that $\mathfrak{N}$ is a filter basis on E. The elements of $\mathfrak{N}$ are absorbing, balanced, convex sets (see Exercise 3.2), and $V \in \mathfrak{N}$ implies $\lambda V \in \mathfrak{N}$ for all $\lambda > 0$. Thus the statement follows from Proposition 5. ▮

COROLLARY. *Let $\mathfrak{B}$ be the collection of all finite intersections of elements of $\mathfrak{S}$ and let $\mathfrak{M}$ be the collection of all sets λV, where $\lambda > 0$ and $V \in \mathfrak{B}$. Then $\mathfrak{M}$ is a fundamental system of neighborhoods of* 0, *equivalent to $\mathfrak{N}$.*

Proof. Clearly, $\mathfrak{M} \subset \mathfrak{N}$. Conversely, if $\lambda = \min_{1 \leqq i \leqq n} \lambda_i$, then

$$\lambda_1 V_1 \cap \lambda_2 V_2 \cap \cdots \cap \lambda_n V_n \supset \lambda(V_1 \cap V_2 \cap \cdots \cap V_n). ▮$$

EXAMPLE 5. In the space $\mathfrak{D}(K)$ of Example 3.4 the sets $V_p = V_{p,1}$ are absorbing, balanced, and convex. Since $V_{p,\epsilon} = \epsilon V_p$, we see again that the finite intersections of the $V_{p,\epsilon}$ form a fundamental system of neighborhoods of 0 for a locally convex topology on $\mathfrak{D}(K)$.

DEFINITION 2. *Given a vector space E, a semi-norm* (pseudo-norm, prenorm) *on E is a map $q: x \mapsto q(x)$ from E into $\mathbf{R}_+$ which satisfies the following axioms:*

(SN 1) $q(\lambda x) = |\lambda| q(x)$ *for all* $\lambda \in \mathbf{K}$ *and* $x \in E$.
(SN 2) $q(x + y) \leqq q(x) + q(y)$.

It follows from (SN 1) that $q(0) = 0$. If, conversely, $q(x) = 0$ implies $x = 0$, then q is a norm.

Let E be a vector space and q a semi-norm on E. The set

$$V = \{x \mid q(x) \leqq 1\}$$

is absorbing, balanced and convex. Indeed, if $q(x) = \alpha \neq 0$, then $q(\alpha^{-1}x) = 1$. Furthermore, $q(\lambda x) \leqq q(x)$ for $|\lambda| \leqq 1$. Finally, if $q(x) \leqq 1$, $q(y) \leqq 1$, and $\alpha \geqq 0$, $\beta \geqq 0$, $\alpha + \beta = 1$, then

$$q(\alpha x + \beta y) \leqq \alpha q(x) + \beta q(y) \leqq \alpha + \beta = 1.$$

Now let $(q_\iota)_{\iota \in I}$ be a family of semi-norms defined on E, and for each $\iota \in I$ let V_ι be the set of all $x \in E$ such that $q_\iota(x) \leqq 1$. It follows from Proposition 6 that the finite intersections of the sets ϵV_ι $(\epsilon > 0)$ form a fundamental system of neighborhoods of 0 for a locally convex topology $\mathcal{T}$ on E. The set $\epsilon V_\iota = V_{\iota,\epsilon}$ is formed by all $x \in E$ such that $q_\iota(x) \leqq \epsilon$ and thus a fundamental system of neighborhoods of 0 for $\mathcal{T}$ is given by the sets

$$V_{\iota_1, \dots, \iota_n, \epsilon_1, \dots, \epsilon_n} = \{x \mid q_{\iota_k}(x) \leqq \epsilon_k \text{ for } 1 \leqq k \leqq n\},$$

where $\iota_1, \dots, \iota_n$ is a finite subset of I and $\epsilon_k > 0$ $(1 \leqq k \leqq n)$. By virtue of the corollary of Proposition 6, an equivalent fundamental system of neighborhoods of 0 for $\mathcal{T}$ is formed by the sets

$$V_{\iota_1, \dots, \iota_n, \epsilon} = \{x \mid q_{\iota_k}(x) \leqq \epsilon \text{ for } 1 \leqq k \leqq n\}. \tag{1}$$

Example 6. On the space $\mathcal{C}(\Omega)$ of Example 3.3 we can define a family of semi-norms in the following way. For every compact subset K of Ω let

$$q_K(f) = \max_{x \in K} |f(x)|.$$

The topology on $\mathcal{C}(\Omega)$ defined by these semi-norms is the same as the one introduced before, since the set V_K defined in Example 4 is given by $V_K = \{f \mid q_K(f) \leqq 1\}$.

Example 7. Let Ω be an open subset of $\mathbf{R}^n$ and m a positive integer. We denote by $\mathcal{E}^m(\Omega)$ the vector space of all functions f defined in Ω and taking their values in $\mathbf{K}$, such that the partial derivatives $\partial^p f$ exist and are continuous for all $p \in \mathbf{N}^n$ with $|p| \leqq m$. For each compact subset K of Ω and multi-index $p \in \mathbf{N}^n$ with $|p| \leqq m$ we define a semi-norm $q_{K,p}$ by

$$q_{K,p}(f) = \max_{x \in K} |\partial^p f(x)|. \tag{2}$$

The family $(q_{K,p})$ of semi-norms defines a locally convex topology on $\mathcal{E}^m(\Omega)$. Observe that for $m = 0$ this space is just $\mathcal{C}(\Omega)$ (Example 6).

Example 8. If Ω is again an open subset of $\mathbf{R}^n$, we denote by $\mathcal{E}(\Omega)$ the vector space of all functions defined in Ω which possess continuous partial derivatives of all orders. We equip $\mathcal{E}(\Omega)$ with the locally convex topology defined by the family $(q_{K,p})$ of semi-norms, where $q_{K,p}$ is defined by (2), K runs through the compact subsets of Ω and p through $\mathbf{N}^n$.

We shall sometimes denote $\mathcal{E}(\Omega)$ also by $\mathcal{E}^\infty(\Omega)$. This will enable us to speak of all the spaces $\mathcal{E}^m(\Omega)$ and $\mathcal{E}(\Omega)$ at once by writing

$$\mathcal{E}^m(\Omega), \quad 0 \leqq m \leqq \infty.$$

EXAMPLE 9. Let K be a compact subset of $\mathbf{R}^n$ and m a positive integer. We denote by $\mathfrak{D}^m(K)$ the vector space of all functions f defined on $\mathbf{R}^n$ whose partial derivatives $\partial^p f$ exist and are continuous for $|p| \leqq m$ and whose support is contained in K. For each $p \in \mathbf{N}^n$ such that $|p| \leqq m$ we define the semi-norm q_p by

$$q_p(f) = \max_{x \in K} |\partial^p f(x)|. \tag{3}$$

Equipped with the family $(q_p)_{|p| \leqq m}$ of semi-norms, $\mathfrak{D}^m(K)$ becomes a locally convex space.

For $m = 0$ we shall also write $\mathcal{K}(K)$ instead of $\mathfrak{D}^0(K)$. Observe that the topology of $\mathcal{K}(K)$ is defined by the *norm* q_0.

EXAMPLE 10. In the space $\mathfrak{D}(K)$ of Example 3.4 the topology is defined by the family of semi-norms (q_p), where q_p is defined by (3) for every $p \in \mathbf{N}^n$. Indeed, the set V_p defined in Example 5 is given by

$$V_p = \{f \mid q_p(f) \leqq 1\}.$$

Similarly as in Example 8, we shall sometimes denote $\mathfrak{D}(K)$ by $\mathfrak{D}^\infty(K)$.

EXAMPLE 11. Let m be a positive integer and k an arbitrary integer. Denote by $\mathcal{S}_k^m$ the vector space of all functions f defined on $\mathbf{R}^n$ whose partial derivatives $\partial^p f$ exist and are continuous for $|p| \leqq m$ and which satisfy the following condition: given $p \in \mathbf{N}^n$ with $|p| \leqq m$ and $\epsilon > 0$, there exists $\rho > 0$ (depending on f, p, and ϵ) such that

$$|(1 + |x|^2)^k \, \partial^p f(x)| \leqq \epsilon \quad \text{for} \quad |x| > \rho.$$

For every p with $|p| \leqq m$ we define the semi-norm

$$q_{k,p}(f) = \max_{x \in \mathbf{R}^n} |(1 + |x|^2)^k \, \partial^p f(x)|. \tag{4}$$

EXAMPLE 12. If k is a fixed integer, we denote by $\mathcal{S}_k$ (or by $\mathcal{S}_k^\infty$) the vector space of all functions defined on $\mathbf{R}^n$ which possess continuous partial derivatives of all orders and which satisfy the condition of the preceding example for every $p \in \mathbf{N}^n$. On $\mathcal{S}_k$ we have the family $(q_{k,p})$ of semi-norms, where $q_{k,p}$ is defined by (4) and p varies in $\mathbf{N}^n$.

EXAMPLE 13. Given a fixed positive integer m, we denote by $\mathcal{S}^m$ the vector space of all functions f defined on $\mathbf{R}^n$ whose partial derivatives $\partial^p f$ exist and are continuous for $|p| \leqq m$ and which satisfy the following condition: for any $k \in \mathbf{Z}$, $p \in \mathbf{N}^n$ with $|p| \leqq m$ and $\epsilon > 0$, there exists $\rho > 0$ such that

$$|(1 + |x|^2)^k \, \partial^p f(x)| \leqq \epsilon \quad \text{if} \quad |x| > \rho.$$

We say that f and its derivatives of order $\leqq m$ are *rapidly decreasing*. We equip $\mathcal{S}^m$ with the family of semi-norms $(q_{k,p})$ defined by (4), where k runs through $\mathbf{N}$ and p through the multi-indices of order $\leqq m$.

EXAMPLE 14. $\mathcal{S}$ (or $\mathcal{S}^\infty$) denotes the vector space of all functions defined on $\mathbf{R}^n$ which possess continuous partial derivatives of all orders and which are rapidly decreasing together with all their derivatives. We equip $\mathcal{S}$ with the locally convex topology defined by the family $(q_{k,p})$ of semi-norms, where $q_{k,p}$ is defined by (4), k runs through $\mathbf{N}$ and p through $\mathbf{N}^n$.

EXAMPLE 15. Let $\mathcal{O}_M$ be the vector space of all functions f defined on $\mathbf{R}^n$ which possess continuous partial derivatives of all orders and are such that for every $\varphi \in \mathcal{S}$ and $p \in \mathbf{N}^n$ the function $\varphi\, \partial^p f$ is bounded on $\mathbf{R}^n$. We say that $\mathcal{O}_M$ is the space of all infinitely differentiable functions which are *slowly increasing* together with all their derivatives. For every $\varphi \in \mathcal{S}$ and $p \in \mathbf{N}^n$ we define the semi-norm $q_{\varphi,p}$ by

$$q_{\varphi,p}(f) = \sup_{x \in \mathbf{R}^n} |\varphi(x)\, \partial^p f(x)|.$$

EXAMPLE 16. Let Ω be an open subset of $\mathbf{R}^n$ and m a positive integer. We denote by $\mathcal{B}_0^m(\Omega)$ the vector space of all functions f defined in Ω whose partial derivatives $\partial^p f$ exist and are continuous for $|p| \leqq m$ and, furthermore, "vanish at the boundary of Ω" in the following sense: for every $f \in \mathcal{B}_0^m$, $p \in \mathbf{N}^n$, with $|p| \leqq m$, and $\epsilon > 0$ there exists a compact subset K of Ω such that $|\partial^p f(x)| \leqq \epsilon$ if $x \in \Omega \cap \complement K$. We define a locally convex topology on $\mathcal{B}_0^m(\Omega)$ with the help of the family of semi-norms (q_p), where q_p is defined by

$$q_p(f) = \max_{x \in \Omega} |\partial^p f(x)| \tag{5}$$

for $|p| \leqq m$.

For $m = 0$ the space $\mathcal{B}_0^0(\Omega)$ is simply the Banach space $\mathcal{C}_0(\Omega)$ introduced in Exercise 1.2.7. For $\Omega = \mathbf{R}^n$ the space $\mathcal{B}_0^m(\mathbf{R}^n)$ is the same as $\mathcal{S}_0^m$, that is, $\mathcal{S}_k^m$ with $k = 0$ (Example 11).

REMARK 1. One could also define the spaces $\mathcal{S}_k^m(\Omega)$ [91].

EXAMPLE 17. Again let Ω be an open subset of $\mathbf{R}^n$. We denote by $\mathcal{B}_0(\Omega)$ or by $\mathcal{B}_0^\infty(\Omega)$ the space of all functions defined in Ω which have continuous partial derivatives of all orders and which, together with all their partial derivatives, vanish at the boundary of Ω. The topology of $\mathcal{B}_0(\Omega)$ is defined by the family (q_p) of semi-norms, where q_p is defined by (5) and p varies in $\mathbf{N}^n$. For $\Omega = \mathbf{R}^n$ we obtain the space $\mathcal{S}_0$.

EXAMPLE 18. If Ω is an open subset of $\mathbf{R}^n$ and m a positive integer, we denote by $\mathcal{B}^m(\Omega)$ the vector space of all functions f defined in Ω which possess continuous and bounded partial derivatives of orders $|p| \leqq m$.

For each $p \in \mathbf{N}^n$ we define the semi-norm q_p by

$$q_p(f) = \sup_{x \in \Omega} |\partial^p f(x)| \tag{6}$$

and equip $\mathcal{B}^m(\Omega)$ with the topology defined by the family (q_p), where $|p| \leqq m$.

EXAMPLE 19. If Ω is an open subset of $\mathbf{R}^n$, we denote by $\mathcal{B}(\Omega)$ or by $\mathcal{B}^\infty(\Omega)$ the vector space of all functions defined in Ω which possess continuous and bounded derivatives of all orders. The topology of $\mathcal{B}(\Omega)$ is defined by the family of all semi-norms q_p defined by (6), where $p \in \mathbf{N}^n$.

The locally convex spaces which figure in Examples 6 through 19 play an important role in the theory of distributions. If $\Omega = \mathbf{R}^n$, we shall write $\mathcal{C}$, $\mathcal{E}^m$, $\mathcal{E}$, $\mathcal{B}_0^m$, $\mathcal{B}_0$, $\mathcal{B}^m$, $\mathcal{B}$ respectively instead of $\mathcal{C}(\mathbf{R}^n)$, $\mathcal{E}^m(\mathbf{R}^n)$, etc.

Let us now turn to some examples of a different type.

EXAMPLE 20. Let E be a normed space and E' its dual (Chapter 1, §7). For every $x' \in E'$ we define the semi-norm q_x on E by

$$q_{x'}(x) = |\langle x, x' \rangle|.$$

The locally convex topology on E defined by the family $(q_{x'})_{x' \in E'}$ of semi-norms is the weak topology mentioned at the end of Chapter 1, §7.

Similarly, for every $x \in E$ we define the semi-norm $q_{x'}$ on E' by

$$q_x(x') = |\langle x, x' \rangle|.$$

The locally convex topology on E' defined by the family $(q_x)_{x \in E}$ of semi-norms is called the topology $\sigma(E', E)$ or the weak* topology (cf. Chapter 3, §2).

EXAMPLE 21. Let E be a Hilbert space and let $\mathcal{L}(E) = \mathcal{L}(E, E)$ be the set of all continuous linear maps of E into itself. Clearly, $\mathcal{L}(E)$ is a vector space (its elements are also called continuous endomorphisms of E or bounded operators). On $\mathcal{L}(E)$ we can define several locally convex topologies [21].

(a) For $A \in \mathcal{L}(E)$ let us define the norm of A by setting

$$\|A\| = \sup_{\substack{x \in E \\ x \neq 0}} \frac{\|Ax\|}{\|x\|}.$$

It is easy to verify that $\|A\|$ is a norm on $\mathcal{L}(E)$ and that $\mathcal{L}(E)$ is indeed a Banach space (see Exercise 1.7.4). The topology defined by this norm is called the *normic* (or uniform) topology on $\mathcal{L}(E)$.

(b) For every $x \in E$ we define the semi-norm q_x on $\mathfrak{L}(E)$ by

$$q_x(A) = \|Ax\|.$$

The family $(q_x)_{x\in E}$ of semi-norms defines a locally convex topology on $\mathfrak{L}(E)$, called the *strong* topology.

(c) For every x, $y \in E$ we define the semi-norm $q_{x,y}$ on $\mathfrak{L}(E)$ by

$$q_{x,y}(A) = |(Ax \mid y)|.$$

The family $(q_{x,y})_{(x,y)\in E\times E}$ of semi-norms defines a locally convex topology, the *weak* topology, on $\mathfrak{L}(E)$.

(d) Let $(x_n)_{n\in\mathbf{N}}$ be a sequence of elements of E such that

$$\sum_{n=0}^{\infty} \|x_n\|^2 < \infty.$$

For $A \in \mathfrak{L}(E)$ we have

$$\sum_{n=0}^{\infty} \|Ax_n\|^2 \leqq \|A\|^2 \sum_{n=0}^{\infty} \|x_n\|^2 < \infty,$$

and thus we can define the semi-norm $q_{(x_n)}$ by

$$q_{(x_n)}(A) = \left(\sum_{n=0}^{\infty} \|Ax_n\|^2\right)^{1/2}.$$

The locally convex topology on $\mathfrak{L}(E)$ defined by the family $(q_{(x_n)})$ of semi-norms on $\mathfrak{L}(E)$ is called the *ultra-strong* (or strongest) topology on $\mathfrak{L}(E)$.

(e) Finally, let (x_n) and (y_n) be two sequences of elements of E such that

$$\sum_{n=0}^{\infty} \|x_n\|^2 < \infty, \qquad \sum_{n=0}^{\infty} \|y_n\|^2 < \infty.$$

For $A \in \mathfrak{L}(E)$ we have

$$\sum_{n=0}^{\infty} (Ax_n \mid y_n) \leqq \sum_{n=0}^{\infty} \|Ax_n\| \cdot \|y_n\| \leqq \|A\| \sum_{n=0}^{\infty} \|x_n\| \cdot \|y_n\|$$

$$\leqq \|A\| \left(\sum_{n=0}^{\infty} \|x_n\|^2\right)^{1/2} \left(\sum_{n=0}^{\infty} \|y_n\|^2\right)^{1/2},$$

and we can define the semi-norm $q_{(x_n),(y_n)}$ by

$$q_{(x_n),(y_n)}(A) = \left|\sum_{n=0}^{\infty} (Ax_n \mid y_n)\right|.$$

The locally convex topology defined by the family $(q_{(x_n),(y_n)})$ is the so-called *ultra-weak* topology on $\mathcal{L}(E)$.

EXAMPLE 22. We finally define the spaces $k^p(\alpha_{\iota\lambda})$ introduced by Köthe, which are useful in constructing various counterexamples in the theory of locally convex spaces. Let $(\alpha_{\iota\lambda})_{(\iota,\lambda)\in I\times L}$ be a double family of positive numbers and let $1 \leqq p < \infty$. We denote by $k^p(\alpha_{\iota\lambda})$ the set of all families $x = (\xi_\iota)_{\iota\in I}$ of real or complex numbers such that the family $(\alpha_{\iota\lambda}|\xi_\iota|^p)_{\iota\in I}$ is summable (Definition 1.3.7) for every $\lambda \in L$. The set $k^p(\alpha_{\iota\lambda})$ clearly forms a vector space under the natural operations, and we obtain a family $(q_\lambda)_{\lambda\in L}$ of semi-norms on $k^p(\alpha_{\iota\lambda})$ by setting for each $\lambda \in L$:

$$q_\lambda(x) = \Big(\sum_{\iota\in I} \alpha_{\iota\lambda}|\xi_\iota|^p\Big)^{1/p}, \qquad x = (\xi_\iota) \in k^p(\alpha_{\iota\lambda}).$$

If $I = \mathbf{N}$, L contains one element, and $\alpha_{\iota\lambda} = 1$ for every (ι, λ), then the space is just l^p (Example 1.1.7).

In the above examples we defined the locally convex topologies with the help of semi-norms. We want to show now that every locally convex topology can be defined with the help of a family of semi-norms. Let E be a vector space and A a subset of E. The *gauge* (or Minkowski functional) g_A of A is a map $x \mapsto g_A(x)$ from E into the extended set $\mathbf{R}_+ \cup \{\infty\}$ of positive real numbers ([2], §3–9, p. 56) defined as follows:

$$g_A(x) = \inf_{\substack{x\in\rho A\\ \rho>0}} \rho \quad \text{if there exist } \rho > 0 \text{ such that } x \in \rho A,$$

$$g_A(x) = +\infty \qquad \text{if } x \notin \rho A \text{ for all } \rho > 0.$$

For $\lambda > 0$ we have $g_A(\lambda x) = \lambda g_A(x)$ since the conditions $x \in \rho A$ and $\lambda x \in \lambda\rho A$ are equivalent. If A is absorbing, then g_A is clearly finite. If 0 belongs to A, then $g_A(0) = 0$.

If A is a convex set, then its gauge $g = g_A$ satisfies the inequality

$$g(x + y) \leqq g(x) + g(y). \tag{7}$$

This is clear if $g(x)$ or $g(y)$ is $+\infty$. Next let us observe that if A is convex, then $\lambda A + \mu A = (\lambda + \mu)A$ for $\lambda > 0$, $\mu > 0$. Indeed,

$$(\lambda + \mu)A \subset \lambda A + \mu A$$

always holds and

$$\frac{\lambda}{\lambda + \mu} A + \frac{\mu}{\lambda + \mu} A \subset A$$

is equivalent to the convexity of A. Thus let us suppose that $g(x) = \xi$ and $g(y) = \eta$ are both finite. For any $\epsilon > 0$ there exist ρ and σ such that

$\xi \leqq \rho < \xi + \epsilon$, $\eta \leqq \sigma < \eta + \epsilon$, $x \in \rho A$, $y \in \sigma A$. But then

$$x + y \in \rho A + \sigma A = (\rho + \sigma)A,$$

i.e.,

$$g(x + y) \leqq \xi + \eta + 2\epsilon.$$

Since ϵ is arbitrary, we have $g(x + y) \leqq \xi + \eta$.

If A is balanced, then its gauge $g = g_A$ satisfies the relation

$$g(\lambda x) = |\lambda| g(x)$$

for all $\lambda \in \mathbf{K}$. Since we know this for $\lambda > 0$, it is sufficient to prove it for $|\lambda| = 1$, but then it follows from the fact that for a balanced set A the relations $\lambda x \in \rho A$ and $x \in \rho A$ are equivalent.

Summing up, we have obtained the following result:

PROPOSITION 7. *In a vector space E the gauge of an absorbing, balanced, convex set is a semi-norm.*

An absorbing, balanced, convex set A in a topological vector space is a neighborhood of 0 if and only if its gauge $g = g_A$ is continuous. Let us first suppose that A is a neighborhood of 0. From (7) we obtain the inequality $|g(x) - g(y)| \leqq g(x - y)$, and thus it is sufficient to prove that g is continuous at 0. But for any $\epsilon > 0$ the relation $x \in \epsilon A$ implies $g(x) \leqq \epsilon$ and ϵA is a neighborhood of 0. Conversely, if g is continuous, then $\{x \mid g(x) < 1\}$ is an open set contained in A and containing 0.

Let V be a balanced, convex neighborhood of 0 in a topological vector space and $g = g_V$ its gauge. Then the closure of V is the set

$$B = \{x \mid g(x) \leqq 1\}.$$

Indeed, the continuity of g implies that B is closed. Since obviously $V \subset B$, we have $\overline{V} \subset B$. To prove that $B \subset \overline{V}$, let $g(x) \leqq 1$ and let W be a neighborhood of x. We have $\rho x \in V$ for $\rho < 1$, and there exists $\epsilon > 0$ such that $\rho x \in W$ for $1 - \epsilon < \rho < 1 + \epsilon$. Hence $\rho x \in V \cap W$ for $1 - \epsilon < \rho < 1$, which proves that $x \in \overline{V}$.

Now let E be a locally convex vector space. We know from Proposition 5 that the balanced, closed, convex neighborhoods of 0 form a fundamental system of neighborhoods of 0. The gauge g_V of each such neighborhood V is a continuous semi-norm on E, and the locally convex topology defined by the family (g_V) of semi-norms is the same as the topology originally given since $V = \{x \mid g_V(x) \leqq 1\}$. Thus we have proved that *a locally convex topology can always be defined by a family of semi-norms, namely, by the family of all semi-norms which are continuous for the given locally convex topology.*

EXAMPLE 23. The locally convex topology defined by the family of all semi-norms on a vector space is the finest locally convex topology introduced in Example 3.

Let q_i $(1 \leqq i \leqq n)$ be a finite family of semi-norms on a vector space E. Then the function $q = \max q_i$ defined by

$$q(x) = \max_{1 \leqq i \leqq n} q_i(x)$$

is also a semi-norm on E, and we have

$$\begin{aligned} \{x \mid q(x) \leqq \epsilon\} &= \bigcap_{i=1}^{n} \{x \mid q_i(x) \leqq \epsilon\} \\ &= \{x \mid q_i(x) \leqq \epsilon, 1 \leqq i \leqq n\}. \end{aligned} \tag{8}$$

If, furthermore, E is a topological vector space and the q_i $(1 \leqq i \leqq n)$ are continuous, then q is also continuous. We say that a family $\mathfrak{F}$ of semi-norms on a vector space is *saturated* if for any finite subfamily (q_i) of $\mathfrak{F}$ the semi-norm $\max q_i$ also belongs to $\mathfrak{F}$. If $(q_\iota)_{\iota \in I}$ is a saturated family of semi-norms defining a locally convex topology $\mathcal{T}$ on E, then it follows from (8) that a fundamental system of neighborhoods of 0 for $\mathcal{T}$ is given by sets $V_{\iota,\epsilon} = \{x \mid q_\iota(x) \leqq \epsilon\}$. A locally convex topology can always be defined by a saturated family of semi-norms, e.g., by the family of all continuous semi-norms.

We say that two families of semi-norms on a vector space E are *equivalent* if they define the same locally convex topology on E.

If $(q_\iota)_{\iota \in I}$ is a family of semi-norms which defines a locally convex topology on a vector space E, then the family obtained by taking for every finite subfamily q_{ι_k} $(1 \leqq k \leqq n)$ of (q_ι) the semi-norm

$$q_{\iota_1, \ldots, \iota_n} = \max_{1 \leqq k \leqq n} q_{\iota_k}$$

is an equivalent saturated family of semi-norms. Indeed, the neighborhood $V_{\iota_1, \ldots, \iota_n, \epsilon}$ defined in (1) is now the set $\{x \mid q_{\iota_1, \ldots, \iota_n}(x) \leqq \epsilon\}$ by virtue of (8).

PROPOSITION 8. *Let the locally convex topology $\mathcal{T}$ on the vector space E be defined by the family $(q_\iota)_{\iota \in I}$ of semi-norms. Then $\mathcal{T}$ is Hausdorff if and only if for every $x \neq 0$ in E there exists an index $\iota \in I$ such that $q_\iota(x) \neq 0$.*

Proof. If $q_\iota(x) = \alpha > 0$, then $\{x \mid q_\iota(x) \leqq \alpha/2\}$ is a neighborhood of 0 in E which does not contain x. Thus $\mathcal{T}$ is Hausdorff by Proposition 3.2. Conversely, let $\mathcal{T}$ be Hausdorff and $x \neq 0$. Then again by Proposition 3.2 there exists a neighborhood W of 0 which does not contain x. The neighborhood W contains a set of the form $\{x \mid q_{\iota_k}(x) \leqq \epsilon, 1 \leqq k \leqq n\}$, and thus $q_{\iota_k}(x) \neq 0$ for some k $(1 \leqq k \leqq n)$. ∎

Exercises

1. (a) Prove that if A is a balanced convex set in a vector space, then for every finite family $(x_i)_{1 \leqq i \leqq n}$ of elements of A and every family $(\lambda_i)_{1 \leqq i \leqq n}$ of scalars such that $\sum_{i=1}^n |\lambda_i| \leqq 1$, we have $\sum_{i=1}^n \lambda_i x_i \in A$.

(b) Let $(A_\iota)_{\iota \in I}$ be a family of balanced convex subsets of a vector space. Prove that the balanced convex hull of $\bigcup_{\iota \in I} A_\iota$ (i.e., the smallest balanced convex set containing $\bigcup_{\iota \in I} A_\iota$) is the set of all linear combinations $\sum_{\iota \in I} \lambda_\iota x_\iota$, where $x_\iota \in A_\iota$, $\sum_{\iota \in I} |\lambda_\iota| \leqq 1$, and $\lambda_\iota = 0$ except for finitely many ι.

(c) Prove that the balanced convex hull of a subset A of a vector space is the set of all linear combinations $\sum_{i=1}^n \lambda_i x_i$, where $(x_i)_{1 \leqq i \leqq n}$ is a finite family of elements of A, and $(\lambda_i)_{1 \leqq i \leqq n}$ is a family of scalars such that $\sum_{i=1}^n |\lambda_i| \leqq 1$.

2. Let A be a convex subset of a vector space over **R**. Show that A contains an absorbing, balanced, convex set if and only if for every straight line L going through the origin the set $A \cap L$ contains an open segment containing the origin (i.e., for every $z \neq 0$ there exists $\alpha > 0$ and $\beta < 0$ such that $\lambda z \in A$ for $\beta < \lambda < \alpha$).

3. Prove that a convex set A is balanced if $\lambda A \subset A$ for all $\lambda \in \mathbf{K}$ such that $|\lambda| = 1$.

4. Prove that the finest locally convex topology defined in Example 3 is Hausdorff. (*Hint:* Given an element $a \neq 0$ of E, consider an algebraic basis $B = (a_\iota)$ of E such that $a \in B$ and observe that the set $\{x \mid x = \sum \lambda_\iota a_\iota, |\lambda_\iota| < 1\}$ is absorbing, balanced and convex.)

5. Let q be a semi-norm on a vector space. Show that q is the gauge of every set A such that $\{x \mid q(x) < 1\} \subset A \subset \{x \mid q(x) \leqq 1\}$.

§5. *Linear maps, subspaces, quotient spaces*

Proposition 1. *A linear map f from a topological vector space E into a topological vector space F is continuous if it is continuous at the origin.*

Proof. Let W be a neighborhood of 0 in F, then there exists a neighborhood V of 0 in E such that $f(V) \subset W$. But then $f(a + V) \subset f(a) + W$ for every $a \in E$. ∎

Proposition 2. *Let E be a locally convex space whose topology is defined by a* saturated *family $(q_\iota)_{\iota \in I}$ of semi-norms and F a locally convex space whose topology is defined by a family $(r_\lambda)_{\lambda \in L}$ of semi-norms. A linear map f from E into F is continuous if and only if for every semi-norm r_λ there exists a semi-norm q_ι and a positive number M such that $r_\lambda(f(x)) \leqq Mq_\iota(x)$ for all $x \in E$.*

Proof. Suppose that for every r_λ such an $M > 0$ and a q_ι can be found. Let W be a neighborhood of 0 in F. Then W contains a set of the form

$$\{x \mid r_{\lambda_k}(x) \leqq \epsilon, 1 \leqq k \leqq n\}.$$

If now $r_{\lambda_k}(f(x)) \leqq M_k q_{\iota_k}(x)$, then let

$$V = \{x \mid q_{\iota_k}(x) \leqq \epsilon/M_k, 1 \leqq k \leqq n\}.$$

It follows that $f(V) \subset W$, and thus by Proposition 1 the map f is continuous.

Conversely, let f be continuous. Since (q_ι) is saturated, for each $\lambda \in L$ there exists an index $\iota \in I$ and a $\gamma > 0$ such that $q_\iota(x) \leqq \gamma$ implies $r_\lambda(f(x)) \leqq 1$. We then have $r_\lambda(f(x)) \leqq (1/\gamma)q_\iota(x)$ for all $x \in E$. Indeed, if $q_\iota(x) = 0$, then also $r_\lambda(f(x)) = 0$, since in this case $q_\iota(\mu x) = 0$ for every $\mu > 0$; that is, $\mu r_\lambda(f(x)) \leqq 1$ for every $\mu > 0$. If, on the other hand, $q_\iota(x) \neq 0$, then

$$q_\iota\left(\frac{\gamma x}{q_\iota(x)}\right) = \gamma,$$

and thus

$$r_\lambda\left(f\left(\frac{\gamma x}{q_\iota(x)}\right)\right) = \frac{\gamma}{q_\iota(x)} r_\lambda(f(x)) \leqq 1. \blacksquare$$

In particular, a linear form f on E is continuous if and only if there exists a semi-norm q_ι and a positive number M such that $|f(x)| \leqq Mq_\iota(x)$ for every $x \in E$.

Let E and F be two locally convex vector spaces whose topologies are defined by the families of semi-norms $(q_\iota)_{\iota \in I}$ and $(r_\lambda)_{\lambda \in L}$ respectively and f a linear map from E into F. It follows from the remarks made at the end of §4 and from Proposition 2 that f is continuous if and only if for every $\lambda \in L$ there exists a finite family $(\iota_1, \ldots, \iota_n)$ of elements of I and a positive number M such that

$$r_\lambda(f(x)) \leqq M \max_{1 \leqq k \leqq n} q_{\iota_k}(x)$$

for all $x \in E$.

Let E be a vector space and let $\mathcal{T}$ and $\mathcal{T}'$ be two locally convex topologies on E, where $\mathcal{T}$ is defined by the family $(q_\iota)_{\iota \in I}$ of semi-norms and $\mathcal{T}'$ is defined by the family $(r_\lambda)_{\lambda \in L}$ of semi-norms. Then $\mathcal{T}$ is finer than $\mathcal{T}'$ if and only if for every r_λ there exists a finite subfamily q_{ι_k} $(1 \leqq k \leqq n)$ of (q_ι) and a number $M \geqq 0$ such that

$$r_\lambda(x) \leqq M \max_{1 \leqq k \leqq n} q_{\iota_k}(x)$$

for all $x \in E$. Indeed, according to the previous remark, this condition means that the identical bijection $x \mapsto x$ from E equipped with the topology $\mathcal{T}$ onto E equipped with the topology $\mathcal{T}'$ is continuous. In particular, $\mathcal{T}$ and $\mathcal{T}'$ coincide—i.e., (q_ι) and (r_λ) are equivalent—if and only if for every $\lambda \in L$ there exists a finite family $(\iota_1, \ldots, \iota_n)$ of elements of I and a positive number M such that

$$r_\lambda(x) \leqq M \max_{1 \leqq k \leqq n} q_{\iota_k}(x)$$

for all $x \in E$ and for every $\iota \in I$ there exists a finite family $(\lambda_1, \ldots, \lambda_m)$ of elements of L and a positive number N such that

$$q_\iota(x) \leqq N \max_{1 \leqq k \leqq m} r_{\lambda_k}(x)$$

for all $x \in E$.

EXAMPLE 1. Let us introduce the following notations, which we shall use frequently in the sequel: if $x = (x_1, \ldots, x_n)$ is a vector in $\mathbf{R}^n$ and $p = (p_1, \ldots, p_n)$ a multi-index in $\mathbf{N}^n$, then we set

$$x^p = x_1^{p_1} x_2^{p_2} \ldots x_n^{p_n}$$

and $p! = p_1! p_2! \cdots p_n!$. With these notations we can write the multinomial theorem of elementary algebra in the convenient form

$$(x_1 + x_2 + \cdots + x_n)^m = \sum_{|p|=m} \frac{m!}{p!} x^p.$$

We want to show that if k is positive, then on the spaces $\mathcal{S}_k^m$ $(0 \leqq m \leqq \infty)$ of Examples 4.11 and 4.12 the family of semi-norms $q_{k,p}$ with $|p| \leqq m$ is equivalent to the family of semi-norms

$$f \mapsto \max_{x \in \mathbf{R}^n} |x^r \, \partial^p f(x)| \tag{1}$$

with $r, p \in \mathbf{N}^n$, $|r| \leqq 2k$, and $|p| \leqq m$. Indeed, by the multinomial theorem, $(1 + |x|^2)^k$ is the linear combination of monomials

$$x^{2r} = x_1^{2r_1} x_2^{2r_2} \cdots x_n^{2r_n}$$

with $|r| \leqq k$, and therefore for every $p \in \mathbf{N}^n$ there exists $M > 0$ such that

$$q_{k,p}(f) \leqq M \max_{|r| \leqq 2k} \max_{x \in \mathbf{R}^n} |x^r \, \partial^p f(x)|.$$

Conversely, $|x^r| \leqq (1 + |x|^2)^k$ if $|r| \leqq 2k$; hence the semi-norm (1) is not greater than $q_{k,p}(f)$.

EXAMPLE 2. On the spaces $\mathcal{S}^m$ $(0 \leqq m \leqq \infty)$ of Examples 4.13 and 4.14 the family of semi-norms $(q_{k,p})$ with $k \in \mathbf{N}$ and $|p| \leqq m$ is equivalent to the family of semi-norms $(q_{k,p})$ with $k \in \mathbf{Z}$ and $|p| \leqq m$. This follows immediately from the inequality $(1 + |x|^2)^k \leqq 1$ which holds for $k \leqq 0$.

Another equivalent family of semi-norms on $\mathcal{S}^m$ is given by (1), where $r, p \in \mathbf{N}^n$ and $|p| \leqq m$.

EXAMPLE 3. Let m be a fixed positive integer or the symbol ∞. We clearly have $\mathcal{S}^m \subset \mathcal{S}_k^m$ for any $k \in \mathbf{Z}$, and the injection $\mathcal{S}^m \hookrightarrow \mathcal{S}_k^m$ is continuous since the family of semi-norms which defines the topology of $\mathcal{S}_k^m$ is a subfamily of the family of semi-norms which defines the topology of $\mathcal{S}^m$.

EXAMPLE 4. If $k \geqq k'$, then $\mathcal{S}_k^m \subset \mathcal{S}_{k'}^m$ and the map $\mathcal{S}_k^m \hookrightarrow \mathcal{S}_{k'}^m$ is continuous, as follows immediately from the inequality $(1 + |x|^2)^{k'} \leqq (1 + |x|^2)^k$.

EXAMPLE 5. If Ω is an open subset of $\mathbf{R}^n$ and m a positive integer or ∞, then $\mathcal{B}^m(\Omega) \subset \mathcal{E}^m(\Omega)$ (cf. Examples 4.7, 4.8, 4.18, and 4.19). The map $\mathcal{B}^m(\Omega) \hookrightarrow \mathcal{E}^m(\Omega)$ is continuous, since if K is a compact subset of Ω and $p \in \mathbf{N}^n$, $|p| \leqq m$, we have

$$q_{K,p}(f) = \max_{x \in K} |\partial^p f(x)| \leqq \sup_{x \in \Omega} |\partial^p f(x)| = q_p(f)$$

for every $f \in \mathcal{B}^m(\Omega)$.

EXAMPLE 6. We can see similarly that the maps $\mathcal{B} \hookrightarrow \mathcal{O}_M \hookrightarrow \mathcal{E}$ are continuous (Exercise 2).

EXAMPLE 7. Let Ω be an open subset of $\mathbf{R}^n$ and m, m' two positive integers such that $m > m'$. We clearly have

$$\mathcal{E}(\Omega) \subset \mathcal{E}^m(\Omega) \subset \mathcal{E}^{m'}(\Omega) \subset \mathcal{C}(\Omega);$$

and the canonical injections are continuous since the family of semi-norms $(q_{K,p})$ with $|p| \leqq m'$ is a subfamily of the family of semi-norms $(q_{k,p})$ with $|p| \leqq m$.

We can see similarly that for $0 \leqq m' \leqq m \leqq \infty$ the maps $\mathcal{D}^m(K) \hookrightarrow \mathcal{D}^{m'}(K)$, $\mathcal{S}_k^m \hookrightarrow \mathcal{S}_k^{m'}$, $\mathcal{S}^m \hookrightarrow \mathcal{S}^{m'}$, $\mathcal{B}_0^m(\Omega) \hookrightarrow \mathcal{B}_0^{m'}(\Omega)$, and $\mathcal{B}^m(\Omega) \hookrightarrow \mathcal{B}^{m'}(\Omega)$ are continuous.

DEFINITION 1. *A bijective continuous linear map f from a topological vector space E onto a topological vector space F is called an isomorphism if the inverse map f^{-1} is continuous* (i.e., if f is a homeomorphism). *An injective continuous linear map f from a topological vector space E into a topological vector space F is a strict morphism* (or topological homomorphism) *if it is an isomorphism from E onto $f(E)$* $=$ Im (f).

Two topological vector spaces E and F are *isomorphic* if there exists an isomorphism from E onto F. An isomorphism from E onto itself is called an *automorphism*.

Before giving an example of an isomorphism between two spaces, we want to prove the Leibniz formula of differential calculus in a convenient form, in which we shall often use it in the sequel. First, let us introduce the following vectorial notations in addition to those already introduced in Example 1. If $p \in \mathbf{N}^n$ and $q \in \mathbf{N}^n$, we write

$$p + q = (p_1 + q_1, p_2 + q_2, \ldots, p_n + q_n);$$

and $p \geqq q$ if $p_1 \geqq q_1$, $p_2 \geqq q_2, \ldots, p_n \geqq q_n$. If $p \geqq q$, we denote by

$p - q$ the multi-index whose components are $p_i - q_i$ $(1 \leqq i \leqq n)$. Furthermore, we set

$$\binom{p}{q} = \binom{p_1}{q_1}\binom{p_2}{q_2}\cdots\binom{p_n}{q_n},$$

where $\binom{p_i}{q_i}$ is the binomial coefficient $\dfrac{p_i!}{q_i!(p_i - q_i)!}$; i.e., with $p!$ defined as in Example 1, we have

$$\binom{p}{q} = \frac{p!}{q!(p-q)!}$$

if $p \geqq q$ and $=0$ otherwise.

PROPOSITION 3. *If φ and ψ are two functions defined in a neighborhood of the point $x \in \mathbf{R}^n$ and whose partial derivatives of orders $\leqq |p|$ $(p \in \mathbf{N}^n)$ exist and are continuous in that neighborhood, then we have*

$$\partial^p(\varphi\psi) = \sum_{q \leqq p} \binom{p}{q} \partial^q\varphi \cdot \partial^{p-q}\psi \tag{2}$$

at x.

Proof. We use induction on the order $|p|$ of p. For $|p| = 0$ the formula (2) reduces to the identity $\varphi\psi = \varphi\psi$. Let us assume that (2) holds for all $p \in \mathbf{N}^n$ such that $|p| = m$, and let $s = (s_1, \ldots, s_n) \in \mathbf{N}^n$ be a multi-index of order $|s| = m + 1$. Since in particular $|s| \geqq 1$, one of the components of s must be $\geqq 1$, and without loss of generality we may assume that $s_1 \geqq 1$. Let $p \in \mathbf{N}^n$ be such that $p_1 = s_1 - 1$, $p_j = s_j$ for $2 \leqq j \leqq n$. Then $|p| = m$ and $\partial^s = \partial_1\partial^p$. By the induction hypothesis and the formula

$$\partial_1(\varphi\psi) = \partial_1\varphi \cdot \psi + \varphi \cdot \partial_1\psi$$

we have

$$\begin{aligned}
\partial^s(\varphi\psi) &= \partial_1 \sum_{q \leqq p} \binom{p}{q} \partial^q\varphi \cdot \partial^{p-q}\psi \\
&= \sum_{q \leqq p} \binom{p}{q} \{\partial_1\partial^q\varphi \cdot \partial^{p-q}\psi + \partial^q\varphi \cdot \partial_1\partial^{p-q}\psi\} \\
&= \sum \binom{s_2}{q_2}\cdots\binom{s_n}{q_n}\left\{\binom{s_1 - 1}{q_1 - 1} + \binom{s_1 - 1}{q_1}\right\} \partial^q\varphi \cdot \partial^{s-q}\psi \\
&= \sum_{q \leqq s} \binom{s}{q} \partial^q\varphi \cdot \partial^{s-q}\psi. \blacksquare
\end{aligned}$$

EXAMPLE 8. Let k be a fixed integer and m a positive integer or ∞. We want to show that the linear map $f \mapsto (1 + |x|^2)^k f$ is an isomorphism from $\mathcal{S}_k^m$ onto $\mathcal{B}_0^m$.

Let us first prove the formula

$$\partial^q(1 + |x|^2)^k = \sum_{|r| \leqq |q|} \gamma_r x^r \cdot (1 + |x|^2)^{k-|q|}, \tag{3}$$

where $q \in \mathbf{N}^n$ and the γ_r are constants which, of course, depend also on k and q. Formula (3) clearly holds for $|q| = 0$. Let us assume that it holds for $|q| = m$ and let $\partial^s = \partial_1 \partial^q$. Then

$$\partial^s(1 + |x|^2)^k = \sum_{|r| \leqq |q|} \partial_1(\gamma_r x^r \cdot (1 + |x|^2)^{k-|q|})$$

$$= \sum \{\gamma_r r_1 x_1^{r_1 - 1} x_2^{r_2} \cdots x_n^{r_n} (1 + |x|^2) + \gamma_r x^r (k - |q|) 2x_1\} (1 + |x|^2)^{k-|q|-1},$$

which is of the same form as (3) since $|q| + 1 = |s|$ and

$$r_1 - 1 + r_2 + \cdots + r_n + 2 \leqq |s|, \qquad |r| + 1 \leqq |s|.$$

If $f \in \mathcal{S}_k^m$, set $g = (1 + |x|^2)^k f$. Using (2) and (3), we see that

$$\begin{aligned} \partial^p g(x) &= \sum_{q \leqq p} \binom{p}{q} \partial^q (1 + |x|^2)^k \cdot \partial^{p-q} f(x) \\ &= \sum_{q \leqq p} \binom{p}{q} \sum_{|r| \leqq |q|} \frac{\gamma_r x^r}{(1 + |x|^2)^{|q|}} \cdot (1 + |x|^2)^k \, \partial^{p-q} f(x), \end{aligned} \tag{4}$$

which shows that $g \in \mathcal{B}_0^m$. Now from (4) follows the existence of an $M > 0$ such that

$$\max_{x \in \mathbf{R}^n} |\partial^p g(x)| \leqq M \max_{q \leqq p} \max_{x \in \mathbf{R}^n} |(1 + |x|^2)^k \, \partial^q f(x)|;$$

hence the map $f \mapsto g$ from $\mathcal{S}_k^m$ into $\mathcal{B}_0^m$ is continuous.

Clearly, the map $f \mapsto g = (1 + |x|^2)^k f$ is injective. But it is also surjective, and the inverse map $g \mapsto f = (1 + |x|^2)^{-k} g$ is continuous, as one sees from the inequality

$$|(1 + |x|^2)^k \, \partial^p f(x)| \leqq \sum_{q \leqq p} \beta_{p,q} |\partial^q g(x)|, \tag{5}$$

where the $\beta_{p,q}$ are positive constants, $\beta_{p,p} = 1$. For $|p| = 0$ formula (5) is evident. Assume that it holds for any multi-index $< p$. From identity (4) we get

$$|(1 + |x|^2)^k \, \partial^p f(x)| \leqq |\partial^p g(x)| + \sum_{q < p} \alpha_{p,q} |(1 + |x|^2)^k \, \partial^{p-q} f(x)|,$$

$\alpha_{p,q} \geqq 0$, and by the induction hypothesis

$$|(1 + |x|^2)^k \, \partial^{p-q} f(x)| \leqq \sum_{r \leqq p-q} \beta_{p-q,r} |\partial^r g(x)|.$$

EXAMPLE 9. If Ω is an open subset of $\mathbf{R}^n$ and K a compact subset of Ω, then the map $\mathfrak{D}^m(K) \hookrightarrow \mathcal{E}^m(\Omega)$ is a strict morphism $(0 \leqq m \leqq \infty)$. In the first place, it is continuous, since if L is an arbitrary compact subset of Ω, we have

$$\max_{x \in L} |\partial^p f(x)| \leqq \max_{x \in K} |\partial^p f(x)|$$

for every $f \in \mathfrak{D}^m(K)$ and $|p| \leqq m$. But, conversely, if V is the neighborhood of 0 in $\mathfrak{D}^m(K)$ defined by the inequality

$$\max_{x \in K} |\partial^p f(x)| \leqq \epsilon,$$

then its image in $\mathcal{E}^m(\Omega)$ is the intersection of $\mathfrak{D}^m(K)$ with the neighborhood of 0 in $\mathcal{E}^m(\Omega)$ defined by the same inequality.

Similarly, it can be seen that the map $\mathfrak{D}^m(K) \hookrightarrow \mathcal{B}_0^m(\Omega)$ is a strict morphism.

EXAMPLE 10. The map $\mathcal{B}_0^m(\Omega) \hookrightarrow \mathcal{B}^m(\Omega)$ $(0 \leqq m \leqq \infty)$ is a strict morphism since the semi-norms which define the topology of $\mathcal{B}_0^m(\Omega)$ are the restrictions of the semi-norms which define the topology of $\mathcal{B}^m(\Omega)$.

REMARK 1. We shall see in Chapter 4 that none of the inclusion maps considered in Examples 3 through 7 is a strict morphism.

If M is a subspace of a topological vector space E, then the topology induced on M by E clearly satisfies the axioms (TVS 1) and (TVS 2), and thus M becomes a topological vector space in its own right. Unless otherwise stated, we shall always consider on M the induced topology. If E is locally convex, then M is locally convex. If the topology of E is defined by a family $(q_\iota)_{\iota \in I}$ of semi-norms, then the topology of M is defined by the restrictions of the q_ι to M.

PROPOSITION 4. *The closure $\overline{M}$ of a linear subspace M of a topological vector space E is a closed subspace of E.*

Proof. We know that $\overline{M}$ is a closed set; we have to prove that it is a linear subspace. Let $a \in \overline{M}$, $b \in \overline{M}$, and let W be a neighborhood of $a + b$. Since the map $(x, y) \mapsto x + y$ is continuous from $E \times E$ into E, there exists a neighborhood U of a and a neighborhood V of b such that $U + V \subset W$. But $U \cap M \neq \emptyset$, $V \cap M \neq \emptyset$; thus $W \cap M \neq \emptyset$ and so $a + b \in \overline{M}$. Similarly, it can be shown that if $a \in \overline{M}$ and $\lambda \in \mathbf{K}$, then $\lambda a \in \overline{M}$. ∎

The definition of the closed subspace generated by a set G, of a total set in a closed subspace (Definition 1.3.1), as well as Definition 1.3.2, can be repeated word for word for any topological vector space.

Let E be a topological vector space and M a linear subspace of E. We have already mentioned in §4 of Chapter 1 that the quotient set E/M

carries a natural vector space structure. We want to prove now that the quotient topology on E/M is compatible with the vector space structure, and to do so we first recall the concept of quotient topology.

Let X be an arbitrary topological space, R an equivalence relation on X, and let φ be the canonical surjection from X onto the quotient set X/R which assigns to each element $x \in X$ its equivalence class $\varphi(x)$ modulo R. The *quotient topology* on X/R is defined as the finest topology for which φ is continuous. In other words, a set A in X/R is open if and only if $\varphi^{-1}(A)$ is open in X.

Let $\bar{f}$ be a map from X/R into a topological space Y, and define the map $f: X \to Y$ by $f = \bar{f} \circ \varphi$. Then $\bar{f}$ is continuous if and only if f is continuous. Indeed, since φ is continuous, we know that if $\bar{f}$ is continuous, then the composite map f is also continuous. Conversely, assume that f is continuous and let A be an open set in Y. Then the set $f^{-1}(A) = \varphi^{-1}(\bar{f}^{-1}(A))$ is open in X, and therefore, by the definition of the quotient topology, $\bar{f}^{-1}(A)$ is open in X/R; i.e., $\bar{f}$ is continuous.

The map f is constant on every equivalence class modulo R. Conversely, if we have a continuous map $f: X \to Y$ which is constant on every equivalence class modulo R, then there exists a unique continuous map $\bar{f}: X/R \to Y$ such that $f = \bar{f} \circ \varphi$.

Let us now consider a topological vector space E, a linear (not necessarily closed) subspace M, and the quotient vector space E/M equipped with the quotient topology. In this case the canonical surjection φ is not only continuous but transforms open sets into open sets. Indeed, let A be an open set in E; then $\varphi^{-1}(\varphi(A))$ is clearly the set $A + M$. Now, the set $A + x$ is open for each $x \in M$, and thus

$$A + M = \bigcup_{x \in M} (A + x)$$

is open, which proves that $\varphi(A)$ is open in E/M. It follows that a set U in E/M is a neighborhood of a point $\dot{x} = \varphi(x)$ if and only if $\varphi^{-1}(U)$ is a neighborhood of x in E. We can now prove that E/M is a topological vector space, i.e., that it satisfies Axioms (TVS 1) and (TVS 2). Let U be a neighborhood of the point $\dot{x} + \dot{y} \in E/M$, and let x and y be two points of E such that $\dot{x} = \varphi(x)$, $\dot{y} = \varphi(y)$. Then $\varphi^{-1}(U)$ is a neighborhood of $x + y$, and there exist neighborhoods V and W of x and y such that $V + W \subset \varphi^{-1}(U)$. But then $\varphi(V)$ and $\varphi(W)$ are neighborhoods of $\dot{x}$ and $\dot{y}$ such that $\varphi(V) + \varphi(W) \subset U$. Similarly, if $\dot{x} = \varphi(x)$ and U is a neighborhood of $\lambda\dot{x}$ in E/M, then $\varphi^{-1}(U)$ is a neighborhood of λx in E and there exists a neighborhood V of λ in $\mathbf{K}$ and a neighborhood W of x in E such that $VW \subset \varphi^{-1}(U)$. But then $\varphi(W)$ is a neighborhood of $\dot{x}$ such that $V\varphi(W) \subset U$.

In the sequel we shall usually consider the quotient vector space E/M equipped with the quotient topology.

PROPOSITION 5. *The quotient space E/M of a topological vector space E modulo a linear subspace M is a Hausdorff space if and only if M is a closed subspace of E.*

Proof. If E/M is a Hausdorff space, then in particular the set $\{\dot{0}\}$ is closed in E/M. Since the canonical surjection φ is continuous, the set $M = \varphi^{-1}(\{\dot{0}\})$ is closed in E.

Conversely, suppose that M is closed and let $\dot{x}$ be an element of E/M different from $\dot{0}$. Then $\dot{x} = \varphi(x)$ for some $x \notin M$. Hence there exists a neighborhood U of x such that $U \cap M = \emptyset$. But then $\dot{x} - \varphi(U)$ is a neighborhood of $\dot{0}$ which does not contain $\dot{x}$ and thus E/M is a Hausdorff space by Proposition 3.2. ▌

Applying Proposition 5 to the case $M = \{0\}$, we see that the topological vector space E is a Hausdorff space if (and only if) the set $\{0\}$ is closed (cf. Exercise 3.1).

If f is a continuous linear map from a topological vector space E into a topological vector Hausdorff space F, then $\text{Ker}(f)$ is a closed subspace of E. In any topological vector space E the set $\overline{\{0\}}$ is a closed subspace of E (Proposition 4). The quotient space $E/\overline{\{0\}}$ is called the *Hausdorff space associated with E*.

Suppose now that E is a locally convex space, and let M be a subspace of E. Then the quotient space E/M is also locally convex. Indeed, if U is a neighborhood of the origin in E/M, then $\varphi^{-1}(U)$ is a neighborhood of the origin in E. By hypothesis $\varphi^{-1}(U)$ contains a convex neighborhood of the origin $V \subset E$ and thus $\varphi(V)$ is a convex neighborhood of the origin in E/M contained in U.

Let q be a semi-norm on a vector space E. If M is a subspace of E, we define the *quotient semi-norm* $\dot{q}$ on E/M by

$$\dot{q}(\dot{x}) = \inf_{x \in \dot{x}} q(x);$$

$\dot{q}$ is a semi-norm on E/M. Indeed,

$$\dot{q}(\lambda\dot{x}) = \inf_{x \in \dot{x}} q(\lambda x) = |\lambda| \inf_{x \in \dot{x}} q(x) = |\lambda| \cdot \dot{q}(\dot{x}).$$

Next let $\epsilon > 0$ and $\dot{x}, \dot{y} \in E/M$. Then there exist $x \in \dot{x}$ and $y \in \dot{y}$ such that

$$q(x) \leqq \dot{q}(\dot{x}) + \epsilon \quad \text{and} \quad q(y) \leqq \dot{q}(\dot{y}) + \epsilon, \quad \text{and thus}$$
$$q(x + y) \leqq q(x) + q(y) \leqq \dot{q}(\dot{x}) + \dot{q}(\dot{y}) + 2\epsilon.$$

Since ϵ is arbitrary,

$$\dot{q}(\dot{x} + \dot{y}) = \inf q(x + y) \leqq \dot{q}(\dot{x}) + \dot{q}(\dot{y}).$$

If q is a semi-norm on E and $U = \{x \mid q(x) < \epsilon\}$, then clearly

$$\varphi(U) = \{\dot{x} \mid \dot{q}(\dot{x}) < \epsilon\}.$$

From here it follows in particular that if q is continuous on the topological vector space E, then $\dot{q}$ is continuous on the quotient space E/M.

Let $(q_\iota)_{\iota \in I}$ be a saturated family of semi-norms defining a locally convex topology on the vector space E. Then $(\dot{q}_\iota)_{\iota \in I}$ is a family of semi-norms defining the quotient topology on the space E/M. Indeed, the sets

$$U_\iota = \{x \mid q_\iota(x) < \epsilon\}$$

form a fundamental system of neighborhoods of 0 in E, and thus the sets $\varphi(U_\iota) = \{\dot{x} \mid \dot{q}_\iota(\dot{x}) < \epsilon\}$ form a fundamental system of neighborhoods of $\dot{0}$ in E/M.

Let f be a continuous linear map from a topological vector space E into a topological vector space F, $\bar{f}: E/\mathrm{Ker}(f) \to F$ the associated injection (cf. Chapter 1, §4), and $\varphi: E \to E/\mathrm{Ker}(f)$ the canonical surjection. Since $f = \bar{f} \circ \varphi$, it follows from the remarks made above concerning the quotient topology that the linear map $\bar{f}$ is continuous.

Definition 2. *A continuous linear map f from a topological vector space E into a topological vector space F is a strict morphism* (or topological homomorphism) *if the associated injection $\bar{f}: E/\mathrm{Ker}(f) \to F$ is a strict morphism* (Definition 1).

In particular, the canonical surjection φ from a topological vector space E onto a quotient space E/M is a strict morphism, since the associated injection is the identity map of E/M onto itself.

Theorem 1. *If f is a continuous linear map from a topological vector space E into a topological vector space F, then the following conditions are equivalent:*

(a) *f is a strict morphism.*
(b) *f maps every neighborhood of 0 in E onto a neighborhood of 0 in $f(E)$.*
(c) *f maps every open set of E onto an open set of $f(E)$.*

Proof. Let N be the kernel of f, φ the canonical surjection of E onto E/N, and $\bar{f}: E/N \to F$ the injection associated with f. We have $f = \bar{f} \circ \varphi$. Denote by g the inverse of the bijection $\bar{f}: E/N \to f(E)$.

Assume that f is a strict morphism. If V is a neighborhood of 0 in E, then $\varphi(V)$ is a neighborhood of 0 in E/N; and since g is continuous, $\bar{f}(\varphi(V)) = f(V)$ is a neighborhood of 0 in $f(E)$. Thus (a) $\Rightarrow$ (b). Similarly, if A is an open set in E, then $\varphi(A)$ is an open set in E/N and $\bar{f}(\varphi(A)) = f(A)$ is an open set in $f(E)$. Thus (a) $\Rightarrow$ (c).

Conversely, suppose that (b) is satisfied. If W is a neighborhood of 0 in E/N, then $\varphi^{-1}(W)$ is a neighborhood of 0 in E; and thus

$$\bar{f}(W) = f(\varphi^{-1}(W))$$

is a neighborhood of 0 in $f(E)$. Hence g is continuous; i.e., f is a strict morphism and (b) $\Rightarrow$ (a). Similarly, suppose that (c) is satisfied. If B is an open set in E/N, then $\varphi^{-1}(B)$ is an open set in E; and thus

$$\bar{f}(B) = f(\varphi^{-1}(B))$$

is an open set in $f(E)$. It follows that g is continuous; i.e., (c) $\Rightarrow$ (a). ▮

PROPOSITION 6. *Let E be a topological vector space over* $\mathbf{K}$. *Assume that E is a Hausdorff space and that its algebraic dimension is* 1. *If a is a fixed nonzero element of E, then the map $\xi \mapsto \xi a$ is an isomorphism from the Banach space* $\mathbf{K}^1$ (cf. Chapter 1, §5) *onto E.*

Proof. Clearly, the map is bijective, linear, and continuous. Let us prove that the inverse map is also continuous. Given $\epsilon > 0$, there exists a balanced neighborhood U of 0 in E which does not contain ϵa, since E is assumed to be Hausdorff. Then $\xi a \in U$ implies $|\xi| < \epsilon$, since otherwise we would have $|\epsilon \xi^{-1}| \leqq 1$, and thus $\epsilon a = (\epsilon\xi^{-1})(\xi a) \in U$. ▮

A hyperplane (cf. Chapter 1, §5) H in a topological vector space E is either closed or everywhere dense. Indeed, by Proposition 4 the set $\overline{H}$ is a closed subspace of E and thus, by the maximality of H, either $\overline{H} = H$ or $\overline{H} = E$.

PROPOSITION 7. *Let E be a topological vector space and H a hyperplane in E with equation $f(x) = 0$. H is closed if and only if the linear form f is continuous.*

Proof. If f is continuous, then $H = f^{-1}(\{0\})$ is closed since $\{0\}$ is closed in $\mathbf{K}$. Conversely, if H is closed, then E/H is a Hausdorff space according to Proposition 5. Let φ be the canonical surjection of E onto E/H. The injection $\bar{f}: E/H \to \mathbf{K}$ associated with f is surjective (and thus bijective) since f is not identically zero. Since E/H is one-dimensional, it follows from Proposition 6 that $\bar{f}$ is an isomorphism and thus $f = \bar{f} \circ \varphi$ is continuous. ▮

EXERCISES

1. Show that the following relations hold between the various topologies on $\mathcal{L}(E)$ introduced in Example 4.21:

$$\begin{array}{ccccc} \text{normic} & \succcurlyeq & \text{ultra-strong} & \succcurlyeq & \text{ultra-weak} \\ & & \curlyvee & & \curlyvee \\ & & \text{strong} & \succcurlyeq & \text{weak} \end{array}$$

where $\succcurlyeq$ means "is finer than."

2. Prove that the maps $\mathcal{B} \hookrightarrow \mathcal{O}_M \hookrightarrow \mathcal{E}$ are continuous (cf. Example 6).

3. Let U be an absorbing, balanced, convex set in a vector space E, and let g be the gauge of U. If M is a subspace of E, show that the quotient semi-norm $\mathring{g}$ is the gauge of the image $\mathring{U}$ of U in E/M. (*Hint:* Let $\bar{g}$ be the gauge of $\mathring{U}$, $\mathring{x} \in U$, and $\epsilon > 0$. Show that if $\mathring{g}(\mathring{x}) = \lambda$, then $\bar{g}(\mathring{x}) \leqq \lambda + 2\epsilon$ and if $\bar{g}(\mathring{x}) = \lambda$, then $\mathring{g}(\mathring{x}) < \lambda + \epsilon$.)

4. Let q be a semi-norm on the vector space E, M a subspace of E, and $\mathring{q}$ the quotient semi-norm on E/M. Show that $\mathring{q}$ is a norm if and only if M is closed in the topology defined by q and contains the set $\{x \mid q(x) = 0\}$.

5. Show that Proposition 6 is false if we do not assume that E is a Hausdorff space. (*Hint:* The chaotic topology on a vector space E is compatible with the vector space structure of E.)

6. Prove that every not identically zero continuous linear form on a topological vector space E is a surjective strict morphism from E onto $\mathbf{K}^1$. (*Hint:* Use Propositions 5, 6, and 7.)

7. Prove that there exists no not identically zero continuous linear form on the space E defined in Example 4.2. (*Hint:* If F is a continuous linear form on E, there exists $V_{\epsilon,\delta}$ such that $f \in V_{\epsilon,\delta}$ implies $|F(f)| \leqq 1$. If f is zero outside an interval of length $< \delta$ then $\lambda f \in V_{\epsilon,\delta}$ for all $\lambda \in \mathbf{R}$, and thus $F(f) = 0$. Finally, by using a "partition of unity" on [0, 1] show that for each $f \in E$ there exist $f_i \in E$ $(1 \leqq i \leqq n)$ such that $f = \sum_{i=1}^{n} f_i$ and that f_i is zero outside an interval of length $< \delta$.)

§6. *Bounded sets, normability, metrizability*

DEFINITION 1. *Let E be a vector space over $\mathbf{K}$ and A, B two subsets of E. We say that A absorbs B if there exists an $\alpha > 0$ such that $B \subset \lambda A$ for all $\lambda \in \mathbf{K}$ such that $|\lambda| \geqq \alpha$.*

With this terminology we can say that $A \subset E$ is absorbing (Definition 3.2) if it absorbs all finite subsets of E. If A is balanced, then it absorbs B if there exists one $\mu \in \mathbf{K}$ such that $B \subset \mu A$. Indeed, if this is the case, then for $|\lambda| \geqq |\mu|$ we have $|\lambda^{-1}\mu| \leqq 1$ and thus

$$B \subset \mu A = \lambda(\lambda^{-1}\mu)A \subset \lambda A.$$

DEFINITION 2. *A set B in a topological vector space is bounded if it is absorbed by every neighborhood of* 0.

This definition, which clearly generalizes the notion of a bounded set in a normed vector space, is due to Kolmogorov and von Neumann and plays a fundamental role in the theory of locally convex spaces. A set is of course bounded if it is absorbed by every neighborhood belonging to a fundamental system of neighborhoods of 0. Bounded sets and neighborhoods of 0 behave in a "contravariant" way. For instance, if E and F are topological vector spaces and f a continuous linear map from E into F, then the image by f of any bounded set of E is a bounded set in F. Indeed, let A be a bounded set in E and W a neighborhood of 0 in F. Then $f^{-1}(W)$ is a neighborhood of 0 in E and thus absorbs A. It follows that W absorbs $f(A)$. In particular, the finer the topology on a vector space E, the fewer the bounded sets. We shall see, however, that different topologies may have the same bounded sets.

A set contained in a bounded set is clearly bounded itself. The union of two bounded sets is a bounded set and hence, by induction, the union of finitely many bounded sets is bounded. Indeed, let A and B be bounded sets. For a balanced neighborhood V of 0 there exist $\alpha > 0$, $\beta > 0$ such that $A \subset \alpha V$ and $B \subset \beta V$. Setting $\gamma = \max(\alpha, \beta)$, we have $A \cup B \subset \gamma V$.

A set consisting of a single point is bounded (because any neighborhood of 0 is absorbing), and therefore any finite set is bounded. It follows from Proposition 3.3 that the closure of any bounded set is bounded.

A collection $\mathfrak{P}$ of bounded sets of E is called a *fundamental system of bounded sets* if for every bounded set B of E there exists a set $P \in \mathfrak{P}$ such that $B \subset P$. In a locally convex space the balanced, closed, convex, bounded sets form a fundamental system of bounded sets. To show this, let B be a bounded set and let P be the intersection of all balanced, closed, convex sets containing B. Then P is a balanced, closed, convex set containing B (i.e., the balanced, closed, convex hull of B), and it is sufficient to prove that P is bounded. But this is clear, since by Proposition 4.4 the balanced, closed, convex neighborhoods of 0 form a fundamental system of neighborhoods of 0; and if V is such a neighborhood and $B \subset \lambda V$, then also $P \subset \lambda V$.

Let E be a locally convex space whose topology is defined by the family $(q_\iota)_{\iota \in I}$ of semi-norms. A set $B \subset E$ is bounded if and only if every q_ι is bounded on B.

Let E be a vector space and q a semi-norm on E. We know that q defines a locally convex topology on E, and in this topology E has a fundamental system of bounded neighborhoods of 0 formed by the sets

$$V_\epsilon = \{x \mid q(x) \leqq \epsilon\}.$$

Conversely, we have the following result:

PROPOSITION 1. *If E is a locally convex space in which there exists a bounded neighborhood of* 0, *then the topology of E can be defined with the help of a single semi-norm.*

Proof. Let V be a bounded neighborhood of 0 in E. Then by Proposition 4.4 there exists a balanced, closed, convex neighborhood W of 0 contained in V. Let q be the gauge of W. By Proposition 4.7 q is a semi-norm.

The sets $W_\epsilon = \{x \mid q(x) \leqq \epsilon\}$ form a fundamental system of neighborhoods of 0 in E. Indeed, if U is any neighborhood of 0 in E, then by the boundedness of V there exists $\epsilon > 0$ such that $\epsilon V \subset U$; but then $W_\epsilon \subset U$. ∎

Let E be a topological vector space whose topology is defined by a single semi-norm q. It follows from Proposition 4.8 that q is a norm if and only if E is a Hausdorff space.

If the topology of E can be defined by the finite family $(q_i)_{1 \leqq i \leqq n}$ of semi-norms, then it can also be defined by the single semi-norm q defined, for instance, by

$$q(x) = \max_{1 \leqq i \leqq n} q_i(x)$$

or by

$$q(x) = \sum_{i=1}^{n} q_i(x)$$

for $x \in E$, since

$$q_i(x) \leqq \max_{1 \leqq i \leqq n} q_i(x) \leqq \sum_{i=1}^{n} q_i(x) \leqq n \max_{1 \leqq i \leqq n} q_i(x).$$

In this case the set $\{x \mid q(x) \leqq 1\}$ is a bounded neighborhood of 0.

If the topology of a topological vector space can be defined by a norm, we say that it is *normable*. As we have seen, for a topology to be normable it is necessary and sufficient that the space be locally convex and Hausdorff, and that it possess a bounded neighborhood of 0.

EXAMPLE 1. If Ω is an open subset of $\mathbf{R}^n$, K a compact subset of $\mathbf{R}^n$, m a positive integer, and k an arbitrary integer, then the spaces $\mathfrak{D}^m(K)$, $\mathcal{S}_k^m$, $\mathcal{B}_0^m(\Omega)$, $\mathcal{B}^m(\Omega)$ (Examples 9, 11, 16, and 18 of §4) are normable since they are clearly all Hausdorff and defined by finite families of semi-norms.

We say that a topological space X is *metrizable* if there exists a metric on X such that the topology defined by it coincides with the topology of X. Clearly, a metrizable space is always Hausdorff, and each point possesses a countable fundamental system of neighborhoods (namely, the balls $B_{1/n}(x)$ with $n \geqq 1$). We shall now prove that in a topological vector space the converse is also true (Theorem 1). The proof of Theorem 1, patterned after the proof of a more general theorem from the theory of uniform spaces ([8], Chapter IX, 2nd ed., §1, no. 4, Proposition 2), is somewhat long and complicated. The reader who is primarily interested in locally convex spaces can skip the proof of Theorem 1 at first reading, since a little further on (Proposition 2) we shall give a simpler proof for locally convex spaces.

A metric δ on a vector space E is said to be *translation-invariant* if $\delta(x + a, y + a) = \delta(x, y)$ for every $x, y, a \in E$. If δ is a translation-invariant metric, then the distance from the origin $|x| = \delta(x, 0)$ has the following properties:

(a) $|x| = 0$ if and only if $x = 0$,
(b) $|x| = |-x|$,
(c) $|x + y| \leqq |x| + |y|$.

Conversely let $x \mapsto |x|$ be a map from the vector space E into $\mathbf{R}_+$ having properties (a), (b), and (c). Then $\delta(x, y) = |x - y|$ defines a translation-invariant metric on E. Clearly, if δ is a translation-invariant metric on a vector space E, then the neighborhoods of a point $x \in E$ are the sets $x + U$, where U is a neighborhood of the origin.

THEOREM 1. *Suppose that the topological vector space E is Hausdorff and that there exists a countable fundamental system of neighborhoods of 0 in E. Then the topology of E can be defined by a translation-invariant metric δ such that the distance from the origin $|x| = \delta(x, 0)$ satisfies the conditions:*

(d) $|\lambda| \leqq 1$ *implies* $|\lambda x| \leqq |x|$,
(e) $\lambda \to 0$ *implies* $|\lambda x| \to 0$ *for every* $x \in E$.

Proof. Let $(V_n)_{n \in \mathbf{N}^*}$ be a fundamental system of neighborhoods of 0 in E such that $\bigcap_{n \in \mathbf{N}^*} V_n = \{0\}$. Let us set $W_1 = V_1$ and define by induction a sequence $(W_n)_{n \in \mathbf{N}^*}$ of balanced neighborhoods of 0 which satisfy the relation

$$W_{n+1} + W_{n+1} + W_{n+1} \subset V_n \cap W_n.$$

Clearly, the W_n also form a fundamental system of neighborhoods of 0 in E; in particular, $\bigcap_{n \in \mathbf{N}^*} W_n = \{0\}$, and furthermore, $W_{n+1} \subset W_n$. Let us define the function γ on E as follows:

$$\gamma(0) = 0,$$

$$\gamma(x) = \frac{1}{2^k} \quad \text{if} \quad x \in W_k \quad \text{but} \quad x \notin W_{k+1},$$

$$\gamma(x) = 1 \quad \text{if} \quad x \notin W_1.$$

For $x \neq 0$ we have $\gamma(x) \neq 0$, and for $|\lambda| = 1$ we have $\gamma(\lambda x) = \gamma(x)$ since the W_n are balanced. For the same reason $|\lambda| \leqq 1$ implies

$$\gamma(\lambda x) \leqq \gamma(x).$$

Define now

$$\delta(x, y) = \inf \sum_{i=1}^{p} \gamma(t_i - t_{i-1}),$$

where the infimum is taken with respect to all finite sequences $(t_i)_{0 \leqq i \leqq p}$ such that $t_0 = x$, $t_p = y$. We have the inequalities

$$\tfrac{1}{2}\gamma(y - x) \leqq \delta(x, y) \leqq \gamma(y - x). \tag{1}$$

The second inequality is obvious. To prove the first one, we will show by induction on p that for every pair of points x, y and every finite sequence $(t_i)_{0 \leqq i \leqq p}$ such that $t_0 = x$, $t_p = y$ we have

$$\tfrac{1}{2}\gamma(y - x) \leqq \sum_{i=1}^{p} \gamma(t_i - t_{i-1}). \tag{2}$$

The statement is trivial for $p = 1$, since then the only sequence (t_i) is $t_0 = x$, $t_1 = y$. Now let $p > 1$ and set

$$\alpha = \sum_{i=1}^{p} \gamma(t_i - t_{i-1}).$$

If $\alpha \geqq \frac{1}{2}$, then (2) is true since $\gamma(x) \leqq 1$ for all $x \in E$. Suppose that $\alpha < \frac{1}{2}$. There exists an integer q such that $1 \leqq q \leqq p$ and

$$\sum_{i=1}^{q-1} \gamma(t_i - t_{i-1}) \leqq \tfrac{1}{2}\alpha, \qquad \sum_{i=1}^{q} \gamma(t_i - t_{i-1}) > \tfrac{1}{2}\alpha.$$

We then necessarily have

$$\sum_{i=q+1}^{p} \gamma(t_i - t_{i-1}) \leqq \tfrac{1}{2}\alpha.$$

Since $q - 1 < p$ and $p - q < p$, we have by the hypothesis of induction

$$\gamma(t_{q-1} - x) \leqq \alpha \qquad \text{and} \qquad \gamma(y - t_q) \leqq \alpha.$$

On the other hand, clearly

$$\gamma(t_q - t_{q-1}) \leqq \alpha.$$

Let k be the smallest integer for which $1/2^k \leqq \alpha$. Then $k \geqq 2$ and

$$t_{q-1} - x \in W_k, \qquad t_q - t_{q-1} \in W_k, \qquad y - t_q \in W_k$$

by the definition of γ. But then

$$y - x = (y - t_q) + (t_q - t_{q-1}) + (t_{q-1} - x) \in W_k + W_k + W_k \subset W_{k-1},$$

and thus

$$\gamma(y - x) \leqq \frac{1}{2^{k-1}} \leqq 2\alpha.$$

Next let us prove that δ is a translation-invariant metric on E. It follows from (1) that $\delta(x, y) = 0$ if and only if $x = y$. The property $\delta(x, y) = \delta(y, x)$ follows from $\gamma(-x) = \gamma(x)$ and the definition of δ. Let $x, y, z \in E$ and $\epsilon > 0$. There exists a sequence $(t_i)_{0 \leqq i \leqq p}$, $t_0 = x$,

$t_p = z$ and a sequence $(u_j)_{0 \leqq j \leqq q}$, $u_0 = z$, $u_q = y$ such that

$$\sum_{i=1}^{p} \gamma(t_i - t_{i-1}) \leqq \delta(x, z) + \epsilon,$$

$$\sum_{j=1}^{q} \gamma(u_j - u_{j-1}) \leqq \delta(z, y) + \epsilon.$$

Then we have

$$\delta(x, y) \leqq \sum_{i=1}^{p} \gamma(t_i - t_{i-1}) + \sum_{j=1}^{q} \gamma(u_j - u_{j-1}) \leqq \delta(x, z) + \delta(z, y) + 2\epsilon,$$

and since ϵ is arbitrary, we obtain the triangle inequality

$$\delta(x, y) \leqq \delta(x, z) + \delta(z, y).$$

The translation-invariance of δ follows from the obvious fact that

$$\gamma(y - x) = \gamma(y + a - (x + a))$$

and from the definition of δ.

Denote by U_ϵ the set of all $x \in E$ such that $\delta(x, 0) = |x| \leqq \epsilon$. It follows from the inequalities (1) that for each integer $k \geqq 1$ we have $U_{1/2^{k+1}} \subset W_k \subset U_{1/2^k}$. Furthermore, it follows from the relation

$$\delta(x, a) = |x - a|$$

that $\{x \mid \delta(x, a) \leqq \epsilon\} = U_\epsilon + a$. Thus the topology defined by δ is the same as the topology of E.

Finally, condition (d) follows from the definition of δ and the analogous property of γ, and condition (e) follows from the fact that if $\lambda \to 0$, then λx tends to 0 in E. ∎

A subspace of a metrizable topological vector space is clearly metrizable. If E is a metrizable topological vector space and M is a closed subspace of E, then the quotient space E/M is metrizable. Indeed, by Proposition 5.5 E/M is a Hausdorff space. Furthermore, if $(V_n)_{n \in \mathbf{N}}$ is a fundamental system of neighborhoods of 0 in E and φ is the canonical surjection from E onto E/M, then $(\varphi(V_n))_{n \in \mathbf{N}}$ is a fundamental system of neighborhoods of 0 in E/M.

A locally convex space is metrizable if and only if it is Hausdorff and its topology can be defined by a countable family of semi-norms. This follows from Theorem 1 and also from Proposition 2 below. If the sequence $(q_n)_{n \in \mathbf{N}}$ defines the topology on the vector space E, then setting

$$r_n(x) = \max_{0 \leqq i \leqq n} q_i(x)$$

we obtain an equivalent family $(r_n)_{n \in \mathbf{N}}$ of semi-norms which satisfies the condition $r_n(x) \leqq r_{n+1}(x)$ for every $x \in E$, and thus the sets

$$V_n = \{x \mid r_n(x) \leqq 1\}$$

satisfy the condition $V_n \supset V_{n+1}$ $(n \in \mathbf{N})$. We say that (r_n) is an increasing sequence of semi-norms.

PROPOSITION 2. *Let E be a locally convex Hausdorff space whose topology $\mathcal{T}$ is defined by an increasing sequence $(q_n)_{n \in \mathbf{N}}$ of semi-norms. The map $x \mapsto |x|$ from E into $\mathbf{R}_+$, where*

$$|x| = \sum_{n=0}^{\infty} \frac{1}{2^n} \frac{q_n(x)}{1 + q_n(x)}, \tag{3}$$

has the following properties:

(a) $|x| = 0$ *if and only if* $x = 0$,
(b) $|x| = |-x|$,
(c) $|x + y| \leqq |x| + |y|$,
(d) $|\lambda| \leqq 1$ *implies* $|\lambda x| \leqq |x|$,
(e) $\lambda \to 0$ *implies* $|\lambda x| \to 0$ *for every* $x \in E$.

Furthermore, the metric $\delta(x, y) = |x - y|$ defines the topology $\mathcal{T}$ and is translation invariant.

Proof. The series with positive terms (3) converges since its n-th term is $\leqq 2^{-n}$. If $x = 0$, then $q_n(x) = 0$ for all $n \in \mathbf{N}$; hence $|x| = 0$. If $x \neq 0$, then by Proposition 4.8 there exists an index n such that $q_n(x) \neq 0$ and therefore $|x| \neq 0$. This proves (a). For each n we have $q_n(x) = q_n(-x)$, which proves (b).

Next we observe that the function $\xi \mapsto \xi/(1 + \xi)$ is increasing for $\xi \neq -1$, as can be seen by taking its derivative. In other words, we have

$$\frac{\xi}{1 + \xi} \leqq \frac{\eta}{1 + \eta} \quad \text{for} \quad -1 < \xi \leqq \eta. \tag{4}$$

If $|\lambda| \leqq 1$, then $q_n(\lambda x) = |\lambda| \cdot q_n(x) \leqq q_n(x)$, and by (4)

$$\frac{q_n(\lambda x)}{1 + q_n(\lambda x)} \leqq \frac{q_n(x)}{1 + q_n(x)},$$

from which (d) follows. Since $q_n(x + y) \leqq q_n(x) + q_n(x)$, we have again by (4)

$$\frac{q_n(x + y)}{1 + q_n(x + y)} \leqq \frac{q_n(x) + q_n(y)}{1 + q_n(x) + q_n(y)} \leqq \frac{q_n(x)}{1 + q_n(x)} + \frac{q_n(y)}{1 + q_n(y)},$$

and (c) follows.

From (a), (b), and (c) we see that $\delta(x, y) = |x - y|$ defines a translation-invariant metric. Let us prove that the topology $\mathcal{T}'$ deduced from it coincides with $\mathcal{T}$. Let $U = \{x \mid |x| \leqq 1/2^k\}$ be a neighborhood of 0 for $\mathcal{T}'$. We want to show that U contains the set

$$V = \left\{x \mid q_{k+1}(x) \leqq \frac{1}{2^{k+2}}\right\},$$

which is a neighborhood of 0 for $\mathcal{T}$. If $x \in V$, then

$$q_0(x) \leqq q_1(x) \leqq \cdots \leqq q_k(x) \leqq q_{k+1}(x) \leqq \frac{1}{2^{k+2}};$$

hence

$$\sum_{n=0}^{k+1} \frac{1}{2^n} \frac{q_n(x)}{1 + q_n(x)} \leqq \frac{1}{2^{k+2}} \sum_{n=0}^{\infty} \frac{1}{2^n} = \frac{1}{2^{k+1}} \tag{5}$$

since

$$\frac{q_n(x)}{1 + q_n(x)} \leqq q_n(x).$$

On the other hand,

$$\frac{q_n(x)}{1 + q_n(x)} \leqq 1$$

and therefore

$$\sum_{n=k+2}^{\infty} \frac{1}{2^n} \frac{q_n(x)}{1 + q_n(x)} \leqq \sum_{n=k+2}^{\infty} \frac{1}{2^n} = \frac{1}{2^{k+1}}. \tag{6}$$

From (5) and (6) we obtain $|x| \leqq 1/2^k$, i.e., $x \in U$.

Conversely, let us show that the neighborhood $W = \{x \mid q_m(x) \leqq 1/2^k\}$ of 0 for $\mathcal{T}$ contains $Z = \{x \mid |x| \leqq 1/2^{m+k+1}\}$, which is a neighborhood of 0 for $\mathcal{T}'$. If $x \in Z$, then we have *a fortiori*

$$\frac{1}{2^m} \frac{q_m(x)}{1 + q_m(x)} \leqq \frac{1}{2^{m+k+1}},$$

that is,

$$\frac{q_m(x)}{1 + q_m(x)} \leqq \frac{1}{2^{k+1}};$$

hence

$$q_m(x)\left(1 - \frac{1}{2^{k+1}}\right) \leqq \frac{1}{2^{k+1}},$$

and therefore

$$q_m(x) \leqq \frac{1}{2^{k+1} - 1} \leqq \frac{1}{2^k},$$

which means that $x \in W$.

Finally, (e) follows now from the fact that if $\lambda \to 0$, then $\lambda x \to 0$ for $\mathcal{T}$. ∎

EXAMPLE 2. The spaces $\mathfrak{D}(K)$, $\mathcal{S}_k$, $\mathcal{S}^m$, $\mathcal{S}$, $\mathcal{B}_0(\Omega)$ and $\mathcal{B}(\Omega)$ are metrizable since they are clearly Hausdorff and their topology is defined by a countable family of semi-norms (see §4, Examples 10, 12, 13, 14, 17, and 19).

EXAMPLE 3. The spaces $\mathcal{C}(\Omega)$, $\mathcal{E}(\Omega)$, and $\mathcal{E}^m(\Omega)$ are metrizable (§4, Examples 6, 7, and 8). Indeed, let the sequence (K_k) of subsets of Ω be defined as follows: $K_0 = \emptyset$, and for $k \geqq 1$ let K_k be the set of all points x such that the distance of x from $\complement\Omega$ is not less than $1/k$ and the distance of x from the origin is not greater than k. Clearly, each K_k is compact and it is easy to see that every compact subset of Ω is contained in some K_k. It follows that the family of semi-norms $(q_{K_k})_{k \in \mathbf{N}}$ on $\mathcal{C}(\Omega)$ is equivalent to the family of semi-norms (q_K) and that the family of semi-norms $(q_{K_k,p})_{k \in \mathbf{N}, |p| \leqq m}$ on $\mathcal{E}^m(\Omega)$ $(1 \leqq m \leqq \infty)$ is equivalent to the family of semi-norms $(q_{K,p})$.

PROPOSITION 3 (Mackey's countability condition). *Let E be a metrizable locally convex space and $(B_k)_{k \in \mathbf{N}}$ a sequence of bounded subsets of E. There exists a bounded subset B of E and a sequence $(\lambda_k)_{k \in \mathbf{N}}$ of positive numbers such that $B_k \subset \lambda_k B$ for all $k \in \mathbf{N}$.*

Proof. Let $(q_j)_{j \in \mathbf{N}}$ be a sequence of semi-norms which defines the topology of E. For each $k \in \mathbf{N}$ there exists a sequence $(\mu_{kj})_{j \in \mathbf{N}}$ of positive numbers such that $q_j(x) \leqq \mu_{kj}$ for $x \in B_k$ and $j \in \mathbf{N}$. Define the sequence $(\mu_j)_{j \in \mathbf{N}}$ by

$$\mu_j = \max(\mu_{1j}, \mu_{2j}, \dots, \mu_{jj}).$$

We have $\mu_{kj} \leqq \mu_j$ for $j \geqq k$, and therefore for each $k \in \mathbf{N}$ there exists $\lambda_k > 0$ such that $\mu_{kj} \leqq \lambda_k \mu_j$ for all $j \in \mathbf{N}$. The set

$$B = \{x \mid q_j(x) \leqq \mu_j, j \in \mathbf{N}\}$$

is bounded. If $x \in B_k$, then $q_j(x) \leqq \mu_{kj} \leqq \lambda_k \mu_j$ for all $j \in \mathbf{N}$; that is, $x \in \lambda_k B$. Hence we also have $B_k \subset \lambda_k B$. ∎

EXERCISES

1. Let E be a vector space and $x \mapsto |x|$ a map from E into $\mathbf{R}_+$ having the properties (a) through (e) above. Prove that the sets $U = \{x \mid |x| \leqq \epsilon\}$ $(\epsilon > 0)$ form a fundamental system of balanced neighborhoods of 0 for a metrizable topology on E which is compatible with the vector space structure on E and for which $|x - y|$ is an invariant metric. (*Hint:* Use Theorem 3.1.)

2. Let K be the interval $[0, 1]$ in $\mathbf{R}$. Prove that the space $\mathfrak{D}(K)$ is not normable.

3. Prove that the spaces $\mathcal{E}^m(\Omega)$ $(0 \leqq m \leqq \infty)$ are not normable. (*Hint:* In the case of $\mathcal{C}(\Omega)$, for instance, if K_1 and K_2 are two disjoint compact subsets of Ω, then, in the notations of Example 4.4, the neighborhood V_{K_1} does not absorb the neighborhood V_{K_2}.)

4. Prove that if A and B are two bounded sets in a topological vector space, then the set $A + B$ is bounded.

§7. *Products and direct sums*

Let $(X_\iota)_{\iota \in I}$ be a family of sets. We recall that the *cartesian product* $X = \prod_{\iota \in I} X_\iota$ of the sets X_ι consists of all families $x = (x_\iota)_{\iota \in I}$, where $x_\iota \in X_\iota$. The element x_ι is called the ι-th *component* (or coordinate) of $x = (x_\iota)$ and the map π_ι from X onto X_ι defined by $\pi_\iota(x) = x_\iota$ is called the ι-th *projection.* If all the sets X_ι are equal to the same set X, we write X^I for $\prod_{\iota \in I} X_\iota$. Given a set Y, there is a one-to-one correspondence between the maps $f: Y \to \prod_{\iota \in I} X_\iota$ and the families of maps $(f_\iota)_{\iota \in I}$, where $f_\iota: Y \to X_\iota$. This correspondence is defined by $f_\iota = \pi_\iota \circ f$ and $f(y) = (f_\iota(y))$. We shall often identify f with the family (f_ι) and write $f = (f_\iota)$.

Given a family $(E_\iota)_{\iota \in I}$ of vector spaces over the same field $\mathbf{K}$, we can define a vector space structure on $E = \prod_{\iota \in I} E_\iota$. Indeed, if $x = (x_\iota) \in E$, $y = (y_\iota) \in E$, we set $x + y = (x_\iota + y_\iota)$; and if $\lambda \in \mathbf{K}$, $x = (x_\iota) \in E$, we set $\lambda x = (\lambda x_\iota)$. It is trivial to check that the operations so defined satisfy all eight axioms of a vector space and that the projections π_ι are linear maps from E onto E_ι. The vector space E is called the cartesian product of the family $(E_\iota)_{\iota \in I}$ of vector spaces. If $f_\iota: F \to E_\iota$ is a family of linear maps from a vector space F into E_ι $(\iota \in I)$, then the map

$$f = (f_\iota): F \to \prod_{\iota \in I} E_\iota$$

is linear.

Now let $(X_\iota)_{\iota \in I}$ be a family of topological spaces. The *product topology* on the cartesian product $X = \prod_{\iota \in I} X_\iota$ is defined as the coarsest topology on X for which all the projections π_ι $(\iota \in I)$ are continuous. We shall call the set X equipped with the product topology the *product space* of the spaces X_ι and the latter will be called the *factor spaces* of X. Let $\mathfrak{O}_\iota$ be the system of all open sets of X_ι; then the collection of all sets $\pi_\iota^{-1}(A)$, where $A \in \mathfrak{O}_\iota$, $\iota \in I$, forms a subbasis of the product topology $\mathcal{T}$ on X. A basis for $\mathcal{T}$ is given by all sets $\prod_{\iota \in I} A_\iota$, where each A_ι is an open set in X_ι and $A_\iota = X_\iota$ except for a finite number of indices. A fundamental system of neighborhoods of a point $x = (x_\iota) \in X$ is given by all sets $\prod_{\iota \in I} V_\iota$, where each V_ι is a neighborhood of x_ι in X_ι and $V_\iota = X_\iota$ except for a finite number of indices. If $f_\iota: Y \to X_\iota$ is a family of maps from a topological space Y into the X_ι $(\iota \in I)$, then the map $f = (f_\iota): Y \to X$ is continuous at the point $y \in Y$ if and only if the maps f_ι are continuous at the point $y \in Y$.

If F_ι is a nonempty subset of X_ι $(\iota \in I)$, then $F = \prod_{\iota \in I} F_\iota$ is closed in X if and only if each F_ι is closed in X_ι. In fact, let F be closed and consider a point $x_\kappa \in \overline{F}_\kappa$. For each $\iota \neq \kappa$ take a point $x_\iota \in F_\iota$ and set $x = (x_\iota)$. Then x adheres to F and therefore belongs to F, which proves that $x_\kappa \in F_\kappa$. Conversely, assume that the sets F_ι are closed, i.e., the sets $\complement_{X_\iota} F_\iota = A_\iota$ are open. Now,

$$F = \prod_{\iota \in I} F_\iota = \bigcap_{\iota \in I} \pi_\iota^{-1}(F_\iota);$$

hence

$$A = \complement_X F = \bigcup_{\iota \in I} \pi_\iota^{-1}(A_\iota).$$

Thus A is open and therefore F is closed.

The product space $X = \prod_{\iota \in I} X_\iota$ is a Hausdorff space if and only if every factor space X_ι ($\iota \in I$) is a Hausdorff space.

Now let $(E_\iota)_{\iota \in I}$ be a family of topological vector spaces over the same field $\mathbf{K}$. *The product topology on the cartesian product* $E = \prod_{\iota \in I} E_\iota$ *is compatible with the vector space structure of* E. Indeed, let $a = (a_\iota) \in E$, $b = (b_\iota) \in E$, and let $W = \prod_{\iota \in I} W_\iota$ be a neighborhood of $a + b$, where $W_\iota = E_\iota$ if ι does not belong to a finite subset H of I. For every $\iota \in H$ there exists a neighborhood U_ι of a_ι and a neighborhood V_ι of b_ι such that $U_\iota + V_\iota \subset W_\iota$. For $\iota \notin H$ let us set $U_\iota = V_\iota = E_\iota$. Then $U = \prod_{\iota \in I} U_\iota$ is a neighborhood of a, $V = \prod_{\iota \in I} V_\iota$ is a neighborhood of b, and $U + V \subset W$. This proves (TVS 1). Next let $\lambda \in \mathbf{K}$, $a = (a_\iota) \in E$, and let $W = \prod_{\iota \in I} W_\iota$ be a neighborhood of λa, where $W_\iota = E_\iota$ if ι does not belong to a finite subset H of I. For every $\iota \in H$ there exists a neighborhood U_ι of λ in $\mathbf{K}$ and a neighborhood V_ι of a_ι in E_ι such that $U_\iota V_\iota \subset W_\iota$. For $\iota \notin H$ we set $V_\iota = E_\iota$. Then $U = \bigcap_{\iota \in H} U_\iota$ is a neighborhood of λ in $\mathbf{K}$ and $V = \prod_{\iota \in I} V_\iota$ is a neighborhood of a in E such that $UV \subset W$. This proves (TVS 2).

The product of a countable family of metrizable topological vector spaces is metrizable. Indeed, let $(E_\iota)_{\iota \in I}$ be a family of metrizable topological vector spaces, where the index set I is countable, and write $E = \prod_{\iota \in I} E_\iota$. Since each E_ι is a Hausdorff space, the same holds for E. Furthermore, let $(V_{\iota,\lambda})_{\lambda \in L_\iota}$ be a fundamental system of neighborhoods of 0 in E_ι, where the index set L_ι ($\iota \in I$) is again countable. Then a fundamental system of neighborhoods of 0 in E is given by all sets $\prod_{\iota \in I} W_\iota$, where for ι belonging to a finite subset H of I the set W_ι is equal to some $V_{\iota,\lambda}$, and $W_\iota = E_\iota$ for $\iota \notin H$. The assertion then follows from Theorem 6.1, if we observe that the collection of all finite subsets of a countable set is itself countable.

Let $E = \prod_{\iota \in I} E_\iota$ be the product of a family $(E_\iota)_{\iota \in I}$ of topological vector spaces. *A set B in E is bounded if and only if $\pi_\iota(B)$ is bounded in E_ι for every $\iota \in I$.* Indeed, suppose that every $B_\iota = \pi_\iota(B)$ is bounded, and let $V = \prod_{\iota \in I} V_\iota$ be a neighborhood of 0 in E, where each V_ι is a balanced neighborhood of 0 in E_ι and $V_\iota = E_\iota$ except when ι belongs to a finite subset H of I. For every $\iota \in I$ there exists a $\lambda_\iota > 0$ such that $B_\iota \subset \lambda_\iota V_\iota$. Set $\lambda = \max_{\iota \in H} \lambda_\iota$. Then $B \subset \lambda V$, i.e., B is bounded. Conversely, let B be bounded in E. Since π_ι is a continuous linear map from E onto E_ι, the set $\pi_\iota(B)$ is bounded in E_ι.

As a corollary of the last result we find again (cf. Exercise 6.4) that if A and B are bounded sets in a topological vector space, then the set $A + B$ is bounded. Indeed, $A \times B$ is then bounded in $E \times E$ and $A + B$ is the image of $A \times B$ under the continuous linear map $(x, y) \mapsto x + y$ from $E \times E$ onto E.

Let $(E_\iota)_{\iota \in I}$ be a family of vector spaces and for each $\iota \in I$ let A_ι be a subset of E_ι. The set $A = \prod_{\iota \in I} A_\iota$ is convex in $E = \prod_{\iota \in I} E_\iota$ if and only if for every $\iota \in I$ the set A_ι is convex in E_ι. This follows from the fact that the projection π_ι from E onto E_ι is a linear map and from the formula $A = \bigcap_{\iota \in I} \pi_\iota^{-1}(A_\iota)$.

If $(E_\iota)_{\iota \in I}$ is a family of locally convex spaces, then clearly their product E is also locally convex. Furthermore, if $(q_{\iota,\lambda})_{\lambda \in L_\iota}$ is a family of semi-norms defining the topology of E_ι, then $(q_{\iota,\lambda} \circ \pi_\iota)$ $(\lambda \in L_\iota, \iota \in I)$ is a family of semi-norms on E defining the product topology.

Let E be a vector space over $\mathbf{K}$. If $(A_\iota)_{\iota \in I}$ is an arbitrary family of subsets of E, then, generalizing our earlier notations, we shall denote by $\sum_{\iota \in I} A_\iota$ the set of all elements $x = \sum x_\iota$ of E, where $x_\iota \in A_\iota$ and $x_\iota = 0$, except for finitely many indices ι.

Now let $(M_\iota)_{\iota \in I}$ be a family of linear subspaces of E and assume that $\sum_{\iota \in I} M_\iota = E$, i.e., that every $x \in E$ can be written in the form $x = \sum_{\iota \in I} x_\iota$ with $x_\iota \in M_\iota$ and $x_\iota = 0$ except for finitely many ι. We say that E is the (algebraic, internal) *direct sum* of the family (M_ι), and we write $E = \bigoplus_{\iota \in I} M_\iota$ if the representation of each element $x \in E$ in the form $x = \sum_{\iota \in I} x_\iota$ is unique. The uniquely determined element $x_\iota \in M_\iota$ is called the *component* of x in M_ι.

If $E = \sum_{\iota \in I} M_\iota$, then the following conditions are equivalent:

(a) $E = \bigoplus_{\iota \in I} M_\iota$,
(b) $\sum_{\iota \in I} x_\iota = 0$, $x_\iota \in M_\iota$ implies $x_\iota = 0$ for all $\iota \in I$,
(c) $M_\iota \cap \sum_{\kappa \neq \iota} M_\kappa = \{0\}$ for all $\iota \in I$.

Proof. (a) $\Rightarrow$ (b): $\sum_{\iota \in I} y_\iota = 0$ with all $y_\iota = 0$ is a representation of 0. Therefore, if the representation of each element of E in the form $\sum_{\iota \in I} x_\iota$, $x_\iota \in M_\iota$, is unique, then $\sum_{\iota \in I} x_\iota = 0$ implies $x_\iota = 0$.

(b) $\Rightarrow$ (c): Let $x \in M_\iota \cap \sum_{\kappa \neq \iota} M_\kappa$. Then $x = x_\iota = \sum_{\kappa \neq \iota} x_\kappa$; hence

$$x_\iota - \sum_{\kappa \neq \iota} x_\kappa = 0.$$

From (b) it follows that $x_\iota = 0$ and therefore $x = 0$.

(c) $\Rightarrow$ (a): Let $\sum x_\iota = \sum y_\iota$. For all $\iota \in I$ we have

$$x_\iota - y_\iota = \sum_{\kappa \neq \iota} (y_\kappa - x_\kappa) \in M_\iota \cap \sum_{\kappa \neq \iota} M_\kappa,$$

i.e., $x_\iota = y_\iota$ by hypothesis. ∎

For example, if $(e_\iota)_{\iota \in I}$ is an algebraic basis (Chapter 1, §2) of the vector space E, then E is the direct sum of the one-dimensional subspaces $\{\lambda e_\iota \mid \lambda \in \mathbf{K}\}$.

A linear map p from a vector space E into itself is called a *projector* if it satisfies the relation $p \circ p = p$. Let E be the algebraic direct sum of the family $(M_\iota)_{\iota \in I}$ of linear subspaces. For each index $\iota \in I$ denote by p_ι the map which assigns to each $x = \sum_{\iota \in I} x_\iota \in E$ its component x_ι in M_ι. Clearly, p_ι is a well-defined linear map from E onto M_ι. Furthermore, p_ι is a projector since

$$p_\iota\left(p_\iota\left(\sum_\kappa x_\kappa\right)\right) = p_\iota(x_\iota) = x_\iota = p_\iota\left(\sum_\kappa x_\kappa\right).$$

If $\iota \neq \kappa$, then $p_\iota \circ p_\kappa = 0$ since

$$p_\iota\left(p_\kappa\left(\sum_\lambda x_\lambda\right)\right) = p_\iota(x_\kappa) = 0.$$

Finally, we have $\sum_{\iota \in I} p_\iota = 1_E$, where 1_E denotes the identity map $x \mapsto x$ of E onto itself. Indeed,

$$\sum_{\iota \in I} p_\iota\left(\sum_{\kappa \in I} x_\kappa\right) = \sum_{\iota \in I} x_\iota.$$

Conversely, let E be a vector space and assume that we have a family (p_ι) of projectors such that $p_\iota \circ p_\kappa = 0$ if $\iota \neq \kappa$ and $\sum_\iota p_\iota = 1_E$. Then E is the algebraic direct sum of the subspaces $M_\iota = \operatorname{Im}(p_\iota)$. Indeed,

$$x = 1_E(x) = \sum_\iota p_\iota(x);$$

that is, $E = \sum_\iota M_\iota$. Next observe that the restriction of any projector p to $\operatorname{Im}(p)$ is the identity map. Indeed, if $x = p(y) \in \operatorname{Im}(p)$, then

$$p(x) = p\big(p(y)\big) = p(y) = x.$$

From $\sum_\iota x_\iota = 0$ it follows therefore that $\sum_\iota p_\iota(x_\iota) = 0$ and thus for any index κ we have

$$0 = p_\kappa\left(\sum_\iota p_\iota(x_\iota)\right) = p_\kappa(p_\kappa(x_\kappa)) = p_\kappa(x_\kappa) = x_\kappa.$$

Hence by condition (b) above we see that $E = \bigoplus_\iota M_\iota$.

Now let E be the algebraic direct sum of the *finite* family $(M_i)_{1 \leqq i \leqq n}$ of subspaces of E. Then the map $(x_i)_{1 \leqq i \leqq n} \mapsto \sum_{i=1}^n x_i$ clearly defines an isomorphism of the product vector space $\prod_{i=1}^n M_i$ onto E, called the *canonical* (algebraic) *isomorphism*.

We say that the linear subspace N of the vector space E is an *algebraic supplement* of the linear subspace M if E is the direct sum of M and N, i.e., if $M + N = E$ and $M \cap N = \{0\}$. Every subspace of a vector space has an algebraic supplement. If $E = M \oplus N$ and p is the corresponding projector from E onto M, then $1_E - p$ is the projector from E onto N, which shows that $N = \operatorname{Ker}(p)$. Conversely, if p is any projector in E and $M = \operatorname{Im}(p)$, then $N = \operatorname{Ker}(p)$ is an algebraic supplement of M and $1_E - p$ is the projector onto

N corresponding to the decomposition $E = M \oplus N$. This follows from the above considerations in view of

$$(1_E - p)\circ(1_E - p) = 1_E - p, \qquad p\circ(1_E - p) = (1_E - p)\circ p = 0$$

and

$$\operatorname{Ker}(p) = \operatorname{Im}(1_E - p).$$

We call p the projector onto M *parallel* to N.

Let $E = M \oplus N$, $q: E \to N$ the corresponding projector, $j: N \hookrightarrow E$ the canonical injection, and $\varphi: E \to E/M = E/\operatorname{Ker}(q)$ the canonical surjection. The injection $\bar{q}$ associated with q is an algebraic isomorphism from E/M onto N, and we have $q = \bar{q}\circ\varphi$, $\bar{q}^{-1} = \varphi\circ j$. In particular, two algebraic supplements of a given subspace M are always isomorphic.

DEFINITION 1. *Let E be a topological vector space and $(M_i)_{1\leqq i\leqq n}$ a finite family of subspaces of E such that $E = \bigoplus_{i=1}^n M_i$. We say that E is the topological direct sum of the M_i if the canonical algebraic isomorphism*

$$(x_i) \mapsto \sum_{i=1}^{n} x_i$$

is a homeomorphism from the product space $\prod_{i=1}^n M_i$ onto E (i.e., an isomorphism of the topological vector space structures in the sense of Definition 5.1).

To make this definition meaningful, let us observe that the map

$$(x_i) \mapsto \sum_{i=1}^{n} x_i$$

from $\prod_{i=1}^n M_i$ onto E is always continuous by Axiom (TVS 1), but the inverse map may fail to be continuous.

PROPOSITION 1. *Let E be a topological vector space and assume that E is the algebraic direct sum of the finite family $(M_i)_{1\leqq i\leqq n}$ of subspaces. Then E is the topological direct sum of the M_i if and only if the projectors $p_i: E \to M_i$ are continuous.*

Proof. The inverse of the bijection $(x_i) \mapsto \sum_{i=1}^n x_i$ is given by the map $x \mapsto (p_i(x))$ which is continuous if and only if the p_i are continuous. ∎

DEFINITION 2. *The linear subspace N of the topological vector space E is a topological supplement of the linear subspace M if E is the topological direct sum of M and N.*

Let E be a topological vector space and assume that N is an algebraic supplement of the linear subspace M. If $p: E \to M$ is the projector corresponding to the decomposition $E = M \oplus N$, then N is a topological supplement of M if and only if p is continuous. This follows from Proposition 1 if we observe that $1_E - p$ is the projector onto N and that it is continuous at the same time as p.

If p is a continuous projector from a topological vector space into itself, then $M = \mathrm{Im}(p)$ has $N = \mathrm{Ker}(p)$ as its topological supplement.

EXAMPLE 1. Let M be a closed subspace of a Hilbert space E. By Proposition 1.3.4 every vector $x \in E$ can be written uniquely in the form $x = y + z$ with $y \in M$, $z \in M^\perp$, which shows that $E = M \oplus M^\perp$ (cf. Chapter 1, §7). We have $\|x\|^2 = \|y\|^2 + \|z\|^2$, and therefore the projector $p: x \mapsto y$ from E onto M satisfies $\|p\| \leqq 1$. In particular, p is continuous and therefore $M^\perp$ is a topological supplement of M.

REMARK 1. In a topological vector space there can exist closed subspaces which have no topological supplement (cf. Exercise 4). Clearly, c_0 can be considered as a closed subspace of m (Examples 6 and 8 of Chapter 1, §1). It can be shown that c_0 does not have a topological supplement in m ([52], §31,2.(5); [110]). It can also be shown that for $p \neq 2$ there exist closed subspaces in l^p (Example 1.1.7) which have no topological supplements ([52], §31,3.(6)).

Let M and N be two algebraic supplements in the topological vector space E, and $q: E \to N$ the corresponding projector. Then M and N are topological supplements if and only if the associated injection $\bar{q}: E/M \to N$ is continuous. Indeed, with the above introduced maps $j: N \hookrightarrow E$ and $\varphi: E \to E/M$, we have $q = \bar{q} \circ \varphi$ and therefore q is continuous if and only if $\bar{q}$ is. Let us observe that $\bar{q}$ is a bijective linear map and its inverse $\bar{q}^{-1} = \varphi \circ j$ is always continuous; hence if $\bar{q}$ is continuous, it is an isomorphism.

It follows that *if M and N are two closed subspaces of a Banach space E which are algebraic supplements of each other, then they are also topological supplements*. Indeed, $\bar{q}^{-1}: N \to E/M$ is then a continuous bijective linear map from the Banach space N onto the Banach space E/M, and therefore by Theorem 1.9.1. its inverse $\bar{q}$ is continuous. In Chapter 3, §17 we shall generalize Theorem 1.9.1 to a larger class of spaces (e.g., complete metrizable spaces) and accordingly we shall obtain a generalization of the last result.

REMARK 2. The last result does not hold for arbitrary topological vector spaces. For instance, let M and N be the two closed subspaces of a Hilbert space E mentioned in Remark 1.7.1 such that $F = M + N$ is not closed. One can see that $F = M \oplus N$, but M and N are not topological supplements, because then F would have to be complete (cf. §9), hence closed. (See also [35].)

PROPOSITION 2. *Let E and F be two topological vector spaces and f a continuous linear map from E into F. There exists a continuous linear map g*

from F into E such that $f \circ g$ shall be the identity map 1_F from F onto itself if and only if f is a surjective strict morphism (Definition 5.2) *and* $\mathrm{Ker}(f)$ *has a topological supplement in E.*

Proof. Suppose that a map g with the indicated properties exists. Then for $y \in F$ we have $f(g(y)) = y$; that is, f is a surjection. Let p be the map $g \circ f$ from E into E. Since $(g \circ f) \circ (g \circ f) = g \circ (f \circ g) \circ f = g \circ f$, the map p is a continuous projector and E is the topological direct sum of $p(E)$ and $\mathrm{Ker}(p)$. Now, $\mathrm{Ker}(p) = \mathrm{Ker}(f)$ since $f(x) = 0$ implies $p(x) = g(f(x)) = 0$ and $p(x) = 0$ implies

$$f(p(x)) = f(g(f(x))) = f(x) = 0.$$

Thus $\mathrm{Ker}(f)$ has $p(E)$ as a topological supplement. The injection $\bar{f}: E/\mathrm{Ker}(f) \to F$ associated with f can be decomposed into $\bar{f} = \tilde{f} \circ \bar{p}$, where $\bar{p}$ is the canonical map from

$$E/\mathrm{Ker}(f) = E/\mathrm{Ker}(p)$$

onto $p(E) = g(F)$ and $\tilde{f}$ is the restriction of f to $g(F)$. Now $\bar{p}$ is an isomorphism by what we have seen above and $\tilde{f}$ is an isomorphism, because its inverse is g. Hence $\bar{f}$ is also an isomorphism and f is a strict morphism.

Conversely, suppose that f is a surjective strict morphism and that $\mathrm{Ker}(f)$ has a topological supplement N in E. Let φ be the canonical surjection from E onto $E/\mathrm{Ker}(f)$ and let j be the canonical injection $N \hookrightarrow E$. Then $\psi = \varphi \circ j$ is an isomorphism from N onto $E/\mathrm{Ker}(f)$. Furthermore, the injection $\bar{f}$ associated with f is an isomorphism from $E/\mathrm{Ker}(f)$ onto F. Thus $\psi^{-1} \circ \bar{f}^{-1}$ is an isomorphism from F onto N and $g = j \circ \psi^{-1} \circ \bar{f}^{-1}$ is a continuous linear map from F into E such that $f \circ g = 1_F$. ∎

Proposition 3. *Let E and F be two topological vector spaces and f a continuous linear map from E into F. There exists a continuous linear map g from F into E such that $g \circ f$ shall be the identity map 1_E from E onto itself if and only if f is an injective strict morphism and $f(E)$ has a topological supplement in F.*

Proof. Suppose that a map g with the indicated properties exists. Then $f(x) = 0$ implies $g(f(x)) = x = 0$ and thus f is injective. Furthermore, the restriction of g to $f(E)$ coincides with f^{-1} and thus f is a strict morphism. Finally, $p = f \circ g$ is a continuous projector from F onto $f(E)$; hence $f(E)$ has a topological supplement in F.

Conversely, suppose that f is an injective strict morphism and that $f(E)$ has a topological supplement in F. Let p be the projector from F onto $f(E)$; then $g = f^{-1} \circ p$ is a continuous linear map from F into E such that $g \circ f = 1_E$. ∎

Exercises

1. Show that every locally convex Hausdorff space is isomorphic to a subspace of a product of Banach spaces. (*Hint:* If $(q_\iota)_{\iota \in I}$ is a family of semi-norms which defines the topology of the space E, let F_ι be the quotient of E modulo the subspace $\{x \mid q_\iota(x) = 0\}$ equipped with the quotient norm $\mathring{q}_\iota$. For each $x \in E$ let x_ι be its canonical image in F_ι. Then $x \mapsto (x_\iota)_{\iota \in I}$ is an injective strict morphism from E into $\prod_{\iota \in I} \hat{F}_\iota$.)

2. Let E be a vector space and $(p_i)_{1 \leqq i \leqq n}$ a finite family of linear maps of E into itself such that $p_i \circ p_k = 0$ for $i \neq k$ and $\sum_{i=1}^{n} p_i = 1_E$. Show that each p_i is a projector.

3. Let E be a Hausdorff topological vector space which is the topological direct sum of the finite family $(M_i)_{1 \leqq i \leqq n}$ of subspaces. Prove that each subspace M_i is closed.

4. Let E be the topological vector space considered in Example 4.2 and Exercise 5.7. Show that a one-dimensional subspace of E has no topological supplement. (*Hint:* Use Exercise 3 and Proposition 5.7.)

5. Let M be the closure of $\{0\}$ in a topological vector space E. Show that E induces on M the chaotic topology. Show that every algebraic supplement N of M is a topological supplement and that E induces on N a Hausdorff topology.

§8. *Convergence of filters*

Definition 1. *Let X be a topological space, x a point of X, and $\mathfrak{V}(x)$ the filter of all neighborhoods of x* (Example 2.2). *We say that the filter $\mathfrak{F}$ on X converges to x if $\mathfrak{F}$ is finer than $\mathfrak{V}(x)$.*

Example 1. Let $(x_n)_{n \in \mathbf{N}}$ be a sequence of elements of a set X. The filter on X generated by the image of the Fréchet filter (Example 2.4) under the map $n \mapsto x_n$ from $\mathbf{N}$ into X is called the *elementary filter* associated with the sequence $(x_n)_{n \in \mathbf{N}}$. In other words, the elementary filter associated with the sequence $(x_n)_{n \in \mathbf{N}}$ has a basis formed by the sets

$$S_m = \{x_n \mid n \geqq m\},$$

where m runs through $\mathbf{N}$.

Now let X be a topological space. Similarly as in a metric space (Chapter 1, §2), we say that a sequence $(x_n)_{n \in \mathbf{N}}$ of points of X converges to x if for every neighborhood V of x there exists an integer $m = m(V)$ such that $x_n \in V$ for every $n \geqq m$. The sequence $(x_n)_{n \in \mathbf{N}}$ converges to the point x if and only if the elementary filter associated with $(x_n)_{n \in \mathbf{N}}$ converges to x. Indeed, both conditions mean that each neighborhood V of x contains a set S_m.

If the filter $\mathfrak{F}$ converges to a point x, we say that x is the *limit* of $\mathfrak{F}$. A filter basis $\mathfrak{B}$ is said to converge to x if the filter generated by $\mathfrak{B}$ converges to x. In this case we also say that x is the limit of $\mathfrak{B}$.

A topological space X is a Hausdorff space if and only if every filter has at most one limit. If X is a Hausdorff space, then a filter which converges to two distinct points $x \neq y$ would have to contain two disjoint neighborhoods of x and y respectively, but this contradicts Axioms (F 2) and (F 3). Conversely, if X is not a Hausdorff space, then there exist two distinct points x and y such that every neighborhood U of x intersects every neighborhood V of y. In this case the sets $U \cap V$ form a filter basis which converges to both x and y.

A point x in a topological space X belongs to the adherence of a set A if and only if there exists a filter basis on A which converges to x. Indeed, if $x \in \overline{A}$, then the sets $U \cap A$, where U is a neighborhood of x, form a filter basis on A which converges to x. Conversely, if there exists a filter basis on A which converges to x, then every neighborhood of x must meet A. In general topological spaces this result takes the place of the proposition concerning metric spaces according to which $x \in \overline{A}$ if and only if there exists a sequence (x_n) with $x_n \in A$ which converges to x. In fact, this last proposition holds not only in a metric space but in any topological space where each point has a countable fundamental system of neighborhoods (cf. Exercise 1) and in particular in a metrizable topological vector space.

Let X and Y be two topological spaces and f a map from X into Y. The map f is continuous at the point $x \in X$ if and only if for every filter basis $\mathfrak{B}$ which converges to x the filter basis $f(\mathfrak{B})$ converges to $f(x) \in Y$. Suppose first that f is continuous at x and let $\mathfrak{B}$ be a filter basis which converges to x. If W is a neighborhood of $f(x)$, then $f^{-1}(W)$ is a neighborhood of x and contains therefore a set $A \in \mathfrak{B}$. Hence W contains $f(A)$ and thus $f(\mathfrak{B})$ converges to $f(x)$. Conversely, if for every filter basis $\mathfrak{B}$ which converges to x the filter basis $f(\mathfrak{B})$ converges to $f(x)$, then in particular $f(\mathfrak{V}(x))$ converges to $f(x)$, where $\mathfrak{V}(x)$ is the filter of all neighborhoods of x. But then each neighborhood W of $f(x)$ contains a set $f(V)$, where V is a neighborhood of x, i.e., f is continuous at x. Again this result is the substitute for the proposition which holds if X is a metric space (or more generally a topological space where each point has a countable fundamental system of neighborhoods; cf. Exercise 1), and according to which f is continuous at x if and only if for every sequence (x_n) in X which converges to x, the sequence $\big(f(x_n)\big)$ converges to $f(x)$.

Let X be a set, Y a topological space, f a map from X into Y, and $\mathfrak{F}$ a filter on X. We say that the point $y \in Y$ is the *limit of f with respect to the filter* $\mathfrak{F}$ if the filter basis $f(\mathfrak{F})$ converges to y. If this is the case, we write $\lim_{x,\mathfrak{F}} f(x) = y$.

If X is also a topological space and we choose for $\mathfrak{F}$ the filter of neighborhoods $\mathfrak{V}(a)$ of some point a belonging to X, then the fact that y is the limit of f with respect to $\mathfrak{V}(a)$ will be expressed by saying that $f(x)$ *tends to* y *as* x *tends to* a, and we write $y = \lim_{x \to a} f(x)$. Using this terminology, we can say that $f: X \to Y$ is continuous at the point $a \in X$ if and only if $f(x)$ tends to $f(a)$ as x tends to a. Indeed, if f is continuous at a and $\mathfrak{F} = \mathfrak{B}(a)$ is the filter of neighborhoods of a, then by our previous result $f(\mathfrak{F})$ converges to $f(a)$. Conversely, suppose that $\lim_{x \to a} f(x) = f(a)$ and let $\mathfrak{B}$ be a filter basis on X which converges to a. Then the filter generated by $\mathfrak{B}$ is finer than the filter of neighborhoods $\mathfrak{B}(a)$ of a and thus $f(\mathfrak{B})$ is finer than $f(\mathfrak{B}(a))$. Since, by assumption, $f(\mathfrak{B}(a))$ converges to $f(a)$, *a fortiori* $f(\mathfrak{B})$ converges to $f(a)$ and therefore f is continuous at a by our previous result.

EXAMPLE 2. If X is the set $\mathbf{N}$, then the image of $\mathbf{N}$ in Y under a map $f: \mathbf{N} \to Y$ is a sequence $(x_n)_{n \in \mathbf{N}}$ of elements of Y. The image of the Fréchet filter on $\mathbf{N}$ under f is the elementary filter associated with (x_n), and therefore it is the same to say that the sequence (x_n) converges to x or that the limit of f with respect to the Fréchet filter is x. If this is the case, we write $x = \lim_{n \to \infty} x_n$ and say that x_n tends to x as n increases.

EXAMPLE 3. Let α be a strictly positive number and $I = \{\epsilon \mid 0 < \epsilon < \alpha\}$ an interval on $\mathbf{R}$. For each $\epsilon \in I$ let S_ϵ denote the set $\{\eta \mid 0 < \eta < \epsilon\}$. Clearly, the sets S_ϵ ($\epsilon \in I$) form a filter basis $\mathfrak{B}$ on I. Now let f be a map from I into a topological space Y and set $x_\epsilon = f(\epsilon)$. If y is the limit of f with respect to the filter generated by $\mathfrak{B}$, then we write $y = \lim_{\epsilon \to 0} x_\epsilon$ and say that x_ϵ tends to y as ϵ tends to 0. This means that for every neighborhood W of y there exists ϵ ($0 < \epsilon < \alpha$) such that $x_\eta \in W$ for all η such that $0 < \eta < \epsilon$.

Let $(X_\iota)_{\iota \in I}$ be a family of topological spaces, $X = \prod_{\iota \in I} X_\iota$ the product space (§7), $\pi_\iota: X \to X_\iota$ the ι-th projection and $\mathfrak{F}$ a filter on X. Then $\mathfrak{F}$ converges to a point $x = (x_\iota)_{\iota \in I}$ if and only if for each $\iota \in I$ the filter basis $\pi_\iota(\mathfrak{F})$ converges to x_ι. Since π_ι is continuous, it follows from a previous result that if $\mathfrak{F}$ converges to x, the filter basis $\pi_\iota(\mathfrak{F})$ converges to x_ι. Conversely, suppose that for each $\iota \in I$ the filter basis $\pi_\iota(\mathfrak{F})$ converges to x_ι. Let $V = \prod_{\iota \in I} V_\iota$ be a neighborhood of x, where V_ι is a neighborhood of x_ι in X_ι and $V_\iota = X_\iota$ for all ι which do not belong to a finite subset H of I. For each $\iota \in H$ let $A_\iota \in \mathfrak{F}$ be such that $\pi_\iota(A_\iota) \subset V_\iota$. Then $\bigcap_{\iota \in H} A_\iota$ is an element of $\mathfrak{F}$ which is contained in V. Thus $\mathfrak{F}$ converges to x.

Let X be a topological space and $\mathfrak{B}$ a filter basis on X. We say that a point $x \in X$ *adheres* to $\mathfrak{B}$ if it adheres to every set $A \in \mathfrak{B}$. Clearly, x adheres to $\mathfrak{B}$ if and only if every neighborhood V of x meets every set $A \in \mathfrak{B}$. It follows that if x adheres to $\mathfrak{B}$, then there exists a filter finer than the filter $\mathfrak{F}$ generated by $\mathfrak{B}$ which converges to x. Indeed, the sets $V \cap A$ form then the basis of a filter which is finer than both $\mathfrak{F}$ and the filter of all neighborhoods of x.

To conclude, we generalize to topological vector spaces the notion of summable families we have defined for normed vector spaces (Definition 1.3.7).

DEFINITION 2. *Let* $(x_\iota)_{\iota \in I}$ *be a family of elements of a topological vector space* E. *With each finite subset* J *of* I *we associate the element* $x_J = \sum_{\iota \in J} x_\iota$ *of* E. *The set* $\Phi(I)$ *of finite subsets of* I *is an ordered set with respect to inclusion and it is a directed set. Let* $\mathfrak{S}$ *be the filter of sections* (Example 2.5) *of* $\Phi(I)$. *We say that* $(x_\iota)_{\iota \in I}$ *is summable to an element* x *of* E *if the filter generated by the image of* $\mathfrak{S}$ *under the map* $J \mapsto x_J$ *converges to* x. *If this is the case, we say that* x *is the sum of* (x_ι) *and we write* $x = \sum_{\iota \in I} x_\iota$.

In other words, $x = \sum_{\iota \in I} x_\iota$ if for every neighborhood V of x there exists a finite subset J of I such that $\sum_{\iota \in H} x_\iota \in V$ for any finite subset H of I such that $H \supset J$.

EXERCISES

1. Let X be a topological space where each point has a countable fundamental system of neighborhoods.

(a) Prove that the filter $\mathfrak{B}(x)$ of neighborhoods of a point $x \in E$ is the intersection of the elementary filters associated with the sequences which converge to x.

(b) Prove that a point $x \in X$ adheres to a set A if and only if there exists a sequence $(x_n)_{n \in \mathbf{N}}$ of elements of A which converges to x.

(c) Let f be a map from X into an arbitrary topological space Y. Prove that f is continuous at a point $x \in X$ if and only if for every sequence $(x_n)_{n \in \mathbf{N}}$ in X which converges to x the sequence $(f(x_n))_{n \in \mathbf{N}}$ converges to $f(x)$ in Y.

2. Give an example of a topological space X, a subset A of X, and a point $x \in \overline{A}$ such that there exists no sequence of elements of A which converges to x. (*Hint:* Take for X the real line and for a fundamental system of neighborhoods of a point $x \in X$ the intervals $]x - \alpha, x + \alpha[$, $\alpha > 0$, from which countably many points $\neq x$ have been removed.)

3. Let X be a set, Y a topological space, f a map from X into Y and $\mathfrak{F}$ a filter on X. We say that the point $y \in Y$ is a *value of adherence of* f *with respect to the filter* $\mathfrak{F}$ if y adheres to the filter basis $f(\mathfrak{F})$.

(a) Prove that y is a value of adherence of f with respect to $\mathfrak{F}$ if and only if there exists on X a filter $\mathfrak{G}$ finer than $\mathfrak{F}$ such that y is the limit of f with respect to $\mathfrak{G}$. (*Hint:* Consider on X the filter basis $f^{-1}(\mathfrak{B})$, where $\mathfrak{B}$ is the filter of neighborhoods of y.)

(b) Let $X = \mathbf{N}$ and $f(n) = x_n$. Prove that the following conditions are equivalent:

(i) $y \in Y$ is a value of adherence of f with respect to the Fréchet filter on $\mathbf{N}$;

(ii) $y \in Y$ adheres to the elementary filter associated with the sequence $(x_n)_{n \in \mathbf{N}}$;

(iii) for every neighborhood W of y and every $m \in \mathbf{N}$ there exists $n \in \mathbf{N}$ such that $n \geqq m$ and $x_n \in W$.

If these conditions are satisfied, we say simply that y is a value of adherence of the sequence (x_n).

(c) Suppose that in Y every point has a countable fundamental system of neighborhoods. Prove that if y is a value of adherence of the sequence (x_n), then there exists a subsequence $(x_{n_k})_{k \in \mathbf{N}}$ of (x_n) which converges to y.

§9. *Completeness*

DEFINITION 1. *Let E be a topological vector space and A a subset of E. A filter $\mathfrak{F}$ on A is said to be a Cauchy filter if for every neighborhood V of 0 in E there exists a set $X \in \mathfrak{F}$ such that $X - X \subset V$; that is, $x - y \in V$ for all $x, y \in X$.*

Suppose that $\mathfrak{F}$ is a filter on a subset A of a topological vector space E and that $\mathfrak{F}$ converges to a point $a \in A$. Let V be a neighborhood of 0 in E and let U be a balanced neighborhood of 0 such that $U + U \subset V$. There exists $X \in \mathfrak{F}$ such that $X \subset U + a$. Hence for $x,y \in X$ we have

$$x - y = (x - a) - (y - a) \in U + U \subset V;$$

i.e., $\mathfrak{F}$ is a Cauchy filter.

PROPOSITION 1. *If the point x adheres to the Cauchy filter $\mathfrak{F}$ on a subset of a topological vector space, then $\mathfrak{F}$ converges to x.*

Proof. Let V be a neighborhood of 0 and let W be a neighborhood of 0 such that $W + W \subset V$. There exists a set $A \in \mathfrak{F}$ such that $A - A \subset W$. On the other hand, $A \cap (x + W) \neq \emptyset$. Let y be a point belonging to $A \cap (x + W)$. If $z \in A$, then $z - y \in W$ and

$$z \in y + W \subset x + W + W \subset x + V.$$

Thus $A \subset x + V$ and $\mathfrak{F}$ converges to x. ∎

DEFINITION 2. *A subset A of a topological vector space is complete if every Cauchy filter on A converges to a point of A. A topological vector space is quasi-complete if every bounded closed subset is complete.*

PROPOSITION 2. *A complete subset A of a Hausdorff topological vector space is closed.*

Proof. Let x be adherent to A. The sets $U \cap A$, where U runs through the neighborhoods of x, form a Cauchy filter $\mathfrak{F}$ on A which converges to a point of A. On the other hand, $\mathfrak{F}$ has x as its only limit and thus $x \in A$. ∎

PROPOSITION 3. *Let A be a complete subset of a topological vector space. Then every closed subset of A is complete.*

Proof. Let B be a closed subset of A and let $\mathfrak{F}$ be a Cauchy filter on B. Then $\mathfrak{F}$ is the basis of a Cauchy filter on A and so converges to a point $x \in A$. But x is clearly adherent to B; hence $x \in B$ and thus the filter $\mathfrak{F}$ converges on B. ▮

DEFINITION 3. *Let E and F be two topological vector spaces. A map f from a subset A of E into F is said to be uniformly continuous if for every neighborhood W of 0 in F there exists a neighborhood V of 0 in E such that $x - y \in V$ implies $f(x) - f(y) \in W$.*

If $\mathfrak{F}$ is the basis of a Cauchy filter on A and f is uniformly continuous, then $f(\mathfrak{F})$ is the basis of a Cauchy filter on F. If f is uniformly continuous on A, then it is clearly continuous on A.

If a linear map f from E into F is continuous at the origin, then it is uniformly continuous, as follows immediately from the relation

$$f(x) - f(y) = f(x - y).$$

Similarly, if the semi-norm q defined on a topological vector space is continuous at the origin, then it is uniformly continuous. This follows from the inequality $|q(x) - q(y)| \leqq q(x - y)$.

PROPOSITION 4. *Let E be a complete topological vector space, F a Hausdorff topological vector space and $f: E \to F$ an injective strict morphism* (Definition 5.1). *Then $f(E)$ is a closed subspace of F.*

Proof. Let $g: f(E) \to E$ be the inverse of the isomorphism $f: E \to f(E)$. If $\mathfrak{F}$ is a Cauchy filter on $f(E)$, then $g(\mathfrak{F})$ is a Cauchy filter on E and converges to some point $x \in E$. Consequently, $\mathfrak{F} = f(g(\mathfrak{F}))$ converges to the point $f(x) \in f(E)$; hence $f(E)$ is complete, and by Proposition 2, also closed in F. ▮

PROPOSITION 5. *Let E be a topological vector space, A a subset of E and F a complete Hausdorff topological vector space. If f is a uniformly continuous map from A into F, then there exists a unique uniformly continuous map $\bar{f}$ from $\bar{A}$ into F such that $\bar{f}(x) = f(x)$ for all $x \in A$.*

Proof. Let $x \in \bar{A}$. The filter of neighborhoods of x induces a Cauchy filter $\mathfrak{F}$ on A and thus $f(\mathfrak{F})$ is the basis of a Cauchy filter on F. Since by our assumptions F is complete and Hausdorff, $f(\mathfrak{F})$ converges to a well-defined point $y \in F$. We set $y = \bar{f}(x)$. If $x \in A$, we clearly have $\bar{f}(x) = f(x)$.

Let us prove that $\bar{f}$ is uniformly continuous on $\bar{A}$. Let W be a *closed* balanced neighborhood of 0 in F. There exists a neighborhood U of 0 in E such that $s \in A$, $t \in A$, $s - t \in U$ imply $f(s) - f(t) \in W$. Let V be a balanced neighborhood of 0 in E such that $V + V + V \subset U$. Let $x \in \bar{A}$, $y \in \bar{A}$ be such that $x - y \in V$. There exist elements $x_1 \in A$,

$y_1 \in A$ such that $x_1 \in x + V$, $y_1 \in y + V$. Let us assume for a moment that we have already proved that

$$\bar{f}(x) \in f(x_1) + W, \qquad \bar{f}(y) \in f(y_1) + W. \tag{1}$$

Then it follows from

$$x_1 - y_1 = (x_1 - x) + (x - y) + (y - y_1) \in V + V + V \subset U$$

that $f(x_1) - f(y_1) \in W$, and this combined with (1) yields

$$\begin{aligned}\bar{f}(x) - \bar{f}(y) &= (\bar{f}(x) - f(x_1)) + (f(x_1) - f(y_1)) + (f(y_1) - \bar{f}(y)) \\ &\in W + W + W.\end{aligned}$$

Now we prove the relations (1), and it is obviously sufficient to prove the first one. Let X be a neighborhood of $\bar{f}(x)$. By the definition of $\bar{f}$ there exists a neighborhood Y of x such that $f(Y \cap A) \subset X$. The set

$$(x + V) \cap Y$$

is a neighborhood of x. Since $x \in \bar{A}$, there exists an element

$$u \in (x + V) \cap Y \cap A.$$

But $u \in x + V$ implies $u - x_1 = (u - x) + (x - x_1) \in V + V \subset U$ and thus $f(u) - f(x_1) \in W$, i.e.,

$$f(u) \in f(x_1) + W.$$

On the other hand, $u \in Y \cap A$ implies $f(u) \in X$. Hence

$$f(u) \in X \cap (f(x_1) + W);$$

and since X was an arbitrary neighborhood of $\bar{f}(x)$, we have

$$\bar{f}(x) \in \overline{f(x_1) + W} = f(x_1) + \overline{W}.$$

But W is closed; thus $\bar{f}(x) \in f(x_1) + W$.

Finally, the uniqueness of $\bar{f}$ follows from the fact that the set on which two continuous maps into a Hausdorff space coincide is necessarily closed. ▮

Let E be a topological vector space, M a linear subspace of E, and f a continuous linear map from M into a complete Hausdorff topological vector space F. The uniformly continuous extension $\bar{f}$ of f to the closure $\overline{M}$ of M, which exists and is unique by Proposition 5, is also a linear map. Indeed, let V be a balanced neighborhood of 0 in F and let U be a balanced neighborhood of 0 in E such that $x, y \in \overline{M}$ and $x - y \in U + U$ imply $\bar{f}(x) - \bar{f}(y) \in V$. Given $x \in \overline{M}$ and $y \in \overline{M}$ there exist $x_1, y_1 \in M$ such that $x - x_1 \in U$, $y - y_1 \in U$, and thus

$$x + y - (x_1 + y_1) \in U + U.$$

Since $\bar{f}$ is linear on M, we have

$$\begin{aligned}\bar{f}(x) + \bar{f}(y) - \bar{f}(x + y) \\ &= \bar{f}(x) + \bar{f}(y) - \bar{f}(x + y) - f(x_1) - f(y_1) + f(x_1 + y_1) \\ &= \bar{f}(x) - f(x_1) + \bar{f}(y) - f(y_1) - \bar{f}(x + y) + f(x_1 + y_1) \\ &\in V + V + V.\end{aligned}$$

Since V is an arbitrary balanced neighborhood of 0 and F is a Hausdorff space, this implies that $\bar{f}(x) + \bar{f}(y) = \bar{f}(x + y)$. We can see similarly that $\bar{f}(\lambda x) = \lambda \bar{f}(x)$ for all $\lambda \in \mathbf{K}$ and $x \in \overline{M}$.

THEOREM 1. *Let E be a Hausdorff topological vector space. Then there exists a complete Hausdorff topological vector space $\hat{E}$ such that E is isomorphic to a dense subspace E_0 of $\hat{E}$. The space $\hat{E}$ is unique up to an isomorphism.*

Proof. Let S be the set of all Cauchy filters on E. We introduce in S an equivalence relation R as follows: two Cauchy filters $\mathfrak{F}$ and $\mathfrak{G}$ are equivalent if for every neighborhood V of 0 in E there exist $A \in \mathfrak{F}$ and $B \in \mathfrak{G}$ such that $x - y \in V$ for all $x \in A$, $y \in B$. The relation R is reflexive by the definition of a Cauchy filter. It is symmetric and transitive by properties (NS 2) and (NS 3) of Theorem 3.1 respectively.

Let $\hat{E}$ be the quotient set S/R. We now define a vector space structure on $\hat{E}$. Let $\mathbf{F}$ and $\mathbf{G}$ be two elements of $\hat{E}$ and let $\mathfrak{F}$ and $\mathfrak{G}$ be two Cauchy filters representing $\mathbf{F}$ and $\mathbf{G}$ respectively. The collection

$$\{A + B \mid A \in \mathfrak{F}, B \in \mathfrak{G}\}$$

is the basis of a Cauchy filter $\mathfrak{H}$ on E, and the class of $\mathfrak{H}$ modulo R will be $\mathbf{F} + \mathbf{G}$ by definition. Similarly, let $\lambda \in \mathbf{K}$, $\mathbf{F} \in \hat{E}$ and $\mathfrak{F} \in \mathbf{F}$. Then the collection $\{\lambda A \mid A \in \mathfrak{F}\}$ is the basis of a Cauchy filter on E whose class modulo R will be $\lambda\mathbf{F}$. It is easy to check that these definitions are independent of the choice of representatives of $\mathbf{F}$ and $\mathbf{G}$.

The axioms (VS 1), (VS 2), (VS 5) through (VS 8) of a vector space are clearly satisfied. The zero vector $\mathbf{0}$ of $\hat{E}$ is formed by all filters tending to 0 in E, and thus (VS 3) is satisfied. Finally, (VS 4) is also true, since the relation $\mathbf{F} - \mathbf{F} = \mathbf{0}$ means precisely that every $\mathfrak{F} \in \mathbf{F}$ is a Cauchy filter.

Let V be a closed balanced neighborhood of 0 in E. The subset $\hat{U}_V$ of $\hat{E}$ will be the collection of all elements $\mathbf{F}$ such that choosing an $\mathfrak{F}$ in $\mathbf{F}$, for every neighborhood W of 0 in E there exists $A \in \mathfrak{F}$ which satisfies $A \subset V + W$. This definition is independent of the choice of $\mathfrak{F}$ in $\mathbf{F}$. Indeed, let $\mathfrak{G}$ be equivalent to $\mathfrak{F}$ modulo R, let W be a neighborhood of 0 in E and let W_1 be another neighborhood of 0 such that $W_1 + W_1 \subset W$. There exist $A \in \mathfrak{F}$ and $B \in \mathfrak{G}$ such that $A \subset V + W_1$ and $B - A \subset W_1$. But then

$$B \subset W_1 + A \subset V + W_1 + W_1 \subset V + W.$$

We can easily check that the following relations hold:

$$\text{if } V \subset W, \text{ then } \hat{U}_V \subset \hat{U}_W, \tag{2}$$

$$\hat{U}_{V \cap W} \subset \hat{U}_V \cap \hat{U}_W, \tag{3}$$

$$\lambda \hat{U}_V = \hat{U}_{\lambda V}, \tag{4}$$

$$\text{if } W + W \subset V, \text{ then } \hat{U}_W + \hat{U}_W \subset \hat{U}_V. \tag{5}$$

Relation (3) shows that the sets $\hat{U}_V$ form a filter basis, (2) and (4) imply that the $\hat{U}_V$ are balanced, and (5) that they satisfy condition (NS 3) of Theorem 3.1. Each set $\hat{U}_V$ is absorbing. Indeed, let W be a balanced neighborhood of 0 such that $W + W \subset V$ and let $\mathbf{F} \in \hat{E}$. Pick $\mathfrak{F} \in \mathbf{F}$. Since $\mathfrak{F}$ is a Cauchy filter, there exists $A \in \mathfrak{F}$ such that $A - A \subset W$. Let $x \in A$; then there exists $\lambda > 0$ such that $x \in \lambda W$, and we may suppose that $\lambda \geqq 1$. Then $A \subset W + x \subset W + \lambda W \subset \lambda V$; i.e., $\mathbf{F} \in \lambda \hat{U}_V$.

By virtue of Theorem 3.1, the sets $\hat{U}_V$ form a fundamental system of neighborhoods of 0 for a topological vector space structure on $\hat{E}$. The topology on $\hat{E}$ is Hausdorff. For let $\mathbf{F} \neq \mathbf{0}$ and choose $\mathfrak{F} \in \mathbf{F}$. Then $\mathfrak{F}$ does not tend to 0 in E; hence there exists a neighborhood V of 0 in E such that $A \not\subset V$ for all $A \in \mathfrak{F}$. Take a closed balanced neighborhood W of 0 in E such that $W + W \subset V$. Then $\mathbf{F} \notin \hat{U}_W$ and it follows from Proposition 3.2 that $\hat{E}$ is Hausdorff.

For every $x \in E$ let $\tilde{x}$ be the collection of all filters tending to x. Clearly, $\tilde{x}$ is exactly one equivalence class of S modulo R, i.e., an element of $\hat{E}$. The map $x \mapsto \tilde{x}$ from E into $\hat{E}$ is linear; it is also injective because E is a Hausdorff space. Let E_0 be the image of E under this injection. If for a balanced closed neighborhood V of 0 in E we denote by V_0 its image in E_0 under $x \mapsto \tilde{x}$, then we have

$$V_0 = \hat{U}_V \cap E_0. \tag{6}$$

Indeed, if $\tilde{x} \in V_0$, then $\tilde{x} \in E_0$. Furthermore, $x \in V$ and if $\mathfrak{F} \in \tilde{x}$, then for every neighborhood W of 0 in E there exists $A \in \mathfrak{F}$ such that

$$A \subset x + W;$$

but then $A \subset V + W$, i.e., $\tilde{x} \in \hat{U}_V$. Conversely, let $\tilde{x} \in \hat{U}_V \cap E_0$. We have to show that $x \in V$. Suppose that $x \notin V$. Then, since V is closed, there exists a neighborhood W of 0 in E such that

$$V \cap (x + W) = \emptyset. \tag{7}$$

Let W_1 be a balanced neighborhood of 0 in E such that $W_1 + W_1 \subset W$. If now $\mathfrak{F} \in \tilde{x}$, then there exists $A \in \mathfrak{F}$ such that $A \subset x + W_1$; and since $\tilde{x} \in \hat{U}_V$, there also exists $B \in \mathfrak{F}$ such that $B \subset V + W_1$. But this is impossible since

$$(x + W_1) \cap (V + W_1) = \emptyset.$$

Indeed, if $x + w_1 = v + w_1'$ ($w_1, w_1' \in W_1, v \in V$), then

$$v = x + w_1 - w_1' \in x + W_1 + W_1 \subset x + W,$$

which contradicts (7). Thus x must belong to V and (6) is proved. It follows from (6) that $x \mapsto \tilde{x}$ is an isomorphism of E onto the subspace E_0 of $\hat{E}$.

The subspace E_0 is dense in $\hat{E}$. Indeed, let $\mathbf{F} \in \hat{E}$ and let $\hat{U}_V$ be a neighborhood of 0 in $\hat{E}$. If $\mathfrak{F} \in \mathbf{F}$, there exists $A \in \mathfrak{F}$ such that $A - A \subset V$. Choose an element $x \in A$; then we have *a fortiori* $x - A \subset V$. But the set $\{x\}$ belongs to the filter formed by all sets which contain x, and this filter belongs to the class $\tilde{x} \in E_0$. It follows that $\tilde{x} - \mathbf{F} \in \hat{U}_V$; that is, $\tilde{x} \in \mathbf{F} + \hat{U}_V$.

To prove that $\hat{E}$ is complete we shall use the following

LEMMA. *Let E be a topological vector space and A a dense subset of E. If every Cauchy filter on A converges to a point in E, then E is complete.*

Assuming the lemma for a moment, let $\mathfrak{F}_0$ be a Cauchy filter on E_0. Then $\mathfrak{F}_0$ is the image of a Cauchy filter $\mathfrak{F}$ on E under the isomorphism $x \mapsto \tilde{x}$. Let $\mathbf{F} \in \hat{E}$ be the class of $\mathfrak{F}$ modulo R. We shall prove that $\mathfrak{F}_0$ converges to $\mathbf{F}$. Let V be a closed balanced neighborhood of 0 in E; then there exists $A \in \mathfrak{F}$ such that $A - A \subset V$. Let $A_0 \in \mathfrak{F}_0$ be the image of A in E_0. We claim that $A_0 \subset \mathbf{F} + \hat{U}_V$, which will prove the assertion. Let $\tilde{x}$ be any element of A_0 and x the corresponding element in A. We have $\{x\} - A \subset V$. Since $\{x\}$ is an element of the filter formed by all subsets of E containing x, the last relation implies $\tilde{x} - \mathbf{F} \in \hat{U}_V$. Since $\tilde{x}$ is arbitrary in A_0, we have $A_0 \subset \mathbf{F} + \hat{U}_V$.

Finally, let $\hat{E}$ and $\hat{F}$ be two complete Hausdorff topological vector spaces. Let E_0 and F_0 be dense subspaces of $\hat{E}$ and $\hat{F}$, respectively, which are both isomorphic to E and consequently to each other. The isomorphism $f: E_0 \to F_0$ can be extended by Proposition 5 into a continuous linear map $\bar{f}$ from $\hat{E}$ into $\hat{F}$. Similarly, the inverse isomorphism $g: F_0 \to E_0$ can be extended into a continuous linear map $\bar{g}$ from $\hat{F}$ into $\hat{E}$. Now $g \circ f$ is the identity map of E_0 onto itself and $f \circ g$ is the identity map of F_0 onto itself. It follows that $\bar{g} \circ \bar{f}$ is the identity map of $\hat{E}$ onto itself and that $\bar{f} \circ \bar{g}$ is the identity map of $\hat{F}$ onto itself; hence $\bar{f}$ is an isomorphism of $\hat{E}$ onto $\hat{F}$. ∎

Proof of the lemma. Let $\mathfrak{F}$ be a Cauchy filter on E. The collection of all sets $X + V$, where $X \in \mathfrak{F}$ and V is a balanced neighborhood of 0 in E, is the basis of a filter $\mathfrak{G}$ on E since if $X \in \mathfrak{F}$, $Y \in \mathfrak{F}$, and V, W are balanced neighborhoods of 0, then $(X + V) \cap (Y + W) \supset (X \cap Y) + (V \cap W)$. Furthermore, $\mathfrak{G}$ is a Cauchy filter on E since for any balanced neighborhood V of 0 there exists $X \in \mathfrak{F}$ such that $X - X \subset V$ and thus

$$(X + V) - (X + V) \subset V + V + V.$$

The sets $(X + V) \cap A$ are not empty since A is dense in E. Thus $\mathfrak{G}$ induces a Cauchy filter $\mathfrak{G}_A$ on A, and by hypothesis $\mathfrak{G}_A$ converges to a point $x \in E$. But then x adheres to $\mathfrak{G}$ and since $\mathfrak{G}$ is a Cauchy filter, by Proposition 1 it converges to x. Consequently, $\mathfrak{F}$, which is finer than $\mathfrak{G}$, also converges to x. ▮

The space $\hat{E}$ constructed in Theorem 1 is called the *completion* of E. We usually identify E with the subspace E_0 of $\hat{E}$ and thus imbed E into $\hat{E}$. With this identification we can say that the closures in $\hat{E}$ of the neighborhoods belonging to a fundamental system $\mathfrak{N}$ of neighborhoods of 0 in E form a fundamental system of neighborhoods of 0 in $\hat{E}$. To prove this let us first observe that if A is an open subset of a topological space T and B is any set in T, then

$$A \cap \overline{B} \subset \overline{A \cap B}. \tag{8}$$

Indeed, if $x \in A \cap \overline{B}$, then for every neighborhood V of x, the set $V \cap A$ is also a neighborhood of x since A is open. Thus $V \cap A \cap B \neq \emptyset$; i.e., $x \in \overline{A \cap B}$. Now let $\hat{U}$ be a closed neighborhood of 0 in $\hat{E}$ and set $U = \hat{U} \cap E$. Then U is a neighborhood of 0 in E and there exists $V \in \mathfrak{N}$ such that $V \subset U$ and consequently $\overline{V} \subset \hat{U}$, where $\overline{V}$ is the closure of V in $\hat{E}$. Let W be an open neighborhood of 0 in E such that $W \subset V$. There exists an open neighborhood $\hat{W}$ of 0 in $\hat{E}$ such that $W = \hat{W} \cap E$. But then by (8), since E is dense in $\hat{E}$, we have $\hat{W} = \hat{W} \cap \overline{E} \subset \overline{W} \subset \overline{V}$; i.e., $\overline{V}$ is a neighborhood of 0 in $\hat{E}$.

It follows from the last remark that the completion of a locally convex space is locally convex since the closure of a convex set is convex (Proposition 4.3). Let $(q_\iota)_{\iota \in I}$ be a family of semi-norms defining a locally convex topology on the vector space E. Since each q_ι is uniformly continuous in E, it can be extended into a uniformly continuous function $\hat{q}_\iota$ on $\hat{E}$ (Proposition 5), and it is clear that $\hat{q}_\iota$ is a semi-norm on $\hat{E}$. The topology of $\hat{E}$ is defined by the family $(\hat{q}_\iota)_{\iota \in I}$. To show this, let $U_{\iota,\epsilon}$ be the open set $\{x \mid x \in E, q_\iota(x) < \epsilon\}$, $\overline{U}_{\iota,\epsilon}$ its closure in $\hat{E}$, and

$$V_{\iota,\epsilon} = \{x \mid x \in \hat{E}, \hat{q}_\iota(x) < \epsilon\}.$$

Then $U_{\iota,\epsilon} \subset V_{\iota,\epsilon}$ and thus $\overline{U}_{\iota,\epsilon} \subset \overline{V}_{\iota,\epsilon}$. Conversely, it follows from the definition of the extended function $\hat{q}_\iota$ that every $x \in \hat{E}$ with $\hat{q}_\iota(x) < \epsilon$ adheres to $U_{\iota,\epsilon}$, i.e., $V_{\iota,\epsilon} \subset \overline{U}_{\iota,\epsilon}$ and thus also $\overline{V}_{\iota,\epsilon} \subset \overline{U}_{\iota,\epsilon}$. Hence

$$\overline{U}_{\iota,\epsilon} = \overline{V}_{\iota,\epsilon},$$

which proves our assertion.

The completion of a metrizable topological vector space is a complete metrizable topological vector space. A sequence in a topological vector space is called a *Cauchy sequence* if the associated elementary filter is a Cauchy filter. In other words, $(x_n)_{n \in \mathbf{N}}$ is a Cauchy sequence if for every neighborhood V of 0 there exists an integer $N = N(V)$ such that $x_n - x_m \in V$ for all $n,m > N$.

Let $(x_n)_{n \in \mathbf{N}}$ be a sequence in a metrizable topological vector space E. Then (x_n) can be a Cauchy sequence with respect to the topological vector space structure of E or with respect to one of the translation-invariant metrics δ (§6) which define the topology of E (cf. Chapter 1, §1). Fortunately the two notions coincide. Indeed, the sets $V_\epsilon = \{x \mid \delta(x, 0) < \epsilon\}$ form a fundamental system of neighborhoods of 0 in E and the condition $x_n - x_m \in V_\epsilon$ is equivalent to $\delta(x_n - x_m, 0) = \delta(x_n, x_m) < \epsilon$.

A metrizable topological vector space E is complete if every Cauchy sequence in E converges. Indeed, let $x \in \hat{E}$. Since E is dense in $\hat{E}$ and $\hat{E}$ is metrizable, there exists a sequence (x_n) of elements of E converging to x. But (x_n) is a Cauchy sequence on E; hence $x \in E$.

In particular, if a metrizable topological vector space is quasi-complete, then it is complete. This follows from the fact that a Cauchy sequence (x_n) in a topological vector space E is always contained in some bounded closed set. Indeed, let V be a balanced neighborhood of 0 in E and let W be a balanced neighborhood of 0 such that $W + W \subset V$. There exists $N \in \mathbf{N}$ such that $x_n - x_m \in W$ for $n,m \geqq N$ and in particular

$$x_n \in x_N + W \qquad \text{for} \qquad n \geqq N.$$

On the other hand, there exist $\lambda_j > 0$ $(0 \leqq j \leqq N)$ such that

$$x_j \in \lambda_j W \quad (0 \leqq j \leqq N).$$

Let $\lambda = \max\ (1, \lambda_0, \lambda_1, \ldots, \lambda_N)$. Then $x_i \in \lambda W + W \subset \lambda V$ for all $i \in \mathbf{N}$; i.e., the set of points x_i is bounded. Consequently the closure of the set of points x_i $(i \in \mathbf{N})$ is a closed, bounded set.

Let us say that a filter $\mathfrak{F}$ in a topological vector space is *bounded* if there exists a bounded set $A \in \mathfrak{F}$. We have just proved that the elementary filter associated to a Cauchy sequence is bounded. In general a Cauchy filter need not be bounded. Thus, for instance, the filter of all neighborhoods of 0 in a nonnormable Hausdorff locally convex space converges to 0 but is not bounded (cf. Proposition 6.1).

DEFINITION 4. *A complete, metrizable, locally convex space is called a Fréchet space.*

EXAMPLE 1. Every Banach space is a Fréchet space.

EXAMPLE 2. The space $\mathcal{C}(\Omega)$ of Examples 3.3 and 4.6 is a Fréchet space. We know from Example 6.3 that it is metrizable. Let $(f_k)_{k\in\mathbf{N}}$ be a Cauchy sequence in $\mathcal{C}(\Omega)$. For every compact subset K of Ω and every $\epsilon > 0$ there exists an integer $N = N(K, \epsilon)$ such that $|f_k(x) - f_l(x)| < \epsilon$ for $x \in K$ and $k,l > N$. In particular, for each $x \in \Omega$ we have a numerical Cauchy sequence $(f_k(x))_{k\in\mathbf{N}}$ which converges to some number $f(x)$. If K is a compact neighborhood of $x_0 \in \Omega$ contained in Ω, we have

$$|f(x) - f_k(x)| \leqq \epsilon$$

for all $x \in K$ and $k > N$; i.e., f_k converges uniformly to f on K. Thus f is continuous at x_0 (cf. Example 1.1.4); and since x_0 was arbitrary in Ω, we have $f \in \mathcal{C}(\Omega)$. Finally, it is clear that (f_k) converges to f in $\mathcal{C}(\Omega)$.

EXAMPLE 3. The spaces $\mathcal{E}(\Omega)$ and $\mathcal{E}^m(\Omega)$ of Examples 4.7 and 4.8 are Fréchet spaces. We know that they are metrizable (Example 6.3). Let $(f_k)_{k\in\mathbf{N}}$ be a Cauchy sequence in $\mathcal{E}^m(\Omega)$ $(1 \leqq m \leqq \infty)$. For every compact subset K of Ω, $p = (p_1, \ldots, p_n) \in \mathbf{N}^n$, $|p| \leqq m$, and $\epsilon > 0$ there exists an integer $N = N(K, p, \epsilon)$ such that

$$|\partial^p f_k(x) - \partial^p f_l(x)| < \epsilon$$

for all $x \in K$ and $k,l > N$. It follows from the previous example that the sequence $(\partial^p f_k)_{k\in\mathbf{N}}$ converges to a continuous function $g^{(p)}$ in Ω, uniformly on every compact subset of Ω. We know from advanced calculus ([2], Theorem 13–13, p. 402) that if a sequence (f_k) tends uniformly to a function f and the sequence of derivatives $(\partial_j f_k)$ tends uniformly to a function h, then $h = \partial_j f$. Thus, setting $f = g^{(0)}$, we have $\partial^p f = g^{(p)}$ and (f_k) converges to f in $\mathcal{E}^m(\Omega)$.

EXAMPLE 4. The spaces $\mathcal{B}^m(\Omega)$ $(0 \leqq m \leqq \infty)$ of Examples 4.18 and 4.19 are complete. The proof is even simpler than in the previous example, since we deal here with uniform convergence on Ω and not only on compact subsets. In fact, let $(f_k)_{k\in\mathbf{N}}$ be a Cauchy sequence in $\mathcal{B}^m(\Omega)$; then for every $p \in \mathbf{N}^n$ with $|p| \leqq m$ we have $|\partial^p f_k(x) - \partial^p f_l(x)| < \epsilon$ for all $x \in \Omega$ and $k,l > N(p, \epsilon)$. Thus for every $p \in \mathbf{N}^n$, $|p| \leqq m$, the sequence $(\partial^p f_k)$ converges uniformly on Ω to a continuous function $g^{(p)}$. Just as in the previous example, we see that $g^{(p)} = \partial^p f$ $(f = g^{(0)})$ and that (f_k) converges to f in $\mathcal{B}^m(\Omega)$.

It follows that $\mathcal{B}(\Omega)$ is a Fréchet space, since we know from Example 6.2 that it is metrizable. For finite m the situation is even better: the space $\mathcal{B}^m(\Omega)$ is *banachizable*, i.e., normable and complete.

EXAMPLE 5. The spaces $\mathcal{K}(K)$, $\mathfrak{D}^m(K)$ $(1 \leqq m < \infty)$ of Example 4.9 are banachizable (cf. Example 6.1), the space $\mathfrak{D}(K)$ of Example 4.10 is a Fréchet space (cf. Example 6.2). Indeed, we know from Example 5.9 that $\mathfrak{D}^m(K)$ can be considered as a subspace of $\mathcal{E}^m(\Omega)$ for $K \subset \Omega$ $(0 \leqq m \leqq \infty)$; and thus to prove that $\mathfrak{D}^m(K)$ is complete, by Proposition 3 it is sufficient to prove that it is closed. If a sequence (f_k) of elements of $\mathfrak{D}^m(K)$ converges to $f \in \mathcal{E}^m(\Omega)$ and $x \notin K$, then

$$|f(x) - f_k(x)| = |f(x)| < \epsilon \qquad \text{for} \qquad k > N(\epsilon);$$

that is, $f(x) = 0$. Hence $\operatorname{Supp} f \subset K$ and $f \in \mathfrak{D}^m(K)$.

EXAMPLE 6. The spaces $\mathcal{B}_0^m(\Omega)$ $(0 \leqq m \leqq \infty)$ of Example 4.16 are banachizable (cf. Example 6.1), the space $\mathcal{B}_0(\Omega)$ of Example 4.17 is a Fréchet space (cf. Example 6.2). Indeed, we know from Example 5.10 that $\mathcal{B}_0^m(\Omega)$ is a subspace of $\mathcal{B}^m(\Omega)$; hence, similarly as in the previous example, it suffices to show that $\mathcal{B}_0^m(\Omega)$ is closed in $\mathcal{B}^m(\Omega)$ $(0 \leqq m \leqq \infty)$. If (f_k) is a sequence of elements of $\mathcal{B}_0^m(\Omega)$ which converges to an element $f \in \mathcal{B}^m(\Omega)$, then for every $\epsilon > 0$ and every $p \in \mathbf{N}^n$ with $|p| \leqq m$ there exists an integer $N > 0$ such that $|\partial^p f(x) - \partial^p f_k(x)| < \frac{1}{2}\epsilon$ for $x \in \Omega$ and $k \geqq N$. On the other hand, there exists a compact subset K of Ω such that $|\partial^p f_N(x)| < \frac{1}{2}\epsilon$ for $x \in \Omega \cap \complement K$. Hence

$$|\partial^p f(x)| < \epsilon \qquad \text{for} \qquad x \in \Omega \cap \complement K;$$

that is, $f \in \mathcal{B}_0^m(\Omega)$.

It follows that the spaces $\mathcal{S}_k^m$ $(m \in \mathbf{N}, k \in \mathbf{Z})$ of Example 4.11 are Fréchet spaces, since by Example 5.8 they are isomorphic to $\mathcal{B}_0^m$.

EXAMPLE 7. The space $\mathcal{O}_M$ of Example 4.15 is complete. Let $\mathfrak{F}$ be a Cauchy filter on $\mathcal{O}_M$. For every function $\varphi \in \mathcal{S}$, $p \in \mathbf{N}^n$, and $\epsilon > 0$ there exists a set $A = A(\varphi, p, \epsilon)$ in $\mathfrak{F}$ such that

$$|\varphi(x)\{\partial^p f(x) - \partial^p g(x)\}| < \epsilon \tag{9}$$

for all $x \in \mathbf{R}^n$ and $f,g \in A$. Given any compact set K in $\mathbf{R}^n$ there exists a $\varphi \in \mathcal{S}$ such that $\varphi(x) \geqq 1$ for $x \in K$ and thus a $B \in \mathfrak{F}$ such that

$$|\partial^p f(x) - \partial^p g(x)| < \epsilon$$

for $x \in K$ and $f,g \in B$. It follows as in Example 3 that there exists a function h defined on $\mathbf{R}^n$ which possesses continuous partial derivatives of all orders and such that $|\partial^p h(x) - \partial^p f(x)| < \epsilon$ for all $x \in K$ and $f \in B$. Given $x \in \mathbf{R}^n$ and $\eta > 0$ there exists $g \in A$ such that

$$|\varphi(x)\{\partial^p h(x) - \partial^p g(x)\}| < \eta$$

and hence by (9)

$$|\varphi(x)\{\partial^p h(x) - \partial^p f(x)\}| < \epsilon + \eta,$$

i.e.,

$$|\varphi(x)\{\partial^p h(x) - \partial^p f(x)\}| \leqq \epsilon$$

for all $x \in \mathbf{R}^n$ and $f \in A$. Thus h belongs to $\mathcal{O}_M$ and $\mathfrak{F}$ converges to h in $\mathcal{O}_M$.

THEOREM 2. *Let E be a complete, metrizable, topological vector space and M a closed subspace of E. Then the quotient space E/M is complete.*

Proof. We know (§6) that E/M is metrizable. Therefore it is enough to prove that every Cauchy sequence is convergent. Let $(V_n)_{n \in \mathbf{N}}$ be a fundamental system of neighborhoods of 0 in E which satisfies

$$V_{n+1} + V_{n+1} \subset V_n$$

for $n \in \mathbf{N}$. If we denote by $\dot{V}_n$ the canonical image of V_n in E/M, then the $\dot{V}_n$ form a fundamental system of neighborhoods of 0 in E/M. Let $(\dot{x}_k)_{k \in \mathbf{N}}$ be a Cauchy sequence in E/M. There exists a subsequence $(\dot{x}_{k_l})_{l \in \mathbf{N}}$ of $(\dot{x}_k)$ such that $\dot{x}_{k_l} - \dot{x}_{k_m} \in \dot{V}_n$ for $l,m \geqq n$. If y and z are elements of E such that $y \in \dot{x}_{k_l}$ and $z \in \dot{x}_{k_m}$, then $y - z \in V_n + M$. It follows that for $z \in \dot{x}_{k_m}$ and $l,m \geqq n$ we have

$$\dot{x}_{k_l} \cap (z + V_n) \neq \emptyset.$$

Indeed, let $y \in \dot{x}_{k_l}$, then $y \in z + V_n + M$, i.e., $y = z + v + u$, where $v \in V_n$ and $u \in M$. Thus $y - u = z + v$ belongs to $z + V_n$ and also to $\dot{x}_{k_l}$ since $y - u \equiv y \pmod{M}$.

We now define inductively a sequence $(x_n)_{n \in \mathbf{N}}$ in E. Take an arbitrary x_0 in $\dot{x}_0$ and suppose that we have already chosen the elements x_i $(1 \leqq i \leqq n)$ so that $x_i \in \dot{x}_{k_i}$, $x_i \in x_{i-1} + V_{i-1}$. Then we pick $x_{n+1} \in \dot{x}_{k_{n+1}}$ so that

$$x_{n+1} \in \dot{x}_{k_{n+1}} \cap (x_n + V_n) \subset x_n + V_n,$$

which is possible by what we have just seen. We have $x_{n+p} \in x_{n+p-1} + V_{n+p-1} \subset x_{n+p-2} + V_{n+p-2} + V_{n+p-1} \subset \cdots \subset x_n + V_n + V_{n+1} + \cdots + V_{n+p-1} \subset x_n + V_{n-1}$ for $p > 0$ and so (x_n) is a Cauchy sequence. Since E is complete, (x_n) converges to some element $x \in E$ and therefore $(\dot{x}_{k_n})$ converges to the image $\dot{x}$ of x in E/M. Hence $\dot{x}$ adheres to the elementary filter associated with the sequence $(\dot{x}_k)$ and by Proposition 1 $(\dot{x}_k)$ converges to $\dot{x}$. ∎

COROLLARY. *Let $f: E \to F$ be a strict morphism* (Definition 5.2) *from a metrizable and complete topological vector space E into a Hausdorff topological vector space F. Then* $\operatorname{Im}(f)$ *is a closed subspace of F.*

Proof. The quotient space $E/\mathrm{Ker}(f)$ is complete by the theorem. Hence, if $\bar{f}: E/\mathrm{Ker}(f) \to F$ is the associated injection, $\mathrm{Im}(\bar{f})$ is a closed subspace of F by Proposition 5. But $\mathrm{Im}(f) = \mathrm{Im}(\bar{f})$. ∎

REMARK 1. There exist complete topological vector spaces E which contain a closed subspace M such that E/M is not complete ([52], §31,6; [9], Chapter IV, §4, Exercise 10(b); [41], p. 92).

Let $(E_\iota)_{\iota \in I}$ be a family of topological vector spaces, $E = \prod_{\iota \in I} E_\iota$ their product (§7) and $\pi_\iota: E \to E_\iota$ the ι-th projection. A filter $\mathfrak{F}$ on E is a Cauchy filter if and only if $\pi_\iota(\mathfrak{F})$ is a Cauchy filter for every $\iota \in I$. Indeed, if $\mathfrak{F}$ is a Cauchy filter, then it follows from the uniform continuity of π_ι that $\pi_\iota(\mathfrak{F})$ is a Cauchy filter on E_ι. Conversely, suppose that each $\pi_\iota(\mathfrak{F})$ is a Cauchy filter. Let $V = \prod_{\iota \in I} V_\iota$ be a neighborhood of 0 in E, where V_ι is a neighborhood of 0 in E_ι and $V_\iota = E_\iota$ for all ι which do not belong to a finite subset H of I. For each $\iota \in H$ let $A_\iota \in \mathfrak{F}$ be such that $\pi_\iota(A_\iota) - \pi_\iota(A_\iota) \subset V_\iota$. Then $A = \bigcap_{\iota \in H} A_\iota \in \mathfrak{F}$ and $A - A \subset V$.

It follows that if for each $\iota \in I$ the set A_ι is a complete subset of E_ι, then the subset $A = \prod_{\iota \in I} A_\iota$ of E is complete. Indeed, if $\mathfrak{F}$ is a Cauchy filter on A, then $\pi_\iota(\mathfrak{F})$ is a Cauchy filter on A_ι and therefore converges to a point $x_\iota \in A_\iota$. But then $\mathfrak{F}$ converges to (x_ι). It is also true that if E is complete, then the E_ι are complete (see Exercise 5).

PROPOSITION 6. *Let $(E_\iota)_{\iota \in I}$ be a family of topological vector spaces and $E = \prod_{\iota \in I} E_\iota$. If the E_ι are quasi-complete* (Definition 2), *then E is quasi-complete.*

Proof. Let B be a closed bounded set in E. For each $\iota \in I$ the set $B_\iota = \pi_\iota(B)$ is bounded; hence $\bar{B}_\iota$ is a closed, bounded set and therefore complete. But then B is a closed subset of the complete set $\prod_{\iota \in I} \bar{B}_\iota$, and thus by Proposition 3 it is complete. ∎

EXERCISES

1. Prove the theorem quoted in Example 3.

2. Prove Theorem 1 for a locally convex Hausdorff space E using Exercise 7.1.

3. (a) Let E be a Hausdorff topological vector space, $\hat{E}$ a completion of E, and j the isomorphism of E onto the dense subspace E_0 of $\hat{E}$. Show that the pair $(j, \hat{E})$ has the following "universal property":

(U) Given a continuous linear map f from E into a complete Hausdorff topological vector space G, there exists a *unique* continuous linear map $\hat{f}: \hat{E} \to G$ such that $f = \hat{f} \circ j$.

(b) Show that (U) characterizes the pair $(j, \hat{E})$ up to an isomorphism in the following sense: If $(j, \hat{E})$ is a pair consisting of a complete Hausdorff topological vector space $\hat{E}$ and of a continuous linear map $j: E \to \hat{E}$ satisfying condition (U), then j is an isomorphism onto a dense subspace of $\hat{E}$. If, furthermore, $(j_1, \hat{E}_1)$

is another such pair, then there exists a unique isomorphism $h: \hat{E} \to \hat{E}_1$ such that $h \circ j = j_1$.

4. Let E be a topological vector space. The set of all Cauchy filters on E is ordered by inclusion. A *minimal Cauchy filter* on E is a minimal element in the set of all Cauchy filters; i.e., $\mathfrak{F}$ is a minimal Cauchy filter if there exists no Cauchy filter coarser than $\mathfrak{F}$ and distinct from $\mathfrak{F}$.

(a) Prove that given a Cauchy filter $\mathfrak{F}$ on E there exists a unique minimal Cauchy filter $\mathfrak{F}_0$ coarser than $\mathfrak{F}$. A basis of $\mathfrak{F}_0$ is given by the sets $A + V$, where $A \in \mathfrak{F}$ and V is a neighborhood of 0 in E. Prove that in particular for every $x \in E$ the filter $\mathfrak{B}(x)$ of all neighborhoods of x is a minimal Cauchy filter.

(b) Suppose that E is a Hausdorff space and denote by $\hat{E}$ the set of all minimal Cauchy filters on E. Define on $\hat{E}$ a structure of complete Hausdorff topological vector space such that E is isomorphic to a dense subspace E_0 of $\hat{E}$. (*Hint:* Establish successively the following steps:

(i) Associating with $x \in E$ the filter $\mathfrak{B}(x) \in \hat{E}$, we obtain a bijection from E onto a subset of E_0 of $\hat{E}$.

(ii) Given $\mathfrak{F}, \mathfrak{G} \in \hat{E}$, define $\mathfrak{F} + \mathfrak{G}$ as the minimal Cauchy filter contained in the Cauchy filter generated by the collection $\{A + B \mid A \in \mathfrak{F}, B \in \mathfrak{G}\}$. Similarly, for $\lambda \in \mathbf{K}$, $\mathfrak{F} \in \hat{E}$, define $\lambda\mathfrak{F}$ as the minimal Cauchy filter contained in the Cauchy filter generated by the collection $\{\lambda A \mid A \in \mathfrak{F}\}$. Prove that these operations satisfy Axioms (VS 1) through (VS 8) of Chapter 1, §1, where $\mathfrak{B}(0)$ is the zero vector.

(iii) Given a balanced neighborhood V of 0 in E, let $\hat{U}_V$ be the collection of those elements $\mathfrak{F}$ of $\hat{E}$ for which there exists an $A \in \mathfrak{F}$ and a $W \in \mathfrak{B}(0)$ such that $A + W \subset V$. Prove that the sets $\hat{U}_V$ form a filter basis on $\hat{E}$ which satisfies conditions (NS 1) through (NS 3) of Theorem 3.1.

(iv) Prove that $\hat{E}$ is a Hausdorff space.

(v) Prove that the map $E \to E_0$ is an isomorphism and that E_0 is dense in $\hat{E}$.

(vi) Prove that $\hat{E}$ is complete.)

5. (a) Let $(E_\iota)_{\iota \in I}$ be a family of sets and $E = \prod_{\iota \in I} E_\iota$ their cartesian product. Suppose that on each set E_ι we are given a filter $\mathfrak{F}_\iota$, and consider on E the collection $\mathfrak{B}$ of all sets of the form $\prod_{\iota \in I} A_\iota$, where $A_\iota \in \mathfrak{F}_\iota$ for all ι and $A_\iota = E_\iota$ for all ι which do not belong to a finite subset H of I. Prove that $\mathfrak{B}$ is a filter basis on E. The filter $\mathfrak{F}$ generated by $\mathfrak{B}$ is called the *product* of the filters $\mathfrak{F}_\iota$ and we write $\mathfrak{F} = \prod_{\iota \in I} \mathfrak{F}_\iota$.

(b) Let $(E_\iota)_{\iota \in I}$ be a family of topological vector spaces and $E = \prod_{\iota \in I} E_\iota$. Let $\mathfrak{F}_\iota$ be a filter on E_ι and $\mathfrak{F} = \prod_{\iota \in I} \mathfrak{F}_\iota$. Prove that $\mathfrak{F}$ is a Cauchy filter if and only if each $\mathfrak{F}_\iota$ is a Cauchy filter.

(c) Prove that if the topological vector space $E = \prod_{\iota \in I} E_\iota$ is complete, then each E_ι is complete. (*Hint:* Given a Cauchy filter $\mathfrak{F}_\iota$ on E_ι, take a Cauchy filter $\mathfrak{F}_\kappa$ on each E_κ with $\kappa \neq \iota$ and consider the product filter $\prod_{\kappa \in I} \mathfrak{F}_\kappa$.)

6. (a) Let E be a topological vector space and $(x_\iota)_{\iota \in I}$ a summable family of elements of E (Definition 8.2). Prove that the following condition is satisfied:

Cauchy's condition: For every neighborhood V of 0 in E there exists a finite subset J of I such that $\sum_{\iota \in H} x_\iota \in V$ for all finite subsets H of I which do not meet J.

(b) Let E be a complete topological vector space and $(x_\iota)_{\iota \in I}$ a family of elements of E. Prove that if (x_ι) satisfies Cauchy's condition, then it is summable. (*Hint:* The image of the filter of sections under the map $J \mapsto x_J$ (see Definition 8.2) is the basis of a Cauchy filter.)

§10. Finite-dimensional and locally compact spaces

LEMMA 1. *Let H be a closed hyperplane in a topological vector space E and let D be a* (one-dimensional) *algebraic supplement of H. Then D is a topological supplement* (Definition 7.2) *of H.*

Proof. Since H is closed, the one-dimensional vector space E/H is a Hausdorff space (Proposition 5.5). Also D is a Hausdorff space since the set $\{0\}$ is closed in D as the intersection of D with the closed set H. If $f: D \to E/H$ is the canonical algebraic isomorphism $f = \varphi \circ j$, where $j: D \hookrightarrow E$ is the canonical injection and $\varphi: E \to E/H$ the canonical surjection, then it follows from Proposition 5.6 that f is an isomorphism of the topological vector spaces and thus D is a topological supplement of H. ▮

THEOREM 1. *Let E be a Hausdorff topological vector space which has finite algebraic dimension n over the field $\mathbf{K}$. Then E is isomorphic to the product $\mathbf{K}^n$ of n spaces $\mathbf{K}^1$* (cf. Chapter 1, §5). *More precisely, if $(e_i)_{1 \leq i \leq n}$ is a basis of E, then the map*

$$(\xi_i)_{1 \leq i \leq n} \mapsto \sum_{i=1}^{n} \xi_i e_i$$

is an isomorphism from $\mathbf{K}^n$ onto E.

Proof. We use induction on n and observe that for $n = 1$ the statement is precisely Proposition 5.6. Suppose therefore that the theorem holds for $n - 1$, and let H be the hyperplane of E spanned by the vectors $e_1, e_2, \ldots, e_{n-1}$. By the induction hypothesis the map

$$(\xi_i)_{1 \leq i \leq n-1} \mapsto \sum_{i=1}^{n-1} \xi_i e_i$$

is an isomorphism from $\mathbf{K}^{n-1}$ onto H. The space $\mathbf{K}^{n-1}$ is complete as the product of the complete spaces $\mathbf{K}^1$. Thus H is complete, and since E is Hausdorff, it follows from Proposition 9.2 that H is closed. By the lemma E is the topological direct sum of H and the $D = \{\lambda e_n \mid \lambda \in \mathbf{K}\}$; i.e., the map

$$(\xi_i)_{1 \leq i \leq n} \mapsto \sum_{i=1}^{n} \xi_i e_i$$

from $\mathbf{K}^{n-1} \times \mathbf{K}^1$ onto E is an isomorphism (Definition 7.1). ▮

Let us now draw some conclusions from the above theorem.

PROPOSITION 1. *A finite-dimensional subspace F of a Hausdorff topological vector space is closed.*

Proof. Since F is a finite-dimensional Hausdorff topological vector space over $\mathbf{K}$, it is isomorphic to the complete space $\mathbf{K}^n$ for some $n \in \mathbf{N}$. Thus F is complete and hence by Proposition 9.2 it is closed. ∎

COROLLARY. *If E is a Hausdorff topological vector space, then every algebraically free finite subset of E is topologically free* (Definition 1.3.2).

PROPOSITION 2. *If E is a finite-dimensional Hausdorff topological vector space and F an arbitrary topological vector space, then every linear map f from E into F is continuous.*

Proof. There exists an isomorphism φ from E onto some space $\mathbf{K}^n$, given by

$$\sum_{i=1}^{n} \xi_i e_i \mapsto (\xi_i)_{1 \leqq i \leqq n},$$

where $(e_i)_{1 \leqq i \leqq n}$ is a basis of E. Let $b_i = f(e_i)$; then the map

$$\psi: (\xi_i)_{1 \leqq i \leqq n} \mapsto \sum_{i=1}^{n} \xi_i b_i$$

from $\mathbf{K}^n$ into F is continuous by the definition of a topological vector space. Hence $f = \psi \circ \varphi$ is continuous. ∎

PROPOSITION 3. *Let E be a topological vector space, M a closed subspace of E, and F a finite-dimensional linear subspace of E. Then the subspace $M + F$ is closed.*

Proof. We know that E/M is a Hausdorff space (Proposition 5.5). Let $\varphi: E \to E/M$ be the canonical surjection. Then $\varphi(F)$ is a finite-dimensional subspace of E/M, and thus by Proposition 1 it is closed. Hence the subspace $M + F = \varphi^{-1}(\varphi(F))$ of E is closed. ∎

REMARK 1. If M and N are two closed subspaces of a topological vector space, then $M + N$ is not necessarily closed (cf. Remark 1.7.1).

Before we pass to the next proposition, let us recall that the (algebraic) *codimension* of a subspace M of a vector space E is the dimension of the quotient space E/M. It is also the dimension of any algebraic supplement of M.

PROPOSITION 4. *If E is a Hausdorff topological vector space and M a closed subspace of E with finite codimension, then every algebraic supplement N of M is a topological supplement* (Definition 7.2) *of M.*

Proof. Let $q: E \to N$ be the projector onto N and $\bar{q}: E/M \to N$ the associated injection. By Proposition 2 the map $\bar{q}$ is continuous since E/M

is a finite-dimensional Hausdorff space. This proves that M and N are topological supplements (p. 122). ∎

REMARK 2. We shall see (Proposition 3.1.3) that if E is a locally convex Hausdorff space and M a finite-dimensional subspace of E, then there exists a topological supplement of M.

In our next theorem we shall prove that a locally compact topological vector space is necessarily finite-dimensional, which will be of fundamental importance in the theory of compact linear maps. First, however, let us recall the definition and basic properties of compact sets.

A topological space X is said to be *compact* if it is a Hausdorff space and if the following equivalent conditions are satisfied:

(C 1) Every filter on X has at least one adherent point.

(C 2) Every ultrafilter on X converges.

(C 3) (*Cantor property*) Every family of closed sets in X whose intersection is empty contains a finite subfamily whose intersection is empty.

(C 4) (*Heine-Borel-Lebesgue property*) Every open cover of X contains a finite subcover.

The last condition means, of course, that given a family of open sets $(U_\iota)_{\iota \in I}$ such that $X = \bigcup_{\iota \in I} U_\iota$, there exists a finite subset J of I such that $X = \bigcup_{\iota \in J} U_\iota$.

Let us prove the equivalence of conditions (C 1) through (C 4).

(C 1) $\Rightarrow$ (C 2): Let $\mathfrak{U}$ be an ultrafilter and $x \in X$ a point which adheres to $\mathfrak{U}$. Then there exists a filter $\mathfrak{G}$, finer than $\mathfrak{U}$, which converges to x (cf. §8). But by the definition of an ultrafilter we have $\mathfrak{G} = \mathfrak{U}$.

(C 2) $\Rightarrow$ (C 1): Let $\mathfrak{F}$ be a filter on X. There exists (cf. §2) an ultrafilter $\mathfrak{U}$ finer than $\mathfrak{F}$. By condition (C 2) the ultrafilter $\mathfrak{U}$ converges to a point $x \in X$. But then x adheres to $\mathfrak{F}$.

(C 1) $\Rightarrow$ (C 3): Let $\mathfrak{S}$ be a family of closed subsets of X whose intersection is empty. Let $\mathfrak{S}'$ be the collection of all finite intersections of sets in $\mathfrak{S}$. If none of the sets in $\mathfrak{S}'$ were empty, then $\mathfrak{S}'$ would generate a filter $\mathfrak{F}$. By (C 1) the filter $\mathfrak{F}$ would have an adherent point x which then would belong to all the sets in $\mathfrak{S}$ in contradiction to the hypothesis.

(C 3) $\Rightarrow$ (C 1): We prove that the negation of (C 1) implies the negation of (C 3). Let $\mathfrak{F}$ be a filter with no adherent point. The collection of all sets $\overline{A}$, where A runs through $\mathfrak{F}$, has an empty intersection (since every point belonging to the intersection would adhere to $\mathfrak{F}$), but by the definition of a filter no finite subcollection has an empty intersection.

Finally, the equivalence of (C 3) and (C 4) is clear by passing to the complements.

A subset A of a topological space X is said to be *compact* if, equipped with the topology induced by X, it is a compact space. In other words, A is compact if the induced topology is Hausdorff and every family $(U_\iota)_{\iota \in I}$ of open sets *in* X

such that $A \subset \bigcup_{\iota \in I} U_\iota$ contains a finite subfamily $(U_\iota)_{\iota \in J}$ such that $A \subset \bigcup_{\iota \in J} U_\iota$. Here are some easy properties of compact sets.

(a) A closed subset of a compact space is compact. This follows immediately from Axiom (C 3) if we observe that if A is a closed subset of the topological space X, then a closed subset of A is also closed in X.

(b) A compact subset of a Hausdorff space is closed. Indeed, let x adhere to the compact subset A of the Hausdorff space X. Then every neighborhood of x meets A, and therefore the filter $\mathfrak{B}(x)$ of neighborhoods of x in X induces a filter $\mathfrak{B}_A$ on A. By Axiom (C 1) there exists a point $y \in A$ which is adherent to $\mathfrak{B}_A$ and also to $\mathfrak{B}(x)$. But $\mathfrak{B}(x)$ converges to x; therefore, since X is a Hausdorff space, we must have $x = y$, i.e., $x \in A$.

(c) If f is a continuous map from a topological space X into a Hausdorff space Y, then the image $f(A)$ of any compact subset A of X is a compact subset of Y. *Proof.* Let $(U_\iota)_{\iota \in I}$ be an open cover of $f(A)$ in Y. Then the sets $(f^{-1}(U_\iota))_{\iota \in I}$ form an open cover of A in X, hence there exists a finite subfamily $(f^{-1}(U_\iota))_{\iota \in J}$ which still covers A. But then $(U_\iota)_{\iota \in J}$ covers A.

A subset A of a topological space X is said to be ***relatively compact*** in X if it is contained in a compact subset of X.

(d) A subset A of a Hausdorff space X is relatively compact if and only if $\overline{A}$ is compact. Since $A \subset \overline{A}$, it follows from the definition that A is relatively compact if $\overline{A}$ is compact. Conversely if A is contained in the compact set B, then also $\overline{A} \subset B$, since by (b) the set B is closed. But then by (a) the set $\overline{A}$ is compact.

(e) If A is a relatively compact subset of the topological space X, then every filter basis on A has at least one adherent point in X. Indeed, if A is contained in the compact subset B of X, then every filter basis on A generates a filter on B, which by Axiom (C 1) has at least one adherent point in B.

The following result is one of the most important theorems of analysis. Its proof is very simple because we can use Axiom (C 2), i.e., the existence of ultrafilters.

TIHONOV'S THEOREM. *If $(X_\iota)_{\iota \in I}$ is a family of compact spaces, then the product space $X = \prod_{\iota \in I} X_\iota$ is compact.*

Proof. We know that X is a Hausdorff space (§7). Let $\pi_\iota: X \to X_\iota$ be the ι-th projection and let $\mathfrak{U}$ be an ultrafilter on X. Then $\pi_\iota(\mathfrak{U})$ is an ultrafilter on X_ι (§2), hence converges to some point x_ι by Axiom (C 2). Consequently, $\mathfrak{U}$ converges to $x = (x_\iota)$. ∎

A topological space X is said to be *locally compact* if it is a Hausdorff space and every point $x \in X$ possesses a compact neighborhood.

A Hausdorff topological vector space is locally compact if 0 possesses a compact neighborhood. It follows from Tihonov's theorem that the product of finitely many locally compact spaces is locally compact. Since the space $\mathbf{K}^1$ is locally compact, it follows in particular that every finite-dimensional Hausdorff space is locally compact (Theorem 1). We shall see in a moment that the converse is also true. First let us, however, establish some results concerning compact sets in topological vector spaces.

LEMMA 2. *Let E be a Hausdorff topological vector space, A a compact subset of E and B a closed subset of E. If $A \cap B = \emptyset$, then there exists a neighborhood V of 0 such that $(A + V) \cap (B + V) = \emptyset$.*

Proof. Since B is closed, for each $x \in A$ there exists a neighborhood W_x of 0 such that $(x + W_x) \cap B = \emptyset$. For each $x \in A$ let V_x be a balanced neighborhood of 0 such that $V_x + V_x + V_x \subset W_x$. Now the sets $x + V_x$ cover A; hence there exists a finite family $(x_i)_{1 \leq i \leq n}$ of points of A such that the sets $x_i + V_{x_i}$ cover A. The set $V = \bigcap_{i=1}^{n} V_{x_i}$ is a neighborhood of 0. Furthermore, $(A + V) \cap (B + V) = \emptyset$. Indeed, if $a + v = b + v'$ with $a \in A$, $b \in B$, $v, v' \in V$, then $a \in x_i + V_{x_i}$ for some index i. Hence

$$a + v - v' \in x_i + V_{x_i} + V_{x_i} + V_{x_i} \subset x_i + W_{x_i}.$$

But $a + v - v' = b \in B$, which contradicts the choice of W_{x_i}. ▮

PROPOSITION 5. *Let E be a Hausdorff topological vector space, A a compact subset of E and B a closed subset of E. Then the set $A + B$ is closed.*

Proof. Let x be a point of E not belonging to $A + B$. Then

$$(x - B) \cap A = \emptyset,$$

and since $x - B$ is closed, by Lemma 2 there exists a neighborhood V of 0 such that $(x - B + V) \cap A = \emptyset$. The set $x + V$ is a neighborhood of x and $(x + V) \cap (A + B) = \emptyset$. This shows that $\complement(A + B)$ is open, i.e., that $A + B$ is closed. ▮

A compact subset A of a topological vector space is always complete. Indeed, let $\mathfrak{F}$ be a Cauchy filter on A. Then by Axiom (C 1) there exists a point $x \in A$ which adheres to $\mathfrak{F}$. But by Proposition 9.1 the filter $\mathfrak{F}$ converges to x.

DEFINITION 1. *A subset A of a Hausdorff topological vector space is said to be precompact* (or totally bounded) *if every ultrafilter on A is a Cauchy filter.*

Clearly every compact subset of a Hausdorff topological vector space is precompact and every complete and precompact subset is compact.

THEOREM 2. *Let E be a Hausdorff topological vector space and A a subset of E. The following conditions are equivalent:*

(a) *A is precompact.*

(b) *If $\hat{E}$ is a completion of E and $j: E \to \hat{E}$ the isomorphism onto the dense subspace E_0 of $\hat{E}$* (cf. Theorem 9.1), *then the closure of $j(A)$ is a compact subset of $\hat{E}$.*

(c) *For every neighborhood V of 0 in E there exists a finite family $(x_i)_{1 \leq i \leq n}$ of points of A such that $A \subset \bigcup_{i=1}^{n} (x_i + V)$.*

Proof. (a) $\Rightarrow$ (b): Suppose that $B = \overline{j(A)}$ is not compact in $\hat{E}$ and let us prove that A is not precompact. There exists on B a filter $\mathfrak{F}$ which has no adherent point. The collection of all sets $X + V$, where $X \in \mathfrak{F}$ and V is a balanced neighborhood of 0 in $\hat{E}$, is the basis of a filter $\mathfrak{G}$ on $\hat{E}$. Since every set $X + V$ meets A, the trace $\mathfrak{G}_A$ of $\mathfrak{G}$ on A is a filter on A. Let $\mathfrak{U}$ be an ultrafilter on A finer than $\mathfrak{G}_A$. If $\mathfrak{U}$ were a Cauchy filter, it would generate a Cauchy filter on B which would converge to some point $x \in B$. But then x would be clearly an adherent point of $\mathfrak{F}$, in contradiction to our original assumption.

(b) $\Rightarrow$ (c): Let W be a closed neighborhood of 0 in E such that $W \subset V$. The closure $\hat{W}$ of $j(W)$ in $\hat{E}$ is a neighborhood of 0 in $\hat{E}$. Since $\overline{j(A)}$ is compact in $\hat{E}$, there exists a finite family $(x_i)_{1 \leqq i \leqq n}$ of points of A such that the sets $j(x_i) + \hat{W}$ cover $\overline{j(A)}$. Since $\hat{W} \cap E = W \subset V$, the sets $x_i + V$ cover A.

(c) $\Rightarrow$ (a): Let $\mathfrak{U}$ be an ultrafilter on A, U a neighborhood of 0 in E, and V a balanced neighborhood of 0 in E such that $V + V \subset U$. Suppose that $A \subset \bigcup_{i=1}^{n} (x_i + V)$. Then $A = \bigcup_{i=1}^{n} (x_i + V) \cap A \in \mathfrak{U}$; hence by a characteristic property of ultrafilters (§2) there exists an index j $(1 \leqq j \leqq n)$ such that the set $X = (x_j + V) \cap A$ belongs to $\mathfrak{U}$. But then $X - X \subset V + V \subset U$, which shows that $\mathfrak{U}$ is a Cauchy filter. ∎

PROPOSITION 6. *Let E and F be two Hausdorff topological vector spaces and $f: E \to F$ a uniformly continuous map* (Definition 9.3). *If A is a precompact subset of E, then $f(A)$ is a precompact subset of F.*

Proof. Let V be a neighborhood of 0 in F and U a neighborhood of 0 in E such that $x - y \in U$ implies $f(x) - f(y) \in V$. There exists a finite family $(x_i)_{1 \leqq i \leqq n}$ of points of A such that the sets $x_i + U$ cover A, and hence the sets $f(x_i) + V$ cover $f(A)$. ∎

PROPOSITION 7. *A precompact subset A of a Hausdorff topological vector space E is bounded.*

Proof. Let V be a balanced neighborhood of 0 in E. By Theorem 2 there exists a finite family $(x_i)_{1 \leqq i \leqq n}$ of points of A such that

$$A \subset \bigcup_{i=1}^{n} (x_i + V).$$

On the other hand, there exists a number $\lambda \geqq 1$ such that $x_i \in \lambda V$ for $1 \leqq i \leqq n$. Thus

$$A \subset \lambda V + V \subset \lambda(V + V). \blacksquare$$

In particular, in a quasi-complete (Definition 9.2) Hausdorff topological vector space every closed, precompact set is compact and every precompact set is relatively compact (cf. Exercise 3(b)).

THEOREM 3 (F. Riesz). *Let E be a Hausdorff topological vector space in which there exists a precompact neighborhood U of 0. Then E has finite algebraic dimension* (and is by Theorem 1 isomorphic to some $\mathbf{K}^n$).

We first prove:

LEMMA 3. *Let E be a topological vector space, F a proper closed subspace of E, U a bounded neighborhood of 0 in E, and V a balanced neighborhood of 0 in E such that $V + V \subset U$. Then U is not contained in $F + V$.*

Proof. Assume that $U \subset F + V$. Then $V + V \subset U \subset F + V$ and *a fortiori* $2V \subset F + V$. We see by induction that $2^n V \subset F + V$ for every integer $n \geqq 1$. Indeed, this holds for $n = 1$; and if we assume it for n, we have

$$2^{n+1}V \subset 2^n V + 2^n V \subset F + V + F + V \subset F + V$$

since $F + F = F$.

Every integer $m \geqq 1$ verifies $m \leqq 2^n$ for some n. Therefore $mV \subset F + V$ for all $m \geqq 1$; i.e.,

$$V \subset \bigcap_{m \geqq 1} \left(F + \frac{1}{m} V\right).$$

Now,

$$F = \bigcap_{m \geqq 1} \left(F + \frac{1}{m} V\right).$$

Indeed, the inclusion

$$F \subset \bigcap_{m \geqq 1} \left(F + \frac{1}{m} V\right)$$

is clear. On the other hand, let

$$x \in \bigcap_{m \geqq 1} \left(F + \frac{1}{m} V\right)$$

and let W be a balanced neighborhood of 0. Since V is bounded, there exists an integer $m \geqq 1$ such that $(1/m)V \subset W$ and therefore $x \in F + W$. Hence $(x + W) \cap F \neq \emptyset$ and $x \in \overline{F}$. The assertion follows from the hypothesis that F is closed.

Thus we have $V \subset F$ and therefore $F = E$ since V is absorbing. This, however, contradicts the assumption that F is a proper subspace of E. ∎

Proof of Theorem 3. Let V be a balanced neighborhood of 0 such that $V + V \subset U$. We assume that E is not finite-dimensional and construct an infinite sequence $(x_n)_{n \in \mathbf{N}}$ of elements of U such that $x_n - x_m \notin V$ for $n \neq m$. This, however, contradicts the assumption that U is precompact. Indeed, if W is a balanced neighborhood of 0 such that $W + W \subset V$, let

(y_i) be a finite sequence of elements of U such that the sets $y_i + W$ cover U. There must exist two distinct indices m and n such that x_n and x_m are in the same set $y_i + W$. But this implies $x_n - x_m \in W + W \subset V$.

It remains for us to construct the sequence (x_n). For x_0 we choose any nonzero element of E. Assume that we have already chosen $x_0, x_1, \ldots, x_n$ and denote by F_n the (necessarily closed) subspace of E generated by these elements. By our hypothesis $E \neq F_n$, and therefore by Lemma 3 there exists $x_{n+1} \in U$ such that $x_{n+1} \notin F_n + V$, and in particular $x_{n+1} \notin x_j + V$ for $0 \leqq j \leqq n$. This defines the sequence (x_n) inductively. ▮

Exercises

1. Let X be a compact space.

(a) Prove that if $\mathfrak{F}$ is a filter on X and A the set of points adherent to $\mathfrak{F}$, then every neighborhood of A belongs to $\mathfrak{F}$. (*Hint:* Let V be a neighborhood of A and assume that every set $B \in \mathfrak{F}$ meets $\complement V$. Consider the filter $\mathfrak{G}$ generated by the sets $B \cap \complement V$. Then $\mathfrak{G}$ has an adherent point $x \in X$, $x \notin A$, which also adheres to $\mathfrak{F}$.)

(b) Prove that a filter on X converges if and only if it has a unique adherent point.

(c) Let x be a point of X. Show that a filter basis $\mathfrak{B}$ formed of closed neighborhoods of x is a fundamental system of neighborhoods of x if and only if the intersection of all the sets belonging to $\mathfrak{B}$ is the point x.

(d) Prove that every compact space is regular (cf. §2).

(e) Show that if A and B are two closed subsets of X such that $A \cap B = \emptyset$, then there exist two open subsets U and V of X such that $A \subset U$, $B \subset V$, and $U \cap V = \emptyset$. (*Hint:* If every neighborhood U of A met every neighborhood V of B, then the sets $U \cap V$ would generate a filter $\mathfrak{F}$. Using (d), show that a point adherent to $\mathfrak{F}$ would belong to $A \cap B$.)

2. Let X be a regular topological space. Prove that if every filter on a subset A of X has at least one adherent point in X, then A is relatively compact. (*Hint:* Let $\mathfrak{F}$ be a filter on $\overline{A}$ and $\mathfrak{G}$ the filter basis on $\overline{A}$ formed by all open sets which contain a set of $\mathfrak{F}$. Consider the trace of $\mathfrak{G}$ on A.)

3. Let E be a Hausdorff topological vector space.

(a) Prove that every relatively compact subset of E is precompact.

(b) Prove that if the subset A of E is precompact, then $\overline{A}$ is also precompact.

(c) Prove that every subset of a precompact set is also precompact.

(d) Prove that the union of two precompact subsets of E is precompact.

4. Let E be a Hausdorff topological vector space.

(a) Prove that the balanced hull of a compact subset A of E is compact. (*Hint:* If B is the set of all $\lambda \in \mathbf{K}$ such that $|\lambda| \leqq 1$, then the balanced hull of A is the image of $B \times A$ under the map $(\lambda, x) \mapsto \lambda x$.)

(b) Prove that the balanced hull of a precompact subset C of E is precompact. (*Hint:* The closure of $j(C)$ is compact in $\hat{E}$, and by part (a) so is its balanced hull.)

5. Give a direct proof of the implication (a) ⇒ (c) in Theorem 2. (*Hint:* If V is a neighborhood of 0 in E such that for no finite family (x_i) of points of A do the sets $x_i + V$ cover A, then the traces on A of the sets $\complement(x + V)$, where $x \in A$, generate a filter $\mathfrak{F}$ on A. If $\mathfrak{U}$ is an ultrafilter on A finer than $\mathfrak{F}$, then for no $B \in \mathfrak{U}$ can we have $B - B \subset V$.)

6. (a) Let X be a locally compact topological space. Denote by X' the set whose points are the points of X and one additional point, "the point at infinity," denoted by ω. Define on X' a topology as follows: the neighborhoods of a point $x \in X$ in the original space X form a fundamental system of neighborhoods of x in X', while a fundamental system of neighborhoods of ω is given by the complements (with respect to X') of the compact subsets of X. Show that the neighborhoods defined in this way satisfy axioms (NB 1) through (NB 4) of §1, that X' induces on X its original topology and that X' is compact. (X' is called the Alexandrov or one-point compactification of X.)

(b) Let X be a locally compact space, X_1' and X_2' two compact spaces. Assume that for $i = 1,2$ there exist maps $f_i: X \to X_i'$ such that $\complement f_i(X)$ is a single point and that f_i is a homeomorphism of X onto $f_i(X)$. Show that there exists a unique homeomorphism $g: X_1' \to X_2'$ such that $f_2 = g \circ f_1$.

(c) We say that a numerical function φ defined on a locally compact space X "vanishes at infinity" (or "vanishes at the boundary of X") if for every $\epsilon > 0$ there exists a compact subset K of X such that $|\varphi(x)| \leqq \epsilon$ for $x \in \complement K$ (cf. Examples 1.1.5, 4.16, and Exercise 1.2.7). Show that φ vanishes at infinity if and only if, setting $\varphi(\omega) = 0$, the extended function is continuous on the Alexandrov compactification X' of X.

§11. *Initial topologies*

Let E be a vector space and $(F_\iota)_{\iota \in I}$ a family of topological vector spaces, all over the same field $\mathbf{K}$, and for each index $\iota \in I$ let f_ι be a linear map from E into F_ι. Let $\mathcal{T}$ be the coarsest topology on E for which all the maps f_ι are continuous (§1), i.e., the initial topology on E for the family (f_ι). If $\mathfrak{N}_\iota$ is a fundamental system of neighborhoods of 0 in F_ι, then a fundamental system of neighborhoods of 0 in E for $\mathcal{T}$ is given by all finite intersections

$$f_{\iota_1}^{-1}(U_1) \cap \cdots \cap f_{\iota_n}^{-1}(U_n), \qquad \text{where} \qquad U_k \in \mathfrak{N}_{\iota_k} \ (1 \leqq k \leqq n).$$

The topology $\mathcal{T}$ is compatible with the vector space structure on E. Indeed, let $a \in E$, $b \in E$ and let $\bigcap_{k=1}^{n} f_{\iota_k}^{-1}(W_k)$ be a neighborhood of 0, where W_k is a neighborhood of 0 in F_{ι_k}. Let $a_k = f_{\iota_k}(a)$ and $b_k = f_{\iota_k}(b)$ $(1 \leqq k \leqq n)$. Then there exist neighborhoods U_k and V_k of 0 in F_{ι_k} $(1 \leqq k \leqq n)$, such that $u \in a_k + U_k$ and $v \in b_k + V_k$ imply

$$u + v \in a_k + b_k + W_k.$$

But then in E the relations

$$x \in a + \bigcap_{k=1}^{n} f_{\iota_k}^{-1}(U_k) \quad \text{and} \quad y \in b + \bigcap_{k=1}^{n} f_{\iota_k}^{-1}(V_k)$$

imply

$$x + y \in a + b + \bigcap_{k=1}^{n} f_{\iota_k}^{-1}(W_k).$$

Next let $\lambda \in \mathbf{K}$, $a \in E$ and let $\bigcap_{k=1}^{n} f_{\iota_k}^{-1}(W_k)$ be a neighborhood of 0 in E, where W_k is a neighborhood of 0 in F_{ι_k}. For each k $(1 \leq k \leq n)$ there exists a neighborhood U_k of 0 in $\mathbf{K}$ and a neighborhood V_k of 0 in F_{ι_k} such that $\xi \in \lambda + U_k$ and $u \in a_k + V_k$ imply $\xi u \in \lambda a_k + W_k$. But then the relations

$$\xi \in \lambda + \bigcap_{k=1}^{n} U_k \quad \text{and} \quad x \in a + \bigcap_{k=1}^{n} f_{\iota_k}^{-1}(V_k)$$

imply

$$\xi x \in \lambda a + \bigcap_{k=1}^{n} f_{\iota_k}^{-1}(W_k).$$

EXAMPLE 1. Let F be a topological vector space, E a linear subspace of the vector space F and $f: E \hookrightarrow F$ the canonical injection. Then the initial topology on E for f is the topology induced on E by F (§5).

EXAMPLE 2. Let $(F_\iota)_{\iota \in I}$ be a family of topological vector spaces and E the product $\prod_{\iota \in I} F_\iota$ of the vector spaces F_ι. For each $\iota \in I$ let π_ι be the ι-th projection from E onto F_ι. Then the initial topology on E for the family (π_ι) is the product topology (§7).

EXAMPLE 3. Let E be a vector space and $(\mathcal{T}_\iota)_{\iota \in I}$ a family of topologies on E compatible with the vector space structure on E. For each ι let F_ι be the vector space E equipped with the topology $\mathcal{T}_\iota$ on it, and let f_ι be the identical bijection from E onto F_ι. Then the initial topology on E for the family (f_ι) is clearly the least upper bound of the topologies $\mathcal{T}_\iota$ (§1). It is compatible with the vector space structure of E.

PROPOSITION 1. *Let E be a vector space, $(G_\iota)_{\iota \in I}$ a family of topological vector spaces, $I = \bigcup_{\lambda \in L} J_\lambda$ a partitioning of the set I, and $(F_\lambda)_{\lambda \in L}$ a family of vector spaces. For each $\lambda \in L$ let h_λ be a linear map from E into F_λ and for each $\iota \in J_\lambda$ let $g_{\iota\lambda}$ be a linear map from F_λ into G_ι. Set $f_\iota = g_{\iota\lambda} \circ h_\lambda$. Equip each F_λ with the initial topology for the family $(g_{\iota\lambda})_{\iota \in J_\lambda}$. Then on E the initial topology for the family $(h_\lambda)_{\lambda \in L}$ coincides with the initial topology for the family $(f_\iota)_{\iota \in I}$.*

Proof. If $W = \bigcap_{k=1}^{n} f_{\iota_k}^{-1}(U_{\iota_k})$ is a neighborhood of 0 in E for the second topology, where U_{ι_k} is a neighborhood of 0 in G_{ι_k}, then $V_{\lambda_k} = g_{\iota_k\lambda_k}^{-1}(U_{\iota_k})$ is a neighborhood of 0 in F_{λ_k} where λ_k is such that $\iota_k \in J_{\lambda_k}$. Since

$$f_{\iota_k}^{-1}(U_{\iota_k}) = h_{\lambda_k}^{-1}(g_{\iota_k\lambda_k}^{-1}(U_{\iota_k})) = h_{\lambda_k}^{-1}(V_{\lambda_k}),$$

it follows that

$$W = \bigcap_{k=1}^{n} h_{\lambda_k}^{-1}(V_{\lambda_k})$$

is also a neighborhood of 0 for the first topology. Conversely, let

$$W = \bigcap_{\lambda \in M} h_\lambda^{-1}(V_\lambda)$$

be a neighborhood of 0 in E for the first topology, where M is a finite subset of L. Then each neighborhood V_λ of 0 in F_λ $(\lambda \in M)$ contains a set $\bigcap_{\iota \in H_\lambda} g_{\iota\lambda}^{-1}(U_\iota)$, where H_λ is a finite subset of J_λ and for each $\iota \in H_\lambda$ the set U_ι is a neighborhood of 0 in G_ι. But then W contains the set

$$\bigcap_{\lambda \in M} \left(\bigcap_{\iota \in H_\lambda} h_\lambda^{-1}(g_{\iota\lambda}^{-1}(U_\iota)) \right) = \bigcap_{\lambda \in M} \bigcap_{\iota \in H_\lambda} f_\iota^{-1}(U_\iota);$$

i.e., W is a neighborhood of 0 for the second topology. ▮

PROPOSITION 2. *Let E be a vector space, $(F_\iota)_{\iota \in I}$ and $(G_\lambda)_{\lambda \in L}$ two families of topological vector spaces. For each $\iota \in I$ let f_ι be a linear map from E into F_ι and for each $\lambda \in L$ let g_λ be a linear map from E into G_λ. Let $\mathfrak{T}$ be the initial topology on E for (f_ι) and $\mathfrak{T}'$ the initial topology on E for (g_λ). Suppose that for each $\iota \in I$ there exists a continuous linear map $u_{\iota\lambda}$ from some G_λ into F_ι so that $f_\iota = u_{\iota\lambda} \circ g_\lambda$. Then $\mathfrak{T}'$ is finer than $\mathfrak{T}$.*

If, furthermore, for every $\lambda \in L$ there exists a continuous linear map $v_{\lambda\iota}$ from some F_ι into G_λ so that $g_\lambda = v_{\lambda\iota} \circ f_\iota$, then the topologies $\mathfrak{T}$ and $\mathfrak{T}'$ are identical.

Proof. The second statement follows obviously from the first one.

To prove the first assertion, it is enough to observe that under the hypotheses, the f_ι are continuous for the topology $\mathfrak{T}'$ on E. ▮

Let E be a vector space, $(F_\iota)_{\iota \in I}$ a family of topological vector spaces, $f_\iota : E \to F_\iota$ a linear map $(\iota \in I)$, and equip E with the initial topology for the family (f_ι). If E is a Hausdorff space, then for each $x \neq 0$ in E there exists an index $\iota \in I$ such that $f_\iota(x) \neq 0$. Indeed, there exists a neighborhood

$$V = \bigcap_{k=1}^{n} f_{\iota_k}^{-1}(U_{\iota_k})$$

of 0 such that $x \notin V$; hence $f_{\iota_k}(x) \notin U_{\iota_k}$ for some index k and *a fortiori* $f_{\iota_k}(x) \neq 0$. It follows that the linear map $x \mapsto (f_\iota(x))_{\iota \in I}$ from E into

$\prod_{\iota \in I} F_\iota$ is injective. It is even a strict morphism (Definition 5.1). In fact, the image of the neighborhood $\bigcap_{k=1}^n f_{\iota_k}^{-1}(U_{\iota_k})$ of 0 in E is the trace on the image of E of the neighborhood $\prod_{\iota \in I} U_\iota$ of 0 in $\prod_{\iota \in I} F_\iota$, where $U_\iota = U_{\iota_k}$ for $\iota = \iota_k$ and $U_\iota = F_\iota$ for $\iota \neq \iota_k$ $(1 \leqq k \leqq n)$. Thus if E is a Hausdorff space, it can be considered as a subspace of the product space $\prod_{\iota \in I} F_\iota$.

Conversely, assume that the F_ι are Hausdorff spaces and that for each $x \neq 0$ in E there exists an index $\iota \in I$ such that $f_\iota(x) \neq 0$. Then E is a Hausdorff space. Indeed, if $f_\iota(x) \neq 0$, then there exists a neighborhood U of 0 in F_ι such that $f_\iota(x) \notin U$. But then $f_\iota^{-1}(U)$ is a neighborhood of 0 in E and $x \notin f_\iota^{-1}(U)$ (cf. Proposition 3.2).

If the F_ι are locally convex spaces, then E is a locally convex space since if U_k $(1 \leqq k \leqq n)$ is a convex neighborhood of 0 in F_{ι_k}, then $\bigcap_{k=\iota}^n f_{\iota_k}^{-1}(U_k)$ is a convex neighborhood of 0 in E. If for each $\iota \in I$ the topology of F_ι is defined by the family $(q_{\iota\lambda})_{\lambda \in L_\iota}$ of semi-norms, then the topology of E is clearly defined by the family of semi-norms $(q_{\iota\lambda} \circ f_\iota)$ $(\lambda \in L_\iota, \iota \in I)$.

EXAMPLE 4. For every positive integer m the space $\mathcal{E}(\Omega)$ defined in Example 4.8 is a linear subspace of the space $\mathcal{E}^m(\Omega)$ of Example 4.7. Let f_m be the canonical injection $\mathcal{E}(\Omega) \hookrightarrow \mathcal{E}^m(\Omega)$ and equip $\mathcal{E}^m(\Omega)$ with the topology defined in Example 4.7. It is clear that the initial topology on $\mathcal{E}(\Omega)$ for (f_m) coincides with the topology defined in Example 4.8.

The situation is entirely analogous for the canonical injections $\mathcal{D}(K) \hookrightarrow \mathcal{D}^m(K)$, $\mathcal{S}_k \hookrightarrow \mathcal{S}_k^m$, $\mathcal{S} \hookrightarrow \mathcal{S}^m$, $\mathcal{B}_0(\Omega) \hookrightarrow \mathcal{B}_0^m(\Omega)$, and $\mathcal{B}(\Omega) \hookrightarrow \mathcal{B}^m(\Omega)$ of the spaces defined in Examples 9 through 14 and 16 through 19 of §4.

EXAMPLE 5. If m is a fixed positive integer and k an arbitrary integer, then the space $\mathcal{S}^m$ of Example 4.13 is a linear subspace of the space $\mathcal{S}_k^m$ of Example 4.11. If we equip $\mathcal{S}_k^m$ with the topology defined in Example 4.11, then the topology of $\mathcal{S}^m$ defined in Example 4.13 coincides with the coarsest topology on $\mathcal{S}^m$ for which all the maps $f^k : \mathcal{S}^m \hookrightarrow \mathcal{S}_k^m$ are continuous, where we may take either all $k \in \mathbf{Z}$ or only $k \in \mathbf{N}$ by virtue of Proposition 2 and Example 5.4 (cf. Example 5.2).

EXAMPLE 6. For any $k \in \mathbf{Z}$ and $m \in \mathbf{N}$ the space $\mathcal{S}$ of Example 4.14 is a linear subspace of the space $\mathcal{S}_k^m$. The topology of $\mathcal{S}$ defined in Example 4.14 is identical with the initial topology for the family of maps

$$f_m^k : \mathcal{S} \hookrightarrow \mathcal{S}_k^m \quad (m \in \mathbf{N}, k \in \mathbf{Z} \quad \text{or} \quad k \in \mathbf{N}).$$

By Proposition 1 the topology of $\mathcal{S}$ is also the initial topology for the family of maps $\mathcal{S} \hookrightarrow \mathcal{S}_k$ or the family of maps $\mathcal{S} \hookrightarrow \mathcal{S}^m$.

With the same notations as above, *a set B is bounded in E if and only if $f_\iota(B)$ is bounded in F_ι for each $\iota \in I$.* Indeed, suppose that each set $B_\iota = f_\iota(B)$

is bounded and let $W = \bigcap_{k=1}^{n} f_{\iota_k}^{-1}(U_k)$ be a neighborhood of 0 in E, where U_k is a balanced neighborhood of 0 in F_{ι_k} $(1 \leqq k \leqq n)$. For each k there exists $\lambda_k > 0$ such that $B_{\iota_k} \subset \lambda_k U_k$. Let $\lambda = \max_{1 \leqq k \leqq n} \lambda_k$. Then $B \subset \lambda W$, i.e., B is bounded. Conversely, if B is bounded, then $f_\iota(B)$ is bounded since f_ι is continuous (§6).

Let $\mathfrak{F}$ be a filter on E. Then $\mathfrak{F}$ is a Cauchy filter if and only if $f_\iota(\mathfrak{F})$ is the basis of a Cauchy filter on F_ι for every $\iota \in I$. Indeed, if $\mathfrak{F}$ is a Cauchy filter, then it follows from the uniform continuity of f_ι that $f_\iota(\mathfrak{F})$ is the basis of a Cauchy filter on F_ι. Conversely, suppose that each $f_\iota(\mathfrak{F})$ is the basis of a Cauchy filter. Let $W = \bigcap_{k=1}^{n} f_{\iota_k}^{-1}(U_k)$ be a neighborhood of 0 in E, where U_k is a neighborhood of 0 in F_{ι_k}. For each k there exists $A_k \in \mathfrak{F}$ such that $f_{\iota_k}(A_k) - f_{\iota_k}(A_k) \subset U_k$. Then $A = \bigcap_{k=1}^{n} A_k \in \mathfrak{F}$ and $A - A \subset W$.

These results are really trivial and their proofs are almost word for word the same as in the case of product spaces. Let us now prove a slightly deeper result.

Proposition 3. *Let $(F_\iota)_{\iota \in I}$ be a family of Hausdorff topological vector spaces, where the index set is ordered and we suppose that for $\iota \leqq \kappa$ we have a continuous linear map $f_{\iota\kappa}: F_\kappa \to F_\iota$. Let E be a vector space and for each $\iota \in I$ let f_ι be a linear map from E into F_ι such that for $\iota \leqq \kappa$ we have $f_\iota = f_{\iota\kappa} \circ f_\kappa$. Assume, furthermore, that if $(x_\iota)_{\iota \in I}$ is a family of elements such that $x_\iota \in F_\iota$ and for each pair (ι, κ) with $\iota \leqq \kappa$ we have $f_{\iota\kappa}(x_\kappa) = x_\iota$, then there exists an $x \in E$ such that $x_\iota = f_\iota(x)$ for all $\iota \in I$. If the spaces F_ι are complete, then E equipped with the initial topology for the family (f_ι) is complete.*

Proof. Let $\mathfrak{F}$ be a Cauchy filter on E, then $\mathfrak{F}_\iota = f_\iota(\mathfrak{F})$ is the basis of a Cauchy filter on F_ι for each $\iota \in I$. By assumption $\mathfrak{F}_\iota$ converges to an element x_ι of F_ι. Let us prove that the family $(x_\iota)_{\iota \in I}$ so obtained satisfies the condition $x_\iota = f_{\iota\kappa}(x_\kappa)$ for $\iota \leqq \kappa$. Since $\mathfrak{F}_\kappa$ tends to x_κ, it follows from the continuity of $f_{\iota\kappa}$ that $f_{\iota\kappa}(\mathfrak{F}_\kappa)$ tends to $f_{\iota\kappa}(x_\kappa)$. On the other hand, $f_{\iota\kappa}(\mathfrak{F}_\kappa) = f_{\iota\kappa}(f_\kappa(\mathfrak{F})) = f_\iota(\mathfrak{F}) = \mathfrak{F}_\iota$ which tends to x_ι. But in a Hausdorff space a filter basis has at most one limit; hence $x_\iota = f_{\iota\kappa}(x_\kappa)$.

Let $x \in E$ be an element such that $f_\iota(x) = x_\iota$. Let $W = \bigcap_{k=1}^{n} f_{\iota_k}^{-1}(U_k)$ be a neighborhood of 0 in E, where U_k is a neighborhood of 0 in F_{ι_k}. For each k there exists $A_k \in \mathfrak{F}$ such that $f_{\iota_k}(A_k) \subset x_{\iota_k} + U_k$. Then

$$A = \bigcap_{k=1}^{n} A_k \in \mathfrak{F} \qquad \text{and} \qquad A \subset x + W;$$

that is, $\mathfrak{F}$ converges to x. ∎

Remark 1. Let $E = \prod_{\iota \in I} F_\iota$, for the f_ι choose the projections π_ι and order I in such a manner that no two different elements of I are comparable. Then every family $(x_\iota)_{\iota \in I}$ satisfies the condition of Proposition 3 and we have proved again that E is complete for the product topology if the F_ι are complete.

COROLLARY. *Let $(F_\iota)_{\iota \in I}$ be a family of Hausdorff topological vector spaces, where the index set is ordered. Suppose that for $\iota \leqq \kappa$ the space F_κ is a linear subspace of F_ι and the topology of F_κ is finer than the topology induced by F_ι on F_κ. Let $E = \bigcap_{\iota \in I} F_\iota$ and equip E with the initial topology for the injections $E \hookrightarrow F_\iota$. If the F_ι are complete, then E is complete.*

EXAMPLE 7. If $m,m' \in \mathbf{N}$ and $m > m'$, then the map $\mathcal{E}^m(\Omega) \hookrightarrow \mathcal{E}^{m'}(\Omega)$ is continuous (Example 5.7) and clearly $\mathcal{E}(\Omega) = \bigcap_{m \in \mathbf{N}} \mathcal{E}^m(\Omega)$. Thus by Example 4 and the corollary the completeness of the $\mathcal{E}^m(\Omega)$ for finite m implies the completeness of $\mathcal{E}(\Omega)$. This is, however, of little interest to us, since in Example 9.3 we have already established the completeness of $\mathcal{E}^m(\Omega)$ with equal ease for both finite and infinite m.

Exactly the same applies to the spaces $\mathcal{B}^m(\Omega)$, $\mathcal{D}^m(K)$, and $\mathcal{B}_0^m(\Omega)$ (cf. Examples 4, 5, and 6 of §9).

EXAMPLE 8. If k is a fixed integer, $m,m' \in \mathbf{N}$ and $m > m'$, then the map $\mathcal{S}_k^m \hookrightarrow \mathcal{S}_k^{m'}$ is continuous (Example 5.7) and $\mathcal{S}_k = \bigcap_{m \in \mathbf{N}} \mathcal{S}_k^m$. It follows from Example 4 and the corollary to Proposition 3 that $\mathcal{S}_k$ is complete since the spaces $\mathcal{S}_k^m$ are complete by Example 9.6.

EXAMPLE 9. If m is a fixed positive integer, $k,k' \in \mathbf{Z}$ and $k > k'$, then the map $\mathcal{S}_k^m \hookrightarrow \mathcal{S}_{k'}^m$ is continuous (Example 5.4) and we have $\mathcal{S}^m = \bigcap_{k \in \mathbf{Z}} \mathcal{S}_k^m$. It follows from Example 5 and the above corollary that $\mathcal{S}^m$ is complete.

EXAMPLE 10. If $m,m' \in \mathbf{N}$, $k,k' \in \mathbf{Z}$, $m > m'$ and $k > k'$, then the map $\mathcal{S}_k^m \hookrightarrow \mathcal{S}_{k'}^{m'}$ is continuous since it is composed of the maps $\mathcal{S}_k^m \hookrightarrow \mathcal{S}_k^{m'} \hookrightarrow \mathcal{S}_{k'}^{m'}$ or $\mathcal{S}_k^m \hookrightarrow \mathcal{S}_{k'}^m \hookrightarrow \mathcal{S}_{k'}^{m'}$. Furthermore, $\mathcal{S} = \bigcap_{m,k} \mathcal{S}_k^m$; hence it follows from Example 6 and the above corollary that the space $\mathcal{S}$ is complete.

Of course this follows also from the completeness of $\mathcal{S}^m$ or $\mathcal{S}_k$ since $\mathcal{S} = \bigcap_m \mathcal{S}^m$ and $\mathcal{S} = \bigcap_k \mathcal{S}_k$ (cf. Example 6).

Let E be a vector space, $(F_n)_{n \in \mathbf{N}}$ a sequence of metrizable topological vector spaces, and for every $n \in \mathbf{N}$ let f_n be a linear map from E into F_n. Then the initial topology on E for the sequence (f_n) is metrizable. Indeed, if $(V_{nm})_{m \in \mathbf{N}}$ is a fundamental system of neighborhoods of 0 in F_n, then the sets $f_{n_1}^{-1}(V_{n_1 m_1}) \cap \cdots \cap f_{n_p}^{-1}(V_{n_p m_p})$ form a fundamental system of neighborhoods of 0 in E. The assertion then follows from Theorem 6.1 if we observe that the collection of all finite subsets of a countable set is itself countable.

EXAMPLE 11. The spaces $\mathcal{S}_k$, $\mathcal{S}^m$, and $\mathcal{S}$ are metrizable, and hence are Fréchet spaces by virtue of Examples 8, 9, and 10.

Exercises

1. Let E be a vector space, $(F_\iota)_{\iota \in I}$ a family of topological vector spaces and for each index $\iota \in I$ let f_ι be a linear map from E into F_ι. For each $x \in E$ denote by $\varphi(x)$ the point $(f_\iota(x))_{\iota \in I}$ of the product space $\prod_{\iota \in I} F_\iota$. Show that the initial topology on E for the family (f_ι) is the coarsest topology on E for which the map $\varphi : E \to \prod_{\iota \in I} F_\iota$ is continuous if we equip $\prod_{\iota \in I} F_\iota$ with the product topology.

2. Let E be a vector space, $(F_\iota)_{\iota \in I}$ a family of Hausdorff topological vector spaces, $f_\iota : E \to F_\iota$ a linear map, and equip E with the initial topology for (f_ι). Show that a set K is precompact (Definition 10.1) in E if and only if $f_\iota(K)$ is precompact in F_ι for each $\iota \in I$.

3. Let I be an ordered set and $(F_\iota)_{\iota \in I}$ a family of topological vector spaces. Suppose that for every pair (ι, κ) of indices such that $\iota \leqq \kappa$ we have a continuous linear map $f_{\iota\kappa} : F_\kappa \to F_\iota$ and that these maps satisfy the following two conditions:

(i) $f_{\iota\iota}$ is the identity map for each $\iota \in I$,

(ii) $f_{\iota\lambda} = f_{\iota\kappa} \circ f_{\kappa\lambda}$ for $\iota \leqq \kappa \leqq \lambda$.

A topological vector space E is said to be an *inverse* (or projective) *limit* of the *inverse* (or projective) *system* $(F_\iota, f_{\iota\kappa})$ if the following conditions are satisfied:

(iii) for every $\iota \in I$ there exists a continuous linear map $f_\iota : E \to F_\iota$ such that $f_\iota = f_{\iota\kappa} \circ f_\kappa$ for $\iota \leqq \kappa$,

(iv) given a topological vector space D and a family of continuous linear maps $g_\iota : D \to F_\iota$ such that $g_\iota = f_{\iota\kappa} \circ g_\kappa$ for $\iota \leqq \kappa$, there exists a *unique* continuous linear map $g : D \to E$ such that $g_\iota = f_\iota \circ g$ for all $\iota \in I$, i.e., such that the diagram

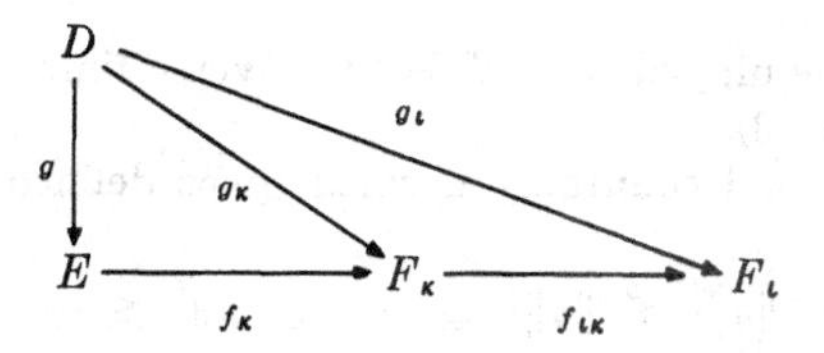

is commutative.

(a) Show that given an inverse system there always exists an inverse limit which is unique up to an isomorphism. (*Hint:* Consider the subspace of $\prod_{\iota \in I} F_\iota$ formed by those vectors $(x_\iota)_{\iota \in I}$ which satisfy $f_{\iota\kappa}(x_\kappa) = x_\iota$ for $\iota \leqq \kappa$.)

(b) Let E be an inverse limit of the inverse system $(F_\iota, f_{\iota\kappa})$. Show that given a family $(x_\iota)_{\iota \in I}$ of elements such that $x_\iota \in F_\iota$ and $f_{\iota\kappa}(x_\kappa) = x_\iota$ for $\iota \leqq \kappa$, there exists a *unique* $x \in E$ such that $f_\iota(x) = x_\iota$ for all $\iota \in I$ and show that this property characterizes E up to an isomorphism. In particular, the map $x \mapsto (f_\iota(x))$ from E into $\prod_{\iota \in I} F_\iota$ is injective, and if the F_ι are complete Hausdorff spaces, then E is complete.

4. Show that there exists a family $(F_\iota)_{\iota \in I}$ of complete, Hausdorff topological vector spaces, a vector space E, and linear maps $f_\iota : E \to F_\iota$ $(\iota \in I)$ such that

E equipped with the initial topology for (f_ι) is not complete. (*Hint:* Take a subspace of $\prod_{\iota \in I} F_\iota$ which is not closed.)

5. Suppose that the hypotheses of Proposition 3 are satisfied. Show that if the spaces F_ι are quasi-complete, then E is quasi-complete for the initial topology for the family (f_ι).

6. Let E be a vector space over $\mathbf{K}$ and suppose that we have on E a sequence of norms $x \mapsto \|x\|_p$ $(p \in \mathbf{N})$ which satisfy the condition

$$\|x\|_0 \leqq \|x\|_1 \leqq \cdots \leqq \|x\|_p \leqq \cdots$$

for every $x \in E$. If E_p is the space E equipped with the norm $\|x\|_p$, then for $p \leqq q$ the identical bijection $E_q \to E_p$ is continuous.

We assume that the norms $\|x\|_p$ are *pairwise coordinated* in the sense of Gelfand and Shilov [31], i.e., that if $p \leqq q$ and (x_n) is a Cauchy sequence in E_q (hence *a fortiori* in E_p) and (x_n) converges to 0 in E_p, then (x_n) converges to 0 in E_q. Denote by $\hat{E}_p$ the completion of E_p (Proposition 1.2.4) and denote the norm of $\hat{E}_p$ also by $\|x\|_p$. Furthermore, let f_{pq} be the continuous linear map from $\hat{E}_q$ into $\hat{E}_p$ which is the unique extension of the identical map $E_q \to E_p$ $(p \leqq q)$.

(a) Show that the map $f_{pq}: \hat{E}_q \to \hat{E}_p$ is injective.

(b) For each index $p \in \mathbf{N}$ let f_p be the canonical injection $E \hookrightarrow \hat{E}_p$. Show that for $p \leqq q$ we have $f_p = f_{pq} \circ f_q$.

(c) Let $\mathcal{T}$ be the coarsest topology on E for which all the maps f_p are continuous. Show that $\mathcal{T}$ is a locally convex metrizable topology defined by the saturated family of norms $x \mapsto \|x\|_p$. (E is a *countably normed space* in the terminology of Gelfand and Shilov.)

(d) Show that E is complete for $\mathcal{T}$ if and only if $E = \bigcap_{p \in \mathbf{N}} \hat{E}_p$. (*Hint:* To show that the condition is necessary observe that E is always dense in $\bigcap_{p \in \mathbf{N}} \hat{E}_p$.)

(e) Show that E equipped with $\mathcal{T}$ is the inverse limit of the inverse system (E_p, f_{pq}) (cf. Exercise 3).

(f) Let F be a second countably normed space defined with the help of the norms

$$\|y\|_0' \leqq \|y\|_1' \leqq \cdots \leqq \|y\|_p' \leqq \cdots$$

Show that a linear map $f: F \to E$ is continuous if and only if for every norm $\|x\|_p$ on E there exists a norm $\|y\|_q'$ on F and a constant $M > 0$ such that

$$\|f(y)\|_p \leqq M\|y\|_q' \qquad \text{for all} \qquad y \in F.$$

(*Hint:* Use (c) and Proposition 5.2.)

(g) Show that $\mathfrak{D}(K)$, equipped with its topology as defined in Example 4.10, is a countably normed space defined with the help of the norms

$$\|f\|_m = \max_{|p| \leqq m} \max_{x \in K} |\partial^p f(x)|.$$

Show that the completion of $\mathfrak{D}(K)$ with respect to the norm $\|f\|_m$ is $\mathfrak{D}^m(K)$ (cf. Chapter 4, §4).

§12. Final topologies

Let F be a vector space and $(E_\iota)_{\iota\in I}$ a family of topological vector spaces, all over the same field $\mathbf{K}$. For each index $\iota \in I$ let f_ι be a linear map from E_ι into F. Let Φ be the set of all locally convex topologies on F for which all the maps f_ι are continuous. The set Φ is not empty since the chaotic topology on F belongs to it. Let $\mathcal{T}$ be the least upper bound of the set Φ of topologies on F. We know from Example 11.3 that $\mathcal{T}$ is compatible with the vector space structure on F.

Let $\mathfrak{N}$ be the filter basis on F formed by all absorbing, balanced, convex sets V such that for every index $\iota \in I$ the set $f_\iota^{-1}(V)$ is a neighborhood of 0 in E_ι. We shall prove that $\mathfrak{N}$ is a fundamental system of neighborhoods of 0 in F for $\mathcal{T}$. It will then follow from Proposition 4.5 that $\mathcal{T}$ is a locally convex topology on F.

Indeed, again by Proposition 4.5, the filter basis $\mathfrak{N}$ is a fundamental system of neighborhoods of 0 for some locally convex topology $\mathcal{T}'$ on F, and clearly all the maps f_ι are continuous for $\mathcal{T}'$. We prove that $\mathcal{T}'$ is finer than $\mathcal{T}$, from which $\mathcal{T} = \mathcal{T}'$ will follow by the definition of $\mathcal{T}$ as the least upper bound of Φ. Let U be the neighborhood of 0 in F for $\mathcal{T}$. There exist a finite family $(\mathcal{T}_k)_{1\leqq k\leqq n}$ of topologies belonging to Φ and for each k a balanced, convex neighborhood V_k of 0 in F for $\mathcal{T}_k$ such that $U \supset \bigcap_{k=1}^n V_k$. Now

$$V = \bigcap_{k=1}^{n} V_k \in \mathfrak{N}$$

since it is absorbing, balanced, convex and for each $\iota \in I$ the set

$$f_\iota^{-1}(V) = \bigcap_{k=1}^{n} f_\iota^{-1}(V_k)$$

is a neighborhood of 0 in E_ι.

We have proved that $\mathcal{T}$ *is the finest locally convex topology on F for which all the maps f_ι are continuous.* We shall also refer to $\mathcal{T}$ as the final locally convex topology on F for the family (f_ι).

Let $\mathfrak{N}_\iota$ be a fundamental system of neighborhoods of 0 in E_ι and assume that the set $\bigcup_{\iota\in I} f_\iota(E_\iota)$ generates the vector space F. Then the balanced, convex hulls U of the sets $\bigcup_{\iota\in I} f_\iota(U_\iota)$, where $U_\iota \in \mathfrak{N}_\iota$, form a fundamental system of neighborhoods of 0 in F for $\mathcal{T}$. Indeed, U is balanced and convex, $f_\iota^{-1}(U) \supset U_\iota$ and by our hypothesis U is absorbing. Thus $U \in \mathfrak{N}$. Conversely, if $V \in \mathfrak{N}$, then $f_\iota^{-1}(V)$ contains some $U_\iota \in \mathfrak{N}_\iota$. Thus $f_\iota(U_\iota) \subset V$, hence $\bigcup_{\iota\in I} f_\iota(U_\iota) \subset V$; and since V is balanced and convex, the balanced, convex hull of $\bigcup_{\iota\in I} f_\iota(U_\iota)$ is also contained in V.

EXAMPLE 1. Let E be a locally convex space and M a subspace of E. Let f be the canonical surjection from E onto the quotient space $F = E/M$.

Then the quotient topology on F (§5) is the finest locally convex topology in F for which the map f is continuous.

Before giving our next example we need another definition from algebra. Let $(E_\iota)_{\iota\in I}$ be a family of vector spaces over the same field $\mathbf{K}$. The *external direct sum* $\coprod_{\iota\in I} E_\iota$ of the vector spaces E_ι is the subspace of $\prod_{\iota\in I} E_\iota$ formed by all those elements $(x_\iota)_{\iota\in I}$ for which $x_\iota = 0$ except for finitely many indices $\iota \in I$. If the index set I is finite, the external direct sum and the product are the same. For every $\iota \in I$ we have an injective linear map j_ι from E_ι into $\coprod_{\iota\in I} E_\iota$, called the ι-th *injection* and defined as follows. Let $y \in E_\iota$. Then $x = j_\iota(y) = (x_\iota)_{\iota\in I}$ is given by $x_\iota = y$ and $x_\kappa = 0$ for $\kappa \neq \iota$. Clearly, $\bigcup_{\iota\in I} j_\iota(E_\iota)$ generates $\coprod_{\iota\in I} E_\iota$. If all spaces E_ι are equal to the same space E, we write $E^{(I)}$ for $\coprod_{\iota\in I} E_\iota$.

Example 2. Let $(E_\iota)_{\iota\in I}$ be a family of locally convex spaces. We shall equip $\coprod_{\iota\in I} E_\iota$ with the finest locally convex topology for which all the injections j_ι are continuous and refer to the locally convex space $\coprod_{\iota\in I} E_\iota$ as the *locally convex direct sum* of the spaces E_ι.

If the index set I is finite, then the product topology and the locally convex direct sum topology coincide on $\prod_{\iota\in I} E_\iota = \coprod_{\iota\in I} E_\iota$. We may assume that I is the interval $[1, n]$ of $\mathbf{N}$. If U_i $(1 \leqq i \leqq n)$ is a balanced convex neighborhood of 0 in E_i, then the balanced, convex hull of $\bigcup_{i=1}^n j_i(U_i)$ is contained in $\prod_{i=1}^n U_i$ and contains $(1/n) \prod_{i=1}^n U_i$.

Example 3. Let F be a vector space and $(\mathcal{T}_\iota)_{\iota\in I}$ a family of locally convex topologies on F. For each ι let E_ι be the vector space F equipped with the topology $\mathcal{T}_\iota$ and let f_ι be the identical bijection from E_ι onto F. Then the finest locally convex topology $\mathcal{T}$ on F for which the f_ι are continuous is the greatest lower bound of the topologies $\mathcal{T}_\iota$ in the set of all locally convex topologies on F. Indeed, $\mathcal{T}$ is coarser than any $\mathcal{T}_\iota$. A balanced, convex set V is a neighborhood of 0 for $\mathcal{T}$ if it is a neighborhood of 0 for all $\mathcal{T}_\iota$. Thus if $\mathcal{T}'$ is a locally convex topology on F which is coarser than every $\mathcal{T}_\iota$ and V is a balanced, convex neighborhood of 0 for $\mathcal{T}'$, then V is a neighborhood of 0 for each $\mathcal{T}_\iota$ and consequently for $\mathcal{T}$. It follows that $\mathcal{T}$ is finer than $\mathcal{T}'$.

Example 4. Let F be a vector space and let $(E_\iota)_{\iota\in I}$ be the family of all finite-dimensional subspaces of F. On each E_ι we consider the unique Hausdorff topology compatible with the vector space structure of E_ι (Theorem 10.1). For each $\iota \in I$ let f_ι be the canonical injection from E_ι into F. Then the finest locally convex topology on F for which all the f_ι are continuous coincides with the finest locally convex topology on F (Examples 4.3 and 4.23). Indeed, if V is an absorbing, balanced, convex set in F, then for each $\iota \in I$ the set $V \cap E_\iota$ is absorbing, balanced, and convex in E_ι, hence a neighborhood of 0 in E_ι.

PROPOSITION 1. *Let F be a vector space, $(E_\iota)_{\iota \in I}$ a family of locally convex spaces, and for each $\iota \in I$ let f_ι be a linear map from E_ι into F. Equip F with the finest locally convex topology $\mathcal{T}$ for which all the maps f_ι are continuous. Let G be a locally convex space and g a linear map from F into G. Then g is continuous for $\mathcal{T}$ if and only if all the maps $g \circ f_\iota$ are continuous.*

Proof. If g is continuous, then by a general theorem from topology all the maps $g \circ f_\iota$ are continuous. Suppose that all the maps $g \circ f_\iota$ are continuous and let W be a balanced, convex neighborhood of 0 in G. Then $f_\iota^{-1}(g^{-1}(W))$ is a neighborhood of 0 in E_ι, and thus the absorbing, balanced, convex set $g^{-1}(W)$ is a neighborhood of 0 in F for $\mathcal{T}$; i.e., g is continuous. ∎

PROPOSITION 2. *Let G be a vector space, $(E_\iota)_{\iota \in I}$ a family of locally convex spaces, $I = \bigcup_{\lambda \in L} J_\lambda$ a partitioning of the set I, and $(F_\lambda)_{\lambda \in L}$ a family of vector spaces. For each $\lambda \in L$ let g_λ be a linear map from F_λ into G and for each $\iota \in J_\lambda$ let $h_{\lambda\iota}$ be a linear map from E_ι into F_λ. Set $f_\iota = g_\lambda \circ h_{\lambda\iota}$. Equip F_λ with the finest locally convex topology for which all the maps $h_{\lambda\iota} (\iota \in J_\lambda)$ are continuous. Then on G the finest locally convex topology for which all the g_λ are continuous coincides with the finest locally convex topology for which all the f_ι are continuous.*

Proof. Let V be an absorbing, balanced, convex set in G. Then $f_\iota^{-1}(V) = h_{\lambda\iota}^{-1}(g_\lambda^{-1}(V))$ is a neighborhood of 0 in E_ι if and only if $g_\lambda^{-1}(V)$ is a neighborhood of 0 in F_λ. ∎

PROPOSITION 3. *Let G be a vector space, $(E_\iota)_{\iota \in I}$ and $(F_\lambda)_{\lambda \in L}$ two families of locally convex spaces. For each $\iota \in I$ let f_ι be a linear map from E_ι into G and for each $\lambda \in L$ let g_λ be a linear map from F_λ into G. Let $\mathcal{T}$ be the finest locally convex topology on G for which the maps f_ι are continuous and $\mathcal{T}'$ the finest locally convex topology on G for which all the maps g_λ are continuous. Suppose that for every $\iota \in I$ there exists a continuous linear map $u_{\lambda\iota}$ from E_ι into some F_λ so that $f_\iota = g_\lambda \circ u_{\lambda\iota}$. Then $\mathcal{T}'$ is coarser than $\mathcal{T}$.*

If, furthermore, for every $\lambda \in L$ there exists a continuous linear map $v_{\iota\lambda}$ from F_λ into some E_ι so that $g_\lambda = f_\iota \circ v_{\iota\lambda}$, then the topologies $\mathcal{T}$ and $\mathcal{T}'$ are identical.

Proof. The second statement follows obviously from the first one.

To prove the first assertion it is enough to observe that under the hypotheses $\mathcal{T}'$ is a locally convex topology on G for which the maps f_ι are continuous. ∎

THEOREM 1 (Dieudonné-Schwartz). *Let F be a vector space and let $(E_n)_{n \in \mathbf{N}}$ be a sequence of linear subspaces of F such that $E_n \subset E_{n+1}$ for all $n \in \mathbf{N}$ and $F = \bigcup_{n \in \mathbf{N}} E_n$. Suppose that each E_n is equipped with a locally convex topology $\mathcal{T}_n$ and for each n the topology induced by $\mathcal{T}_{n+1}$ on E_n is $\mathcal{T}_n$* (i.e., the map $E_n \hookrightarrow E_{n+1}$ is a strict morphism). *Let $\mathcal{T}$ be the finest locally convex topology on F for which all the canonical injections $f_n : E_n \hookrightarrow F$ are continuous. Then $\mathcal{T}$ induces on each E_n the topology $\mathcal{T}_n$.*

REMARK 1. If the hypotheses of Theorem 1 are satisfied, then F is often said to be the *strict inductive limit* of the sequence (E_n).

Proof. Let $\mathcal{T}'_n$ be the topology induced by $\mathcal{T}$ on E_n. Since f_n is continuous, $\mathcal{T}'_n$ is coarser than $\mathcal{T}_n$. We have to prove that $\mathcal{T}'_n$ is finer than $\mathcal{T}_n$. Let V_n be a convex neighborhood of 0 in E_n for $\mathcal{T}_n$. We shall construct a sequence $(V_{n+p})_{p \in \mathbf{N}}$, where V_{n+p} is a convex neighborhood of 0 in E_{n+p} for $\mathcal{T}_{n+p}$, $V_{n+p} \subset V_{n+p+1}$ and $V_{n+p} \cap E_n = V_n$ for all $p \geqq 1$. Assuming that such a sequence has been constructed, set

$$V = \bigcup_{p \in \mathbf{N}} V_{n+p}.$$

Then V is a convex set in F such that for each $k \in \mathbf{N}$ the set $V \cap E_k$ is a neighborhood of 0 in E_k for $\mathcal{T}_k$; that is, V is a neighborhood of 0 in F for $\mathcal{T}$. Since $V \cap E_n = V_n$, the set V_n is a neighborhood of 0 in E_n for $\mathcal{T}'_n$, and thus $\mathcal{T}'_n$ is finer than $\mathcal{T}_n$. The possibility of constructing the sequence $(V_{n+p})_{p \in \mathbf{N}}$ results by induction on p from part (i) of the following lemma. ∎

LEMMA 1. *Let F be a locally convex space, M a subspace of F and V a convex neighborhood of 0 in M.*

(i) *There exists a convex neighborhood W of 0 in F such that $W \cap M = V$.*

(ii) *If M is closed, then for every x_0 not belonging to M there exists a convex neighborhood W_0 of 0 in F such that $W_0 \cap M = V$ and $x_0 \notin W_0$.*

Proof. (i) By the definition of the induced topology there exists a convex neighborhood U of 0 in F such that $U \cap M \subset V$. Let W be the convex hull of $U \cup V$. Since $W \supset U$, the set W is a neighborhood of 0 in F. Also $W \cap M \supset V$; hence we must prove that $W \cap M \subset V$. By Proposition 4.2 an element $z \in W$ can be written in the form $z = \alpha x + \beta y$, where $x \in V$, $y \in U$, $0 \leqq \alpha \leqq 1$, $\alpha + \beta = 1$. Let us now assume that z also belongs to M. If $\beta \neq 0$, then $y = \beta^{-1}(z - \alpha x) \in M$ and thus $y \in U \cap M \subset V$, i.e., $z \in V$. If $\beta = 0$, i.e., $\alpha = 1$, then $z = x \in V$.

(ii) If M is closed, then by Proposition 5.5 the quotient space F/M is a Hausdorff space. Hence there exists a convex neighborhood U_0 of 0 in F such that $U_0 \cap M \subset V$ and $U_0 \cap (x_0 + M) = \emptyset$. Let W_0 be the convex hull of $U_0 \cup V$. Again W_0 is a neighborhood of 0 in F for which $W_0 \cap M = V$. Suppose that $x_0 \in W_0$. Then $x_0 = \alpha x + \beta y$, where $x \in V$, $y \in U_0$, $0 \leqq \alpha \leqq 1$, $\alpha + \beta = 1$. Hence $\beta y = x_0 - \alpha x$ would belong to $U_0 \cap (x_0 + M)$ since $\beta y = \beta y + \alpha 0 \in U_0$ and $x_0 - \alpha x \in x_0 + M$. But this is impossible. ∎

COROLLARY 1. *Under the conditions of Theorem* 1, *if each* $\mathcal{T}_n$ *is a Hausdorff topology, then* $\mathcal{T}$ *is a Hausdorff topology.*

Proof. Let $x \in F$, $x \neq 0$. There exists an index $n \in \mathbf{N}$ such that $x \in E_n$. Since the topology $\mathcal{T}_n$ is Hausdorff, there exists a neighborhood V_n of 0 in E_n for $\mathcal{T}_n$ such that $x \notin V_n$. By Theorem 1 there exists a neighborhood V of 0 in F for $\mathcal{T}$ such that $V \cap E_n = V_n$. Then $x \notin V$ and it follows from Proposition 3.2 that $\mathcal{T}$ is a Hausdorff topology. ∎

COROLLARY 2. *Under the conditions of Theorem* 1, *if each subspace* E_n *is closed in* E_{n+1} *for the topology* $\mathcal{T}_{n+1}$, *then* E_n *is closed in* F *for* $\mathcal{T}$.

Proof. It is clear by mathematical induction that E_n is closed in each E_{n+p} for $\mathcal{T}_{n+p}$ $(p \geqq 1)$. Let $x \in F$, $x \notin E_n$. There exists $p \in \mathbf{N}$ such that $x \in E_{n+p}$. Since E_n is closed in E_{n+p}, there exists a neighborhood V_{n+p} of 0 in E_{n+p} for $\mathcal{T}_{n+p}$ such that $(x + V_{n+p}) \cap E_n = \emptyset$. By Theorem 1 there exists a neighborhood V of 0 in F for $\mathcal{T}$ such that $V \cap E_{n+p} = V_{n+p}$. But then

$$(x + V) \cap E_n = \emptyset;$$

i.e., $\complement E_n$ is open in F for $\mathcal{T}$. ∎

LEMMA 2. *A set* B *in a topological vector space* F *is bounded* (Definition 6.2) *if and only if for every sequence* $(x_n)_{n \in \mathbf{N}}$ *of elements of* B *and every sequence* $(\lambda_n)_{n \in \mathbf{N}}$ *of positive scalars converging to zero the sequence* $(\lambda_n x_n)_{n \in \mathbf{N}}$ *tends to* 0 *in* F.

Proof. Suppose that B is bounded and let V be a balanced neighborhood of 0 in F. There exists an $\alpha > 0$ such that $\lambda x_n \in V$ for $|\lambda| \leqq \alpha$ and $n \in \mathbf{N}$. By hypothesis there exists an $n_0 \in \mathbf{N}$ such that $\lambda_n \leqq \alpha$ for $n \geqq n_0$. Hence $\lambda_n x_n \in V$ for $n \geqq n_0$.

Suppose that B is not bounded. There exists a balanced neighborhood U of 0 in F such that $B \not\subset nU$ for all $n \in \mathbf{N}$. Choose $x_n \in B \cap \complement(nU)$ and $\lambda_n = 1/n$. Then

$$\lambda_n x_n = (1/n)x_n \notin U$$

for all $n \geqq 1$ and $(\lambda_n x_n)$ does not tend to 0 in F. ∎

THEOREM 2 (Dieudonné-Schwartz). *Let* F *be a vector space and let* $(E_n)_{n \in \mathbf{N}}$ *be a sequence of linear subspaces of* F *such that* $E_n \subset E_{n+1}$ *for all* $n \in \mathbf{N}$ *and* $F = \bigcup_{n \in \mathbf{N}} E_n$. *Suppose that each* E_n *is equipped with a locally convex topology* $\mathcal{T}_n$, *that* $\mathcal{T}_{n+1}$ *induces the topology* $\mathcal{T}_n$ *on* E_n, *and that* E_n *is closed in* E_{n+1} *for* $\mathcal{T}_{n+1}$. *Equip* F *with the finest locally convex topology* $\mathcal{T}$ *for which all the canonical injections* $f_n: E_n \hookrightarrow F$ *are continuous. Then a set* B *in* F *is bounded if and only if* B *is contained in some* E_n *and is bounded there.*

Proof. If B is a bounded set in E_n, then it is also bounded in F since the map $f_n: E_n \hookrightarrow F$ is continuous.

Next suppose that B is not contained in any subspace E_n and choose a sequence (x_n) such that $x_n \in B \cap \complement E_n$. There exists a subsequence (y_k) of (x_n) and a strictly increasing sequence (n_k) of positive integers such that for all $k \geqq 1$ we have $y_k \notin E_{n_k}$, $y_k \in E_{n_{k+1}}$. By part (ii) of Lemma 1 there exists an increasing sequence (V_k) of convex sets such that V_k is a neighborhood of 0 in E_{n_k}, $V_{k+1} \cap E_{n_k} = V_k$ and $y_k/k \notin V_{k+1}$. Then $V = \bigcup_{k \in \mathbf{N}} V_k$ is a neighborhood of 0 in F for $\mathcal{T}$ and $y_k/k \notin V$ for all $k \geqq 1$; i.e., the sequence y_k/k does not tend to 0 in F. It follows from Lemma 2 that B cannot be bounded.

Thus if B is bounded in F, it must be contained in some E_n. Let us show that it must be bounded there. If V_n is a balanced neighborhood of 0 in E_n for $\mathcal{T}_n$, then by Theorem 1 there exists a balanced neighborhood V of 0 in F for $\mathcal{T}$ such that $V \cap E_n \subset V_n$. We have $B \subset \lambda V$ for some $\lambda \in \mathbf{K}$; hence $B \subset \lambda V_n$. ▮

REMARK 2. Douady [22] proved that Theorem 2 is false if the sequence (E_n) is replaced by an uncountable family of subspaces.

THEOREM 3 (Köthe). *Let F be a vector space and let $(E_n)_{n \in \mathbf{N}}$ be a sequence of linear subspaces of F such that $E_n \subset E_{n+1}$ for all $n \in \mathbf{N}$ and $F = \bigcup_{n \in \mathbf{N}} E_n$. Suppose that each E_n is equipped with a locally convex topology $\mathcal{T}_n$ and let $\mathcal{T}$ be the finest locally convex topology on F for which all the canonical injections $f_n : E_n \hookrightarrow F$ are continuous.*

Then F is complete with respect to the topology $\mathcal{T}$ if and only if for every Cauchy filter $\mathfrak{F}$ on E_n with respect to the topology induced by $\mathcal{T}$ on E_n there exists an index $p \geqq n$ such that $\mathfrak{F}$ is convergent in E_p with respect to the topology induced by $\mathcal{T}$ on E_p.

Proof. Suppose that F is complete with respect to $\mathcal{T}$ and let $\mathfrak{F}$ be a Cauchy filter on E_n with respect to the topology induced by $\mathcal{T}$ on E_n. Then $\mathfrak{F}$ is the basis of a Cauchy filter on F and so converges to some point $x \in F$. But $x \in E_p$ for some $p \geqq n$ and $\mathfrak{F}$ converges to x in E_p for the topology induced by $\mathcal{T}$ on E_p.

Conversely, suppose that the condition is satisfied and let $\mathfrak{F}$ be a Cauchy filter on F. We shall use the following:

LEMMA 3. *Let $\mathfrak{G}$ be the Cauchy filter whose basis is formed by all the sets $M + V$, where M runs through $\mathfrak{F}$ and V through the filter of neighborhoods of 0 in F. Then there exists an integer n such that $\mathfrak{G}$ induces a Cauchy filter on E_n.*

Let us assume the lemma for a moment and conclude the proof of the theorem. Let $\mathfrak{G}_n$ be the Cauchy filter induced by $\mathfrak{G}$ on E_n. Then $\mathfrak{G}_n$ converges in some E_p to a point $x \in E_p$ which therefore adheres to $\mathfrak{G}$. It follows from Proposition 9.1 that $\mathfrak{G}$ converges to x in F and *a fortiori* $\mathfrak{F}$, which is finer than $\mathfrak{G}$, converges to x. ▮

Proof of Lemma 3. If there exists an index $n \in \mathbf{N}$ such that the sets $(M + V) \cap E_n$ are all nonempty, then the lemma is proved. We assume therefore that this is not the case and obtain a contradiction. Let $(V_n)_{n \in \mathbf{N}}$ be a decreasing sequence of balanced, convex neighborhoods of 0 in F and let $(M_n)_{n \in \mathbf{N}}$ be a sequence of sets $M_n \in \mathfrak{F}$ such that $M_n - M_n \subset V_n$ and $(M_n + V_n) \cap E_n = \emptyset$ for $n \in \mathbf{N}$. If we denote by W the convex hull in F of the set $\bigcup_{n \in \mathbf{N}} \frac{1}{2}(V_n \cap E_n)$, then W is a neighborhood of 0 in F. We shall show that there exists no set $Q \in \mathfrak{F}$ such that $Q - Q \subset W$. This, however, will contradict the fact that $\mathfrak{F}$ is a Cauchy filter; it is the desired contradiction.

Let W_n be the convex hull in F of the set

$$\tfrac{1}{2}(V_1 \cap E_1) \cup \tfrac{1}{2}(V_2 \cap E_2) \cup \cdots \cup \tfrac{1}{2}(V_{n-1} \cap E_{n-1}) \cup \tfrac{1}{2}V_n.$$

Then W_n is a neighborhood of 0 in F such that $W \subset W_n$. Let $P_n \in \mathfrak{F}$ be such that $P_n - P_n \subset W_n$. Then $(P_n + W_n) \cap E_n = \emptyset$. Indeed, let $x_0 \in P_n \cap M_n \in \mathfrak{F}$. By Proposition 4.2 the elements $y \in P_n$ have the form

$$y = x_0 + \sum_{i=1}^{n} \alpha_i x_i,$$

where $x_i \in \frac{1}{2}(V_i \cap E_i)$ for $1 \leqq i \leqq n - 1$, $x_n \in \frac{1}{2}V_n$, $\alpha_i \geqq 0$,

$$\sum_{i=1}^{n} \alpha_i = 1,$$

and therefore the elements $z \in P_n + W_n$ have the form

$$z = x_0 + \sum_{i=1}^{n} \alpha_i x_i + \sum_{i=1}^{n} \beta_i t_i,$$

where $t_i \in \frac{1}{2}(V_i \cap E_i)$ for $1 \leqq i \leqq n - 1$, $t_n \in \frac{1}{2}V_n$, $\beta_i \geqq 0$,

$$\sum_{i=1}^{n} \beta_i = 1.$$

Now

$$\alpha_n x_n + \beta_n t_n \in \tfrac{1}{2}V_n + \tfrac{1}{2}V_n = V_n.$$

Since $x_0 \in M_n$ and $(M_n + V_n) \cap E_n = \emptyset$, the element $x_0 + \alpha_n x_n + \beta_n t_n$ is not in E_n. On the other hand,

$$\sum_{i=1}^{n-1} \alpha_i x_i + \sum_{i=1}^{n-1} \beta_i t_i \in E_{n-1} \subset E_n;$$

hence z does not belong to E_n, i.e., $(P_n + W_n) \cap E_n = \emptyset$.

Suppose now that there exists a set $Q \in \mathfrak{F}$ such that $Q - Q \subset W$. Let $y_0 \in Q$. Then there exists an index $n \in \mathbf{N}$ such that $y_0 \in E_n$. We shall prove that for this index n we have $Q \cap P_n = \emptyset$, which contradicts the fact that $\mathfrak{F}$ is a filter. Let $y \in P_n$. Then $(y + W_n) \cap E_n = \emptyset$ and in particular $y_0 \notin y + W_n$; i.e., $y_0 - y \notin W_n$ and *a fortiori* $y_0 - y \notin W$. But then y cannot belong to Q since $Q - Q \subset W$. ∎

COROLLARY. *Let F be a vector space and let $(E_n)_{n \in \mathbf{N}}$ be a sequence of linear subspaces of F such that $E_n \subset E_{n+1}$ for all $n \in \mathbf{N}$ and $F = \bigcup_{n \in \mathbf{N}} E_n$. Suppose that each E_n is equipped with a locally convex topology $\mathcal{T}_n$ such that for each n the topology induced by $\mathcal{T}_{n+1}$ on E_n is $\mathcal{T}_n$ and E_n is closed in E_{n+1} for $\mathcal{T}_{n+1}$. Let $\mathcal{T}$ be the finest locally convex topology on F for which all the canonical injections $f_n: E_n \hookrightarrow F$ are continuous. Then F is complete if and only if all the E_n are complete.*

Proof. By Theorem 1 the topology $\mathcal{T}$ induces $\mathcal{T}_n$ on E_n. If E_n is complete, then every Cauchy filter on E_n (for $\mathcal{T}$ or for $\mathcal{T}_n$) converges in E_n. Hence by Theorem 3 the space F is complete. Conversely, it follows from Corollary 2 of Theorem 1 that E_n is closed in F. Hence if F is complete, it follows from Proposition 9.3 that E_n is complete. ∎

EXAMPLE 5. Let Ω be an open subset of $\mathbf{R}^n$ and let $\mathcal{K}(\Omega)$ be the vector space of all continuous functions on Ω whose support (cf. Example 3.4) is contained in some compact subset of Ω. The space $\mathcal{K}(\Omega)$ is the union of the linear subspaces $\mathcal{K}(K)$ introduced in Example 4.9, where K runs through the compact subsets of Ω. We equip $\mathcal{K}(\Omega)$ with the finest locally convex topology $\mathcal{T}$ for which all the canonical injections $\mathcal{K}(K) \hookrightarrow \mathcal{K}(\Omega)$ are continuous.

If $K \subset K'$, then $\mathcal{K}(K) \subset \mathcal{K}(K')$, and for $f \in \mathcal{K}(K)$ the norm of f in $\mathcal{K}(K)$ is the same as its norm in $\mathcal{K}(K')$. Hence in particular $\mathcal{K}(K')$ induces on $\mathcal{K}(K)$ its own topology; i.e., the injection $\mathcal{K}(K) \hookrightarrow \mathcal{K}(K')$ is a strict morphism. Let $(K_k)_{k \in \mathbf{N}}$ be the sequence of compact subsets of Ω constructed in Example 6.3 such that $K_k \subset \mathring{K}_{k+1}$ for all $k \in \mathbf{N}$ and every compact subset of Ω is contained in some K_k. Then $\mathcal{K}(K_k) \subset \mathcal{K}(K_{k+1})$ for all $k \in \mathbf{N}$ and

$$\mathcal{K}(\Omega) = \bigcup_{k \in \mathbf{N}} \mathcal{K}(K_k).$$

It follows from Proposition 3 that the topology $\mathcal{T}$ on $\mathcal{K}(\Omega)$ can also be defined as the finest locally convex topology for which the canonical injections $f_k: \mathcal{K}(K_k) \hookrightarrow \mathcal{K}(\Omega)$ $(k \in \mathbf{N})$ are continuous.

By Example 9.5 the space $\mathcal{K}(K_k)$ is complete, hence it is a closed subspace of $\mathcal{K}(K_{k+1})$. It follows therefore from all the above results that $\mathcal{K}(\Omega)$ is a complete Hausdorff space which induces on each $\mathcal{K}(K)$ its own topology.

Each subspace $\mathcal{K}(K)$ is closed in $\mathcal{K}(\Omega)$, and a set B is bounded in $\mathcal{K}(\Omega)$ if and only if there exists a compact subset K of Ω and a number $\mu > 0$ such that all $f \in B$ have their support in K and $|f(x)| \leqq \mu$ for all $x \in \Omega$ and $f \in B$.

The topology $\mathcal{T}$ on $\mathcal{K}(\Omega)$ is strictly finer than the topology $\mathcal{T}_u$ of uniform convergence on Ω; i.e., the injection $\mathcal{K}(\Omega) \hookrightarrow \mathcal{C}_0(\Omega)$ (cf. Example 4.16 or Exercise 1.2.7) is continuous but not a strict morphism. Indeed, let h be a continuous, strictly positive function on Ω and denote by V_h the set of all $f \in \mathcal{K}(\Omega)$ such that $|f(x)| \leqq h(x)$ for all x. The set V_h is absorbing, balanced, and convex. Furthermore, if $\mu_K > 0$ is the minimum of h on the compact subset K of Ω, then $V_h \cap \mathcal{K}(K)$ contains the ball $\{f \mid \max_{x \in K} |f(x)| \leqq \mu_K\}$ in $\mathcal{K}(K)$; hence V_h is a neighborhood of 0 in $\mathcal{K}(\Omega)$ for $\mathcal{T}$. However, if h is not bounded away from zero in Ω, then V_h is not a neighborhood of 0 for $\mathcal{T}_u$.

We find that $\mathcal{T}$ is *a fortiori* finer than the topology of uniform convergence on compact subsets of Ω (Examples 3.3 and 4.6). Finally, let us recall that not only $\mathcal{T}$ but also $\mathcal{T}_u$ induces on each $\mathcal{K}(K)$ its own Banach space topology (cf. Example 5.9).

The space $\mathcal{K}(\Omega)$ will have an important role in measure theory: we shall define a measure on Ω as a continuous linear form on $\mathcal{K}(\Omega)$.

Example 6. Let Ω be an open subset of $\mathbf{R}^n$ and let $\mathfrak{D}(\Omega)$ be the vector space of all functions defined on Ω whose partial derivatives of all orders exist and are continuous and whose support is contained in some compact subset of Ω. For each compact subset K of Ω let $\mathfrak{D}(K)$ be the linear subspace of $\mathfrak{D}(\Omega)$ formed by those functions whose support is contained in K. Then the space $\mathfrak{D}(\Omega)$ is the union of all the subspaces $\mathfrak{D}(K)$; it plays a fundamental role in the theory of distributions.

We have equipped each space $\mathfrak{D}(K)$ with a locally convex topology (Examples 3.4 and 4.10) for which it is a Fréchet space (Example 9.5). We equip $\mathfrak{D}(\Omega)$ with the finest locally convex topology $\mathcal{T}$ for which all the canonical injections $\mathfrak{D}(K) \hookrightarrow \mathfrak{D}(\Omega)$ are continuous.

If $K \subset K'$, then $\mathfrak{D}(K) \subset \mathfrak{D}(K')$ and it follows immediately from the definition of the topologies of these two spaces that $\mathfrak{D}(K')$ induces on $\mathfrak{D}(K)$ its own topology. Let $(K_k)_{k \in \mathbf{N}}$ be an increasing sequence of compact subsets of Ω such that $K_k \subset \mathring{K}_{k+1}$ and every compact subset of Ω is contained in some K_k (cf. Example 6.3). Then $\mathfrak{D}(K_k) \subset \mathfrak{D}(K_{k+1})$ for all $k \in \mathbf{N}$ and $\mathfrak{D}(\Omega) = \bigcup_{k \in \mathbf{N}} \mathfrak{D}(K_k)$. It follows from Proposition 3 that $\mathcal{T}$ is also the finest locally convex topology on $\mathfrak{D}(\Omega)$ for which the canonical injections $f_k \colon \mathfrak{D}(K_k) \hookrightarrow \mathfrak{D}(\Omega)$ $(k \in \mathbf{N})$ are continuous.

The subspace $\mathfrak{D}(K_k)$ is complete and thus closed in $\mathfrak{D}(K_{k+1})$. Hence, by the above results, $\mathfrak{D}(\Omega)$ is a complete Hausdorff space which induces on each $\mathfrak{D}(K)$ its own topology. Each subspace $\mathfrak{D}(K)$ is closed in $\mathfrak{D}(\Omega)$, and a set $B \subset \mathfrak{D}(\Omega)$ is bounded if and only if there exists a compact subset K of Ω and positive numbers μ_p, where $p = (p_1, \ldots, p_n) \in \mathbf{N}^n$, such that all $f \in B$ have their supports contained in K and for any $p \in \mathbf{N}^n$ we have $|\partial^p f(x)| \leqq \mu_p$ for all $x \in \Omega, f \in B$.

Before proving other properties of the spaces $\mathfrak{D}(\Omega)$ we have to consider in detail the behavior of infinitely differentiable functions with compact support. It is not at all trivial that such functions exist. To construct an example, one observes with Cauchy that the function defined on $\mathbf{R}$ whose value for $\xi > 0$ is e^{-1/ξ^2} and which vanishes for $\xi \leqq 0$ is infinitely differentiable ([2], Exercise 13–30, p. 427). Similarly, the function φ defined by

$$\begin{aligned} \varphi(x) &= \exp\left(-\frac{1}{1-|x|^2}\right) \quad &\text{if} \quad |x| < 1, \\ \varphi(x) &= 0 \quad &\text{if} \quad |x| \geqq 1 \end{aligned} \tag{1}$$

is infinitely differentiable on $\mathbf{R}^n$ and its support is the unit ball

$$B_1 = \{x \mid |x| \leqq 1\}$$

of $\mathbf{R}^n$. In particular, the function

$$x \mapsto \varphi\left(\frac{x - x_0}{\epsilon}\right)$$

has the ball $B_\epsilon(x_0) = \{x \mid |x - x_0| \leqq \epsilon\}$ for its support, which shows that if $\Omega \neq \emptyset$, then $\mathfrak{D}(\Omega)$ contains nonzero elements. If we divide φ by the positive constant $\int_{\mathbf{R}^n} \varphi(x)\, dx$, we obtain a function ρ which has the following properties:

(i) ρ is infinitely differentiable in $\mathbf{R}^n$,
(ii) Supp $\rho = B_1$,
(iii) $\rho(x) > 0$ for $|x| < 1$,
(iv) $\rho(x) \geqq 0$ for all $x \in \mathbf{R}^n$,
(v) $\int_{\mathbf{R}^n} \rho(x)\, dx = 1$.

For $\epsilon > 0$ the function ρ_ϵ defined by

$$\rho_\epsilon(x) = \frac{1}{\epsilon^n} \rho\left(\frac{x}{\epsilon}\right) \tag{2}$$

has properties (i), (iv), (v) and furthermore,

(ii′) Supp $\rho_\epsilon = B_\epsilon$,
(iii′) $\rho_\epsilon(x) > 0$ for $|x| < \epsilon$.

Let X be a topological space and $(A_\iota)_{\iota \in I}$ a family of subsets of X. We say that (A_ι) is *locally finite* if every point $x \in X$ has a neighborhood V such that $V \cap A_\iota = \emptyset$ except for finitely many indices $\iota \in I$.

LEMMA 4. *Let Ω be an open subset of $\mathbf{R}^n$ and $(U_k)_{k\in\mathbf{N}}$ a countable locally finite open cover of Ω. There exists an open cover $(V_k)_{k\in\mathbf{N}}$ of Ω such that $\overline{V}_k \subset U_k$ for each $k \in \mathbf{N}$.*

REMARK 3. A topological space X is said to be *normal* if it is a Hausdorff space and if the following axiom is satisfied:

(N) Given a closed subset A of X and an open set V containing A, there exists an open set W containing A such that $\overline{W} \subset V$.

An open subset of $\mathbf{R}^n$, and more generally, any metric space is normal. Indeed, for every $x \in A$ let $B(x)$ be a ball with center x, radius $\sigma(x)$ contained in V. If $C(x)$ is the open ball with center x and radius $\frac{1}{2}\sigma(x)$, we can take $W = \bigcup_{x\in A} C(x)$. By Exercise 10.1(e) any compact space is normal. We shall prove Lemma 4 under the sole assumption that Ω is a normal space.

Proof. We shall define by induction a sequence (V_k) of open sets such that

(a) $\overline{V}_k \subset U_k$ for every $k \in \mathbf{N}$.

(b) For every $m \in \mathbf{N}$ the sets V_k with $k \leqq m$ and the sets U_k with $k > m$ cover Ω.

This will prove the lemma since each $x \in \Omega$ is contained only in finitely many sets U_k, say $U_{k_1}, \ldots, U_{k_l}$. Taking $m > k_i$ $(1 \leqq i \leqq l)$, it follows from (b) that (V_k) is a covering of Ω.

Suppose that we have defined the sets V_k with $k < p$ such that (a) is satisfied for $k < p$ and (b) for $m < p$ (with an obvious interpretation we may also suppose $p = 0$ to start the induction). Write

$$W = \left(\bigcup_{k<p} V_k\right) \cup \left(\bigcup_{k>p} U_k\right).$$

The set W is open and taking $m = p - 1$, it follows from (b) that $\complement U_p \subset W$. Since Ω is a normal space, there exists an open subset Z of Ω such that $\complement U_p \subset Z \subset \overline{Z} \subset W$. If we set $V_p = \complement\overline{Z}$, we have $V_p \subset \complement Z \subset U_p$, and since $\complement Z$ is closed, also $\overline{V}_p \subset U_p$. Furthermore, $V_p \cup W = \Omega$; i.e., (b) is satisfied for $m = p$. ∎

PROPOSITION 4. *Let Ω be an open subset of $\mathbf{R}^n$ and $(U_k)_{k\in\mathbf{N}}$ a countable locally finite open cover of Ω by relatively compact sets. Then there exists a family $(\beta_k)_{k\in\mathbf{N}}$ of functions belonging to $\mathfrak{D}(\Omega)$ such that*

(a) *$\beta_k(x) \geqq 0$ for all $x \in \Omega$,*

(b) *$\operatorname{Supp} \beta_k \subset U_k$ for each $k \in \mathbf{N}$,*

(c) *$\sum_{k\in\mathbf{N}} \beta_k(x) = 1$ for all $x \in \Omega$.*

Proof. By Lemma 4 there exists an open covering (V_k) of Ω such that $\overline{V}_k \subset U_k$ for all $k \in \mathbf{N}$.

Consider a fixed index $k \in \mathbf{N}$ and for each $x \in \overline{V}_k$ let $B(x)$ be a ball with center x contained in U_k. Since $\overline{V}_k$ is compact, it can be covered by the interiors of a finite number of these balls, say $B_1, \ldots, B_l$. Let x_i be the center and ϵ_i the radius of B_i $(1 \leqq i \leqq l)$ and define the function ρ_i by $x \mapsto \rho_{\epsilon_i}(x - x_i)$, where ρ_ϵ is the function introduced in Eq. (2). Then $\rho_i(x) > 0$ in $\mathring{B}_i$, Supp $\rho_i = B_i$ and $\rho_i \in \mathfrak{D}(\Omega)$. Consequently, if we set

$$\gamma_k = \sum_{i=1}^{l} \rho_i,$$

then $\gamma_k(x) > 0$ for $x \in \overline{V}_k$, Supp $\gamma_k \subset U_k$ and $\gamma_k \in \mathfrak{D}(\Omega)$.

The sum $\gamma = \sum_{k \in \mathbf{N}} \gamma_k$ is well-defined since the cover (U_k) is locally finite, and γ is infinitely differentiable for the same reason. Furthermore, $\gamma(x) > 0$ for all $x \in \Omega$. Setting $\beta_k = \gamma_k/\gamma$, we obtain the required functions. ∎

LEMMA 5. *Let Ω be an open subset of $\mathbf{R}^n$ and $(\Omega_\iota)_{\iota \in I}$ an open cover of Ω. There exists a countable locally finite open cover $(U_k)_{k \in \mathbf{N}}$ of Ω by relatively compact sets U_k, which is a refinement of (Ω_ι); i.e., for each $k \in \mathbf{N}$ there exists an $\iota \in I$ such that $U_k \subset \Omega_\iota$.*

Proof. Let $(K_i)_{i \in \mathbf{N}}$ be the sequence of compact subsets of Ω constructed in Example 6.3 such that $K_i \subset \mathring{K}_{i+1}$, $\Omega = \bigcup_{i \in \mathbf{N}} K_i$, and set $K_i = \emptyset$ for $i \leqq 0$. The set $L_i = K_i \cap \complement \mathring{K}_{i-1}$ is compact and $V_i = \mathring{K}_{i+1} \cap \complement K_{i-2}$ is an open neighborhood of L_i. For each point $x \in L_i$ there exists an open neighborhood $W(x)$ contained in V_i and in some set Ω_ι. Since L_i is compact, it is covered by a finite number of these sets $W(x)$, say

$$W_{i,1}, \ldots, W_{i,l(i)}.$$

The family of all sets W_{ij} $(1 \leqq j \leqq l(i), i \in \mathbf{N})$ form a countable open cover of Ω by relatively compact sets, which is a refinement of (Ω_ι). Let us show that it is locally finite. Let $z \in \Omega$ and let i be the smallest integer for which $z \in \mathring{K}_i$. Then $z \notin \mathring{K}_{i-1}$ and there exists a neighborhood T of z such that $T \subset \mathring{K}_i$ and $T \cap \mathring{K}_{i-2} = \emptyset$. Consequently, T meets at most those sets W_{mj} for which $i - 2 \leqq m \leqq i + 1$, which are finite in number. ∎

THEOREM 4. *Let Ω be an open subset of $\mathbf{R}^n$ and $(\Omega_\iota)_{\iota \in I}$ an open cover of Ω. There exists a family $(\alpha_\iota)_{\iota \in I}$ of functions possessing continuous partial derivatives of all orders and having the following properties:*

(a) $\alpha_\iota(x) \geqq 0$ *for all* $x \in \Omega$,
(b) $A_\iota = \text{Supp } \alpha_\iota \subset \Omega_\iota$ *for each* $\iota \in I$,
(c) *the family* $(A_\iota)_{\iota \in I}$ *is locally finite,*
(d) $\sum_{\iota \in I} \alpha_\iota(x) = 1$ *for all* $x \in \Omega$.

Proof. By Lemma 5 there exists a countable locally finite open refinement $(U_k)_{k \in \mathbf{N}}$ of (Ω_ι) composed of relatively compact sets U_k. Consider a sequence (β_k) of functions belonging to $\mathfrak{D}(\Omega)$ and having the properties (a) through (c) of Proposition 4.

There exists a map $\kappa: \mathbf{N} \to I$ such that $U_k \subset \Omega_{\kappa(k)}$ for each $k \in \mathbf{N}$. Let us set

$$\alpha_\iota = \sum_{\kappa(k)=\iota} \beta_k$$

for $\iota \in I$. This sum is well-defined since the family $(\operatorname{Supp} \beta_k)_{k \in \mathbf{N}}$ is locally finite, and the functions α_ι have continuous partial derivatives of all orders for the same reason. Clearly $\alpha_\iota(x) \geqq 0$ for all $x \in \Omega$.

Next let C_ι be the union of all the sets $\operatorname{Supp} \beta_k$ with $\kappa(k) = \iota$. Clearly $C_\iota \subset \Omega_\iota$. On the other hand, let $x \in A_\iota$. Then every neighborhood of x meets some set $\operatorname{Supp} \beta_k$ with $\kappa(k) = \iota$, and there exists a neighborhood of x which meets only finitely many sets U_k and *a fortiori* only finitely many sets $\operatorname{Supp} \beta_k$. Therefore there must exist an index k_0 with $\kappa(k_0) = \iota$ such that $x \in \operatorname{Supp} \beta_{k_0}$. Hence $x \in C_\iota$, that is, $A_\iota \subset C_\iota$ and we have proved that $A_\iota \subset \Omega_\iota$.

The family $(C_\iota)_{\iota \in I}$ is locally finite. Indeed, for each $x \in \Omega$ there exists a neighborhood V of x and a finite subset H of $\mathbf{N}$ such that $V \cap U_k = \emptyset$ for $k \notin H$. But then $V \cap C_\iota = \emptyset$ for all ι which do not belong to the finite set $\kappa(H)$.

The family $(A_\iota)_{\iota \in I}$ is *a fortiori* locally finite. Finally, for every $x \in \Omega$ we have

$$1 = \sum_{k \in \mathbf{N}} \beta_k(x) = \sum_{\iota \in I} \Big(\sum_{\kappa(k)=\iota} \beta_k(x) \Big) = \sum_{\iota \in I} \alpha_\iota(x). \blacksquare$$

REMARK 4. The family (α_ι) constructed in Theorem 4 is called a *locally finite, infinitely differentiable partition of unity subordinated to the cover* (Ω_ι). In Theorem 4 the sum in (d) is well defined because of condition (c). It also follows from condition (c) that each compact subset of Ω meets only finitely many sets A_ι. Finally, if the sets Ω_ι are relatively compact, then (b) implies that $\alpha_\iota \in \mathfrak{D}(\Omega)$ for all $\iota \in I$.

PROPOSITION 5. *Let Ω be an open subset of $\mathbf{R}^n$, A a closed subset of Ω, and V an open subset of Ω containing A. There exists an infinitely differentiable function φ such that $0 \leqq \varphi(x) \leqq 1$ for $x \in \Omega$, $\varphi(x) = 1$ for $x \in A$, and $\varphi(x) = 0$ for $x \in \complement V$.*

Proof. Apply Theorem 4 to the cover of Ω formed by the two sets $\Omega_1 = V$ and $\Omega_2 = \complement A$. If (α_1, α_2) is the partition of unity subordinated to (Ω_1, Ω_2), then we can choose $\varphi = \alpha_1$ since $\alpha_1 + \alpha_2 = 1$ and $\operatorname{Supp} \alpha_2 \subset \complement A$ imply that $\alpha_1(x) = 1$ for $x \in A$. $\blacksquare$

We can now return to the study of the spaces $\mathfrak{D}(\Omega)$.

EXAMPLE 7. It is useful to indicate explicitly a fundamental system of neighborhoods of 0 in $\mathfrak{D}(\Omega)$. Let $m = (m_k)_{k \geqq 1}$ be an increasing sequence of positive integers and $\epsilon = (\epsilon_k)_{k \geqq 1}$ a decreasing sequence of strictly positive numbers. Denote by $V_{m,\epsilon}$ the set of all functions $f \in \mathfrak{D}(\Omega)$ which for every $k \geqq 1$ satisfy $|\partial^p f(x)| \leqq \epsilon_k$ if $|p| \leqq m_k$ and $x \in \Omega \cap \complement K_{k-1}$, where $(K_k)_{k \in \mathbf{N}}$ is the sequence of compact subsets of Ω considered e.g. in Example 6. The set $V_{m,\epsilon}$ is balanced and convex and for each $k \geqq 1$ the intersection $V_{m,\epsilon} \cap \mathfrak{D}(K_k)$ contains a neighborhood of 0 in $\mathfrak{D}(K_k)$, to wit, the set of those functions $f \in \mathfrak{D}(K_k)$ which satisfy $|\partial^p f(x)| \leqq \epsilon_k$ for $|p| \leqq m_k$ and $x \in \Omega$. Thus $V_{m,\epsilon}$ is a neighborhood of 0 in $\mathfrak{D}(\Omega)$. Conversely, let V be a balanced, convex neighborhood of 0 in $\mathfrak{D}(\Omega)$. Then for each $k \in \mathbf{N}$ the set $V \cap \mathfrak{D}(K_{k+1})$ is a neighborhood of 0 in $\mathfrak{D}(K_{k+1})$; hence there exists an integer l_k and a number $\eta_k > 0$ such that $V \cap \mathfrak{D}(K_{k+1})$ contains the set of those functions $f \in \mathfrak{D}(K_{k+1})$ which satisfy $|\partial^p f(x)| \leqq \eta_k$ for $x \in K_{k+1}$ and $|p| \leqq l_k$. Choose an infinitely differentiable partition of unity $(\alpha_k)_{k \geqq 1}$ subordinated to the cover

$$(\complement K_{k-1} \cap \mathring{K}_{k+1})_{k \geqq 1}$$

of Ω. For $f \in \mathfrak{D}(\Omega)$ we can write

$$f = \sum_{k \geqq 1} \frac{1}{2^k} (2^k \alpha_k f),$$

where the sum is actually finite since the support of f is contained in some K_k. If all the functions $2^k \alpha_k f$ belong to V, then f belongs to V since it is balanced and convex.

By virtue of the Leibniz formula (Proposition 5.3), $\partial^p(\alpha_k f)$ is a finite linear combination of products of the form $\partial^q \alpha_k \cdot \partial^r f$, where $|q| \leqq |p|$ and $|r| \leqq |p|$. Since the sequence (α_k) has been fixed once and for all and since only the values taken by f in $\complement K_{k-1} \cap K_{k+1}$ intervene, it follows that there exists a constant λ_k such that

$$|\partial^p f(x)| \leqq \epsilon_k$$

for $|p| \leqq m_k$ and $x \in \Omega \cap \complement K_{k-1}$ imply

$$|\partial^p (2^k \alpha_k f)| \leqq \lambda_k \epsilon_k$$

for $|p| \leqq m_k$ and $x \in \Omega$. Suppose now that m_i and ϵ_i have already been chosen for $0 \leqq i \leqq k - 1$ and determine m_k and ϵ_k so that $m_k \geqq m_{k-1}$, $m_k \geqq l_k$, $\epsilon_k \leqq \epsilon_{k-1}$, $\lambda_k \epsilon_k \leqq \eta_k$. Let $V_{m,\epsilon}$ be the neighborhood of 0 defined by the sequences $m = (m_k)$ and $\epsilon = (\epsilon_k)$ just constructed. If $f \in V_{m,\epsilon}$ then for every $k \geqq 1$ the function $2^k \alpha_k f$, whose support is contained in $\complement K_{k-1} \cap \mathring{K}_{k+1} \subset K_{k+1}$, belongs to $V \cap \mathfrak{D}(K_{k+1}) \subset V$. Hence $f \in V$; i.e., V contains $V_{m,\epsilon}$.

We can also indicate a family of semi-norms which determines the topology of $\mathfrak{D}(\Omega)$. We consider families $\theta = (\theta_p)_{p\in\mathbf{N}^n}$ of continuous functions θ_p defined in Ω and such that the family of sets $(\operatorname{Supp}\theta_p)_{p\in\mathbf{N}^n}$ is locally finite. For each θ and $f \in \mathfrak{D}(\Omega)$

$$q_\theta(f) = \max_{p\in\mathbf{N}^n} \max_{x\in\Omega} |\theta_p(x)\, \partial^p f(x)|, \tag{3}$$

where the expression on the right-hand side exists since $\operatorname{Supp} f$ meets $\operatorname{Supp}\theta_p$ only for finitely many multi-indices p. Let us show that the family of semi-norms (q_θ), where θ runs through all possible families, has the required property.

Conserving our earlier notations, let $V_{m,\epsilon}$ be a neighborhood of 0 in $\mathfrak{D}(\Omega)$, where we may obviously assume that the sequence $m = (m_k)$ is strictly increasing. If $p \in \mathbf{N}^n$, $m_{k-1} < |p| \leqq m_k$, we define

$$\theta_p(x) = \sum_{l\geqq k-1} \epsilon_{l+1}^{-1}\alpha_l(x)$$

(with $\alpha_0(x) \equiv 0$). Suppose that $q_\theta(f) \leqq 1$ and consider a point

$$x \in \mathbf{C}K_{k-1} \cap K_k.$$

Then for $|p| \leqq m_k$ we have

$$\{\epsilon_k^{-1}\alpha_{k-1}(x) + \epsilon_{k+1}^{-1}\alpha_k(x)\}|\partial^p f(x)| \leqq 1$$

and *a fortiori* $|\partial^p f(x)| \leqq \epsilon_k$ since $\epsilon_k \geqq \epsilon_{k+1}$ and $\alpha_{k-1}(x) + \alpha_k(x) = 1$. Hence $f \in V_{m,\epsilon}$.

Conversely, let $\theta = (\theta_p)$ be the family which defines q_θ of (3). Determine m_k inductively so that $m_k \geqq m_{k-1}$ and $K_k \cap \operatorname{Supp}\theta_p = \emptyset$ for $|p| > m_k$. Set

$$\mu_k = \max_{|p|\leqq m_k} \max_{x\in K_k} |\theta_p(x)|$$

and determine $\epsilon_k > 0$ inductively so that $\epsilon_k \leqq \epsilon_{k-1}$ and $\epsilon_k\mu_k \leqq 1$. If $|\partial^p f(x)| \leqq \epsilon_k$ for $|p| \leqq m_k$ and $x \in \mathbf{C}K_{k-1} \cap K_k$ $(k \geqq 1)$, then $q_\theta(f) \leqq 1$.

Obviously instead of (3) we can consider the semi-norms defined by

$$\sum_{p\in\mathbf{N}^n} \max_{x\in\Omega} |\theta_p(x)\, \partial^p f(x)|.$$

EXAMPLE 8. For a fixed positive integer m denote by $\mathfrak{D}^m(\Omega)$ the vector space of all functions f defined in the open subset Ω of $\mathbf{R}^n$, whose partial derivatives $\partial^p f$ exist and are continuous for $|p| \leqq m$ and whose support is contained in some compact subset of Ω. For $m = 0$ we obtain the space $\mathcal{K}(\Omega)$ of Example 5.

Analogously as in Examples 5 and 6, the space $\mathfrak{D}^m(\Omega)$ is the union of all spaces $\mathfrak{D}^m(K)$ of Example 4.9, where K is a compact subset of Ω. We equip $\mathfrak{D}^m(\Omega)$ with the finest locally convex topology for which the maps $\mathfrak{D}^m(K) \hookrightarrow \mathfrak{D}^m(\Omega)$, or equivalently the maps $\mathfrak{D}^m(K_k) \hookrightarrow \mathfrak{D}^m(\Omega)$, are continuous, where (K_k) is the often-mentioned sequence of compact subsets of Ω.

Again $\mathfrak{D}^m(\Omega)$ is a complete Hausdorff space which induces on each $\mathfrak{D}^m(K)$ its own topology. Each subspace $\mathfrak{D}^m(K)$ is closed in $\mathfrak{D}^m(\Omega)$. A set $B \subset \mathfrak{D}^m(\Omega)$ is bounded if and only if there exists a compact set $K \subset \Omega$ and for each $p \in \mathbf{N}^n$ with $|p| \leqq m$ a number $\mu_p > 0$ such that $\operatorname{Supp} f \subset K$ and $|\partial^p f(x)| \leqq \mu_p$ for all $f \in B$, $x \in \Omega$, and $|p| \leqq m$.

As before, we shall also write $\mathfrak{D}^\infty(\Omega)$ for $\mathfrak{D}(\Omega)$. Instead of $\mathfrak{D}^m(\mathbf{R}^n)$ we write $\mathfrak{D}^m$ $(0 \leqq m \leqq \infty)$.

Let us prove that for each m $(0 \leqq m \leqq \infty)$ the injection $\mathfrak{D}^m \hookrightarrow \mathcal{S}^m$ (cf. Examples 4.13 and 4.14) is continuous. By virtue of Proposition 1 we have to prove that for every compact set $K \subset \mathbf{R}^n$ the map $\mathfrak{D}^m(K) \hookrightarrow \mathcal{S}^m$ is continuous. Let $k \in \mathbf{N}$ and denote by M the maximum of $(1 + |x|^2)^k$ on K. Then for $f \in \mathfrak{D}^m(K)$ and $|p| \leqq m$ we have

$$\max_{x \in \mathbf{R}^n} |(1 + |x|^2)^k \, \partial^p f(x)| \leqq M \max_{x \in K} |\partial^p f(x)|,$$

from which the conclusion follows (Proposition 5.2).

For any open set $\Omega \subset \mathbf{R}^n$ and $0 \leqq m \leqq \infty$ the map $\mathfrak{D}^m(\Omega) \hookrightarrow \mathcal{B}_0^m(\Omega)$ is continuous. This follows from Proposition 1 and from the fact that for every compact set $K \subset \Omega$ the map $\mathfrak{D}^m(K) \hookrightarrow \mathcal{B}_0^m(\Omega)$ is continuous (Example 5.9).

Let $m > m'$, then the maps

$$\mathfrak{D}(\Omega) \hookrightarrow \mathfrak{D}^m(\Omega) \hookrightarrow \mathfrak{D}^{m'}(\Omega) \hookrightarrow \mathcal{K}(\Omega)$$

are continuous. Indeed, for every compact set $K \subset \Omega$ the map

$$\mathfrak{D}^m(K) \hookrightarrow \mathfrak{D}^{m'}(K)$$

is continuous by Example 5.7 and $\mathfrak{D}^{m'}(K) \hookrightarrow \mathfrak{D}^{m'}(\Omega)$ by definition. Hence by Proposition 1 $\mathfrak{D}^m(\Omega) \hookrightarrow \mathfrak{D}^{m'}(\Omega)$ is also continuous, where we may take $m = \infty$ and $m' = 0$.

The vector space $\mathfrak{D}(\Omega)$ is clearly the intersection of all the spaces $\mathfrak{D}^m(\Omega)$ with $m \in \mathbf{N}$. For reasons which will appear in Chapter 4, §4, we denote by $\mathfrak{D}^F(\Omega)$ the vector space $\mathfrak{D}(\Omega)$ equipped with the coarsest topology for which all the maps $\mathfrak{D}(\Omega) \hookrightarrow \mathfrak{D}^m(\Omega)$ are continuous. $\mathfrak{D}^F(\Omega)$ is a locally convex Hausdorff space and by the corollary to Proposition 11.3 it is complete. The topology of $\mathfrak{D}^F(\Omega)$ is coarser than the topology of $\mathfrak{D}(\Omega)$ defined in Example 6, i.e., the identity map $\mathfrak{D}(\Omega) \hookrightarrow \mathfrak{D}^F(\Omega)$ is continuous. Indeed, we have the diagram:

$$\begin{array}{ccccc} \mathfrak{D}(K) & \overset{f}{\hookrightarrow} & \mathfrak{D}(\Omega) & \overset{g}{\hookrightarrow} & \mathfrak{D}^F(\Omega) \\ h \downarrow & & & & \nearrow k \\ \mathfrak{D}^m(K) & & \overset{l}{\hookrightarrow} & & \mathfrak{D}^m(\Omega). \end{array}$$

We know from Example 5.7 that h is continuous. Since l is continuous by definition, we see that $l \circ h$ is continuous. But $k \circ g \circ f = l \circ h$; hence $g \circ f$ is continuous by a property of the initial topology recalled in §1. Hence g is continuous by Proposition 1. We shall see in Chapter 4, §4 that the topology of $\mathfrak{D}^F(\Omega)$ is strictly coarser than the topology of $\mathfrak{D}(\Omega)$.

EXAMPLE 9. Let m be a positive integer or the symbol ∞. We denote by $\mathfrak{O}_C^m$ the union of the spaces $\mathcal{S}_k^m$ (Examples 4.11 and 4.12) as k varies in $\mathbf{Z}$. Instead of $\mathfrak{O}_C^\infty$ we also write $\mathfrak{O}_C$.

A function f belongs to $\mathfrak{O}_C^m$ if and only if it is of the form $f = Pg$, where P is a polynomial and $g \in \mathfrak{B}_0^m$. Indeed, if $f \in \mathfrak{O}_C^m$, then f belongs to some $\mathcal{S}_k^m$; and since $\mathcal{S}_k^m \subset \mathcal{S}_{k'}^m$ for $k \geqq k'$, we may suppose that $k = -l < 0$. But by Example 5.8 the function $g = (1 + |x|^2)^{-l} f$ belongs to $\mathfrak{B}_0^m$; hence $f = (1 + |x|^2)^l g$. Conversely, if $f = Pg$, where $g \in \mathfrak{B}_0^m$ and P is a polynomial of degree l, then clearly $f \in \mathcal{S}_k^m$ for $k = -l$.

We equip $\mathfrak{O}_C^m$ with the finest locally convex topology for which all the maps $\mathcal{S}_k^m \hookrightarrow \mathfrak{O}_C^m$ are continuous. The map $\mathfrak{O}_C^m \hookrightarrow \mathcal{E}^m$ is continuous. By virtue of Proposition 1 we have to prove that for each $k \in \mathbf{Z}$ the map $\mathcal{S}_k^m \hookrightarrow \mathcal{E}^m$ is continuous. Let K be a compact subset of $\mathbf{R}^n$. Then we have

$$\max_{x \in K} |\partial^p f(x)| \leqq \max_{x \in K} (1 + |x|^2)^{-k} \max_{x \in \mathbf{R}^n} |(1 + |x|^2)^k \partial^p f(x)|,$$

from which the conclusion follows by Proposition 5.2.

In particular, $\mathfrak{O}_C^m$ is a Hausdorff space since it has a topology finer than a Hausdorff topology. Observe that we cannot apply the corollary of Theorem 1. In fact, as we shall see in Chapter 4, §11, the linear subspace $\mathfrak{D}$ is dense in $\mathcal{S}_k^m$ and *a fortiori* $\mathcal{S}_{k+1}^m$ is dense in $\mathcal{S}_k^m$. Now since $\mathcal{S}_{k+1}^m$ is complete (Examples 9.6 and 11.8), if $\mathcal{S}_{k+1}^m \hookrightarrow \mathcal{S}_k^m$ were a strict morphism, we would have $\mathcal{S}_{k+1}^m = \mathcal{S}_k^m$, which is obviously not the case.

If $0 \leqq m' < m \leqq \infty$, then the map $\mathfrak{O}_C^m \hookrightarrow \mathfrak{O}_C^{m'}$ is continuous. Again by Proposition 1 we have to show that for every $k \in \mathbf{Z}$ the map $\mathcal{S}_k^m \hookrightarrow \mathfrak{O}_C^{m'}$ is continuous. Now this map is composed of the maps $\mathcal{S}_k^m \hookrightarrow \mathcal{S}_k^{m'}$ and $\mathcal{S}_k^{m'} \hookrightarrow \mathfrak{O}_C^{m'}$, of which the first is continuous by Example 5.7 and the second by the definition of the topology of $\mathfrak{O}_C^{m'}$.

Exercises

1. (a) Let $(E_\iota)_{\iota \in I}$ be a family of locally convex spaces. On $\prod_{\iota \in I} E_\iota$ define the topology $\mathcal{T}^*$, for which a fundamental system of neighborhoods of a point (x_ι) is given by the sets $\prod_{\iota \in I} V_\iota$, where for each $\iota \in I$ the set V_ι is a neighborhood of x_ι. Show that $\mathcal{T}^*$ induces on $\coprod_{\iota \in I} E_\iota$ a locally convex topology $\mathcal{T}_0$ which is coarser than the locally convex direct sum topology $\mathcal{T}$ on $\coprod_{\iota \in I} E_\iota$.

(b) Show that if the index set I is countable, then the topologies $\mathcal{T}_0$ and $\mathcal{T}$ coincide on $\coprod_{\iota \in I} E_\iota$. (*Hint:* Assuming that $I = \mathbf{N}^*$, the balanced, convex hull of $\bigcup_{n \geq 1} j_n(V_n)$ contains $\coprod_{n \geq 1} (1/2^n) V_n$ and is contained in $\coprod_{n \geq 1} V_n$.)

2. Let F be a vector space and let $(E_\iota)_{\iota \in I}$ be a family of linear subspaces of F. Suppose that each E_ι is equipped with a locally convex topology $\mathcal{T}_\iota$ and let $\mathcal{T}$ be the finest locally convex topology on F for which the canonical injections $f_\iota: E_\iota \hookrightarrow F$ are continuous. Show that $\mathcal{T}$ is defined by those semi-norms on F whose restrictions to each E_ι are continuous for $\mathcal{T}_\iota$.

3. Let F be a vector space, $(E_\iota)_{\iota \in I}$ a family of locally convex spaces, and for each index $\iota \in I$ let f_ι be a linear map from E_ι into F. Define a linear map φ from the locally convex direct sum $\coprod_{\iota \in I} E_\iota$ into F by

$$\varphi\left(\sum_{\iota \in I} x_\iota\right) = \sum_{\iota \in I} f_\iota(x_\iota).$$

Show that the finest locally convex topology on F for which the maps f_ι are continuous coincides with the finest locally convex topology on F for which the map $\varphi: \coprod_{\iota \in I} E_\iota \to F$ is continuous.

4. Let I be a directed set and $(E_\iota)_{\iota \in I}$ a family of locally convex spaces. Suppose that for every pair (ι, κ) of indices such that $\iota \leq \kappa$ we have a continuous linear map $f_{\kappa\iota}: E_\iota \to E_\kappa$ and that these maps satisfy the following two conditions:

(i) $f_{\iota\iota}$ is the identity map for each $\iota \in I$,

(ii) $f_{\lambda\iota} = f_{\lambda\kappa} \circ f_{\kappa\iota}$ for $\iota \leq \kappa \leq \lambda$.

A locally convex space F is said to be a *direct* (or inductive) *limit* of the *direct* (or inductive) *system* $(E_\iota, f_{\kappa\iota})$ if the following conditions are satisfied:

(iii) For every $\iota \in I$ there exists a continuous linear map $f_\iota: E_\iota \to F$ such that $f_\iota = f_\kappa \circ f_{\kappa\iota}$ for $\iota \leq \kappa$.

(iv) Given a locally convex space G and a family of continuous linear maps $g_\iota: E_\iota \to G$ such that $g_\iota = g_\kappa \circ f_{\kappa\iota}$ for $\iota \leq \kappa$, there exists a *unique* continuous linear map $g: F \to G$s uch that $g_\iota = g \circ f_\iota$ for all $\iota \in I$, i.e., such that the diagram

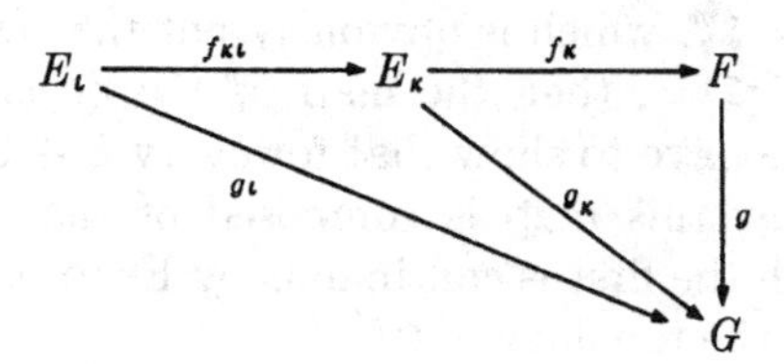

is commutative.

(a) Show that given a direct system there always exists a direct limit which is unique up to an isomorphism. (*Hint:* Consider the quotient space of $\coprod_{\iota \in I} E_\iota$ modulo the subspace generated by the vectors $x_\iota - f_{\kappa\iota}(x_\iota)$, where $x_\iota \in E_\iota$ and $\iota \leqq \kappa$.)

(b) Let F be the direct limit of the direct system $(E_\iota, f_{\kappa\iota})$. Show that if $f_\iota(x_\iota) = 0$ for some $x_\iota \in E_\iota$, then there exists some $\kappa \geqq \iota$ such that $f_{\kappa\iota}(x_\iota) = 0$.

5. Let I be a directed set, F a vector space, and $(E_\iota)_{\iota \in I}$ a family of subspaces of F such that $E_\iota \subset E_\kappa$ for $\iota \leqq \kappa$ and $F = \bigcup_{\iota \in I} E_\iota$. Suppose that each E_ι is equipped with a locally convex topology $\mathcal{T}_\iota$ and for $\iota \leqq \kappa$ the topology $\mathcal{T}_\kappa$ induces $\mathcal{T}_\iota$ on E_ι. Let $\mathcal{T}$ be the finest locally convex topology on F for which all the canonical injections $f_\iota: E_\iota \hookrightarrow F$ are continuous. Suppose that there exists on F a locally convex topology $\mathcal{T}'$ which induces on each E_ι the topology $\mathcal{T}_\iota$. Show that $\mathcal{T}$ induces on each E_ι the topology $\mathcal{T}_\iota$. Apply this result to the space $\mathcal{K}(\Omega)$, taking for $\mathcal{T}'$ the topology $\mathcal{T}_u$ (cf. Example 5).

6. Let F be a vector space and let $(E_n)_{n \in \mathbf{N}}$ be a sequence of subspaces of F such that $E_n \subset E_{n+1}$, $E_n \neq E_{n+1}$ for all $n \in \mathbf{N}$ and $F = \bigcup_{n \in \mathbf{N}} E_n$. Suppose that each E_n is equipped with a locally convex Hausdorff topology $\mathcal{T}_n$ such that for each n the topology induced by $\mathcal{T}_{n+1}$ on E_n is $\mathcal{T}_n$ and the linear subspace E_n is closed in E_{n+1} for $\mathcal{T}_{n+1}$. Equip F with the finest locally convex topology $\mathcal{T}$ for which all the canonical injections $f_n: E_n \hookrightarrow F$ are continuous.

(a) Show that if each E_n is complete for $\mathcal{T}_n$, then F is not metrizable. (*Hint:* Use Baire's theorem, Chapter 1, §8.)

(b) Show that there exists a sequence of bounded sets $(B_k)_{k \in \mathbf{N}}$ in F such that for no choice of the sequence $(\lambda_k)_{k \in \mathbf{N}}$ of strictly positive numbers is the set $\bigcup_{k \in \mathbf{N}} \lambda_k B_k$ bounded (cf. Proposition 6.3).

7. Prove that the function φ defined by formula (1) has all the properties claimed in the text.

8. (a) Prove that the union of a locally finite collection of closed subsets of a topological space is closed.

(b) Show that the sets C_ι which figure in the proof of Theorem 4 are closed, and using this, give a shorter proof of the inclusion $A_\iota \subset C_\iota$.

9. (a) Show that the sets V_h introduced in Example 5 form a fundamental system of neighborhoods of 0 in $\mathcal{K}(\Omega)$.

(b) Again let h be a continuous, strictly positive function defined on the open set $\Omega \subset \mathbf{R}^n$. Let $m \in \mathbf{N}$ and denote by V_h the set of all functions $f \in \mathfrak{D}^m(\Omega)$ such that $|\partial^p f(x)| \leqq h(x)$ for $|p| \leqq m$ and $x \in \Omega$. Show that the sets V_h form a fundamental system of neighborhoods of 0 in $\mathfrak{D}^m(\Omega)$.

10. Let $(E_\iota)_{\iota \in I}$ be a family of locally convex Hausdorff spaces and let E be their locally convex direct sum (Example 2). Show that a subset B of E is bounded in E if and only if there exists a finite subset H of I such that B is contained in the subspace $F = \prod_{\iota \in H} E_\iota$ and B is bounded in F. (*Hint:* To show that the condition is necessary, use an argument similar to the one used in the proof of Theorem 2.)

CHAPTER 3

Duality

§1. The Hahn-Banach theorem

Theorem 1.6.1 can be generalized in the following way.

THEOREM 1. *Let E be a vector space over $\mathbf{K}$, q a semi-norm on E, and M a linear subspace of E. Let f be a linear form defined on M which satisfies $|f(x)| \leqq q(x)$ for all $x \in M$. Then there exists a linear form g on E such that $g(x) = f(x)$ for all $x \in M$ and $|g(x)| \leqq q(x)$ for all $x \in E$.*

The theorem quoted above follows from this result taking

$$q(x) = \|f\| \cdot \|x\|.$$

It is possible to give a proof of Theorem 1 following the lines of the proof given in Chapter 1, §6. We shall leave this, however, as an exercise to the reader. Instead we shall give a different, more geometric proof.

A subset C of a vector space E is called a *cone with vertex at x_0* if for every $x \in C$ the half-line $\{x_0 + \lambda(x - x_0) \mid \lambda > 0\}$ belongs to C. If $x_0 \in C$, the cone is said to be *pointed;* if $x_0 \notin C$, the cone is *blunt* (épointé in French). In what follows we shall mainly consider cones with vertex at 0. Any cone can be obtained from a cone with vertex at 0 by translation.

A subset C of E is a convex cone with vertex at 0 if and only if $\lambda C \subset C$ for $\lambda > 0$ and $C + C \subset C$. Indeed, the first condition means that C is a cone with vertex at 0. If C is convex, then for $x, y \in C$ we have

$$x + y = 2(\tfrac{1}{2}x + \tfrac{1}{2}y) \in C,$$

i.e., $C + C \subset C$. Conversely, if $C + C \subset C$, then for $\alpha + \beta = 1$, $\alpha > 0$, $\beta > 0$ and $x, y \in C$ we have $\alpha x \in C$, $\beta y \in C$, and $\alpha x + \beta y \in C$.

If C is a convex pointed cone with vertex at 0 in a real vector space E, then $M = C \cap (-C)$ is the greatest linear subspace contained in C. Indeed, if $x, y \in M$, then $x + y \in M$, since clearly $-C$ is also a convex cone with vertex at 0. Furthermore, if $x \in M$, then $0 \cdot x = 0 \in M$ since C is pointed and $\lambda x \in M$ for $\lambda \neq 0$; thus M is a subspace of E. Finally, it is clear that any subspace contained in C must be contained in M.

The translate of a linear subspace of a vector space E is called a *linear manifold* (or affine subspace). In other words, the subset M of E is a linear manifold if there exists a linear subspace M_0 of E and a vector $x_0 \in E$ such that $M = M_0 + x_0$. Clearly, $x_0 \in M$ and for every $y_0 \in M$ we have $M = M_0 + y_0$. We also say that M is the linear manifold through x_0, parallel to M_0.

If H_0 is a hyperplane of E, the linear manifold $H = H_0 + x_0$ is said to be an *affine hyperplane*. If $f(x) = 0$ is the equation of H_0 and $f(x_0) = \alpha$, then clearly $H = \{x \mid f(x) = \alpha\}$. We say that H has equation $f(x) = \alpha$.

Let E be a vector space over the field $\mathbf{R}$ and let H be an affine hyperplane with equation $f(x) = \alpha$. The sets defined by $f(x) \geqq \alpha$ and $f(x) \leqq \alpha$ are called the two algebraically closed *half-spaces* determined by H. The sets defined by $f(x) > \alpha$ and $f(x) < \alpha$ are called the two algebraically open half-spaces determined by H. Each of these four sets is convex.

If E is a real topological vector space and H a closed affine hyperplane with equation $f(x) = \alpha$, then f is continuous (Proposition 2.5.7), and thus the two algebraically closed (open) half-spaces determined by H are closed (open) sets in E.

If H is an affine hyperplane in a real vector space E and A a nonempty convex set in E such that $A \cap H = \emptyset$, then A is contained in one of the two algebraically open half-spaces determined by H. Indeed, let $f(x) = \alpha$ be the equation of H. The set $f(A)$ is convex in $\mathbf{R}$ and $\alpha \notin f(A)$. Thus $f(A)$ can contain either only numbers $< \alpha$ or only numbers $> \alpha$.

We can now prove the following theorem due to Mazur, which is also referred to as the geometric form of the Hahn-Banach theorem:

THEOREM 2. *Let E be a topological vector space, A an open convex non-empty set in E, and M a linear manifold in E which does not meet A. There exists a closed affine hyperplane H which contains M and does not meet A.*

Proof. Performing, if necessary, a translation, we may suppose that $0 \in M$, i.e., that M is a linear subspace.

We shall consider first the case of real scalars. We can assume that $M = \{0\}$. Indeed, if φ is the canonical surjection of E onto the quotient space E/M, then $\varphi(A)$ is a convex open nonempty set in E/M which does not contain 0. If H_1 is a closed hyperplane in E/M such that

$$H_1 \cap \varphi(A) = \emptyset,$$

then $\varphi^{-1}(H_1)$ is a closed hyperplane in E which contains M and does not meet A.

Let a be an interior point of A; then $-a \notin A$ since otherwise 0 would belong to A. Let $\mathfrak{C}$ be the collection of all convex pointed cones with vertex at 0 which contain A but not $-a$. The set $B = \bigcup_{\lambda \geqq 0} \lambda A$ belongs to $\mathfrak{C}$

and in particular $\mathfrak{C}$ is not empty. Indeed, $B + B \subset B$ follows from the relation $\lambda A + \mu A = (\lambda + \mu)A$ for $\lambda \geqq 0$, $\mu \geqq 0$, equivalent to the convexity of A, and the other properties are evident.

If ordered by inclusion, $\mathfrak{C}$ is inductive. In fact, let $(C_\iota)_{\iota \in I}$ be a totally ordered family in $\mathfrak{C}$ and let $C = \bigcup_{\iota \in I} C_\iota$. We have $C + C \subset C$ since if $x \in C$, $y \in C$, then $x \in C_\iota$, $y \in C_\kappa$; and if for instance $C_\iota \subset C_\kappa$, then $x \in C_\kappa$ and $x + y \in C_\kappa \subset C$. It is clear now that C belongs to $\mathfrak{C}$ and is the least upper bound of the family (C_ι).

Let C be a maximal element of $\mathfrak{C}$. We shall prove that $H = C \cap (-C)$ is a closed hyperplane which does not meet A.

Let us first prove that $C \cup (-C) = E$. Suppose that $C \cup (-C) \neq E$. Then there exists a vector x such that $\lambda x \notin C$ for $\lambda \neq 0$. The set $C_1 = \{\lambda x + y \mid \lambda \geqq 0, y \in C\}$ is clearly an element of $\mathfrak{C}$ and $C \subset C_1$, but $C \neq C_1$ since $x \in C_1$.

We say that $x \in C$ is an *internal* point of C if for every straight line $z(\tau) = x + \tau y$ through x there exists $\tau_0 > 0$ such that $z(\tau) \in C$ for $|\tau| \leqq \tau_0$. If $x \in C$ is not an internal point of C, then $x \in H$, i.e., $x \in -C$. Indeed, since C is convex, there exists $y \in E$ such that $x + \tau y \notin C$ for $\tau > 0$. Suppose that $x \notin -C$, then the set $C_1 = \{w - \lambda x \mid w \in C, \lambda \geqq 0\}$ is a convex pointed cone with vertex at 0 and $C \subset C_1$, but $C \neq C_1$ since $-x \in C_1$. We have $y \notin C_1$, i.e., $C_1 \neq E$. In the first place, $y \notin C$ since $y \in C$ would imply $x + \tau y \in C$ for all $\tau \geqq 0$. Next $y \in C_1$ would mean $y = w - \lambda x$, where necessarily $\lambda > 0$. But then it would follow that

$$x + \frac{1}{\lambda} y = \frac{1}{\lambda} w \in C,$$

which again contradicts the choice of y. Now we can prove that $-a \notin C_1$. Let $b \in E$. Then since a is an interior point of A, we have

$$a + \lambda(b - a) \in A \subset C_1 \qquad \text{for} \qquad |\lambda| \leqq \lambda_0.$$

If $-a$ belonged to C_1, then it would follow that $\lambda(b - a) \in C_1$, i.e., that $b \in C_1$, but this contradicts the relation $C_1 \neq E$. Thus $C_1 \in \mathfrak{C}$, which is a contradiction of the fact that C is a maximal element of $\mathfrak{C}$. Hence we necessarily have $x \in -C$.

If $y \in C$ and $z \in -C$, then on the segment $x(\tau) = (1 - \tau)y + \tau z$ $(0 \leqq \tau \leqq 1)$ there exists a point $x(\sigma) \in H$. To show this let σ be the least upper bound of all values τ for which $x(\tau) \in C$. Since C is convex, we have $x(\tau) \in C$ for $\tau < \sigma$ and $x(\tau) \notin C$ for $\tau > \sigma$. If $x(\sigma) \in C$, then $x(\sigma)$ is not an internal point of C and thus $x(\sigma) \in H$. If $x(\sigma) \in -C$, then either there exist points $x(\tau) \in -C$ with $\tau < \sigma$, and then for these values of τ we have $x(\tau) \in H$, or $x(\tau) \notin -C$ for $\tau < \sigma$, in which case $x(\sigma)$ is not an internal point of $-C$, that is, $-x(\sigma)$ is not an internal point of C, and thus $x(\sigma) \in H$.

We are now in the position to prove that H is a hyperplane. In the first place, $H \neq E$ since $-a \notin H$. Let $x \in C$. Then the segment

$$x(\tau) = \tau x - (1 - \tau)a$$

contains a point $y = x(\sigma) \in H$, and $\sigma \neq 0$ since $-a \notin C$. Thus

$$x = \frac{1}{\sigma} y + \frac{1 - \sigma}{\sigma} a = w + \lambda a, \qquad w \in H, \quad \lambda \geqq 0.$$

Similarly it can be shown that if $x \in -C$, then $x = w + \lambda a$ with $w \in H$, $\lambda \leqq 0$.

Next we prove that $A \cap H = \emptyset$. Every point of A is an interior point of C and thus evidently an internal point of C; hence it is sufficient to prove that H contains no internal points of C. Let $x \in H$; then $x - \lambda a \notin C$ for $\lambda > 0$ since $x - \lambda a = y \in C$ would imply

$$-a = \frac{1}{\lambda}(-x + y) \in C.$$

Finally, H is closed. Indeed, it cannot be dense since $\complement H$ contains the open set A.

We still have to consider the case of complex scalars. Let E_0 be the real vector space underlying the complex vector space E. Clearly, E_0 is a real topological vector space and M is also a linear subspace of E_0. Next let H_0 be a closed hyperplane in the real space E_0 which contains M and does not meet A. Then $H = H_0 \cap (iH_0)$ is a closed hyperplane in the complex space E (cf. Exercise 5) which contains $M = M \cap (iM)$ and *a fortiori* does not meet A. ∎

Proof of Theorem 1. If $f(x) = 0$ for every $x \in M$, then $g = 0$ gives the required extension. Thus we may suppose that f is not identically zero on M.

We first consider the case $\mathbf{K} = \mathbf{R}$. Let us equip E with the locally convex topology defined by the semi-norm q. Then the set $A = \{x \mid q(x) < 1\}$ is convex, nonempty, and open in E. Let N be the affine hyperplane of M defined by $f(x) = 1$. Then N is a linear manifold in E and does not meet A since on A we have $|f(x)| \leqq q(x) < 1$. By Theorem 2 there exists a closed affine hyperplane H which contains N and does not meet A. Let the equation of H be $g(x) = 1$, where g is a linear form on E. For $y \in N$ we have $g(y) = f(y) = 1$. Also $f(z) = 0$ implies $g(z) = 0$ since

$$\begin{aligned} \{x \mid x \in M, f(x) = 0\} &= N - x_0 \subset H - x_0 \\ &= \{x \mid x \in E, g(x) = 0\}, \end{aligned}$$

where $x_0 \in N \subset H$. Every $x \in M$ can be written in the form $x = \xi y + z$, where $y \in N$ and $f(z) = 0$. It follows that $g(x) = f(x)$ for all $x \in M$. Finally, since $0 \in A$ and H does not meet A, we have $g(x) < 1$ for all $x \in A$, i.e., for $q(x) < 1$. In other words, $g(x) = 1$ implies $q(x) \geqq 1$. Let x be an arbitrary element of E. If $g(x) = 0$, then clearly $|g(x)| \leqq q(x)$. If $g(x) = \rho \neq 0$, then $x = \rho y$ with $g(y) = 1$ and thus

$$g(x) = \rho g(y) \leqq |\rho| q(y) = q(x).$$

Since also $-g(x) = g(-x) \leqq q(-x) = q(x)$ we have $|g(x)| \leqq q(x)$ for every $x \in E$.

Let us now pass to the complex case. Let E_0 be the real vector space underlying the complex vector space E and let M_0 be the set M considered as subspace of E_0. If we set $f_1(x) = \mathcal{R}e\, f(x)$ for each $x \in M$, then $f_1(x)$ is a real linear form on M_0, and by Eq. (2) of Chapter 1, §6 we have

$$f(x) = f_1(x) - i f_1(ix).$$

Furthermore, we obviously have $|f_1(x)| \leqq q(x)$ for $x \in M_0$. By the first part of the proof there exists a real linear form g_1 on E_0 such that $g_1(x) = f_1(x)$ on M_0 and $|g_1(x)| \leqq q(x)$ for $x \in E_0$. Set

$$g(x) = g_1(x) - i g_1(ix)$$

for $x \in E$. Clearly, $g(x) = f(x)$ on M and g is a linear form on E since $g(ix) = g_1(ix) - i g_1(-x) = i g(x)$. Finally, for $x \in E$ set $g(x) = \rho e^{i\theta}$, where $\rho \geqq 0$. Then

$$|g(x)| = \rho = e^{-i\theta} g(x) = g(e^{-i\theta}x) = g_1(e^{-i\theta}x) \leqq q(e^{-i\theta}x) = q(x). \blacksquare$$

We now draw some conclusions from the above theorems.

PROPOSITION 1. *Let E be a locally convex space, M a linear subspace of E, and f a continuous linear form defined on M. Then there exists a continuous linear form g defined on E such that $g(x) = f(x)$ for all $x \in M$.*

Proof. There exists a continuous semi-norm q on E such that $|f(x)| \leqq q(x)$ for all $x \in M$ (cf. Proposition 2.5.2). By Theorem 1 there exists a linear form g on E such that $g(x) = f(x)$ for $x \in M$ and $|g(x)| \leqq q(x)$ for all $x \in E$. But then g is continuous on E. $\blacksquare$

PROPOSITION 2. *Let E be a locally convex space, M a closed subspace of E, and $z \in E$ a vector which does not belong to M. Then there exists a continuous linear form f on E such that $f(z) = 1$ and $f(x) = 0$ for $x \in M$.*

Proof. Let φ be the canonical surjection of E onto the quotient space E/M and set $\dot{z} = \varphi(z)$. Since $z \notin M$, we have $\dot{z} \neq 0$; and since E/M is a Hausdorff space (Proposition 2.5.5), there exists a continuous semi-norm $\dot{q}$ on E/M such that $\dot{q}(\dot{z}) \neq 0$ (Proposition 2.4.8). Define the linear form

g on the line $\{\lambda\dot{z} \mid \lambda \in \mathbf{K}\}$ by setting $g(\lambda\dot{z}) = \lambda$. We have

$$|g(\lambda\dot{z})| = |\lambda| = \frac{\dot{q}(\lambda\dot{z})}{\dot{q}(\dot{z})}.$$

By Theorem 1 there exists a linear form h defined on E/M which satisfies

$$|h(\dot{x})| \leqq \frac{1}{\dot{q}(\dot{z})}\,\dot{q}(\dot{x})$$

for all $\dot{x} \in E/M$ and for which $h(\dot{z}) = 1$. The linear form $f = h \circ \varphi$ defined on E satisfies all the requirements. ∎

From this proposition, which is a straightforward generalization of Proposition 1.6.1, we can deduce conclusions generalizing Propositions 2, 3, and 4 of Chapter 1, §6. This will be left as an exercise.

PROPOSITION 3. *Let E be a locally convex Hausdorff space and M a finite-dimensional linear subspace of E. Then there exists a closed subspace N of E which is a topological supplement* (Definition 2.7.2) *of M.*

Proof. We know that M has a topological supplement if there exists a continuous projector p from E onto M. Since M is a Hausdorff space, it is isomorphic to $\mathbf{K}^n$ (Theorem 2.10.1). Let $(e_i)_{1 \leqq i \leqq n}$ be a basis of M and consider the continuous linear form f_i $(1 \leqq i \leqq n)$ defined on M by

$$f_i\left(\sum_{k=1}^{n} \xi_k e_k\right) = \xi_i.$$

By Proposition 1 there exists a continuous linear form g_i $(1 \leqq i \leqq n)$ defined on E which coincides with f_i on M. Then p can be defined by

$$p(x) = \sum_{i=1}^{n} g_i(x)e_i$$

for $x \in E$. Clearly, p is a continuous projector from E onto M and $N = \operatorname{Ker}(p)$ is a closed topological supplement of M. ∎

PROPOSITION 4. *In a locally convex space every closed linear manifold M is the intersection of the closed affine hyperplanes which contain it.*

Proof. If $x \notin M$ there exists a convex open neighborhood V of x which does not meet M. Hence by Theorem 2 there exists a closed hyperplane H containing M which does not meet V. *A fortiori* H does not contain x. ∎

Let E be a real topological vector space, H a closed affine hyperplane with equation $f(x) = \alpha$ in E, and V an open set contained in the closed half-space $f(x) \leqq \alpha$. Then V is contained in the open half-space $f(x) < \alpha$. Indeed, suppose that $V \cap H \neq \emptyset$. Then $\alpha \in f(V)$. On the other hand, f is a surjective strict morphism of E onto $\mathbf{R}^1$ (Exercise 2.5.6). Hence

$f(V)$ is an open set in $\mathbf{R}$ (Theorem 2.5.1). Thus V contains points with $f(x) > \alpha$, which is a contradiction.

PROPOSITION 5. *Let E be a locally convex space over $\mathbf{R}$, A a closed convex nonempty set in E, and a a point which does not belong to A. Then there exists a continuous linear form f defined on E and a real number α such that $f(x) > \alpha$ for all $x \in A$ and $f(a) < \alpha$.*

Proof. There exists an open convex neighborhood V of a which does not meet A. The set $A - V = \bigcup_{x \in A} (x - V)$ is open, convex, nonempty and does not contain the origin. By Theorem 2 there exists a closed hyperplane H which does not meet $A - V$, and we can find an equation $f(x) = 0$ of H such that $f(z) \geqq 0$ for all $z \in A - V$, i.e., $f(x) \geqq f(v)$ for $x \in A$, $v \in V$. Set $\beta = \inf_{x \in A} f(x)$; then by the preceding remark V is contained in the open half-space $f(v) < \beta$, and in particular $f(a) < \beta$. The number $\alpha = \frac{1}{2}(f(a) + \beta)$ will satisfy the required conditions. ∎

COROLLARY. *In a real locally convex space every closed convex set A is the intersection of the closed half-spaces which contain it.*

Proof. If $x \notin A$, there exists a closed affine hyperplane H such that x and A belong to different open half-spaces determined by H. ∎

EXERCISES

1. Prove Theorem 1 by the method of Chapter 1, §6. (*Hint:* Prove successively the following statements:

(a) Consider on E the locally convex topology defined by q. Prove that M can be supposed to be closed, since otherwise f can be extended to $\overline{M}$ by continuity (Proposition 2.9.5).

(b) If E is a topological vector space, M a closed subspace of E, and $a \notin M$, then every x of the closed subspace generated by M and a can be written uniquely in the form $x = y + \lambda a$, $y \in M$, $\lambda \in \mathbf{K}$.

(c) If E is a real vector space, q a semi-norm, M a closed subspace (for the topology defined by q), f a linear form defined on M which satisfies

$$|f(x)| \leqq q(x)$$

for every $x \in M$, and $a \notin M$, then there exists a linear form g defined on the closed subspace N generated by M and a such that $g(x) = f(x)$ for all $x \in M$ and $|g(x)| \leqq q(x)$ for all $x \in N$. Show that

$$\xi = \sup_{y \in M} \big(-q(y + a) - f(y)\big) \leq \Xi = \inf_{y \in M} \big(q(y + a) - f(y)\big).$$

Then choose μ such that $\xi \leqq \mu \leqq \Xi$ and prove that the linear form g defined by $g(y + \lambda a) = f(y) + \lambda\mu$ $(y \in M, \lambda \in \mathbf{R})$ satisfies the requirements.

(d) Prove Theorem 1 for the case of real scalars using Zorn's lemma.

(e) Extend to the case of complex scalars.)

2. A function $q: E \to \mathbf{R}$ defined on a vector space E is said to be positively homogeneous if $q(\lambda x) = \lambda q(x)$ for all $\lambda \geqq 0$, $x \in E$. It is said to be subadditive if $q(x + y) \leqq q(x) + q(y)$ for all $x, y \in E$.

(a) Prove the following generalization of Theorem 1: Let E be a vector space over $\mathbf{R}$; let q be a positively homogeneous, subadditive function on E. Let f be a linear form defined on a subspace M of E which satisfies $f(x) \leqq q(x)$ for all $x \in M$. Then there exists a linear form g on E such that $g(x) = f(x)$ for all $x \in M$ and $g(x) \leqq q(x)$ for all $x \in E$. (*Hint:* Use the method of Exercise 1.)

(b) Deduce Theorem 2 from (a). (*Hint:* Suppose that $0 \in A$ and let q be the gauge of A. Set $M = x_0 + F$, where F is a linear subspace of E. Then $q(x_0 + y) \geqq 1$ for $y \in F$. Define $f(\lambda x_0 + y) = \lambda$, $\lambda \in \mathbf{R}$, $y \in F$ and prove that $f(\lambda x_0 + y) \leqq q(\lambda x_0 + y)$. By (a) there exists g such that $g(x) = f(x)$ for $x \in M$ and $g(x) \leqq q(x)$. Then $A \subset \{x \mid g(x) < 1\}$ and

$$M \subset H = \{x \mid g(x) = 1\}.)$$

3. Deduce Proposition 1 directly from Theorem 2. (*Hint:* Suppose that M is closed and $f \neq 0$. Then $N = \{x \mid f(x) = 1\}$ is a closed linear manifold in E and $0 \notin N$. If H is a closed affine hyperplane of E such that $N \subset H$, $0 \notin H$ (cf. Proposition 4), and H has equation $g(x) = 1$, then g is the required linear form (cf. the proof of Theorem 1).)

4. Using Proposition 2, generalize Propositions 2, 3, and 4 of Chapter 1, §6 to locally convex spaces.

5. Let E be a complex vector space and E_0 the underlying real vector space.

(a) Let H be a hyperplane of E with equation $f(x) = 0$ and let $f_1(x) = \mathcal{R}e\, f(x)$. Prove that $f_1(x) = 0$ defines a hyperplane H_0 of E_0 and that $H = H_0 \cap iH_0$.

(b) Prove that if H is an affine hyperplane of E with equation $f(x) = \alpha + i\beta$ ($\alpha, \beta \in \mathbf{R}$), then $f_1(x) = \alpha$ and $f_1(ix) = -\beta$ define two affine hyperplanes H_1 and H_2 of E_0 such that $H = H_1 \cap H_2$.

(c) Let H_0 be a hyperplane of E_0 with equation $f_1(x) = 0$. Prove that $H = H_0 \cap iH_0$ is a hyperplane of E with equation $f(x) = f_1(x) - if_1(ix) = 0$.

(d) Prove that a linear subspace M of E_0 is also a linear subspace of E if and only if $M = iM$.

§2. *Pairings*

DEFINITION 1. *Let F and G be two vector spaces over the same field $\mathbf{K}$. If a bilinear form $(x, y) \mapsto B(x, y)$ is given on $F \times G$, then we say that the spaces F and G are paired or form a pairing with respect to the bilinear form B. The pairing or the bilinear form separates points of F if for $x \in F$, $x \neq 0$, there exists $y \in G$ such that $B(x, y) \neq 0$. Similarly, the pairing or the bilinear form separates points of G if for $y \in G$, $y \neq 0$, there exists $x \in F$ such that $B(x, y) \neq 0$. If the pairing separates points of both F and G, then we say that it is separated or that (F, G) is a dual system with respect to the bilinear form B.*

If F and G form a dual system with respect to the bilinear form $(x, y) \mapsto B(x, y)$, then the underlying real vector spaces F_0 and G_0 form a dual system with respect to the bilinear form $(x, y) \mapsto B_1(x, y)$, where $B_1(x, y) = \mathcal{R}e(B(x, y))$. We see exactly as in the proof of formula (2) of Chapter 1, §6 that $B(x, y) = B_1(x, y) - iB_1(ix, y)$.

Example 1. Let E be a vector space and let E^* be the set of all linear forms on E. We shall denote by the letters $x, y, \ldots$ the elements of E and by the primed letters $x', y', \ldots$ the elements of E^*. If we define addition and multiplication by a scalar on E^* by the formulas

$$(x' + y')(x) = x'(x) + y'(x), \qquad (\lambda x')(x) = x'(\lambda x), \tag{1}$$

then it is trivial to verify that Axioms (VS 1) through (VS 8) of a vector space are satisfied and in particular the zero vector of E^* is the map which assigns 0 to each vector $x \in E$. The vector space E^* will be referred to as the *algebraic dual* of E. If we set $\langle x, x' \rangle = \langle x', x \rangle = x'(x)$ for $x \in E$, $x' \in E^*$, then for each $x' \in E^*$ the map $x \mapsto \langle x, x' \rangle$ is linear since x' is a linear form on E and for each $x \in E$ the map $x' \mapsto \langle x, x' \rangle$ is linear by the definitions (1) of the operations on E^*. The map $(x, x') \mapsto \langle x, x' \rangle$ is called the *canonical bilinear form*. It follows from the definition of the zero vector in E^* that the canonical bilinear form separates points of E^*. But it also separates points of E. Indeed, let x be a nonzero element of E and let $(e_\iota)_{\iota \in I}$ be an algebraic basis of E. Then

$$x = \sum_{\iota \in I} \xi_\iota e_\iota,$$

where $\xi_\kappa \neq 0$ for some index κ. Let $x' \in E^*$ be the linear form on E which associates with each vector $y = \sum_{\iota \in I} \eta_\iota e_\iota$ its κ-th coordinate η_κ. Then $\langle x, x' \rangle = \xi_\kappa \neq 0$.

Example 2. Our second example will be of fundamental importance in the remainder of this book and is in fact one of the most important notions of mathematics.

Let E be a topological vector space and denote by E' the set of all *continuous* linear forms on E. Then E' is a linear subspace of the algebraic dual E^*. Indeed, let $x', y' \in E'$. Given $\epsilon > 0$, there exist neighborhoods U and V of 0 in E such that $|\langle x, x' \rangle| < \frac{1}{2}\epsilon$ for $x \in U$ and $|\langle x, y' \rangle| < \frac{1}{2}\epsilon$ for $x \in V$. But then $|\langle x, x' + y' \rangle| < \epsilon$ for $x \in U \cap V$, and thus by Proposition 2.5.1 we have $x' + y' \in E'$. Similarly, let $\lambda \in \mathbf{K}$, $\lambda \neq 0$, and $x' \in E'$. Given $\epsilon > 0$, there exists a neighborhood U of 0 in E such that $|\langle x, x' \rangle| < \epsilon/|\lambda|$ for $x \in U$. But then

$$|\langle x, \lambda x' \rangle| = |\langle \lambda x, x' \rangle| < \epsilon$$

for $x \in U$; i.e., $\lambda x' \in E'$. The vector space E' will be called the *dual* (adjoint, conjugate, or topological dual) space of E. In the particular case when E is a normed space we have introduced the dual space E' in Chapter 1, §7.

We also denote by $(x, x') \mapsto \langle x, x' \rangle$ the restriction to $E \times E'$ of the canonical bilinear form defined on $E \times E^*$ in Example 1. It clearly separates points of E'. If E is locally convex and Hausdorff, then it also separates points of E by virtue of Proposition 1.2 applied to the subspace $M = \{0\}$, which is closed because E is a Hausdorff space.

In the sequel the dual systems (E, E^*) and (E, E') will always be considered with respect to the canonical bilinear form.

Let F and G form a pairing with respect to the bilinear form

$$(x, y) \mapsto B(x, y).$$

For every $y \in G$ the map $x \mapsto B(x, y)$ is a linear form on F, i.e., an element y^* of the algebraic dual F^* of F, and the map $\Psi\colon y \mapsto y^*$ from G into F^* is clearly linear. If B separates points of G, then this map is injective; hence in this case we can consider G as the linear subspace $\operatorname{Im}(\Psi)$ of F^*, and if we do so we shall say that we identify G canonically with a linear subspace of F^*. If this identification is made, then the bilinear form $(x, y) \mapsto B(x, y)$ is the restriction to $F \times G$ of the canonical bilinear form $(x, y) \mapsto \langle x, y \rangle$ on $F \times F^*$, and therefore in such a situation we shall usually write $\langle x, y \rangle$ instead of $B(x, y)$. Similarly, if the pairing separates points of F, then the map $\Phi\colon x \mapsto x^*$ from F into G^* is injective and F can be identified canonically with the linear subspace $\operatorname{Im}(\Phi)$ of the algebraic dual G^* of G.

Given a pairing of F and G with respect to the bilinear form $(x, y) \mapsto B(x, y)$, we can define a locally convex topology $\sigma(F, G)$ on F and a locally convex topology $\sigma(G, F)$ on G. These are called the *weak topologies* defined by the pairing of F and G. In the special case when F is a normed space E, and G its dual E', the weak topologies were already introduced in Example 2.4.20. Since the definitions of the two topologies are perfectly symmetric, we shall consider only $\sigma(F, G)$.

Let $y \in G$, then the map $x \mapsto |B(x, y)|$ is clearly a semi-norm on F. We denote this semi-norm by q_y, and $\sigma(F, G)$ will be the topology on F defined by the family of all semi-norms q_y as y runs through G. A fundamental system of neighborhoods of 0 in F for the topology $\sigma(F, G)$ is given by the sets

$$U_{y_1, \ldots, y_n, \epsilon} = \{x \mid |B(x, y_k)| \leqq \epsilon\},$$

where $\epsilon > 0$ and $(y_k)_{1 \leqq k \leqq n}$ is a finite family of elements of G.

Proposition 1. *Let F and G be two vector spaces paired with respect to the bilinear form $(x, y) \mapsto B(x, y)$. The pairing separates points of F if and only if $\sigma(F, G)$ is a Hausdorff topology on F.*

Proof. Assume that the pairing separates points of F and let $x \in F$, $x \neq 0$. There exists $y \in G$ such that $q_y(x) = |B(x, y)| \neq 0$; hence by Proposition 2.4.8 the topology $\sigma(F, G)$ is Hausdorff.

Conversely, if $\sigma(F, G)$ is a Hausdorff topology and $x \in F$, $x \neq 0$, then again by Proposition 2.4.8 there exists a semi-norm q_y such that $q_y(x) \neq 0$, i.e., $B(x, y) \neq 0$. ∎

Let us observe that if B separates points of F and we identify F canonically with a linear subspace of G^*, then $\sigma(G^*, G)$ induces on F the topology $\sigma(F, G)$.

We know that each element $y \in G$ defines a linear form $y^* = \Psi(y) \in F^*$ on F according to the formula $y^*(x) = B(x, y)$. Furthermore, each such y^* is continuous for the topology $\sigma(F, G)$ by virtue of the relation

$$|y^*(x)| = |B(x, y)| = q_y(x)$$

and Proposition 2.5.2. It is, however, also true that conversely every linear form on F which is continuous for $\sigma(F, G)$ is of the form $x \mapsto B(x, y)$ for some $y \in G$. To prove this, we shall need the following result from algebra.

LEMMA. *Let E be a vector space and $u_1, u_2, \ldots, u_n$ a finite family of linear forms on E. If u is a linear form on E such that*

$$\operatorname{Ker}(u) \supset \bigcap_{k=1}^{n} \operatorname{Ker}(u_k),$$

then there exists a family of scalars $(\lambda_k)_{1 \leq k \leq n}$ such that

$$u = \sum_{k=1}^{n} \lambda_k u_k.$$

Proof. We may obviously assume that none of the forms u_i is zero. Then each $\operatorname{Ker}(u_i)$ is a hyperplane H_i. For $n = 1$ the lemma follows from the considerations of Chapter 1, §5. Indeed, if $u = 0$, then $u = 0 \cdot u_1$; and if $u \neq 0$, then necessarily $\operatorname{Ker}(u) = \operatorname{Ker}(u_1)$; i.e., $u = \lambda u_1$.

Assume that the lemma holds for $n - 1$. The kernel of the restriction of u to H_n contains the intersection of the kernels of the restrictions of the u_i to H_n. Hence there exist scalars $\lambda_1, \ldots, \lambda_{n-1}$ such that

$$u(x) - \sum_{k=1}^{n-1} \lambda_k u_k(x) = 0$$

for all $x \in H_n$. But then, by the result used already at the beginning of the proof, there exists a scalar λ_n such that

$$u - \sum_{k=1}^{n-1} \lambda_k u_k = \lambda_n u_n. \blacksquare$$

Now let f be a linear form on F which is continuous for the topology $\sigma(F, G)$. By Proposition 2.5.2 there exists a finite family $(y_k)_{1 \leqq k \leqq n}$ of elements of G such that

$$|f(x)| \leqq \max_{1 \leqq k \leqq n} |B(x, y_k)|$$

for all $x \in F$. In particular, $f(x) = 0$ for all $x \in F$ such that $B(x, y_k) = 0$, $1 \leqq k \leqq n$. It follows from the lemma that

$$f = \sum_{k=1}^{n} \lambda_k y_k^*$$

for some family $(\lambda_k)_{1 \leqq k \leqq n}$ of scalars, where $y_k^* = \Psi(y_k)$. If we set

$$y = \sum_{k=1}^{n} \lambda_k y_k,$$

then $f(x) = B(x, y)$ for all $x \in F$. We have thus proved the following result:

PROPOSITION 2. *Let the vector spaces F and G form a pairing with respect to the bilinear form $(x, y) \mapsto B(x, y)$ and equip F with the topology $\sigma(F, G)$. For every $y \in G$ the map $x \mapsto B(x, y)$ is a continuous linear form on F, and conversely, if f is a continuous linear form on F, then there exists $y \in G$ such that $f(x) = B(x, y)$ for every $x \in F$.*

In other words, the dual of F equipped with the topology $\sigma(F, G)$ can be identified with $G/\mathrm{Ker}(\Psi)$. If B separates points of G, then Ψ is injective and the dual of F can be identified with G.

Let us observe that $\sigma(F, G)$ is the coarsest topology on F for which all the maps $x \mapsto B(x, y)$ are continuous as y runs through G. Indeed, if for some topology $\mathcal{T}$ all the maps $x \mapsto B(x, y)$ are continuous, then given $x_0 \in F$, $\epsilon > 0$, and a finite family $(y_k)_{1 \leqq k \leqq n}$ of elements of G, there exists a neighborhood V of x_0 for $\mathcal{T}$ such that $x \in V$ implies $|B(x - x_0, y_k)| < \epsilon$ for $1 \leqq k \leqq n$. In other words, $V \subset x_0 + U_{y_1, \ldots, y_n, \epsilon}$ and thus $\mathcal{T}$ is finer than $\sigma(F, G)$. In particular, if B separates points of G, then $\sigma(F, G)$ is the coarsest locally convex topology on F for which G is the dual of F. We shall determine in §5 all the locally convex topologies on F for which G is the dual of F (cf. Definition 4.1, Example 4.2, and Proposition 5.4).

PROPOSITION 3. *Let F and G be two vector spaces which form a pairing separating points of F. Identify F canonically with a linear subspace of G^* and equip G^* with the topology $\sigma(G^*, G)$. The pairing separates points of G if and only if F is dense in G^*.*

Proof. Assume that the pairing separates points of G. Let f be a continuous linear form on G^* such that $f(x) = 0$ for all $x \in F$, and thus also $f(x) = 0$ for all x belonging to the closure $\overline{F}$ of F. By Proposition 2 there exists $y \in G$ such that $f(x) = \langle x, y\rangle$ for all $x \in G^*$, and in particular $\langle x, y\rangle = 0$ for all $x \in F$. By hypothesis we have $y = 0$, that is, $f = 0$, and it follows from Proposition 1.2 that $\overline{F} = G^*$.

Conversely, assume that $\overline{F} = G^*$. Let $y \in G$ be such that $\langle x, y\rangle = 0$ for all $x \in F$. Since $x \mapsto \langle x, y\rangle$ is a continuous linear form on G^*, we also have $\langle x, y\rangle = 0$ for all $x \in G^*$. It follows from Example 1 that $y = 0$; i.e., the pairing separates points of G. ∎

PROPOSITION 4. *Let F and G be two vector spaces paired with respect to the bilinear form B which separates points of G and let N be a subspace of G, distinct from G. Then the topology $\sigma(F, G)$ on F is strictly finer than the topology $\sigma(F, N)$.*

Proof. If $y \in G$ but $y \notin N$, then the linear form $x \mapsto B(x, y)$ on F is continuous for $\sigma(F, G)$. On the other hand, it is not continuous for $\sigma(F, N)$ since otherwise by Proposition 2 there would exist an element $z \in N$ such that $B(x, y) = B(x, z)$, i.e., $B(x, y - z) = 0$ for all $x \in F$, which by hypothesis would imply $y = z \in N$. ∎

EXAMPLE 3. Let E be a vector space over the field $\mathbf{K}$ and let $(e_\iota)_{\iota \in I}$ be an algebraic basis of E. The map

$$\sum_{\iota \in I} \xi_\iota e_\iota \mapsto (\xi_\iota)_{\iota \in I}$$

is clearly an isomorphism from E onto the external direct sum $\mathbf{K}^{(I)}$. The algebraic dual E^* of E is then isomorphic to the product $\mathbf{K}^I$. Indeed, the map which assigns to $(\xi'_\iota)_{\iota \in I}$ the linear map

$$\sum_{\iota \in I} \xi_\iota e_\iota \mapsto \sum_{\iota \in I} \xi_\iota \xi'_\iota$$

is clearly an injective linear map from $\mathbf{K}^I$ into E^*. Furthermore, it is also surjective since if $x' \in E^*$, then we set $\xi'_\iota = \langle e_\iota, x'\rangle$ and obtain

$$\langle x', \sum_{\iota \in I} \xi_\iota e_\iota\rangle = \sum_{\iota \in I} \xi_\iota \langle x', e_\iota\rangle = \sum_{\iota \in I} \xi_\iota \xi'_\iota;$$

that is, x' is the image of $(\xi'_\iota)_{\iota \in I}$ under the above map.

Let us identify E with $\mathbf{K}^{(I)}$ and E^* with $\mathbf{K}^I$. The topology $\sigma(E, E^*)$ on E is coarser than the locally convex direct sum topology $\mathcal{T}$ on $E = \mathbf{K}^{(I)}$ (Example 2.12.2). It is enough to show that each set

$$U_{x',\epsilon} = \{x \mid |\langle x, x'\rangle| < \epsilon\}$$

contains a neighborhood of 0 for $\mathcal{T}$. Let $x' = (\xi'_\iota)_{\iota \in I}$; then the set

$$V = \left\{x \mid x = \sum_{\iota \in I} \lambda_\iota \xi_\iota e_\iota, \sum_{\iota \in I} |\lambda_\iota| \leq 1, |\xi_\iota \xi'_\iota| < \epsilon\right\}$$

forms a neighborhood of 0 for $\mathcal{T}$ in E since for each $\iota \in I$ the set

$$\{\xi \mid |\xi\xi'_\iota| < \epsilon\}$$

is a neighborhood of 0 in $\mathbf{K}$. Clearly $V \subset U_{x',\epsilon}$. If I is infinite, then it can be shown that $\sigma(E, E^*)$ is strictly coarser than $\mathcal{T}$ (cf. Exercise 1).

On the other hand, the topology $\sigma(E^*, E)$ coincides with the product topology $\mathcal{T}^*$ on $E^* = \mathbf{K}^I$. Indeed, for every $x \in E$ the linear map $x' \mapsto \langle x, x'\rangle$ is continuous on E^* for $\mathcal{T}^*$. For, let $x = \sum_{\iota\in I} \xi_\iota e_\iota$ and $\xi_\iota = 0$ for all ι which do not belong to the finite subset H of I having n elements. Given $\epsilon > 0$, the set

$$V = \left\{(\xi'_\iota)_{\iota\in I} \mid |\xi'_\iota| < \frac{\epsilon}{n|\xi_\iota|}, \quad \iota \in H\right\}$$

is a neighborhood of 0 in E^* for $\mathcal{T}^*$ and $x' \in V$ implies

$$|\langle x, x'\rangle| \leqq \sum_{\iota\in I} |\xi_\iota\xi'_\iota| < \epsilon.$$

Thus $\sigma(E^*, E)$ is coarser than $\mathcal{T}^*$. Conversely, if

$$V = \{(\xi'_\iota)_{\iota\in I} \mid |\xi'_{\iota_k}| < \epsilon_k, \quad 1 \leqq k \leqq n\}$$

is a neighborhood of 0 in E^* for $\mathcal{T}^*$, then

$$U_{e_{\iota_1},\ldots,e_{\iota_k},\epsilon} \subset V,$$

where $\epsilon = \min_{1\leqq k\leqq n} \epsilon_k$. Hence $\sigma(E^*, E)$ is also finer than $\mathcal{T}^*$.

It follows that a set in E^* is relatively compact for the topology $\sigma(E^*, E)$ if and only if it is bounded, i.e., the analogue of the Bolzano-Weierstrass theorem holds. By Proposition 2.10.7 any relatively compact set is bounded. Conversely, let B be a bounded set in $E^* = \mathbf{K}^I$. Then the projections $B_\iota = \pi_\iota(B)$ are bounded in $\mathbf{K}$, hence relatively compact by the Bolzano-Weierstrass theorem. Thus B is contained in $\prod_{\iota\in I} \overline{B}_\iota$, which is compact by Tihonov's theorem; i.e., B is relatively compact.

Finally, E^* is complete for the topology $\sigma(E^*, E)$ as the product of complete spaces. We have the even stronger result:

PROPOSITION 5. *Let F and G be two vector spaces which form a pairing separating points of F. Identify F canonically with a linear subspace of G^*, equip G^* with the topology $\sigma(G^*, G)$ and F with the induced topology $\sigma(F, G)$.*

(a) *The completion of F is its closure in G^*.*

(b) *If the pairing also separates points of G, then G^* is the completion of F.*

Proof. (a) We have just seen that G^* is complete. Hence the closure of F in G^* is also complete (Proposition 2.9.3).

(b) follows from (a) and Proposition 3. ▮

Exercises

1. Let E be a vector space with algebraic basis $(e_\iota)_{\iota \in I}$ over the field $\mathbf{K}$ and let E^* be the algebraic dual of E. As in Example 3, identify E with $\mathbf{K}^{(I)}$.

(a) Show that the locally convex direct sum topology $\mathcal{T}$ on E is the finest locally convex topology on E. (*Hint:* See Example 2.12.4.)

(b) Show that if I is infinite, then the topology $\sigma(E, E^*)$ is strictly coarser than $\mathcal{T}$. (*Hint:* The $\mathcal{T}$-neighborhood

$$V = \{x \mid x = \sum_{\iota \in I} \xi_\iota e_\iota, |\xi_\iota| \leqq 1, \iota \in I\}$$

of 0 contains no subspace of E, but every $\sigma(E, E^*)$-neighborhood of 0 contains a subspace of finite codimension.)

2. Let (F, G) be a dual system. Show that every subset of F which is bounded for the topology $\sigma(F, G)$ is precompact (Definition 2.10.1) for this topology. (*Hint:* Use Proposition 3 and Example 3.)

3. Let E be a locally convex Hausdorff space and $\hat{E}$ its completion. Show that the dual of E can be identified with the dual of $\hat{E}$. (*Hint:* Use Proposition 2.9.5.)

4. Show that if E is a topological vector space and the pairing (E, E') separates points of E, then E is a Hausdorff space.

§3. *Polarity*

Definition 1. *Let F and G be two vector spaces over the same field $\mathbf{K}$, paired with respect to the bilinear form $(x, y) \mapsto B(x, y)$. If A is a subset of F, then the polar of A is the subset A° of G, formed by those elements $y \in G$ which satisfy $|B(x, y)| \leqq 1$ for all $x \in A$.*

We define in a similar way the polar of a subset of G.

Remark 1. In the literature ([9], Chapter IV, §1, no. 3, Definition 2–3; [52], §20, 8) A° is often called the *absolute polar* of A, while the term polar is used for the set of those $y \in G$ which satisfy $\mathcal{R}e\, B(x, y) \leqq 1$ for all $x \in A$. If A is balanced, the two coincide (see Exercise 2).

We shall list the trivial properties of the polar in the following:

Proposition 1. *Suppose that the vector spaces F and G are paired with respect to the bilinear form $(x, y) \mapsto B(x, y)$.*

(a) *If $A_1 \subset A_2 \subset F$, then $A_1^\circ \supset A_2^\circ$.*

(b) *If $A \subset F$ and D is the balanced hull of A, then $A^\circ = D^\circ$.*

(c) $A \subset A^{\circ\circ} = (A^\circ)^\circ$.

(d) $A^\circ = A^{\circ\circ\circ}$.

(e) *If $A \subset F$, then A° is a balanced, convex set in G closed for $\sigma(G, F)$.*

(f) $(\lambda A)^{\circ} = (1/\lambda)A^{\circ}$ *for* $\lambda \in \mathbf{K}$, $\lambda \neq 0$. *In particular,* A° *is absorbing if and only if* A *is bounded for* $\sigma(F, G)$.

(g) *If* $(A_\iota)_{\iota \in I}$ *is a family of subsets of* F, *then*

$$\Big(\bigcup_{\iota \in I} A_\iota\Big)^{\circ} = \bigcap_{\iota \in I} A_\iota^{\circ}.$$

Proof. (a) If $y \in A_2^{\circ}$, then $|B(x, y)| \leqq 1$ for all $x \in A_2$ and *a fortiori* for all $x \in A_1$; hence $y \in A_1^{\circ}$.

(b) Since $A \subset D$, we have $A^{\circ} \supset D^{\circ}$ by (a). Conversely, let $y \in A^{\circ}$. Then $|B(\lambda x, y)| \leqq 1$ for all $x \in A$ and $|\lambda| \leqq 1$. Since

$$D = \{\lambda x \mid x \in A, |\lambda| \leqq 1\}$$

(Exercise 2.3.3(a)), we have $y \in D^{\circ}$, that is, $A^{\circ} \subset D^{\circ}$.

(c) If $x \in A$, then $|B(x, y)| \leqq 1$ for all $y \in A^{\circ}$, that is, $x \in A^{\circ\circ}$.

(d) From (c) we obtain $A \subset A^{\circ\circ}$; hence by (a) $A^{\circ} \supset A^{\circ\circ\circ}$. On the other hand, from (c) we get $A^{\circ} \subset (A^{\circ})^{\circ\circ} = A^{\circ\circ\circ}$.

(e) Let $y \in A^{\circ}$, $z \in A^{\circ}$, λ and μ two scalars such that $|\lambda| + |\mu| \leqq 1$. Then for any $x \in A$ we have

$$\begin{aligned} |B(x, \lambda y + \mu z)| &\leqq |\lambda| \cdot |B(x, y)| + |\mu| \cdot |B(x, z)| \\ &\leqq |\lambda| + |\mu| \leqq 1; \end{aligned}$$

that is, $\lambda y + \mu z \in A^{\circ}$ and thus A° is balanced and convex.

For each $x \in F$ the map $y \mapsto B(x, y)$ is continuous on G for $\sigma(G, F)$ (Proposition 2.2). Hence the set $A_x = \{y \mid |B(x, y)| \leqq 1\}$ is closed in G as the inverse image of the closed set $\{\xi \mid |\xi| \leqq 1\}$ in $\mathbf{K}$. Thus

$$A^{\circ} = \bigcap_{x \in A} A_x$$

is closed for $\sigma(G, F)$.

(f) $y \in (\lambda A)^{\circ}$ is equivalent to $|B(\lambda x, y)| = |B(x, \lambda y)| \leqq 1$ for all $x \in A$, i.e., to $\lambda y \in A^{\circ}$.

Suppose that A° is absorbing and let $y \in G$. There exists $\mu > 0$ such that $\mu y \in A^{\circ}$, that is, $y \in (\mu A)^{\circ}$. Thus

$$q_y(x) = |B(x, y)| \leqq \frac{1}{\mu}$$

for all $x \in A$; i.e., A is $\sigma(F, G)$-bounded. Conversely, if A is $\sigma(F, G)$-bounded, then for every $y \in G$ there exists a $\mu > 0$ such that

$$|B(x, y)| \leqq \frac{1}{\mu}$$

for all $x \in A$. Hence $\mu y \in A^\circ$, and so A° is absorbing.

(g) $y \in (\bigcup_{\iota \in I} A_\iota)^\circ \Leftrightarrow |B(x, y)| \leqq 1$, for all $x \in \bigcup_{\iota \in I} A_\iota$

$\Leftrightarrow |B(x, y)| \leqq 1$ for all $x \in A_\iota$ and $\iota \in I$

$\Leftrightarrow y \in \bigcap_{\iota \in I} A_\iota^\circ$. ∎

Our next result on polars is much deeper since it uses the Hahn-Banach theorem.

THEOREM 1 (Theorem of the bipolars). *Suppose that the vector spaces F and G form a pairing. If A is a nonempty subset of F, then $A^{\circ\circ}$ is the balanced, convex, $\sigma(F, G)$-closed hull of A, i.e., the smallest balanced, convex set containing A which is closed for the topology $\sigma(F, G)$.*

Proof. By Proposition 1(e) the set $A^{\circ\circ}$ is balanced, convex, $\sigma(F, G)$-closed, and by Proposition 1(c) it contains A. Thus we have only to show that if D is a balanced, convex, $\sigma(F, G)$-closed set containing A, then $A^{\circ\circ} \subset D$. Let $a \notin D$. By Proposition 1.5 there exists a continuous linear form f on the real vector space F_0 underlying F (taking, of course, $F_0 = F$ if F is real) and a real number α such that $f(x) < \alpha$ for $x \in D$ and $f(a) > \alpha$. Since $0 \in D$, we can choose $\alpha = 1$. In the case when F is a complex vector space then, as we saw in the proof of Theorem 1.1 (cf. Chapter 1, §6, (2)), the map $x \mapsto f(x) - if(ix)$ is a linear form on F, continuous for $\sigma(F, G)$. By Proposition 2.2 there exists $y \in G$ such that $f(x) = \mathcal{R}e\, B(x, y)$ for all $x \in F$. Since D is balanced, we have $|B(x, y)| \leqq 1$ for all $x \in D$; i.e., $y \in D^\circ$ and therefore by Proposition 1(b) $y \in A^\circ$. On the other hand, $|B(a, y)| \geqq f(a) > 1$, i.e., $a \notin A^{\circ\circ}$. ∎

PROPOSITION 2. *Let $(A_\iota)_{\iota \in I}$ be a family of balanced, convex, nonempty subsets of F which are closed for the topology $\sigma(F, G)$. Then the polar of $\bigcap_{\iota \in I} A_\iota$ is the balanced, convex, $\sigma(G, F)$-closed hull of the union of the sets A_ι°.*

Proof. By Theorem 1 we have $A_\iota = A_\iota^{\circ\circ}$ and therefore by Proposition 1(g)

$$\bigcap_{\iota \in I} A_\iota = \bigcap_{\iota \in I} A_\iota^{\circ\circ} = \Big(\bigcup_{\iota \in I} A_\iota^\circ\Big)^\circ;$$

hence

$$\Big(\bigcap_{\iota \in I} A_\iota\Big)^\circ = \Big(\bigcup_{\iota \in I} A_\iota^\circ\Big)^{\circ\circ}.$$

Again by Theorem 1 the right-hand side is the balanced, convex, $\sigma(G, F)$-closed hull of $\bigcup_{\iota \in I} A_\iota^\circ$. ∎

If M is a linear subspace of F, then $|B(x, y)| \leqq 1$ for all $x \in M$ is possible only if $B(x, y) = 0$ for all $x \in M$. Thus in this case M° is the linear subspace $M^\perp$ of G, orthogonal to M, which we considered briefly earlier in a special case (Chapter 1, §7). In general we state:

Definition 2. *Let F and G be two vector spaces over the same field $\mathbf{K}$, paired with respect to the bilinear form $(x, y) \mapsto B(x, y)$. If M is a subset of F, the elements $y \in G$ which are orthogonal to M, i.e., which satisfy $B(x, y) = 0$ for all $x \in M$, form a linear subspace of G, called the subspace orthogonal to M and denoted by $M^{\perp}$.*

We define in a similar way the linear subspace $N^{\perp}$ of F when N is a subset of G. If $(M_\iota)_{\iota \in I}$ is a family of subsets of F, let us denote by $\bigvee_{\iota \in I} M_\iota$ the $\sigma(F, G)$-closed subspace generated by $\bigcup_{\iota \in I} M_\iota$, i.e., the smallest linear subspace of F which contains all the sets M_ι and which is closed for the topology $\sigma(F, G)$ (cf. Exercise 1.7.2). Clearly, $\bigvee_{\iota \in I} M_\iota$ is the $\sigma(F, G)$-closure of the linear subspace of F formed by all the sums $\sum_{\iota \in I} \lambda_\iota x_\iota$, where $\lambda_\iota \in \mathbf{K}$, $x_\iota \in M_\iota$, and $\lambda_\iota = 0$ except for finitely many indices ι (Proposition 2.5.4).

Proposition 3. *Let F and G be two vector spaces paired with respect to the bilinear form $(x, y) \mapsto B(x, y)$.*

(a) *If $M_1 \subset M_2 \subset F$, then $M_1^{\perp} \supset M_2^{\perp}$.*

(b) $M \subset M^{\perp\perp} = (M^{\perp})^{\perp}$.

(c) $M^{\perp} = M^{\perp\perp\perp}$.

(d) *$M^{\perp}$ is closed for the topology $\sigma(G, F)$.*

(e) *$M^{\perp\perp}$ is the $\sigma(F, G)$-closed subspace of F generated by M.*

(f) *Let $(M_\iota)_{\iota \in I}$ be a family of subsets of F. Then*

$$\Big(\bigcup_{\iota \in I} M_\iota\Big)^{\perp} = \Big(\bigvee_{\iota \in I} M_\iota\Big)^{\perp} = \bigcap_{\iota \in I} M_\iota^{\perp}.$$

(g) *Let $(M_\iota)_{\iota \in I}$ be a family of $\sigma(F, G)$-closed subspaces of F. Then*

$$\Big(\bigcap_{\iota \in I} M_\iota\Big)^{\perp} = \bigvee_{\iota \in I} M_\iota^{\perp}.$$

Proof. (a) If $y \in M_2^{\perp}$, then $B(x, y) = 0$ for all $x \in M_2$ and *a fortiori* for all $x \in M_1$; hence $y \in M_1^{\perp}$.

(b) If $x \in M$, then $B(x, y) = 0$ for all $y \in M^{\perp}$; i.e., $x \in M^{\perp\perp}$.

(c) From (b) we obtain $M \subset M^{\perp\perp}$; hence by (a) $M^{\perp} \supset M^{\perp\perp\perp}$. On the other hand, from (b) we get $M^{\perp} \subset (M^{\perp})^{\perp\perp} = M^{\perp\perp\perp}$.

(d) For each $x \in F$ the map $y \mapsto B(x, y)$ is continuous on G for $\sigma(G, F)$; hence the set $M_x = \{y \mid B(x, y) = 0\}$ is closed as the inverse image of the closed set $\{0\}$ in $\mathbf{K}$. Thus $M^{\perp} = \bigcap_{x \in M} M_x$ is also closed for $\sigma(G, F)$.

(e) By (d) $M^{\perp\perp}$ is a $\sigma(F, G)$-closed subspace of F and by (b) $M \subset M^{\perp\perp}$. Let L be the $\sigma(F, G)$-closed subspace of F generated by M and assume that $L \neq M^{\perp\perp}$. Then by Propositions 1.2 and 2.2 there exists an element $y \in G$ such that $B(x, y) = 0$ for all $x \in L$, and in particular for all $x \in M$, but $B(a, y) = 1$ for some $a \in M^{\perp\perp}$. Thus we would have $y \in M^{\perp}$ but $y \notin M^{\perp\perp\perp}$ in contradiction to (c).

(f) Clearly

$$\Big(\bigcup_{\iota \in I} M_\iota\Big)^{\perp} = \bigcap_{\iota \in I} M_\iota^{\perp}.$$

On the other hand,

$$\Big(\bigcup_{\iota \in I} M_\iota\Big)^{\perp} = \Big(\bigcup_{\iota \in I} M_\iota\Big)^{\perp\perp\perp}$$

by (c) and

$$\Big(\bigcup_{\iota \in I} M_\iota\Big)^{\perp\perp} = \bigvee_{\iota \in I} M_\iota$$

by (e).

(g) By (e) we have $M_\iota = M_\iota^{\perp\perp}$ and therefore by (f)

$$\bigcap_{\iota \in I} M_\iota = \bigcap_{\iota \in I} M_\iota^{\perp\perp} = \Big(\bigvee_{\iota \in I} M_\iota^{\perp}\Big)^{\perp}.$$

Using (e) again, we obtain

$$\Big(\bigcap_{\iota \in I} M_\iota\Big)^{\perp} = \bigvee_{\iota \in I} M_\iota^{\perp}. \blacksquare$$

Parts (d), (e), and (f) of Proposition 3 show a striking similarity to Propositions 1.7.1, 1.7.2, and Exercise 1.7.2. Actually these earlier results are special cases of the proposition just proved, since it will soon turn out that in a normed space E the linear subspaces which are closed for the topology defined by the norm are exactly those which are closed for $\sigma(E, E')$ (Proposition 4.3), and that in E' every subspace which is closed for $\sigma(E', E)$ is also closed for the topology defined by the norm (Example 4.5). However, in Chapter 1 we were unable to prove the analogue of part (g) of Proposition 3 (cf. Exercise 1.7.3(c)). This is due to the fact that if E is a nonreflexive normed space, then there exist in E' subspaces which are closed for the norm topology but not for $\sigma(E', E)$. We shall return to these questions in greater detail in §8.

Exercises

1. Why do we have to assume in Theorem 1 that A is not empty?

2. Let F and G be two vector spaces over the same field $\mathbf{K}$, paired with respect to the bilinear form $(x, y) \mapsto B(x, y)$. If A is a subset of F, we denote by A^p the subset of G formed by those elements y which satisfy $\mathcal{R}e\, B(x, y) \leqq 1$ (or simply $B(x, y) \leqq 1$ if $\mathbf{K} = \mathbf{R}$) for all $x \in A$ (cf. Remark 1). We define similarly A^p for a set $A \subset G$.

(a) If A is a subset of F, show that $A^{\circ} \subset A^p$ and that $A^{\circ} = A^p$ if A is balanced. More generally, if D is the balanced hull of A, show that $A^{\circ} = D^p$. Give an example in which A^p is not balanced and in particular $A^{\circ} \neq A^p$.

(b) Prove the following properties of A^p:

(i) If $A_1 \subset A_2 \subset F$, then $A_1^p \supset A_2^p$.

(ii) $A \subset A^{pp} = (A^p)^p$.

(iii) $A^p = A^{ppp}$.

(iv) If $A \subset F$, then A^p is a convex, $\sigma(G, F)$-closed subset of G.

(v) $(\lambda A)^p = (1/\lambda)A^p$ for $\lambda \in \mathbf{K}$, $\lambda \neq 0$.

(vi) A^p is absorbing if and only if A is bounded for $\sigma(F, G)$.

(vii) If $(A_\iota)_{\iota \in I}$ is a family of subsets of F, then

$$\left(\bigcup_{\iota \in I} A_\iota\right)^p = \bigcap_{\iota \in I} A_\iota^p.$$

(c) Show that A^{pp} is the convex, $\sigma(F, G)$-closed hull of $A \cup \{0\}$.

(d) Show that if A is a cone with vertex at 0 (§1), then A^p consists of those $y \in G$ for which $\mathcal{R}e\, B(x, y) \leqq 0$ for all $x \in A$ and that A^p is a cone (the dual cone).

3. Let F and G be two paired vector spaces, A a subset of F, and M the linear subspace of F generated by A. Show that $A^\perp = M^\perp$.

4. (a) Prove that the convex hull of the balanced hull of a set A is the balanced, convex hull of A. (*Hint:* Using the expression for the convex hull C of a set A given in the corollary to Proposition 2.4.2, show that C is balanced if A is.)

(b) Show that the balanced hull of the convex hull of a set is not necessarily convex. (*Hint:* Consider the points $(0, 0)$, $(0, 1)$ and $(1, 0)$ in $\mathbf{R}^2$.)

§4. $\mathfrak{S}$-topologies

Let F and G be two vector spaces which form a pairing with respect to the bilinear form $(x, y) \mapsto B(x, y)$. The concept of polarity allows us to define all kinds of important locally convex topologies on F and G.

Let $\mathfrak{S}$ be a collection of $\sigma(F, G)$-bounded subsets of F. Then by Proposition 3.1 the polars A° of the sets $A \in \mathfrak{S}$ form a collection of absorbing, balanced, convex sets in G and thus define a locally convex topology on G according to Proposition 2.4.6 or its corollary. We shall call it the $\mathfrak{S}$-*topology* or the topology of uniform convergence on sets belonging to $\mathfrak{S}$. As we know, the finite intersections of the sets λA°, where $\lambda > 0$ and $A \in \mathfrak{S}$, form a fundamental system of neighborhoods of 0 for the $\mathfrak{S}$-topology.

The $\mathfrak{S}$-topology on G is defined by the family of semi-norms q_A, where $A \in \mathfrak{S}$, which are given by

$$q_A(y) = \sup_{x \in A} |B(x, y)|,$$

since $q_A(y) \leqq \epsilon$ is equivalent to $y \in \epsilon A^\circ$.

PROPOSITION 1. *Let F and G be paired and $\mathfrak{S}$ a collection of $\sigma(F, G)$-bounded sets in F. The $\mathfrak{S}$-topology on G is Hausdorff if and only if $\bigcup_{A\in\mathfrak{S}} A$ is total in F for $\sigma(F, G)$ and the pairing separates points of G.*

Proof. By Proposition 2.4.8 the $\mathfrak{S}$-topology is Hausdorff if and only if $q_A(y) = 0$ for all $A \in \mathfrak{S}$ implies $y = 0$. But $q_A(y) = 0$ for all $A \in \mathfrak{S}$ is equivalent to

$$y \in \bigcap_{A\in\mathfrak{S}} A^{\perp} = \Big(\bigvee_{A\in\mathfrak{S}} A\Big)^{\perp}$$

(Proposition 3.3(f)). If $\bigcup_{A\in\mathfrak{S}} A$ is total in F, i.e., if $\bigvee_{A\in\mathfrak{S}} A = F$, then $q_A(y) = 0$ for all $A \in \mathfrak{S}$ implies $B(x, y) = 0$ for all $x \in F$, which in turn implies $y = 0$ if the pairing separates points of G. Conversely, if $\bigvee_{A\in\mathfrak{S}} A \neq F$ or if B does not separate points of G, then there exists $y \in G$, $y \neq 0$, such that $y \in A^{\perp}$ for all $A \in \mathfrak{S}$, i.e., such that $q_A(y) = 0$ for all $A \in \mathfrak{S}$. ▮

PROPOSITION 2. *Let F and G be paired and $\mathfrak{S}$ a collection of $\sigma(F, G)$-bounded sets in F. The $\mathfrak{S}$-topology on G will not be changed if we replace $\mathfrak{S}$ by one of the following collections of subsets of F:*

(a) *all the subsets of the sets in $\mathfrak{S}$;*
(b) *the finite unions of the sets in $\mathfrak{S}$;*
(c) *the sets λA, where $\lambda \in \mathbf{K}$ and $A \in \mathfrak{S}$;*
(d) *the balanced hulls of the sets in $\mathfrak{S}$;*
(e) *the $\sigma(F, G)$-closures of the sets in $\mathfrak{S}$;*
(f) *the balanced, convex, $\sigma(F, G)$-closed hulls of the sets in $\mathfrak{S}$.*

Proof. (a) if $A_1 \subset A$ and $A \in \mathfrak{S}$, then $A^{\circ} \subset A_1^{\circ}$ and so A_1° is a neighborhood of 0 for the $\mathfrak{S}$-topology.

(b) The sets

$$\Big(\bigcup_{i=1}^{n} A_i\Big)^{\circ} = \bigcap_{i=1}^{n} A_i^{\circ},$$

where $A_i \in \mathfrak{S}$ $(1 \leqq i \leqq n)$, are neighborhoods of 0 for the $\mathfrak{S}$-topology.

(c) If $\lambda \neq 0$, then the set $(\lambda A)^{\circ} = (1/\lambda)A^{\circ}$ is a neighborhood of 0 for the $\mathfrak{S}$-topology. If $\lambda = 0$, then $(\lambda A)^{\circ} = G$.

(d) If D is the balanced hull of A, we have $D^{\circ} = A^{\circ}$.

(e) We have $A \subset \overline{A} \subset A^{\circ\circ}$, hence $A^{\circ} \supset \overline{A}^{\circ} \supset A^{\circ\circ\circ}$. But $A^{\circ} = A^{\circ\circ\circ}$; thus $\overline{A}^{\circ} = A^{\circ}$.

(f) We have $A^{\circ} = A^{\circ\circ\circ}$, where $A^{\circ\circ}$ is the balanced, convex, $\sigma(F, G)$-closed hull of A (Theorem 3.1). ▮

We may thus always assume that $\mathfrak{S}$ is a collection of balanced, $\sigma(F, G)$-bounded (or even of balanced, convex, $\sigma(F, G)$-closed and $\sigma(F, G)$-bounded) sets. We shall need in §5 the fact that we may even suppose that $\mathfrak{S}$ is a collection of balanced, convex, $\sigma(F, G)$-closed, $\sigma(F, G)$-bounded subsets of F which has the following two stability properties:

(i) Given a finite family $(A_i)_{1 \leqq i \leqq n}$ of sets in $\mathfrak{S}$, the balanced, convex, $\sigma(F, G)$-closed hull $(\bigcup_{i=1}^n A_i)^{\circ\circ}$ of the union of the A_i belongs to $\mathfrak{S}$.

(ii) If $A \in \mathfrak{S}$, then $\lambda A \in \mathfrak{S}$ for $\lambda \neq 0$.

Indeed, if $\mathfrak{S}$ is an arbitrary collection of $\sigma(F, G)$-bounded subsets of F, let $\mathfrak{S}_1$ be the collection of all the sets λA, where $\lambda \neq 0$, $A \in \mathfrak{S}$, let $\mathfrak{S}_2$ be the collection of the finite unions of the sets in $\mathfrak{S}_1$, and finally, let $\mathfrak{S}_3$ be the collection of the balanced, convex, $\sigma(F, G)$-closed hulls of the sets in $\mathfrak{S}_2$. It follows from Proposition 2(c), (b), and (f) that the $\mathfrak{S}$-topology coincides with the $\mathfrak{S}_3$-topology. It is also clear that the sets in $\mathfrak{S}_3$ are balanced, convex, $\sigma(F, G)$-closed and $\sigma(F, G)$-bounded. Thus we have only to show that $\mathfrak{S}_3$ has the stability properties (i) and (ii). Now every set in $\mathfrak{S}_3$ is of the form $(\bigcup \lambda_j S_j)^{\circ\circ}$, with $S_j \in \mathfrak{S}$. Hence (i) follows from

$$\left(\bigcup C_i^{\circ\circ}\right)^{\circ\circ} = \left(\bigcup C_i\right)^{\circ\circ}. \tag{1}$$

Indeed, if $A_i = (\bigcup_j \lambda_{ij} S_{ij})^{\circ\circ} \in \mathfrak{S}_3$, then setting $C_i = \bigcup_j \lambda_{ij} S_{ij}$ we have

$$\Big(\bigcup_i A_i\Big)^{\circ\circ} = \Big(\bigcup_i C_i^{\circ\circ}\Big)^{\circ\circ} = \Big(\bigcup_i C_i\Big)^{\circ\circ} = \Big(\bigcup_{i,j} \lambda_{ij} S_{ij}\Big)^{\circ\circ} \in \mathfrak{S}_3.$$

Formula (1) itself is a consequence of

$$\left(\bigcup C_i\right)^{\circ} = \bigcap C_i^{\circ} = \bigcap C_i^{\circ\circ\circ} = \left(\bigcup C_i^{\circ\circ}\right)^{\circ}.$$

Property (ii) follows from

$$\lambda \left(\bigcup \lambda_j S_j\right)^{\circ\circ} = \left(\bigcup \lambda\lambda_j S_j\right)^{\circ\circ},$$

which is an immediate consequence of Proposition 3.1(f).

Example 1. The weak topology $\sigma(G, F)$ is the $\mathfrak{S}_i$-topology $(1 \leqq i \leqq 5)$ for each of the following collections of subsets of F:

$\mathfrak{S}_1$ = all subsets of F consisting of a single point.

$\mathfrak{S}_2$ = all subsets of F of the form $\{\lambda x \mid |\lambda| \leqq 1\}$, where x is an arbitrary element of F. These sets are the balanced, convex, $\sigma(F, G)$-closed hulls of the sets in $\mathfrak{S}_1$.

$\mathfrak{S}_3$ = all subsets of F of the form $\{\sum_{i=1}^n \lambda_i x_i \mid \sum|\lambda_i| \leqq \lambda\}$, where $(x_i)_{1 \leqq i \leqq n}$ is an arbitrary finite family of elements of F. This collection $\mathfrak{S}_3$ has properties (i) and (ii).

$\mathfrak{S}_4$ = all balanced, convex, $\sigma(F, G)$-closed, $\sigma(F, G)$-bounded, finite-dimensional subsets of F.

$\mathfrak{S}_5$ = all $\sigma(F, G)$-bounded, finite-dimensional subsets of F. (This collection is saturated; cf. Exercise 2.)

We know that if F and G form a pairing which separates points of G, then G can be considered canonically as a subspace of F^*. Thus we can pose the following:

DEFINITION 1. *Suppose that the vector spaces F and G form a pairing which separates points of G. We say that a locally convex topology $\mathcal{T}$ on F is compatible with the pairing of F and G if G is the dual of F for the topology $\mathcal{T}$.*

In other words, $\mathcal{T}$ is compatible with the pairing of F and G if the continuous linear forms on F are precisely those defined by elements of G, i.e., are of the form $x \mapsto \langle x, y\rangle$ for some $y \in G$.

EXAMPLE 2. We have shown that if the pairing of F and G separates points of G, then the topology $\sigma(F, G)$ on F is compatible with the pairing of F and G (Proposition 2.2). We have even shown that $\sigma(F, G)$ is the coarsest such topology on F.

EXAMPLE 3. Let E be a topological vector space with topology $\mathcal{T}$ and let E' be its dual (Example 2.2). Then $\mathcal{T}$ is compatible with the pairing of E and E' by the very definition of E'.

EXAMPLE 4. Let E be a Banach space and E' its dual. If E is not reflexive then the topology defined by the norm on E' is not compatible with the pairing of E' and E since the dual of E' is strictly larger than E. The topology $\sigma(E', E)$ is of course compatible with the pairing of E' and E.

PROPOSITION 3. *Let the vector spaces F and G form a pairing which separates points of G. The closed convex sets of F are the same for all locally convex topologies on F which are compatible with the pairing of F and G.*

Proof. We may suppose that F and G are real vector spaces since a set is closed and convex in F if and only if it is closed and convex in the underlying real vector space F_0. By the corollary to Proposition 1.5 a closed convex set is the intersection of the closed half-spaces which contain it. Now a closed half-space is defined by an equation $f(x) \leqq \alpha$, where f is a continuous linear form on F. But the continuous linear forms on F are the same for all topologies compatible with the pairing of F and G. ∎

Our next goal is to show that every locally convex topology can be obtained as an $\mathfrak{S}$-topology for an appropriate collection $\mathfrak{S}$. Let E be a topological space, F a topological vector space, and $\mathcal{H}$ a set of maps from E into F. We say that $\mathcal{H}$ is *equicontinuous* at the point $a \in E$ if for every neighborhood W of 0 in F there exists a neighborhood V of a in E such that $f(x) - f(a) \in W$ for all $x \in V$ and $f \in \mathcal{H}$. If $\mathcal{H}$ is equicontinuous at a, then in particular every function $f \in \mathcal{H}$ is continuous at a. We say that $\mathcal{H}$ is equicontinuous on E if it is equicontinuous at every point of E.

Suppose now that E is also a topological vector space. We say that $\mathcal{H}$ is *uniformly equicontinuous* if for every neighborhood W of 0 in F there exists a neighborhood V of 0 in E such that $x - y \in V$ implies

$$f(x) - f(y) \in W$$

for all $f \in \mathcal{H}$. If $\mathcal{H}$ is uniformly equicontinuous, then it is equicontinuous and each $f \in \mathcal{H}$ is uniformly continuous.

PROPOSITION 4. *A set $\mathcal{H}$ of linear maps from a topological vector space E into a topological vector space F is uniformly equicontinuous if it is equicontinuous at the origin.*

Proof. Let W be a neighborhood of 0 in F. There exists a neighborhood V of 0 in E such that $f(x) \in W$ for all $x \in V$ and $f \in \mathcal{H}$. But then $x - y \in V$ implies that $f(x) - f(y) = f(x - y) \in W$ for all $f \in \mathcal{H}$. ∎

In particular, a set $\mathcal{H}$ of linear maps from a topological vector space E into a topological vector space F is equicontinuous if for each neighborhood W of 0 in F, the set $\bigcap_{h \in \mathcal{H}} h^{-1}(W)$ is a neighborhood of 0 in E.

PROPOSITION 5. *Let F be a vector space and $(E_\iota)_{\iota \in I}$ a family of locally convex spaces. For each $\iota \in I$ let f_ι be a linear map from E_ι into F and assume that $\bigcup_{\iota \in I} f_\iota(E_\iota)$ generates F. Equip F with the finest locally convex topology $\mathfrak{T}$ for which all the maps f_ι are continuous. Let G be a locally convex space and $\mathcal{H}$ a set of linear maps from F into G. Then $\mathcal{H}$ is equicontinuous if and only if for each $\iota \in I$ the set $\mathcal{H} \circ f_\iota = \{h \circ f_\iota \mid h \in \mathcal{H}\}$ of maps from E_ι into G is equicontinuous.*

Proof. Assume that $\mathcal{H}$ is equicontinuous and let W be a neighborhood of 0 in G. Then

$$V = \bigcap_{h \in \mathcal{H}} h^{-1}(W)$$

is a neighborhood of 0 in F and

$$f_\iota^{-1}(V) = \bigcap_{h \in \mathcal{H}} f_\iota^{-1}(h^{-1}(W))$$

is a neighborhood of 0 in E_ι, which proves that $\mathcal{H} \circ f_\iota$ is equicontinuous. Conversely, suppose that all the sets $\mathcal{H} \circ f_\iota$ are equicontinuous, and let W be a balanced, convex neighborhood of 0 in G. Then for each $\iota \in I$ the set

$$U_\iota = \bigcap_{h \in \mathcal{H}} f_\iota^{-1}(h^{-1}(W))$$

is a neighborhood of 0 in E_ι. The subset $V = \bigcap_{h \in \mathcal{H}} h^{-1}(W)$ of F is clearly balanced and convex. Since it contains $\bigcup_{\iota \in I} f_\iota(U_\iota)$, we conclude from an observation made in Chapter 2, §12 that V is a neighborhood of 0 in F. ∎

Let E and F be two locally convex vector spaces whose topologies are defined by the families of semi-norms $(q_\iota)_{\iota \in I}$ and $(r_\lambda)_{\lambda \in L}$ respectively. A set $\mathcal{H}$ of linear maps from E into F is equicontinuous if and only if for every $\lambda \in L$ there exists a finite family $(\iota_1, \ldots, \iota_n)$ of elements of I and a positive number μ such that

$$r_\lambda(h(x)) \leqq \mu \max_{1 \leqq k \leqq n} q_{\iota_k}(x)$$

for all $x \in E$ and $h \in \mathcal{H}$. The proof is similar to that of Proposition 2.5.2.

Continuous linear forms on the topological vector space E are continuous linear maps from E into the topological vector space $\mathbf{K}^1$ and therefore we can speak about equicontinuous subsets of the dual E'. If the space E is locally convex, then $A \subset E'$ is equicontinuous if and only if there exists a continuous semi-norm q on E such that $|u(x)| \leqq q(x)$ for all $x \in E$ and $u \in A$.

Let E be a locally convex space, M a linear subspace of E and $(f_\iota)_{\iota \in I}$ an equicontinuous family of linear forms defined on M. One sees, exactly as in the proof of Proposition 1.1, that there exists an equicontinuous family $(g_\iota)_{\iota \in I}$ of linear forms defined on E such that $g_\iota(x) = f_\iota(x)$ for all $x \in M$ and $\iota \in I$.

Proposition 6. *A subset M of the dual E' of a topological vector space E is equicontinuous if and only if it is contained in the polar V° of a neighborhood V of 0 in E.*

Proof. Suppose that M is equicontinuous. Then there exists a neighborhood V of 0 in E such that $|\langle x, u \rangle| \leqq 1$ for all $x \in V$ and $u \in M$. Thus $M \subset V^\circ$. Conversely, let $M \subset V^\circ$, where V is a neighborhood of 0 in E. Then $|\langle x, u \rangle| \leqq \epsilon$ for $x \in \epsilon V$ and $u \in M$; hence M is equicontinuous at 0 and therefore equicontinuous by Proposition 4. ∎

It follows from this proposition that every equicontinuous subset M of E' is bounded for $\sigma(E', E)$. We do not state it as a formal proposition, since we shall soon have a better result (Proposition 6.1). The proof is easy. We have $M \subset V^\circ$, where V is some neighborhood of 0 in E. On the other hand, a fundamental system of neighborhoods of 0 in E' for $\sigma(E', E)$ is formed by the polars A° of finite subsets A of E. Now A is bounded; hence $A \subset \lambda V$ for some $\lambda > 0$. But then $M \subset V^\circ \subset \lambda A^\circ$; i.e., M is bounded for $\sigma(E', E)$.

Proposition 7. *Let E be a locally convex space with topology $\mathcal{T}$. Then $\mathcal{T}$ coincides with the $\mathfrak{S}$-topology, where $\mathfrak{S}$ is the collection of all equicontinuous subsets of E'.*

Proof. By Proposition 2.4.4 the balanced, closed, convex neighborhoods V of 0 form a fundamental system of neighborhoods of 0 for $\mathcal{T}$. By Proposition 3 each of these sets V is also closed for the topology $\sigma(E, E')$. Hence by the theorem of bipolars (Theorem 3.1) we have $V = V^{\circ\circ}$. But by Proposition 6 the set V° is equicontinuous in E'; hence V is a neighborhood of 0 for the $\mathfrak{S}$-topology. Thus the $\mathfrak{S}$-topology is finer than the topology $\mathcal{T}$. Conversely, let M be an equicontinuous set in E'. Then $M \subset V^\circ$, where V is a balanced, convex, $\sigma(E, E')$-closed neighborhood of 0 for $\mathcal{T}$ in E. It follows that $M^\circ \supset V^{\circ\circ} = V$; i.e., M° is a neighborhood of 0 for the topology $\mathcal{T}$. Thus $\mathcal{T}$ is finer than the $\mathfrak{S}$-topology. ∎

THEOREM 1 (Alaoglu-Bourbaki). *Let E be a topological vector space. Then any equicontinuous subset of E' is relatively compact for the topology $\sigma(E', E)$.*

Proof. By Proposition 6 it is sufficient to prove that if V is a neighborhood of the origin in E, then V° is compact for $\sigma(E', E)$. Now E' is a subspace of the algebraic dual E^* of E and the topology $\sigma(E^*, E)$ induces $\sigma(E', E)$ on E'. Denote by $V^\bullet$ the polar of V in E^*. Since $V^\bullet$ is a bounded, closed set for $\sigma(E^*, E)$, it is compact in E^* (Example 2.3). But if $u \in V^\bullet$, then $|u(x)| \leqq \epsilon$ for $x \in \epsilon V$; i.e., u is a continuous linear form on E and therefore $u \in V^\circ$. Thus $V^\bullet = V^\circ \subset E'$ and V° is a compact subset of E'. ∎

COROLLARY. *Let E be a normed vector space and E' its dual. The closed unit ball $B_1' = \{x' \mid \|x'\| \leqq 1\}$ of E' is compact for the topology $\sigma(E', E)$.*

Proof. By formula (1) of Chapter 1, §5 the norm in E' is defined by

$$\|x'\| = \sup_{\|x\| \leqq 1} |\langle x, x' \rangle|.$$

In other words, the closed unit ball B_1' of E' is the polar of the closed unit ball B_1 in E. The conclusion then follows from Theorem 1. ∎

If E is a reflexive Banach space (in particular, if it is a Hilbert space), then $E = E''$ and therefore its closed unit ball is compact for the topology $\sigma(E, E')$, as we already mentioned in Chapter 1, §7. We shall see in §8 that this necessary condition is also sufficient for the reflexivity of E.

Since a collection $\mathfrak{S}$ of subsets on F which defines the $\mathfrak{S}$-topology on G must consist of $\sigma(F, G)$-bounded sets, a special interest is attached to the topology defined by all bounded sets of F. Accordingly, we introduce the following:

DEFINITION 2. *Let F and G be two vector spaces which form a pairing. If $\mathfrak{S}$ is the collection of all $\sigma(F, G)$-bounded sets in F, then the corresponding $\mathfrak{S}$-topology on G will be called the strong topology and denoted by $\beta(G, F)$.*

We also refer to $\beta(G, F)$ as the topology of the uniform convergence on bounded subsets of F.

EXAMPLE 5. Let E be a normed vector space and E' its dual. A set $A \subset E$ is $\sigma(E, E')$-bounded if for each $x' \in E'$ there exists a scalar

$$\lambda = \lambda(x') > 0$$

such that $|\langle x, x' \rangle| \leqq \lambda$ for all $x \in A$. Therefore if A is bounded in the sense of the norm, then it is bounded for $\sigma(E, E')$ since $\|x\| \leqq M$ implies $|\langle x, x' \rangle| \leqq M \cdot \|x'\|$. Conversely, we can consider a set $A \subset E$ as a collection of linear forms on the Banach space E' defined by $x' \mapsto \langle x, x' \rangle$. All these linear forms are continuous for the norm topology on E' since $|\langle x, x' \rangle| \leqq \|x\| \cdot \|x'\|$. Suppose that A is bounded for $\sigma(E, E')$, i.e., that for each $x' \in E'$ the set $\{|\langle x, x' \rangle| \mid x \in A\}$ is bounded. Then we can apply the Banach-Steinhaus theorem (Theorem 1.8.1) to the Banach space E' to find that $|\langle x, x' \rangle| \leqq M\|x'\|$ for all $x \in A$; i.e.,

$$\|x\| = \sup_{x' \in E'} \frac{|\langle x, x' \rangle|}{\|x'\|} \leqq M$$

for all $x \in A$, which means that A is bounded in the sense of the norm on E. Therefore in E the sets bounded for $\sigma(E, E')$ are exactly the sets bounded for the norm (cf. Theorem 5.3).

Now in E a set is bounded in the sense of the norm if and only if it is contained in some set λB_1, where $\lambda > 0$ and B_1 is the closed unit ball in E. Therefore the topology $\beta(E', E)$ is also the $\mathfrak{S}$-topology, where $\mathfrak{S}$ consists of the unique element B_1. But the polar of B_1 is the unit ball B'_1 of E'. Hence we have proved that $\beta(E', E)$ is the topology defined by the norm on E'.

In particular, the topology $\beta(G, F)$ is not necessarily compatible with the pairing of the spaces F and G (see Example 4).

EXERCISES

1. Suppose that the vector spaces F and G form a pairing and let $\mathfrak{S}$ be a collection of balanced, convex, $\sigma(F, G)$-closed, $\sigma(F, G)$-bounded subsets of F. Suppose furthermore that given a finite family of sets in $\mathfrak{S}$, the balanced, convex, $\sigma(F, G)$-closed hull of their union belongs to $\mathfrak{S}$. Prove that the family of seminorms $(q_A)_{A \in \mathfrak{S}}$ is saturated (cf. Chapter 2, §4). (*Hint:* If $(\bigcup_{i=1}^n A_i)^{\circ\circ}$, then $q_A(y) = \max_{1 \leqq i \leqq n} q_{A_i}(y)$ for every $y \in G$.)

2. Let F and G be paired and $\mathfrak{S}$ a collection of $\sigma(F, G)$-bounded subsets of F. We say that $\mathfrak{S}$ is *saturated* if the following conditions are satisfied:

(i) Every subset of a set $A \in \mathfrak{S}$ belongs to $\mathfrak{S}$.
(ii) The union of a finite number of sets in $\mathfrak{S}$ belongs to $\mathfrak{S}$.
(iii) If $A \in \mathfrak{S}$, then $\lambda A \in \mathfrak{S}$ for all $\lambda \neq 0$.
(iv) The balanced, convex, $\sigma(F, G)$-closed hull of every set in $\mathfrak{S}$ belongs to $\mathfrak{S}$.

(a) Show that given a collection $\mathfrak{S}$ of $\sigma(F, G)$-bounded subsets of F, there exists a smallest saturated collection $\widehat{\mathfrak{S}}$ of $\sigma(F, G)$-bounded subsets containing

$\mathfrak{S}$. The collection $\widehat{\mathfrak{S}}$ is called the saturated hull of $\mathfrak{S}$. Show that the $\mathfrak{S}$-topology on G coincides with the $\widehat{\mathfrak{S}}$-topology.

(b) Let $\mathfrak{S}_1$ be a saturated collection of $\sigma(F, G)$-bounded subsets of F and let $\mathfrak{S}_2$ be a collection of $\sigma(F, G)$-bounded subsets of F containing $\mathfrak{S}_1$ but distinct from $\mathfrak{S}_1$. Show that the $\mathfrak{S}_2$-topology is strictly finer than the $\mathfrak{S}_1$-topology.

(c) Deduce from (b) that if $\mathfrak{S}_1$ and $\mathfrak{S}_2$ are two collections of $\sigma(F, G)$-bounded subsets of F, then the $\mathfrak{S}_1$-topology coincides with the $\mathfrak{S}_2$-topology on G if and only if $\widehat{\mathfrak{S}}_1 = \widehat{\mathfrak{S}}_2$.

3. Let F and G form a pairing which separates points of G and let M be a subspace of F. Show that $M = M^{\perp\perp}$ if and only if M is closed with respect to some topology on F, compatible with the duality between F and G.

4. Give an example of a family of continuous real-valued functions defined on a closed interval of the real line which is not equicontinuous.

5. Show by an example that Proposition 5 does not necessarily hold if the $f_\iota(E_\iota)$ do not generate the space F. (*Hint:* Let I consist of one element and take $E = \mathbf{R}$, $F = \mathbf{R}^2$, $G = \mathbf{R}^2$, $f: x \mapsto (x, 0)$, $h_n(x, y) = ny$.)

6. Show that if E is a Banach space and E' its dual, then the topology $\beta(E, E')$ coincides with the original normed topology of E. (*Hint:* Use the Banach-Steinhaus theorem similarly as in Example 5.)

7. Let E be the subspace of l^2 formed by those elements $x = (\xi_n)_{n\in\mathbf{N}}$ such that $\xi_n = 0$ for large n.

(a) Show that E is dense in l^2 and that the dual E' of E can be identified with l^2 (cf. Example 1.7.1).

(b) Show that a subset $A \subset E'$ is bounded for $\sigma(E', E)$ if and only if there exists a sequence $(m_n)_{n\in\mathbf{N}}$ of positive numbers such that $|\xi'_n| \leqq m_n$ for $n \in \mathbf{N}$ and $x' = (\xi'_n) \in A$.

(c) Show that a subset $A \subset E'$ is bounded for $\beta(E', E)$ if and only if there exists a positive number M such that $\|x'\| \leqq M$ for all $x' \in A$.

8. Let E be a locally convex Hausdorff space and $\hat{E}$ its completion. Show that the equicontinuous subsets of E' are the same whether we consider E' as the dual of E or as the dual of $\hat{E}$ (cf. Exercise 2.3).

§ 5. *The Mackey topology*

Let F and G be two vector spaces forming a pairing which separates points of F (Definition 2.1). We know that F can be considered as a linear subspace of the algebraic dual G^* of G. Let $\mathfrak{S}$ be a collection of $\sigma(F, G)$-bounded subsets of F and equip G with the $\mathfrak{S}$-topology $\mathcal{T}$. We want to find conditions which will ensure that the continuous linear forms on G for $\mathcal{T}$ be precisely the elements of F, i.e., that F be the dual G' of G. In the terminology of Definition 4.1 we want to find the $\mathfrak{S}$-topologies on G which are compatible with the pairing of G and F. More generally, we shall determine the dual of G for any $\mathfrak{S}$-topology. We start with a special case when $\mathfrak{S}$ possesses the stability conditions mentioned in §4.

PROPOSITION 1. *Let F and G be two vector spaces which form a pairing separating points of F. Let $\mathfrak{S}$ be a collection of balanced, convex, $\sigma(F, G)$-closed, $\sigma(F, G)$-bounded subsets of F which has the following two properties:*

(i) *Given a finite family of sets in $\mathfrak{S}$, the balanced, convex, $\sigma(F, G)$-closed hull of their union belongs to $\mathfrak{S}$.*

(ii) *If $A \in \mathfrak{S}$, then $\lambda A \in \mathfrak{S}$ for $\lambda \neq 0$.*

The dual of the space G equipped with the $\mathfrak{S}$-topology is the linear subspace of G^* formed by the $\sigma(G^*, G)$-closures of the sets $A \in \mathfrak{S}$.

Proof. Let u be an $\mathfrak{S}$-continuous linear form on G. By conditions (i) and (ii) there exists $A \in \mathfrak{S}$ such that $|u(x)| \leqq 1$ for $x \in A^{\circ}$; hence $u \in A^{\circ\circ}$. But $A^{\circ\circ}$ is the $\sigma(G^*, G)$-closure of A in G^* (Theorem 3.1). Conversely, let $u \in G^*$ belong to some set $A^{\circ\circ}$, where $A \in \mathfrak{S}$. Then $|u(x)| \leqq \epsilon$ for $x \in \epsilon A^{\circ} = [(1/\epsilon)A]^{\circ}$ and thus u is continuous for the $\mathfrak{S}$-topology. ∎

In order to pass to arbitrary collections of $\sigma(F, G)$-bounded sets we shall need the following result, which is also of independent interest:

PROPOSITION 2. *Let E be a locally convex Hausdorff space and $(A_i)_{1 \leqq i \leqq n}$ a finite family of convex, compact subsets of E. Then the convex hull and the balanced convex hull of the union of the sets A_i are both compact.*

Proof. Let L be the compact subset of $\mathbf{K}^n$ formed by all points $\lambda = (\lambda_1, \ldots, \lambda_n) \in \mathbf{K}^n$ such that $\lambda_i \geqq 0$ $(1 \leqq i \leqq n)$ and $\sum_{i=1}^{n} \lambda_i = 1$. By Tihonov's theorem (Chapter 2, §10) the set $K = L \times A_1 \times \cdots \times A_n$ is a compact subset of $\mathbf{K}^n \times E^n$. Define the continuous map f from K into E by

$$f(\lambda_1, \ldots, \lambda_n, x_1, \ldots, x_n) = \sum_{i=1}^{n} \lambda_i x_i.$$

By Proposition 2.4.2 the image $f(K)$ of K is the convex hull of $\bigcup_{i=1}^{n} A_i$. Since K is compact and f continuous, $f(K)$ is also compact.

If we define L as the compact subset of $\mathbf{K}^n$ formed by all points $\lambda = (\lambda_1, \ldots, \lambda_n) \in \mathbf{K}^n$ such that $\sum_{i=1}^{n} |\lambda_i| \leqq 1$, then by Exercise 2.4.1(b) the set $f(K)$ will be the balanced convex hull of $\bigcup_{i=1}^{n} A_i$. By the same reasoning as above, we see that $f(K)$ is compact. ∎

PROPOSITION 3. *Suppose that the vector spaces F and G form a pairing which separates points of F, and let $\mathfrak{S}_0$ be a collection of $\sigma(F, G)$-bounded subsets of F. Equip G with the $\mathfrak{S}_0$-topology. Then the dual of G is the linear subspace of G^* generated by the balanced, convex, $\sigma(G^*, G)$-closed hulls of the sets $B \in \mathfrak{S}_0$.*

Proof. Let $\mathfrak{S}$ be the collection of all the balanced, convex, $\sigma(F, G)$-closed hulls in F of the sets $\bigcup_{i=1}^{n} \lambda_i B_i$, where $B_i \in \mathfrak{S}_0$ $(1 \leqq i \leqq n)$. We have seen in §4 that $\mathfrak{S}$ has the properties (i) and (ii) stated in Proposition 1 and that the $\mathfrak{S}_0$-topology on G coincides with the $\mathfrak{S}$-topology. Thus by Proposition 1 the dual G' of G is the union of the $\sigma(G^*, G)$-closures of the sets in $\mathfrak{S}$. In particular, the balanced, convex, $\sigma(G^*, G)$-closed hulls in G^* of the sets belonging to $\mathfrak{S}_0$ are contained in G' and therefore the subspace H of G^* generated by these hulls is also contained in G'.

Thus we have only to show that conversely G' is contained in the subspace H of G^*. Let $(B_i)_{1 \leqq i \leqq n}$ be a finite family of elements of $\mathfrak{S}_0$ and $(\lambda_i)_{1 \leqq i \leqq n}$ a finite family of nonzero scalars. Let A_i be the balanced, convex, $\sigma(G^*, G)$-closed hull in G^* of B_i $(1 \leqq i \leqq n)$. Since A_i is the polar in G^* of the $\mathfrak{S}_0$-neighborhood B_i° of 0 in G, it is $\sigma(G^*, G)$-compact (Theorem 4.1). Thus by Proposition 2 the balanced convex hull A of the union of the sets $\lambda_i A_i$ is $\sigma(G^*, G)$-compact and hence $\sigma(G^*, G)$-closed in G^*. But $\lambda_i B_i \subset A$ for $1 \leqq i \leqq n$; hence the $\sigma(G^*, G)$-closure of $(\bigcup \lambda_i B_i)^{\circ\circ}$ is also a subset of A. Since A is a subset of H, we have proved that the $\sigma(G^*, G)$-closure of every set in $\mathfrak{S}$ is a subset of H, hence that $G' \subset H$. ∎

Corollary. *Suppose that the vector spaces F and G form a pairing which separates points of F, and let $\mathfrak{S}$ be a collection of $\sigma(F, G)$-bounded subsets of F. Equip G with the $\mathfrak{S}$-topology and let G' be the dual of G for that topology. Then:*

(a) *G' will be contained in F if and only if the balanced, convex, $\sigma(F, G)$-closed hulls in F of the sets $B \in \mathfrak{S}$ are $\sigma(F, G)$-compact.*

(b) *G' will contain F if and only if the subspace of F generated by the balanced, convex, $\sigma(F, G)$-closed hulls of the sets $B \in \mathfrak{S}$ is equal to F.*

Proof. (a) If the balanced, convex, $\sigma(F, G)$-closed hulls of the sets $B \in \mathfrak{S}$ are $\sigma(F, G)$-compact, then they are also $\sigma(G^*, G)$-closed. Thus by Proposition 3 the dual G' of G is contained in F.

Conversely, if $G' \subset F$, then the balanced, convex, $\sigma(F, G)$-closed hulls of the sets $B \in \mathfrak{S}$ are also $\sigma(G^*, G)$-closed, hence of the form $B^{\circ\circ}$. Since B° is a neighborhood of 0 in G for the $\mathfrak{S}$-topology, $B^{\circ\circ}$ is $\sigma(G^*, G)$-compact.

(b) If the subspace of F generated by the balanced, convex, $\sigma(F, G)$-closed hulls of the sets $B \in \mathfrak{S}$ is equal to F, then the subspace of G^* generated by the balanced, convex, $\sigma(G^*, G)$-closed hulls of the sets $B \in \mathfrak{S}$, i.e., the space G', must contain F.

Conversely the subspace N of F generated by the balanced, convex, $\sigma(F, G)$-closed hulls of the sets $B \in \mathfrak{S}$ is the intersection of G' with F. Thus if $G' \supset F$, then $N = F$. ∎

Theorem 1 (Mackey-Arens). *Suppose that the vector spaces F and G form a pairing which separates points of G. A locally convex topology $\mathcal{T}$ on F is compatible with the pairing of F and G* (Definition 4.1) *if and only if $\mathcal{T}$ is an $\mathfrak{S}$-topology, where $\mathfrak{S}$ is some collection of balanced, convex, $\sigma(G, F)$-compact subsets of G which cover G.*

Proof. Let $\mathcal{T}$ be compatible with the pairing of F and G and let $\mathfrak{N}$ be the collection of all balanced, convex, $\mathcal{T}$-closed, and thus $\sigma(F, G)$-closed (Proposition 4.3) $\mathcal{T}$-neighborhoods of 0 in F. Let us denote by $\mathfrak{S}$ the collection of the polars V° of the sets $V \in \mathfrak{N}$. Then $\mathcal{T}$ is the $\mathfrak{S}$-topology since $V = V^{\circ\circ}$ for each $V \in \mathfrak{N}$. By part (a) of the preceding corollary (with the roles of F and G interchanged) it follows that the sets V° are $\sigma(G, F)$-compact. One could prove from part (b) of the corollary that the sets V° cover G (cf. Exercise 2), but a direct proof is simpler. Thus let $y \in G$. Then the linear form $x \mapsto \langle x, y\rangle$ is continuous on F for the topology $\mathcal{T}$, hence there exists a neighborhood $V \in \mathfrak{N}$ such that $|\langle x, y\rangle| \leqq 1$ for $x \in V$, that is, $y \in V^\circ$.

Conversely, if $\mathcal{T}$ is an $\mathfrak{S}$-topology, where $\mathfrak{S}$ is a collection of balanced, convex, $\sigma(G, F)$-compact subsets of G which cover G, then by the preceding corollary (again interchanging the roles of F and G) the dual of F is G; i.e., $\mathcal{T}$ is compatible with the pairing of F and G. ∎

The theorem just proved shows that a special interest is attached to the $\mathfrak{S}$-topology on F defined by the collection of *all* balanced, convex, $\sigma(G, F)$-compact subsets of G. Thus we are led to the following:

DEFINITION 1. *Let F and G be two vector spaces which form a pairing which separates points of G. If $\mathfrak{S}$ is the collection of all balanced, convex, $\sigma(G, F)$-compact subsets of G, then the corresponding $\mathfrak{S}$-topology on F will be called the Mackey topology and will be denoted by $\tau(F, G)$.*

With this definition we can restate Theorem 1 as follows.

PROPOSITION 4. *Suppose that the vector spaces F and G form a pairing which separates points of G. A locally convex topology $\mathcal{T}$ on F is compatible with the pairing of F and G if and only if $\mathcal{T}$ is finer than $\sigma(F, G)$ and coarser than $\tau(F, G)$.*

Proof. If $\mathcal{T}$ is compatible with the duality between F and G, then by Theorem 1 it is an $\mathfrak{S}$-topology, where $\mathfrak{S}$ is *some* collection of balanced, convex, $\sigma(G, F)$-compact subsets of G. Thus $\mathcal{T}$ is coarser than the topology of the uniform convergence on *all* balanced, convex, $\sigma(G, F)$-compact subsets of G, i.e., coarser than $\tau(F, G)$. On the other hand, we know (Example 4.2) that $\sigma(F, G)$ is coarser than any topology compatible with the duality between F and G.

Conversely, if $\mathcal{T}$ is finer than $\sigma(F, G)$, then all elements of G define linear forms on F which are continuous for $\mathcal{T}$ (Proposition 2.2) and thus the dual of F for the topology $\mathcal{T}$ contains G. If, on the other hand, $\mathcal{T}$ is coarser than $\tau(F, G)$, then the dual of F for the topology $\mathcal{T}$ is contained in the dual of F for the topology $\tau(F, G)$, i.e., in G. ∎

Our next goal is to prove that the bounded subsets of a locally convex Hausdorff space E are the same for all locally convex topologies which are compatible with the duality between E and E'. To do this we must first prove a chain of propositions which will also be useful in other connections. In the proof of Theorem 2 we shall use the Banach-Steinhaus theorem (Theorem 1.8.1). For the benefit of the reader who has skipped Chapter 1 we note that the Banach-Steinhaus theorem will be proved again in the next section (Corollary to Proposition 6.3) and that the proof does not use the last two theorems of this section. Therefore, if the reader wishes, he may just read Definition 2, Proposition 7, Definition 3, and Proposition 8 below and then go on to §6, returning to the present section after the corollary to Proposition 6.3.

PROPOSITION 5. *Let E be a vector space and let $\mathcal{T}$ and $\mathcal{T}'$ be two Hausdorff topologies on E compatible with the vector space structure of E. Suppose that $\mathcal{T}$ is finer than $\mathcal{T}'$ and that for $\mathcal{T}$ there exists a fundamental system of neighborhoods $\mathfrak{V}$ of 0 which are complete for $\mathcal{T}'$. Then E is complete for $\mathcal{T}$.*

Proof. We first prove that each $V \in \mathfrak{V}$ is complete for $\mathcal{T}$. Let $\mathfrak{F}$ be a Cauchy filter on V for $\mathcal{T}$. *A fortiori* $\mathfrak{F}$ is a Cauchy filter for $\mathcal{T}'$, and thus by our assumption $\mathfrak{F}$ converges in the topology $\mathcal{T}'$ to some point $x_0 \in V$. Let W be a balanced neighborhood of 0 for the topology $\mathcal{T}$, which is complete and thus by Proposition 2.9.2 also closed for $\mathcal{T}'$. By definition $\mathfrak{F}$ contains a set A such that $A - A \subset W$. Let x_1 be a point of A, then $A \subset x_1 + W$. But $x_1 + W$ is closed for $\mathcal{T}'$ and x_0 belongs to the $\mathcal{T}'$-closure of A; hence $x_0 \in x_1 + W$ and $A \subset x_0 + W + W$. Thus $\mathfrak{F}$ converges to x_0 for the topology $\mathcal{T}$, and V is $\mathcal{T}$-complete.

Now let $\mathfrak{F}$ be a Cauchy filter on E for the topology $\mathcal{T}$. Let $V \in \mathfrak{V}$ and $A \in \mathfrak{F}$ such that $A - A \subset V$. Let x_1 be a point of A. Then we have $A \subset x_1 + V$. But the set $x_1 + V$ is complete for the topology $\mathcal{T}$ by the first part of the proof, and the filter induced on $x_1 + V$ by $\mathfrak{F}$ is a Cauchy filter on $x_1 + V$, which is therefore convergent. Thus $\mathfrak{F}$ is also convergent, and E is complete for $\mathcal{T}$. ∎

Let us now introduce a very useful notation due to Grothendieck. Let E be a locally convex space and A a nonempty, balanced, convex subset of E. We know that the linear subspace of E generated by A is simply $\bigcup_{n=1}^{\infty} nA$ and that on this subspace the gauge q_A of A is a semi-norm (Proposition 2.4.7). We denote by E_A the linear subspace of E generated by A and equipped with the semi-norm q_A. If A is absorbing (in particular, if A is a neighborhood of 0 in E), then the subspace generated by A is the whole of E.

PROPOSITION 6. *Let E be a locally convex space and A a nonempty balanced, convex, bounded subset of E.*

(a) *If E is a Hausdorff space, then so is E_A; i.e., E_A is a normed space.*

(b) *If furthermore A is a complete subset of E, then E_A is complete, i.e., a Banach space.*

Proof. (a) Let x be a nonzero vector in E_A. There exists a neighborhood U of 0 in E such that $x \notin U$. Since A is bounded, there exists $\lambda > 0$ such that $\lambda A \subset U$. Then $x \notin \lambda A$, that is, $q_A(x) \geqq \lambda$. Thus q_A is a norm on E_A.

(b) The sets λA with $\lambda > 0$ form a fundamental system of neighborhoods of 0 in the normed space E_A. Since A is bounded in E, every neighborhood of 0 in E contains a set λA, $\lambda > 0$; i.e., the normed topology $\mathcal{T}$ on E_A is finer than the topology $\mathcal{T}'$ induced by E on E_A. Furthermore, each set λA is complete for the topology $\mathcal{T}'$. Hence Proposition 5 shows that the space E_A is complete for the topology $\mathcal{T}$. ▮

If now V is an *absorbing*, balanced, convex subset of E, then the gauge q_V of V is a semi-norm on E (Proposition 2.4.7). If N is the linear subspace of E formed by the vectors $x \in E$ such that $q_V(x) = 0$, then $x - y \in N$ implies $q_V(x) = q_V(y)$. Thus we can define a norm on the quotient space E/N by setting $\|\dot{x}\|_V = q_V(x)$, where x is any representative of the equivalence class $\dot{x}$. We denote by E_V the normed space obtained this way.

The similarity of the two notations E_A and E_V does not in general lead to any confusion, since if E is a locally convex Hausdorff space and $A = V$ is an absorbing, balanced, convex, bounded subset of E, then E_A and E_V coincide.

DEFINITION 2. *Let E be a locally convex space. An absorbing, balanced, convex, and closed subset of E is said to be a barrel* (keg; tonneau in French).

PROPOSITION 7. *Let E be a locally convex space and E' its dual. A subset M of E' is bounded for the topology $\sigma(E', E)$ if and only if it is contained in the polar T° of a barrel T in E.*

Proof. If the set M is $\sigma(E', E)$-bounded, then by Proposition 3.1 its polar $T = M^{\circ}$ is balanced, convex, $\sigma(E, E')$-closed, and absorbing. By Proposition 4.3 the set T is also closed for the original topology of E. Thus T is a barrel in E and $T^{\circ} = M^{\circ\circ} \supset M$. Conversely, if T is a barrel in E, then $T = T^{\circ\circ}$ by Theorem 3.1, and by Proposition 3.1(f) its polar T° is $\sigma(E', E)$-bounded. Therefore if $M \subset T^{\circ}$, then M is also $\sigma(E', E)$-bounded. ▮

THEOREM 2 (Banach-Mackey). *Let E be a locally convex Hausdorff space and T a barrel in E. Then T absorbs* (Definition 2.6.1) *every balanced, convex, bounded and complete subset A of E.*

Proof. By Proposition 6 the space E_A is a Banach space; the set $S = T \cap E_A$ is a barrel in E_A since it is obviously absorbing, balanced, convex, and the topology of E_A is finer than the topology induced by the original topology of E. By Proposition 7 the set S° is bounded for the topology $\sigma((E_A)', E_A)$ in the dual space $(E_A)'$ of E_A. In other words, for every $x \in E_A$ there exists $\lambda = \lambda(x) > 0$ such that $|\langle x, x'\rangle| < \lambda$ for all

$x' \in S^{\circ}$. But then by the Banach-Steinhaus theorem (Theorem 1.8.1) there exists $\mu > 0$ such that $|\langle x, x' \rangle| < \mu q_A(x)$ for all $x \in E_A$ and $x' \in S^{\circ}$, where q_A is the norm in E_A. Consequently the ball $\{x \mid q_A(x) \leqq 1/\mu\}$ is contained in $S^{\circ\circ} = S$. By the definition of the gauge of A we thus have $(1/\mu)A \subset S \subset T$; i.e., T absorbs A. ▮

Now we are in the position to prove the result announced earlier.

THEOREM 3 (Mackey). *Suppose that the vector spaces F and G form a dual system. The bounded subsets of F are the same for all locally convex topologies compatible with the pairing of F and G.*

Proof. If a subset of F is bounded for some topology on F, then it is bounded for every coarser topology. Thus by Proposition 4 it is enough to prove that if a set is bounded for the topology $\sigma(F, G)$, then it is bounded for the topology $\tau(F, G)$.

Let M be a subset of F which is bounded for the topology $\sigma(F, G)$. Since the pairing of F and G separates points of F, we can apply Proposition 7 to the case when $E = G$, $E' = F$, and the topology on G is $\sigma(G, F)$ and see that $M \subset T^{\circ}$, where T is a barrel in G for the topology $\sigma(G, F)$. Since the pairing also separates points of G, by Proposition 2.1 the topology $\sigma(G, F)$ is Hausdorff and we can consider the collection $\mathfrak{S}$ of all balanced, convex, $\sigma(G, F)$-compact subsets of G. Since each such set is $\sigma(G, F)$-complete, by Theorem 2 for every $A \in \mathfrak{S}$ there exists $\lambda > 0$ such that $A \subset \lambda T$. It follows that $M \subset T^{\circ} \subset \lambda A^{\circ}$. Since the sets A°, where $A \in \mathfrak{S}$, form a fundamental system of neighborhoods of 0 for the topology $\tau(F, G)$, the set M is bounded for $\tau(F, G)$. ▮

In particular, if E is a locally convex Hausdorff space with topology $\mathcal{T}$, and E' its dual, then the subsets of E which are bounded for $\mathcal{T}$ are exactly those which are bounded for $\sigma(E, E')$. Furthermore, the topology $\beta(E', E)$ on E' is the topology of the uniform convergence on subsets of E which are bounded in the sense of the topology $\mathcal{T}$. For the case of a normed vector space E we proved this result in Example 4.5. Observe that in both the general and the particular cases the proof depended essentially on the Banach-Steinhaus theorem for Banach spaces.

By virtue of the last theorem we sometimes speak of the bounded sets of a locally convex Hausdorff space without specifying for which topology they are bounded.

To conclude this section we want to use the techniques just introduced to show that if E is a quasi-complete (Definition 2.9.2) locally convex Hausdorff space, then every subset of its dual E' which is bounded for $\sigma(E', E)$ is also bounded for $\beta(E', E)$. This does not follow from Theorem 3, since, as we have observed above (Example 4.5), $\beta(E', E)$ is not necessarily compatible with the duality between E' and E. Also if E is not quasi-complete, the assertion is not necessarily true (Exercise 4.7).

Definition 3. *A set S in a topological vector space E is said to be bornivorous if it absorbs every bounded subset of E.*

By the definition of bounded sets (Definition 2.6.2) every neighborhood of 0 in E is bornivorous.

Proposition 8. *Let E be a locally convex space and E' its dual. A subset M of E' is bounded for the topology $\beta(E', E)$ if and only if it is contained in the polar T° of a bornivorous barrel T of E.*

Proof. Let M be a $\beta(E', E)$-bounded subset of E'. By Proposition 3.1(e) its polar $T = M^{\circ}$ is balanced, convex, $\sigma(E, E')$-closed, and by Proposition 4.3 also closed for the original topology of E. Furthermore, T is bornivorous since given a bounded set A in E, there exists $\lambda = \lambda(A) > 0$ such that $M \subset \lambda A^{\circ}$ and therefore $A \subset A^{\circ\circ} \subset \lambda M^{\circ} = \lambda T$. Thus T is a bornivorous barrel in E and $T^{\circ} = M^{\circ\circ} \supset M$. Conversely, if T is a bornivorous barrel and A a bounded set in E, then there exists $\lambda > 0$ such that $A \subset \lambda T$ and thus $T^{\circ} \subset \lambda A^{\circ}$; that is, T° is $\beta(E', E)$-bounded. Therefore if $M \subset T^{\circ}$, then M is also $\beta(E', E)$-bounded. ∎

Theorem 4. *Let E be a quasi-complete* (Definition 2.9.2), *locally convex Hausdorff space and E' its dual. Then every subset of E' which is bounded for the topology $\sigma(E', E)$ is also bounded for the topology $\beta(E', E)$.*

Proof. Let M be a subset of E' which is bounded for the topology $\sigma(E', E)$. By Proposition 7 there exists a barrel T in E such that $M \subset T^{\circ}$. Every bounded set in E is contained in a balanced, convex, closed bounded set A (p. 109) and by our assumption such a set is complete. Hence T is bornivorous by Theorem 2 and thus M is bounded for $\beta(E', E)$ by Proposition 8. ∎

Exercises

1. Let F and G be two vector spaces which form a pairing separating points of F. Let $\mathfrak{S}$ be a collection of balanced, convex, $\sigma(F, G)$-closed, and $\sigma(F, G)$-bounded subsets of F which satisfies the stability properties (i) and (ii) of Proposition 1. Equip G with the $\mathfrak{S}$-topology. Show that a subset of G^* is an equicontinuous set of linear forms on G if and only if it is contained in the $\sigma(G^*, G)$-closure $\bar{A}$ of some set $A \in \mathfrak{S}$.

2. In the first part of the proof of Theorem 1 show that the sets in $\mathfrak{S}$ cover G using the corollary to Proposition 3. (*Hint:* If $x \in V^{\circ}$ and $y \in W^{\circ}$, then $x + y \in 2(V \cap W)^{\circ}$.)

3. Let E be a vector space and let $\mathcal{T}_1$ and $\mathcal{T}_2$ be two Hausdorff topologies on E compatible with the vector space structure of E. Denote by E_k the space E equipped with the topology $\mathcal{T}_k$ ($k = 1, 2$). Suppose that $\mathcal{T}_1$ is finer than $\mathcal{T}_2$, i.e., that the identity map $i: E_1 \to E_2$ is continuous. By Proposition 2.9.5 there

exists a continuous extension $\hat{\imath}$ of i which maps the completion $\hat{E}_1$ of E_1 into the completion $\hat{E}_2$ of E_2. Suppose that for $\mathcal{T}_1$ there exists a fundamental system of neighborhoods of 0 which are closed for $\mathcal{T}_2$. Show that the map $\hat{\imath}$ is injective. (*Hint:* If $z \in \hat{E}_1$ such that $\hat{\imath}(z) = 0$ and $\mathfrak{F}_1$ is a Cauchy filter on E_1 which converges to z, construct a Cauchy filter $\mathfrak{F}_2$ on E_2 which converges to 0 and is coarser than $\mathfrak{F}_1$. Then show by an argument similar to the one used in the proof of Proposition 5 that $\mathfrak{F}_2$ converges to 0 in the topology $\mathcal{T}_1$.)

4. Show that if A is a bounded, balanced, convex subset of the locally convex Hausdorff space E, then $\overline{A}$ is the closed unit ball of the normed space E_A.

5. Let F and G be two vector spaces which form a pairing and let $\mathfrak{S}$ be a collection of $\sigma(F, G)$-bounded subsets of F. A set S in F is said to be $\mathfrak{S}$-bornivorous if it absorbs every set $A \in \mathfrak{S}$.

(a) Show that a set $M \subset G$ is bounded for the $\mathfrak{S}$-topology in G if and only if it is contained in the polar of an $\mathfrak{S}$-bornivorous subset of F. (*Hint:* See the proofs of Propositions 7 and 8.)

(b) Suppose that the pairing of F and G separates points of F and that $\mathfrak{S}$ is formed by balanced, convex, $\sigma(F, G)$-bounded, and $\sigma(F, G)$-complete subsets of F. Show that every subset of G which is bounded for the topology $\sigma(G, F)$ is also bounded for the $\mathfrak{S}$-topology. (*Hint:* Apply Theorem 2.)

§6. *Barrelled spaces*

PROPOSITION 1. *Let E be a locally convex space and E' its dual.*

(a) *Every equicontinuous subset of E' is bounded for the topology $\beta(E', E)$.*

(b) *Every subset of E' which is bounded for the topology $\beta(E', E)$ is also bounded for the topology $\sigma(E', E)$.*

Proof. (a) Let M be an equicontinuous subset of E'. By Proposition 2.4.4 and Proposition 4.6 we have $M \subset V^{\circ}$, where V is a balanced, convex, closed neighborhood of 0 in E. Since V is also bornivorous, it is a bornivorous barrel. Hence by Proposition 5.8 the set M is bounded for $\beta(E', E)$.

(b) Let M be a $\beta(E', E)$-bounded subset of E'. Then by Proposition 5.8 we have $M \subset T^{\circ}$, where T is a bornivorous barrel in E. It follows from Proposition 5.7 that M is bounded for $\sigma(E', E)$. ∎

REMARK 1. Proposition 1(b) also follows from the fact that the topology $\beta(E', E)$ is finer than $\sigma(E', E)$.

EXAMPLE 1. Let E be a normed vector space and E' its dual. The balls $B_\rho = \rho B_1 = \{x \mid \|x\| \leqq \rho\}$ in E constitute at the same time a fundamental system of neighborhoods of 0 and a fundamental system of bounded sets.

A subset $M \subset E'$ is equicontinuous if and only if it is contained in the polar $\rho^{-1}B_1^{\circ} = (\rho B_1)^{\circ}$ of a ball in E, i.e., if and only if it is bounded in the sense of the norm in E'.

A subset $M \subset E'$ is bounded for $\beta(E', E)$ if and only if the numbers $\|u\| = \sup_{\|x\| \leqq 1} |u(x)|$ are bounded for $u \in M$ (Example 4.5), i.e., if and only if M is bounded in the sense of the norm in E'.

Thus in the dual of a normed vector space the equicontinuous sets are exactly the strongly bounded sets.

EXAMPLE 2. Let E be a Banach space and E' its dual. A subset $M \subset E'$ is bounded for $\sigma(E', E)$ if for each $x \in E$ there exists $\lambda_x > 0$ such that $|u(x)| \leqq \lambda_x$ for all $u \in M$. By the Banach-Steinhaus theorem (Theorem 1.8.1) there exists then a $\mu > 0$ such that $|u(x)| \leqq \mu\|x\|$ for all $u \in M$ and thus

$$\|u\| = \sup_{x \neq 0} \frac{|u(x)|}{\|x\|} \leqq \mu \qquad \text{for all} \qquad u \in M.$$

Thus M is bounded in the sense of the norm of E', and consequently in the dual of a Banach space equicontinuous, $\beta(E', E)$-bounded and $\sigma(E', E)$-bounded subsets coincide.

We shall now characterize those locally convex spaces in whose dual every weakly bounded subset, resp. every strongly bounded subset, is equicontinuous.

DEFINITION 1. *A locally convex space E is said to be barrelled* (kegly; tonnelé in French) *if every barrel* (Definition 5.2) *in E is a neighborhood of* 0.

Let us observe that by Proposition 2.4.4 there exists in a locally convex space a fundamental system of neighborhoods of 0 formed by barrels.

PROPOSITION 2. *A locally convex space E is barrelled if and only if every subset of its dual E' which is bounded for the topology $\sigma(E', E)$ is also equicontinuous.*

Proof. Suppose that E is barrelled. Let M be a $\sigma(E', E)$-bounded subset of E'. By Proposition 5.7 there exists a barrel T in E such that $M \subset T^{\circ}$. But T is a neighborhood of 0, hence M is equicontinuous by Proposition 4.6.

Conversely, suppose that every $\sigma(E', E)$-bounded subset of E' is equicontinuous and let T be a barrel in E. Then by Proposition 5.7 the set T° is $\sigma(E', E)$-bounded in E' and consequently equicontinuous. But then by Proposition 4.7 the set $T = T^{\circ\circ}$ is a neighborhood of 0 in E. ∎

COROLLARY. *Let E be a barrelled locally convex space. Then the following collections of subsets of its dual E' are identical:*

(a) *the equicontinuous sets,*
(b) *the relatively $\sigma(E', E)$-compact sets,*
(c) *the $\beta(E', E)$-bounded sets,*
(d) *the $\sigma(E', E)$-bounded sets.*

Proof. We have (a) $\Rightarrow$ (c) $\Rightarrow$ (d) by Proposition 1, (d) $\Rightarrow$ (a) by Proposition 2, (a) $\Rightarrow$ (b) by Theorem 4.1, and (b) $\Rightarrow$ (d) by Proposition 2.10.7. ∎

In spite of its trivial appearance, Proposition 2 will enable us to find large classes of spaces in whose duals every weakly bounded subset is equicontinuous. First, let us reformulate Baire's theorem (Chapter 1, §8) in the context of topological spaces.

A subset A of a topological space X is *rare* (or nowhere dense) if the closure of A has a void interior, i.e., if $\mathring{\overline{A}} = \emptyset$. Clearly A is rare in X if and only if $\complement\overline{A}$ is everywhere dense in X. We say that the subset A of X is *meager* (of first category) if there exists a sequence $(A_n)_{n\in\mathbf{N}}$ of rare subsets of X such that $A = \bigcup_{n\in\mathbf{N}} A_n$. Finally, we say that a topological space X is a *Baire space* if no nonempty open subset of X is meager. If X is a Baire space, then X itself is not the union of countably many rare subsets; and in particular, if X is the union of a sequence $(F_n)_{n\in\mathbf{N}}$ of closed sets, then at least one set F_n has an interior point.

The topological space X is a Baire space if and only if the intersection of an arbitrary countable family $(G_n)_{n\in\mathbf{N}}$ of everywhere dense open subsets of X is everywhere dense. Indeed, suppose that this condition is satisfied and let $A = \bigcup_{n\in\mathbf{N}} A_n$, where the A_n are rare subsets of X. Then each $\complement\overline{A}_n$ is an everywhere dense open subset of X and thus

$$\bigcap_{n\in\mathbf{N}} \complement\overline{A}_n = \complement \bigcup_{n\in\mathbf{N}} \overline{A}_n$$

is everywhere dense. Hence A cannot be a nonempty open set. Conversely, let X be a Baire space and let $(G_n)_{n\in\mathbf{N}}$ be a sequence of everywhere dense open subsets of E. Then each $A_n = \complement G_n$ is a closed rare subset of X, and therefore $A = \bigcup_{n\in\mathbf{N}} A_n$ has no interior points (since if A contained the open neighborhood V of one of its points, then V would be the union of the rare sets $V \cap A_n$). Thus

$$\complement A = \complement \bigcup_{n\in\mathbf{N}} A_n = \bigcap_{n\in\mathbf{N}} \complement A_n = \bigcap_{n\in\mathbf{N}} G_n$$

is everywhere dense.

BAIRE'S THEOREM. *A complete metrizable topological space is a Baire space.*

Proof. Let X be a complete metrizable space and let δ be a metric on X which defines the topology of X. Let $(G_n)_{n\in\mathbf{N}}$ be a sequence of everywhere dense open subsets of X and let H be an arbitrary nonempty open subset of X. We have to show that $H \cap (\bigcap_{n\in\mathbf{N}} G_n)$ is nonempty.

We shall determine inductively a sequence $(B_{\rho_n}(a_n))_{n\in\mathbf{N}}$ of balls

$$B_{\rho_n}(a_n) = \{x \mid \delta(x, a_n) \leq \rho_n\}$$

such that $B_{\rho_0}(a_0) \subset H$ and

$$B_{\rho_n}(a_n) \subset B_{\rho_{n-1}}(a_{n-1}) \cap G_{n-1}, \qquad \rho_n \leqq \frac{1}{n} \tag{1}$$

for $n \geqq 1$. Suppose that we have already defined the balls $B_{\rho_i}(a_i)$ with the required properties for $0 \leqq i \leqq n-1$. The set $\mathring{B}_{\rho_{n-1}}(a_{n-1}) \cap G_{n-1}$ is open and nonempty. Hence there exists a ball $B_{\rho_n}(a_n)$ satisfying conditions (1).

The sequence $(a_n)_{n\in\mathbf{N}}$ is a Cauchy sequence since for $n, m \geqq k$ we have $a_n, a_m \in B_{\rho_k}(a_k)$ and thus $\delta(a_n, a_m) \leqq 2/k$. By hypothesis (a_n) converges to a point a which belongs to each closed ball $B_{\rho_n}(a_n)$. Hence

$$a \in \bigcap_{n\in\mathbf{N}} B_{\rho_n}(a_n) \subset H \cap \Big(\bigcap_{n\in\mathbf{N}} G_n\Big) \cdot \blacksquare$$

PROPOSITION 3. *A locally convex space E which is a Baire space is barrelled.*

Proof. Let T be a barrel in E. Since T is absorbing, we have

$$\bigcup_{n\geqq 1} nT = E.$$

By the definition of a Baire space, one of the closed sets nT has an interior point. Since $x \mapsto nx$ is a homeomorphism, the set T itself has an interior point x_0. Let V be a balanced neighborhood of 0 in E such that $x_0 + V \subset T$. Since T is balanced, we have $-x_0 + V \subset T$. But then $V \subset T$ since if $x \in V$, then

$$x = \tfrac{1}{2}(x_0 + x) + \tfrac{1}{2}(-x_0 + x) \in T,$$

because T is convex. Hence T is a neighborhood of 0. ∎

From this proposition and Baire's theorem we obtain the following:

COROLLARY. *Every Fréchet space* (Definition 2.9.4) *is barrelled.*

This corollary, combined with Proposition 2, contains the Banach-Steinhaus theorem (Theorem 1.8.1) as a particular case, which has thus been proved anew.

EXAMPLE 3. The spaces $\mathcal{E}^m(\Omega)$, $\mathcal{D}^m(K)$, $\mathcal{B}_0^m(\Omega)$, $\mathcal{B}^m(\Omega)$, $\mathcal{S}_k^m$, $\mathcal{S}^m$ $(0 \leqq m \leqq \infty)$ are barrelled since they are Fréchet spaces (Examples 2.9.2 through 2.9.6 and 2.11.8 through 2.11.10).

PROPOSITION 4. *Let F be a vector space, $(E_\iota)_{\iota\in I}$ a family of barrelled locally convex spaces, and for each $\iota \in I$ let f_ι be a linear map from E_ι into F. Then the space F equipped with the finest locally convex topology for which all the maps f_ι are continuous* (Chapter 2, §12) *is barrelled.*

Proof. Let T be a barrel in F. For each index ι the set $f_\iota^{-1}(T)$ is a barrel in E_ι. Indeed, $f_\iota^{-1}(T)$ is balanced and convex since f_ι is linear, and $f_\iota^{-1}(T)$ is closed since f_ι is continuous. Furthermore, $f_\iota^{-1}(T)$ is absorbing since given $x \in E_\iota$, there exists $\lambda > 0$ such that $\lambda f_\iota(x) \in T$, that is, $\lambda x \in f_\iota^{-1}(T)$. Since E_ι is barrelled, $f_\iota^{-1}(T)$ is a neighborhood of 0 in E_ι. Thus T is a neighborhood of 0 in F. ∎

From Examples 1 and 2 of Chapter 2, §12 we obtain the following:

COROLLARY. (a) *Let E be a barrelled locally convex space and M a subspace of E. Then the quotient space E/M is barrelled.*

(b) *Let $(E_\iota)_{\iota \in I}$ be a family of barrelled locally convex spaces. Then the locally convex direct sum $\coprod_{\iota \in I} E_\iota$ is barrelled.*

It can be shown that the product of barrelled spaces is barrelled (see Exercise 14.2). On the other hand, a closed subspace of a barrelled space is not necessarily barrelled ([52], §31, 5, p. 437 and §27, 2, p. 373).

EXAMPLE 4. The spaces $\mathfrak{D}^m(\Omega)$ and $\mathcal{O}_C^m$ $(0 \leqq m \leqq \infty)$ (Examples 2.12.5 through 2.12.9) are barrelled.

As an application of Proposition 2, we want to generalize Proposition 1.8.1 to barrelled spaces. We first prove two lemmas which will also be useful later.

LEMMA 1. *Let E be a locally convex space over the field $\mathbf{K}$ and E' its dual. Let $\mathfrak{F}$ be a filter basis on E'. For a fixed $x \in E$ and $A \in \mathfrak{F}$ denote by $A(x)$ the subset of $\mathbf{K}$ formed by all elements $x'(x)$, where x' runs through A. Then the collection $\mathfrak{F}(x)$ of all sets $A(x)$, $A \in \mathfrak{F}$, is a filter basis on $\mathbf{K}$. Suppose furthermore that for each x the filter basis $\mathfrak{F}(x)$ converges to some number $u(x)$. Then the map $u: x \mapsto u(x)$ is a linear form on E.*

Proof. The collection $\mathfrak{F}(x)$ is indeed a filter basis since if $A, B, C \in \mathfrak{F}$ and $A \subset B \cap C$, then $A(x) \subset B(x) \cap C(x)$.

Given $x, y \in E$ and $\epsilon > 0$, there exist sets $A, B, C \in \mathfrak{F}$ such that

$$|u(x) - x'(x)| < \frac{\epsilon}{3} \quad \text{for all} \quad x' \in A,$$

$$|u(y) - x'(y)| < \frac{\epsilon}{3} \quad \text{for all} \quad x' \in B$$

and

$$|u(x + y) - x'(x + y)| < \frac{\epsilon}{3} \quad \text{for all} \quad x' \in C.$$

Choosing $x' \in A \cap B \cap C$, we have therefore

$$\begin{aligned} &|u(x + y) - u(x) - u(y)| \\ &\quad = |u(x + y) - u(x) - u(y) - \{x'(x + y) - x'(x) - x'(y)\}| \\ &\quad \leqq |u(x + y) - x'(x + y)| + |u(x) - x'(x)| + |u(y) - x'(y)| < \epsilon. \end{aligned}$$

Since ϵ is arbitrary, we obtain $u(x + y) = u(x) + u(y)$. We can prove in an entirely analogous fashion that $u(\lambda x) = \lambda u(x)$ for $\lambda \in \mathbf{K}$ and $x \in E$. ∎

LEMMA 2. *Let E be a locally convex space and E' its dual. Let M be an equicontinuous subset of E', and let $\mathfrak{F}$ be a filter basis on M. Suppose that for each $x \in E$ the filter basis $\mathfrak{F}(x)$* (cf. Lemma 1) *converges to some number $u(x)$. Then the map $u: x \mapsto u(x)$ is a continuous linear form on E.*

Proof. It follows from Lemma 1 that u is a linear form on E. Given $\epsilon > 0$, there exists a neighborhood V of 0 in E such that $|x'(x)| \leqq \frac{1}{2}\epsilon$ for all $x \in V$ and $x' \in M$. For each $x \in V$ there exists $x' \in M$ such that $|u(x) - x'(x)| \leqq \frac{1}{2}\epsilon$. Thus

$$|u(x)| \leqq |u(x) - x'(x)| + |x'(x)| \leqq \epsilon \quad \text{for} \quad x \in V.$$

Hence u is continuous by Proposition 2.5.1. ∎

Now we are ready to prove the announced result, which is also often called the Banach-Steinhaus theorem.

PROPOSITION 5. *Let E be a barrelled locally convex space and E' its dual. Let $\mathfrak{F}$ be a filter basis on E' which contains a $\sigma(E', E)$-bounded set* [i.e., $\mathfrak{F}$ generates a filter which is bounded for the topology $\sigma(E', E)$ (see Chapter 2, §9)]. *Suppose that for every $x \in E$ the filter basis $\mathfrak{F}(x)$* (cf. Lemma 1) *converges to some number $f(x)$. Then $f: x \mapsto f(x)$ is a continuous linear form on E.*

Proof. Let A be a $\sigma(E', E)$-bounded set belonging to $\mathfrak{F}$. Then A is equicontinuous by Proposition 2. The filter basis $\mathfrak{F}_A = \{A \cap B \mid B \in \mathfrak{F}\}$ induced by $\mathfrak{F}$ on A is such that for every $x \in E$ the filter basis $\mathfrak{F}_A(x)$ converges to $f(x)$. Hence the conclusion follows from Lemma 2. ∎

COROLLARY. *Let E be a barrelled space and suppose that one of the following conditions is satisfied:*

(a) *$(f_n)_{n \in \mathbf{N}}$ is a sequence of continuous linear forms on E such that for every $x \in E$ the sequence $(f_n(x))$ converges to some number $f(x)$.*

(b) *$(f_\epsilon)_{0<\epsilon<\alpha}$ is a family of continuous linear forms on E such that $f_\epsilon(x)$ converges to some number $f(x)$ as $\epsilon \to 0$ for every $x \in E$.*

Then $f: x \mapsto f(x)$ is a continuous linear form on E.

Proof. (1) The elementary filter associated with the sequence (f_n) is bounded for $\sigma(E', E)$. Hence the conclusion holds under condition (a).

(2) Let $\mathfrak{F}$ be the filter basis formed by the sets $\{f_\epsilon \mid \epsilon \leqq \eta\}$, where $0 < \eta < \alpha$. The fact that $f_\epsilon(x)$ converges to $f(x)$ as $\epsilon \to 0$ means precisely that $\mathfrak{F}(x)$ converges to $f(x)$. Now if we set $\epsilon_n = 1/n$ for $n \geqq 1$, then the sequence $f_{\epsilon_n}(x)$ converges to $f(x)$ as $n \to \infty$. Thus by part (1) the conclusion also holds under condition (b). ∎

DEFINITION 2. *A locally convex space E is said to be infrabarrelled* (quasi-barrelled, evaluable) *if every bornivorous* (Definition 5.3) *barrel in E is a neighborhood of* 0.

Clearly, every barrelled space is infrabarrelled. Conversely, there exist infrabarrelled spaces which are not barrelled (see Exercise 7.1).

PROPOSITION 6. *A locally convex space E is infrabarrelled if and only if every subset of its dual E' which is bounded for the topology $\beta(E', E)$ is equicontinuous.*

Proof. Suppose that E is infrabarrelled and let M be a $\beta(E', E)$-bounded subset of E'. By Proposition 5.8 there exists a bornivorous barrel T in E such that $M \subset T^{\circ}$. But T is a neighborhood of 0; hence M is equicontinuous by Proposition 4.6.

Conversely, suppose that every $\beta(E', E)$-bounded subset of E' is equicontinuous and let T be a bornivorous barrel in E. Then by Proposition 5.8 the set T° is $\beta(E', E)$-bounded in E' and consequently equicontinuous. But then by Proposition 4.7 the set $T = T^{\circ\circ}$ is a neighborhood of 0 in E. ∎

It follows from this proposition and from Theorem 5.4 that if a quasi-complete locally convex Hausdorff space is infrabarrelled, then it is barrelled. The following diagram helps to remember some of the relations proved in this and the previous section:

A balanced, convex, closed

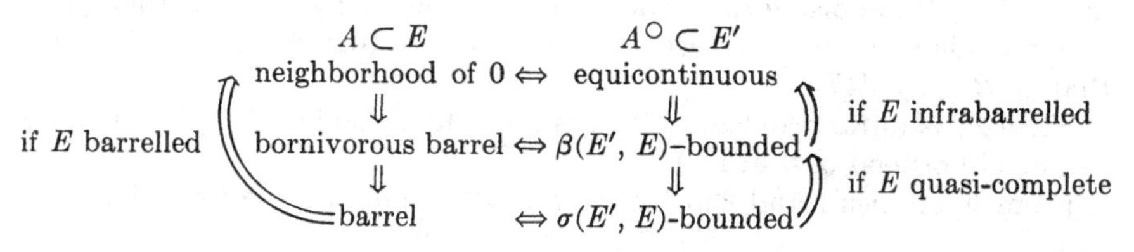

PROPOSITION 7. *Let E be a locally convex Hausdorff space and E' its dual. Every convex subset of E' which is relatively compact for the topology $\sigma(E', E)$ is bounded for the topology $\beta(E', E)$.*

Proof. Let M be a convex, relatively $\sigma(E', E)$-compact subset of E'. By Proposition 5.2 the balanced convex hull N of the closure of M is a balanced, convex, $\sigma(E', E)$-compact subset of E'. Its polar N° is a neighborhood of 0 in E for the Mackey topology $\tau(E, E')$. Let B be an arbitrary bounded subset of E. Then B is also bounded for the Mackey topology (Theorem 5.3) and thus there exists $\lambda > 0$ such that $B \subset \lambda N^{\circ}$. But then $M \subset N = N^{\circ\circ} \subset \lambda B^{\circ}$. Hence by the definition of the topology $\beta(E', E)$ the set M is $\beta(E', E)$-bounded. ∎

REMARK 1. If we do not assume that M is convex, then the conclusion is not necessarily true (see Exercise 7).

PROPOSITION 8. *Let E be an infrabarrelled locally convex Hausdorff space with topology $\mathcal{T}$. Then the topology $\mathcal{T}$ coincides with the Mackey topology $\tau(E, E')$.*

Proof. By Proposition 4.7 and the definition of $\tau(E, E')$ we have to show that the collection of balanced, convex, $\sigma(E', E)$-closed, equicontinuous subsets of E' is identical with the collection of all balanced, convex $\sigma(E', E)$-compact subsets of E'.

By Theorem 4.1 a $\sigma(E', E)$-closed, equicontinuous subset of E' is $\sigma(E', E)$-compact. Conversely, a convex, $\sigma(E', E)$-compact subset of E' is $\sigma(E', E)$-closed. It is also $\beta(E', E)$-bounded by Proposition 7. Hence it is equicontinuous by Proposition 6 and the assumption that E is infrabarrelled. ∎

We shall see in the next section that every metrizable locally convex space is infrabarrelled.

PROPOSITION 9. *Let F be a vector space, $(E_\iota)_{\iota \in I}$ a family of infrabarrelled locally convex spaces, and for each $\iota \in I$ let f_ι be a linear map from E_ι into F. Then the space F equipped with the finest locally convex topology for which all the maps f_ι are continuous* (Chapter 2, §12) *is infrabarrelled.*

Proof. Let T be a bornivorous barrel in F. For each index ι the set $f_\iota^{-1}(T)$ is a bornivorous barrel in E_ι. Indeed, $f_\iota^{-1}(T)$ is balanced and convex since f_ι is linear, and $f_\iota^{-1}(T)$ is closed since f_ι is continuous. Furthermore, $f_\iota^{-1}(T)$ is bornivorous since if B is a bounded set in E_ι, then $f_\iota(B)$ is a bounded set in F. Hence there exists $\lambda > 0$ such that $f_\iota(B) \subset \lambda T$, that is, $B \subset \lambda f_\iota^{-1}(T)$.

Since E_ι is infrabarrelled, $f_\iota^{-1}(T)$ is a neighborhood of 0 in E_ι. Thus T is a neighborhood of 0 in F. ∎

From Examples 1 and 2 of Chapter 2, §12 we obtain the following:

COROLLARY. (a) *Let E be an infrabarrelled locally convex space and M a subspace of E. Then the quotient space E/M is infrabarrelled.*

(b) *Let $(E_\iota)_{\iota \in I}$ be a family of infrabarrelled locally convex spaces. Then the locally convex direct sum $\coprod_{\iota \in I} E_\iota$ is infrabarrelled.*

It can be shown that the product of infrabarrelled spaces is infrabarrelled (see Exercise 14.2). On the other hand, a closed subspace of a barrelled space is not even necessarily infrabarrelled ([52], §27, 1, pp. 371–372 and §31, 5, p. 437).

If E is a barrelled space, then its dual E' is quasi-complete for the topology $\sigma(E', E)$ since a bounded, closed subset of E' is compact by the Corollary of Proposition 2 and thus complete. A much stronger result is:

THEOREM 1. *Let E be a barrelled locally convex space and E' its dual. Let $\mathfrak{S}$ be a collection of bounded subsets of E which cover E. Then E' is quasi-complete for the $\mathfrak{S}$-topology.*

Proof. Let M be a bounded, closed subset of E' for the $\mathfrak{S}$-topology and let $\mathfrak{F}$ be a Cauchy filter on M. Then M is bounded for the topology $\sigma(E', E)$ since given $x \in E$, there exists $B \in \mathfrak{S}$ such that $x \in B$ and

$$|\langle x, x'\rangle| \leqq \sup_{y \in B} |\langle y, x'\rangle| < \infty$$

for all $x' \in M$. Thus M is equicontinuous by Proposition 2.

For each $x \in E$ the collection $\mathfrak{F}(x)$ (cf. Lemma 1) is the basis of a Cauchy filter on $\mathbf{K}$. Indeed, given $\epsilon > 0$ and $x \in B \in \mathfrak{S}$, there exists $A \in \mathfrak{F}$ such that $A - A \subset \epsilon B^{\circ}$ and in particular $|x'(x) - y'(x)| \leqq \epsilon$ for $x', y' \in A$. Since $\mathbf{K}$ is complete, $\mathfrak{F}(x)$ converges to an element $u(x) \in \mathbf{K}$ and by Lemma 2 the map $u: x \mapsto u(x)$ is a continuous linear form on E.

Let us prove that $\mathfrak{F}$ converges to u for the $\mathfrak{S}$-topology. Since $\mathfrak{F}$ is a Cauchy filter, for every $B \in \mathfrak{S}$ and $\epsilon > 0$ there exists $A \in \mathfrak{F}$ such that $|x'(x) - y'(x)| < \frac{1}{2}\epsilon$ for all $x \in B$ and $x', y' \in A$. On the other hand, for every $x \in B$ there exists $x' \in A$ such that $|u(x) - x'(x)| < \frac{1}{2}\epsilon$. Hence

$$|u(x) - y'(x)| \leqq |u(x) - x'(x)| + |x'(x) - y'(x)| < \epsilon$$

for every $x \in B$ and $y' \in A$, that is, $A \subset u + \epsilon B^{\circ}$. If now $B_i \in \mathfrak{S}$ for $1 \leqq i \leqq n$, and $\epsilon > 0$, then there exists $A \in \mathfrak{F}$ such that

$$A \subset u + \epsilon \bigcap_{i=1}^{n} B_i^{\circ},$$

which proves our assertion. ∎

Exercises

1. Let E be a topological space. A function $f: E \to \mathbf{R}$ is said to be *lower semi-continuous* at the point $a \in E$ if for every $\epsilon > 0$ there exists a neighborhood V of a such that $f(x) > f(a) - \epsilon$ for every $x \in V$. We say that f is lower semi-continuous on E if it is lower semi-continuous at every point on E.

(a) Prove that f is lower semi-continuous on E if and only if for every $\lambda \in \mathbf{R}$ the set $\{x \mid f(x) \leqq \lambda\}$ is closed in E.

(b) Let q be a semi-norm on a locally convex space. Prove that the set $\{x \mid q(x) \leqq 1\}$ is a barrel if and only if q is lower semi-continuous.

(c) Prove that a locally convex space E is barrelled if and only if every lower semi-continuous semi-norm on E is continuous.

2. Prove that the completion of an infrabarrelled locally convex Hausdorff space is barrelled. (*Hint:* Use Exercise 4.8.)

3. (a) Prove that if a filter $\mathfrak{F}$ has a countable basis, then it is the intersection of all the elementary filters finer than $\mathfrak{F}$. (*Hint:* If $(A_n)_{n \in \mathbf{N}}$ is a basis of $\mathfrak{F}$, set $B_n = \bigcap_{p=0}^{n} A_p$. Then $(B_n)_{n \in \mathbf{N}}$ is a basis of $\mathfrak{F}$ and $B_{n+1} \subset B_n$. If $a_n \in B_n$, then $\mathfrak{F}$ is coarser than the elementary filter associated with (a_n). The intersection of the elementary filters finer than $\mathfrak{F}$ is a filter $\mathfrak{I}$ finer than $\mathfrak{F}$. If $\mathfrak{I}$ were strictly finer than $\mathfrak{F}$, there would exist $M \in \mathfrak{I}$ such that $B_n \cap \complement M \neq \emptyset$ for all $n \in \mathbf{N}$.

Picking $b_n \in B_n \cap \complement M$, the elementary filter associated with (b_n) would be finer than $\mathfrak{F}$, but M would not belong to it.)

(b) Let E be a barrelled Hausdorff space, E' its dual, and $\mathfrak{F}$ a filter on E' which has a countable basis. Suppose that for every $x \in E$ the filter basis $\mathfrak{F}(x)$ (cf. Lemma 1) converges to some number $f(x)$. Prove that $f: x \mapsto f(x)$ is a continuous linear form on E. (*Hint:* Use (a) and the corollary to Proposition 5.)

4. Give an example of a locally convex Hausdorff space which is quasi-complete but not complete. (*Hint:* Consider the space l^1 with the topology $\sigma(l^1, c_0)$ (cf. Example 1.7.2). Use Theorem 1 and Proposition 2.5.)

5. Let E be a locally convex Hausdorff space with topology $\mathcal{T}$ and let E' be its dual. Denote by $\beta(E, E')$ the topology of the uniform convergence on $\sigma(E', E)$-bounded subsets of E' on E and by $\beta^*(E, E')$ the topology of the uniform convergence on $\beta(E', E)$-bounded subsets of E' on E.

(a) Prove that

$$\beta(E, E') \succcurlyeq \beta^*(E, E') \succcurlyeq \tau(E, E') \succcurlyeq \mathcal{T},$$

where $\succcurlyeq$ means "is finer than."

(b) Prove that E is barrelled if and only if $\beta(E, E') = \mathcal{T}$.

(c) Prove that E is infrabarrelled if and only if $\beta^*(E, E') = \mathcal{T}$.

6. Show that if in a Fréchet space E there exists a countable fundamental system of bounded sets (cf. Chapter 2, §6), then E is normable. (*Hint:* Use Baire's theorem and Proposition 2.6.1.)

7. Give an example of a $\sigma(E', E)$-compact set which is not $\beta(E', E)$-bounded. (*Hint:* Let E be the subspace of l^2 considered in Exercise 4.7 and consider in E' the sequence nu_n, where the linear form u_n is given by $(\xi_i) \mapsto \xi_n$.)

§7. *Bornological spaces*

We pointed out in Exercise 1.4.3 that if f is a linear map from a normed vector space E into a normed vector space F which maps bounded sets into bounded sets, then f is continuous. We shall now characterize those locally convex spaces E which possess this property.

DEFINITION 1. *A locally convex space E is said to be bornological if every balanced, convex, bornivorous* (Definition 5.3) *subset of E is a neighborhood of* 0.

Since a bornivorous barrel is in particular a balanced, convex, bornivorous set, it follows that every bornological space is infrabarrelled.

PROPOSITION 1. (a) *Let E be a bornological locally convex space and F an arbitrary locally convex space. If f is a linear map from E into F which maps bounded sets into bounded sets, then f is continuous.*

(b) *If E is a locally convex space such that, given a normed space F and a linear map f from E into F which maps bounded sets into bounded sets, f is necessarily continuous, then E is bornological.*

Proof. (a) Suppose that E is bornological and let $f: E \to F$ be a linear map into the locally convex space F which maps bounded sets into bounded sets. Let V be a balanced, convex neighborhood of 0 in F and let A be a bounded set in E. Then $f(A)$ is a bounded set in F; hence there exists $\lambda > 0$ such that $f(A) \subset \lambda V$, that is, $A \subset \lambda f^{-1}(V)$. Thus $f^{-1}(V)$ is balanced, convex, and bornivorous and therefore a neighborhood of 0 in E. Hence f is continuous.

(b) Suppose that E has the property indicated in part (b) of the proposition, and let V be a balanced, convex, bornivorous subset of E. Let E_V be the normed space introduced in §5 (following Proposition 5.6) and f the canonical surjection from E onto E_V. If A is a bounded set in E, then there exists $\lambda > 0$ such that $A \subset \lambda V$, that is, $q_V(x) \leqq \lambda$ for all $x \in A$. Hence $f(A)$ is bounded in E_V and thus by our assumption f is continuous. But if B_ρ is the ball with center 0 and radius ρ in E_V, then $V \supset f^{-1}(B_\rho)$ for $\rho < 1$ and thus V is a neighborhood of 0 in E. ∎

We shall now give a characterization of bornological spaces in terms of linear forms instead of linear maps.

Proposition 2. *A locally convex Hausdorff space E is bornological if and only if the following two conditions are satisfied:*

(i) *The topology $\mathcal{T}$ of E coincides with the Mackey topology $\tau(E, E')$ on E.*

(ii) *If a linear form f transforms bounded sets of E into bounded sets of* $\mathbf{K}$, *then f is continuous.*

Proof. (a) Suppose that E is bornological. Then E is infrabarrelled and (i) follows from Proposition 6.8. Setting $F = \mathbf{K}$ in Proposition 1(a) we obtain (ii).

(b) Suppose that conditions (i) and (ii) are satisfied. By Proposition 2.4.5 the collection $\mathfrak{N}$ of all balanced, convex, bornivorous subsets form a fundamental system of neighborhoods of 0 for a locally convex topology $\mathcal{T}'$ on E. The topology $\mathcal{T}'$ is finer than $\mathcal{T}$ since every balanced, convex neighborhood of 0 is in particular bornivorous.

On the other hand, $\mathcal{T}'$ is coarser than $\tau(E, E')$. Indeed, let f be a $\mathcal{T}'$-continuous linear form on E. There exists $V \in \mathfrak{N}$ such that $|f(x)| \leqq 1$ for $x \in V$. If A is a $\mathcal{T}$-bounded subset of E, then there exists $\lambda > 0$ such that $A \subset \lambda V$. Hence $|f(x)| \leqq \lambda$ for $x \in A$. Thus f maps bounded subsets of E into bounded subsets of $\mathbf{K}$ and is therefore $\mathcal{T}$-continuous by condition (ii). Since conversely every $\mathcal{T}$-continuous linear form on E is $\mathcal{T}'$-continuous, the dual space of E for the topology $\mathcal{T}'$ is the same as for the topology $\mathcal{T}$. Hence by Proposition 5.4 the topology $\mathcal{T}'$ is indeed coarser than $\tau(E, E')$.

But then by condition (i) the topology $\mathcal{T}'$ must coincide with both $\mathcal{T}$ and $\tau(E, E')$ and in particular every balanced, convex, $\mathcal{T}$-bornivorous set in E is a neighborhood of 0 for $\mathcal{T}$; i.e., E is bornological. ∎

Proposition 3. *Every locally convex space in which the origin has a countable fundamental system of neighborhoods is bornological. In particular, every metrizable locally convex space is bornological.*

Proof. Let $(V_n)_{n \geqq 1}$ be a countable, decreasing (i.e., $V_n \supset V_{n+1}$) fundamental system of balanced neighborhoods of 0 in E, and let U be a balanced, convex, bornivorous subset of E. Suppose that U contains no set of the form $(1/n)V_n$. For each $n \geqq 1$ let x_n be a point of $(1/n)V_n$ which does not belong to U. The sequence $(nx_n)_{n \geqq 1}$ converges to 0, and thus the set $A = \{nx_n \mid n \geqq 1\}$ is bounded. However, U does not absorb A since if $nx_n \in \lambda U$, $\lambda > 0$, for all $n \geqq 1$, then $x_n \in (\lambda/n)U \subset U$ for $n \geqq \lambda$, contradicting the choice of x_n. But this is impossible, since U is bornivorous and so we have proved that U contains a set $(1/n)V_n$ and is therefore itself a neighborhood of 0. ▮

More particularly, a metrizable locally convex space is always infrabarrelled, as we already mentioned in the previous section.

Example 1. The spaces $\mathcal{E}^m(\Omega)$, $\mathcal{D}^m(K)$, $\mathcal{B}_0^m(\Omega)$, $\mathcal{B}^m(\Omega)$, $\mathcal{S}_k^m$, $\mathcal{S}^m$, where $0 \leqq m \leqq \infty$, are bornological and infrabarrelled since they are metrizable (Examples 2.6.1 through 2.6.3).

Proposition 4. *Let F be a vector space, $(E_\iota)_{\iota \in I}$ a family of bornological locally convex spaces, and for each $\iota \in I$ let f_ι be a linear map from E_ι into F. Then the space F equipped with the finest locally convex topology for which all the maps f_ι are continuous* (Chapter 2, §12) *is bornological.*

Proof. Let U be a balanced, convex, bornivorous set in F. For each index ι the set $f_\iota^{-1}(U)$ is balanced and convex in E_ι since f_ι is linear. Furthermore, if B is a bounded set in E_ι, then $f_\iota(B)$ is bounded in F since f_ι is continuous. Hence there exists $\lambda > 0$ such that $f_\iota(B) \subset \lambda U$, that is, $B \subset \lambda f_\iota^{-1}(U)$. Thus we have proved that $f_\iota^{-1}(U)$ is a balanced, convex, bornivorous set in E_ι.

Since each E_ι is bornological, each set $f_\iota^{-1}(U)$ is a neighborhood of 0. Hence U is a neighborhood of 0 in F. ▮

From Examples 2.12.1 and 2 we obtain the following:

Corollary. (a) *Let E be a bornological locally convex space and M a subspace of E. Then the quotient space E/M is bornological.*

(b) *Let $(E_\iota)_{\iota \in I}$ be a family of bornological locally convex spaces. Then the locally convex direct sum $\coprod_{\iota \in I} E_\iota$ is bornological.*

Example 2. The spaces $\mathcal{D}^m(\Omega)$ and $\mathcal{O}_C^m$ $(0 \leqq m \leqq \infty)$ (Examples 2.12.5 through 2.12.9) are bornological and infrabarrelled.

It can be shown that the product of a countable family of bornological spaces is bornological ([52], §28,4.(4), p. 387). The problem, whether the product of an arbitrary family of bornological spaces is bornological, seems to be unsolved and is closely related to some problems posed by Ulam and Mackey in the theory of infinite cardinals and in measure theory. (For an account and references, see [52], §28,8, pp. 393–395 and [43], pp. 279–281.) A closed subspace of a bornological space is not necessarily bornological ([52], §28,4, p. 387).

Propositions 3 and 4 have a kind of converse. Let E be a bornological locally convex space and let A be a bounded, balanced, closed, convex subset of E. As in §5 (cf. Proposition 5.6), we denote by E_A the subspace of E generated by A and equipped with the semi-norm $q_A =$ the gauge of A. Denote by f_A the canonical injection $E_A \hookrightarrow E$. Let $\mathcal{T}$ be the original topology of E and let $\mathcal{T}'$ be the finest locally convex topology on E for which all the maps f_A are continuous. We want to show that the topologies $\mathcal{T}$ and $\mathcal{T}'$ are identical. Let V be a balanced, convex neighborhood of 0 in E for $\mathcal{T}$. Then $\lambda A \subset V$ for some $\lambda > 0$; that is, $\lambda A \subset f_A^{-1}(V)$. Hence for every A the set $f_A^{-1}(V)$ is a neighborhood of 0 in E_A, and thus V is a neighborhood of 0 in E for $\mathcal{T}'$. Conversely, let V be a balanced, convex neighborhood of 0 in E for $\mathcal{T}'$. Then for each A there exists $\lambda > 0$ such that $\lambda A \subset f_A^{-1}(V)$ and *a fortiori* $\lambda A \subset V$. Thus V is a bornivorous set and therefore a neighborhood of 0 for the bornological topology $\mathcal{T}$. We have therefore proved the first part of the following result:

PROPOSITION 5. *Let F be a bornological locally convex space with topology $\mathcal{T}$. Then there exists a family $(E_\iota)_{\iota \in I}$ of locally convex vector spaces, whose topology can be defined with the help of a single semi-norm, and a family $(f_\iota)_{\iota \in I}$ of linear maps $f_\iota : E_\iota \to F$, so that $\mathcal{T}$ is the finest locally convex topology for which the maps f_ι are continuous and furthermore $F = \bigcup_{\iota \in I} f_\iota(E_\iota)$.*

If F is a Hausdorff space, then the E_ι can be taken to be normed spaces. If furthermore F is quasi-complete, then the E_ι can be taken to be Banach spaces.

The last two statements follow from Proposition 5.6.

In Chapter 1, §7 we showed that the dual of a normed space is always complete for the normed topology. We shall now generalize this fact.

PROPOSITION 6. *Let E be a bornological space and E' its dual. Then E' equipped with the topology $\beta(E', E)$ is complete.*

Proof. Let $\mathfrak{F}$ be a Cauchy filter on E'. For each $x \in E$ the filter basis $\mathfrak{F}(x)$ introduced in Lemma 6.1 is the basis of a Cauchy filter on $\mathbf{K}$ since given $\epsilon > 0$ and a bounded set B in E containing x, there exists $A \in \mathfrak{F}$ such that $A - A \subset \epsilon B^\circ$, and in particular $|x'(x) - y'(x)| \leqq \epsilon$ for $x', y' \in A$.

Since $\mathbf{K}$ is complete, $\mathfrak{F}(x)$ converges to an element $u(x) \in \mathbf{K}$, and by Lemma 6.1 the map $u : x \mapsto u(x)$ is a linear form on E. Let us show that u is continuous. Let B be a bounded subset of E. There exists $A \in \mathfrak{F}$ such that $A - A \subset B^\circ$; i.e., $|x'(x) - y'(x)| \leqq 1$ for $x \in B$ and $x', y' \in A$. In particular, there exists $\alpha > 0$ such that $|x'(x)| \leqq \alpha$ for $x \in B$, $x' \in A$.

On the other hand, given $x \in B$, there exists $x' \in A$ such that

$$|u(x) - x'(x)| \leqq \alpha.$$

Hence

$$|u(x)| \leqq |u(x) - x'(x)| + |x'(x)| \leqq 2\alpha$$

for all $x \in B$; that is, u maps B into a bounded subset of $\mathbf{K}$. Since E is bornological, it follows from Proposition 1 that u is continuous.

Finally, $\mathfrak{F}$ converges to u for the topology $\beta(E', E)$. Indeed, since $\mathfrak{F}$ is a Cauchy filter, for every bounded subset B of E and $\epsilon > 0$ there exists $A \in \mathfrak{F}$ such that $|x'(x) - y'(x)| < \frac{1}{2}\epsilon$ for all $x \in B$ and $x', y' \in A$. On the other hand, for every $x \in B$ there exists $x' \in A$ such that

$$|u(x) - x'(x)| < \tfrac{1}{2}\epsilon.$$

Hence

$$|u(x) - y'(x)| \leqq |u(x) - x'(x)| + |x'(x) - y'(x)| < \epsilon$$

for every $x \in B$ and $y' \in A$; that is, $A \subset u + \epsilon B^{\circ}$. If now $(B_i)_{1 \leqq i \leqq n}$ is a finite family of bounded subsets of E and $\epsilon > 0$, then there exists $A \in \mathfrak{F}$ such that

$$A \subset u + \epsilon \bigcap_{i=1}^{n} B_i^{\circ},$$

which proves our assertion. ▮

Exercises

1. Give an example of a locally convex space which is bornological, and thus also infrabarrelled, but which is not barrelled. (*Hint:* Use Exercise 4.7, Proposition 3, and the corollary to Proposition 6.2.)

Remark 1. There exist barrelled spaces which are not bornological [67, 95].

2. Prove that a locally convex space E is bornological if and only if every semi-norm on E which is bounded on every bounded subset of E is continuous.

3. Let E be a locally convex space. Prove that the following conditions are equivalent:

(i) E is bornological.

(ii) If M is a set of linear maps from E into a locally convex space F such that for every bounded set $A \subset E$ the set $M(A) = \bigcup_{u \in M} \bigcup_{x \in A} u(x)$ is bounded in F, then M is equicontinuous.

(*Hint:* Prove (i) $\Rightarrow$ (ii) as in Proposition 1(a). Then prove (ii) $\Rightarrow$ (b) of Proposition 1.)

4. Let E be a locally convex Hausdorff space. Prove that the following conditions are equivalent:

(α) E is bornological.

(β) If M is a set of linear forms on E which is uniformly bounded on every bounded subset of E, then M is equicontinuous.

(*Hint:* (α) $\Rightarrow$ (ii) of Exercise 3 $\Rightarrow$ (β). To prove (β) $\Rightarrow$ (α), use Proposition 2. To show that (β) $\Rightarrow$ (i) of Proposition 2, observe that every balanced, convex, $\sigma(E', E)$-compact set in E' is equicontinuous.)

5. (a) Let E be a locally convex space and suppose that given a normed space F and a linear map $f: E \to F$ which maps sequences converging to 0 into bounded sequences, f is necessarily continuous. Prove that E is bornological. (*Hint:* Use Proposition 1(b).)

(b) Let E be a metrizable locally convex space. Prove that every balanced subset of E which absorbs all sequences converging to 0 is a neighborhood of 0. (*Hint:* Use an argument similar to that employed in the proof of Proposition 3.)

(c) Using (a) and (b), prove that a metrizable locally convex space is bornological.

6. (a) Let E and F be two topological vector spaces and let $f: E \to F$ be a linear map which transforms sequences converging to 0 into bounded sequences. Prove that f transforms bounded sets into bounded sets. (*Hint:* Use an argument similar to that employed in the proof of Proposition 3.)

(b) Let E be a bornological space, F a locally convex space, and $f: E \to F$ a linear map which transforms sequences converging to 0 into bounded sequences. Prove that f is continuous. (*Hint:* Use (a) and Proposition 1(a).)

7. Let E be a locally convex Hausdorff space. A sequence (x_n) of elements of E is said to *converge* to $x_0 \in E$ *in the Mackey sense* (or locally) if there exists a bounded, balanced, convex subset A of E so that (x_n) converges to x_0 in the normed space E_A (cf. Proposition 5.6).

(a) Prove that (x_n)converges to x_0 in the Mackey sense if and only if $(x_n - x_0)$ converges to 0 in the Mackey sense.

(b) Prove that (x_n) converges to 0 in the Mackey sense if and only if there exists a sequence (λ_n) of strictly positive numbers tending to $+\infty$ such that $(\lambda_n x_n)$ converges to 0 in the ordinary sense.

(c) Prove that if a sequence converges to 0 in the Mackey sense, then it converges to 0 in the ordinary sense.

(d) Prove that if E is metrizable, then every sequence which converges to 0 in the ordinary sense also converges to 0 in the Mackey sense.

(e) Prove that E is bornological if and only if every balanced, convex set which absorbs all sequences which converge to zero in the Mackey sense is a neighborhood of 0. (*Hint:* If M does not absorb the bounded set B, pick $x_n \in B$ such that $x_n/n^3 \notin M$ and consider the sequence x_n/n^2.)

(f) Let E and F be two locally convex Hausdorff spaces and $u: E \to F$ a linear map. Prove that the following conditions are equivalent:

(i) u transforms bounded sets into bounded sets.

(ii) u transforms sequences which converge to 0 in the Mackey sense into sequences which converge to 0 in the Mackey sense.

(iii) u transforms sequences which converge to 0 in the Mackey sense into sequences which converge to 0 in the ordinary sense.

(iv) u transforms sequences which converge to 0 in the Mackey sense into bounded sequences.

(*Hint:* To prove (iv) $\Rightarrow$ (i), suppose that B is bounded in E, but that there exists a neighborhood V of 0 in F such that $u(B) \not\subset n^3V$ for all $n \geqq 1$. Pick $x_n \in B$ such that $u(x_n)/n^3 \notin V$. Then x_n/n^2 tends to 0 in the Mackey sense, but $u(x_n/n^2)$ is not bounded.)

(g) Using (d) and (f), prove that a metrizable locally convex space is bornological. (*Hint:* If $u: E \to F$ satisfies (i) of (f), then it satisfies (iii) of (f). If E is metrizable, then by (d) the map u transforms sequences tending to 0 into sequences tending to 0; that is, u is continuous. Apply Proposition 1(b).)

(h) Let E be a bornological Hausdorff space, F a locally convex Hausdorff space, and $u: E \to F$ a linear map. Prove that (i) through (iv) of (f) are equivalent to the following conditions:

(v) u transforms sequences which converge to 0 in the ordinary sense into sequences which converge to 0 in the ordinary sense.

(vi) u is continuous.

8. Let E be a locally convex space with topology $\mathcal{T}$.

(a) Prove that among all the locally convex topologies on E for which the bounded sets are the same as for $\mathcal{T}$ there exists a finest one $\mathcal{T}'$, which is the only one among these topologies for which E is bornological. The vector space E equipped with the topology $\mathcal{T}'$ is called the bornological space associated with E. (*Hint:* The collection $\mathfrak{N}$ of all balanced, convex, bornivorous subsets of E forms a fundamental system of neighborhoods of 0 for the topology $\mathcal{T}'$.)

(b) Let F be a locally convex space. Prove that a linear map $u: E \to F$ transforms every $\mathcal{T}$-bounded set of E into a bounded set of F if and only if it is continuous for the topology $\mathcal{T}'$.

(c) For every bounded, balanced, closed, convex subset A of E let E_A be the space introduced in §5 and let f_A be the canonical injection $E_A \hookrightarrow E$. Prove that $\mathcal{T}'$ is the finest locally convex topology on E for which the maps f_A are continuous.

§8. *Reflexivity*

DEFINITION 1. *Let E be a locally convex Hausdorff space and E' its dual. The dual E'' of E' equipped with the strong topology $\beta(E', E)$ is called the* bidual (second conjugate, second adjoint) *of E.*

If E is a normed space, this definition coincides with the one given in Chapter 1, §7. Let x be an element of E and define the linear form $\tilde{x}$ on E' by writing

$$\tilde{x}(x') = \langle x, x' \rangle \quad \text{for all} \quad x' \in E'.$$

We know from Proposition 2.2 that $\tilde{x}$ is continuous on E' for the topology $\sigma(E', E)$, and thus $\tilde{x}$ is also continuous on E' for the finer topology $\beta(E', E)$; that is, $\tilde{x}$ is an element of the bidual E''. The map $x \mapsto \tilde{x}$ from E into E'' is clearly linear. Since we suppose that E is a locally convex Hausdorff space, the pairing of E and E' separates points of E (Example 2.2), and therefore $\tilde{x} = 0$ implies $x = 0$; i.e., the map $x \mapsto \tilde{x}$ is also injective. We shall call this map $x \mapsto \tilde{x}$ the *canonical imbedding* of the (algebraic) vector space E into the (algebraic) vector space E''. It is possible to consider E as a linear subspace of E''. If we do so, we say that E is identified canonically with a subspace of E''. We will then have the nested sequence of spaces $E \subset E'' \subset (E')^*$ (cf. §2). Our goal is to investigate under what conditions this canonical imbedding is surjective or is an isomorphism for the topology $\beta(E'', E')$ on E''.

DEFINITION 2. *A locally convex Hausdorff space E is said to be semi-reflexive if the canonical imbedding from E into its bidual E'' is surjective.*

The fact that E is semi-reflexive is clearly equivalent to any one of the following statements:

(a) Every linear form on E' which is continuous for the topology $\beta(E', E)$ is of the form $x' \mapsto \langle x, x' \rangle$ for some $x \in E$.

(b) Every linear form on E' which is continuous for the topology $\beta(E', E)$ is also continuous for $\sigma(E', E)$.

(c) The topology $\beta(E', E)$ on E' is compatible with the duality between E and E'.

In general, the strong topology $\beta(E', E)$ on the dual E' of a locally convex Hausdorff space E is finer than the Mackey topology $\tau(E', E)$ (Definition 5.1) since every $\sigma(E, E')$-compact set is bounded (Proposition 2.10.7). Therefore it follows from the characterization (c) above and from Proposition 5.4 that E is semi-reflexive if and only if the topologies $\beta(E', E)$ and $\tau(E', E)$ are identical. This can be expressed also in several other forms.

PROPOSITION 1. *Let E be a locally convex Hausdorff space and E' its dual. The space E is semi-reflexive if and only if every bounded, $\sigma(E, E')$-closed subset of E is compact for the topology $\sigma(E, E')$.*

Proof. If E is semi-reflexive, then in particular the bidual E'' of E is contained in E. Hence it follows from the corollary to Proposition 5.3, applied to $F = E$, $G = E'$, and $\mathfrak{S} =$ the collection of $\sigma(E, E')$-bounded subsets of E, that the balanced, convex, $\sigma(E, E')$-closed hull of every $\sigma(E, E')$-bounded set is $\sigma(E, E')$-compact. Therefore every bounded, $\sigma(E, E')$-closed set is also $\sigma(E, E')$-compact.

Conversely, suppose that every bounded, $\sigma(E, E')$-closed subset of E is $\sigma(E, E')$-compact. By Proposition 4.2, we can consider the strong topology $\beta(E', E)$ as the topology of the uniform convergence on all balanced, convex, $\sigma(E, E')$-closed, and $\sigma(E, E')$-bounded subsets of E. But these sets are then precisely the balanced, convex, $\sigma(E, E')$-compact subsets of E; that is, $\beta(E', E)$ coincides with the Mackey topology $\tau(E', E)$. ∎

Proposition 2. *Let E be a locally convex Hausdorff space and E' its dual. The space E is semi-reflexive if and only if it is quasi-complete for the topology $\sigma(E, E')$.*

Proof. Suppose that E is semi-reflexive and let A be a bounded, closed subset of E with respect to the topology $\sigma(E, E')$. By Proposition 1 the set A is compact and hence also complete.

Conversely, suppose that E is quasi-complete for $\sigma(E, E')$ and let A be a bounded, $\sigma(E, E')$-closed subset of E. Since $E \subset E'^*$ and $\sigma(E'^*, E')$ induces on E the topology $\sigma(E, E')$, the set A is bounded in E'^* for $\sigma(E'^*, E')$. It follows from Example 2.3 that A is relatively compact in E'^* for $\sigma(E'^*, E')$. On the other hand, by virtue of Proposition 2.5, the space E'^* equipped with the topology $\sigma(E'^*, E')$ is the completion of E equipped with the topology $\sigma(E, E')$, and therefore A is a precompact subset of E (Theorem 2.10.2). By our hypothesis A is complete; hence it is compact for $\sigma(E, E')$. It follows from Proposition 1 that E is semi-reflexive. ∎

Proposition 3. *Let E be a locally convex Hausdorff space and E' its dual. Then E is semi-reflexive if and only if every convex subset of E' which is closed for the topology $\beta(E', E)$ is also closed for the topology $\sigma(E', E)$.*

Proof. If E is semi-reflexive, then $\beta(E', E)$ is compatible with the duality between E and E', and thus by Proposition 4.3 every $\beta(E', E)$-closed convex set is also $\sigma(E', E)$-closed.

Conversely, if every $\beta(E', E)$-closed convex set is also $\sigma(E', E)$-closed, then in particular every $\beta(E', E)$-closed hyperplane in E' is also $\sigma(E', E)$-closed, which, by Proposition 2.5.7, means that every $\beta(E', E)$-continuous linear form on E' is also $\sigma(E', E)$-continuous. ∎

We shall see in §§14 and 15 that a closed subspace of a semi-reflexive space is semi-reflexive (Proposition 15.2), and that the product (Proposition 14.4) and the locally convex direct sum of a family of semi-reflexive spaces are semi-reflexive (Exercise 14.4). On the other hand, the quotient space of a semi-reflexive space modulo a closed subspace is not necessarily semi-reflexive ([52], §23,5, p. 307).

Proposition 4. *The dual E' of a semi-reflexive locally convex Hausdorff space E equipped with the topology $\beta(E', E)$ is barrelled.*

Proof. Let T be a barrel in E'. Then its polar T° is a bounded set for the topology $\sigma(E, E') = \sigma(E'', E')$ on $E = E''$. Thus $T^{\circ\circ} = T$ is a neighborhood of 0 in E' for the topology $\beta(E', E)$. ∎

Remark 1. The converse of Proposition 4 is false. Indeed, the dual of a nonreflexive Banach space (see Example 1.7.4) is a Banach space and in particular a barrelled space (cf. Proposition 16.1).

Let E be a locally convex Hausdorff space, E' its dual and E'' its bidual. By the definition of E'' and by Example 2.2 the two spaces E' and E'' form a dual system. An equicontinuous subset of E' is bounded for the topology $\beta(E', E)$ (Proposition 6.1) and therefore also for the topology $\sigma(E', E'')$ (Theorem 5.3). Let us denote by $\epsilon(E'', E')$ the topology of the uniform convergence on all equicontinuous subsets of E'. By Proposition 4.7 the topology $\epsilon(E'', E')$ induces on E its original topology.

The topology $\beta(E'', E')$ on E'' is finer than $\epsilon(E'', E')$, and therefore the canonical imbedding $E \rightarrow E''$ is not necessarily continuous if we equip E'' with the topology $\beta(E'', E')$. We have the following result:

PROPOSITION 5. *Let E be a locally convex Hausdorff space, E' its dual, and E'' its bidual. The topologies $\beta(E'', E')$ and $\epsilon(E'', E')$ on E'' coincide if and only if E is infrabarrelled. If this condition is satisfied, then the canonical imbedding $E \rightarrow E''$ is an injective strict morphism if we equip E'' with the topology $\beta(E'', E')$.*

Proof. The equality of the topologies $\beta(E'', E')$ and $\epsilon(E'', E')$ means that every $\beta(E', E)$-bounded subset of E' is equicontinuous. By Proposition 6.6 this is the case if and only if E is infrabarrelled. The last statement follows from the considerations preceding the proposition. ▮

DEFINITION 3. *A locally convex Hausdorff space E is said to be reflexive if the canonical imbedding from E into its bidual E'' is an isomorphism when we equip E'' with the topology $\beta(E'', E')$.*

PROPOSITION 6. *A locally convex Hausdorff space is reflexive if and only if it is semi-reflexive and infrabarrelled.*

Proof. If E is reflexive, then the canonical imbedding $E \rightarrow E''$ is surjective; i.e., E is semi-reflexive. Furthermore, $\beta(E'', E')$ induces on E its original topology, i.e., coincides with $\epsilon(E'', E')$. Thus E is infrabarrelled by Proposition 5.

Conversely, if E is semi-reflexive, then $E \rightarrow E''$ is surjective, and if furthermore E is infrabarrelled, then $E \rightarrow E''$ is an isomorphism by Proposition 5. ▮

COROLLARY. *A reflexive space is always barrelled.*

Proof. Let M be a $\sigma(E', E)$-bounded subset of E'. If E is semi-reflexive, then the topology $\beta(E', E)$ is compatible with the duality between E and E'. Hence M is also bounded for $\beta(E', E)$ (Theorem 5.3). But E is infrabarrelled. Hence M is equicontinuous by Proposition 6.6. Consequently, E is barrelled by Proposition 6.2. ▮

PROPOSITION 7. *The dual E' of a reflexive space E equipped with the strong topology $\beta(E', E)$ is reflexive.*

Proof. In the first place, E' is barrelled by virtue of Proposition 4. Next, let M be a subset of E' which is bounded and closed for the topology $\sigma(E', E'')$. Since E is reflexive, M is also bounded and closed for $\sigma(E', E)$. Since E is barrelled by virtue of the preceding corollary, it follows from the corollary to Proposition 6.2 that M is compact for $\sigma(E', E)$. Thus E' is also semi-reflexive. ∎

Since a normed space E is in particular infrabarrelled, the canonical imbedding $E \to E''$ is always an injective strict morphism. Actually we have proved much more in Chapter 1, §7; namely, that the canonical imbedding is an isometric isomorphism from E onto a subspace of E''. Thus if E is a semi-reflexive normed space, then it is reflexive and the isomorphism $E \to E''$ is an isometry. In particular, E is then a Banach space. The following theorem tells us when this is the case.

PROPOSITION 8 (Banach-Bourbaki). *A normed vector space E is reflexive if and only if its closed unit ball $\{x \mid \|x\| \leqq 1\}$ is compact for the topology $\sigma(E, E')$.*

Proof. If E is reflexive, then in particular it is semi-reflexive, and therefore by Proposition 1 every bounded, $\sigma(E, E')$-closed subset of E is compact for the topology $\sigma(E, E')$. Now, the closed unit ball is bounded; and since it is convex, it is $\sigma(E, E')$-closed by Proposition 4.3.

Conversely, if the closed unit ball is $\sigma(E, E')$-compact, then every ball $B_\rho = \{x \mid \|x\| \leqq \rho\}$ is $\sigma(E, E')$-compact. A bounded, $\sigma(E, E')$-closed subset of E is contained in some ball B_ρ and is therefore $\sigma(E, E')$-compact. Thus E is semi-reflexive by Proposition 1 and hence reflexive by the above discussion. ∎

REMARK 2. The first half of the proposition also follows from the corollary to Theorem 4.1. Indeed, if E is reflexive, then it is the dual of E'.

EXERCISES

1. Show that a normed space E is reflexive if and only if its closed unit ball is complete for the topology $\sigma(E, E')$. (*Hint:* Use Proposition 2.)

2. Show that a reflexive space is always quasi-complete. (*Hint:* Use Proposition 7, the corollary to Proposition 6, and Theorem 6.1.)

3. Let E be a locally convex space, E' its dual equipped with the topology $\beta(E', E)$, and E'' the dual of E'. For each $x \in E$ define $\tilde{x} \in E''$ by

$$\tilde{x}(x') = \langle x, x' \rangle$$

for all $x' \in E'$. Show that the kernel N of the linear map $x \mapsto \tilde{x}$ from E into E'' is the adherence of $\{0\}$. Show that E' can be identified canonically with $(E/N)'$ and that $\beta(E', E) = \beta(E', E/N)$. Finally, show that the map $x \mapsto \tilde{x}$ is surjective if and only if E/N is semi-reflexive.

§9. *Montel spaces*

Let E be a locally convex Hausdorff space with topology $\mathcal{T}$. In the preceding section we examined those spaces in which every bounded set is relatively compact for the topology $\sigma(E, E')$. This leads us to inquire about the spaces in which every bounded set is relatively compact for the topology $\mathcal{T}$. If E is a normed space, then this property implies that E is finite-dimensional. Indeed, in this case the balls B_ρ are compact; and since they form a fundamental system of neighborhoods of 0, the space E is locally compact, and hence finite-dimensional by Theorem 2.10.3. On the other hand, in a nonnormable space a bounded set is never a neighborhood of 0 (Proposition 2.6.1); in fact, some of the most important spaces have the property just described. It is convenient to introduce the following definitions:

DEFINITION 1. *A locally convex Hausdorff space E is said to be a semi-Montel* (perfect) *space if every bounded subset of E is relatively compact. An infrabarrelled semi-Montel space is said to be a Montel space.*

Clearly every semi-Montel space is quasi-complete.

PROPOSITION 1. *Every semi-Montel space is semi-reflexive.*

Proof. Every bounded subset of a semi-Montel space E is relatively compact, hence *a fortiori* relatively compact for the coarser topology $\sigma(E, E')$. Thus by Proposition 8.1 the space E is semi-reflexive. ∎

COROLLARY. *Every Montel space is reflexive.*

In particular, every Montel space is barrelled (Corollary to Proposition 8.6).

For our next proposition we need a property of compact sets. Let X be a compact space and Y a Hausdorff space. Then every continuous bijection $f: X \mapsto Y$ is necessarily a homeomorphism (in the case of metric spaces, we already observed this in Chapter 1, §9). Indeed, let A be a closed subset of X. Then A is compact in X; hence $f(A)$ is compact in Y. Thus f^{-1} is continuous.

In particular, let X be a compact space with topology $\mathcal{T}_1$. Then any Hausdorff topology $\mathcal{T}_2$ on X which is coarser than $\mathcal{T}_1$ is necessarily identical with $\mathcal{T}_2$. Indeed, let X_i be the set X equipped with the topology $\mathcal{T}_i$ ($i = 1, 2$). Then the identical bijection $X_1 \rightarrow X_2$ is continuous and therefore a homeomorphism.

PROPOSITION 2. *Let E be a semi-Montel space with topology $\mathcal{T}$. If B is a bounded subset of E, then the topology induced on B by $\mathcal{T}$ is the same as the topology induced by $\sigma(E, E')$.*

Proof. We can suppose that B is closed for the topology $\mathcal{T}$ since if $\mathcal{T}$ and $\sigma(E, E')$ induce the same topology on the $\mathcal{T}$-closure $\overline{B}$ of B, then they also induce the same topology on B.

If B is a bounded, $\mathcal{T}$-closed subset of E, then it is compact for the topology induced by $\mathcal{T}$. The topology induced by $\sigma(E, E')$ on B is Hausdorff and coarser than the topology induced by $\mathcal{T}$, with which it must therefore coincide by what we have said above. ▌

COROLLARY. *Let E be a semi-Montel space with topology $\mathcal{T}$.*

(a) *Let $(x_n)_{n\in\mathbf{N}}$ be a sequence of elements of E which converges to a point $x \in E$ for the topology $\sigma(E, E')$. Then (x_n) converges to x also for the topology $\mathcal{T}$.*

(b) *Let $(x_\epsilon)_{0<\epsilon<\alpha}$ be a family of elements of E which converges to a point $x \in E$ for the topology $\sigma(E, E')$ as $\epsilon \to 0$. Then (x_ϵ) converges to x also for the topology $\mathcal{T}$.*

Proof. (a) The subset of E formed by the points x_n $(n \in \mathbf{N})$ and x is bounded for the topology $\sigma(E, E')$ and therefore also for the topology $\mathcal{T}$ (Theorem 5.3). Thus the topologies induced by $\mathcal{T}$ and by $\sigma(E, E')$ on this set coincide.

(b) If (x_ϵ) does not converge to x for the topology $\mathcal{T}$, then there exists a sequence (ϵ_n) of positive numbers tending to zero such that the sequence (x_{ϵ_n}) does not tend to x. But according to our hypothesis the sequence (x_{ϵ_n}) tends to x for the topology $\sigma(E, E')$ and therefore by part (a) also for the topology $\mathcal{T}$. ▌

PROPOSITION 3. *A closed subspace M of a semi-Montel space E is a semi-Montel space.*

Proof. Let A be a bounded subset of M. Then A is a bounded subset of E and therefore relatively compact by hypothesis. Since M is closed, A is also relatively compact in M. Hence M is a semi-Montel space. ▌

PROPOSITION 4. *Let $(E_\iota)_{\iota\in I}$ be a family of locally convex Hausdorff spaces. If each E_ι is a semi-Montel space, then the product $E = \prod_{\iota\in I} E_\iota$ is a semi-Montel space.*

Proof. Let A be a bounded subset of E. If π_ι denotes the projection $E \to E_\iota$, then $\pi_\iota(A)$ is a bounded subset of E_ι. Hence if E_ι is a semi-Montel space, then $\pi_\iota(A)$ is relatively compact. But in this case the product set $B = \prod_{\iota\in I} \pi_\iota(A)$ is relatively compact by Tihonov's theorem. Hence the set A, which is contained in B, is also relatively compact. ▌

REMARK 1. Since the product of infrabarrelled spaces is infrabarrelled (Exercise 14.2), the product of Montel spaces is a Montel space.

For our next result we shall need the following slight variant of Proposition 2.11.3:

PROPOSITION 5. *Let $(F_\iota)_{\iota\in I}$ be a family of Hausdorff topological vector spaces where the index set is ordered, and we suppose that for $\iota \leqq \kappa$ we have a* *continuous linear map* $f_{\iota\kappa}: F_\kappa \to F_\iota$. *Let* E *be a vector space and for each* $\iota \in I$ *let* f_ι *be a linear map from* E *into* F_ι *such that for* $\iota \leqq \kappa$ *we have* $f_\iota = f_{\iota\kappa} \circ f_\kappa$. *Assume that if* $(x_\iota)_{\iota \in I}$ *is an element of* $\prod_{\iota \in I} F_\iota$ *such that for* $\iota \leqq \kappa$ *we have* $f_{\iota\kappa}(x_\kappa) = x_\iota$, *then there exists an element* $x \in E$ *such that* $x_\iota = f_\iota(x)$. *If the linear map* $\varphi: E \to \prod_{\iota \in I} F_\iota$ *is defined by* $\varphi(x) = (f_\iota(x))_{\iota \in I}$ (cf. Exercise 2.11.1), *then* $\operatorname{Im}(\varphi)$ *is closed.*

Proof. Let $(x_\iota) \in \overline{\varphi(E)}$. Consider two indices ι and κ such that $\iota \leqq \kappa$ and let U be a balanced neighborhood of 0 in F_ι. There exist a balanced neighborhood V of 0 in F_κ such that $f_{\iota\kappa}(V) \subset U$, and an element $(y_\iota)_{\iota \in I} \in \varphi(E)$ such that $y_\iota \in U + x_\iota$ and $y_\kappa \in V + x_\kappa$. Let $y \in E$ be such that $(y_\iota) = \varphi(y)$. Then

$$y_\kappa = f_\kappa(y) \qquad \text{and} \qquad y_\iota = f_\iota(y) = f_{\iota\kappa}(f_\kappa(y)) = f_{\iota\kappa}(y_\kappa).$$

Hence

$$y_\iota = f_{\iota\kappa}(y_\kappa) \in f_{\iota\kappa}(V + x_\kappa) \subset U + f_{\iota\kappa}(x_\kappa)$$

and

$$x_\iota - f_{\iota\kappa}(x_\kappa) \in U + U.$$

Since U is arbitrary and F_ι is a Hausdorff space, we have $x_\iota = f_{\iota\kappa}(x_\kappa)$. Thus there exists $x \in E$ such that $(x_\iota) = \varphi(x)$, that is, $(x_\iota) \in \varphi(E)$. ∎

Proposition 6. *Let* I *be an ordered set and for each* $\iota \in I$ *let* F_ι *be a semi-Montel space. Suppose that for* $\iota \leqq \kappa$ *we have a continuous linear map* $f_{\iota\kappa}: F_\kappa \to F_\iota$. *Let* E *be a vector space and for each* $\iota \in I$ *let* f_ι *be a linear map from* E *into* F_ι *such that for* $\iota \leqq \kappa$ *we have* $f_\iota = f_{\iota\kappa} \circ f_\kappa$. *Assume that the following condition is satisfied:*

(P) *If* $(x_\iota)_{\iota \in I}$ *is an element of* $\prod_{\iota \in I} F_\iota$ *such that for* $\iota \leqq \kappa$ *we have* $f_{\iota\kappa}(x_\kappa) = x_\iota$, *then there exists a* unique *element* $x \in E$ *such that* $x_\iota = f_\iota(x)$ *for all* $\iota \in I$.

Then E, *equipped with the coarsest topology for which the* f_ι *are continuous, is a semi-Montel space.*

Proof. The uniqueness requirement in condition (P) implies in particular that if $x \in E$ and $x \neq 0$, then there exists an index $\iota \in I$ such that $f_\iota(x) \neq 0$; i.e, the linear map $\varphi: E \to \prod_{\iota \in I} F_\iota$ defined by $\varphi(x) = (f_\iota(x))_{\iota \in I}$ is injective and that E is a Hausdorff space. By the preceding proposition, φ is an isomorphism of E onto a closed subspace of $\prod_{\iota \in I} F_\iota$ (cf. Exercise 2.11.1), and by Proposition 4, this product is a semi-Montel space. Hence E is a semi-Montel space by Proposition 3. ∎

Corollary. *Let* $(F_\iota)_{\iota \in I}$ *be a family of locally convex Hausdorff spaces, where the index set is ordered. Suppose that for* $\iota \leqq \kappa$ *the space* F_κ *is a linear subspace of* F_ι *and the topology of* F_κ *is finer than the topology induced* *by F_ι on F_κ. Let $E = \bigcap_{\iota \in I} F_\iota$ and equip E with the coarsest topology for which the injections $E \hookrightarrow F_\iota$ are continuous. If the F_ι are semi-Montel spaces, then E is a semi-Montel space.*

Our next aim is to prove that the strong dual of a Montel space is a Montel space. In order to do this we have to introduce a definition which is also of independent interest.

DEFINITION 2. *Let E be a locally convex Hausdorff space and let $\mathfrak{S}$ be the collection of all precompact* (Definition 2.10.1) *subsets of E. The corresponding $\mathfrak{S}$-topology on the dual E' of E will be denoted by $\lambda(E', E)$, and is called the topology of the uniform convergence on precompact subsets of E.*

The reason why one considers precompact sets rather than compact ones lies in the following stability theorem:

PROPOSITION 7. *In a locally convex Hausdorff space E, the balanced convex hull of a precompact set is precompact.*

Proof. Let A be a precompact subset of E. The balanced convex hull B of A consists of all points $x = \sum_{i=1}^n \lambda_i x_i$, where $n \in \mathbf{N}^*$, $\sum_{i=1}^n |\lambda_i| \leqq 1$, and $x_i \in A$ $(1 \leqq i \leqq n)$ (Exercise 2.4.1(c)). Let V be a balanced, convex neighborhood of 0 in E. We have to show that there exist finitely many points $a_j \in B$ $(1 \leqq j \leqq p)$ such that $B \subset \bigcup_{j=1}^p (a_j + V)$ (Theorem 2.10.2).

Since A is precompact, there exists a finite family of points $b_k \in A$ $(1 \leqq k \leqq q)$ such that $A \subset \bigcup_{k=1}^q (b_k + \frac{1}{2}V)$. Let Λ be the subset of $\mathbf{K}^q$ formed by those points $\lambda = (\lambda_1, \ldots, \lambda_q)$ for which $\sum_{k=1}^q |\lambda_k| \leqq 1$. Since Λ is compact, for every $\delta > 0$ there exist finitely many points $\mu^{(j)} = (\mu_1^{(j)}, \ldots, \mu_q^{(j)})$ $(1 \leqq j \leqq p)$ such that for every $\lambda \in \Lambda$ we can find a $\mu^{(j)}$ which satisfies $\sum_{k=1}^q |\lambda_k - \mu_k^{(j)}| < \delta$. Let us choose $\delta > 0$ so small that $\delta A \subset \frac{1}{2}V$. Since V is balanced and convex, we also have $\delta B \subset \frac{1}{2}V$.

We will show that

$$B \subset \bigcup_{j=1}^p \left(\sum_{k=1}^q \mu_k^{(j)} b_k + V \right). \tag{1}$$

Indeed, if $x = \sum_{i=1}^n \lambda_i x_i \in B$, then for each x_i choose b_{k_i} so that $x_i = b_{k_i} + y_i$ with $y_i \in \frac{1}{2}V$ $(1 \leqq i \leqq n)$. Hence

$$x = \sum_{i=1}^n \lambda_i x_i = \sum_{i=1}^n \lambda_i b_{k_i} + \sum_{i=1}^n \lambda_i y_i = \sum_{k=1}^q \nu_k b_k + y$$

with $\sum_{k=1}^q |\nu_k| \leqq 1$ and $y \in \frac{1}{2}V$.

For an appropriate $\mu^{(j)}$ we have

$$x = \sum_{k=1}^q \mu_k^{(j)} b_k + \sum_{k=1}^q (\nu_k - \mu_k^{(j)}) b_k + y = \sum_{k=1}^q \mu_k^{(j)} b_k + z + y,$$

where $z \in \delta B \subset \frac{1}{2}V$. Consequently,

$$x \in \sum_{k=1}^{q} \mu_k^{(j)} b_k + V;$$

and since x was arbitrary in B, we have proved relation (1). ∎

The balanced convex hull of a compact set is not necessarily compact (see Exercise 5). If, however, the space E is quasi-complete, then every precompact set is relatively compact since it is bounded (Proposition 2.10.7). Hence in a quasi-complete, locally convex Hausdorff space the balanced, convex, closed hull of a compact set is compact; and by Proposition 4.2(f) the topology $\lambda(E', E)$ coincides with *the topology* $\kappa(E', E)$ *of the uniform convergence on balanced, convex, compact sets.* In general, $\kappa(E', E)$ is coarser than $\lambda(E', E)$. Since each compact subset of E is *a fortiori* $\sigma(E, E')$-compact, we see that $\kappa(E', E)$ is coarser than the Mackey topology $\tau(E', E)$. Also $\kappa(E', E)$ is clearly finer than $\sigma(E', E)$. Hence the dual of E' equipped with the topology $\kappa(E', E)$ is the space E (Proposition 5.4).

If E is a semi-Montel space, then $\kappa(E', E)$ and $\beta(E', E)$ coincide on E', and we have, of course, $\kappa(E', E) = \lambda(E', E) = \tau(E', E) = \beta(E', E)$. Conversely, if E is a quasi-complete, locally convex Hausdorff space and $\kappa(E', E) = \lambda(E', E) = \beta(E', E)$, then E is a semi-Montel space.

PROPOSITION 8. *Let E be a locally convex Hausdorff space and E' its dual. On every equicontinuous subset M of E' the topology induced by $\lambda(E', E)$ coincides with the topology induced by $\sigma(E', E)$.*

Proof. The topology induced by $\lambda(E', E)$ on M is clearly finer than the topology induced by $\sigma(E', E)$. Thus to prove that the two topologies coincide, it is sufficient to show that given $x'_0 \in M$ and a precompact subset A of E, there exists a finite family $(x_i)_{1 \leqq i \leqq n}$ of points of E such that $\sup_{1 \leqq i \leqq n} |\langle x' - x'_0, x_i \rangle| \leqq \frac{1}{2}$ implies $\sup_{x \in A} |\langle x' - x'_0, x \rangle| \leqq 1$ for x' belonging to M, i.e., such that $(x'_0 + \frac{1}{2}F^{\circ}) \cap M \subset (x'_0 + A^{\circ}) \cap M$, where F is the finite set $\{x_i \mid 1 \leqq i \leqq n\}$.

Since M, and therefore $M - x'_0$, is equicontinuous, there exists a neighborhood V of 0 in E such that $\sup_{x' \in M, x \in V} |\langle x' - x'_0, x \rangle| \leqq \frac{1}{2}$. On the other hand, A is precompact. Hence there exist a finite number of points x_i $(1 \leqq i \leqq n)$ such that $A \subset \bigcup_{i=1}^{n} (x_i + V)$; i.e., every $x \in A$ has the form $x = x_i + y$, $1 \leqq i \leqq n$, $y \in V$. Suppose now that $x' \in M$ is such that $\sup_{1 \leqq i \leqq n} |\langle x' - x'_0, x_i \rangle| \leqq \frac{1}{2}$. Then

$$\sup_{x \in A} |\langle x' - x'_0, x \rangle| \leqq \sup_{1 \leqq i \leqq n} |\langle x' - x'_0, x_i \rangle| + \sup_{y \in V} |\langle x' - x'_0, y \rangle| \leqq 1. \blacksquare$$

We shall return to the topology $\lambda(E', E)$ in the next section, where we shall prove that if E is metrizable, then $\lambda(E', E)$ is the finest topology which induces on equicontinuous subsets the same topology as $\sigma(E', E)$.

Now we are in the position to prove the result announced earlier.

PROPOSITION 9. *If E is a Montel space, then its dual E' equipped with the topology $\beta(E', E)$ is also a Montel space.*

Proof. E is reflexive by the corollary to Proposition 1; hence E' is barrelled by Proposition 8.4.

Let M be a subset of E' which is bounded for the topology $\beta(E', E)$. The set M is contained in a balanced, convex, $\beta(E', E)$-closed, $\beta(E', E)$-bounded set N. Since E is infrabarrelled, it follows from Proposition 6.6 that N is equicontinuous. Now, $\beta(E', E)$ is compatible with the duality between E and E'; hence N is also $\sigma(E', E)$-closed by Proposition 4.3, and thus N is compact for the topology $\sigma(E', E)$ by the Bourbaki-Alaoglu theorem (Theorem 4.1). By Proposition 8 the topologies induced by $\sigma(E', E)$ and $\lambda(E', E)$ on N coincide. Hence N is compact for the topology $\lambda(E', E)$. Finally, since E is a Montel space, the topologies $\lambda(E', E)$ and $\beta(E', E)$ on E' are identical. Therefore N is compact for the topology $\beta(E', E)$, and M is relatively compact for that topology. ▮

EXAMPLE 1. Every finite-dimensional Hausdorff topological vector space is a Montel space by the Bolzano-Weierstrass theorem since every such space is isomorphic to some space $\mathbf{K}^n$ (Theorem 2.10.1).

Before we give further examples, we must recall a classical theorem of analysis. Let X be a locally compact space and denote by $\mathcal{C}(X)$ the vector space of all numerical functions defined and continuous on X. For each compact subset K of X define the semi-norm q_K by

$$q_K(f) = \max_{x \in K} |f(x)|.$$

It follows from Proposition 2.4.8 that, equipped with the family (q_K) of semi-norms, the space $\mathcal{C}(X)$ becomes a locally convex Hausdorff space, and we refer to the topology $\mathcal{T}_c$ defined by these semi-norms as the topology of compact convergence. If X is a compact space, then the family $(q_K)_K$ of semi-norms on $\mathcal{C}(X)$ is equivalent to the unique norm q_X, and $\mathcal{C}(X)$ is a normed space. We have already met these spaces in the particular cases where X is an open subset Ω of $\mathbf{R}^n$ (Examples 2.3.3 and 2.4.6) and where X is a compact subset K of $\mathbf{R}^n$ (Exercise 1.2.6). The present generalization enables us to deal with these two particular cases simultaneously. Since every point of X possesses a compact neighborhood, we can prove similarly as in the special case mentioned above (Example 2.9.2) that $\mathcal{C}(X)$ is complete.

On $\mathcal{C}(X)$ we can define the topology of pointwise convergence $\mathcal{T}_s$ by the semi-norms $q_x = |f(x)|$, where $x \in X$. Clearly, $\mathcal{T}_s$ is coarser than $\mathcal{T}_c$. Consider the family of spaces $(\mathbf{K}_x)_{x \in X}$, where each $\mathbf{K}_x$ is equal to $\mathbf{K}$, and let $\mathbf{K}^X = \prod_{x \in X} \mathbf{K}_x$ be their product. The space $\mathcal{C}(X)$ can be considered as a subset of $\mathbf{K}^X$ if we associate with every $f \in \mathcal{C}(X)$ the family $(f(x))_{x \in X}$. The product topology on $\mathbf{K}^X$ induces on $\mathcal{C}(X)$ the topology $\mathcal{T}_s$. We have the following result which is closely related to Proposition 8:

Lemma 1. *On every equicontinuous subset $\mathcal{H}$ of $\mathcal{C}(X)$ the topology induced by $\mathcal{T}_c$ coincides with the topology induced by $\mathcal{T}_s$.*

Proof. Clearly, the topology induced by $\mathcal{T}_c$ on $\mathcal{H}$ is finer than the topology induced by $\mathcal{T}_s$. Thus to prove that the two topologies coincide, we have to show that given $f_0 \in \mathcal{H}$, $\epsilon > 0$, and a compact subset K of X, there exists a finite family $(x_i)_{1 \leqq i \leqq n}$ of points of X such that

$$\max_{1 \leqq i \leqq n} |f(x_i) - f_0(x_i)| \leqq \tfrac{1}{2}\epsilon$$

implies

$$\max_{x \in K} |f(x) - f_0(x)| \leqq \epsilon$$

for all $f \in \mathcal{H}$.

Since $\mathcal{H}$ is equicontinuous, each point $x \in X$ has a neighborhood V_x such that $y \in V_x$ implies $|f(y) - f(x)| \leqq \frac{1}{4}\epsilon$ for $f \in \mathcal{H}$. The compact set K can be covered by finitely many such neighborhoods V_x, say $V_{x_1}, \ldots, V_{x_n}$. Assume now that $f \in \mathcal{H}$ and $|f(x_i) - f_0(x_i)| \leqq \frac{1}{2}\epsilon$ for $1 \leqq i \leqq n$. Let x be an arbitrary point in K. There exists an index i $(1 \leqq i \leqq n)$ such that $x \in V_{x_i}$. Hence we have

$$|f(x) - f_0(x)| \leqq |f(x) - f(x_i)| + |f(x_i) - f_0(x_i)| + |f_0(x_i) - f_0(x)| \leqq \epsilon. \blacksquare$$

The next result is a variant of Lemma 6.2.

Lemma 2. *The closure in $\mathbf{K}^X$ of an equicontinuous subset $\mathcal{H}$ of $\mathcal{C}(X)$ is itself an equicontinuous subset of $\mathcal{C}(X)$.*

Proof. Let $x \in X$ and $\epsilon > 0$. There exists a neighborhood V of x such that $y \in V$ implies $|f(y) - f(x)| < \frac{1}{3}\epsilon$ for all $f \in \mathcal{H}$. Now let $g \in \overline{\mathcal{H}}$. Given the points $x, y \in X$, there exists $f \in \mathcal{H}$ such that $|f(x) - g(x)| < \frac{1}{3}\epsilon$ and $|f(y) - g(y)| < \frac{1}{3}\epsilon$. Thus if $y \in V$, we have

$$|g(y) - g(x)| \leqq |g(y) - f(y)| + |f(y) - f(x)| + |f(x) - g(x)| < \epsilon. \blacksquare$$

Ascoli's theorem. *A subset $\mathcal{H}$ of $\mathcal{C}(X)$ is relatively compact for the topology of compact convergence if and only if it is equicontinuous and for each $x \in X$ the set $\mathcal{H}(x) = \{f(x) \mid f \in \mathcal{H}\}$ is bounded in* $\mathbf{K}$.

Proof. Assume first that the two conditions of the theorem are satisfied. By Lemma 2 the $\mathcal{T}_s$-closure $\overline{\mathcal{H}}$ of $\mathcal{H}$ is also an equicontinuous subset of $\mathcal{C}(X)$, and clearly $\overline{\mathcal{H}}(x)$ is bounded in $\mathbf{K}$ for each $x \in X$. By the Bolzano-

Weierstrass theorem each set $\overline{\mathcal{H}}(x)$ is relatively compact in $\mathbf{K}$, and thus by Tihonov's theorem the set $\overline{\mathcal{H}}$ is compact in $\mathcal{C}(X)$ for $\mathcal{T}_s$. But by Lemma 1 the topologies $\mathcal{T}_c$ and $\mathcal{T}_s$ coincide on $\overline{\mathcal{H}}$. Hence $\overline{\mathcal{H}}$ is also compact for $\mathcal{T}_c$. Thus $\mathcal{H}$ is indeed relatively compact.

Conversely, assume that $\mathcal{H}$ is a relatively compact subset of $\mathcal{C}(X)$ for the topology $\mathcal{T}_c$. *A fortiori* $\mathcal{H}$ is relatively compact in $\mathcal{C}(X)$ for the topology $\mathcal{T}_s$ induced by $\mathbf{K}^X$. For each $x \in X$ the projection $\pi_x: f \mapsto f(x)$ from $\mathbf{K}^X$ onto the factor $\mathbf{K}_x$ is continuous, and therefore the set $\pi_x(\mathcal{H}) = \mathcal{H}(x)$ is relatively compact in $\mathbf{K}$.

On the other hand, let x be a point of X, K a compact neighborhood of x, and $\epsilon > 0$. Since $\mathcal{H}$ is relatively compact, there exists a finite family $(f_i)_{1 \leqq i \leqq n}$ of elements of $\mathcal{H}$ such that for each $f \in \mathcal{H}$ there exists an index i $(1 \leqq i \leqq n)$ such that $|f(y) - f_i(y)| \leqq \frac{1}{3}\epsilon$ for all $y \in K$. Furthermore, for each i there exists a neighborhood U_i of x such that

$$|f_i(x) - f_i(y)| \leqq \tfrac{1}{3}\epsilon$$

for all $y \in U_i$. Hence for $f \in \mathcal{H}$ and $y \in K \cap U_1 \cap \cdots \cap U_n$ we have

$$\begin{aligned} |f(x) - f(y)| &\leqq |f(x) - f_i(x)| + |f_i(x) - f_i(y)| + |f_i(y) - f(y)| \\ &\leqq \tfrac{1}{3}\epsilon + \tfrac{1}{3}\epsilon + \tfrac{1}{3}\epsilon = \epsilon; \end{aligned}$$

i.e., $\mathcal{H}$ is equicontinuous. ∎

Now we are ready to give further examples of Montel spaces.

Example 2. Let us explain where the name "Montel space" comes from. In its classical form, Montel's theorem from complex function theory ([1], Chapter IV, Section 4.1, Theorem 9, p. 171) can be stated as follows. Let Ω be a domain in $\mathbf{C}$, $\mathcal{H}$ a collection of holomorphic functions defined on Ω, and suppose that for every compact subset K of Ω there exists a $\mu_K > 0$ such that $|f(z)| \leqq \mu_K$ for all $f \in \mathcal{H}$ and $z \in K$. Then, given a sequence (f_n) of elements of $\mathcal{H}$, there exists a subsequence (f_{n_k}) of (f_n) which converges to a holomorphic function g, uniformly on every compact subset of Ω.

We shall translate this statement into the language of topological vector spaces. Let Ω be a domain of $\mathbf{C}$, and let $H(\Omega)$ be the vector space of all holomorphic functions defined in Ω. Since Ω can also be considered as a domain in $\mathbf{R}^2$, the space $H(\Omega)$ is a subspace of $\mathcal{C}(\Omega)$. If we consider the topology of compact convergence on $\mathcal{C}(\Omega)$, then $H(\Omega)$ is a closed subspace of $\mathcal{C}(\Omega)$ since, by a theorem of Weierstrass ([1], Chapter IV, Section 1.4, Theorem 1, p. 138), if a sequence of holomorphic functions converges in Ω to some function g, uniformly on every compact subset of Ω, then g is a holomorphic function in Ω. Thus $H(\Omega)$, being the closed subspace of a Fréchet space, is itself a Fréchet space and is, in particular, barrelled.

Montel's theorem is now clearly equivalent to the assertion that every bounded subset of $H(\Omega)$ is relatively compact. It then follows that $H(\Omega)$ is a Montel space. In complex function theory a relatively compact subset of $H(\Omega)$ is usually called a "normal family."

To prove Montel's theorem it is sufficient by Ascoli's theorem to show that every bounded subset $\mathcal{H}$ of $H(\Omega)$ is equicontinuous. Suppose that $\mathcal{H}$ is a bounded subset of $H(\Omega)$, i.e., that for each compact subset K of Ω there exists $\mu_K > 0$ such that $|f(z)| \leqq \mu_K$ for $f \in \mathcal{H}$ and $z \in K$. Let K be a compact subset of Ω. There exists a $\delta > 0$ such that the compact set $L = \{\zeta \mid |\zeta - z| \leqq \delta \text{ for some } z \in K\}$ is contained in Ω. If $z \in K$ and γ is the circle with center z and radius δ, then we have by Cauchy's formula

$$|f'(z)| = \left|\frac{1}{2\pi i}\int_\gamma \frac{f(\zeta)}{(\zeta - z)^2}\,d\zeta\right| \leqq \frac{1}{2\pi}\int_\gamma \frac{|f(\zeta)|}{\delta^2}\,|d\zeta| \leqq \frac{\mu_L}{\delta} = \nu_K.$$

Thus $|f(z_1) - f(z_2)| \leqq |f'(z)| \cdot |z_1 - z_2| \leqq \nu_K |z_1 - z_2|$ if $f \in \mathcal{H}$, z_1, $z_2 \in K$, $|z_1 - z_2| < \delta$, and in particular $\mathcal{H}$ is equicontinuous.

The modern proof of Riemann's mapping theorem ([1], Chapter V, Section 4.2, Theorem 10, p. 172) is based on the fact that $H(\Omega)$ is a Montel space.

Example 3. Let Ω be an open subset of $\mathbf{R}^n$. The space $\mathcal{E}(\Omega)$ (Example 2.4.8) is a Montel space. Indeed, it is a Fréchet space (Example 2.9.3) and thus in particular it is barrelled.

For each $p \in \mathbf{N}^n$ let $\mathcal{C}_p(\Omega)$ be a space equal to $\mathcal{C}(\Omega)$. If we associate with each $f \in \mathcal{E}(\Omega)$ the family $(\partial^p f)_{p \in \mathbf{N}^n}$, we obtain an imbedding of $\mathcal{E}(\Omega)$ into the product space $\prod_{p \in \mathbf{N}^n} \mathcal{C}_p(\Omega)$, and clearly the topology induced by this product space on $\mathcal{E}(\Omega)$ is its own topology. Furthermore, $\mathcal{E}(\Omega)$ is a closed subspace of $\prod_p \mathcal{C}_p(\Omega)$ since it is complete. Now let $\mathcal{H}$ be a bounded subset of $\mathcal{E}(\Omega)$. This means that for every compact subset K of Ω and for every $p \in \mathbf{N}^n$ there exists $\mu_{K,p} > 0$ such that $|\partial^p f(x)| \leqq \mu_{K,p}$ for all $f \in \mathcal{H}$ and $x \in K$. The projection $\partial^p \mathcal{H} = \{\partial^p f \mid f \in \mathcal{H}\}$ of $\mathcal{H}$ in $\mathcal{C}_p(\Omega)$ is equicontinuous. Indeed, given a compact subset K of Ω there exists $\delta > 0$ such that the compact set $L = \{x \mid |x - y| \leqq \delta \text{ for some } y \in K\}$ is contained in Ω. If x and y are two points of K such that $|x - y| < \delta$, then the line segment joining them lies in L, and we have

$$|\partial^p f(x) - \partial^p f(y)| = \left|\sum_{i=1}^n \int_x^y \partial_i \partial^p f(z)\,dz_i\right| \leqq n \cdot |x - y| \cdot \max_r \mu_{L,r},$$

where $r = (r_1, \ldots, r_n) \in \mathbf{N}^n$ runs through the multi-indices such that $|r| = r_1 + \cdots + r_n = |p| + 1$.

By Ascoli's theorem each $\partial^p \mathcal{H}$ is relatively compact in $\mathcal{C}(\Omega)$, and thus by Tihonov's theorem $\mathcal{H}$ is relatively compact in $\mathcal{E}(\Omega)$.

EXAMPLE 4. Let K be a compact subset of $\mathbf{R}^n$. The space $\mathfrak{D}(K)$ (Examples 2.3.4 and 2.4.10) is a Montel space. Indeed, it is barrelled (Example 6.3) and it is a semi-Montel space since by Example 2.9.5 it is a closed subspace of $\mathcal{E}$ (Proposition 3).

EXAMPLE 5. The space $\mathcal{S}$ (Example 2.4.14) is a Montel space. We know from Example 6.3 that it is barrelled.

Let $\mathcal{H}$ be a bounded subset of $\mathcal{S}$. For every $k \in \mathbf{N}$ and $p \in \mathbf{N}^n$ there exists $\mu_{k,p} > 0$ such that $|(1 + |x|^2)^k \partial^p f(x)| \leqq \mu_{k,p}$ for all $f \in \mathcal{H}$ and $x \in \mathbf{R}^n$. We have *a fortiori* $|\partial^p f(x)| \leqq \mu_{k,p}$ for all $f \in \mathcal{H}$ and $x \in \mathbf{R}^n$. Since $\mathcal{S} \subset \mathcal{E}$, the set $\mathcal{H}$ is bounded in $\mathcal{E}$, and by Example 3 it is relatively compact in $\mathcal{E}$. To show that $\mathcal{H}$ is relatively compact in $\mathcal{S}$ it is clearly enough to show that if (f_j) is a sequence of elements of $\mathcal{H}$ which converges in $\mathcal{E}$ to some function $g \in \mathcal{E}$, then $g \in \mathcal{S}$ and (f_j) converges to g in $\mathcal{S}$. We have

$$|(1 + |x|^2)^{k+1} \partial^p f(x)| \leqq \mu_{k+1,p}$$

for all $f \in \mathcal{H}$ and $x \in \mathbf{R}^n$, and given $\epsilon > 0$, we have

$$|(1 + |x|^2)^k \partial^p f(x)| \leqq \mu_{k+1,p}(1 + |x|^2)^{-1} \leqq \tfrac{1}{2}\epsilon$$

for $|x| > \rho$. Since $(\partial^p f_j)$ converges to $\partial^p g$ pointwise, we have

$$|(1 + |x|^2)^k \partial^p g(x)| \leqq \tfrac{1}{2}\epsilon$$

for $|x| > \rho$ and in particular g belongs to $\mathcal{S}$. On the compact set $|x| \leqq \rho$ the sequence $(\partial^p f_j)$ converges uniformly to $\partial^p g$. Hence we have

$$|\partial^p f_j(x) - \partial^p g(x)| \leqq \epsilon(1 + \rho^2)^{-k}$$

for $j \geqq J$ and *a fortiori* $(1 + |x|^2)^k|\partial^p f_j(x) - \partial^p g(x)| \leqq \epsilon$ for $|x| \leqq \rho$ and $j \geqq J$. Collecting these inequalities, we obtain

$$q_{k,p}(f_j - g) = \max_{x \in \mathbf{R}^n} (1 + |x|^2)^k|\partial^p f_j(x) - \partial^p g(x)| \leqq \epsilon$$

for $j \geqq J$.

PROPOSITION 10. *Let F be a vector space and let $(E_n)_{n \in \mathbf{N}}$ be a sequence of linear subspaces of F such that $E_n \subset E_{n+1}$ for all $n \in \mathbf{N}$ and*

$$F = \bigcup_{n \in \mathbf{N}} E_n.$$

Suppose that each E_n is equipped with a locally convex Hausdorff topology $\mathcal{T}_n$, that $\mathcal{T}_{n+1}$ induces the topology $\mathcal{T}_n$ on E_n, and that E_n is closed in E_{n+1} for $\mathcal{T}_{n+1}$. We equip F with the finest locally convex topology for which all the injections $E_n \hookrightarrow F$ are continuous.

If the E_n are semi-Montel spaces, then F is a semi-Montel space. If the E_n are Montel spaces, then F is a Montel space.

Proof. F is a Hausdorff space by Corollary 1 of Theorem 2.12.1. Assume that the E_n are semi-Montel spaces and let B be a bounded, closed subset of F. By Theorem 2.12.2 the set B is a bounded, closed subset of one of the subspaces E_n, and therefore B is compact in E_n. Consequently, B is also compact in F. This proves that F is a semi-Montel space. On the other hand, if the spaces E_n are infrabarrelled, then it follows from Proposition 6.9 that F is infrabarrelled. ▮

EXAMPLE 6. Let Ω be an open subset of $\mathbf{R}^n$. The space $\mathfrak{D}(\Omega)$ of Example 2.12.6 is a Montel space by virtue of Example 4 and Proposition 10.

We conclude this section with conditions for compactness in the spaces $\mathcal{E}^m(\Omega)$ (Example 2.4.7) and $\mathfrak{D}^m(\Omega)$ (Example 2.12.8) for finite m.

PROPOSITION 11. *Let Ω be an open subset of $\mathbf{R}^n$ and m a positive integer. A subset $\mathcal{H}$ of $\mathcal{E}^m(\Omega)$ is relatively compact if and only if it satisfies the following two conditions:*

(1) *For every $p \in \mathbf{N}^n$ such that $|p| \leqq m$ the set $\{\partial^p f \mid f \in \mathcal{H}\}$ is equicontinuous.*

(2) *For every $x \in \Omega$ and every $p \in \mathbf{N}^n$ such that $|p| \leqq m$ the set*

$$\{\partial^p f(x) \mid f \in \mathcal{H}\}$$

is bounded in $\mathbf{K}$.

Proof. For $m = 0$ this is precisely Ascoli's theorem. In the general case, for each $p \in \mathbf{N}^n$ such that $|p| \leqq m$ denote by $\mathcal{C}_p(\Omega)$ a space identical with $\mathcal{C}(\Omega)$. The map $f \mapsto (\partial^p f)_{|p| \leqq m}$ is an injective strict morphism from $\mathcal{E}^m(\Omega)$ into $\prod_{|p| \leqq m} \mathcal{C}_p(\Omega)$ and the image of $\mathcal{E}^m(\Omega)$ is closed in the product space since $\mathcal{E}^m(\Omega)$ is complete. By Tihonov's theorem $\mathcal{H}$ is a relatively compact subset of $\mathcal{E}^m(\Omega)$ if and only if for each p the set

$$\partial^p \mathcal{H} = \{\partial^p f \mid f \in \mathcal{H}\}$$

is relatively compact in $\mathcal{C}_p(\Omega)$. Thus the proposition follows immediately from the special case $m = 0$. ▮

PROPOSITION 12. *Let Ω be an open subset of $\mathbf{R}^n$ and m a positive integer. A subset $\mathcal{H}$ of $\mathfrak{D}^m(\Omega)$ is relatively compact if and only if it satisfies the following three conditions:*

(1) *There exists a compact subset K of Ω such that* Supp $f \subset K$ *for all* $f \in \mathcal{H}$.

(2) *For every $p \in \mathbf{N}^n$ such that $|p| \leqq m$ the set $\{\partial^p f \mid f \in \mathcal{H}\}$ is equicontinuous.*

(3) *For every $x \in \Omega$ and every $p \in \mathbf{N}^n$ such that $|p| \leqq m$ the set $\{\partial^p f(x) \mid f \in \mathcal{H}\}$ is bounded in* $\mathbf{K}$.

Proof. In the first place, if $\mathcal{H}$ is relatively compact in $\mathfrak{D}^m(\Omega)$, then it is bounded there, and thus by Theorem 2.12.2 there exists a compact subset K of Ω such that $\mathcal{H}$ is contained in $\mathfrak{D}^m(K)$; i.e., condition (1) is satisfied. Furthermore, by Theorem 2.12.1 the set $\mathcal{H}$ is relatively compact in $\mathfrak{D}^m(K)$. Conversely, if $\mathcal{H}$ is a relatively compact subset of $\mathfrak{D}^m(K)$, then it is also a relatively compact subset of $\mathfrak{D}^m(\Omega)$.

For each p such that $|p| \leqq m$ let $\mathcal{K}_p(K)$ be a space equal to $\mathcal{K}(K)$ (Example 2.12.5). The map $f \mapsto (\partial^p f)_{|p| \leqq m}$ from $\mathfrak{D}^m(K)$ into $\prod_{|p| \leqq m} \mathcal{K}_p(K)$ is an injective strict morphism, and the image of $\mathfrak{D}^m(K)$ is closed in the product space, since $\mathfrak{D}^m(K)$ is complete. By Tihonov's theorem a set $\mathcal{H}$ is relatively compact in $\mathfrak{D}^m(K)$ if and only if for each p such that $|p| \leqq m$ the set $\partial^p \mathcal{H} = \{\partial^p f \mid f \in \mathcal{H}\}$ is relatively compact in $\mathcal{K}(K)$. Now $\mathcal{K}(K)$ is a closed subspace of $\mathcal{C}(K)$, and therefore by Ascoli's theorem the sets $\partial^p \mathcal{H}$ are relatively compact in $\mathcal{K}(K)$ if and only if conditions (2) and (3) are satisfied. ∎

Exercises

1. Let E be a semi-Montel space with topology $\mathcal{T}$, and let $\mathfrak{F}$ be a filter with countable basis on E which converges to an element $x \in E$ for the topology $\sigma(E, E')$. Prove that $\mathfrak{F}$ converges to x for the topology $\mathcal{T}$. (*Hint:* Use Exercise 6.3(a) and Corollary (a) to Proposition 2.)

2. Prove that the locally convex direct sum (Example 2.12.2) of a family of semi-Montel spaces is a semi-Montel space. (*Hint:* Use Exercise 2.12.10 and an argument similar to that employed in the proof of Proposition 4.)

3. Prove that the inverse limit (Exercise 2.11.3) of an inverse system of semi-Montel spaces is a semi-Montel space.

4. Show that Proposition 2.11.3 follows from Proposition 5 and from the fact that a product of complete spaces is complete.

5. Show by an example that the balanced convex hull of a compact set in a topological vector space is not necessarily compact. (*Hint:* Let E be the subspace of l^2 considered in Exercise 4.7. Consider in E the subset A formed by 0 and by the vectors $(1/n)e_n$, where e_n $(n \geqq 1)$ is the n-th unit vector of l^2. Let $\lambda_n > 0$, $\sum_{n \geqq 1} \lambda_n = 1$. The vectors

$$\left(\sum_{n=1}^{m} \lambda_n\right)^{-1} \sum_{n=1}^{m} \frac{\lambda_n}{n} e_n$$

form a Cauchy sequence in the balanced convex hull of A which converges to no point of E.)

6. Let E be a locally convex Hausdorff space. Prove that the collection of precompact subsets of E is saturated in the sense of Exercise 4.2.

7. Let E be a locally convex Hausdorff space and E' its dual.

(a) Show that every equicontinuous subset of E' is relatively compact for the topology $\lambda(E', E)$. (*Hint:* Use Proposition 8 and Theorem 4.1.)

(b) Show that if E is bornological, then E' equipped with the topology $\lambda(E', E)$ is complete. (*Hint:* Using Exercise 7.7(f), prove first that if the restriction of a linear form u to every compact subset of E is continuous, then u is continuous.)

(c) Deduce Proposition 7.6 from (b). (*Hint:* Use Proposition 5.5.)

8. Let E be a locally convex Hausdorff space, E' its dual, equip E' with the topology $\kappa(E', E)$, and identify E with the dual of E'. Denote by $\gamma(E, E')$ the topology of uniform convergence on balanced, convex, compact subsets of E'.

(a) Show that $\gamma(E, E')$ is finer than the original topology $\mathcal{T}$ of E and coarser than $\tau(E, E')$. (*Hint:* Use Proposition 4.7.)

(b) Show that E has the topology $\gamma(E, E')$ if and only if E is isomorphic to the dual F' of a locally convex Hausdorff space F, equipped with $\kappa(F', F)$. (*Hint:* $F = E'$.)

§10. *The Banach-Dieudonné theorem*

Let E be a locally convex Hausdorff space and E' its dual. Denote by $\nu(E', E)$ the finest topology on E' which induces on every equicontinuous subset of E' the same topology as $\sigma(E', E)$. We know from Proposition 9.8 that $\nu(E', E)$ is finer than $\lambda(E', E)$ and we want to show that if E is metrizable, then $\nu(E', E) = \lambda(E', E)$.

It was recently proved by Y. Kōmura [51] that the topology $\nu(E', E)$ is not necessarily compatible with the vector space structure of E' since there can exist a neighborhood V of 0 such that for no neighborhood U of 0 does $U + U \subset V$ hold. We shall show, however, that it exists, that it is translation-invariant and that the origin possesses a fundamental system of absorbing, balanced neighborhoods. This follows from:

PROPOSITION 1. *Let $\mathfrak{U}$ be the collection of all subsets A of E' such that for every equicontinuous subset M of E' the intersection $A \cap M$ is open for the topology induced on M by $\sigma(E', E)$. Then $\mathfrak{U}$ is the collection of all open sets for the topology $\nu(E', E)$.*

Proof. Let us first show that $\mathfrak{U}$ satisfies Axioms (O 1) and (O 2) of open sets.

(O 1) Let $(A_\iota)_{\iota \in I}$ be a family of sets of $\mathfrak{U}$ and let M be an equicontinuous subset of E'. Then each set $A_\iota \cap M$ is open in M for $\sigma(E', E)$ and thus

$$\Big(\bigcup_{\iota \in I} A_\iota\Big) \cap M = \bigcup_{\iota \in I} (A_\iota \cap M)$$

is open in M. Consequently,

$$\bigcup_{\iota \in I} A_\iota \in \mathfrak{U}.$$

(O 2) Let $(A_i)_{1 \leqq i \leqq n}$ be a finite family of sets of $\mathfrak{U}$ and let M be an equicontinuous subset of E'. Then each set $A_i \cap M$ is open in M for $\sigma(E', E)$, and thus

$$\left(\bigcap_{i=1}^{n} A_i\right) \cap M = \bigcap_{i=1}^{n} (A_i \cap M)$$

is open in M. Therefore $\bigcap_{i=1}^{n} A_i \in \mathfrak{U}$.

Clearly, the topology for which the sets belonging to $\mathfrak{U}$ are the open sets has the property that on every equicontinuous subset of E' it induces the same topology as $\sigma(E', E)$. Conversely, if a topology is such that on every equicontinuous subset of E' it induces the same topology as $\sigma(E', E)$, then its open sets belong to the collection $\mathfrak{U}$. ∎

COROLLARY. *$\nu(E', E)$ is a translation-invariant Hausdorff topology, and the origin possesses a fundamental system of neighborhoods consisting of absorbing, balanced sets.*

Proof. A neighborhood of a point $a \in E'$ for $\nu(E', E)$ is a set U such that for every equicontinuous subset M of E' such that $a \in M$ the set $U \cap M$ is a neighborhood of a for the topology induced on M by $\sigma(E', E)$. It follows that the neighborhoods of a point $a \in E'$ for $\nu(E', E)$ are the sets $U + a$, where U is a neighborhood of 0, since M is an equicontinuous subset of E' containing 0 if and only if $M + a$ is an equicontinuous subset of E' containing a and $(U + a) \cap (M + a) = (U \cap M) + a$ is a $\sigma(E', E)$-neighborhood of a in $M + a$ if and only if $U \cap M$ is a $\sigma(E', E)$-neighborhood of 0 in M. Thus $\nu(E', E)$ is translation-invariant.

Since the topology $\nu(E', E)$ is finer than the Hausdorff topology $\sigma(E', E)$, it is also a Hausdorff topology.

The balanced core V of a $\nu(E', E)$-neighborhood U of 0 is a $\nu(E', E)$-neighborhood of 0. Indeed, if M is a balanced, equicontinuous subset of E' containing 0, then there exists a balanced $\sigma(E', E)$-neighborhood W_M of 0 such that $W_M \cap M \subset U \cap M \subset U$. Since $W_M \cap M$ is balanced, we have $W_M \cap M \subset V$ and thus $W_M \cap M \subset V \cap M$. If N is an arbitrary equicontinuous subset of E', then its balanced hull M is equicontinuous (cf. Proposition 4.6) and $W_M \cap N \subset V \cap N$. Hence V is indeed a $\nu(E', E)$-neighborhood of 0.

Finally, let V be a balanced $\nu(E', E)$-neighborhood of 0 and $x' \in E'$. There exist a balanced equicontinuous subset M of E' containing x' and a $\sigma(E', E)$-neighborhood W of 0 such that $W \cap M \subset V \cap M$. There exists $\alpha > 0$ such that $\lambda x' \in W$ for $|\lambda| \leqq \alpha$ and

$$\lambda x' \in W \cap M \subset V \cap M \subset V$$

for $|\lambda| \leqq \min (\alpha, 1)$. Thus V is absorbing. ∎

THEOREM 1 (Banach-Dieudonné). *Let E be a metrizable locally convex space and E' its dual. The finest topology on E', which induces on every equicontinuous subset of E' the same topology as $\sigma(E', E)$, is the topology $\lambda(E', E)$ of the uniform convergence on precompact subsets of E.*

Proof. Since $\nu(E', E)$ is finer than $\lambda(E', E)$ and since $\nu(E', E)$ is translation invariant, it is sufficient to show that given an open $\nu(E', E)$-neighborhood U of 0, there exists a precompact subset A of E such that $A^\circ \subset U$.

Let $(V_n)_{n\in\mathbf{N}}$ be a countable fundamental system of balanced, convex, closed neighborhoods of 0 in E such that $V_0 = E$ and $V_n \supset V_{n+1}$ $(n \in \mathbf{N})$. The theorem is then a consequence of the following

LEMMA. *For every $n \in \mathbf{N}$ there exists a finite subset F_n of V_n so that if we set $A_n = \bigcup_{k=0}^{n-1} F_k$, then the set $V_n^\circ \cap A_n^\circ$ is contained in U.*

Let us assume for a moment that we have proved the lemma. Let $A = \bigcup_{k\in\mathbf{N}} F_k$. The set A is precompact in E since every sequence of elements of A has a subsequence which is either stationary or converges to 0. On the other hand, $A^\circ \subset A_n^\circ$ and therefore $A^\circ \cap V_n^\circ \subset U$ for every $n \in \mathbf{N}$. Now, $\bigcap_{n\in\mathbf{N}} V_n = \{0\}$. Hence $\bigcup_{n\in\mathbf{N}} V_n^\circ = E'$, and therefore $A^\circ \subset U$. ∎

Proof of the lemma. We use mathematical induction. Suppose that for $n \geqq 0$ we have constructed the sets F_k $(0 \leqq k \leqq n - 1)$ with the required properties (i.e., for $n = 0$ we suppose nothing). We have to show that there exists a finite subset F_n of V_n so that $V_{n+1}^\circ \cap (A_n \cup F_n)^\circ \subset U$. Let us assume that no such set F_n exists and write $B_n = V_{n+1}^\circ \cap \complement U$. Then for every finite subset F of V_n the set $(A_n \cup F)^\circ$ meets B_n; i.e., $A_n^\circ \cap F^\circ \cap B_n \neq \emptyset$. The collection of all sets $A_n^\circ \cap F^\circ \cap B_n$, where F runs through the finite subsets of V_n, form a filter basis on B_n. Now, the set V_{n+1}° is compact for the topology $\sigma(E', E)$ by Theorem 4.1 and therefore also for the topology $\nu(E', E)$, which by hypothesis coincides with $\sigma(E', E)$ on V_{n+1}°. Since U is open for $\nu(E', E)$, $\complement U$ is closed and B_n is compact. The sets $A_n^\circ \cap F^\circ \cap B_n$ are closed for $\sigma(E', E)$ and $\nu(E', E)$, and therefore there exists a point x' which is contained in all of them. Now V_n is the union of all its finite subsets F. Hence V_n° is the intersection of all the sets F°, and thus $x' \in A_n^\circ \cap V_n^\circ \cap B_n$. But this contradicts the fact that $A_n^\circ \cap V_n^\circ \subset U$, and the contradiction proves the lemma. ∎

COROLLARY. *Let E be a metrizable locally convex space and E' its dual. A subset A of E' is closed for the topology $\lambda(E', E)$ if and only if for every balanced, convex, $\sigma(E', E)$-closed, equicontinuous subset M of E' the set $A \cap M$ is closed for $\sigma(E', E)$.*

Proof. If A is closed for $\lambda(E', E)$, then its intersection with M is closed for the topology induced on M by $\lambda(E', E)$. Now $\lambda(E', E)$ and $\sigma(E', E)$

induce the same topology on M. Hence $A \cap M$ is closed for the topology induced on M by $\sigma(E', E)$. Since M is $\sigma(E', E)$-closed, $A \cap M$ is $\sigma(E', E)$-closed in E'.

Conversely, assume that $A \cap M$ is $\sigma(E', E)$-closed for every balanced, convex, $\sigma(E', E)$-closed, equicontinuous subset M of E'. Let N be an arbitrary equicontinuous subset. The balanced, convex, $\sigma(E', E)$-closed hull $N^{\circ\circ} = M$ of N is equicontinuous (Proposition 4.6), and it follows from the relation $A \cap N = (A \cap M) \cap N$ that $A \cap N$ is closed in N for the topology induced on N by $\sigma(E', E)$. By Proposition 1 the set A is closed for $\nu(E', E)$. But by Theorem 1 we have $\nu(E', E) = \lambda(E', E)$. ∎

THEOREM 2 (Krein-Šmulian). *Let E be a Fréchet space and E' its dual. A convex subset A of E' is closed for the topology $\sigma(E', E)$ if and only if for every balanced, convex, $\sigma(E', E)$-closed, equicontinuous subset M of E' the set $A \cap M$ is closed for $\sigma(E', E)$.*

Proof. If A is closed for $\sigma(E', E)$, then $A \cap M$ is closed for $\sigma(E', E)$. Conversely, if the condition is satisfied, then A is closed for the topology $\lambda(E', E)$ by the previous corollary. On the other hand, E is complete. Hence $\lambda(E', E) = \kappa(E', E)$ is compatible with the duality between E and E' by the observation made following Proposition 9.7. Thus A is closed for $\sigma(E', E)$ by Proposition 4.3. ∎

COROLLARY (Banach). *Let E be a Banach space and E' its dual. A linear subspace F of E' is closed for the topology $\sigma(E', E)$ if and only if its intersection with the closed unit ball $B'_1 = \{x' \mid \|x'\| \leqq 1\}$ is closed for $\sigma(E', E)$.*

Proof. Any equicontinuous subset of E' is contained in a closed ball $B'_\rho = \{x' \mid \|x'\| \leqq \rho\}$. Thus if $F \cap B'_1$ is $\sigma(E', E)$-closed, then $F \cap M$ is $\sigma(E', E)$-closed for any $\sigma(E', E)$-closed equicontinuous subset M of E'. The corollary now follows from Theorem 2. ∎

We shall encounter in the remainder of this chapter several situations which are similar to the one in Theorem 2. For the orientation of the reader we shall present here a short list of these situations. Let E be a locally convex Hausdorff space and denote by E' its dual. Consider the following collections of subsets of E':

$\mathbf{C}$ = all convex subsets of E',
$\mathbf{A}$ = all balanced, convex subsets of E',
$\mathbf{S}$ = all linear subspaces of E',
$\mathbf{H}$ = all hyperplanes of E',
$\mathbf{D}$ = all linear subspaces of E' which are dense in E' for the topology $\sigma(E', E)$.

Clearly, $\mathbf{H} \subset \mathbf{S} \subset \mathbf{A} \subset \mathbf{C}$ and $\mathbf{D} \subset \mathbf{S}$. Let $\mathbf{X}$ be any of the collections just introduced and denote by $(\mathbf{X})$ the following statement.

"If $A \in \mathbf{X}$ and for every balanced, convex, $\sigma(E', E)$-closed, equicontinuous subset M of E' the set $A \cap M$ is $\sigma(E', E)$-closed, then A is closed for $\sigma(E', E)$."

Clearly, (**C**) $\Rightarrow$ (**A**) $\Rightarrow$ (**S**) $\Rightarrow$ (**H**) and (**S**) $\Rightarrow$ (**D**). Now the following can be said about these statements:

(1) We just saw that if E is metrizable and complete, then (**C**) holds.

(2) (**H**) holds if and only if E is complete, as we shall see in the following section (Corollary 5 to Theorem 11.1 and Exercise 11.5).

(3) If (**S**) holds, we shall say that E is a Pták space (Definition 17.2). We owe to Pták the deep and beautiful result that every continuous, "almost open" (Definition 17.1) linear map from E into any locally convex Hausdorff space is a strict morphism if and only if (**S**) holds for E (Theorem 17.2). This enables us to generalize the Banach homomorphism theorem (Theorem 1.9.1) and the closed-graph theorem (Proposition 1.9.3), as we shall see in §17.

(4) If (**D**) holds, we say that E is an infra-Pták space (cf. Exercise 17.5).

(5) (**A**) means that E is a hypercomplete space in the sense of Kelley ([49], 13F, p. 116 and 18G, p. 177).

We shall see that (**D**) $\Rightarrow$ (**H**) (Exercise 17.6). It seems to be an open problem whether there are spaces which satisfy (**D**) but not (**S**). On the negative side, we have the following results: (**H**) $\not\Rightarrow$ (**S**) (Exercise 17.4), (**H**) $\not\Rightarrow$ (**D**) ([75], pp. 72–73), and (**A**) $\not\Rightarrow$ (**C**) ([49], 18H, p. 178).

Exercises

1. Let E be a metrizable locally convex space and E' its dual.

(a) Prove that every precompact subset of E lies in the balanced, closed, convex hull of a sequence of points of E converging to 0. (*Hint:* Examine the proof of Theorem 1.)

(b) Let $\mathfrak{S}$ be the collection of subsets of E consisting of points of a sequence which converges to 0. Prove that the $\mathfrak{S}$-topology on E' coincides with $\lambda(E', E)$ and with $\nu(E', E)$.

§11. *Grothendieck's completeness theorem*

In this section we shall characterize the completion of the dual of a locally convex Hausdorff space equipped with an $\mathfrak{S}$-topology as a space of linear forms. Among the numerous consequences of this characterization will figure one of the results announced at the end of the preceding section.

We begin with the following approximation lemma:

PROPOSITION 1. *Let F be a locally convex Hausdorff space, F' its dual, and A a balanced, closed, convex subset of F. Let u be a linear form on F whose restriction to A is continuous on A. For every $\epsilon > 0$ there exists a linear form $x' \in F'$ such that $|u(x) - \langle x', x\rangle| \leqq \epsilon$ for all $x \in A$.*

Proof. We consider F' as a linear subspace of the algebraic dual F^* of F. The topology $\sigma(F^*, F)$ induces on F' the topology $\sigma(F', F)$.

Since the restriction of u to A is continuous at 0, there exists a balanced, closed, convex neighborhood U of 0 in F such that $x \in U \cap A$ implies $|u(x)| \leqq \epsilon$. The polar U° of U in F^* is formed by all the linear forms v on F such that $|v(x)| \leqq 1$ for all $x \in U$. Thus U° is a $\sigma(F', F)$-closed, equicontinuous subset of F', and therefore by Theorem 4.1 it is compact in F' and also in F^*. On the other hand, A° is a balanced, $\sigma(F^*, F)$-closed, convex subset of F^*, and by Proposition 2.10.5 the set $U^\circ + A^\circ$ is $\sigma(F^*, F)$-closed in F^*. By the theorem of bipolars (Theorem 3.1) we have $U = U^{\circ\circ}$, $A = A^{\circ\circ}$; and since $U^\circ + A^\circ$ is balanced and convex, we also have $U^\circ + A^\circ = (U^\circ + A^\circ)^{\circ\circ}$. Since $0 \in U^\circ$ and $0 \in A^\circ$, we have $U^\circ \cup A^\circ \subset U^\circ + A^\circ$ and (Proposition 3.2)

$$(U \cap A)^\circ = (U^\circ \cup A^\circ)^{\circ\circ} \subset (U^\circ + A^\circ)^{\circ\circ} = U^\circ + A^\circ.$$

Now $(1/\epsilon)u \in (U \cap A)^\circ$, and therefore there exist $v \in U^\circ$ and $w \in A^\circ$ such that $u = \epsilon v + \epsilon w$; i.e., $|u(x) - \epsilon v(x)| = |\epsilon w(x)| \leqq \epsilon$ for all $x \in A$. Setting $\epsilon v = x' \in F'$, we obtain the required linear form. ∎

We shall also need the following

LEMMA 1. *Let F be a locally convex space, A a balanced, convex subset of F, and u a linear map from F into a locally convex space G. Then the restriction of u to A is uniformly continuous on A if it is continuous at 0.*

Proof. Suppose that the restriction of u to A is continuous at 0. Let V be a balanced, convex neighborhood of 0 in G. There exists a balanced, convex neighborhood U of 0 in F such that $u(U \cap A) \subset \frac{1}{2}V$. Let x and y be two elements of A such that $x - y \in U$. Then $x - y \in 2A$ since A is balanced and convex. It follows that $x - y \in U \cap 2A = 2(\frac{1}{2}U \cap A)$. Hence $\frac{1}{2}(x - y) \in \frac{1}{2}U \cap A \subset U \cap A$. Therefore $u(\frac{1}{2}(x - y)) \in \frac{1}{2}V$; that is, $u(x - y) \in V$. Thus $x - y \in U$ implies $u(x) - u(y) \in V$, which means that the restriction of u to A is uniformly continuous. ∎

Now we are ready to state the main result of this section.

THEOREM 1 (Grothendieck). *Let F be a locally convex Hausdorff space with topology $\mathfrak{T}$, and F' its dual. Let $\mathfrak{S}$ be a covering of F by balanced, bounded, closed, convex sets such that $A \in \mathfrak{S}$, $B \in \mathfrak{S}$ implies the existence of $C \in \mathfrak{S}$ for which $A \cup B \subset C$. Consider F' equipped with the X-topology. Furthermore, let R be the vector space formed by all linear forms on F whose* *restrictions to all sets $A \in \mathfrak{S}$ are continuous for the topology induced by $\mathcal{T}$ and equip R with the $\mathfrak{S}$-topology. Then R is the completion of F'.*

Proof. Observe first that $F' \subset R \subset F^*$ and therefore the spaces F and R form a dual system. Next we observe that the sets $A \in \mathfrak{S}$ are $\sigma(F, R)$-bounded. Indeed, let $A \in \mathfrak{S}$ and $u \in R$. There exists a $\mathcal{T}$-neighborhood U of 0 such that $x \in A \cap U$ implies $|u(x)| \leqq 1$. Since A is bounded, there exists $0 < \lambda < 1$ such that $\lambda A \subset U$. But then $x \in A$ implies $\lambda x \in A \cap U$, and thus $|u(x)| \leqq 1/\lambda$, which shows that A is $\sigma(F, R)$-bounded. Thus the $\mathfrak{S}$-topology is a locally convex Hausdorff topology on R (Proposition 4.1). It follows from Proposition 1 that F' is dense in R for the $\mathfrak{S}$-topology.

Thus we have only to prove that R is complete. Let $\mathfrak{F}$ be a Cauchy filter on R for the $\mathfrak{S}$-topology. The topology $\sigma(F^*, F)$ induces on R the topology $\sigma(R, F)$, and $\sigma(R, F)$ is coarser than the $\mathfrak{S}$-topology on R since the sets $A \in \mathfrak{S}$ cover F. Consequently, $\mathfrak{F}$ is the basis of a Cauchy filter on F^* for the topology $\sigma(F^*, F)$. But F^* is complete for $\sigma(F^*, F)$ (Example 2.3). Thus $\mathfrak{F}$ converges to an element $u_0 \in F^*$; i.e., given $x_1, \ldots, x_n$ in F and $\epsilon > 0$, there exists $X \in \mathfrak{F}$ such that $|\langle u - u_0, x_i\rangle| < \epsilon$ for all $u \in X$ and $1 \leqq i \leqq n$.

$\mathfrak{F}$ converges to u_0 in the $\mathfrak{S}$-topology on F^*. Indeed, let $A \in \mathfrak{S}$ and $\epsilon > 0$. Since $\mathfrak{F}$ is the basis of a Cauchy filter, there exists $X \in \mathfrak{F}$ such that $|\langle u - v, x\rangle| < \frac{1}{2}\epsilon$ for all $u, v \in X$ and $x \in A$. Now for each $x \in A$ there exists $Y_x \in \mathfrak{F}$ such that $|\langle u_0 - u, x\rangle| < \frac{1}{2}\epsilon$ for all $u \in Y_x$. For a given $v \in X$ and $x \in A$ we pick u in $X \cap Y_x$ and obtain

$$|\langle u_0 - v, x\rangle| \leqq |\langle u_0 - u, x\rangle| + |\langle u - v, x\rangle| < \epsilon;$$

i.e., $|\langle u_0 - v, x\rangle| < \epsilon$ for all $v \in X$ and $x \in A$.

The linear form u_0 belongs to R. Indeed, let $A \in \mathfrak{S}$ and $\epsilon > 0$. There exists $X \in \mathfrak{F}$ such that $|\langle u_0 - u, x\rangle| < \frac{1}{2}\epsilon$ for all $u \in X$ and $x \in A$. On the other hand, for each $u \in R$ there exists a neighborhood U of 0 in F so that $|\langle u, x\rangle| < \frac{1}{2}\epsilon$ for all $x \in U \cap A$. Picking u in X, we have therefore

$$|\langle u_0, x\rangle| \leqq |\langle u_0 - u, x\rangle| + |\langle u, x\rangle| < \epsilon$$

for all $x \in U \cap A$; i.e., the restriction of u_0 to A is continuous at 0. But then this restriction is continuous everywhere on A by Lemma 1; i.e., $u_0 \in R$. This concludes the proof of the fact that R is complete. ∎

Remark 1. The fact that R is complete follows also from a theorem in general topology. Let X be a topological space, $\mathfrak{S}$ a collection of subsets of X, and Y a complete uniform space. Then the space of all maps from X into Y whose restrictions to the sets of $\mathfrak{S}$ are continuous, equipped with the topology of uniform convergence on the sets of $\mathfrak{S}$, is complete ([8], Chapter X, 2nd ed., §1, no. 6, Corollary 2 to Theorem 2; [49], 8.11, p. 73).

Corollary 1. *Let F and G be two vector spaces which form a dual system (Definition 2.1). Let $\mathfrak{S}$ be a covering of F by balanced, convex, $\sigma(F, G)$-closed and $\sigma(F, G)$-bounded sets. Equip G with the $\mathfrak{S}$-topology. Then the completion of G can be identified with the vector space of all linear forms on F whose restrictions to the sets $A \in \mathfrak{S}$ are continuous for the topology induced by $\sigma(F, G)$.*

Proof. For the topology $\mathcal{T}$ in Theorem 1 take $\sigma(F, G)$. Then $F' = G$ (Proposition 2.2). ▮

Corollary 2. *Let E be a locally convex Hausdorff space and E' its dual. Then the completion $\hat{E}$ of E can be identified with the vector space of all linear forms on E' whose restrictions to the equicontinuous subsets of E' are continuous for $\sigma(E', E)$.*

Proof. We take E' equipped with the topology $\sigma(E', E)$ for the space F of Theorem 1 and denote by $\mathfrak{S}$ the collection of all balanced, closed, convex, equicontinuous subsets of E'. Then the dual of E' is the space E, and by Proposition 4.7 the $\mathfrak{S}$-topology on E coincides with its original topology. Thus by Theorem 1 the completion of E is the vector space formed by all linear forms on E' whose restrictions to the sets $A \in \mathfrak{S}$ are $\sigma(E', E)$-continuous. ▮

Of course the topology of $\hat{E}$ is the $\mathfrak{S}$-topology, where $\mathfrak{S}$ has the same meaning as in the preceding proof.

Corollary 3. *Let E be a locally convex Hausdorff space and E' its dual. Then E is complete if and only if every linear form on E', whose restrictions to the equicontinuous subsets of E' are continuous for $\sigma(E', E)$, is continuous for $\sigma(E', E)$, i.e., is an element of E.*

Let us note the special case of a Banach space.

Corollary 4 (Banach). *Let E be a Banach space and E' its dual. A linear form u on E' is continuous for $\sigma(E', E)$ if and only if its restriction to the closed unit ball $B'_1 = \{x' \mid \|x'\| \leqq 1\}$ is continuous for $\sigma(E', E)$.*

Proof. This is an immediate consequence of Corollary 3, considering that every equicontinuous subset of E' is contained in some ball

$$B'_\rho = \{x' \mid \|x'\| \leqq \rho\}. \ ▮$$

Corollary 5. *Let E be a locally convex Hausdorff space and E' its dual. Then E is complete if the following condition is satisfied:*

(H) *Let H be a hyperplane in E' and suppose that for any balanced, convex, $\sigma(E', E)$-closed, equicontinuous subset M of E' the intersection $H \cap M$ is $\sigma(E', E)$-closed. Then H is $\sigma(E', E)$-closed.*

Proof. Assume that condition (**H**) is satisfied and let u be a linear form on E' whose restriction to any equicontinuous subset of E' is continuous for the topology induced by $\sigma(E', E)$. Let H be the hyperplane in E' defined by the equation $u(x') = 0$. For every balanced, convex, $\sigma(E', E)$-closed, equicontinuous subset M of E the set $H \cap M$ is the inverse image of 0 by the restriction of u to M. Hence $H \cap M$ is $\sigma(E', E)$-closed, and therefore by condition (**H**) the hyperplane H is $\sigma(E', E)$-closed. It follows from Proposition 2.5.7 that u is $\sigma(E', E)$-continuous, and thus by Corollary 3 the space E is complete. ▮

In order to give another useful formulation of Corollary 2 to Theorem 1, we introduce still another topology on the dual of a locally convex space. Let E be a locally convex Hausdorff space and E' its dual. We denote by $\mu(E', E)$ the finest locally convex topology on E' which induces on every equicontinuous subset of E' the same topology as $\sigma(E', E)$. Since it is not obvious that the least upper bound of all locally convex topologies on E', which induce on equicontinuous subsets the same topology as $\sigma(E', E)$, is itself a locally convex topology, we have to prove that $\mu(E', E)$ exists.

PROPOSITION 2. *Denote by* $\mathfrak{U}$ *the collection of all absorbing, balanced, convex subsets of* E' *whose intersection with any equicontinuous set* M *containing* 0 *is a neighborhood of* 0 *in* M *for the topology induced by* $\sigma(E', E)$. *Then* $\mathfrak{U}$ *is a fundamental system of neighborhoods of* 0 *for* $\mu(E', E)$.

Proof. If U and V belong to $\mathfrak{U}$, then $U \cap V$ also belongs to $\mathfrak{U}$, and for every $\lambda > 0$ the set λU belongs to $\mathfrak{U}$. Thus by Proposition 2.4.5 $\mathfrak{U}$ is a fundamental system of neighborhoods of 0 for a locally convex topology $\mathcal{T}$ on E'. The topology $\mathcal{T}$ is finer than $\sigma(E', E)$, hence induces on every equicontinuous subset of E' a finer topology than does $\sigma(E', E)$. Conversely, let M be an equicontinuous subset of E', $a \in M$, and $U \in \mathfrak{U}$. Then $0 \in M - a$. Hence there exists a $\sigma(E', E)$-neighborhood W of 0 such that $U \cap (M - a) \supset W \cap (M - a)$, i.e., that

$$(a + U) \cap M \supset (a + W) \cap M.$$

This proves that $\mathcal{T}$ induces on M a coarser topology than $\sigma(E', E)$. Finally, if U is a balanced, convex neighborhood of 0 for any locally convex topology on E' which induces on every equicontinuous subset the same topology as $\sigma(E', E)$, then $U \in \mathfrak{U}$. ▮

Clearly,

$$\sigma(E', E) \preccurlyeq \kappa(E', E) \preccurlyeq \lambda(E', E) \preccurlyeq \mu(E', E) \preccurlyeq \nu(E', E),$$

where $\preccurlyeq$ means "coarser than." We know that if E is metrizable, then $\lambda(E', E) = \nu(E', E)$ (Theorem 10.1), and that $\kappa(E', E) = \lambda(E', E)$ if E

is quasi-complete. If E is complete, then

$$\kappa(E', E) = \lambda(E', E) = \mu(E', E)$$

(cf. Exercise 4). On the other hand, examples can be given where $\lambda(E', E) \neq \mu(E', E)$ and $\mu(E', E) \neq \nu(E', E)$ ([52], §21,8.(2)). We can now give the following formulation of Corollary 2 to Theorem 1.

PROPOSITION 3. *Let E be a locally convex Hausdorff space and E' its dual. Equip E' with the topology $\mu(E', E)$. Then the dual of E' is the completion $\hat{E}$ of E.*

Proof. By Corollary 2 of Theorem 1, if $u \in \hat{E}$, then given $\epsilon > 0$, for every balanced, convex, equicontinuous subset M of E' there exists a balanced, convex $\sigma(E', E)$-neighborhood U_M of 0 so that $x \in U_M \cap M$ implies $|u(x)| \leqq \epsilon$. The balanced convex hull U of $\bigcup_M (U_M \cap M)$ is a neighborhood of 0 in E' for the topology $\mu(E', E)$. Every $x \in U$ is of the form $\sum_M \lambda_M x_M$ with $x_M \in U_M \cap M$ and $\sum_M |\lambda_M| \leqq 1$ (Exercise 2.4.1(b)). Hence $x \in U$ implies $|u(x)| \leqq \sum |\lambda_M| \cdot |u(x_M)| \leqq \epsilon$. Thus u belongs to the dual of E'. Conversely, let u be a continuous linear form on E'. Then the restriction of u to any equicontinuous subset of E' is continuous for $\mu(E', E)$ and thus also for $\sigma(E', E)$. Again by Corollary 2 to Theorem 1 we have $u \in \hat{E}$. ∎

Corollary 3 to Theorem 1 has an interesting application to the spaces in which there exists a countable everywhere dense subset. Let us first prove a property of the dual of such a space.

Let E be a locally convex Hausdorff space and E' its dual. Let A be a subset of E which is total for the topology $\sigma(E, E')$ and denote by $\mathcal{T}_A$ the $\mathfrak{S}$-topology on E', where $\mathfrak{S}$ is the collection of all sets $\{x\}$ with $x \in A$. We refer to $\mathcal{T}_A$ as the topology of pointwise convergence on A. By Proposition 4.1 the topology $\mathcal{T}_A$ is Hausdorff and it is clearly coarser than $\sigma(E', E)$.

LEMMA 2. *On every equicontinuous subset M of E' the topology induced by $\mathcal{T}_A$ coincides with the topology induced by $\sigma(E', E)$.*

Proof. The $\sigma(E', E)$-closure of M is equicontinuous (cf. Lemmas 6.2 and 9.2) and thus $\sigma(E', E)$-compact by Theorem 4.1. Hence the conclusion follows from the remarks preceding Proposition 9.2. ∎

PROPOSITION 4. *Let E be a locally convex Hausdorff space and E' its dual. Suppose that there exists in E a countable total subset A. Then every equicontinuous subset M of E' is metrizable for the topology induced on it by $\sigma(E', E)$.*

Proof. The set A is *a fortiori* total in E for $\sigma(E, E')$, and thus $\mathcal{T}_A$ is a Hausdorff topology. On the other hand, since A is countable, there exists a countable fundamental system of neighborhoods of 0 for the topology $\mathcal{T}_A$. By Theorem 2.6.1 the topology $\mathcal{T}_A$ is metrizable. But by Lemma 2 the topologies $\mathcal{T}_A$ and $\sigma(E', E)$ induce the same topology on M. ▮

Now we come to the announced application of Corollary 3 of Theorem 1.

PROPOSITION 5. *Let E be a locally convex, complete Hausdorff space and E' its dual. Suppose furthermore that there exists in E a countable total subset. Then a linear form u on E' is continuous for the topology $\sigma(E', E)$ if and only if for every sequence $(x'_n)_{n \in \mathbf{N}}$ in E' which converges for $\sigma(E', E)$ to a point $x' \in E'$ we have $\lim_{n\to\infty} u(x'_n) = u(x')$.*

Proof. The condition is clearly necessary. Conversely if the condition is satisfied, then the restriction of u to every equicontinuous subset of E' is continuous for $\sigma(E', E)$, since by Proposition 3 every such set is metrizable. It follows from Corollary 3 to Theorem 1 that u is continuous on E' for $\sigma(E', E)$. ▮

EXERCISES

1. Deduce Proposition 7.6 from Theorem 1. (*Hint:* Let $\mathfrak{S}$ be the collection of all balanced, bounded, convex, closed subsets of the bornological Hausdorff space E and let u be a linear form on E whose restrictions to every $A \in \mathfrak{S}$ are continuous. Show that u is bounded on every $A \in \mathfrak{S}$.)

2. Deduce Proposition 2.5(b) from Theorem 1.

3. Let F and G be two vector spaces which form a dual system. Let $\mathfrak{S}_F$ be a collection of balanced, convex, $\sigma(F, G)$-bounded subsets of F which covers F, and $\mathfrak{S}_G$ be a collection of balanced, convex, $\sigma(G, F)$-bounded subsets of G which covers G.

(a) Show that the following statements are equivalent:

(α) Every set $A \in \mathfrak{S}_F$ is precompact for the $\mathfrak{S}_G$-topology on F.

(β) Every set $B \in \mathfrak{S}_G$ is precompact for the $\mathfrak{S}_F$-topology on G.

(γ) On every subset $A \in \mathfrak{S}_F$ the topology induced by the $\mathfrak{S}_G$-topology coincides with the topology induced by $\sigma(F, G)$.

(δ) On every subset $B \in \mathfrak{S}_G$ the topology induced by the $\mathfrak{S}_F$-topology coincides with the topology induced by $\sigma(G, F)$.

(*Hint:* It is sufficient to prove that $(\alpha) \Rightarrow (\delta)$ and $(\delta) \Rightarrow (\beta)$ since then by symmetry $(\beta) \Rightarrow (\gamma)$ and $(\gamma) \Rightarrow (\alpha)$ will follow. To prove $(\alpha) \Rightarrow (\delta)$, observe that the sets B°, where $B \in \mathfrak{S}_G$, form a fundamental system of neighborhoods of 0 for the $\mathfrak{S}_G$-topology; hence the sets $B^{\circ\circ}$ and thus also the sets B are equicontinuous in G for the $\mathfrak{S}_G$-topology on F. By Proposition 9.8 the topology induced on B by $\sigma(G, F)$ coincides with the topology induced by the topology of uniform convergence on precompact subsets of F for the $\mathfrak{S}_G$-topology. If (α) holds, each $A \in \mathfrak{S}_F$ is precompact for the $\mathfrak{S}_G$-topology. Hence the $\mathfrak{S}_F$-topology on G is coarser than the topology of uniform convergence on precompact subsets of F for the $\mathfrak{S}_G$-topology. Since $F = \bigcup_{A \in \mathfrak{S}_F} A$, the $\mathfrak{S}_F$-topology is finer than $\sigma(G, F)$. Hence (δ) follows.

To prove $(\delta) \Rightarrow (\beta)$, observe that every $B \in \mathfrak{S}_G$ is equicontinuous for the $\mathfrak{S}_G$-topology on F, hence relatively compact for $\sigma(G, F)$ and precompact for the topology induced by $\sigma(G, F)$. Therefore if (δ) holds, B is precompact for the topology induced by the $\mathfrak{S}_F$-topology.)

(b) Denote by $\mathcal{T}_F$ the $\mathfrak{S}_G$-topology on F and by $\mathcal{T}_F^\lambda$ the topology of uniform convergence on $\mathcal{T}_F$-precompact subsets of F on G. Show that $\mathcal{T}_F^\lambda$ is the finest locally convex topology on G which induces on every $\mathcal{T}_F$-equicontinuous subset of G the same topology as $\sigma(G, F)$. (*Hint:* By Proposition 9.8 $\mathcal{T}_F^\lambda$ induces the same topology as $\sigma(G, F)$ on every $\mathcal{T}_F$-equicontinuous subset of G. If $\mathcal{T}$ is some locally convex topology on G which induces on every $\mathcal{T}_F$-equicontinuous subset of G the same topology as $\sigma(G, F)$, then $\mathcal{T}$ is an $\mathfrak{S}$-topology, where $\mathfrak{S}$ is the collection of $\mathcal{T}$-equicontinuous subsets of F; and by (α) of part (a) each set $M \in \mathfrak{S}$ is precompact for $\mathcal{T}_F$. Thus $\mathcal{T}$ is coarser than $\mathcal{T}_F^\lambda$.)

4. Show that if the locally convex Hausdorff space E is complete, then the topology $\mu(E', E)$ coincides with the topology $\lambda(E', E)$. (*Hint:* By Proposition 3 the space E is the dual of E' equipped with the topology $\mu(E', E)$, and $\mu(E', E)$ is the $\mathfrak{S}$-topology for some collection $\mathfrak{S}$ of balanced, convex, $\sigma(E, E')$-bounded subsets of E. By Exercise 3(a) each set $A \in \mathfrak{S}$ is precompact for the topology of E. Thus $\mu(E', E)$ is coarser than $\lambda(E', E)$.)

5. Show that if the locally convex Hausdorff space E is complete, then condition (**H**) of Corollary 5 to Theorem 1 is satisfied.

§12. *The transpose of a linear map*

Let us begin by recalling from algebra the definition of the transpose of a linear map. Let E and F be two vector spaces over the same field $\mathbf{K}$ and let E^* and F^* be their respective algebraic duals. If u is a linear map from E into F, then there exists a unique linear map ${}^tu$ from F^* into E^* which satisfies

$$\langle u(x), y' \rangle = \langle x, {}^tu(y') \rangle$$

for every $x \in E$ and $y' \in F^*$. Indeed, for every $y' \in F^*$ the map $x \mapsto \langle u(x), y' \rangle$ is clearly a linear form $x' \in E^*$; we set ${}^tu(y') = x'$. The map ${}^tu$ is linear since for $y', z' \in F^*$ and $\lambda \in \mathbf{K}$ we have

$$\begin{aligned}\langle x, {}^tu(y' + z') \rangle &= \langle u(x), y' + z' \rangle = \langle u(x), y' \rangle + \langle u(x), z' \rangle \\ &= \langle x, {}^tu(y') \rangle + \langle x, {}^tu(z') \rangle = \langle x, {}^tu(y') + {}^tu(z') \rangle\end{aligned}$$

and

$$\langle x, {}^tu(\lambda y') \rangle = \langle u(x), \lambda y' \rangle = \langle \lambda u(x), y' \rangle = \langle u(\lambda x), y' \rangle = \langle \lambda x, {}^tu(y') \rangle = \langle x, \lambda {}^tu(y') \rangle.$$

We call ${}^tu$ the *transpose* of u.

Next, let us consider the transpose of a weakly continuous linear map.

PROPOSITION 1. *Let (F_1, G_1) and (F_2, G_2) be two pairs of vector spaces forming dual systems. We consider G_i imbedded canonically into F_i^* $(i = 1, 2)$. Let u be a linear map from F_1 into F_2. Then the restriction of ${}^tu$ to G_2 maps G_2 into G_1 if and only if u is continuous for the topologies $\sigma(F_1, G_1)$ and $\sigma(F_2, G_2)$.*

Proof. (a) Suppose that ${}^t u(G_2) \subset G_1$ and let W be a $\sigma(F_2, G_2)$-neighborhood of 0 in F_2 defined by $|\langle y, y_i' \rangle| \leqq \epsilon$, where $y \in F_2$, $y_i' \in G_2$ $(1 \leqq i \leqq n)$. Set $x_i' = {}^t u(y_i') \in G_1$ for $1 \leqq i \leqq n$ and let V be the $\sigma(F_1, G_1)$-neighborhood of 0 in F_1 defined by $|\langle x, x_i' \rangle| \leqq \epsilon$ for $1 \leqq i \leqq n$. If $x \in V$, then

$$|\langle u(x), y_i' \rangle| = |\langle x, {}^t u(y_i') \rangle| = |\langle x, x_i' \rangle| \leqq \epsilon \qquad (1 \leqq i \leqq n);$$

i.e., $u(x) \in W$. Thus u is continuous.

(b) Suppose that u is continuous for the weak topologies. Then for each $y' \in G_2$ the linear form $x \mapsto \langle u(x), y' \rangle$ is continuous on F_1 for $\sigma(F_1, G_1)$. By Proposition 2.2 there exists $x' \in G_1$ such that

$$\langle u(x), y' \rangle = \langle x, x' \rangle$$

for all $x \in F_1$. This shows that ${}^t u(y') = x'$, i.e., that ${}^t u$ maps G_2 into G_1. ∎

If the conditions of Proposition 1 are satisfied, we call the restriction of ${}^t u$ to G_2 the transpose of u and go on denoting it by ${}^t u$.

COROLLARY. *If u is a linear map from F_1 into F_2, continuous for the topologies $\sigma(F_1, G_1)$ and $\sigma(F_2, G_2)$, then the map ${}^t u\colon G_2 \to G_1$ is continuous for the topologies $\sigma(G_2, F_2)$ and $\sigma(G_1, F_1)$. We have furthermore ${}^t({}^t u) = u$.*

Proof. It follows from the identity $\langle u(x), y' \rangle = \langle x, {}^t u(y') \rangle$, $x \in F_1$, $y' \in G_2$ that ${}^t({}^t u) = u$. Now u maps F_1 into F_2; hence by Proposition 1 the map ${}^t u$ is weakly continuous. ∎

PROPOSITION 2. *Let (F_1, G_1) and (F_2, G_2) be two pairs of vector spaces forming dual systems and let $u\colon F_1 \to F_2$ be a linear map which is continuous for the topologies $\sigma(F_1, G_1)$ and $\sigma(F_2, G_2)$. Let A be a subset of F_1 and B a subset of F_2. Then:*

(a) $(u(A))^\circ = {}^t u^{-1}(A^\circ)$.

(b) *$u(A) \subset B$ implies ${}^t u(B^\circ) \subset A^\circ$.*

(c) *If A and B are nonempty, balanced, convex and closed for the topologies $\sigma(F_1, G_1)$ and $\sigma(F_2, G_2)$ respectively, then the relations $u(A) \subset B$ and ${}^t u(B^\circ) \subset A^\circ$ are equivalent.*

Proof. (a) The relation $y' \in (u(A))^\circ$ means that

$$|\langle u(x), y' \rangle| = |\langle x, {}^t u(y') \rangle| \leqq 1$$

for all $x \in A$, which is equivalent to ${}^t u(y') \in A^\circ$, i.e., to $y' \in {}^t u^{-1}(A^\circ)$.

(b) If $u(A) \subset B$, then by Proposition 3.1(a) we have

$$B^\circ \subset (u(A))^\circ = {}^t u^{-1}(A^\circ).$$

(c) Since ${}^t u\colon G_2 \to G_1$ is continuous for $\sigma(G_2, F_2)$ and $\sigma(G_1, F_1)$ (Corollary to Proposition 1), by part (b) the relation ${}^t u(B^{\circ}) \subset A^{\circ}$ implies that ${}^t({}^t u)(A^{\circ\circ}) \subset B^{\circ\circ}$. But ${}^t({}^t u) = u$ by the preceding corollary, and $A^{\circ\circ} = A$, $B^{\circ\circ} = B$ by Theorem 3.1. ∎

COROLLARY 1. *Let $u\colon F_1 \to F_2$ be a linear map which is continuous for $\sigma(F_1, G_1)$ and $\sigma(F_2, G_2)$. Then the kernel of ${}^t u$ is the subspace $u(F_1)^{\perp}$ of G_2, i.e.,* $\operatorname{Ker}({}^t u) = (\operatorname{Im} u)^{\perp}$.

Proof. If $A = F_1$, then $A^{\circ} = \{0\}$, and ${}^t u^{-1}(A^{\circ})$ is the kernel of ${}^t u$. Now, by part (a) of Proposition 2 ${}^t u^{-1}(A^{\circ}) = (u(A))^{\circ} = (u(F_1))^{\perp}$. ∎

COROLLARY 2. *Let $u\colon F_1 \to F_2$ be a linear map which is continuous for $\sigma(F_1, G_1)$ and $\sigma(F_2, G_2)$. Then $u(F_1)$ is everywhere dense in F_2 for the topology $\sigma(F_2, G_2)$ if and only if the map ${}^t u$ is injective.*

Proof. This is a consequence of the preceding corollary since $\overline{u(F_1)} = F_2$ means that $u(F_1)^{\perp} = \{0\}$ and ${}^t u$ is injective if and only if its kernel is $\{0\}$. ∎

PROPOSITION 3. *Let E and F be two locally convex Hausdorff spaces with topologies $\mathcal{T}_E$ and $\mathcal{T}_F$, and let E' and F' be their duals.*

(a) *If the linear map $u\colon E \to F$ is continuous for the topologies $\mathcal{T}_E$ and $\mathcal{T}_F$, then it is continuous for the topologies $\sigma(E, E')$ and $\sigma(F, F')$.*

(b) *If the linear map $u'\colon F' \to E'$ is continuous for the topologies $\sigma(F', F)$ and $\sigma(E', E)$, then it is continuous for the topologies $\beta(F', F)$ and $\beta(E', E)$.*

Proof. (a) Let $u\colon E \to F$ be continuous for the topologies $\mathcal{T}_E$ and $\mathcal{T}_F$. Then for each $y' \in F'$ the linear form $x \mapsto \langle u(x), y' \rangle$ is continuous on E for $\mathcal{T}_E$; hence it is also continuous for $\sigma(E, E')$ (Example 4.3). By the definition of the topology $\sigma(F, F')$ this means that u is continuous for the topologies $\sigma(E, E')$ and $\sigma(F, F')$.

(b) Let $u'\colon F' \to E'$ be continuous for the topologies $\sigma(F', F)$ and $\sigma(E', E)$. Then by the corollary to Proposition 1 we can set $u' = {}^t u$, where $u\colon E \to F$ is a linear map, continuous for $\sigma(E, E')$ and $\sigma(F, F')$. Let A be a $\sigma(E, E')$-bounded subset of E. Then $B = u(A)$ is a $\sigma(F, F')$-bounded subset of F, and by Proposition 2(b) we have $u'(B^{\circ}) \subset A^{\circ}$, which proves that u' is continuous for the topologies $\beta(F', F)$ and $\beta(E', E)$. ∎

COROLLARY. *Let E and F be two locally convex Hausdorff spaces with topologies $\mathcal{T}_E$ and $\mathcal{T}_F$, and let E' and F' be their duals. If the linear map $u\colon E \to F$ is continuous for $\mathcal{T}_E$ and $\mathcal{T}_F$, then it is continuous for $\sigma(E, E')$ and $\sigma(F, F')$, and its transpose ${}^t u\colon F' \to E'$ is continuous for $\sigma(F', F)$ and $\sigma(E', E)$ and also for $\beta(F', F)$ and $\beta(E', E)$.*

Proof. By Proposition 3(a) the map u is continuous for $\sigma(E, E')$ and $\sigma(F, F')$. By the corollary to Proposition 1 the map ${}^t u$ is continuous for $\sigma(F', F)$ and $\sigma(E', E)$. Finally, by Proposition 3(b) the map ${}^t u$ is continuous for $\beta(F', F)$ and $\beta(E', E)$. ∎

For normed spaces we have the more precise result:

Proposition 4. *Let E and F be two normed spaces and $u: E \to F$ a continuous linear map. Then*

$$\|{}^t u\| = \|u\|.$$

Proof. By the definition of the norm of a continuous linear map (Exercise 1.7.4) we have

$$\|{}^t u\| = \sup_{\|y'\| \leqq 1} \|{}^t u(y')\|, \tag{1}$$

where y' varies in F'. By the definition of the norm in E' (cf. formula (1) of Chapter 1, §5) we have

$$\sup_{\|y'\| \leqq 1} \|{}^t u(y')\| = \sup_{\|y'\| \leqq 1} \sup_{\|x\| \leqq 1} |\langle x, {}^t u(y')\rangle|, \tag{2}$$

where x varies in E. By the definition of the transpose we have

$$\sup_{\|y'\| \leqq 1} \sup_{\|x\| \leqq 1} |\langle x, {}^t u(y')\rangle| = \sup_{\|x\| \leqq 1} \sup_{\|y'\| \leqq 1} |\langle u(x), y'\rangle|. \tag{3}$$

Since the canonical imbedding of F into its bidual is an isometry, we have

$$\sup_{\|x\| \leqq 1} \sup_{\|y'\| \leqq 1} |\langle u(x), y'\rangle| = \sup_{\|x\| \leqq 1} \|u(x)\|. \tag{4}$$

Finally, again using the definition of the norm of a linear map, we have

$$\sup_{\|x\| \leqq 1} \|u(x)\| = \|u\|. \tag{5}$$

The assertion of the proposition follows by combining the equalities (1) through (5). ∎

If a map $u: E \to F$ is continuous for the topologies $\sigma(E, E')$ and $\sigma(F, F')$, then it is not necessarily continuous for the topologies $\mathcal{T}_E$ and $\mathcal{T}_F$. For instance, let G and H be two vector spaces which form a dual system and suppose that $\sigma(G, H) \neq \tau(G, H)$. Take for E the space G equipped with the topology $\mathcal{T}_E = \sigma(G, H)$, for F the space G equipped with the topology $\mathcal{T}_F = \tau(G, H)$, and for u the identical map $G \to G$. Then u is continuous for $\sigma(E, E') = \sigma(G, H)$ and $\sigma(F, F') = \sigma(G, H)$ but not for $\mathcal{T}_E$ and $\mathcal{T}_F$.

We have, however, the following result:

Proposition 5. *Let E and F be two locally convex Hausdorff spaces and let E' and F' be their duals. Every linear map $u: E \to F$ which is continuous for the topologies $\sigma(E, E')$ and $\sigma(F, F')$ is also continuous for the Mackey topologies $\tau(E, E')$ and $\tau(F, F')$.*

Proof. Let $W = B^{\circ}$ be a neighborhood of 0 in F for the topology $\tau(F, F')$, where B is a balanced, convex, $\sigma(F', F)$-compact subset of F'. Since ${}^t u: F' \to E'$ is continuous for the topologies $\sigma(F', F)$ and $\sigma(E', E)$

(Corollary to Proposition 1), the set $A = {}^tu(B)$ is a balanced, convex, $\sigma(E', E)$-compact subset of E', and thus $V = A^{\circ}$ is a neighborhood of 0 in E for the topology $\tau(E, E')$. By Proposition 2(b) we have ${}^t({}^tu)(A^{\circ}) \subset B^{\circ}$, that is, $u(V) \subset W$, which proves that u is continuous for $\tau(E, E')$ and $\tau(F, F')$. ▮

Corollary. *If the topology $\mathcal{T}_E$ of E coincides with the Mackey topology $\tau(E, E')$, then every linear map $u\colon E \to F$ which is continuous for the topologies $\sigma(E, E')$ and $\sigma(F, F')$ is also continuous for the topologies $\mathcal{T}_E$ and $\mathcal{T}_F$.*

Proof. By the assumption and by Proposition 5 the map u is continuous for the topologies $\mathcal{T}_E = \tau(E, E')$ and $\tau(F, F')$. But $\tau(F, F')$ is finer than $\mathcal{T}_F$ (Proposition 5.4). ▮

Let us recall that the hypothesis of this corollary is satisfied if E is an infrabarrelled space (Proposition 6.8).

Returning to Proposition 3, if $u'\colon F' \to E'$ is continuous for the strong topologies $\beta(F', F)$ and $\beta(E', E)$, then it is not necessarily continuous for the topologies $\sigma(F', F)$ and $\sigma(E', E)$. For instance, let F be a non semi-reflexive space and $E = E' = \mathbf{K}^1$. Then there exist linear forms on F' which are continuous for $\beta(F', F)$ (i.e., belong to F'') but are not continuous for $\sigma(F', F)$ (i.e., do not belong to F). Here we have the following result:

Proposition 6. *Let E and F be two locally convex Hausdorff spaces and let E' and F' be their duals. Suppose that F is semi-reflexive. If the linear map $u'\colon F' \to E'$ is continuous for the topologies $\beta(F', F)$ and $\beta(E', E)$, then it is also continuous for the topologies $\sigma(F', F)$ and $\sigma(E', E)$.*

Proof. It follows from Proposition 3(a) that u' is continuous for the topologies $\sigma(F', F'')$ and $\sigma(E', E'')$. By our hypothesis

$$\sigma(F', F'') = \sigma(F', F);$$

on the other hand, $\sigma(E', E'')$ is finer than $\sigma(E', E)$. ▮

Proposition 7. *Let E and F be two locally convex Hausdorff spaces and $u\colon E \to F$ a continuous linear map. If we equip the duals E' and F' with the topologies $\kappa(E', E)$ and $\kappa(F', F)$ or with the topologies $\lambda(E', E)$ and $\lambda(F', F)$, then the map ${}^tu\colon F' \to E'$ is continuous.*

Proof. Let K be a balanced, convex, compact subset of E. Then $L = u(K)$ is a balanced, convex, compact subset of F. If $x \in K$ and $y' \in L^{\circ}$, then

$$|\langle {}^tu(y'), x\rangle| = |\langle y', u(x)\rangle| \leqq 1;$$

i.e., ${}^tu(y') \subset K^{\circ}$. Thus ${}^tu$ is continuous for the topologies $\kappa(F', F)$ and $\kappa(E', E)$.

To prove the statement concerning the topologies λ one considers, similarly as above, a precompact subset K of E and uses Proposition 2.10.6. ▮

We conclude this section with the description of a situation which we shall encounter frequently in Chapter 4 (cf. Definition 4.2.3). Let D be a reflexive locally convex Hausdorff space and D' its dual equipped with the topology $\beta(D', D)$. Let E be another locally convex Hausdorff space and assume that we have two continuous injective linear maps $i: D \to E$ and $j: E \to D'$. Let us also assume that $j(i(D))$ is dense in D'. If $i(D)$ is dense in E, we say that (D, i, E, j, D') is a *normal system*. It will be often convenient to identify D with a linear subspace of E with the help of the injection i and to identify E with a linear subspace of D' with the help of the injection j. Then we can say that D' induces on E a topology coarser than its own and E induces on D a topology coarser than its own.

SCHOLIUM 1. *Let (D, i, E, j, D') be a normal system. The restriction of an element $x' \in E'$ to D is a continuous linear form on D, i.e., an element $y' \in D'$. We have $y' = {}^ti(x')$; and if we equip E' with the topology $\beta(E', E)$, then the map ${}^ti: E' \to D'$ is continuous and injective. Similarly, the restriction of an element $u \in D = D''$ to E is a continuous linear form on E, i.e., an element $v' \in E'$. We have $v' = {}^tj(u)$, and the map ${}^tj: D \to E'$ is continuous and injective.*

If D is a Montel space, then the same results hold if we equip E' with the topology $\kappa(E', E)$. Furthermore, $(D, {}^tj, E', {}^ti, D')$ is then a normal system.

Proof. For $y \in D$ we have $\langle y, y' \rangle = \langle i(y), x' \rangle = \langle y, {}^ti(x') \rangle$, hence $y' = {}^ti(x')$. Similarly, for $v \in E$ we have $\langle v, v' \rangle = \langle j(v), u \rangle = \langle v, {}^tj(u) \rangle$, hence $v' = {}^tj(u)$. By the corollary to Proposition 3 the maps ${}^ti$ and ${}^tj$ are continuous. Since $i(D)$ is dense in E, it is *a fortiori* dense for the coarser topology $\sigma(E, E')$, and therefore by Corollary 2 of Proposition 2 the map ${}^ti$ is injective (this is of course also directly obvious). Now, $j(i(D))$ and *a fortiori* $j(E)$ are dense in D'; hence, just like before, ${}^tj$ is injective.

Next $j \circ i$ is injective. Hence ${}^t(j \circ i)(D) = {}^ti({}^tj(D))$ is dense in D' for any topology compatible with the duality between D and D' by Corollary 2 to Proposition 2 and Proposition 4.3. For the same reason, ${}^tj(D)$ is dense in E' equipped with the topology $\kappa(E', E)$. Finally, if D is a Montel space, the injections ${}^tj$ and ${}^ti$ are continuous for the topologies

$$\beta(D', D) = \kappa(D', D)$$

and $\kappa(E', E)$ by Proposition 7. ▮

Observe that if E' is equipped with $\beta(E', E)$, then ${}^tj(D)$ is not necessarily dense in E' (take $D = \mathfrak{D}$, $E = \mathcal{B}'_0$ and $E' = \mathcal{B}$; cf. Chapter 4, §5).

Exercises

1. Let E be a finite-dimensional vector space with basis $(e_i)_{1 \leqq i \leqq n}$ and let $(e'_i)_{1 \leqq i \leqq n}$ be the basis of the dual space E^* defined by

$$\langle e_i, e'_j \rangle = \begin{cases} 1 & \text{if } i = j, \\ 0 & \text{if } i \neq j. \end{cases}$$

Let $u: E \to E$ be a linear map whose matrix (α_{ij}) is defined by

$$u(e_i) = \sum_{k=1}^{n} \alpha_{ik} e_k.$$

Show that if the matrix (β_{ij}) of ${}^t u: E^* \to E^*$ is defined by ${}^t u(e'_j) = \sum_{k=1}^{n} \beta_{jk} e'_k$, then $\beta_{ij} = \alpha_{ji}$ for $1 \leqq i, j \leqq n$; i.e., (β_{ij}) is the "transpose" of (α_{ij}).

2. If E and F are two vector spaces, M a set of linear maps $u: E \to F$, A a subset of E, and B a subset of F, we write

$$M(A) = \bigcup_{u \in M} u(A) \quad \text{and} \quad M^{-1}(B) = \bigcap_{u \in M} u^{-1}(B).$$

Let (F_1, G_1) and (F_2, G_2) be two pairs of vector spaces which form dual systems and let M be a collection of linear maps $u: F_1 \to F_2$ which are continuous for the topologies $\sigma(F_1, G_1)$ and $\sigma(F_2, G_2)$. Denote by ${}^t M$ the set of all transpose maps ${}^t u: G_2 \to G_1$, where u runs through M. Let A be a subset of F_1 and B a subset of G_2.

(a) Prove that $(M(A))^{\circ} = {}^t M^{-1}(A^{\circ})$ and $M^{-1}(B^{\circ}) = ({}^t M(B))^{\circ}$.

(b) Prove that the balanced, convex, $\sigma(F_2, G_2)$-closed hull of $M(A)$ is $({}^t M^{-1}(A^{\circ}))^{\circ}$.

3. Let E and F be two normed vector spaces over the same field and let $B: (x, y) \mapsto B(x, y)$ be a bilinear form on $E \times F$. We define the norm $\|B\|$ of B by

$$\|B\| = \sup_{\|x\| \leqq 1, \|y\| \leqq 1} |B(x, y)|.$$

Let $u: E \to F$ be a continuous linear map. Prove that the norm of the bilinear form $(x, y') \mapsto \langle u(x), y' \rangle = \langle x, {}^t u(y') \rangle$ on $E \times F'$ is $\|u\|$. (*Hint:* Proof of Proposition 4.)

§13. Duals of subspaces and quotient spaces

Let F and G be two vector spaces which form a dual system, M a linear subspace of F, and $j: M \hookrightarrow F$ the canonical injection. Let us consider the transpose ${}^t j$ of j as a map from G into the algebraic dual M^* of M defined by

$$\langle x, {}^t j(y') \rangle = \langle j(x), y' \rangle$$

for $x \in M$, $y' \in G$. Let us prove that

(a) $x' \in M^*$ belongs to ${}^t j(G)$ if and only if x' is continuous for the topology induced by $\sigma(F, G)$ on M,

(b) the kernel of ${}^t j$ is $M^{\perp}$.

Indeed, if x' is continuous on M for the topology induced by $\sigma(F, G)$ on M, then by Proposition 1.1 there exists $y' \in G$ such that

$$\langle j(x), y' \rangle = \langle x, x' \rangle.$$

Hence $x' = {}^t j(y')$; i.e., $x' \in {}^t j(G)$. Conversely, observe that ${}^t j(y') = y' \circ j$ is the restriction of y' to M. Thus ${}^t j(y')$ is continuous on M since y' is continuous on F for $\sigma(F, G)$. Finally, (b) follows from Corollary 1 to Proposition 12.2, but it is also directly clear that the restriction of $y' \in G$ to M is zero if and only if $y' \in M^{\perp}$.

Let us denote by N the subspace of M^* formed by all linear forms on M which are continuous for the topology induced by $\sigma(F, G)$. Then the map ${}^t j$ can be factored into the maps

$$\psi: G \to G/M^{\perp}, \qquad \alpha: G/M^{\perp} \to N, \qquad \iota: N \hookrightarrow M^*,$$

where ψ is the canonical surjection, α is a bijective linear map, and ι is the canonical injection.

PROPOSITION 1. *The vector spaces M and N form a dual system, and the topology $\sigma(M, N)$ coincides with the topology induced on M by $\sigma(F, G)$.*

Proof. If $x' \in N$ is such that $\langle x, x' \rangle = 0$ for all $x \in M$, then $x' = {}^t j(y')$ for some $y' \in G$ and $\langle x, {}^t j(y') \rangle = \langle j(x), y' \rangle = 0$ for all $x \in M$, i.e., $y' \in M^{\perp}$. Thus $x' = 0$ by (b) above.

If $x \in M$ is such that $\langle x, x' \rangle = 0$ for all $x' \in N$, then

$$\langle j(x), y' \rangle = \langle x, {}^t j(y') \rangle = 0$$

for all $y' \in G$. Thus $j(x) = 0$ and also $x = 0$, since j is injective. This proves that M and N form a dual system.

Let V be the neighborhood of 0 in F defined by the inequalities

$$|\langle y, y'_k \rangle| \leqq \epsilon,$$

where $y \in F$ and $y'_k \in G$ for $1 \leqq k \leqq n$. Set $x'_k = {}^t j(y'_k) \in N$. Then the trace of V on M is the set W defined by the inequalities $|\langle x, x'_k \rangle| \leqq \epsilon$ $(1 \leqq k \leqq n)$ since $\langle x, x'_k \rangle = \langle j(x), y'_k \rangle$ for all $x \in M$. Conversely, if

$$W = \{x \mid |\langle x, x'_k \rangle| \leqq \epsilon, 1 \leqq k \leqq n\}$$

is a neighborhood of 0 in M for $\sigma(M, N)$, then choose $y'_k \in G$ such that ${}^t j(y'_k) = x'_k$ for $1 \leqq k \leqq n$. Clearly, W is the trace on M of the neighborhood V of 0 in F defined by the inequalities $|\langle y, y'_k \rangle| \leqq \epsilon$ $(1 \leqq k \leqq n)$. ∎

Since we now know that the vector spaces M and N form a dual system, it is legitimate to consider ${}^t j$ as a linear map from G onto N. Then

$\alpha: G/M^{\perp} \to N$ is the injection associated with ${}^t j$, and therefore it is continuous if we equip $G/M^{\perp}$ with the quotient topology of $\sigma(G, F)$ modulo the subspace $M^{\perp}$ and N with the topology $\sigma(N, M)$. We shall sometimes identify the spaces $G/M^{\perp}$ and N with the help of the bijection α.

COROLLARY. *Let E be a locally convex Hausdorff space, M a subspace of E, and E' the dual of E. Then there exists a canonical bijective linear map from $E'/M^{\perp}$ onto the dual M' of M, and the topology $\sigma(M, M')$ coincides with the topology induced by $\sigma(E, E')$ on M.*

Proof. This follows from the previous considerations if we take into account that a linear form u on M is continuous for the topology $\mathcal{T}_M$ induced by the original topology $\mathcal{T}_E$ of E if and only if it is continuous for the topology σ_M induced by $\sigma(E, E')$. Indeed, if u is continuous for σ_M, it is *a fortiori* continuous for $\mathcal{T}_M$. Conversely, if u is continuous for $\mathcal{T}_M$, then by Proposition 1.1 it is the restriction to M of a linear form v on E, continuous for $\mathcal{T}_E$. But v is continuous for $\sigma(E, E')$; hence u is continuous for σ_M. ∎

PROPOSITION 2. *Let us equip $G/M^{\perp}$ with the quotient topology of $\sigma(G, F)$ modulo the subspace $M^{\perp}$ and N with the topology $\sigma(N, M)$. The continuous bijective linear map $\alpha: G/M^{\perp} \to N$ is an isomorphism if and only if M is closed in F for $\sigma(F, G)$.*

Proof. (a) Suppose that M is closed for $\sigma(F, G)$. Let

$$V = \{y' \mid |\langle y_k, y'\rangle| \leqq \epsilon, \quad 1 \leqq k \leqq n\}$$

be a neighborhood of 0 in G for $\sigma(G, F)$. Let L be the subspace of F generated by M and the vectors y_k $(1 \leqq k \leqq n)$. There exists in L a supplement P of M having finite dimension m. Let $(z_l)_{1 \leqq l \leqq m}$ be a basis of P. The restrictions to $M^{\perp}$ of the linear forms $y' \mapsto \langle z_l, y'\rangle$ are linearly independent, since otherwise there would exist $z = \sum_{l=1}^{m} \mu_l z_l$, where not all μ_l are zero, such that $\langle z, y'\rangle = 0$ for all $y' \in M^{\perp}$. This would mean that $z \in M^{\perp\perp}$, i.e., $z \in M$ since M is closed (Proposition 3.3(e)) in contradiction to the definition of P.

Each y_k $(1 \leqq k \leqq n)$ can be written uniquely in the form

$$y_k = x_k + \sum_{l=1}^{m} \lambda_{kl} z_l,$$

where $x_k \in M$. By the linear independence of the linear forms

$$y' \mapsto \langle z_l, y'\rangle$$

on $M^{\perp}$, for every $y' \in G$ there exists $s' \in M^{\perp}$ such that $\langle z_l, y'\rangle = \langle z_l, s'\rangle$

for $1 \leqq l \leqq m$. Hence

$$\langle y_k, y' - s' \rangle = \langle j(x_k), y' - s' \rangle = \langle j(x_k), y' \rangle$$

for $1 \leqq k \leqq n$.

Now let U be the neighborhood of 0 in N for the topology $\sigma(N, M)$ defined by the inequalities $|\langle x_k, x' \rangle| \leqq \epsilon, 1 \leqq k \leqq n$. We have ${}^t j(V) \supset U$. Indeed, let $x' = {}^t j(y') \in U$. For some $s' \in M^\perp$ we have

$$|\langle y_k, y' - s' \rangle| = |\langle j(x_k), y' \rangle| = |\langle x_k, {}^t j(y') \rangle| = |\langle x_k, x' \rangle| \leqq \epsilon$$

for $1 \leqq k \leqq n$, whence $y' - s' \in V$. But ${}^t j(y' - s') = {}^t j(y') = x'$ since s' belongs to the kernel $M^\perp$ of ${}^t j$.

It follows now from Theorem 2.5.1 that the surjective map ${}^t j$ is a strict morphism; hence α is indeed an isomorphism.

(b) Suppose that M is not closed for $\sigma(F, G)$ and let $\overline{M}$ be its closure. Then N is also the space of all continuous linear forms on $\overline{M}$ (Proposition 2.9.5), and furthermore $\overline{M}^\perp = (M^{\perp\perp})^\perp = M^\perp$ by Proposition 3.3. Thus by part (a) the canonical bijection $\alpha\colon G/M^\perp \to N$ is an isomorphism for the topology $\sigma(N, \overline{M})$ on N. Since $M \neq \overline{M}$, it follows from Proposition 2.4 that the topology $\sigma(N, \overline{M})$ is strictly finer than $\sigma(N, M)$, and thus α cannot be an isomorphism for $\sigma(N, M)$. ▮

Corollary. *Let E be a locally convex Hausdorff space, L a closed subspace of E, and E' the dual of E. Then there exists a canonical bijective linear map from $(E/L)'$ onto $L^\perp$, and the topology $\sigma(E/L, (E/L)')$ coincides with the quotient of the topology $\sigma(E, E')$ modulo L.*

Proof. Let $\varphi\colon E \to E/L$ be the canonical surjection. Its transpose ${}^t\varphi\colon (E/L)' \to E'$ associates with every continuous linear form y' on E/L the linear form ${}^t\varphi(y') = y' \circ \varphi$ on E. Clearly, ${}^t\varphi$ is injective. Applying Corollary 1 of Proposition 12.2 to $u = {}^t\varphi$ and ${}^t u = \varphi$, we see that

$${}^t\varphi((E/L)') = L^\perp.$$

The assertion concerning the topologies follows from Proposition 2, taking $F = E'$, $G = E$, $M = L^\perp$, and for j the injection $L^\perp \hookrightarrow E'$. We have $M^\perp = L^{\perp\perp} = L$ because L is assumed to be closed and thus also closed for $\sigma(E, E')$. Thus if we identify $(E/L)'$ with $L^\perp$, then we have indeed $N = E/L$ and $\sigma(N, M) = \sigma(E/L, (E/L)')$. ▮

Let us now give some applications of the results proved so far in this section.

Proposition 3. *Let (F_1, G_1) and (F_2, G_2) be two pairs of vector spaces forming dual systems, and let $u\colon F_1 \to F_2$ be a linear map which is continuous for the topologies $\sigma(F_1, G_1)$ and $\sigma(F_2, G_2)$. Then u is a strict morphism* (Definition 2.5.2) *if and only if ${}^t u(G_2)$ is closed in G_1 for $\sigma(G_1, F_1)$.*

Proof. Set $N = {}^t u(G_2) \subset G_1$. Then $N^\perp = u^{-1}(0)$ by Corollary 1 of Proposition 12.2 applied to ${}^t u$. Let $\varphi: F_1 \to F_1/N^\perp$ be the canonical surjection and set $u = \overline{u} \circ \varphi$, where $\overline{u}: F_1/N^\perp \to F_2$ is the injection associated with u. The spaces $F_1/N^\perp$ and N form a dual system. Let $\dot{x} \in F_1/N^\perp$, $y' \in G_2$, and choose $x \in F_1$ so that $\varphi(x) = \dot{x}$. Then we have

$$\langle \overline{u}(\dot{x}), y' \rangle = \langle u(x), y' \rangle = \langle x, {}^t u(y') \rangle = \langle \dot{x}, {}^t u(y') \rangle$$

if we consider ${}^t u$ as a map from G_2 onto N. This relation shows that $\overline{u}$ is an injective strict morphism of $F_1/N^\perp$ equipped with the topology $\sigma(F_1/N^\perp, N)$ into F_2 equipped with the topology $\sigma(F_2, G_2)$. By Proposition 2 the topology $\sigma(F_1/N^\perp, N)$ coincides with the quotient topology of $\sigma(F_1, G_1)$ modulo $N^\perp$ if and only if N is closed in G_1 for $\sigma(G_1, F_1)$. ∎

COROLLARY. *Let $u: F_1 \to F_2$ be a linear map which is continuous for the topologies $\sigma(F_1, G_1)$ and $\sigma(F_2, G_2)$. Then u is surjective if and only if ${}^t u$ is an injective strict morphism for the topologies $\sigma(G_2, F_2)$ and $\sigma(G_1, F_1)$.*

Proof. $u(F_1) = F_2$ if and only if $u(F_1)$ is both closed and everywhere dense in F_2. Hence the corollary follows from Proposition 3 (applied to ${}^t u$) and from Corollary 2 of Proposition 12.2. ∎

PROPOSITION 4. *Let E and F be two locally convex Hausdorff spaces with topologies $\mathcal{T}_E$ and $\mathcal{T}_F$, and let E' and F' be their duals. If $u: E \to F$ is a strict morphism for the topologies $\mathcal{T}_E$ and $\mathcal{T}_F$, then u is also a strict morphism for the topologies $\sigma(E, E')$ and $\sigma(F, F')$.*

Proof. Let N be the kernel of u, H the quotient space E/N equipped with the quotient topology $\mathcal{T}_H$ of $\mathcal{T}_E$ modulo N, $\varphi: E \to H$ the canonical surjection, and $\overline{u}: H \to F$ the injection associated with u. If H' is the dual of H, then by the Corollary of Proposition 2 the map φ is a surjective strict morphism for the topologies $\sigma(E, E')$ and $\sigma(H, H')$. On the other hand, $\overline{u}$ is an injective strict morphism for the topologies $\sigma(H, H')$ and $\sigma(F, F')$ by the corollary of Proposition 1, since by assumption it is an injective strict morphism for the topologies $\mathcal{T}_H$ and $\mathcal{T}_F$. Thus $u = \overline{u} \circ \varphi$ is a strict morphism for the topologies $\sigma(E, E')$ and $\sigma(F, F')$. ∎

Conversely, it is not necessarily true that if $u: E \to F$ is a strict morphism for the topologies $\sigma(E, E')$ and $\sigma(F, F')$, then it is also a strict morphism for the topologies $\mathcal{T}_E$ and $\mathcal{T}_F$ (see Exercise 2 and [9], Chapter IV, §4, Exercises 3 and 4). We have, however, the following result:

PROPOSITION 5. *Let E be a locally convex Hausdorff space with topology $\mathcal{T}_E$, F a metrizable locally convex space with topology $\mathcal{T}_F$, and let E' and F' be their duals. Let $u: E \to F$ be a linear map which is continuous for the topologies $\mathcal{T}_E$ and $\mathcal{T}_F$ and which is a strict morphism for the topologies $\sigma(E, E')$ and $\sigma(F, F')$. Then u is a strict morphism for $\mathcal{T}_E$ and $\mathcal{T}_F$.*

Proof. (a) First let us assume that u is injective. Let $v: u(E) \to E$ be the inverse map of $u: E \to u(E)$. Since $u(E)$ is metrizable, it follows from the corollary of Proposition 12.5 that v is continuous for the topology induced by $\mathcal{T}_F$ on $u(E)$ and for $\mathcal{T}_E$; i.e., u is a strict morphism for the topologies $\mathcal{T}_E$ and $\mathcal{T}_F$.

(b) In the general case, let N be the kernel of u, H the quotient space E/N, $\mathcal{T}_H$ the quotient of $\mathcal{T}_E$ modulo N, $\varphi: E \to H$ the canonical surjection, and $\bar{u}: H \to F$ the injection associated with u. The map $\bar{u}$ is continuous for $\mathcal{T}_H$ and $\mathcal{T}_F$, and it is a strict morphism for $\sigma(H, H')$ and $\sigma(F, F')$ since $\sigma(H, H')$ is the quotient of $\sigma(E, E')$ modulo N (Corollary of Proposition 2). It follows from part (a) of the proof that $\bar{u}$ is a strict morphism for $\mathcal{T}_H$ and $\mathcal{T}_F$. Thus $u = \bar{u} \circ \varphi$ is a strict morphism for $\mathcal{T}_E$ and $\mathcal{T}_F$. ∎

COROLLARY. *Let E be a locally convex Hausdorff space whose topology $\mathcal{T}_E$ coincides with the Mackey topology $\tau(E, E')$, and let F be a metrizable locally convex space with topology $\mathcal{T}_F$. If the linear map $u: E \to F$ is a strict morphism for the topologies $\sigma(E, E')$ and $\sigma(F, F')$, then it is a strict morphism for the topologies $\mathcal{T}_E$ and $\mathcal{T}_F$.*

Proof. By the corollary of Proposition 12.5 the map u is continuous for the topologies $\mathcal{T}_E$ and $\mathcal{T}_F$, and thus by the preceding proposition u is a strict morphism for $\mathcal{T}_E$ and $\mathcal{T}_F$. ∎

EXERCISES

1. Let (F_1, G_1) and (F_2, G_2) be two pairs of vector spaces forming dual systems, and let $u: F_1 \to F_2$ be a strict morphism for the topologies $\sigma(F_1, G_1)$ and $\sigma(F_2, G_2)$. Show that the dual of $u(F_1)$ equipped with the topology induced by $\sigma(F_2, G_2)$ is isomorphic to the subspace $N = {}^t u(G_2)$ of G_1 and the topology $\sigma(N, u(F_1))$ coincides with the topology induced by $\sigma(G_1, F_1)$ on N. (*Hint:* By Proposition 3 the subspace N is closed in G_1 for $\sigma(G_1, F_1)$. Therefore, by Proposition 2 the dual of N is $F_1/N^{\perp}$. But since u is a strict morphism, $u(F_1)$ is isomorphic to $F_1/N^{\perp}$. Finally, the two topologies on N coincide by Proposition 1.)

2. Let E and F be two locally convex Hausdorff spaces with topologies $\mathcal{T}_E$ and $\mathcal{T}_F$ and let E' and F' be their duals. Show that a continuous linear map $u: E \to F$ is a strict morphism for the topologies $\mathcal{T}_E$ and $\mathcal{T}_F$ if and only if the following two conditions are satisfied:

(α) u is a strict morphism for the topologies $\sigma(E, E')$ and $\sigma(F, F')$.

(β) Every equicontinuous subset of E' contained in ${}^t u(F')$ is the image by ${}^t u$ of some equicontinuous subset of F'.

(*Hint:* (α) is necessary by Proposition 5. To prove sufficiency, consider first the case where u is injective.)

3. Let E be a locally convex Hausdorff space. For each closed subspace F of finite codimension we equip E/F with the unique Hausdorff topology compatible with its vector space structure (Theorem 2.10.1). For $F \supset G$ we have a natural map $f_{FG}: E/G \to E/F$. Show that the spaces E/F and the maps f_{FG} form an inverse system (Exercise 2.11.3) whose inverse limit L is canonically isomorphic to the space E'^* (cf. Proposition 2.5) equipped with the topology $\sigma(E'^*, E')$. (*Hint:* For $x \in L$ define $\tilde{x} \in E'^*$ by setting $\langle \tilde{x}, y' \rangle = \langle x_F, y' \rangle$ if $y' \in F^\perp \subset E'$, and x_F is the image of x in E/F.)

§14. Duals of products and direct sums

Let $(E_\iota)_{\iota \in I}$ be a family of locally convex Hausdorff spaces, for each index $\iota \in I$ let E'_ι be the dual of E_ι and let $E = \prod_{\iota \in I} E_\iota$ be the product of the spaces E_ι equipped with the product topology. Let us define a canonical map χ from the external direct sum $\coprod_{\iota \in I} E'_\iota$ of the spaces E'_ι onto the dual E' of E as follows. Let $x' = (x'_\iota) \in \coprod_{\iota \in I} E'_\iota$, then $\chi(x')$ is the linear form on E defined by

$$\langle \chi(x'), x \rangle = \sum_{\iota \in I} \langle x'_\iota, x_\iota \rangle,$$

where $x = (x_\iota) \in E$. The sum has a meaning since $x'_\iota = 0$ except for finitely many indices ι. The linear form $\chi(x')$ is continuous. Indeed, let H be the finite subset of I such that $x'_\iota \neq 0$ for $\iota \in H$. Let $\epsilon > 0$ and suppose that H has n elements. For each $\iota \in H$ let U_ι be a neighborhood of 0 such that $x_\iota \in U_\iota$ implies $|\langle x'_\iota, x_\iota \rangle| < \epsilon/n$. Set $U_\iota = E_\iota$ for $\iota \notin H$. Then $x \in \prod_{\iota \in I} U_\iota$ implies $|\langle \chi(x'), x \rangle| < \epsilon$.

It is clear that χ is linear and injective. Let us show that it is also surjective. Let y' be a continuous linear form on E. There exists a neighborhood $V = \prod_{\iota \in I} V_\iota$ of 0 in E, where $V_\iota = E_\iota$ except for those indices ι which belong to a finite subset H of I, such that $x \in V$ implies $|\langle y', x \rangle| \leqq 1$. Define the element $x'_\iota \in E'_\iota$ by $\langle x'_\iota, x_\iota \rangle = \langle y', x \rangle$, where x_ι is an arbitrary element in E_ι and $x = (x_\kappa) \in E$ is such that $x_\kappa = x_\iota$ for $\kappa = \iota$ and $x_\kappa = 0$ for $\kappa \neq \iota$. We have $x'_\iota = 0$ for all ι which are not in H. Indeed, if $\iota \notin H$, then $|\langle x'_\iota, \lambda x_\iota \rangle| = |\langle y', \lambda x \rangle| \leqq 1$ for all $x_\iota \neq 0$ in E_ι and $\lambda \in \mathbf{K}$; hence $x'_\iota = 0$. It follows that the element $x' = (x'_\iota)$ lies in $\coprod_{\iota \in I} E'_\iota$, and clearly we have $y' = \chi(x')$.

We can summarize our result as follows:

PROPOSITION 1. *Let $(E_\iota)_{\iota \in I}$ be a family of locally convex Hausdorff spaces. Then the external direct sum $\coprod_{\iota \in I} E'_\iota$ of the dual spaces E'_ι can be identified canonically with the dual of the product space $\prod_{\iota \in I} E_\iota$. The canonical bilinear form is given by*

$$\langle x, x' \rangle = \sum_{\iota \in I} \langle x_\iota, x'_\iota \rangle$$

for $x = (x_\iota)_{\iota \in I} \in \prod_{\iota \in I} E_\iota$ and $x' = (x'_\iota)_{\iota \in I} \in \coprod_{\iota \in I} E'_\iota$.

Now let $F = \coprod_{\iota \in I} E_\iota$ be the external direct sum of the spaces E_ι equipped with the locally convex direct sum topology (Example 2.12.2). We define a canonical map θ from the product $\prod_{\iota \in I} E'_\iota$ of the dual spaces E'_ι onto the dual F' of F as follows. Let $x' = (x'_\iota) \in \prod_{\iota \in I} E'_\iota$. Then $\theta(x')$ is the linear form on F defined by

$$\langle \theta(x'), x \rangle = \sum_{\iota \in I} \langle x'_\iota, x_\iota \rangle,$$

where $x = (x_\iota) \in F$. Again this sum has a meaning, since $x_\iota = 0$ except for finitely many indices ι. The linear form $\theta(x')$ is continuous. Indeed, let $\epsilon > 0$ and for each $\iota \in I$ let U_ι be a neighborhood of 0 in E_ι such that $x_\iota \in U_\iota$ implies $|\langle x'_\iota, x_\iota \rangle| < \epsilon$. Denote by j_ι the ι-th injection and let U be the balanced, convex hull of $\bigcup_{\iota \in I} j_\iota(U_\iota)$. Then U is a neighborhood of 0 in F. Furthermore, if $x = \sum_{\iota \in I} \lambda_\iota x_\iota \in U$, $x_\iota \in U_\iota$, $\sum_{\iota \in I} |\lambda_\iota| \leqq 1$, then

$$|\langle \theta(x'), x \rangle| \leqq \sum_{\iota \in I} |\lambda_\iota| \cdot |\langle x'_\iota, x_\iota \rangle| < \epsilon.$$

It is clear that the map θ is linear and injective. It is also surjective, since if $y' \in F'$, define $x'_\iota \in E'_\iota$ by $\langle x'_\iota, x_\iota \rangle = \langle y', j_\iota(x_\iota) \rangle$ for $x_\iota \in E_\iota$. Setting $x' = (x'_\iota)$ we have $\theta(x') = y'$. We can again summarize our results.

Proposition 2. *Let $(E_\iota)_{\iota \in I}$ be a family of locally convex Hausdorff spaces. The product $\prod_{\iota \in I} E'_\iota$ of the dual spaces E'_ι can be identified canonically with the dual of the locally convex direct sum $\coprod_{\iota \in I} E_\iota$. The canonical bilinear form is given by*

$$\langle x, x' \rangle = \sum_{\iota \in I} \langle x_\iota, x'_\iota \rangle$$

for $x = (x_\iota) \in \coprod_{\iota \in I} E_\iota$, $x' = (x'_\iota) \in \prod_{\iota \in I} E'_\iota$.

Consider next a family $(F_\iota, G_\iota)_{\iota \in I}$ of pairs of vector spaces which form dual systems. Set $F = \prod_{\iota \in I} F_\iota$, $G = \coprod_{\iota \in I} G_\iota$, and define a bilinear form on $F \times G$ by

$$\langle x, y \rangle = \sum_{\iota \in I} \langle x_\iota, y_\iota \rangle$$

for $x = (x_\iota) \in F$ and $y = (y_\iota) \in G$. The spaces F and G form a dual system, since if for instance $x \neq 0$, then there exists an index $\iota \in I$ such that $x_\iota \neq 0$. Choose y_ι such that $\langle x_\iota, y_\iota \rangle \neq 0$ and $y_\kappa = 0$ for $\kappa \neq \iota$. Then $\langle x, y \rangle = \langle x_\iota, y_\iota \rangle \neq 0$. It follows from the preceding two propositions that G can be considered as the dual of F equipped with the product of the topologies $\sigma(F_\iota, G_\iota)$, and F can be considered as the dual of G equipped with the locally convex direct sum of the topologies $\sigma(G_\iota, F_\iota)$.

Proposition 3. *The topology* $\sigma(F, G)$ *on* $F = \prod_{\iota \in I} F_\iota$ *coincides with the product of the topologies* $\sigma(F_\iota, G_\iota)$.

Proof (cf. Example 2.3). Let us denote by $\mathcal{T}$ the product of the topologies $\sigma(F_\iota, G_\iota)$. As we have just mentioned, each element of G defines a linear form on F which is continuous for $\mathcal{T}$. It follows that $\mathcal{T}$ is finer than $\sigma(F, G)$. Conversely, let $V = \prod_{\iota \in I} V_\iota$ be a neighborhood of 0 in F for $\mathcal{T}$, where $V_\iota = F_\iota$ for those indices ι which do not belong to a finite subset H of I and let

$$V_\iota = \{x_\iota \mid x_\iota \in F_\iota, |\langle x_\iota, y_{\iota,k}\rangle| < \epsilon, k = 1, \ldots, m_\iota\}$$

for $\iota \in H$. Then

$$U = \{x \mid x \in F, |\langle x, j_\iota(y_{\iota,k})\rangle| < \epsilon, k = 1, \ldots, m_\iota, \iota \in H\}$$

is a neighborhood of 0 for $\sigma(F, G)$ and $U \subset V$. Thus $\sigma(F, G)$ is finer than $\mathcal{T}$. ∎

Remark 1. In general, if I is infinite, the topology $\sigma(G, F)$ on G is strictly coarser than the locally convex direct sum of the topologies $\sigma(G_\iota, F_\iota)$ (see Exercise 2.1(b)).

Proposition 4. *Let* $(E_\iota)_{\iota \in I}$ *be a family of locally convex Hausdorff spaces. If each* E_ι *is semi-reflexive* (Definition 8.2), *then the product* $E = \prod_{\iota \in I} E_\iota$ *is semi-reflexive.*

Proof. Let A be a bounded, $\sigma(E, E')$-closed subset of E. If π_ι denotes the projection $E \to E_\iota$, then by Proposition 3 the set $\pi_\iota(A)$ is $\sigma(E_\iota, E'_\iota)$-bounded in E_ι. If E_ι is a semi-reflexive space, then by Proposition 8.1 the set $\pi_\iota(A)$ is relatively compact for the topology $\sigma(E_\iota, E'_\iota)$. It follows from Proposition 3 and from Tihonov's theorem that the set

$$B = \prod_{\iota \in I} \pi_\iota(A)$$

is relatively compact in E for the topology $\sigma(E, E')$. Thus the set A, which is contained in B, is compact for $\sigma(E, E')$ and therefore by Proposition 8.1 the space E is semi-reflexive. ∎

Remark 2. It follows from Proposition 4 and Exercise 2 that the product of reflexive (Definition 8.3) spaces is reflexive (cf. Proposition 8.6).

Proposition 5. *Let* $(F_\iota)_{\iota \in I}$ *be a family of locally convex Hausdorff spaces,* E *a vector space, and for each* $\iota \in I$ *let* $f_\iota \colon E \to F_\iota$ *be a linear map. Assume that for each* $x \in E$, $x \neq 0$, *there exists an index* $\iota \in I$ *such that* $f_\iota(x) \neq 0$.

(a) *If we equip* E *with the coarsest topology for which all the maps* f_ι *are continuous, then for every continuous linear form* x' *on* E *there exists a finite family* $(y'_{\iota_k})_{1 \leqq k \leqq n}$, *where* $y'_{\iota_k} \in F'_{\iota_k}$ $(1 \leqq k \leqq n)$, *such that*

$$x' = \sum_{k=1}^{n} {}^t f_{\iota_k}(y'_{\iota_k}) = \sum_{k=1}^{n} y'_{\iota_k} \circ f_{\iota_k}.$$

(b) *If we equip each space* F_ι *with the topology* $\sigma(F_\iota, F'_\iota)$, *then* $\sigma(E, E')$ *is the coarsest topology on* E *for which the maps* f_ι *are continuous.*

Proof. (a) We know (Exercise 2.11.1) that the linear map

$$\varphi: E \to \prod_{\iota \in I} F_\iota,$$

defined by $\varphi(x) = (f_\iota(x))_{\iota \in I}$, is an injective strict morphism. Define the linear map $\psi: \coprod_{\iota \in I} F'_\iota \to E'$ by

$$\psi((y'_\iota)) = \sum_{\iota \in I} {}^t f_\iota(y'_\iota).$$

If we identify the dual of $\prod_{\iota \in I} F_\iota$ with $\coprod_{\iota \in I} F'_\iota$ according to Proposition 1, then $\psi = {}^t\varphi$ since for $x \in E$ and $(y'_\iota) \in \coprod_{\iota \in I} F'_\iota$ we have

$$\begin{aligned} \langle \varphi(x), (y'_\iota) \rangle &= \sum_{\iota \in I} \langle f_\iota(x), y'_\iota \rangle \\ &= \sum_{\iota \in I} \langle x, {}^t f_\iota(y'_\iota) \rangle = \langle x, \psi((y'_\iota)) \rangle. \end{aligned}$$

Now it follows from the Corollary of Proposition 13.3 and Proposition 13.4 (or from Proposition 1.1) that the map ψ is surjective. Hence for $x' \in E'$ there exists (y'_ι) such that $x' = \psi((y'_\iota)) = \sum_{\iota \in I} {}^t f_\iota(y'_\iota)$, with $y'_\iota = 0$ except for finitely many indices ι.

(b) The maps f_ι are continuous for the topologies $\sigma(E, E')$ and $\sigma(F_\iota, F'_\iota)$. Indeed, let V_ι be the neighborhood of 0 in F_ι defined by the inequalities $|\langle y, y'_k \rangle| \leqq \epsilon$, where $y'_k \in F'_\iota$ $(1 \leqq k \leqq n)$, and set $x'_k = {}^t f_\iota(y'_k)$. Then

$$U = \{x \mid |\langle x, x'_k \rangle| \leqq \epsilon, 1 \leqq k \leqq n\}$$

is a $\sigma(E, E')$-neighborhood of 0 in E, and for $x \in U$ we have

$$|\langle f_\iota(x), y'_k \rangle| = |\langle x, {}^t f_\iota(y'_k) \rangle| = |\langle x, x'_k \rangle| \leqq \epsilon;$$

i.e., $f_\iota(U) \subset V_\iota$.

Conversely, let $U = \{x \mid |\langle x, x'_k \rangle| \leqq \epsilon, 1 \leqq k \leqq n\}$ be a $\sigma(E, E')$-neighborhood of 0 in E. By part (a) there exists for each x'_k a family $(y'_{k,\iota})_{\iota \in I}$ such that $x'_k = \sum_{\iota \in I} {}^t f_\iota(y'_{k,\iota})$, $y'_{k,\iota} \in F'_\iota$, and $y'_{k,\iota} = 0$ except for finitely many indices ι. For each k $(1 \leqq k \leqq n)$ let $p(k)$ be the number of $y'_{k,\iota}$ which are different from zero, and define the $\sigma(F_\iota, F'_\iota)$-neighborhood V_ι of 0 in F_ι by the inequalities $|\langle y, y'_{k,\iota} \rangle| \leqq \epsilon/p(k)$, $(1 \leqq k \leqq n)$, where, of course, $V_\iota = F_\iota$, except for finitely many indices ι. If $x \in \bigcap_{\iota \in I} f_\iota^{-1}(V_\iota)$, then

$$|\langle x, x'_k \rangle| = \left| \left\langle x, \sum_{\iota \in I} {}^t f_\iota(y'_{k,\iota}) \right\rangle \right| = \sum_{\iota \in I} |\langle f_\iota(x), y'_{k,\iota} \rangle| \leqq \epsilon;$$

hence $x \in U$, i.e., $\bigcap_{\iota \in I} f_\iota^{-1}(V_\iota) \subset U$. ∎

COROLLARY. *Let I be a directed set and for each $\iota \in I$ let F_ι be a locally convex space. Assume that for $\iota \leqq \kappa$ the space F_κ is a linear subspace of F_ι and that the map $f_{\iota\kappa}: F_\kappa \hookrightarrow F_\iota$ is continuous. Equip $E = \bigcap_{\iota \in I} F_\iota$ with the coarsest topology for which all the injections $f_\iota: E \hookrightarrow F_\iota$ are continuous. Then every linear form on E is the restriction to E of a continuous linear form on some F_κ.*

Proof. Let

$$x' = \sum_{k=1}^{n} {}^t f_{\iota_k}(y'_{\iota_k}) \in E'.$$

There exists $\kappa \in I$ such that $\iota_k \leqq \kappa$ for $1 \leqq k \leqq n$. We have $x' = {}^t f_\kappa(y')$, where

$$y' = \sum_{k=1}^{n} {}^t f_{\iota_k \kappa}(y'_{\iota_k}) \in F'_\kappa. \ ∎$$

If $f_\iota(E)$ is dense in each F_ι, then the maps

$${}^t f_\iota: F'_\iota \to E' \qquad \text{and} \qquad {}^t f_{\iota\kappa}: F'_\iota \to F'_\kappa$$

are injective. Identifying each $y' \in F'_\iota$ with its restrictions to the spaces $F_\kappa \subset F_\iota$ and $E \subset F_\iota$ (i.e., identifying F'_ι with ${}^t f_{\iota\kappa}(F'_\iota)$ and with ${}^t f_\iota(F'_\iota)$), we can then say that $E' = \bigcup_{\iota \in I} F'_\iota$.

EXERCISES

1. Let $(F_\iota, G_\iota)_{\iota \in I}$ be a family of vector spaces which form dual systems, and set $F = \prod_{\iota \in I} F_\iota$, $G = \coprod_{\iota \in I} G_\iota$. Then the pair F, G forms a dual system with respect to the bilinear form on $F \times G$ defined by

$$\langle x, y \rangle = \sum_{\iota \in I} \langle x_\iota, y_\iota \rangle$$

for $x = (x_\iota) \in F$, $y = (y_\iota) \in G$.

(a) Show that $\beta(F, G)$ is the product of the topologies $\beta(F_\iota, G_\iota)$ and that $\beta(G, F)$ is the locally convex direct sum of the topologies $\beta(G_\iota, F_\iota)$. (*Hint:* Use Exercise 2.12.10.)

(b) Show that $\beta^*(F, G)$ (Exercise 6.5) is the product of the topologies $\beta^*(F_\iota, G_\iota)$ and that $\beta^*(G, F)$ is the locally convex direct sum of the topologies $\beta^*(G_\iota, F_\iota)$.

2. (a) Let $(E_\iota)_{\iota \in I}$ be a family of barrelled Hausdorff spaces. Prove that $\prod_{\iota \in I} E_\iota$ is barrelled. (*Hint:* Use Exercise 1(a) and Exercise 6.5(b).)

(b) Let $(E_\iota)_{\iota \in I}$ be a family of infrabarrelled Hausdorff spaces. Prove that $\prod_{\iota \in I} E_\iota$ is infrabarrelled. (*Hint:* Use Exercise 1(b) and Exercise 6.5(c).)

3. Let F and G be two vector spaces forming a dual system and suppose that F equipped with the topology $\sigma(F, G)$ is the topological direct sum (Definition 2.7.1) of the two subspaces M and N. Show that G equipped with the topology $\sigma(G, F)$ is the topological direct sum of the subspaces $M^\perp$ and $N^\perp$. (*Hint:* F can be identified with $M \times N$, where, by Proposition 13.1, the space M is equipped with $\sigma(M, G/M^\perp)$ and N with $\sigma(N, G/N^\perp)$. Hence by Proposition 3 the space G can be identified with $G/M^\perp \times G/N^\perp$, and finally $G/M^\perp = N^\perp$, $G/N^\perp = M^\perp$.)

4. Prove that the locally convex direct sum (Example 2.12.2) of a family of semi-reflexive spaces is semi-reflexive. (*Hint:* Use Exercise 2.12.10 and an argument similar to that employed in the proof of Proposition 4.)

5. Assume that the hypotheses of Proposition 5 are satisfied and that E is equipped with the coarsest topology for which the f_ι are continuous. Prove that every equicontinuous subset of E' is contained in the $\sigma(E', E)$-closure of a set of the form

$$\sum_{k=1}^{n} {}^t f_{\iota_k}(M_{\iota_k}),$$

where M_{ι_k} is an equicontinuous subset of F'_{ι_k} $(1 \leqq k \leqq n)$.

§15. *Schwartz spaces*

Let E be a locally convex Hausdorff space and E' its dual. Let M be a closed subspace of E and $\varphi: E \to E/M$ the canonical surjection. Then it follows from Proposition 13.1 that ${}^t\varphi: (E/M)' \to E'$ is an injective linear map with image $M^\perp$ and furthermore ${}^t\varphi$ is a strict morphism if we equip $(E/M)'$ with the topology $\sigma\big((E/M)', E/M\big)$ and E' with $\sigma(E', E)$ (see Exercise 1). In general, it is not true that ${}^t\varphi$ is a strict morphism for the topologies $\beta\big((E/M)', E/M\big)$ and $\beta(E', E)$.

Similarly, if $j: M \hookrightarrow E$ is the canonical injection, then it follows from Proposition 13.2 that ${}^tj: E' \to M'$ is a surjective linear map with kernel $M^\perp$, and furthermore ${}^tj$ is a strict morphism if we equip E' with the topology $\sigma(E', E)$ and M' with $\sigma(M', M)$ (see Exercise 1). In general, ${}^tj$ will not be a strict morphism for the topologies $\beta(E', E)$ and $\beta(M', M)$.

For normed spaces, however, the situation is more favorable since we have the following result:

PROPOSITION 1. *Let E be a normed space. Then:*

(a) *The map ${}^t\varphi$ is an isometry from $(E/M)'$ onto the subspace $M^\perp$ of E'.*

(b) *The injection $\alpha: E'/M^\perp \to M'$ associated with ${}^tj$ is an isometry (and a bijection, of course).*

Proof. (a) For $y' \in (E/M)'$ we have

$$\|y'\| = \sup_{\substack{\|y\|<1 \\ y \in E/M}} |\langle y, y' \rangle| = \sup_{\substack{\|\varphi(x)\|<1 \\ x \in E}} |\langle \varphi(x), y' \rangle|.$$

If $\|y\| < 1$, then by the definition of the quotient norm there exists $x \in E$ such that $\varphi(x) = y$ and $\|x\| < 1$. Thus

$$\|y'\| = \sup_{\substack{\|x\|<1 \\ x \in E}} |\langle \varphi(x), y' \rangle| = \sup_{\substack{\|x\|<1 \\ x \in E}} |\langle x, {}^t\varphi(y') \rangle| = \|{}^t\varphi(y')\|.$$

(b) We have to prove that for every $y' \in M'$ we have

$$\|y'\| = \inf_{\substack{{}^tj(x')=y' \\ x' \in E'}} \|x'\|.$$

Now ${}^tj(x')$ is the restriction of x' to M. Hence if ${}^tj(x') = y'$, we have

$$\|y'\| = \|{}^tj(x')\| = \sup_{\substack{\|x\| \leqq 1 \\ x \in M}} |\langle x, x' \rangle| \leqq \sup_{\substack{\|x\| \leqq 1 \\ x \in E}} |\langle x, x' \rangle| = \|x'\|.$$

Thus

$$\|y'\| \leqq \inf_{\substack{{}^tj(x')=y' \\ x' \in E'}} \|x'\|.$$

On the other hand, by the Hahn-Banach theorem (Theorem 1.6.1) there exists $z' \in E'$ such that $\|z'\| = \|y'\|$ and such that the restriction of z' to M is y', i.e., ${}^tj(z') = y'$. This proves the assertion. ∎

In what follows we shall identify the space $(E/M)'$ with $M^\perp$ and the space M' with $E'/M^\perp$. We shall also denote the canonical surjection ${}^tj: E' \to E'/M^\perp = M'$ by ψ. Then we can say that in the case of the dual E' of a normed space E the strong topology $\beta(M^\perp, E/M)$ on $M^\perp$ coincides with the topology induced by $\beta(E', E)$ on $M^\perp$ and that the strong topology $\beta(E'/M^\perp, M)$ on $E'/M^\perp$ coincides with the quotient of the topology $\beta(E', E)$ modulo $M^\perp$. We shall now see some other classes of spaces where the same is true.

PROPOSITION 2 (Dieudonné-Schwartz). *Let E be a semi-reflexive* (Definition 8.2) *locally convex Hausdorff space, M a closed subspace of E, and E' the dual of E. Then:*

(a) *The strong topology $\beta(E'/M^\perp, M)$ on $E'/M^\perp$ coincides with the quotient of the topology $\beta(E', E)$ modulo $M^\perp$.*

(b) *M is semi-reflexive.*

Proof. (a) By the corollary of Proposition 12.3 the map $\psi: E' \to M'$ is continuous for the topologies $\beta(E', E)$ and $\beta(M', M)$. Hence $\beta(M', M)$ is coarser than the quotient topology $\mathfrak{Q}$ of $\beta(E', E)$ modulo $M^\perp$.

Conversely, let U be a balanced, closed, convex neighborhood of 0 in $E'/M^\perp$ for the topology $\mathfrak{Q}$. Then $\psi^{-1}(U)$ is a neighborhood of 0 in E' for the topology $\beta(E', E)$; i.e., there exists a balanced, bounded, closed, convex subset B of E such that $B^\circ \cup M^\perp \subset \psi^{-1}(U)$. Since $\psi^{-1}(U)$ is balanced, $\beta(E', E)$-closed and convex, it also contains the balanced, $\beta(E', E)$-closed, convex hull H of $B^\circ \cup M^\perp$. Now, E is semi-reflexive, hence by Proposition 8.3 the set H is also the balanced, $\sigma(E', E)$-closed, convex hull of $B^\circ \cup M^\perp$; i.e., $H = (B \cap M)^\circ$ (Proposition 3.2). But $B \cap M$ is a bounded, balanced, closed, convex set in M, and U contains the polar of $B \cap M$ with respect to the duality between M and M'. Thus $\beta(M', M)$ is finer than $\mathfrak{Q}$.

(b) The topology $\sigma(M, M')$ is the topology induced by $\sigma(E, E')$ on M (Corollary of Proposition 13.1). A bounded subset B of M is also a bounded subset of E and thus relatively compact for the topology $\sigma(E, E')$ by Proposition 8.1. But then B is also relatively compact for the topology $\sigma(M, M')$, and therefore, again by Proposition 8.1, the space M is semi-reflexive. ▮

Corollary. *Let I be an ordered set and for each $\iota \in I$ let F_ι be a semi-reflexive space. Suppose that for $\iota \leqq \kappa$ we have a continuous linear map $f_{\iota\kappa}: F_\kappa \to F_\iota$. Let E be a vector space, and for each $\iota \in I$ let f_ι be a linear map from E into F_ι such that for $\iota \leqq \kappa$ we have $f_\iota = f_{\iota\kappa} \circ f_\kappa$. Assume that the following condition is satisfied:*

(P) *If $(x_\iota)_{\iota \in I}$ is an element of $\prod_{\iota \in I} F_\iota$ such that for $\iota \leqq \kappa$ we have $f_{\iota\kappa}(x_\kappa) = x_\iota$, then there exists a* unique *element $x \in E$ such that $x_\iota = f_\iota(x)$ for all $\iota \in I$.*

Then E, equipped with the coarsest topology for which the f_ι are continuous, is a semi-reflexive space.

Proof (cf. Proposition 9.6). The uniqueness requirement in condition (P) implies in particular that if $x \in E$ and $x \neq 0$, then there exists an index $\iota \in I$ such that $f_\iota(x) \neq 0$; i.e., the linear map $\varphi: E \to \prod_{\iota \in I} F_\iota$ defined by $\varphi(x) = (f_\iota(x))_{\iota \in I}$ is injective and that E is a Hausdorff space. By Proposition 9.5 and Exercise 2.11.1 the map φ is an isomorphism of E onto a closed subspace of $\prod_{\iota \in I} F_\iota$. By Proposition 14.4 this product space is semi-reflexive; hence by part (b) of the preceding proposition so is E. ▮

Consider now the topology $\beta(M^\perp, E/M)$ and the topology $\mathfrak{J}$ induced by $\beta(E', E)$ on $M^\perp$. It follows from the corollary of Proposition 12.3 that ${}^t\varphi: M^\perp \to E'$ is continuous for the topologies $\beta(M^\perp, E/M)$ and $\beta(E', E)$; i.e., $\beta(M^\perp, E/M)$ is finer than $\mathfrak{J}$. Conversely, let U be a neighborhood of 0 in $M^\perp$ for the topology $\beta(M^\perp, E/M)$. Then U contains a set of the form

B°, where B is a bounded subset of E/M. *Suppose now that there exists a bounded subset* C *of* E *such that* $B \subset \varphi(C)$. In that case, $B \subset \varphi(C + M)$ and $U \supset B^{\circ} \supset (\varphi(C + M))^{\circ} = {}^t\varphi^{-1}(C^{\circ} \cap M^{\perp}) = C^{\circ} \cap M^{\perp}$. Since C° is a neighborhood of 0 in E' for $\beta(E', E)$, it follows that $\mathcal{G}$ is finer than $\beta(M^{\perp}, E/M)$.

In general, we do not know whether every bounded set in E/M is contained in the canonical image of a bounded subset of E. The following lemma belonging to general topology will help us to single out a class of spaces in which this is the case.

Before we can state our lemma we need a definition. Let X be a topological space, R an equivalence relation on X, X/R the quotient space equipped with the quotient topology, and $\varphi: X \to X/R$ the canonical surjection. We say that the equivalence relation R is *open* if the image $\varphi(A)$ of any open subset A of X is open in the quotient space X/R. An excellent example of an open equivalence relation is given by the congruence modulo a subspace M of a topological vector space E.

LEMMA. *Let* X *be a complete metric space and* R *an open equivalence relation on* X *such that the quotient space* X/R *is Hausdorff. Let* $\varphi: X \to X/R$ *be the canonical surjection. Then for every compact subset* K *of* X/R *there exists a compact subset* L *of* X *such that* $\varphi(L) = K$.

Proof. Let $\mathfrak{B}_1$ be the collection of all open balls with radius $\frac{1}{2}$ in X. Since R is open, the sets $\varphi(B)$, where B runs through $\mathfrak{B}_1$, form an open covering of K. Hence there exists a finite set $H_1 = \{x_1, \ldots, x_m\}$ of points of X such that if B_i $(1 \leqq i \leqq m)$ is the open ball in X with radius $\frac{1}{2}$ and center at x_i $(1 \leqq i \leqq m)$, then the sets $\varphi(B_i)$ $(1 \leqq i \leqq m)$ cover K.

Suppose now that for all indices i with $1 \leqq i \leqq n$ we have constructed a finite subset H_i of X with the following properties:

(α) $H_i \subset H_{i+1}$, and each point of H_{i+1} is at a distance $\leqq 1/2^i$ from H_i $(1 \leqq i \leqq n - 1)$.

(β) If B runs through the open balls in X with radius $1/2^i$ and center at a point in H_i, then the sets $\varphi(B)$ cover K $(1 \leqq i \leqq n)$.

Let $\mathfrak{B}_{n+1}$ be the collection of all open balls with radius $1/2^{n+1}$ and whose center x satisfies $\delta(x, H_n) < 1/2^n$, where δ denotes the distance in X. By the properties of H_n, the sets $\varphi(B)$, where B runs through $\mathfrak{B}_{n+1}$, form an open covering of K. Since K is compact, there exists a finite set $G_{n+1} \subset X$ such that if B runs through the open balls in X with radius $1/2^{n+1}$ and center at a point of G_{n+1}, then the sets $\varphi(B)$ cover K. If we take $H_{n+1} = H_n \cup G_{n+1}$, we see that we can define inductively an infinite sequence $(H_n)_{n \in \mathbf{N}}$ of finite subsets of X which has the properties (α) and (β).

Set $H = \bigcup_{n \in \mathbf{N}} H_n$ and let us prove that H is precompact. For every $p > 0$ and every point $y_{n+p} \in H_{n+p}$ there exists a sequence of points $y_{n+i} \in H_{n+i}$ $(0 \leqq i \leqq p - 1)$ such that

$$\delta(y_{n+i}, y_{n+i+1}) < \frac{1}{2^{n+i}}$$

for $0 \leqq i \leqq p - 1$. Thus

$$\delta(y_n, y_{n+p}) < \sum_{i=0}^{p-1} \frac{1}{2^{n+i}} \leqq \frac{1}{2^{n-1}},$$

and therefore for any $y \in H$ and any $n \geqq 1$ we have $\delta(y, H_n) \leqq 1/2^{n-1}$, which proves that H is precompact.

Since by assumption X is complete, the set $\overline{H}$ is compact and hence $\varphi(\overline{H})$ also. Let us show that $K \subset \varphi(\overline{H})$. If $z \in K$, then by the definition of H_n we have $\delta(\varphi^{-1}(z), H_n) \leqq 1/2^n$ for every n; hence $\delta(\varphi^{-1}(z), \overline{H}) = 0$. Since $\varphi^{-1}(z)$ is closed and $\overline{H}$ compact, this implies that $\overline{H} \cap \varphi^{-1}(z) \neq \emptyset$ (see Exercise 4) and therefore $z \in \varphi(\overline{H})$.

If we set $L = \overline{H} \cap \varphi^{-1}(K)$, then L is compact as a closed subset of the compact set $\overline{H}$, and we have $\varphi(L) = K$. ▮

In order to apply this lemma we introduce the following:

Definition 1 (Grothendieck). *A locally convex Hausdorff space E is said to be a Schwartz space if for every balanced, closed, convex neighborhood U of 0 in E there exists a neighborhood V of 0 such that for every $\alpha > 0$ the set V can be covered by finitely many translates of αU.*

Let us recall a notation introduced in §5. If U is a balanced, convex neighborhood of 0 in a locally convex space E, and q_U the gauge of U, then we denote by E_U the quotient space of E modulo the subspace $N = \{x \mid q_U(x) = 0\}$, equipped with the quotient norm of q_U. Using this notation, we can reformulate the above definition.

Proposition 3. *A locally convex Hausdorff space E is a Schwartz space if and only if for every balanced, closed, convex neighborhood U of 0 in E there exists a neighborhood V of 0 in E whose canonical image in the space E_U is precompact.*

Proof. (a) Suppose that E is a Schwartz space and let U be a balanced, closed, convex neighborhood of 0 in E. Let V be a neighborhood of 0 in E with the property stated in Definition 1. If $\varphi: E \to E_U$ is the canonical surjection, set $\dot{U} = \varphi(U)$ and $\dot{V} = \varphi(V)$. For every $\alpha > 0$ the set $\dot{V}$ can be covered by finitely many translates of $\alpha\dot{U}$. Since the sets $\alpha\dot{U}$ $(\alpha > 0)$ form a fundamental system of neighborhoods of 0 in E_U, it follows that $\dot{V}$ is precompact in E_U.

(b) Suppose that the condition of the proposition is satisfied, i.e., that, with the notations introduced previously, we have

$$\dot{V} \subset \bigcup_{i=1}^{n} (\dot{x}_i + \tfrac{1}{2}\alpha\dot{U}),$$

where $(\dot{x}_i)_{1 \leqq i \leqq n}$ is a finite family of points in $E_U = E/N$. If $\varphi(x_i) = \dot{x}_i$, then we have

$$V \subset \bigcup_{i=1}^{n} (x_i + \tfrac{1}{2}\alpha U) + N.$$

Now $N \subset \lambda U$ for every $\lambda > 0$, and since U is convex, we have

$$\tfrac{1}{2}\alpha U + N \subset \tfrac{1}{2}\alpha U + \tfrac{1}{2}\alpha U \subset \alpha U;$$

hence

$$V \subset \bigcup_{i=1}^{n} (x_i + \alpha U).$$

Thus E is a Schwartz space. ∎

Here is another useful characterization of Schwartz spaces.

PROPOSITION 4. *A locally convex Hausdorff space E is a Schwartz space if and only if the following two conditions are satisfied:*

(i) *Every bounded subset of E is precompact.*

(ii) *For every balanced, closed, convex neighborhood U of 0 there exists a neighborhood V of 0 such that for every $\alpha > 0$ we can find a bounded subset A of E which satisfies the relation*

$$V \subset \alpha U + A.$$

Proof. (a) Suppose that E is a Schwartz space. Let B be a bounded subset of E and U a balanced, closed, convex neighborhood of 0 in E. Let V be the neighborhood of 0 associated with U in Definition 1. We have $B \subset \lambda V$ for some $\lambda > 0$. On the other hand,

$$V \subset \bigcup_{i=1}^{n} \left(x_i + \frac{1}{\lambda} U\right)$$

for some finite family $(x_i)_{1 \leqq i \leqq n}$ of points of E. Hence

$$B \subset \bigcup_{i=1}^{n} (\lambda x_i + U),$$

and thus B is precompact (Theorem 2.10.2).

Let U be a balanced, closed, convex neighborhood of 0. If

$$V \subset \bigcup_{i=1}^{n} (x_i + \alpha U),$$

set $A = \{x_i \mid 1 \leqq i \leqq n\}$. Then A is bounded, and $V \subset \alpha U + A$.

(b) Suppose that conditions (i) and (ii) are satisfied. Let U be a balanced, closed, convex neighborhood of 0 in E and let V be the neighborhood of 0 whose existence is postulated in (ii). For a given $\alpha > 0$ determine a bounded subset A of E such that $V \subset \frac{1}{2}\alpha U + A$. By (i) the set A is precompact; hence there exists a finite family $(x_i)_{1 \leqq i \leqq n}$ of elements of E such that $A \subset \bigcup_{i=1}^{n} (x_i + \frac{1}{2}\alpha U)$. Hence $V \subset \bigcup_{i=1}^{n} (x_i + \alpha U)$ since U is convex. This proves that E is a Schwartz space. ▮

For obvious reasons, we shall call a locally convex Hausdorff space a *Fréchet-Schwartz space* if it is a Fréchet space (Definition 2.9.4) and a Schwartz space.

Corollary. *Every quasi-complete Schwartz space is a semi-Montel space* (Definition 9.1). *In particular, every Fréchet-Schwartz space is a Montel space.*

Finally, we want to characterize Schwartz spaces by a property of their duals, but before doing so we have to prove a result which will be of great importance later.

Theorem 1 (Schauder). *Let E and F be two normed spaces and $u: E \to F$ a linear map which maps the unit ball of E into a precompact subset of F. Then the transpose ${}^t u: F' \to E'$ maps the unit ball of F' onto a compact subset of E'* (equipped with the topology defined by the norm).

Proof. Let K be the image in F of the unit ball B of E. Then ${}^t u(K^{\circ}) \subset B^{\circ}$, which proves that ${}^t u$ is continuous if we equip F' with the topology $\lambda(F', F)$ (Definition 9.2) and E' with the topology $\beta(E', E)$. But by virtue of Proposition 9.8 and the corollary of Theorem 4.1 the unit ball of F' is compact for the topology $\lambda(F', F)$; hence its image under the map ${}^t u$ is compact in E' for $\beta(E', E)$. ▮

Let us observe that if E is a locally convex space and U a balanced convex neighborhood of 0 in E, then the dual of the normed space E_U is the normed space $(E')_{U^{\circ}}$. Indeed, a linear form on E will define a continuous linear form on E_U if and only if it is bounded on U, i.e., belongs to the subspace of E' generated by U°. Furthermore, it is clear that U° is the polar of the unit ball of E_U in the duality between E_U and $(E')_{U^{\circ}}$.

Proposition 5. *Let E be a locally convex Hausdorff space and E' its dual. Then E is a Schwartz space if and only if for every equicontinuous subset B of E' there exists a neighborhood V of 0 in E such that B is relatively compact in the normed space $(E')_{V^{\circ}}$.*

Proof. (a) Assume that E is a Schwartz space; let B be an equicontinuous subset of E' and $U = B^{\circ}$. By Proposition 3 there exists a neighborhood V of 0 in E whose image in E_U is precompact. We may assume that V is balanced, convex and contained in U. In this case the kernel of the canonical surjection $E \to E_V$ is contained in the kernel of the canonical surjection $E \to E_U$, and therefore we have a continuous linear map $u: E_V \to E_U$ which maps the unit ball of E_V into a precompact subset of E_U. Now $V^{\circ} \supset U^{\circ}$, and ${}^t u$ is clearly the canonical injection

$$(E')_{U^{\circ}} \hookrightarrow (E')_{V^{\circ}}.$$

By Theorem 1 the map ${}^t u$ maps U° onto a compact subset of $(E')_{V^\circ}$ and B onto a relatively compact subset, since $B \subset B^{\circ\circ} = U^{\circ}$.

(b) Conversely, assume that the condition of the proposition is satisfied and let U be a balanced, closed, convex neighborhood of 0 in E. Then $B = U^{\circ}$ is an equicontinuous subset of E', and therefore there exists a neighborhood V of 0 in E such that B is relatively compact in $(E')_{V^\circ}$; i.e., for given $\alpha > 0$ there exists a finite subset C of $(E')_{V^\circ}$ such that

$$B \subset C + \tfrac{1}{3}\alpha V^{\circ}.$$

On the other hand, the canonical image of V in $E_{V^{\circ\circ}}$ is precompact for the topology $\sigma(E_{V^{\circ\circ}}, (E')_{V^\circ})$, and C° contains the kernel of the canonical surjection $E \to E_{V^{\circ\circ}}$. Hence there exists a finite family $(x_i)_{1 \leqq i \leqq n}$ of points of V such that $V \subset \bigcup_{i=1}^n (x_i + \frac{1}{3}\alpha C^{\circ})$. If $x \in V$ and $y' \in U^{\circ}$, there exists an index i $(1 \leqq i \leqq n)$ such that $x - x_i \in \frac{1}{3}\alpha C^{\circ}$ and elements $z' \in C$ and $w' \in \frac{1}{3}\alpha V^{\circ}$ such that $y' = z' + w'$. Thus

$$\begin{aligned} |\langle x - x_i, y'\rangle| &\leqq |\langle x - x_i, z'\rangle| + |\langle x, w'\rangle| + |\langle x_i, w'\rangle| \\ &\leqq \tfrac{1}{3}\alpha + \tfrac{1}{3}\alpha + \tfrac{1}{3}\alpha = \alpha; \end{aligned}$$

i.e., $x - x_i \in \alpha U^{\circ\circ} = \alpha U$, and therefore V is covered by finitely many translates of αU. Thus E is a Schwartz space (Definition 1). ∎

Schwartz spaces have remarkable stability properties which we list in the following three propositions.

Proposition 6. (a) *A subspace M of a Schwartz space E is a Schwartz space.*

(b) *Let $(E_\iota)_{\iota \in I}$ be a family of Schwartz spaces. Then the product $\prod_{\iota \in I} E_\iota$ is a Schwartz space.*

(c) *Let E be a vector space, $(F_\iota)_{\iota \in I}$ a family of Schwartz spaces and for each index $\iota \in I$ let $f_\iota: E \to F_\iota$ be a linear map. Equip E with the coarsest topology $\mathcal{T}$ for which all the maps $f_\iota: E \to F_\iota$ are continuous, and suppose that E is a Hausdorff space. Then E is a Schwartz space.*

Proof. (a) Let U be a balanced, convex, closed neighborhood of 0 in M. There exists a balanced, convex, closed neighborhood U_1 of 0 in E such that $U \supset U_1 \cap M$. Let V_1 be a neighborhood of 0 in E such that for every $\alpha > 0$ the set V_1 can be covered by finitely many translates of αU_1. Set $V = V_1 \cap M$.

Given $\alpha > 0$, there exists a finite family $(x_i)_{1 \leqq i \leqq n}$ of elements of E such that $V_1 \subset \bigcup_{i=1}^n (x_i + \frac{1}{2}\alpha U_1)$. Let H be the set of those indices i $(1 \leqq i \leqq n)$ for which $(x_i + \frac{1}{2}\alpha U_1) \cap M \neq \emptyset$, and for each $i \in H$ select a point $y_i \in (x_i + \frac{1}{2}\alpha U_1) \cap M$. Then we have $V \subset \bigcup_{i \in H} (y_i + \alpha U)$, and therefore M is a Schwartz space.

(b) Let U be a balanced, convex, closed neighborhood of 0 in

$$E = \prod_{\iota \in I} E_\iota.$$

There exist a finite subset H of I and for each $\iota \in I$ a balanced, convex, closed neighborhood U_ι of 0 in E_ι such that $U_\iota = E_\iota$ for $\iota \notin H$ and $U \supset \prod_{\iota \in I} U_\iota$. For each $\iota \in H$ choose a neighborhood V_ι of 0 in E_ι according to Definition 1, for each $\iota \in \complement H$ set $V_\iota = E_\iota$ and define

$$V = \prod_{\iota \in I} V_\iota.$$

Given $\alpha > 0$, for every $\iota \in H$ there exists a finite family $(x_{\iota,k})_{1 \leqq k \leqq n(\iota)}$ of points in E_ι such that $V_\iota \subset \bigcup_{k=1}^{n(\iota)} (x_{\iota,k} + \alpha U_\iota)$. Denote by $(y_l)_{1 \leqq l \leqq n}$ the finite family of points $y_l = (y_{l,\iota})$ of E, where for $\iota \in H$ the component $y_{l,\iota}$ equals some $x_{\iota,k}$ $(1 \leqq k \leqq n(\iota))$ and $y_{l,\iota} = 0$ for $\iota \in \complement H$. Then we have $V \subset \bigcup_{l=1}^{n} (y_l + \alpha U)$, and thus E is a Schwartz space.

(c) Since $\mathcal{T}$ is a Hausdorff topology, the map $x \mapsto (f_\iota(x))_{\iota \in I}$ from E into $\prod_{\iota \in I} F_\iota$ is injective; and if we identify E with its image, then $\mathcal{T}$ is the topology induced on E by the product topology on $\prod_{\iota \in I} F_\iota$. Thus the conclusion follows from parts (a) and (b) of the proposition. ∎

Proposition 7. *Let E be a Schwartz space and M a closed subspace of E. Then the quotient space E/M is a Schwartz space.*

Proof. First of all, E/M is a Hausdorff space since M is closed (Proposition 2.5.5).

Let $\dot{U}$ be a balanced, convex, closed neighborhood of 0 in E/M. If $\varphi: E \to E/M$ is the canonical surjection, then $U = \varphi^{-1}(\dot{U})$ is a balanced, convex, closed neighborhood of 0 in E. Choose a neighborhood V of 0 in E which satisfies the requirements of Definition 1 and set $\dot{V} = \varphi(V)$. Then $\dot{V}$ is a neighborhood of 0 in E/M. If $\alpha > 0$ and $V \subset \bigcup_{i=1}^{n} (x_i + \alpha U)$, where $(x_i)_{1 \leqq i \leqq n}$ is a finite family of elements of E, then setting $y_i = \varphi(x_i)$ for $1 \leqq i \leqq n$ we have $\dot{V} \subset \bigcup_{i=1}^{n} (y_i + \alpha \dot{U})$. ∎

Corollary. *Let E be a Fréchet-Schwartz space and M a closed subspace of E. Then the quotient E/M is a Fréchet-Schwartz space.*

Remark 1. In particular, E/M is a Montel space. This is of interest because the quotient space of a Fréchet-Montel space modulo a closed subspace need not be a Montel space or even a reflexive space ([52], §27,2, p. 373, §31,5, p. 437; [9], Chapter IV, §5, Exercise 21).

Proof. We have just seen that E/M is a Schwartz space. It is also metrizable and complete (Theorem 2.9.2). ∎

PROPOSITION 8. *Let F be a vector space and let $(E_n)_{n \in \mathbf{N}}$ be a sequence of linear subspaces of F such that $E_n \subset E_{n+1}$ for all $n \in \mathbf{N}$ and $F = \bigcup_{n \in \mathbf{N}} E_n$. Suppose that each E_n is equipped with a locally convex topology $\mathcal{T}_n$ for which E_n is a Schwartz space. Suppose furthermore that $\mathcal{T}_{n+1}$ induces on E_n the topology $\mathcal{T}_n$ and that E_n is closed in E_{n+1} for $\mathcal{T}_{n+1}$.*

Let $\mathcal{T}$ be the finest locally convex topology on F for which the canonical injections $f_n: E_n \hookrightarrow F$ are continuous. Then F equipped with the topology $\mathcal{T}$ is a Schwartz space.

Proof. F is a Hausdorff space by Corollary 1 of Theorem 2.12.1. We shall show that the two conditions of Proposition 4 are satisfied.

(i) Let B be a bounded subset of F. By Theorem 2.12.2 the set B is contained in some subspace E_n and is bounded there. Since E_n is a Schwartz space, B is precompact in E_n. Hence by Proposition 2.10.6 the set B is precompact in F.

(ii) Let U be a balanced, convex, closed neighborhood of 0 in F. Then $U_n = f_n^{-1}(U) = U \cap E_n$ is a balanced, convex, closed neighborhood of 0 in E_n, and therefore there exists a neighborhood V_n of 0 in E_n such that $V_n \subset U_n$ and that V_n satisfies condition (ii) of Proposition 4 with respect to U_n. Let V be the balanced convex hull of $\bigcup_{n \in \mathbf{N}} [1/(n+1)]V_n$. Then V is a neighborhood of 0 in F. Let us show that it satisfies condition (ii) of Proposition 4 with respect to U.

Let $\alpha > 0$. Then

$$\frac{1}{n+1} V_n \subset \alpha U_n \subset \alpha U \qquad \text{for} \qquad n + 1 \geqq \frac{1}{\alpha}\cdot$$

For each n such that $0 \leqq n \leqq 1/\alpha$ let A_n be a bounded set in E_n such that

$$\frac{1}{n+1} V_n \subset A_n + \alpha U_n$$

and let A be the balanced convex hull of the union of the sets A_n $(0 \leqq n \leqq 1/\alpha)$ in F. The set A is bounded and we have

$$\frac{1}{n+1} V_n \subset A + \alpha U \qquad \text{for} \qquad n \in \mathbf{N}.$$

Finally, since $A + \alpha U$ is balanced and convex, we have $V \subset A + \alpha U$, which proves our assertion. ∎

EXAMPLE 1. Let F and G be two vector spaces which form a dual system. Then F equipped with the topology $\sigma(F, G)$ is a Schwartz space. Indeed, every bounded subset of F is relatively compact in G^* for $\sigma(G^*, G)$ and thus precompact in F by Proposition 2.5(b). Hence condition (i) of Proposition 4 is satisfied.

To prove that condition (ii) is also satisfied, let

$$U = \{x \mid |\langle x, y_k\rangle| \leqq \epsilon, 1 \leqq k \leqq n\}$$

be a neighborhood of 0, where we may suppose that the vectors y_k are linearly independent. Let M be the subspace of G generated by the y_k ($1 \leqq k \leqq n$), let N be an algebraic supplement of M and let $(z_\iota)_{\iota\in I}$ be a basis of N. We choose $V = U$ and

$$A = \{x \mid |\langle x, y_k\rangle| \leqq \epsilon, 1 \leqq k \leqq n; |\langle x, z_\iota\rangle| \leqq \epsilon, \iota \in I\}$$

for the set A which figures in condition (ii), independently of α. In the first place, A is bounded since if

$$y = \sum_{k=1}^{n} \eta_k y_k + \sum_{\iota\in I} \zeta_\iota z_\iota \in G,$$

then

$$|\langle x, y\rangle| \leqq \epsilon\left(\sum_{k=1}^{n} |\eta_k| + \sum_{\iota\in I} |\zeta_\iota|\right).$$

Next we see from an elementary algebraic consideration (cf. Exercise 10) or from Proposition 1.2 that there exists a basis $(x_k)_{1\leqq k\leqq n}$ in $N^\perp$ for which

$$\langle x_k, y_l\rangle = 0 \quad \text{if} \quad k \neq l,$$
$$\langle x_k, y_k\rangle = 1.$$

For each $x \in U$ set $u = \sum_{k=1}^{n} \langle x, y_k\rangle x_k$. Then $u \in A$ since

$$|\langle u, y_k\rangle| = |\langle x, y_k\rangle| \leqq \epsilon \quad \text{and} \quad \langle u, z_\iota\rangle = 0.$$

Finally, $x - u \in \alpha U$ since $\langle x - u, y_k\rangle = \langle x, y_k\rangle - \langle x, y_k\rangle = 0$. Thus $U \subset \alpha U + A$, and therefore condition (ii) is satisfied.

Example 2. Let Ω be an open subset of $\mathbf{R}^n$. The space $E = \mathcal{E}(\Omega)$ (Example 2.4.8) is a Schwartz space. To prove this, consider the neighborhood

$$U = \{f \mid |\partial^p f(x)| \leqq \epsilon, |p| \leqq m, x \in K\}$$

of 0 in E, where $m \in \mathbf{N}$ and K is a compact subset of Ω. The space E_U consists of all equivalence classes, where two functions are equivalent if their partial derivatives up to order m coincide on the compact set K. On E_U the topology is defined by the norm

$$q(f) = \max_{\substack{|p|\leqq m\\ x\in K}} |\partial^p f(x)|,$$

where f actually stands for the equivalence class it represents. Choose $\delta > 0$ so that the compact set

$$L = \{x \mid |x - y| \leqq \delta \text{ for some } y \in K\}$$

is contained in Ω. Set

$$V = \{f \mid |\partial^p f(x)| \leqq 1, |p| \leqq m + 1, x \in L\}.$$

We shall show that the canonical image $\dot{V}$ of V in E_U is precompact.

Indeed, for each $p \in \mathbf{N}^n$ with $|p| \leqq m$ let $\mathcal{C}_p(K)$ be a space isomorphic to the Banach space $\mathcal{C}(K)$. The map $f \mapsto (\partial^p f)_{|p| \leqq m}$ is an injection from E_U into $\prod_{|p| \leqq m} \mathcal{C}_p(K)$, and E_U can be identified with a subspace of the Banach space $\prod_{|p| \leqq m} \mathcal{C}_p(K)$. The projection $\partial^p \dot{V} = \{\partial^p f \mid f \in \dot{V}\}$ of $\dot{V}$ in $\mathcal{C}_p(K)$ is equicontinuous since if x and y are two points of K such that $|x - y| \leqq \delta$, then the line segment joining them lies in L and we have

$$|\partial^p f(x) - \partial^p f(y)| \leqq \left| \sum_{i=1}^{n} \int_x^y \partial_i \partial^p f(z)\, dz_i \right| \leqq n \cdot |x - y|.$$

By Ascoli's theorem each set $\partial^p \dot{V}$ is relatively compact in $\mathcal{C}_p(K)$. Hence by Tihonov's theorem $\dot{V}$ is relatively compact in $\prod_{|p| \leqq m} \mathcal{C}_p(K)$ and therefore precompact in E_U.

Since $\mathcal{E}(\Omega)$ is a Fréchet space (Example 2.9.3), we obtain from the corollary of Proposition 4 that $\mathcal{E}(\Omega)$ is a Montel space. We have seen this already in Example 9.3 with a proof very similar to the present one.

Example 3. If K denotes a compact subset of $\mathbf{R}^n$, then the space $\mathfrak{D}(K)$ introduced in Examples 2.3.4 and 2.4.10 is a Schwartz space. This follows from Proposition 6(a), since by Example 2.5.9 if Ω is an open subset of $\mathbf{R}^n$ containing K, then $\mathfrak{D}(K)$ is a subspace of $\mathcal{E}(\Omega)$.

Since $\mathfrak{D}(K)$ is a Fréchet space (Example 2.9.5), it follows from the corollary of Proposition 4 that it is a Montel space, as we already know from Example 9.4.

Applying Proposition 8 to the space $\mathfrak{D}(\Omega)$ of Example 2.12.6, we see that it is a Schwartz space. Since it is barrelled (Example 6.4) and complete (Example 2.12.6), it is a Montel space (cf. Example 9.6), but it is not a Fréchet space (cf. Exercise 2.12.6).

In the next example we shall use the following result:

Proposition 9 (Sebastião e Silva). *Let I be an ordered set and for each index $\iota \in I$ let F_ι be a normed space. Suppose that for $\kappa \geqq \iota$ we have a continuous linear map $f_{\iota\kappa}: F_\kappa \to F_\iota$, and assume furthermore that for each index $\iota \in I$ there exists an index $\kappa \geqq \iota$ such that $f_{\iota\kappa}$ maps the unit ball B_κ of F_κ onto a precompact subset of F_ι.*

Let E be a vector space and for each $\kappa \in I$ let $f_\kappa: E \to F_\kappa$ be a linear map such that $f_\iota = f_{\iota\kappa} \circ f_\kappa$ for $\iota \leqq \kappa$. Equip E with the coarsest topology for which the f_ι are continuous and assume that E is a Hausdorff space. Then E is a Schwartz space.

REMARK 1. The proof is similar to that of Proposition 6. It is possible to give a statement which generalizes simultaneously Propositions 6 and 9 (see Exercise 11(b)).

Proof. Let U be a balanced, closed, convex neighborhood of 0 in E. There exists a finite subfamily $(\iota_l)_{1 \leqq l \leqq r}$ of I and for each ι_l a neighborhood U_{ι_l} of 0 in F_{ι_l} such that $U \supset \bigcap_{l=1}^r f_{\iota_l}^{-1}(U_{\iota_l})$. For each ι_l $(1 \leqq l \leqq r)$ let $\kappa_l \in I$ be such that $f_{\iota_l\kappa_l}(B_{\kappa_l})$ is precompact in F_{ι_l}. The set $V = \bigcap_{l=1}^r f_{\kappa_l}^{-1}(B_{\kappa_l})$ is a neighborhood of 0 in E. Let us prove that it satisfies the condition of Definition 1.

Given $\alpha > 0$, for each l $(1 \leqq l \leqq r)$ there exists a *finite* family $(x_j^{(l)})_{1 \leqq j \leqq n(l)}$ of points of F_{ι_l} such that

$$f_{\iota_l\kappa_l}(B_{\kappa_l}) \subset \bigcup_{j=1}^{n(l)} (x_j^{(l)} + \tfrac{1}{2}\alpha U_{\iota_l}).$$

Let P be the finite set of multi-indices $p = (p_1, \ldots, p_r)$ such that $1 \leqq p_l \leqq n(l)$ and the set

$$W_p = \bigcap_{l=1}^r f_{\iota_l}^{-1}(x_{p_l}^{(l)} + \tfrac{1}{2}\alpha U_{\iota_l})$$

is not empty. For each $p \in P$ choose a point $z_p \in W_p$. If $z \in V$, then for every l $(1 \leqq l \leqq r)$ there exists an index q_l $(1 \leqq q_l \leqq n(l))$ such that

$$f_{\iota_l}(z) \in f_{\iota_l\kappa_l}(B_{\kappa_l}) \in x_{q_l}^{(l)} + \tfrac{1}{2}\alpha U_{\iota_l}.$$

Choosing $p \in P$ such that $p_l = q_l$, we have $f_{\iota_l}(z - z_p) \in \alpha U_{\iota_l}$ for each l, i.e.,

$$z \in z_p + \alpha \bigcap_{l=1}^r f_{\iota_l}^{-1}(U_{\iota_l}) \subset z_p + \alpha U.$$

Thus $V \subset \bigcup_{p \in P} (z_p + \alpha U)$. ∎

EXAMPLE 4. The space $\mathcal{S}$ introduced in Example 2.4.14 is a Schwartz space. As we observed in Example 2.6.1, for $k \in \mathbf{N}$ and $m \in \mathbf{N}$ the topology of the space $\mathcal{S}_k^m$ (Example 2.4.11) is defined with the help of the norm

$$\|f\|_k^m = \max_{|p| \leqq m} \max_{x \in \mathbf{R}^n} |(1 + |x|^2)^k \, \partial^p f(x)|.$$

In view of Proposition 9 and Example 2.11.6 it is sufficient to prove that the unit ball B_{k+1}^{m+1} of $\mathcal{S}_{k+1}^{m+1}$ is precompact in $\mathcal{S}_k^m$. We shall prove that it is relatively compact, using a method similar to that of Example 9.5.

In the first place, B_{k+1}^{m+1} is compact in $\mathcal{E}^m$ since it satisfies the conditions of Proposition 9.11. Indeed, for $|p| \leqq m + 1$ and $f \in B_{k+1}^{m+1}$ we have $|\partial^p f(x)| \leqq 1$ for all $x \in \mathbf{R}^n$, and the inequality

$$|\partial^p f(x) - \partial^p(y)| = \left| \sum_{i=1}^{n} \int_x^y \partial_i \partial^p f(z)\, dz \right| \leqq n \cdot |x - y|$$

proves that for $|p| \leqq m$ the set $\{\partial^p f \mid f \in B_{k+1}^{m+1}\}$ is equicontinuous

It is now clearly enough to prove that if a sequence (f_j) of elements of B_{k+1}^{m+1} converges in $\mathcal{E}^m$ to some function g, then $g \in \mathcal{S}_k^m$ and (f_j) converges to g in $\mathcal{S}_k^m$. We have

$$\max_{|p| \leqq m} |(1 + |x|^2)^{k+1} \partial^p f(x)| \leqq 1$$

for all $f \in B_{k+1}^{m+1}$, $x \in \mathbf{R}^n$. Hence given $\epsilon > 0$, there exists $\rho > 0$, such that

$$\max_{|p| \leqq m} |(1 + |x|^2)^k \partial^p f(x)| \leqq (1 + |x|^2)^{-1} \leqq \tfrac{1}{2}\epsilon \tag{1}$$

for $|x| > \rho$. Since in particular $(\partial^p f_j)$ converges to $(\partial^p g)$ pointwise, we have

$$\max_{|p| \leqq m} |(1 + |x|^2)^k \partial^p g(x)| \leqq \tfrac{1}{2}\epsilon \tag{2}$$

for $|x| > \rho$, which proves already that $g \in \mathcal{S}_k^m$. For $|p| \leqq m$ each sequence $(\partial^p f_j)$ converges uniformly to $\partial^p g$ on the compact set $|x| \leqq \rho$. We have therefore

$$\max_{|p| \leqq m} |\partial^p f_j(x) - \partial^p g(x)| \leqq \epsilon(1 + \rho^2)^{-k}$$

and *a fortiori*

$$\max_{|p| \leqq m} |(1 + |x|^2)^k \{\partial^p f_j(x) - \partial^p g(x)\}| \leqq \epsilon \tag{3}$$

for $|x| \leqq \rho$ and $j \geqq J$. Collecting the inequalities (1), (2) and (3), we obtain

$$\max_{|p| \leqq m} \max_{x \in \mathbf{R}^n} |(1 + |x|^2)^k \{\partial^p f_j(x) - \partial^p g(x)\}| \leqq \epsilon$$

for $j \geqq J$, which proves our assertion.

Since $\mathcal{S}$ is a Fréchet space by Example 2.11.11, it is a Montel space, as we already know.

PROPOSITION 10. *Let E be a Fréchet-Schwartz space and M a closed subspace of E. Then every bounded subset of the quotient space E/M is contained in the image of some compact subset of E.*

Proof. Since E/M is a Schwartz space (Proposition 7), its bounded subsets are precompact (Proposition 4). But E/M is also a Fréchet space (Theorem 2.9.2), and therefore its bounded subsets are relatively compact. Hence the statement follows from the lemma proved above. ∎

From the discussion preceding the lemma we obtain:

COROLLARY. *Let E be a Fréchet-Schwartz space, E' its dual, and M a closed subspace of E. Then the strong topology $\beta(M^{\perp}, E/M)$ on $M^{\perp}$ coincides with the topology induced by $\beta(E', E)$ on $M^{\perp}$.*

EXERCISES

1. Prove the statements made at the beginning of this section. (*Hint:* To prove that ${}^t\varphi: (E/M)' \to E'$ is an injective strict morphism for the topologies $\sigma((E/M)', E/M)$ and $\sigma(E', E)$, replace F by E', G by E, M by $M^{\perp} \simeq (E/M)'$, N by E/M, and j by ${}^t\varphi$ in Proposition 13.1. To prove that ${}^tj: E' \to M'$ is a surjective strict morphism for the topologies $\sigma(E', E)$ and $\sigma(M', M)$, replace F by E, G by E', M by M, and N by M' in Proposition 13.2.)

2. Let E be a reflexive Banach space and M a closed subspace of E. Prove that E/M is a reflexive Banach space. (*Hint:* E' is reflexive and by Proposition 2(b) $M^{\perp}$ is also reflexive. Furthermore, $(E/M)'' \simeq (M^{\perp})' \simeq E/M$ by Proposition 1.)

3. Let X be a set, R an equivalence relation on X, and A a subset of X. We say that A is *saturated* for R if $x \in A$ and $x \equiv y \bmod R$ imply $y \in A$. If A is saturated, then it is the union of equivalence classes modulo R.

(a) Prove that, given any subset B of X, there exists a smallest saturated subset A of X containing B. We call A the saturate of B.

(b) Suppose that X is a topological space. Prove that R is open if and only if for every open set B in X the saturate of B is open.

(c) Still suppose that X is a topological space and let $\varphi: X \to X/R$ be the canonical surjection. Prove that R is open if and only if for every point $x \in X$ the filter of neighborhoods of $\varphi(x)$ in X/R coincides with the image under φ of the filter of neighborhoods of x in X.

4. (a) Let X be a metric space with metric δ, A a closed subset of X, and B a compact subset of X. Show that if $\delta(A, B) = 0$, then $A \cap B \neq \emptyset$. (*Hint:* The function $x \mapsto \delta(A, x)$ attains its minimum on B.)

(b) Show by an example in $\mathbf{R}^2$ that if A and B are two closed sets, the condition $\delta(A, B) = 0$ does not imply $A \cap B \neq \emptyset$.

5. Let Ω be a domain in $\mathbf{C}$. Show that the space $H(\Omega)$ introduced in Example 9.2 is a Schwartz space. (*Hint:* Using Cauchy's formula for $f^{(p)}(z)$, show that the injection $H(\Omega) \hookrightarrow \mathcal{E}(\Omega)$ is a strict morphism.)

6. Let E be a locally convex Hausdorff space. We say that E satisfies the *Mackey convergence condition* if every sequence which converges to 0 also converges to 0 in the Mackey sense (Exercise 7.7). Thus every metrizable locally convex space satisfies the Mackey convergence condition (Exercise 7.7(d)).

We say that E satisfies the *strict Mackey convergence condition* if for every bounded subset A of E there exists a bounded, balanced, closed, convex subset $B \supset A$ such that the topology induced by E on A coincides with the topology induced on A by the normed space E_B.

(a) Prove that if E satisfies the strict Mackey convergence condition, then it also satisfies the Mackey convergence condition.

(b) Let E be an infrabarrelled locally convex Hausdorff space and E' its dual equipped with the strong topology $\beta(E', E)$. Prove that E' satisfies the Mackey convergence condition if and only if every sequence in E' which converges to 0 for $\beta(E', E)$, converges to 0 uniformly on some neighborhood V of 0 in E. Prove that E' satisfies the strict Mackey convergence condition if and only if for every equicontinuous subset A of E' there exists a neighborhood V of 0 in E such that on A the topology induced by $\beta(E', E)$ coincides with the topology of uniform convergence on V.

(c) Following Grothendieck ([41], p. 106), we say that a locally convex Hausdorff space E is *quasi-normable* if for every equicontinuous subset A of E' there exists a neighborhood V of 0 in E such that on A the topology induced by $\beta(E', E)$ coincides with the topology of uniform convergence on V. Prove that if E is an infrabarrelled quasi-normable space, then its dual E' equipped with the strong topology $\beta(E', E)$ satisfies the strict Mackey convergence condition.

(d) Prove that a locally convex Hausdorff space E is quasi-normable if and only if for every balanced, closed, convex neighborhood U of 0 there exists a neighborhood V of 0 such that for every $\alpha > 0$ we can find a bounded subset M of E which satisfies the relation $V \subset \alpha U + M$. In particular, every Schwartz space is quasi-normable by Proposition 4. More precisely, a locally convex Hausdorff space is a Schwartz space if and only if it is quasi-normable and its bounded subsets are precompact. (*Hint:* If $A^{\circ} = U$, then $\lambda V \subset U + M$ is equivalent to $\mu V^{\circ} \supset A \cap M^{\circ}$, where M° is a neighborhood of 0 for $\beta(E', E)$.)

7. Prove the following converse of Theorem 1. Let E and F be two normed spaces and $u: E \to F$ a continuous linear map. If ${}^t u: F' \to E'$ maps the unit ball of F' onto a compact subset of E' equipped with the topology $\beta(E', E)$, then u maps the unit ball of E onto a precompact subset of F. (*Hint:* Consider the map ${}^{tt}u$, or use an argument similar to part (b) of the proof of Proposition 5.)

8. (a) Let E be a locally convex Hausdorff space, M a subspace of E and (x'_n) an equicontinuous sequence of linear forms on M which converges to 0 for the topology $\beta(M', M)$. Prove that if M is quasi-normable, then every x'_n has an extension $\tilde{x}'_n$ to E such that the sequence $(\tilde{x}'_n)$ of linear forms is equicontinuous on E and converges to 0 for the topology $\beta(E', E)$. (*Hint:* There exists a balanced, closed, convex neighborhood V of 0 in E such that $\sup_{x \in V \cap M} |\langle x, x'_n \rangle| = \lambda_n \to 0$. By the Hahn-Banach theorem there exist $\tilde{x}'_n \in E'$ such that $|\langle x, \tilde{x}'_n \rangle| \leqq \lambda_n$ for $x \in V$.)

(b) Let E be a vector space, $(F_\iota)_{\iota \in I}$ a family of locally convex Hausdorff spaces, and for each $\iota \in I$ let $f_\iota: E \to F_\iota$ be a linear map. Equip E with the coarsest topology for which all the maps f_ι are continuous. Suppose that E is quasi-normable and let (x'_n) be an equicontinuous sequence of linear forms on E which converges to 0 for the topology $\beta(E', E)$. Prove that there exist a finite subset H of I and for each $\iota \in H$ an equicontinuous sequence $(x'_{\iota,n})_{n \in \mathbf{N}}$ of linear forms on F_ι which converges to 0 for $\beta(F'_\iota, F_\iota)$ such that $x'_n = \sum_{\iota \in H} x'_{\iota,n} \circ f_\iota$ for every n. (*Hint:* Consider E as a subset of $\prod_{\iota \in I} F_\iota$; then apply Proposition 14.1 and part (a).)

9. A locally convex Hausdorff space F is said to be *ultrabornological* if there exist a family $(E_\iota)_{\iota \in I}$ of Banach spaces and for each $\iota \in I$ a linear map $f_\iota: E_\iota \to F$ such that the topology of F is the finest locally convex topology on F for which all the maps f_ι are continuous.

(a) Prove that an ultrabornological space is barrelled and bornological. (*Hint:* Use Propositions 6.4 and 7.4.)

(b) Prove that a quasi-complete bornological Hausdorff space is ultrabornological. (*Hint:* Proposition 7.5.)

(c) Prove that the dual E' of a complete Schwartz space E equipped with the topology $\beta(E', E)$ is ultrabornological. (*Hint:* Every balanced, convex, $\sigma(E', E)$-closed, equicontinuous subset B of E' is $\sigma(E', E)$-compact, hence $\sigma(E', E)$-complete, and so $(E')_B$ is a Banach space by Proposition 5.6. Let $\mathcal{T}$ be the finest locally convex topology on E' for which the injections $f_B: (E')_B \to E'$ are continuous. $\mathcal{T}$ is finer than $\lambda(E', E)$ since the f_B are continuous if we equip E' with $\lambda(E', E)$. To prove that $\mathcal{T} = \lambda(E', E)$ we have to show by Exercise 11.4 that the two topologies induce the same topology on every balanced, convex, $\sigma(E', E)$-closed, equicontinuous subset A of E'. Let B be a balanced, convex, $\sigma(E', E)$-closed, equicontinuous subset of E' such that A is compact in $(E')_B$ (Proposition 5). Then on $(E')_B$ and *a fortiori* on A the topologies $\mathcal{T}$ and $\lambda(E', E)$ are coarser than the Banach space topology of $(E')_B$. But A is compact. Hence both these topologies coincide on A. Finally, $\beta(E', E) = \lambda(E', E)$ since E is a complete Schwartz space.)

10. Let E be a vector space, E^* its algebraic dual and $(x_k)_{1 \leqq k \leqq n}$ a finite family of linearly independent elements of E. Prove that there exists a family $(x'_k)_{1 \leqq k \leqq n}$ of elements of E^* such that $\langle x_k, x'_k \rangle = 1$ for $1 \leqq k \leqq n$ and $\langle x_k, x'_l \rangle = 0$ for $k \neq l$, $1 \leqq k, l \leqq n$. (*Hint:* Use induction. If the elements x'_i for $1 \leqq i \leqq k-1$ have been constructed, then we can write every $x' \in E$ in the form

$$x' = \sum_{i=1}^{k-1} \langle x_i, x' \rangle \cdot x'_i + y',$$

where $\langle x_i, y' \rangle = 0$ for $1 \leqq i \leqq k-1$. We cannot have $\langle x_k, y' \rangle = 0$ for every x' since this would imply $x_k = \sum_{i=1}^{k-1} \langle x_k, x'_i \rangle x_i$. There exists $z'_k \in E^*$ such that $\langle x_i, z'_k \rangle = 0$ for $1 \leqq i \leqq k-1$ and $\langle x_k, z'_k \rangle = 1$. Write $z'_i = x'_i - \langle x_k, x'_i \rangle x'_k$ for $1 \leqq i \leqq k-1$. The family $(z'_i)_{1 \leqq i \leqq k}$ satisfies the conditions $\langle x_i, z'_i \rangle = 1$ for $1 \leqq i \leqq k$ and $\langle x_i, z'_j \rangle = 0$ for $i \neq j$, $1 \leqq i, j \leqq k$.)

11. (a) Prove the following converse of Proposition 9. Let E be a Schwartz space. Then there exists a family $(F_\iota)_{\iota \in I}$ of normed spaces, where I is an ordered set, such that:

(i) for $\kappa \geqq \iota$ there exists a continuous linear map $f_{\iota\kappa}: F_\kappa \to F_\iota$,

(ii) for each $\kappa \in I$ there exists a linear map $f_\kappa: E \to F_\kappa$ such that $f_\iota = f_{\iota\kappa} \circ f_\kappa$ for $\kappa \geqq \iota$,

(iii) for each $\iota \in I$ there exists a $\kappa \geqq \iota$ such that $f_{\iota\kappa}$ maps the unit ball B_κ of F_κ onto a precompact subset of F_ι,

(iv) the topology of E is the coarsest topology for which the f_κ are continuous. (*Hint:* Consider the spaces E_U.)

(b) Let $(F_\iota)_{\iota\in I}$ be a family of locally convex Hausdorff spaces, where I is an ordered set. Suppose that for $\kappa \geqq \iota$ we have a continuous linear map $f_{\iota\kappa}: F_\kappa \to F_\iota$ and assume furthermore that given an index $\iota \in I$ and a balanced, closed, convex neighborhood U of 0 in F_ι, there exists an index $\kappa \geqq \iota$ and a neighborhood V of 0 in F_κ such that the canonical image of $f_{\iota\kappa}(V)$ in $(F_\iota)_U$ is precompact. Let E be a vector space and for each $\kappa \in I$ let $f_\kappa: E \to F_\kappa$ be a linear map such that $f_\iota = f_{\iota\kappa}\circ f_\kappa$ if $\iota \leqq \kappa$. Equip E with the coarsest topology for which all the maps f_κ are continuous, and assume that this topology is Hausdorff. Prove that E is a Schwartz space. Show that both Propositions 6 and 9 are special cases of this result. Finally, verify that in Example 2 we made implicit use of this result.

§16. *Distinguished spaces*

In subsequent chapters the strong duals of the spaces $\mathcal{E}(\Omega)$, $\mathcal{S}$ and $\mathcal{D}(\Omega)$ will play an essential role, and it will be useful to know that these duals are bornological. In this section we shall prove some results from which the desired conclusion will follow (see also Exercise 15.9(c)).

Let E be a locally convex Hausdorff space, E' its dual, and E'' its bidual (Definition 8.1). *In this section we shall use the symbol* $^\circ$ *for polarity in the duality between the spaces E and E' and the symbol* $^\bullet$ *for polarity in the duality between the spaces E' and E''.* If A is a bounded set in E, then A° is a neighborhood of 0 in E' for the topology $\beta(E', E)$ and $A^{\circ\bullet}$ is an equicontinuous, balanced, convex, $\sigma(E'', E')$-closed, and in particular $\sigma(E'', E')$-bounded subset of E''. If we consider E canonically imbedded into E'', then it follows from the theorem of bipolars (Theorem 3.1) that $A^{\circ\bullet}$ is the $\sigma(E'', E')$-closure of A. Conversely, it is not always true that every bounded subset of E'' is contained in the $\sigma(E'', E')$-closure of a bounded subset of E ([52], §31,7, p. 438; [41], p. 88). Thus we are led to

Definition 1. *A locally convex Hausdorff space E is said to be distinguished if every $\sigma(E'', E')$-bounded subset of its bidual E'' is contained in the $\sigma(E'', E')$-closure of some bounded subset of E.*

In other words, E is distinguished if for every $\sigma(E'', E')$-bounded subset B of E'' there exists a bounded subset A of E such that $B \subset A^{\circ\bullet}$. Clearly, a semi-reflexive space is always distinguished. Therefore the following result generalizes Proposition 8.4.

Proposition 1. *A locally convex Hausdorff space E is distinguished if and only if E', equipped with the topology $\beta(E', E)$, is barrelled.*

Proof. (a) Suppose that E is distinguished and let B be a $\sigma(E'', E')$-bounded subset of E''. Then there exists a bounded subset A of E such that $B \subset A^{\circ\bullet}$. Since A° is a neighborhood of 0 in E' for $\beta(E', E)$, it follows from Proposition 4.6 that $A^{\bullet\circ}$, and hence also B, is equicontinuous in E''. Consequently (Proposition 6.2), E' is barrelled for the topology $\beta(E', E)$.

(b) Suppose now that E' is barrelled. Let B be a $\sigma(E'', E')$-bounded subset of E''. Then B is equicontinuous. Hence there exists a neighborhood V of 0 in E' such that $B \subset V^{\bullet}$ (Proposition 4.6). On the other hand, by the definition of the topology $\beta(E', E)$ there exists a bounded set A in E such that $A^{\circ} \subset V$. Thus $B \subset V^{\bullet} \subset A^{\circ\bullet}$; hence E is distinguished. ∎

In particular, every normed space E is distinguished since E' equipped with $\beta(E', E)$ is a Banach space and therefore barrelled. A Fréchet space is not, however, necessarily distinguished (*loc. cit.* before Definition 1).

THEOREM 1 (Grothendieck). *Let E be a metrizable locally convex space. Then E is distinguished if and only if E', equipped with the topology $\beta(E', E)$, is bornological.*

The proof will be preceded by two lemmas.

LEMMA 1. *Let E be a metrizable locally convex space and suppose that its dual E' equipped with the strong topology $\beta(E', E)$ is infrabarrelled. Let $(A_n)_{n\in\mathbf{N}}$ be a countable fundamental system of bounded, balanced, $\sigma(E', E)$-closed, convex subsets of E'. Let $(\lambda_n)_{n\in\mathbf{N}}$ be an arbitrary sequence of strictly positive numbers and let A be the balanced convex hull of $\bigcup_{n\in\mathbf{N}} \lambda_n A_n$. Then the $\beta(E', E)$-closure $\overline{A}$ of A coincides with its "algebraic closure" $\widetilde{A}$ defined by*

$$\widetilde{A} = \{y \mid y \in E', \lambda y \in A \text{ for } 0 < \lambda < 1\}.$$

Proof. It is clear that $\widetilde{A} \subset \overline{A}$ since λy converges to y as $\lambda \to 1$.

To establish the inclusion $\overline{A} \subset \widetilde{A}$ we shall prove $\mathbf{C}\widetilde{A} \subset \mathbf{C}\overline{A}$, and we do this by showing that for each $x \in \mathbf{C}\widetilde{A}$ there exists $x' \in A^{\bullet} \subset E''$ such that $|\langle x, x'\rangle| > 1$, since in this case $x \notin A^{\bullet\bullet} = \overline{A}$, i.e., $x \in \mathbf{C}\overline{A}$.

If $x \in \mathbf{C}\widetilde{A}$, then there exists some λ with $0 < \lambda < 1$ such that $\lambda x \in \mathbf{C}A$; i.e., there exists $\mu > 1$ such that $x \in \mathbf{C}\mu A$. We shall find $x' \in A^{\bullet}$ such that $\langle x, x'\rangle = \mu$.

Let B_n be the balanced convex hull of $\bigcup_{i=0}^{n} \lambda_i A_i$. Then $B_n \subset B_{n+1}$ for every $n \in \mathbf{N}$ and $A = \bigcup_{n\in\mathbf{N}} B_n$. The sets A_n are equicontinuous, hence $\sigma(E', E)$-compact (Theorem 4.1), and it follows from Proposition 5.2 that B_n is $\sigma(E', E)$-compact, hence $\sigma(E', E)$-closed, and *a fortiori* $\sigma(E', E'')$-closed. We have $x \in \mathbf{C}\mu B_n = \mathbf{C}\mu B_n^{\bullet\bullet}$, and therefore there exists $x'_n \in B_n^{\bullet}$ such that $\langle x, x'_n\rangle = \mu$.

The sequence $(x'_n)_{n\in\mathbf{N}}$ is bounded for the topology $\beta(E'', E')$. Indeed, the sets $\lambda_n B_n^{\bullet}$ form a fundamental system of neighborhoods of 0 for $\beta(E'', E')$. Since $B_n^{\bullet} \supset B_{n+1}^{\bullet}$ for $n \in \mathbf{N}$, given an index $n \in \mathbf{N}$, we have $x'_k \in B_n^{\bullet}$ for $k \geqq n$, and the finite set $\{x'_k \mid 0 \leqq k < n\}$ is certainly bounded.

Since E' is infrabarrelled, the set $\{x'_n \mid n \in \mathbf{N}\}$ is equicontinuous, hence relatively $\sigma(E'', E')$-compact by Theorem 4.1. It follows that the sequence (x'_n) has a value of adherence $x' \in E''$ for the topology $\sigma(E'', E')$ (Exercise 2.8.3(a)). Since the linear form $y' \mapsto \langle x, y' \rangle$ is continuous on E'' for $\sigma(E'', E')$, we have $\langle x, x' \rangle = \mu$. Furthermore,

$$x' \in \bigcap_{n \in \mathbf{N}} B_n^{\bullet} = \Big(\bigcup_{n \in \mathbf{N}} B_n\Big)^{\bullet} = A^{\bullet}. \blacksquare$$

LEMMA 2. *Let E be a metrizable locally convex space and suppose that its dual E', equipped with the strong topology $\beta(E', E)$, is infrabarrelled. Then every balanced, convex, bornivorous subset of E' contains a balanced, convex, bornivorous,* closed *subset of E'.*

Proof. Let T be a balanced, convex, bornivorous subset of E'. Then $\frac{1}{2}T$ has the same properties. Let $(A_n)_{n \in \mathbf{N}}$ be a countable fundamental system of bounded, balanced, convex, closed subsets of E'. For each $n \in \mathbf{N}$ there exists $\lambda_n > 0$ such that $\lambda_n A_n \subset \frac{1}{2}T$. Hence $\frac{1}{2}T$ contains the union $\bigcup_{n \in \mathbf{N}} \lambda_n A_n$ and also its balanced convex hull A. Now the "algebraic closure" of A is clearly contained in T, and therefore by the previous lemma we have $\overline{A} \subset T$. But $\overline{A}$ is a balanced, convex, bornivorous, closed set. ∎

Proof of Theorem 1. (a) Suppose that E is distinguished. Then by Proposition 1 the space E' is barrelled. Let T be a balanced, convex, bornivorous set in E'. By Lemma 2 the set T contains a bornivorous barrel S. Since E' is barrelled, S, and *a fortiori* T, is a neighborhood of 0; i.e., E' is bornological.

(b) Suppose that E' is bornological. Then E' is infrabarrelled. But E is bornological (Proposition 7.3), therefore E' is complete (Proposition 7.6) and thus also barrelled. Hence E is distinguished by Proposition 1. ∎

EXAMPLE 1. We know (Examples 3, 4, and 5 of §9) that $\mathcal{E}(\Omega)$, $\mathfrak{D}(K)$ and $\mathcal{S}$ are Montel spaces. Hence they are reflexive and more particularly distinguished. On the other hand, they are Fréchet spaces (Examples 2.9.3, 2.9.5, and 2.11.11); hence their duals equipped with the strong topology are bornological and barrelled.

THEOREM 2 (Grothendieck). *Let F be a vector space and let $(E_n)_{n \in \mathbf{N}}$ be a sequence of subspaces of F such that $E_n \subset E_{n+1}$ for all $n \in \mathbf{N}$ and $F = \bigcup_{n \in \mathbf{N}} E_n$. Suppose that each E_n is equipped with a locally convex topology $\mathfrak{T}_n$ and that $\mathfrak{T}_{n+1}$ induces $\mathfrak{T}_n$ on E_n.*

Assume that each space E_n is metrizable and distinguished. Equip F with the finest locally convex topology $\mathfrak{T}$ for which all the injections $E_n \hookrightarrow F$ are continuous.

Then F is distinguished and its dual F' equipped with the strong topology $\beta(F', F)$ is barrelled and bornological.

Proof. Suppose that we know already that F' is bornological. Then it is infrabarrelled. On the other hand, the spaces E_n are bornological since they are metrizable (Proposition 7.3), and therefore F is bornological (Proposition 7.4). Hence F' is complete for $\beta(F', F)$ (Proposition 7.6), and consequently it is barrelled. But then by Proposition 1 the space F is distinguished. Thus it suffices to show that F' is bornological.

We shall need the following lemma.

LEMMA 3. *Let F be a vector space and let $(E_n)_{n \in \mathbf{N}}$ be a sequence of subspaces of F such that $E_n \subset E_{n+1}$ for all $n \in \mathbf{N}$ and $F = \bigcup_{n \in \mathbf{N}} E_n$. Suppose that each E_n is equipped with a locally convex topology $\mathcal{T}_n$ and that $\mathcal{T}_{n+1}$ induces $\mathcal{T}_n$ on E_n.*

Equip F with the finest locally convex topology $\mathcal{T}$ for which all the injections $f_n: E_n \hookrightarrow F$ are continuous.

(a) *If S is a balanced, convex subset of F' which absorbs every equicontinuous subset of F', then there exists an index $m \in \mathbf{N}$ such that S contains the subspace $E_m^\perp$ of F' orthogonal to E_m* [or, more precisely, to $f_m(E_m)$].

(b) *If, furthermore, ${}^t f_m: F' \to E_m'$ is the transpose of f_m, then*

$$S_m = \{x' \mid x' \in E_m', {}^t f_m^{-1}(\{x'\}) \subset S\}$$

is a balanced, convex subset of E_m' which absorbs every equicontinuous subset of E_m'.

Let us assume the lemma for a moment and conclude the proof of the theorem. Let S be a balanced, convex, bornivorous subset of F'. Since every equicontinuous subset of F' is bounded for the topology $\beta(F', F)$ (Proposition 6.1), it follows that S absorbs every equicontinuous subset of F'. Hence the set S_m of Lemma 3 is balanced, convex and absorbs every equicontinuous subset of E_m'. But E_m is infrabarrelled (Proposition 7.3), hence every $\beta(E_m', E_m)$-bounded subset of E_m' is equicontinuous. Thus S_m is a balanced, convex, bornivorous set in E_m'. By Theorem 1 the space E_m' is bornological, and therefore S_m is a neighborhood of 0 in E_m'. By the corollary of Proposition 12.3 the map ${}^t f_m$ is continuous, and therefore ${}^t f_m^{-1}(S_m)$ is a neighborhood of 0 in F'. But $S \supset {}^t f_m^{-1}(S_m)$ and therefore S is also a neighborhood of 0 in F', which proves that F' is bornological. ∎

Proof of Lemma 3. (a) Suppose that for no $n \in \mathbf{N}$ is the subspace $E_n^\perp$ contained in S, and choose $x_n' \in E_n^\perp \cap \complement S$ for each n. Since $E_n^\perp \supset E_{n+1}^\perp$ for all $n \in \mathbf{N}$, we have $x_n' \in E_k^\perp$ for $n \geqq k$, i.e., $x_n' \circ f_k = 0$ for $n \geqq k$. In particular, for each $k \in \mathbf{N}$ the sequence $(n x_n' \circ f_k)_{n \in \mathbf{N}}$ is equicontinuous on E_k. It follows, similarly as in the proof of Proposition 2.12.1, that the sequence $(n x_n')$ is an equicontinuous subset of F'.

On the other hand, S does not absorb the sequence $(n x_n')$ since $n x_n' \in \lambda S$ would imply $x_n' \in S$ for $n \geqq \lambda$, which is impossible by the choice of x_n'.

Since S is assumed to absorb all equicontinuous subsets of F', we have arrived at a contradiction.

(b) Let T be the subset of S formed by those elements y' for which $y' + E_m^\perp \subset S$. By part (a) we have $0 \in T$. The set T is balanced and convex. Indeed, let $y' \in T$, $z' \in T$, $\alpha, \beta \in \mathbf{K}$, $0 \leqq \alpha \leqq 1$, and $\alpha + \beta = 1$. We have $y' + u' \in S$ and $z' + v' \in S$ for every $u', v' \in E_m^\perp$. But then

$$\alpha y' + \beta z' + w' = \alpha(y' + w') + \beta(z' + w') \in S$$

for every $w' \in E_m^\perp$. Hence $\alpha y' + \beta z' \in T$. Also, if $\lambda \in \mathbf{K}$, $0 < |\lambda| \leqq 1$, then

$$\lambda y' + w' = \lambda(y' + \lambda^{-1}w') \in S$$

for every $w' \in E_m^\perp$; hence $\lambda y' \in T$.

The set T absorbs every equicontinuous subset B of F'. Indeed, we have $B \subset \lambda S$ for some $\lambda > 0$. Since $E_m^\perp \subset \lambda S$, we have $B + E_m^\perp \subset 2\lambda S$, i.e., $B \subset 2\lambda T$.

Next we prove that $S_m = {}^t f_m(T)$ and in particular S_m is a balanced, convex subset of E'_m. Let us first recall that ${}^t f_m$ is the canonical surjection from F' onto $F'/E_m^\perp = E'_m$ (§13). Observe also that by the definition of S_m we have $x' \in S_m$ if and only if every $y' \in F'$ for which ${}^t f_m(y') = x'$ belongs to S. Consider now $x' \in S_m$ and let y' be such that ${}^t f_m(y') = x'$. Then for every $u' \in E_m^\perp$ we have

$${}^t f_m(y' + u') = {}^t f_m(y') + {}^t f_m(u') = x' + 0 = x',$$

hence $y' + u' \in S$ and $y' \in T$. Thus $S_m \subset {}^t f_m(T)$. Conversely, let $x' = {}^t f_m(z')$ for some $z' \in T$ and let $y' \in F'$ be such that ${}^t f_m(y') = x'$. Then ${}^t f_m(y' - z') = 0$; i.e., $y' = z' + u'$ for some $u' \in E_m^\perp$. But $z' + u' \in S$ by hypothesis, i.e., $y' \in S$ and thus $x' \in S_m$. Consequently, we also have ${}^t f_m(T) \subset S_m$.

Finally, we show that S_m absorbs every equicontinuous subset of E'_m. Let A be an equicontinuous subset of E'_m. By Theorem 2.12.1 the topology induced on E_m by $\mathcal{T}$ is precisely $\mathcal{T}_m$. Thus by Proposition 4.6 there exists a balanced, closed, convex neighborhood V of 0 in F such that $A \subset (V \cap E_m)^\circ$. Let q be the gauge of V. If $x' \in A$, then $|\langle x', x\rangle| \leqq q(x)$ for every $x \in E_m$, and by the Hahn-Banach theorem (Theorem 1.1) there exists an element $y' \in F'$ such that $|\langle y', x\rangle| \leqq q(x)$ for all $x \in F$, i.e., $y' \in V^\circ$, and $\langle y', f_m(x)\rangle = \langle x', x\rangle$ for all $x \in E_m$, i.e., $x' = {}^t f_m(y')$. The set $B = V^\circ$ is equicontinuous in F' and $A \subset {}^t f_m(B)$ by what we have just shown. There exists $\lambda > 0$ such that $B \subset \lambda T$; hence

$$A \subset {}^t f_m(B) \subset {}^t f_m(\lambda T) = \lambda S_m. \blacksquare$$

EXAMPLE 2. The dual of the space $\mathfrak{D}(\Omega)$ (Example 2.12.6) equipped with the strong topology is barrelled and bornological since the $\mathfrak{D}(K)$ are distinguished Fréchet spaces by Example 1. Similarly, for finite $m \in \mathbf{N}$ the duals of the spaces $\mathfrak{D}^m(\Omega)$ (Example 2.12.8) equipped with the strong topology are barrelled and bornological since the spaces $\mathfrak{D}^m(K)$ are normable (Example 2.6.1).

Exercises

1. (a) Let E be a metrizable locally convex space and E' its dual equipped with the strong topology $\beta(E', E)$. Let $(M_n)_{n \geqq 1}$ be a sequence of equicontinuous subsets of the bidual E''. Show that if $M = \bigcup_{n \geqq 1} M_n$ is bounded for the topology $\sigma(E'', E')$, then it is equicontinuous.

REMARK 1. The conclusion would be immediate if we knew that E' is infrabarrelled. However, if E' were infrabarrelled, then it would be barrelled since it is complete (Proposition 7.6), and thus E would be distinguished by Proposition 1. But as we mentioned earlier, there even exist Fréchet spaces which are not distinguished. The property expressed above serves as a substitute for the infrabarrelledness of E' and has been used by Grothendieck to define his $(\mathfrak{DF})$ spaces ([41], p. 63; [52], §29,3, p. 399).

(*Hint:* (α) Observe that a fundamental system of neighborhoods of 0 for $\beta(E', E)$ in E' is given by all absorbing, balanced, convex, $\sigma(E', E)$-closed sets since these are the polars of the bounded subsets of E (Proposition 3.1).

(β) Set $U_n = M_n^\bullet$ and $U = \bigcap_{n \geqq 1} U_n = (\bigcup_{n \geqq 1} M_n)^\bullet$. Each U_n is a balanced, convex, $\beta(E', E)$-closed neighborhood of 0 in E' (Proposition 4.6), and U is an absorbing, balanced, convex, $\beta(E', E)$-closed subset of E' (Proposition 5.7). Show that it suffices to find an absorbing, balanced, convex, $\sigma(E', E)$-closed set $V \subset U$.

(γ) Let $(A_n)_{n \geqq 1}$ be a countable fundamental system of $\beta(E', E)$-bounded sets in E' obtained by taking the polars of a fundamental system of neighborhoods of 0 in E. The sets A_n are balanced, convex, and $\sigma(E', E)$-compact (Theorem 4.1). Suppose that a sequence (λ_n) of positive numbers and a sequence (V_n) of balanced, convex, $\sigma(E', E)$-closed $\beta(E', E)$-neighborhoods of 0 in E' has been constructed such that

(i) $\lambda_i A_i \subset (1/2^{i+1}) U$ for all i,

(ii) $\lambda_i A_i \subset V_j$ for all i and j,

(iii) $V_i \subset U_i$ for all i.

Show that the balanced, convex, $\sigma(E', E)$-closed set $V = \bigcap_{i \geqq 1} V_i$ is absorbing and $V \subset U$.

(δ) Construct the sequences (λ_i) and (V_i) by induction. Suppose that λ_i and V_i satisfy (i) through (iii) for $1 \leqq i \leqq n$. By Theorem 5.2 there exists $\lambda_{n+1} > 0$ such that

$$\lambda_{n+1} A_{n+1} \subset \frac{1}{2^{n+2}} U \cap \left(\bigcap_{i=1}^{n} V_i \right), \tag{1}$$

since A_{n+1} is $\beta(E', E)$-complete (Proposition 7.6) and the set on the right-hand side is a $\beta(E', E)$-barrel. $B_{n+1} = \lambda_1 A_1 + \cdots + \lambda_{n+1} A_{n+1}$ is a balanced, convex $\sigma(E', E)$-compact set (see the proof of Proposition 5.2) and

$$B_{n+1} \subset \tfrac{1}{4}U + \cdots + (1/2^{n+2})U \subset \tfrac{1}{2}U \subset \tfrac{1}{2}U_{n+1}.$$

If W_{n+1} is a balanced, convex, $\sigma(E', E)$-closed $\beta(E', E)$-neighborhood of 0 contained in $\frac{1}{2}U_{n+1}$, then the balanced, convex set $V_{n+1} = B_{n+1} + W_{n+1}$ is a $\beta(E', E)$-neighborhood of 0 since $V_{n+1} \supset W_{n+1}$, and it is $\sigma(E', E)$-closed by Proposition 2.10.5. Finally, $\lambda_{n+1}A_{n+1} \subset V_j$ for $1 \leqq j \leqq n$ by (1) and $\lambda_i A_i \subset B_{n+1} \subset V_{n+1}$ for $1 \leqq i \leqq n + 1$.)

(b) Prove that Lemmas 1 and 2 hold without the assumption that E' is infrabarrelled for the topology $\beta(E', E)$.

2. Let E be a metrizable locally convex space, E' its dual, and E'' its bidual. Prove that $\beta(E', E'')$ is the bornological topology $\mathfrak{T}'$ associated with $\beta(E', E)$ (see Exercise 7.8(a)). (*Hint:* A fundamental system $\mathfrak{V}$ of neighborhoods of 0 for $\beta(E', E'')$ is given by all absorbing, balanced, convex, $\beta(E', E)$-closed sets, i.e., $\beta(E', E)$-barrels. A fundamental system $\mathfrak{W}$ of neighborhoods of 0 for $\mathfrak{T}'$ is given by all $\beta(E', E)$-bornivorous, balanced, convex sets. By Lemma 2 and Exercise 1(b) each $T \in \mathfrak{W}$ contains an $S \in \mathfrak{V}$; i.e., $\beta(E', E'')$ is finer than $\mathfrak{T}'$. On the other hand, E' is complete for $\beta(E', E)$. Thus every barrel is bornivorous (Theorem 5.2); hence $\mathfrak{V} \subset \mathfrak{W}$, i.e., $\mathfrak{T}'$ is finer than $\beta(E', E'')$.)

§17. *The homomorphism theorem and the closed-graph theorem*

The three main results of the first chapter were the Hahn-Banach theorem, the Banach-Steinhaus theorem and Banach's homomorphism theorem (or the related closed-graph theorem). In the preceding sections we extended the validity of the first two of these theorems. We saw that the Hahn-Banach theorem holds for all locally convex spaces (Theorem 1.1 and Proposition 1.1) and that the Banach-Steinhaus theorem holds if and only if the space is barrelled (Example 6.2, Definition 6.1, and Proposition 6.2). No such simple characterization of the spaces for which the Banach homomorphism theorem holds can be expected, and we shall try to find some large classes of spaces for which it does hold.

To begin with, let us show that the proof of Theorem 1.9.1 carries over to metrizable complete spaces which are not even necessarily locally convex.

THEOREM 1 (Banach's homomorphism theorem or the open-mapping theorem). *Let E and F be two metrizable complete topological vector spaces and f a continuous surjective linear map from E onto F. Then f is a strict morphism* (Definition 2.5.2).

We shall need two lemmas.

LEMMA 1. *Let E and F be two topological vector spaces and let f be a continuous linear map from E onto F. If F is a Baire space, then for every neighborhood U of 0 in E the set $\overline{f(U)}$ is a neighborhood of 0 in F.*

Proof. Let V be a balanced neighborhood of 0 in E such that $V + V \subset U$ (Theorem 2.3.1). Since V is absorbing, we have $E = \bigcup_{n \geqq 1} nV$ and therefore $F = \bigcup_{n \geqq 1} nf(V)$ since f is surjective. Now F is a Baire space; hence one of the sets $\overline{nf(V)} = n\overline{f(V)}$ has an interior point. It follows that $\overline{f(V)}$ also has an interior point x.

Since V is balanced, so is $f(V)$ and also $\overline{f(V)}$. Thus $-\overline{f(V)} = \overline{f(V)}$, and therefore $0 = x + (-x)$ is an interior point of $\overline{f(V)} + \overline{f(V)}$. Now the map $(y, z) \mapsto y + z$ from $F \times F$ into F is continuous, and therefore $\overline{f(V)} + \overline{f(V)} \subset \overline{f(V) + f(V)}$. But $f(V) + f(V) = f(V + V) \subset f(U)$, and therefore $\overline{f(V)} + \overline{f(V)} \subset \overline{f(U)}$. Thus 0 is an interior point of $\overline{f(U)}$; i.e., $\overline{f(U)}$ is a neighborhood of 0. ∎

REMARK 1. The proof just given is strikingly similar to the proof of the fact that a locally convex Baire space is barrelled (Proposition 6.3). The reason for this similarity should become clear soon (Proposition 1).

LEMMA 2. *Let E and F be two metric spaces and assume that E is complete. Denote the metric on either space by δ. Suppose that f is a continuous map from E into F which has the following property: for every $r > 0$ there exists $\rho > 0$ such that for every $x \in E$ the image $f(B_r(x))$ of the ball*

$$B_r(x) = \{z \mid z \in E,\ \delta(x, z) \leqq r\}$$

is dense in the ball

$$B_\rho(f(x)) = \{y \mid y \in F,\ \delta(y, f(x)) \leqq \rho\}.$$

Then for every $a > r$ the set $f(B_a(x))$ contains $B_\rho(f(x))$.

Proof. Let $(r_n)_{n \geqq 1}$ be a sequence of strictly positive numbers such that $r_1 = r$ and $a = \sum_{n=1}^{\infty} r_n$. For every index n there exists a strictly positive number ρ_n (with $\rho_1 = \rho$) such that $f(B_{r_n}(x))$ is dense in $B_{\rho_n}(f(x))$, and we may suppose that $\lim_{n \to \infty} \rho_n = 0$.

Let x_0 be a point of E and let y be a point of $B_\rho(f(x_0))$. We want to show that $y = f(x)$ for some $x \in B_a(x_0)$, and we shall obtain x as the limit of a sequence $(x_n)_{n \in \mathbf{N}}$ whose first term is x_0.

We construct inductively the sequence $(x_n)_{n \in \mathbf{N}}$ so that

$$x_n \in B_{r_n}(x_{n-1}) \qquad \text{and} \qquad f(x_n) \in B_{\rho_{n+1}}(y). \tag{1}$$

For $n = 0$ the first condition is vacuous, and the second is satisfied since $f(x_0) \in B_\rho(y)$ is equivalent to $y \in B_\rho(f(x_0))$. Suppose that we have already constructed the points x_i with $0 \leqq i \leqq n - 1$ which satisfy these conditions. Then we have $y \in B_{\rho_n}(f(x_{n-1}))$. Since the image of $B_{r_n}(x_{n-1})$ is dense in $B_{\rho_n}(f(x_{n-1}))$, there exists a point $x_n \in B_{r_n}(x_{n-1})$ such that $f(x_n) \in B_{\rho_{n+1}}(y)$. This proves the existence of a sequence $(x_n)_{n \in \mathbf{N}}$ satisfying (1).

The sequence (x_n) is a Cauchy sequence since

$$\delta(x_n, x_{n+p}) \leqq r_{n+1} + \cdots + r_{n+p},$$

and this sum is arbitrarily small for large n. Since E is complete, (x_n) converges to some point x, and $\delta(x, x_0) \leqq \sum_{n=1}^{\infty} r_n = a$; i.e., $x \in B_a(x_0)$. Also f is continuous; hence $f(x_n)$ converges to $f(x)$. Now $f(x_n) \in B_{\rho_{n+1}}(y)$ for all n; thus $f(x) = y$. ∎

Proof of Theorem 1. Let us equip each of the two metrizable spaces E and F with a translation-invariant metric compatible with their topologies (Theorem 2.6.1). Since F is a Baire space, it follows from Lemma 1 that for every $r > 0$ the set $\overline{f(B_r(0))}$ is a neighborhood of 0 in F. Hence there exists $\rho > 0$ such that $f(B_r(0))$ is dense in $B_\rho(0) \subset F$. By translation we see that for every $x \in E$ the set $f(B_r(x))$ is dense in $B_\rho(f(x))$. Since E is complete, it follows from Lemma 2 that for $a > r$ the set $f(B_a(0))$ is a neighborhood of 0 in F. Since $r > 0$ is arbitrary, Theorem 2.5.1 implies that f is a strict morphism. ∎

We shall now generalize Theorem 1, at least for the case where E and F are locally convex. Lemma 1 motivates the following:

DEFINITION 1 (Pták). *Let E and F be two topological vector spaces and f a linear map from E into F. We say that f is almost* (nearly) *open if for every neighborhood U of* 0 *in E the closure $\overline{f(U)}$ of $f(U)$ is a neighborhood of* 0 *in the subspace $f(E)$ of F.*

The following result generalizes Lemma 1 for the case where the spaces involved are locally convex.

PROPOSITION 1. *Let E be a locally convex space and F a barrelled space. Then every surjective linear map $f\colon E \to F$ is almost open.*

Proof. Let U be a balanced, closed, convex neighborhood of 0 in E. Then $\overline{f(U)}$ is a barrel in F. Indeed, $f(U)$ is balanced and convex; hence so is $\overline{f(U)}$ (Exercise 2.3.5 and Proposition 2.4.3). Clearly, $\overline{f(U)}$ is closed. Finally, $f(U)$ is absorbing since f is surjective and U is absorbing.

It follows from the definition of a barrelled space (Definition 6.1) that $\overline{f(U)}$ is a neighborhood of 0 in F. ∎

Lemma 2 will be replaced by the following result.

THEOREM 2 (Pták). *Let E be a locally convex Hausdorff space and E' its dual. The following two conditions are equivalent:*

(a) *Let F be an arbitrary locally convex Hausdorff space. Every continuous, almost open linear map $f\colon E \to F$ is a strict morphism.*

(b) *If L is a linear subspace of E' such that for every balanced, convex, $\sigma(E', E)$-closed, equicontinuous subset M of E' the intersection $L \cap M$ is $\sigma(E', E)$-closed, then L is $\sigma(E', E)$-closed.*

Proof. (a) ⇒ (b): Suppose that condition (a) is satisfied and let L be a subspace of E' such that for every balanced, convex, $\sigma(E', E)$-closed, equicontinuous subset M of E' the intersection $L \cap M$ is $\sigma(E', E)$-closed. By Proposition 4.6 for each such set M there exists a balanced, closed, convex neighborhood U of 0 in E such that $M = U^{\circ}$.

Set $F = E/L^{\perp}$. The canonical surjection $f: E \to F$ is the transpose of the injection $j: L \hookrightarrow E'$, and the spaces F and L form a dual system (Proposition 13.1). Let us denote by the symbol $^{\circ}$ the polarity for the dual pair (E, E') and by $^{\bullet}$ the polarity for the pair (F, L). We equip F with the locally convex Hausdorff topology $\mathcal{T}_1$ which has as a fundamental system of neighborhoods of 0 the sets

$$(U^{\circ} \cap L)^{\bullet} = f((U^{\circ} \cap L)^{\circ}),$$

where U runs through the balanced, closed, convex neighborhoods of 0 in E.

The dual of F is then L. Indeed, let x' be a linear form on F which is continuous for $\mathcal{T}_1$. Then there exists a balanced, closed, convex neighborhood U of 0 in E such that $x \in (L \cap U^{\circ})^{\circ}$ implies $|\langle x, x' \rangle| \leqq 1$, i.e., $x' \in (L \cap U^{\circ})^{\circ\circ}$. By our assumption concerning L, the set $L \cap U^{\circ}$ is $\sigma(E', E)$-closed, i.e., $(L \cap U^{\circ})^{\circ\circ} = L \cap U^{\circ}$, and therefore

$$x' \in L \cap U^{\circ} \subset L.$$

Conversely, if $x' \in L$, then $x' \in L \cap U^{\circ}$ for some U. But then

$$x \in (L \cap (\epsilon U)^{\circ})^{\circ}$$

implies $|\langle x, x' \rangle| \leqq \epsilon$; hence x' is a continuous linear form on F.

The map $f: E \to F$ is continuous since

$$f^{-1}(f((U^{\circ} \cap L)^{\circ})) \supset (U^{\circ} \cap L)^{\circ} \supset U.$$

Furthermore, f is almost open since

$$f(U)^{\bullet} = U^{\circ} \cap L \qquad \text{and} \qquad \overline{f(U)} = f(U)^{\bullet\bullet} = (U^{\circ} \cap L)^{\bullet}$$

for each balanced, closed, convex neighborhood U of 0 in E.

By condition (a) the map f is a strict morphism; i.e., $\mathcal{T}_1$ coincides with the quotient topology $\mathcal{T}$ of E modulo $L^{\perp}$. Now, the dual of F for the topology $\mathcal{T}$ is the subspace $L^{\perp\perp}$ of E' by the corollary of Proposition 13.2. Thus $L = L^{\perp\perp}$, and therefore L is $\sigma(E', E)$-closed.

(b) ⇒ (a): We first prove

LEMMA 3. *Let E and F be two locally convex Hausdorff spaces and E' and F' their duals. Let $f: E \to F$ be a surjective, continuous, almost open linear map and set $L = {}^tf(F')$. Then for every balanced, closed, convex neighborhood U of 0 in E the set $L \cap U^{\circ}$ is $\sigma(E', E)$-closed.*

Proof of Lemma 3. For any balanced, closed, convex neighborhood U of 0 in E we have

$$L \cap U^{\circ} = {}^t f(f(U)^{\circ}). \tag{1}$$

Indeed, if $x' \in L \cap U^{\circ}$, then $x' = {}^t f(y')$ for some $y' \in F'$ and

$$|\langle x', x\rangle| \leqq 1$$

for all $x \in U$. It follows that

$$|\langle x', x\rangle| = |\langle {}^t f(y'), x\rangle| = |\langle y', f(x)\rangle| \leqq 1 \tag{2}$$

for all $x \in U$, i.e., $y' \in f(U)^{\circ}$, and therefore $x' \in {}^t f(f(U)^{\circ})$. Conversely, if $x' \in {}^t f(f(U)^{\circ})$, then clearly $x' \in L$, and setting $x' = {}^t f(y')$ with $y' \in f(U)^{\circ}$, we obtain from (2) that $x' \in U^{\circ}$.

Since $f(U)^{\circ} = (\overline{f(U)})^{\circ}$, and $\overline{f(U)}$ is by assumption a neighborhood of 0 in F, it follows from Theorem 4.1 that $f(U)^{\circ}$ is $\sigma(F', F)$-compact. By the corollary of Proposition 12.3 the map ${}^t f$ is continuous for the topologies $\sigma(F', F)$ and $\sigma(E', E)$, and therefore ${}^t f(f(U)^{\circ})$ is $\sigma(E', E)$-compact and hence also $\sigma(E', E)$-closed in F'. The conclusion now follows from (1). ∎

Let us now suppose that condition (b) is satisfied, let F be a vector space equipped with a locally convex Hausdorff topology $\mathcal{T}_1$ and let $f: E \to F$ be a continuous, almost open linear map. We may assume that f is *surjective*. It follows then from Lemma 3 and condition (b) that $L = {}^t f(F')$ is a $\sigma(E', E)$-closed subspace of E'.

Denote by $\mathcal{T}_2$ the locally convex Hausdorff topology on F for which a fundamental system of neighborhoods of 0 is given by the sets $f(U)$, where U is a balanced, closed, convex neighborhood of 0 in E. Since f is continuous, $\mathcal{T}_2$ is finer than $\mathcal{T}_1$.

Let us denote by $(F_2)'$ the dual of the vector space F equipped with $\mathcal{T}_2$. Clearly $F' \subset (F_2)'$, and we shall show that $F' = (F_2)'$. Let $u \in (F_2)'$. There exists a balanced, closed, convex neighborhood U of 0 such that $|u(y)| \leqq 1$ for all $y \in f(U)$; i.e., $|u(f(x))| \leqq 1$ for all $x \in U$. Thus the map $x \mapsto u(f(x))$ is continuous on E, and therefore there exists $x' \in E'$ such that

$$\langle x, x'\rangle = u(f(x)) \qquad \text{for all} \qquad x \in E. \tag{3}$$

If $x' \in L$, then we have $u(f(x)) = \langle x, {}^t f(y')\rangle = \langle f(x), y'\rangle$ for some $y' \in F'$ and all $x \in E$; i.e., $u = y' \in F'$. Thus to prove our assertion we must show that $x' \in L$. Suppose that $x' \notin L$. Since L is a $\sigma(E', E)$-closed subspace of E', by Proposition 1.2 there exists $x_0 \in E$ such that $\langle x_0, z'\rangle = 0$ for all $z' \in L$ and $\langle x_0, x'\rangle = 1$. It follows that

$$\langle x_0, {}^t f(y')\rangle = \langle f(x_0), y'\rangle = 0$$

for all $y' \in F'$ so that $f(x_0) = 0$. On the other hand, by (3) we have $u(f(x_0)) = \langle x_0, x'\rangle = 1$.

Take now a balanced, convex, $\mathcal{T}_2$-closed $\mathcal{T}_2$-neighborhood V of 0 in F. Then V contains a set $f(U)$, where U is a balanced, closed, convex neighborhood of 0 in E. But V is also $\mathcal{T}_1$-closed since $F' = (F_2)'$, and therefore V also contains the $\mathcal{T}_1$-closure $\overline{f(U)}$ of $f(U)$. By our assumption $\overline{f(U)}$ is a $\mathcal{T}_1$-neighborhood of 0 in F and therefore $\mathcal{T}_1 = \mathcal{T}_2$. Thus for every balanced, closed, convex neighborhood U of 0 in E the set $f(U)$ is a neighborhood of 0 in F for $\mathcal{T}_1$, which by Theorem 2.5.1 implies that f is a strict morphism. ∎

DEFINITION 2. *A locally convex Hausdorff space which satisfies the two equivalent conditions of Theorem 2 is called a Pták space* (fully complete, B-complete).

Combining the last two results, we obtain:

PROPOSITION 2. *Let E be a Pták space, F a barrelled Hausdorff space, and f a continuous surjective linear map from E onto F. Then f is a strict morphism.*

Proof. By Proposition 1 the map f is almost open, and so by Theorem 2 it is a strict morphism. ∎

Part (a) of the following result shows that Proposition 2 indeed generalizes Theorem 1, at least for locally convex spaces.

PROPOSITION 3. (a) *Every Fréchet space is a Pták space.*
(b) *Every Pták space is complete.*

Proof. (a) This follows immediately from Theorem 10.2 since a linear subspace of E' is in particular a convex set.
(b) This follows immediately from Corollary 5 of Theorem 11.1. ∎

PROPOSITION 4 (Collins). *A closed linear subspace of a Pták space is a Pták space.*

Proof. Let E be a Pták space, M a closed subspace of E, and $j: M \hookrightarrow E$ the canonical injection. By the corollary of Proposition 13.1 the dual M' of M can be identified with the quotient space $E'/M^{\perp}$ and

$$\psi = {}^t j: E' \to E'/M^{\perp}$$

is the canonical surjection.

Let L be a linear subspace of $E'/M^{\perp}$ such that for every balanced, closed, convex neighborhood U of 0 in M the set $U^{\circ} \cap L$ is $\sigma(M', M)$-closed in $E'/M^{\perp}$. If V is a balanced, closed, convex neighborhood of 0 in E, then $j^{-1}(V) = V \cap M = U$ is a balanced, closed, convex neighborhood of 0 in M, and therefore $U^{\circ} \cap L$ is $\sigma(M', M)$-closed. Now,

$$\psi^{-1}(U^{\circ} \cap L) = \psi^{-1}(U^{\circ}) \cap \psi^{-1}(L),$$

and by Proposition 12.2(a) we have $\psi^{-1}(U^{\circ}) = j(U)^{\circ}$. Hence

$$j(U)^{\circ} \cap \psi^{-1}(L)$$

is $\sigma(E', E)$-closed in E'. But $j(U) \subset V$; hence $j(U)^{\circ} \supset V^{\circ}$, and therefore $V^{\circ} \cap \psi^{-1}(L) = V^{\circ} \cap j(U)^{\circ} \cap \psi^{-1}(L)$, which shows that $V^{\circ} \cap \psi^{-1}(L)$ is $\sigma(E', E)$-closed. Since E is a Pták space, it follows that $\psi^{-1}(L)$ is a $\sigma(E', E)$-closed subspace of E'. The topology $\sigma(M', M)$ coincides with the quotient of the topology $\sigma(E', E)$ modulo $M^{\perp}$ (Exercise 15.1), and so L is a $\sigma(M', M)$-closed subspace of M'. Thus M is a Pták space. ∎

Proposition 5. *Let E be a Pták space and M a closed subspace of E. Then the quotient space E/M is a Pták space.*

Proof. Let F be a locally convex Hausdorff space and $f: E/M \to F$ a continuous, almost open linear map. Let $\varphi: E \to E/M$ be the canonical surjection and set $g = f \circ \varphi$. Then $g: E \to F$ is a continuous linear map. It is also almost open since if U is a neighborhood of 0 in E, then $\varphi(U)$ is a neighborhood of 0 in E/M, and therefore $\overline{g(U)} = \overline{f(\varphi(U))}$ is a neighborhood of 0 in $g(E)$. Since E is a Pták space, g is a strict morphism.

Let V be a neighborhood of 0 in E/M. Then $U = \varphi^{-1}(V)$ is a neighborhood of 0 in E, and thus $f(V) = g(U)$ is a neighborhood of 0 in $f(E/M) = g(E)$. This means (Theorem 2.5.1) that f is a strict morphism and proves that E/M is a Pták space. ∎

Proposition 6. *Let E be a reflexive Fréchet space. Then its dual E' equipped with the strong topology $\beta(E', E)$ is a Pták space.*

Proof. The polars of the balanced, convex, closed $\beta(E', E)$-neighborhoods of 0 in E' are the balanced, convex, closed, bounded subsets B of E. Let L be a subspace of E and suppose that for every such set B the intersection $B \cap L$ is closed. If x adheres to L, then there exists a sequence (x_n) of elements of L which converges to x. But (x_n) is contained in a balanced, convex, closed, bounded set, and thus $x \in L$; i.e., L is closed and hence also $\sigma(E, E')$-closed (Proposition 4.3). ∎

In Chapter 1 we deduced the closed-graph theorem (Proposition 1.9.3) for Banach spaces from the homomorphism theorem. Let us do the same now in the more general context of Theorem 1. First, let us point out, however, that the graph of any continuous map is closed.

Let X be a topological space, Y a Hausdorff space, and $f: X \to Y$ a continuous map. Then the graph $G = G(f) = \{(x, f(x)) \mid x \in X\}$ is a closed subset of the product space $X \times Y$. Indeed, suppose that the point $(x, y) \in X \times Y$ adheres to G but $y \neq f(x)$. There exists a neighborhood V of y and a neighborhood W of $f(x)$ such that $V \cap W = \emptyset$. Since f is continuous, there exists a neighborhood U of x such that $z \in U$ implies $f(z) \in W$. But then the neighborhood $U \times V$ of (x, y) does not meet G, in contradiction to $(x, y) \in \overline{G}$.

THEOREM 3 (Closed-graph theorem). *Let E and F be two metrizable, complete topological vector spaces, $f: E \to F$ a linear map, and let us assume that the graph G of f is closed in the product space $E \times F$. Then f is continuous.*

Proof. By our assumption G is a closed linear subspace of $E \times F$, and therefore it is both metrizable and complete (Proposition 2.9.3). The projection $\pi_1: (x, y) \mapsto x$ is a continuous, bijective linear map from G onto E, and therefore by Theorem 1 it is an isomorphism. If $\pi_2: (x, y) \mapsto y$ is the projection from G into F, then $f = \pi_2 \circ \pi_1^{-1}$, whence f is continuous. ∎

Similarly as in §9 of Chapter 1, the theorem can be rephrased in the following way:

COROLLARY. *Let E and F be two metrizable, complete topological vector spaces and $f: E \to F$ a linear map. Suppose that for every sequence (x_n) of points of E which tends to 0 and for which $(f(x_n))$ tends to some vector $y \in F$, we have necessarily $y = 0$. Then f is continuous.*

Proof. Suppose that the point (x, z) adheres to the graph G of f. Then there exists a sequence (x_n) of points of E such that x_n tends to x and $f(x_n)$ tends to z. But then $x_n - x$ tends to 0, and $f(x_n - x) = f(x_n) - f(x)$ tends to $z - f(x)$. By our assumption $z = f(x)$, and thus $(x, z) \in G$. This means that G is closed, and thus by Theorem 3 the map f is continuous. ∎

We now generalize the closed-graph theorem, at least for locally convex spaces, in the same spirit as we have done for the homomorphism theorem.

THEOREM 4 (Robertson-Robertson). *Let E be a barrelled Hausdorff space, F a Pták space, $f: E \to F$ a linear map, and suppose that the graph G of f is closed in the product space $E \times F$. Then f is continuous.*

The proof of this theorem will follow after some preparation.

DEFINITION 3 (Pták). *Let E and F be two topological vector spaces and f a linear map from E into F. We say that f is almost* (nearly) *continuous if for every neighborhood V of 0 in F the closure $\overline{f^{-1}(V)}$ of $f^{-1}(V)$ is a neighborhood of 0 in E.*

PROPOSITION 7. *Let E be a barrelled space and F a locally convex space. Then every linear map $f: E \to F$ is almost continuous.*

Proof. Let V be a balanced, convex neighborhood of 0 in F. Then $f^{-1}(V)$ is a balanced, convex set, and $\overline{f^{-1}(V)}$ a balanced, convex, closed set in E (Exercise 2.3.5 and Proposition 2.4.3). But $f^{-1}(V)$ is also absorbing since V is.

Thus $\overline{f^{-1}(V)}$ is a barrel in E, and therefore by the definition of a barrelled space (Definition 6.1) it is a neighborhood of 0. ∎

LEMMA 4. *Let E and F be two locally convex Hausdorff spaces and $f: E \to F$ a linear map. The graph G of f is closed in $E \times F$ if and only if the subspace Q of F', consisting of those $y' \in F'$ for which the map*

$$x \mapsto \langle f(x), y' \rangle$$

is continuous on E, is dense in F' for $\sigma(F', F)$.

Proof. Let $z_0 \in F$ be such that $\langle z_0, y' \rangle = 0$ for all $y' \in Q$. We show that $(0, z_0) \in \overline{G}$. Indeed, if $(0, z_0) \notin \overline{G}$, then by Proposition 1.2 there exists a point $(x', z') \in E' \times F'$ (cf. Proposition 14.1) such that

$$\langle (x, f(x)), (x', z') \rangle = \langle x, x' \rangle + \langle f(x), z' \rangle = 0 \tag{4}$$

for all $x \in E$ and

$$\langle (0, z_0), (x', z') \rangle = \langle z_0, z' \rangle = 1. \tag{5}$$

By (4) the map $x \mapsto \langle f(x), z' \rangle$ is continuous; i.e., $z' \in Q$. But then $\langle z_0, z' \rangle = 0$ in contradiction to (5).

Suppose now that $G = \overline{G}$. Then $(0, z_0) \in G$; i.e., $z_0 = f(0) = 0$. Thus $Q^\perp = \{0\}$, and therefore (Proposition 3.3(e)) Q is dense in F' for $\sigma(F', F)$.

Conversely, suppose that Q is dense in F' for $\sigma(F', F)$. Let (x_0, y_0) be a point of $E \times F$ which does not belong to G. We want to show that (x_0, y_0) does not belong to $\overline{G}$ either. We have $y_0 - f(x_0) \neq 0$. There exists $y' \in Q$ such that $\langle y_0 - f(x_0), y' \rangle \neq 0$, and by the definition of Q there exists $x' \in E'$ such that $\langle x, x' \rangle + \langle f(x), y' \rangle = 0$ for all $x \in E$. We have thus $\langle (x, f(x)), (x', y') \rangle = 0$ for all $x \in E$ and also

$$\langle (x, y), (x', y') \rangle = 0$$

for all $(x, y) \in \overline{G}$, but

$$\begin{aligned} \langle (x_0, y_0), (x', y') \rangle &= \langle x_0, x' \rangle + \langle y_0, y' \rangle \\ &= \langle x_0, x' \rangle + \langle f(x_0), y' \rangle + \langle y_0 - f(x_0), y' \rangle \\ &= \langle y_0 - f(x_0), y' \rangle \neq 0, \end{aligned}$$

which proves that $(x_0, y_0) \notin \overline{G}$. ∎

PROPOSITION 8 (Pták). *Let E be a locally convex Hausdorff space, F a Pták space, and $f: E \to F$ an almost continuous linear map. Suppose that the graph G of f is closed in $E \times F$. Then f is continuous.*

Proof. It is sufficient to show that f is continuous for the topology $\sigma(F, F')$ on F. Indeed, suppose this has been proved already, and let V be a balanced, convex, closed neighborhood of 0 in F. Then V is $\sigma(F, F')$-closed (Proposition 4.3), and so $f^{-1}(V)$ is closed in E; i.e.,

$$f^{-1}(V) = \overline{f^{-1}(V)}.$$

By our assumption $\overline{f^{-1}(V)}$ is a neighborhood of 0 in E; hence f is continuous.

Let Q be the subspace of F' consisting of those $y' \in F'$ for which the map $x \mapsto \langle f(x), y' \rangle$ is continuous on E. We have to show that for every balanced, closed, convex neighborhood V of 0 in F the set $Q \cap V^{\circ}$ is $\sigma(F', F)$-closed in F'. It will then follow that Q is $\sigma(F', F)$-closed in F', since F is a Pták space, and therefore that $Q = F'$ since, by Lemma 4 and by the hypothesis that G is closed, Q is dense in F'. But then $x \mapsto \langle f(x), y' \rangle$ is continuous for every $y' \in F'$; i.e., f is continuous for $\sigma(F, F')$.

Let us observe that if U is a balanced neighborhood of 0 in E, then $f(U)^{\circ} \subset Q$. Indeed, if $y' \in f(U)^{\circ}$, then $|\langle f(x), y' \rangle| \leqq 1$ for all $x \in U$; i.e., $|\langle f(x), y' \rangle| \leqq \epsilon$ for $x \in \epsilon U$.

Now let V be a balanced, convex, closed neighborhood of 0 in F. Then $U = \overline{f^{-1}(V)}$ is, by assumption, a balanced, convex neighborhood of 0 in E. By the remark just made we have $f(U)^{\circ} \subset Q$. On the other hand, $Q \cap V^{\circ} \subset f(U)^{\circ}$. Indeed, if $y' \in Q \cap V^{\circ}$, then the function

$$x \mapsto \langle f(x), y' \rangle$$

is continuous on E, and $|\langle y, y' \rangle| \leqq 1$ for all $y \in V$. Hence if $x \in f^{-1}(V)$, we have $|\langle f(x), y' \rangle| \leqq 1$, and therefore by continuity we also have

$$|\langle f(x), y' \rangle| \leqq 1$$

for $x \in U$. Therefore $y' \in f(U)^{\circ}$.

Thus we have the inclusions $Q \cap V^{\circ} \subset f(U)^{\circ} \subset Q$, whence

$$Q \cap V^{\circ} = f(U)^{\circ} \cap V^{\circ},$$

and this last set is $\sigma(F', F)$-closed. ▮

Proof of Theorem 4. By Proposition 7 the map f is almost continuous and so by Proposition 8 it is continuous. ▮

Before we state other variants of Theorem 4, let us show that in a certain sense it is the best possible.

PROPOSITION 9 (Mahowald). *Let E be a locally convex Hausdorff space. Suppose that for every Banach space F every linear map $f: E \to F$ whose graph is closed is necessarily continuous. Then E is barrelled.*

We shall need a lemma, but first let us recall a notation introduced in §5. Let E be a locally convex Hausdorff space and T a barrel in E. If q_T is the gauge of T, then we denote by E_T the quotient space of E modulo the subspace $N = \{x \mid q_T(x) = 0\}$ equipped with the quotient norm of q_T. The completion $\hat{E}_T$ of E_T is then a Banach space. With these notations we can state the following lemma.

LEMMA 5. *Let E be a locally convex Hausdorff space, T a barrel in E, and $\varphi: E \to E_T$ the canonical surjection. Then the graph of the linear map $\varphi: E \to \hat{E}_T$ is closed.*

Proof. The subspace $N = \{x \mid q_T(x) = 0\}$ is also given by

$$N = \bigcap_{\lambda > 0} \lambda T$$

and is therefore closed in E. Let us denote by M the quotient space E/N equipped with the quotient of the topology of E modulo N. The dual of M can be identified with the subspace $N^{\perp}$ of E' (Corollary of Proposition 13.2).

By Lemma 4 we have to prove that the subspace Q of $(\hat{E}_T)'$ consisting of those $y' \in (\hat{E}_T)'$ for which the map $x \mapsto \langle \varphi(x), y' \rangle$ is continuous on E, is dense in $(\hat{E}_T)'$ for the topology $\sigma((\hat{E}_T)', \hat{E}_T)$. By the definition of the quotient topology, $y' \in (\hat{E}_T)'$ belongs to Q if and only if it is continuous on M.

We show that $\bigcup_{\lambda > 0} \lambda T^{\circ} \subset Q$. In the first place, $\bigcup_{\lambda > 0} \lambda T^{\circ} \subset N^{\perp}$, and therefore each $x' \in \bigcup_{\lambda > 0} \lambda T^{\circ}$ is a continuous linear form on M; and we need only show that $x' \in (\hat{E}_T)'$, i.e., that $x' \in (E_T)'$ since $(\hat{E}_T)' = (E_T)'$ (Exercise 2.3). If $x' \in \lambda T^{\circ}$, then $|\langle x, x' \rangle| \leqq \epsilon$ for all $x \in (\epsilon/\lambda)T$; i.e., $|\langle \varphi(x), x' \rangle| \leqq \epsilon$ for all $\varphi(x) \in (\epsilon/\lambda)\varphi(T)$, and the set $(\epsilon/\lambda)\varphi(T)$ is a neighborhood of 0 in E_T.

Thus it is enough to prove that $\bigcup_{\lambda > 0} \lambda T^{\circ}$ is dense in $(E_T)'$ for $\sigma((E_T)', \hat{E}_T)$. Let us denote by the symbol $^{\bullet}$ the polarity for the pair of spaces $\hat{E}_T$ and $(E_T)'$ which form a dual system. The closure of $\varphi(T)$ in $\hat{E}_T$ is the set $T^{\circ\bullet}$, and the sets $\lambda T^{\circ\bullet}$ $(\lambda > 0)$ form a fundamental system of neighborhoods of 0 in $\hat{E}_T$. Hence the sets $\lambda T^{\circ\bullet\bullet}$ $(\lambda > 0)$ form a fundamental system of equicontinuous subsets of $(E_T)'$, and so

$$\bigcup_{\lambda > 0} \lambda T^{\circ\bullet\bullet} = (E_T)'.$$

But $T^{\circ\bullet\bullet}$ is the $\sigma((E_T)', \hat{E}_T)$-closure of T°, and therefore the $\sigma((E_T)', \hat{E}_T)$-closure of $\bigcup_{\lambda > 0} \lambda T^{\circ}$ contains $\bigcup_{\lambda > 0} \lambda T^{\circ\bullet\bullet}$. ∎

Proof of Proposition 9. Let T be a barrel in E. Using the notations of Lemma 5, it follows from our assumption that the map $\varphi: E \to \hat{E}_T$ is continuous. Since $\overline{\varphi(T)}$ is a neighborhood of 0 in $\hat{E}_T$ and $T \supset \varphi^{-1}\overline{(\varphi(T))}$, it follows that T is a neighborhood of 0 in E. ∎

PROPOSITION 10 (Robertson-Robertson). *Let E be a vector space, $(E_{\iota})_{\iota \in I}$ a family of locally convex Baire spaces, and for each $\iota \in I$ let u_{ι} be a linear map from E_{ι} into E. Equip E with the finest locally convex topology for which all the maps u_{ι} are continuous.*

Let F be a vector space, $(F_n)_{n\in\mathbf{N}}$ a sequence of Pták spaces, and for each $n \in \mathbf{N}$ let v_n be a linear map from F_n into F. Suppose that $F = \bigcup_{n\in\mathbf{N}} v_n(F_n)$. Equip F with the finest locally convex topology for which all the maps v_n are continuous, and suppose that F is a Hausdorff space for this topology.

Then every linear map $f: E \to F$, whose graph G is closed in the product space $E \times F$, is continuous.

REMARK 2. The conditions on E are certainly fulfilled if E is a quasi-complete, bornological Hausdorff space (Proposition 7.5) or more generally an ultrabornological space (Exercise 15.9).

We first prove

LEMMA 6. *Let E be a locally convex space and L a non-meager subspace of E. Then L is barrelled.*

Proof (cf. proof of Proposition 6.3). Let T be a barrel in L and $\overline{T}$ the closure of T in E. Since T is absorbing in L, we have $L \subset \bigcup_{n\geqq 1} n\overline{T}$. Since L is not meager, one of the closed sets $n\overline{T}$ has an interior point, and therefore $\overline{T}$ has an interior point x_0. Let V be a balanced neighborhood of 0 in E such that $x_0 + V \subset \overline{T}$. Since $\overline{T}$ is balanced (Exercise 2.3.5), we have $-x_0 + V \subset \overline{T}$. But then $V \subset \overline{T}$, since if $x \in V$, then

$$\tfrac{1}{2}(x_0 + x) + \tfrac{1}{2}(-x_0 + x) \in \overline{T},$$

because $\overline{T}$ is convex (Proposition 2.4.3). Hence $\overline{T}$ is a neighborhood of 0 in E. Now $T = \overline{T} \cap L$ since T is closed in L, and therefore T is a neighborhood of 0 in L for the topology induced on L by E. ∎

Proof of Proposition 10. (a) Let us first assume that E itself is a locally convex Baire space. Then

$$E = f^{-1}(F) = f^{-1}\Big(\bigcup_{n\in\mathbf{N}} v_n(F_n)\Big) = \bigcup_{n\in\mathbf{N}} f^{-1}(v_n(F)).$$

Hence there must exist an index m such that $L = f^{-1}(v_m(F))$ is not meager in E since otherwise E itself would be meager. In particular, L is not rare and so $\overline{L}$ contains an interior point x_0. Since $\overline{L}$ is a subspace of E (Proposition 2.5.4), it is in particular a balanced, convex subset in E, and therefore we see by repeating the argument used in the proof of Lemma 6 that $\overline{L}$ is a neighborhood of 0 in E. In particular, $\overline{L}$ is absorbing, and therefore $\overline{L} = E$.

Since F is a Hausdorff space, $N = v_m^{-1}(0)$ is a closed subspace of F_m, and so by Proposition 5 the quotient space $M = F_m/N$ is a Pták space. Denote by $w: M \to F$ the injection associated with v_m.

If f_L is the restriction of f to L, then the graph of the map

$$g = w^{-1} \circ f_L: L \to M$$

is closed in $L \times M$ since it is the inverse image of $G \subset E \times F$ by the continuous map $(x, y) \mapsto (x, w(y))$. By Lemma 6 the space L is barrelled, and so by Theorem 4 the map g is continuous.

By Proposition 3(b) the space M is complete. On the other hand, L is dense in E. Hence by Proposition 2.9.5 the map g has a unique continuous extension $\bar{g}: E \to M$. For each $x \in L$ we have $w(\bar{g}(x)) = f(x)$. Let us prove that the same holds for all $x \in E$.

Suppose that the relation does not hold, i.e., that there exists $x_1 \in E$ such that $(x_1, w(\bar{g}(x_1))) \notin G$. Since G is assumed to be closed, there exists a neighborhood U of 0 in E and a neighborhood V of 0 in F such that

$$\{(x_1 + U) \times (w(\bar{g}(x_1)) + V)\} \cap G = \emptyset.$$

Since $\bar{g}$ is continuous, there exists a neighborhood U_1 of 0 in E such that $U_1 \subset U$ and $w(\bar{g}(U_1)) \subset V$. But L is dense in E. Hence there exists a point $x_2 \in L \cap (x_1 + U_1)$. We have

$$(x_2, f(x_2)) = (x_2, w(\bar{g}(x_2))) \in (x_1 + U_1) \times (w(\bar{g}(x_1)) + V),$$

and this last set does not meet G. We have arrived at a contradiction.

Thus we have $f = w \circ \bar{g}$, and since $\bar{g}$ and w are both continuous, we see that f is continuous.

(b) Consider now the general case where E is equipped with the final locally convex topology for the maps u_ι. For each $\iota \in I$ let G_ι be the graph of the map $f_\iota = f \circ u_\iota: E_\iota \to F$. Then G_ι is a closed subset of $E_\iota \times F$ since it is the inverse image of G under the continuous map

$$(x, y) \mapsto (u_\iota(x), y).$$

It follows from part (a) of the proof that each f_ι is continuous. Hence by Proposition 2.12.1 the map f is continuous. ▮

The preceding result can be used to prove a variant of the homomorphism theorem.

Proposition 11. *Let E be a vector space, $(E_n)_{n \in \mathbf{N}}$ a sequence of Pták spaces, and for each $n \in \mathbf{N}$ let u_n be a linear map from E_n into E. Suppose that $E = \bigcup_{n \in \mathbf{N}} u_n(E_n)$, and equip E with the finest locally convex topology for which all the maps u_n are continuous.*

Let F be a vector space, $(F_\iota)_{\iota \in I}$ a family of locally convex Baire spaces, and for each $\iota \in I$ let v_ι be a linear map from F_ι into F. Equip F with the finest locally convex topology for which all the maps v_ι are continuous, and suppose that F is a Hausdorff space.

Every continuous surjective linear map from E onto F is a strict morphism.

Proof. Let $f: E \to F$ be a continuous surjective linear map. Since F is a Hausdorff space, the subspace $N = f^{-1}(0)$ is closed in E, and the subspace $N_n = u_n^{-1}(f^{-1}(0))$ is closed in E_n. Set $L = E/N$ and $L_n = E_n/N_n$. Since $u_n(N_n) \subset N$, the map u_n defines a linear map $w_n\colon L_n \to L$, and it is easy to check that the quotient topology on L is the finest locally convex topology for which all the maps w_n are continuous. L is a Hausdorff space, and the L_n are Pták spaces by Proposition 5. Furthermore,

$$L = \bigcup_{n \in \mathbf{N}} w_n(L_n).$$

Let $\bar{f}\colon L \to F$ be the injection associated with f. The bijective map $\bar{f}$ is continuous; hence its graph G is closed in $L \times F$. The graph $\bar{f}^{-1}$ is the image of G in $F \times L$ under the homeomorphism $(x, y) \mapsto (y, x)$, and therefore it is also closed. It follows from Proposition 10 that $\bar{f}^{-1}$ is continuous, i.e., that $\bar{f}$ is an isomorphism. But this means precisely that f is a strict morphism (Definition 2.5.2). ▮

To conclude, we shall draw some consequences of the results proved so far in this section.

PROPOSITION 12. *Let E and F be two metrizable and complete topological vector spaces and $f\colon E \to F$ a continuous linear map. Then f is a strict morphism if and only if $f(E)$ is closed in F.*

Proof. If $f(E)$ is closed in F, then it is metrizable and complete (Proposition 2.9.3). Hence by Theorem 1 the map f is a strict morphism from E onto $f(E)$. The converse follows from the corollary of Theorem 2.9.2. ▮

The condition of the proposition is satisfied in particular if $f(E)$ has finite codimension, since then it has a topological supplement (Proposition 2.10.4) and is therefore closed (Exercise 2.7.3). More specifically:

PROPOSITION 13. *Let E and F be two metrizable and complete topological vector spaces. Then every continuous bijective linear map from E onto F is an isomorphism.*

Similarly, if E is a Pták space and F a barrelled Hausdorff space, then every continuous bijective linear map from E onto F is an isomorphism (Proposition 2). As a consequence, we have the following result:

PROPOSITION 14. *Let E be a vector space and $\mathcal{T}_1$ and $\mathcal{T}_2$ two locally convex topologies on E such that E is a Pták space for $\mathcal{T}_1$ and barrelled for $\mathcal{T}_2$. If $\mathcal{T}_1$ is finer than $\mathcal{T}_2$, then $\mathcal{T}_1 = \mathcal{T}_2$.*

Proof. Let E_i be the space E equipped with the topology $\mathcal{T}_i$ $(i = 1, 2)$ and $f\colon E_1 \to E_2$ the identity map. Then f is a bijective, continuous, linear map and therefore an isomorphism. ▮

By virtue of Proposition 13 the same conclusion holds if $\mathcal{T}_1$ and $\mathcal{T}_2$ are complete metrizable topologies compatible with the vector space structure of E.

PROPOSITION 15. *Let E be a metrizable and complete topological vector space. If M and N are two closed subspaces of E which are algebraic supplements of each other, then they are also topological supplements* (Definition 2.7.2).

Proof. The product space $M \times N$ is metrizable and complete, and the map $(x, y) \mapsto x + y$ is continuous, bijective and linear from $M \times N$ onto E, hence an isomorphism by Proposition 13. ∎

PROPOSITION 16. *Let E be a barrelled Hausdorff space, F a Pták space, and denote by $\mathcal{T}_0$ the topology of F. Let $\mathcal{T}$ be a Hausdorff topology on F which is coarser than $\mathcal{T}_0$. If a linear map $f: E \to F$ is continuous for the topology $\mathcal{T}$ on F, then it is also continuous for the topology $\mathcal{T}_0$ on F.*

Proof. The graph of f is closed in $E \times F$ if F is equipped with the topology $\mathcal{T}$, and therefore *a fortiori* if F is equipped with $\mathcal{T}_0$. The conclusion follows from Theorem 4. ∎

PROPOSITION 17 (Dieudonné-Schwartz). *Let E and F be two Fréchet spaces with topologies $\mathcal{T}_E$ and $\mathcal{T}_F$ respectively, E' and F' their duals, and let $u: E \to F$ be a linear map. Then the following five conditions are equivalent:*

(α) *u is a strict morphism for $\mathcal{T}_E$ and $\mathcal{T}_F$;*
(β) *u is a strict morphism for $\sigma(E, E')$ and $\sigma(F, F')$;*
(γ) *$u(E)$ is closed in F;*
(δ) *${}^tu$ is a strict morphism for $\sigma(F', F)$ and $\sigma(E', E)$;*
(ϵ) *${}^tu(F')$ is closed in E' for $\sigma(E', E)$.*

Proof. We shall prove the proposition according to the following logical scheme:

$$(\delta) \Leftrightarrow (\gamma) \Leftrightarrow (\alpha) \Leftrightarrow (\beta) \Leftrightarrow (\epsilon).$$

(β) $\Leftrightarrow$ (ϵ) by Propositions 12.1 and 13.3.
(γ) $\Leftrightarrow$ (δ) by Propositions 12.1 and 13.3.
(α) $\Rightarrow$ (β) by Proposition 13.4.
(β) $\Rightarrow$ (α) by the corollary of Proposition 13.5.
(α) $\Rightarrow$ (γ) by Proposition 12.
(γ) $\Rightarrow$ (α) since by Proposition 12.1 the map ${}^tu$ is continuous for $\sigma(F', F)$ and $\sigma(E', E)$, by its corollary u is continuous for $\sigma(E, E')$ and $\sigma(F, F')$, by the corollary of Proposition 12.5 the map u is continuous for $\mathcal{T}_E$ and $\mathcal{T}_F$, and so (α) is true by Proposition 12. ∎

COROLLARY. *Let E and F be two Fréchet spaces, E' and F' their duals, and $u: E \to F$ a continuous linear map.*

(i) *u is an injective strict morphism if and only if ${}^tu(F') = E'$.*

(ii) *u is a surjective strict morphism if and only if ${}^tu(F')$ is closed in E' for $\sigma(E', E)$ and ${}^tu$ is injective.*

(iii) *u is an isomorphism if and only if ${}^tu$ is an isomorphism for the topologies $\sigma(F', F)$ and $\sigma(E', E)$.*

Proof. (i) If u is an injective strict morphism, then ${}^tu(F')$ is closed in E' for $\sigma(E', E)$ by Proposition 17, and ${}^tu(F')$ is dense in E' for $\sigma(E', E)$ by Corollary 2 of Proposition 12.2. Hence ${}^tu(F') = E'$.

Conversely, if ${}^tu(F') = E'$, then ${}^tu(F')$ is both closed and dense in E' for $\sigma(E', E)$. Hence by the results just quoted, u is an injective strict morphism.

(ii) If u is a surjective strict morphism, then ${}^tu(F')$ is closed in E' for $\sigma(E', E)$ by Proposition 17, and ${}^tu$ is injective by Corollary 2 of Proposition 12.2.

Conversely, if ${}^tu(F')$ is closed in E' for $\sigma(E', E)$, then u is a strict morphism, and in particular $u(E)$ is closed in F. If, furthermore, ${}^tu$ is injective, then by Corollary 2 of Proposition 12.2 the set $u(E)$ is dense in F; hence $u(E) = F$.

(iii) If u is an isomorphism, then ${}^tu$ is a strict morphism for the topologies $\sigma(F', F)$ and $\sigma(E', E)$ by Proposition 17; it is surjective by (i) and injective by (ii).

Conversely, if ${}^tu$ is an isomorphism for the topologies $\sigma(F', F)$ and $\sigma(E', E)$, then u is a strict morphism by Proposition 17; it is injective by (i) and surjective by (ii). ▮

In general, the conditions (α) through (ϵ) of Proposition 17 do not imply that ${}^tu$ is a strict morphism for $\beta(F', F)$ and $\beta(E', E)$ ([9], Chapter IV, §4, Exercise 5(b)). We have, however, the following result (see also Exercises 8 and 11):

PROPOSITION 18. *Let E be a Fréchet-Schwartz space, F a reflexive Fréchet space, E' and F' their duals, and $u: E \to F$ a strict morphism. Then ${}^tu: F' \to E'$ is a strict morphism for the topologies $\beta(F', F)$ and $\beta(E', E)$.*

Proof. By Proposition 17 the subspace $M = {}^tu(F')$ is closed in E' for $\sigma(E', E)$ and *a fortiori* for $\beta(E', E)$. By the corollary of Proposition 15.10 the topology induced on M by $\beta(E', E)$ coincides with the topology $\beta(M, E/M^{\perp})$ since $M = M^{\perp\perp}$. But $E/M^{\perp}$ is a Fréchet-Schwartz space (Corollary of Proposition 15.7), and therefore by Proposition 16.1 the space M equipped with the topology $\beta(M, E/M^{\perp})$ is a barrelled Hausdorff space. On the other hand, F' equipped with $\beta(F', F)$ is a Pták space by Proposition 6. Finally, ${}^tu$ is continuous for the topologies $\beta(F', F)$ and $\beta(E', E)$ by the corollary of Proposition 12.3. It follows from Proposition 2 that ${}^tu: F' \to M$ is a surjective strict morphism, i.e., that ${}^tu: F' \to E'$ is a strict morphism. ▮

In the opposite direction we have the following result (see also Exercise 9).

PROPOSITION 19 (Dieudonné-Schwartz). *Let E and F be two Fréchet spaces, E' and F' their duals, and $u: E \to F$ a continuous linear map. If ${}^tu: F' \to E'$ is an injective strict morphism for the topologies $\beta(F', F)$ and $\beta(E', E)$, then u is a surjective strict morphism.*

Proof. By part (ii) of the corollary of Proposition 17 we have to show that $A = {}^tu(F')$ is closed in E' for $\sigma(E', E)$, and we shall do this with the help of Theorem 10.2.

Let M be a balanced, convex, $\sigma(E', E)$-closed, equicontinuous subset of E'. Since ${}^tu$ is an injective strict morphism for the topologies $\beta(F', F)$ and $\beta(E', E)$, the set ${}^tu^{-1}(M)$ is bounded in F'. On the other hand, ${}^tu$ is continuous for the topologies $\sigma(F', F)$ and $\sigma(E', E)$ by the corollary of Proposition 12.3, and therefore ${}^tu^{-1}(M)$ is closed for $\sigma(F', F)$. It follows (Corollary of Proposition 6.2) that ${}^tu^{-1}(M)$ is compact in F' for $\sigma(F', F)$. Again by the continuity of ${}^tu$, the set ${}^tu({}^tu^{-1}(M)) = A \cap M$ is compact and in particular closed for $\sigma(E', E)$. Hence A is closed for $\sigma(E', E)$ by Theorem 10.2. ∎

EXERCISES

1. Deduce Theorem 1 from Theorem 3. (*Hint:* Assume that $f: E \to F$ is bijective and prove that the graph of $f^{-1}: F \to E$ is closed.)

2. Let E be a Fréchet space and E' its dual. Prove that E' equipped with a topology finer than the topology $\lambda(E', E)$ of uniform convergence on compact sets but coarser than the Mackey topology $\tau(E', E)$ is a Pták space. (*Hint:* See the proof of Proposition 6.)

3. Let E be a vector space and E^* its algebraic dual. Prove that E^* equipped with the topology $\sigma(E^*, E)$ is a Pták space. (*Hint:* Every linear subspace of E is $\sigma(E, E^*)$-closed.)

4. Give an example of a complete locally convex Hausdorff space which is not a Pták space. (*Hint:* Consider an infinite-dimensional Banach space E equipped with the finest locally convex topology. One can also use Proposition 3(b), Proposition 5 and Remark 2.9.1.)

5. (a) Let E be a locally convex Hausdorff space with topology $\mathcal{T}$ and let E' be its dual. Show that the following three conditions are equivalent:

(i) Let F be an arbitrary locally convex Hausdorff space. Every injective, continuous, almost open linear map $f: E \to F$ is a strict morphism.

(ii) If $\mathcal{T}_1$ is a locally convex Hausdorff topology on E, coarser than $\mathcal{T}$, such that the $\mathcal{T}_1$-closure of any $\mathcal{T}$-neighborhood of 0 is a $\mathcal{T}_1$-neighborhood of 0, then $\mathcal{T} = \mathcal{T}_1$.

(iii) If L is a dense subspace of E' for the topology $\sigma(E', E)$ such that for every balanced, convex, $\sigma(E', E)$-closed, equicontinuous subset M of E' the intersection $L \cap M$ is $\sigma(E', E)$-closed, then $L = E'$.

A space E which satisfies conditions (i), (ii), and (iii) is called an *infra-Pták* (or B_r-complete) space. (*Hint:* To prove (ii) $\Rightarrow$ (iii), let L be as in (iii). The sets $(L \cap U^{\circ})^{\circ}$, where U runs through the $\mathcal{T}$-neighborhoods of 0, form a fundamental system of neighborhoods of 0 for a locally convex topology $\mathcal{T}_1$ on E. The topology $\mathcal{T}_1$ is Hausdorff since if $x \in (L \cap U^{\circ})^{\circ}$ for all U, then $|\langle x, x' \rangle| \leqq 1$ for all $x' \in L \cap U^{\circ}$ and all U, i.e., $\langle x, x' \rangle = 0$ for all $x' \in L$. Since L is dense in E', we have $x = 0$. Also, $\mathcal{T}_1$ is coarser than $\mathcal{T}$ since $U \subset (L \cap U^{\circ})^{\circ}$. If E_1 is E equipped with the topology $\mathcal{T}_1$, then $(E_1)' = L$. Clearly, $L \subset (E_1)'$. If $y' \in (E_1)'$, there exists U such that $|\langle x, y' \rangle| \leqq 1$ for $x \in (L \cap U^{\circ})^{\circ}$, i.e., $y' \in (L \cap U^{\circ})^{\circ\circ}$. But by the assumption in (iii) we have $(L \cap U^{\circ})^{\circ\circ} = L \cap U^{\circ}$ and thus $y' \in L$. For every U the $\mathcal{T}_1$-closure of U is $(U^{\circ} \cap L)^{\circ}$, i.e., a $\mathcal{T}_1$-neighborhood of 0. It follows from (ii) that $\mathcal{T} = \mathcal{T}_1$, and therefore $E' = L$. To prove (iii) $\Rightarrow$ (ii), let L be the dual of E equipped with $\mathcal{T}_1$. Then $L \subset E'$, and L is dense in E' for $\sigma(E', E)$ by Proposition 2.3. If U is a balanced, convex, $\mathcal{T}$-closed $\mathcal{T}$-neighborhood of 0, then the $\mathcal{T}_1$-closure $\overline{U}$ of U is a $\mathcal{T}_1$-neighborhood of 0, so that U and $\overline{U}$ have the same polar $L \cap U^{\circ}$ in L. But $L \cap U^{\circ}$ is $\sigma(L, E)$-compact, hence $\sigma(E', E)$-closed. By (iii) this implies that L is $\sigma(E', E)$-closed, and since L is dense, we have $L = E'$. Therefore the balanced, convex set U has the same closure for $\mathcal{T}$ and $\mathcal{T}_1$; i.e., $U = \overline{U}$, which proves that $\mathcal{T} = \mathcal{T}_1$.)

(b) Prove that every Pták space is an infra-Pták space.

(c) Prove that Proposition 8 and Theorem 4 hold if we only suppose that F is an infra-Pták space.

6. Prove that every infra-Pták space E is complete. (*Hint:* A hyperplane H in E', such that for every balanced, closed, convex, equicontinuous subset M of E' the set $H \cap M$ is $\sigma(E', E)$-closed, cannot be dense in E'. Apply Corollary 5 of Theorem 11.1.)

7. (a) Let E be an infra-Pták space and F a locally convex Hausdorff space. Let f be an injective linear map from a subspace E_0 of E into F. Prove that if the graph of f is closed in $E \times F$ and f is almost open, then f maps open sets of E_0 onto open sets of $f(E_0)$. (*Hint:* Let $g: f(E_0) \to E_0$ be the inverse of the map $f: E_0 \to f(E_0)$. Then g is almost continuous and the graph of g is closed in $f(E_0) \times E$. Hence by Proposition 8 and Exercise 5(c) the map g is continuous.)

(b) Let E and F be two locally convex Hausdorff spaces and f a linear map from a subspace E_0 of E into F. Suppose that the graph of f is closed in $E \times F$. Show that the kernel $N = \{x \mid x \in E_0, f(x) = 0\}$ is a closed subset of E. (*Hint:* If $x_0 \in \overline{N}$, then $(x_0, 0)$ belongs to the closure of the graph of f.)

(c) Let E be a Pták space and F a locally convex Hausdorff space. Let f be a linear map from a subspace E_0 of E into F whose graph is closed in $E \times F$. Prove that if f is almost open, then it maps open sets of E_0 onto open sets of $f(E_0)$. (*Hint:* By part (b) and Proposition 5 the quotient E/N is a Pták space. The injection $\bar{f}: E_0/N \to F$ associated with f is almost open and its graph is closed in $E/N \times F$. Apply part (a). For another proof see [75], pp. 57–58.)

(d) Let E be a Pták space and F a barrelled Hausdorff space. Let f be a surjective linear map from a subspace E_0 of E onto F, and suppose that the graph of f is closed in $E \times F$. Prove that f maps open sets of E_0 onto open sets of F.

8. Let E and F be two normed vector spaces, E' and F' their duals equipped with the normed topologies $\beta(E', E)$ and $\beta(F', F)$. Show that if $u: E \to F$ is a strict morphism, then ${}^t u: F' \to E'$ is a strict morphism.

9. Let E and F be two Banach spaces, E' and F' their duals equipped with the normed topologies $\beta(E', E)$ and $\beta(F', F)$, and $u: E \to F$ a continuous linear map. Show that if ${}^t u: F' \to E'$ is a strict morphism, then $u: E \to F$ is a strict morphism. (*Hint:* Use Proposition 13.3 and the corollary of Theorem 10.2.)

10. Let E be a locally convex Hausdorff space. Show that E is infrabarrelled if and only if for every Banach space F every linear map $f: E \to F$ such that

(i) f takes bounded sets into bounded sets,
(ii) the graph of f is closed,

is necessarily continuous. (*Hint:* If E is infrabarrelled, then (i) implies that f is almost continuous. Apply Proposition 8. Conversely, assume that the condition holds. If T is a bornivorous barrel in E, then by Lemma 5 the map φ has a closed graph in $E \times \hat{E}_T$, and φ sends bounded sets into bounded sets since T is bornivorous. Hence φ is continuous; i.e., T is a neighborhood of 0 in E.)

11. Let E and F be two Fréchet spaces, E' and F' their duals equipped with the topologies $\kappa(E', E)$ and $\kappa(F', F)$. Show that if $u: E \to F$ is a strict morphism, then ${}^t u: F' \to E'$ is a strict morphism. (*Hint:* Let $U = C^{\circ}$, where C is a compact subset of F. By Proposition 17 and the lemma in §15 there exists a compact subset A of E such that $u(A) = C \cap \operatorname{Im}(u)$. Show that ${}^t u(U) = A^{\circ}$ and apply Theorem 2.5.1.)

CHAPTER 4

Distributions

§1. The definition of distributions

DEFINITION 1 (Laurent Schwartz). *Let Ω be an open subset of $\mathbf{R}^n$ and $\mathfrak{D}(\Omega)$ the space of infinitely differentiable functions with compact support defined in Ω, equipped with the topology introduced in Example* 2.12.6. *A continuous linear form on $\mathfrak{D}(\Omega)$ is called a distribution defined in Ω.*

The dual $E' = \mathfrak{D}'(\Omega)$ of $E = \mathfrak{D}(\Omega)$ is thus the space of all distributions defined in Ω. Usually we consider E' equipped with the strong topology $\beta(E', E)$ (Definition 3.4.2) which coincides with the topologies $\kappa(E', E)$ and $\lambda(E', E)$ (Definition 3.9.2) since $\mathfrak{D}(\Omega)$ is a Montel space. If $\Omega = \mathbf{R}^n$, we shall write simply $\mathfrak{D}'$ instead of $\mathfrak{D}'(\mathbf{R}^n)$.

Let K be a compact subset of Ω and $i_K\colon \mathfrak{D}(K) \hookrightarrow \mathfrak{D}(\Omega)$ the canonical injection. By Proposition 2.12.1 a linear form T on $\mathfrak{D}(\Omega)$ is continuous if and only if $T \circ i_K$ is continuous on $\mathfrak{D}(K)$ for every K. Using Proposition 2.5.2 and the definition of the semi-norms on $\mathfrak{D}(K)$ (Example 2.4.10), we obtain:

PROPOSITION 1. *A linear form T on $\mathfrak{D}(\Omega)$ is a distribution if and only if for every compact subset K of Ω there exists a positive number M and a positive integer m such that for all $\varphi \in \mathfrak{D}(K)$*

$$|\langle T, \varphi\rangle| \leqq M \max_{|p| \leqq m} \max_{x} |\partial^p \varphi(x)|. \tag{1}$$

Clearly, condition (1) can be replaced by

$$|\langle T, \varphi\rangle| \leqq M \sum_{|p| \leqq m} \max_{x} |\partial^p \varphi(x)|.$$

EXAMPLE 1. Let Ω be an open subset of $\mathbf{R}^n$ and let f be an element of the space $\mathcal{C}(\Omega)$ of continuous functions defined on Ω (Example 2.3.3). For each $\varphi \in \mathfrak{D}(\Omega)$, the function $f\varphi$ is continuous and vanishes outside the support of φ. If we define $f\varphi$ to be equal to zero outside Ω, the extended function is continuous in $\mathbf{R}^n$ and has compact support. It follows that the integral $\int_{\mathbf{R}^n} f(x)\varphi(x)\,dx$ exists. The map $T_f\colon \varphi \mapsto \int_{\mathbf{R}^n} f(x)\varphi(x)\,dx$ is a linear form on $\mathfrak{D}(\Omega)$, and it is also continuous since if $\varphi \in \mathfrak{D}(K)$, where K is contained in the cube $\{x \mid |x_i| \leqq \frac{1}{2}a, 1 \leqq i \leqq n\}$, and

$$b = \max_{x \in K} |f(x)|,$$

then

$$|\langle T_f, \varphi\rangle| \leqq \int_{\mathbf{R}^n} |f(x)\varphi(x)|\, dx \leqq a^n b \cdot \max_x |\varphi(x)|.$$

The map $f \mapsto T_f$ from $\mathfrak{C}(\Omega)$ into $\mathfrak{D}'(\Omega)$ is clearly linear. We want to show that it is also injective. Let $f \not\equiv 0$. There exists a point $x_0 \in \Omega$ such that $f(x_0) \neq 0$. We can write $f = f_1 + if_2$, where f_1 and f_2 are real-valued continuous functions. There exists $\alpha > 0$ and a ball $B_r(x_0)$ such that one of the inequalities $f_i(x) > \alpha$ and $f_i(x) < -\alpha$ holds for $x \in B_r(x_0)$ at least for one of the indices $i = 1, 2$. Let $\chi \in \mathfrak{D}(\Omega)$ be such that $\chi(x) \geqq 0$, $\chi(x) = 1$ for $x \in B_{(1/2)r}(x_0)$, and $\chi(x) = 0$ for $x \in \complement B_r(x_0)$ (Proposition 2.12.5). Then

$$\langle T_f, \chi\rangle = \int_{B_r(x_0)} f_1(x)\chi(x)\, dx + i\int_{B_r(x_0)} f_2(x)\chi(x)\, dx \neq 0$$

since if, for instance, $f_1(x) > \alpha$ in $B_r(x_0)$, then

$$\int_{B_r(x_0)} f_1(x)\chi(x)\, dx \geqq \int_{B_{(1/2)r}(x_0)} f_1(x)\, dx \geqq \alpha|B_{(1/2)r}(x_0)|,$$

where $|B|$ is the volume of the set B. Thus $T_f \not\equiv 0$.

We shall often identify the function f with the distribution T_f and write $\langle f, \varphi\rangle$ for $\langle T_f, \varphi\rangle$. Thus the notion of distribution generalizes that of continuous function. More generally, we shall see in the next chapter that distributions generalize equivalence classes of locally integrable functions. For this reason some authors call distributions "generalized functions."

EXAMPLE 2. Let Ω again be an open subset of $\mathbf{R}^n$ and a a point of Ω. The linear form $\varphi \mapsto \varphi(a)$ is continuous on $\mathfrak{D}(\Omega)$ since if K is a compact subset of Ω we have $|\varphi(a)| \leqq \max_x |\varphi(x)|$ for all $\varphi \in \mathfrak{D}(K)$. Thus we obtain a distribution δ_a such that $\langle \delta_a, \varphi\rangle = \varphi(a)$. This distribution belongs to a class of distributions called measures (Definition 4.1); if $a = 0$, we call $\delta = \delta_0$ the *Dirac measure*. By physical analogy we also say that δ_a is "the unit mass (or charge) placed at the point a."

As we know, $\mathfrak{D}(\Omega)$ is complete (Example 2.12.6), barrelled (Example 3.6.4), bornological (Example 3.7.2), a Montel space (Example 3.9.6), and a Schwartz space (Example 3.15.3). The space $\mathfrak{D}'(\Omega)$ is complete (Proposition 3.7.6), barrelled (Proposition 3.8.4), bornological (Theorem 3.16.2), and a Montel space (Proposition 3.9.9). We can apply the results obtained in Chapter 3, for instance to obtain Proposition 2.

PROPOSITION 2. *Let $(T_n)_{n\in\mathbf{N}}$ be a sequence of distributions [resp. let $(T_\epsilon)_{0<\epsilon<\alpha}$ be a family of distributions] and suppose that for every $\varphi \in \mathfrak{D}(\Omega)$ the limit $T(\varphi) = \lim_{n\to\infty} T_n(\varphi)$ [resp. $T(\varphi) = \lim_{\epsilon\to 0} T_\epsilon(\varphi)$] exists. Then $T\colon \varphi \mapsto T(\varphi)$ is a distribution and (T_n) [resp. (T_ϵ)] converges to T strongly in $\mathfrak{D}'(\Omega)$.*

Proof. T is a distribution by the corollary of Proposition 3.6.5. Since $\mathfrak{D}'(\Omega)$ is a Montel space, (T_n) [resp. (T_ϵ)] converges strongly to T by the corollary of Proposition 3.9.2. ∎

EXAMPLE 3. For each $\varphi \in \mathfrak{D}(\mathbf{R})$ the limit

$$\lim_{\epsilon\to 0} \int_{|x|>\epsilon} \frac{\varphi(x)}{x}\,dx \tag{2}$$

exists. Indeed, let $\varphi(x) = \varphi(0) + x\psi(x)$, where ψ is a continuous function for which $\psi(0) = \varphi'(0)$. Suppose that $\varphi(x) = 0$ for $|x| \geqq a$. Then

$$\begin{aligned}\int_{|x|>\epsilon} \frac{\varphi(x)}{x}\,dx &= \varphi(0)\int_{\epsilon<|x|<a} \frac{dx}{x} + \int_{\epsilon<|x|<a} \psi(x)\,dx \\ &= \varphi(0)\left\{\log\frac{a}{\epsilon} - \log\frac{a}{\epsilon}\right\} + \int_{\epsilon<|x|<a} \psi(x)\,dx \\ &= \int_{\epsilon<|x|<a} \psi(x)\,dx,\end{aligned}$$

and this last expression tends to $\int_{-a}^{a} \psi(x)\,dx$ as $\epsilon \to 0$.

On the other hand, for each $\epsilon > 0$ the linear map

$$\varphi \mapsto \int_{|x|>\epsilon} \frac{\varphi(x)}{x}\,dx$$

is a distribution defined on $\mathbf{R}$. In fact, if $\varphi \in \mathfrak{D}(K)$ and K is contained in the interval $-a \leqq x \leqq a$, then

$$\left|\int_{|x|>\epsilon} \frac{\varphi(x)}{x}\,dx\right| \leqq 2\log\frac{a}{\epsilon}\cdot\max_x |\varphi(x)|.$$

It follows from Proposition 2 that the limit (2) defines a distribution on $\mathbf{R}$. Since in classical analysis the limit (2) is called the "Cauchy principal value" of the integral $\int_{-\infty}^{\infty} [\varphi(x)/x]\,dx$, we denote this distribution by v.p. $1/x$, i.e.,

$$\left\langle \text{v.p.}\,\frac{1}{x}, \varphi\right\rangle = \lim_{\epsilon\to 0}\int_{|x|>\epsilon} \frac{\varphi(x)}{x}\,dx.$$

We shall sometimes omit the letters v.p. and simply write x^{-1} or $1/x$ for v.p. $1/x$.

EXAMPLE 4. Let ρ be a continuous function defined on $\mathbf{R}^n$ which has the following properties:

(a) $\operatorname{Supp} \rho \subset B_1$,

(b) $\rho(x) \geqq 0$ for all $x \in \mathbf{R}^n$,

(c) $\int_{\mathbf{R}^n} \rho(x)\, dx = 1$,

where $B_1 = \{x \mid |x| \leqq 1\}$ is the unit ball of $\mathbf{R}^n$. We saw in Chapter 2, §12 that such functions exist even in $\mathfrak{D}$. For every $\epsilon > 0$ define the function ρ_ϵ by

$$\rho_\epsilon(x) = \frac{1}{\epsilon^n} \rho\left(\frac{x}{\epsilon}\right).$$

As we observed in Chapter 2, §12, the function ρ_ϵ satisfies the conditions

(a′) $\operatorname{Supp} \rho_\epsilon \subset B_\epsilon = \{x \mid |x| \leqq \epsilon\}$,

(b′) $\rho_\epsilon(x) \geqq 0$ for all $x \in \mathbf{R}^n$,

(c′) $\int_{\mathbf{R}^n} \rho_\epsilon(x)\, dx = 1$.

We want to prove that ρ_ϵ converges in $\mathfrak{D}'$ to the Dirac measure δ as $\epsilon \to 0$. For $\varphi \in \mathfrak{D}$ we have by (c′)

$$\int_{\mathbf{R}^n} \rho_\epsilon(x)\varphi(x)\, dx - \varphi(0) = \int_{\mathbf{R}^n} \rho_\epsilon(x)\{\varphi(x) - \varphi(0)\}\, dx,$$

and using (a′), (b′), (c′), we obtain

$$\left| \int_{\mathbf{R}^n} \rho_\epsilon(x)\varphi(x)\, dx - \varphi(0) \right| \leqq \max_{|x| \leqq \epsilon} |\varphi(x) - \varphi(0)|.$$

Since the right-hand side can be made arbitrarily small by taking ϵ sufficiently small, we have

$$\lim_{\epsilon \to 0} \int_{\mathbf{R}^n} \rho_\epsilon(x)\varphi(x)\, dx = \varphi(0);$$

i.e., ρ_ϵ converges to δ weakly and thus by Proposition 2 also strongly.

We conclude this section with two results whose importance will appear in §2. Let us recall that in Example 1 we defined an injection $\varphi \mapsto T_\varphi$ from $\mathfrak{D}(\Omega)$ into $\mathfrak{D}'(\Omega)$.

PROPOSITION 3. *The image M of $\mathfrak{D}(\Omega)$ in $\mathfrak{D}'(\Omega)$ is everywhere dense.*

Proof. Assume that $\overline{M} \neq \mathfrak{D}'(\Omega)$. By Proposition 3.1.2 there exists a continuous linear form u on $\mathfrak{D}'(\Omega)$ such that $\langle u, T_\varphi \rangle = 0$ for all $\varphi \in \mathfrak{D}(\Omega)$ but $u \neq 0$. By the corollary of Proposition 3.9.1 the space $\mathfrak{D}(\Omega)$ is reflexive. Hence there exists $\psi \in \mathfrak{D}(\Omega)$ such that $\langle u, T \rangle = \langle T, \psi \rangle$ for every $T \in \mathfrak{D}'(\Omega)$, and in particular

$$\int_\Omega \varphi(x)\psi(x)\, dx = 0$$

for all $\varphi \in \mathfrak{D}(\Omega)$. In other words, we have $\langle T_\psi, \varphi\rangle = 0$ for all $\varphi \in \mathfrak{D}(\Omega)$; i.e., $T_\psi = 0$. Since the map $\psi \mapsto T_\psi$ is injective, we have $\psi = 0$, i.e., $u = 0$, and this contradiction proves that $\overline{M} = \mathfrak{D}'(\Omega)$. ∎

PROPOSITION 4. *The injection $f \mapsto T_f$ from $\mathcal{C}(\Omega)$ into $\mathfrak{D}'(\Omega)$ is continuous.*

Proof. Since $\mathcal{C}(\Omega)$ is metrizable, it is sufficient to prove that if (f_n) is a sequence which tends to 0 in $\mathcal{C}(\Omega)$, then the corresponding sequence (T_{f_n}) tends to 0 in $\mathfrak{D}'(\Omega)$. Let $\varphi \in \mathfrak{D}(\Omega)$ and $K = \operatorname{Supp} \varphi$. Then

$$\max_{x\in K} |f_n(x)| \leqq \epsilon \qquad \text{for} \qquad n \geqq N(\epsilon),$$

and

$$\left|\int_\Omega f_n(x)\varphi(x)\,dx\right| \leqq \epsilon\int_\Omega |\varphi(x)|\,dx;$$

i.e., T_{f_n} tends to 0 weakly in $\mathfrak{D}'(\Omega)$. But by Proposition 2 the sequence (T_{f_n}) also tends strongly to 0. ∎

EXERCISES

1. Let T be a linear form on the space $\mathfrak{D}(\Omega)$. Prove that T is continuous (i.e., a distribution) if and only if the following condition is satisfied:

Suppose that $(\varphi_k)_{k\in\mathbf{N}}$ is a sequence of elements of $\mathfrak{D}(\Omega)$ such that

(1) the sets $\operatorname{Supp} \varphi_k$ are contained in a fixed compact subset K of Ω,

(2) for each $p \in \mathbf{N}^n$ the sequence $(\partial^p\varphi_k)_{k\in\mathbf{N}}$ tends uniformly to zero as $k \to \infty$.

Then $\langle T, \varphi_k\rangle$ tends to zero as $k \to \infty$.

(*Hint:* That the condition is necessary follows from Proposition 1. Conversely, if the condition of Proposition 1 is not verified, then there exists a compact subset K of Ω and a sequence (φ_k) of elements of $\mathfrak{D}(K)$ such that

$$\max_x |\partial^p\varphi_k(x)| \leqq \frac{1}{k}$$

for $|p| \leqq k$ but $\langle T, \varphi_k\rangle = 1$.)

§2. *Support*

Let Ω be an open subset of $\mathbf{R}^n$ and U an open subset of Ω. Every function belonging to $\mathfrak{D}(U)$ can be considered as a function belonging to $\mathfrak{D}(\Omega)$. If $T \in \mathfrak{D}'(\Omega)$, then its restriction to $\mathfrak{D}(U)$ is the distribution $T_U \in \mathfrak{D}'(U)$ defined by $\langle T_U, \varphi\rangle = \langle T, \varphi\rangle$ for every $\varphi \in \mathfrak{D}(U)$. In other words, the map $T \mapsto T_U$ is the transpose of the injection $\mathfrak{D}(U) \hookrightarrow \mathfrak{D}(\Omega)$. We shall call T_U the *restriction* of T to U, or the distribution *induced* on U by T. If $T_U = 0$, we shall say that T is zero in U. Similarly, if Ω_1 and Ω_2 are two open subsets of $\mathbf{R}^n$, U an open subset of $\Omega_1 \cap \Omega_2$, $S \in \mathfrak{D}'(\Omega_1)$, and $T \in \mathfrak{D}'(\Omega_2)$, we shall say that S and T are equal in U if $S_U = T_U$.

Proposition 1. *Let Ω be an open subset of $\mathbf{R}^n$, $T \in \mathfrak{D}'(\Omega)$, and U an open subset of Ω. If for every $x \in U$ there exists an open neighborhood V_x such that T is zero in V_x, then T is zero in U.*

Proof. Let (α_x) be a locally finite infinitely differentiable partition of unity subordinated to the cover (V_x) of U (Remark 2.12.4). If $\varphi \in \mathfrak{D}(U)$, then $\varphi = \sum \alpha_x \varphi$, and the sum has only finitely many non-zero terms since $\operatorname{Supp} \alpha_x \cap \operatorname{Supp} \varphi = \emptyset$ except for finitely many x. Also $\operatorname{Supp} \alpha_x \varphi \subset V_x$, and therefore we have

$$\langle T, \varphi \rangle = \langle T, \textstyle\sum \alpha_x \varphi \rangle = \sum \langle T, \alpha_x \varphi \rangle = 0;$$

i.e., $T = 0$. ∎

It follows in particular from Proposition 1 that if $(U_\iota)_{\iota \in I}$ is a family of open subsets of Ω and $T = 0$ in each U_ι, then $T = 0$ in $\bigcup_{\iota \in I} U_\iota$. Therefore the following definition is meaningful.

Definition 1. *Let Ω be an open subset of $\mathbf{R}^n$ and $T \in \mathfrak{D}'(\Omega)$. If U is the largest open subset of Ω in which T is zero, then the set $\complement_\Omega U$ is called the support of T and denoted by* Supp T.

The set Supp T is closed in Ω by definition. A point $x \in \Omega$ belongs to Supp T if and only if for every neighborhood V of x there exists $\varphi \in \mathfrak{D}(\Omega)$ with $\operatorname{Supp} \varphi \subset V$ and such that $\langle T, \varphi \rangle \neq 0$.

Example 1. If $f \in \mathcal{C}(\Omega)$ and T_f is the corresponding distribution (Example 1.1), then $\operatorname{Supp} T_f = \operatorname{Supp} f$.

Example 2. $\operatorname{Supp} \delta_a = \{a\}$.

Proposition 2. *If S and T belong to $\mathfrak{D}'(\Omega)$, then*

$$\operatorname{Supp}(S + T) \subset \operatorname{Supp} S \cup \operatorname{Supp} T,$$

and $\operatorname{Supp}(\lambda T) = \operatorname{Supp} T$ *for* $\lambda \neq 0$, $\lambda \in \mathbf{K}$.

Proof. If $\varphi \in \mathfrak{D}(\Omega)$ is such that

$$\operatorname{Supp} \varphi \subset \complement (\operatorname{Supp} S \cup \operatorname{Supp} T) = (\complement \operatorname{Supp} S) \cap (\complement \operatorname{Supp} T),$$

then $\langle S + T, \varphi \rangle = \langle S, \varphi \rangle + \langle T, \varphi \rangle = 0$, and therefore

$$\complement (\operatorname{Supp} S \cup \operatorname{Supp} T) \subset \complement \operatorname{Supp}(S + T).$$

The second assertion follows immediately from the relations

$$\operatorname{Supp}(\lambda \varphi) = \operatorname{Supp} \varphi \qquad (\lambda \neq 0, \lambda \in \mathbf{K})$$

and

$$\langle \lambda T, \varphi \rangle = \langle T, \lambda \varphi \rangle. \ ∎$$

It follows from Proposition 2 that the distributions belonging to $\mathfrak{D}'(\Omega)$ whose support is contained in a fixed subset of Ω form a vector space. Similarly, all distributions with compact support form a vector space. In Proposition 3 we shall characterize this vector space, but first we must introduce a few notions which will play an important role in the sequel.

DEFINITION 2. *Let Ω be an open subset of $\mathbf{R}^n$. A pair (E, j), consisting of a locally convex Hausdorff space E and a continuous injective linear map $j: E \to \mathfrak{D}'(\Omega)$, will be called an injective pair. The image $j(E)$ will be called a space of distributions on Ω.*

We shall most often identify E with its image $j(E)$ and say simply that E is a space of distributions. It is to be emphasized, however, that even after this identification, E keeps its own topology, which by assumption is finer than the one induced by $\mathfrak{D}'(\Omega)$.

DEFINITION 3. *Let Ω be an open subset of $\mathbf{R}^n$. A triple (i, E, j), consisting of a locally convex Hausdorff space E, a continuous injective linear map $i: \mathfrak{D}(\Omega) \to E$, and a continuous injective linear map $j: E \to \mathfrak{D}'(\Omega)$, will be called a normal triple if the following conditions are satisfied:*

(a) $\text{Im}(i)$ *is everywhere dense,*
(b) $j \circ i$ *is the map* $\varphi \mapsto T_\varphi$ *of Example* 1.1.

The image $j(E)$ will then be called a normal space of distributions.

By Proposition 1.3 the image of $\mathfrak{D}(\Omega)$ in $\mathfrak{D}'(\Omega)$ is everywhere dense. Therefore if (i, E, j) is a normal triple, we are precisely in the situation described in Scholium 3.12.1, and in particular $(E', {}^t i)$ is then an injective pair provided that we equip E' with either the topology $\beta(E', E)$ or the topology $\kappa(E', E)$. If we identify $\mathfrak{D}(\Omega)$ with $\text{Im}(i)$, we can say that a distribution $T \in \mathfrak{D}'(\Omega)$ belongs to the space of distributions ${}^t i(E')$ if and only if it is continuous on $\mathfrak{D}(\Omega)$ for the (coarser) topology induced on it by E.

EXAMPLE 3. Take for i the identity map $1_{\mathfrak{D}(\Omega)}: \mathfrak{D}(\Omega) \to \mathfrak{D}(\Omega)$ and for $j: \mathfrak{D}(\Omega) \to \mathfrak{D}'(\Omega)$ the map $\varphi \mapsto T_\varphi$ of Example 1.1. Then $(i, \mathfrak{D}(\Omega), j)$ is a normal triple. Indeed, the only point to establish is the continuity of j. But j is composed of the injections $\mathfrak{D}(\Omega) \hookrightarrow \mathfrak{B}_0(\Omega) \hookrightarrow \mathfrak{B}(\Omega) \hookrightarrow \mathcal{E}(\Omega) \hookrightarrow \mathfrak{C}(\Omega)$ and of the map $f \mapsto T_f$ from $\mathfrak{C}(\Omega)$ into $\mathfrak{D}'(\Omega)$, which are continuous by Examples 2.12.8, 2.5.10, 2.5.5, 2.5.7, and Proposition 1.4 respectively.

EXAMPLE 4. This time let $i: \mathfrak{D}(\Omega) \to \mathfrak{D}'(\Omega)$ be the map $\varphi \mapsto T_\varphi$ and $j = 1_{\mathfrak{D}'(\Omega)}$. Then $(i, \mathfrak{D}'(\Omega), j)$ is a normal triple. Indeed, we saw in the previous example that i is continuous, and we know from Proposition 1.3 that $\text{Im}(i)$ is everywhere dense.

EXAMPLE 5. If $i: \mathfrak{D}(\Omega) \hookrightarrow \mathcal{E}(\Omega)$ and $j: \mathcal{E}(\Omega) \to \mathfrak{D}'(\Omega)$ is the map $f \mapsto T_f$, then $(i, \mathcal{E}(\Omega), j)$ is a normal triple. In fact, it follows from Example 3 that both i and j are continuous. Thus it remains for us to prove that $\mathrm{Im}(i)$ is dense in $\mathcal{E}(\Omega)$. Let $\varphi_0 \in \mathcal{E}(\Omega)$ and let V be a neighborhood of 0 in $\mathcal{E}(\Omega)$ which we may assume to consist of all functions $\varphi \in \mathcal{E}(\Omega)$ which satisfy $|\partial^p\varphi(x)| \leqq \epsilon$ for all $|p| \leqq m$ and $x \in K$, where $m \in \mathbf{N}$ and K is a compact subset of Ω. By Proposition 2.12.5 there exists $\chi \in \mathfrak{D}(\Omega)$ such that $\chi(x) = 1$ for all x belonging to a neighborhood of K. Then

$$\chi\varphi_0 \in \mathfrak{D}(\Omega) \qquad \text{and} \qquad \chi\varphi_0 \in \varphi_0 + V,$$

which proves our assertion.

It follows that the dual $\mathcal{E}'(\Omega)$ of $\mathcal{E}(\Omega)$ is a space of distributions. We shall now characterize the distributions belonging to $\mathcal{E}'(\Omega)$.

PROPOSITION 3. *Let Ω be an open subset of $\mathbf{R}^n$. A distribution $T \in \mathfrak{D}'(\Omega)$ belongs to $\mathcal{E}'(\Omega)$ if and only if* Supp T *is a compact subset of* Ω.

Proof. (a) Suppose that $K = \mathrm{Supp}\, T$ is a compact subset of Ω. Let $\chi \in \mathfrak{D}(\Omega)$ be such that $\chi(x) = 1$ for all x belonging to a neighborhood of K. For $\varphi \in \mathcal{E}(\Omega)$ we have $\chi\varphi \in \mathfrak{D}(\Omega)$, and we define

$$\langle S, \varphi\rangle = \langle T, \chi\varphi\rangle.$$

Clearly, S is a linear form on $\mathcal{E}(\Omega)$. Also, S coincides with T on $\mathfrak{D}(\Omega)$. Indeed, if $\varphi \in \mathfrak{D}(\Omega)$, then $\varphi = \chi\varphi + (1 - \chi)\varphi$, and since

$$\mathrm{Supp}(1 - \chi)\varphi \cap K = \emptyset,$$

we have $\langle T, \varphi\rangle = \langle T, \chi\varphi\rangle = \langle S, \varphi\rangle$.

Thus we have only to prove that S is continuous on $\mathcal{E}(\Omega)$. Let (φ_k) be a sequence of elements of $\mathcal{E}(\Omega)$ which tends to zero in $\mathcal{E}(\Omega)$. Then

$$\mathrm{Supp}(\chi\varphi_k) \subset \mathrm{Supp}\, \chi$$

for every k, and the sequence $(\chi\varphi_k)$ tends to zero in $\mathfrak{D}(\Omega)$ since by the Leibniz formula (Proposition 2.5.3) $\partial^p(\chi\varphi_k)$ is a linear combination of terms $\partial^q\chi \cdot \partial^r\varphi_k$ with $q \leqq p$, $r \leqq p$. Since T is continuous on $\mathfrak{D}(\Omega)$, the expression $\langle S, \varphi_k\rangle = \langle T, \chi\varphi_k\rangle$ tends to zero as $k \to \infty$.

(b) Let $T \in \mathcal{E}'(\Omega)$ and assume that Supp T is not compact. Let $(K_k)_{k\in\mathbf{N}}$ be a sequence of compact subsets of Ω such that $K_k \subset K_{k+1}$ and that every compact subset of Ω is contained in some K_k (Example 2.6.3). Then Supp T meets every set $\complement_\Omega K_k$, and thus for every $k \in \mathbf{N}$ there exists $\varphi_k \in \mathfrak{D}(\Omega)$ such that $\langle T, \varphi_k\rangle = 1$ and Supp $\varphi_k \subset \complement_\Omega K_k$. But the sequence (φ_k) converges to zero in $\mathcal{E}(\Omega)$ since given any compact subset K of Ω, we have $\varphi_k(x) = 0$ for $x \in K$, provided k is such that $K \subset K_k$. This contradicts the assumption that T is a continuous linear form on $\mathcal{E}(\Omega)$. ∎

We know (Example 2.5.9) that the space $\mathcal{E}(\Omega)$ induces on each space $\mathcal{D}(K)$ its own topology ($K \subset \Omega$). Furthermore, the topology of $\mathcal{D}(\Omega)$, which we have introduced in Example 2.12.6 as the finest locally convex topology for which the injections $\mathcal{D}(K) \hookrightarrow \mathcal{D}(\Omega)$ are continuous, coincides with the topology $\beta(\mathcal{D}(\Omega), \mathcal{D}'(\Omega))$ since $\mathcal{D}(\Omega)$ is barrelled (Exercise 3.6.5(b)). The situation is analogous for distributions:

THEOREM 1. *Let Ω be an open subset of $\mathbf{R}^n$, and for each compact subset K of Ω denote by $\mathcal{E}'(K)$ the vector space of all distributions on Ω whose support is contained in K. Equip $\mathcal{E}'(K)$ with the topology induced by $\mathcal{D}'(\Omega)$. Then the strong topology $\beta(\mathcal{E}'(\Omega), \mathcal{E}(\Omega))$ on $\mathcal{E}'(\Omega)$ is the finest locally convex topology $\mathcal{T}$ for which the injections $\mathcal{E}'(K) \hookrightarrow \mathcal{E}'(\Omega)$ are continuous. Furthermore, $\mathcal{D}'(\Omega)$ and $\mathcal{E}'(\Omega)$ induce on $\mathcal{E}'(K)$ the same topology.*

Proof. Denote by $\mathcal{E}'_\beta$ the space $\mathcal{E}'(\Omega)$ equipped with the strong topology and by $\mathcal{E}'_\mathcal{T}$ the space $\mathcal{E}'(\Omega)$ equipped with the topology $\mathcal{T}$. Let us prove first that $\mathcal{E}'_\beta$ induces on $\mathcal{E}'(K)$ the same topology as $\mathcal{D}'(\Omega)$. Since the injection $\mathcal{D}(\Omega) \hookrightarrow \mathcal{E}(\Omega)$ is continuous, its transpose $\mathcal{E}'_\beta \to \mathcal{D}'(\Omega)$ is also continuous, and $\mathcal{E}'_\beta$ induces on $\mathcal{E}'(K)$ a finer topology than $\mathcal{D}'(\Omega)$. Conversely, let $\chi \in \mathcal{D}(\Omega)$ be such that $\chi(x) = 1$ in a neighborhood of K. If B is a bounded subset of $\mathcal{E}(\Omega)$, then χB is a bounded subset of $\mathcal{D}(\Omega)$. Indeed, for all $\varphi \in B$ we have $\operatorname{Supp}(\chi\varphi) \subset \operatorname{Supp}\chi$, and it follows from the Leibniz formula (Proposition 2.5.3) that for every $p \in \mathbf{N}^n$ the expressions $\max_x |\partial^p(\chi\varphi)(x)|$ are bounded as φ varies in B. We have

$$(\chi B)^\circ \cap \mathcal{E}'(K) \subset B^\circ \cap \mathcal{E}'(K)$$

since if $T \in (\chi B)^\circ \cap \mathcal{E}'(K)$ and $\varphi \in B$, then

$$|\langle T, \varphi\rangle| = |\langle T, \chi\varphi\rangle| \leqq 1.$$

Now, an arbitrary neighborhood of 0 in $\mathcal{E}'_\beta$ contains a set of the form B°, and $(\chi B)^\circ$ is a neighborhood of 0 in $\mathcal{D}'(\Omega)$, which proves that the topology induced by $\mathcal{E}'_\beta$ on $\mathcal{E}'(K)$ is also coarser than the one induced by $\mathcal{D}'(\Omega)$.

It follows that the injections $\mathcal{E}'(K) \hookrightarrow \mathcal{E}'_\beta$ are continuous; hence by Proposition 2.12.1 the bijection $\mathcal{E}'_\mathcal{T} \to \mathcal{E}'_\beta$ is continuous. Since the space $\mathcal{E}'_\beta$ is bornological (Example 3.16.1), in order to show that the bijection $\mathcal{E}'_\beta \to \mathcal{E}'_\mathcal{T}$ is also continuous, it is enough to show that every bounded subset of $\mathcal{E}'_\beta$ is also a bounded subset of $\mathcal{E}'_\mathcal{T}$ (Proposition 3.7.1(a)). Let A be a bounded subset of $\mathcal{E}'_\beta$. Since $\mathcal{E}(\Omega)$ is barrelled, the set A is equicontinuous; i.e., there exists $\epsilon > 0$, $m \in \mathbf{N}$ and a compact subset K of Ω such that

$$|\partial^p\varphi(x)| \leqq \epsilon \quad \text{for} \quad |p| \leqq m \quad \text{and} \quad x \in K \tag{1}$$

implies $|\langle T, \varphi\rangle| \leqq 1$ for all $T \in A$. Consequently, $\operatorname{Supp} T \subset K$ for all $T \in A$, since if $\operatorname{Supp}\varphi \subset \complement K$, then $\lambda\varphi$ satisfies conditions (1) for every

$\lambda \in \mathbf{K}$, and therefore $\langle T, \varphi \rangle = 0$. Thus $A \subset \mathcal{E}'(K)$ and A is bounded in $\mathcal{E}'(K)$ since $\mathcal{E}'(K) \hookrightarrow \mathcal{E}'_\beta$ is a strict morphism. It follows from the trivial half of Theorem 2.12.2 that A is bounded in $\mathcal{E}'_T$. ∎

We conclude this section with the following result which permits us to "piece together" distributions.

PROPOSITION 4. *Let Ω be an open subset of $\mathbf{R}^n$ and $(\Omega_\iota)_{\iota \in I}$ an open cover of Ω. Suppose that for each index $\iota \in I$ we are given a distribution $T_\iota \in \mathfrak{D}'(\Omega_\iota)$ and that if $\Omega_\iota \cap \Omega_\kappa \neq \emptyset$, then T_ι is equal to T_κ in $\Omega_\iota \cap \Omega_\kappa$. Then there exists a unique distribution $T \in \mathfrak{D}'(\Omega)$ such that T is equal to T_ι in Ω_ι for each $\iota \in I$.*

REMARK 1. In terms of the theory of sheaves, the proposition expresses the fact that the presheaf $\mathfrak{F}$ of germs of distributions on Ω is a sheaf. Furthermore we have

$$H^p(V, \mathfrak{F}) = 0$$

for $p \geqq 1$ and any paracompact differentiable manifold V. This follows from the fact that $\mathfrak{F}$ is a fine sheaf, which in turn is a consequence (just like Proposition 4) of the existence of partitions of unity subordinated to a given open cover [87, 45, 36].

Proof. The uniqueness of T follows from Proposition 1. To prove its existence, let $(\alpha_\iota)_{\iota \in I}$ be a locally finite infinitely differentiable partition of unity subordinated to the cover (Ω_ι) (Remark 2.12.4). For $\varphi \in \mathfrak{D}(\Omega)$ define

$$\langle T, \varphi \rangle = \sum_{\iota \in I} \langle T_\iota, \alpha_\iota \varphi \rangle.$$

The sum on the right-hand side has only finitely many non-zero terms since the compact set Supp φ meets only finitely many of the sets Supp α_ι. Clearly, T is a linear form on $\mathfrak{D}(\Omega)$. To prove that T is continuous on $\mathfrak{D}(\Omega)$, it is sufficient to prove that for each compact subset K of Ω the map $\varphi \mapsto \langle T, \varphi \rangle$ is continuous on $\mathfrak{D}(K)$ (Proposition 2.12.1). Let H be the finite subset of I consisting of those indices ι for which Supp α_ι meets K. It follows from the Leibniz formula (Proposition 2.5.3), similarly as in part (a) of the proof of Proposition 3, that for each $\iota \in I$ the map $\varphi \mapsto \alpha_\iota \varphi$ from $\mathfrak{D}(K)$ into $\mathfrak{D}(\Omega_\iota)$ is continuous. Let h be the number of elements in H and $\epsilon > 0$ arbitrary. Since T_ι is continuous on $\mathfrak{D}(\Omega_\iota)$, for each $\iota \in H$ there exists a neighborhood V_ι of 0 in $\mathfrak{D}(K)$ such that $|\langle T_\iota, \alpha_\iota \varphi \rangle| < \epsilon h^{-1}$ for $\varphi \in V_\iota$. Thus $\varphi \in \bigcap_{\iota \in H} V_\iota$ implies

$$|\langle T, \varphi \rangle| \leqq \sum_{\iota \in H} |\langle T_\iota, \alpha_\iota \varphi \rangle| < \epsilon,$$

which proves the continuity of T.

Finally, let φ be an element of $\mathfrak{D}(\Omega)$ which has its support in Ω_ι. Then $\text{Supp}(\alpha_\kappa\varphi) \subset \Omega_\iota \cap \Omega_\kappa$, and thus by hypothesis $\langle T_\kappa, \alpha_\kappa\varphi\rangle = \langle T_\iota, \alpha_\kappa\varphi\rangle$. Consequently, we have

$$\langle T, \varphi\rangle = \sum_{\kappa\in I} \langle T_\kappa, \alpha_\kappa\varphi\rangle = \sum_{\kappa\in I} \langle T_\iota, \alpha_\kappa\varphi\rangle = \left\langle T_\iota, \sum_{\kappa\in I} \alpha_\kappa\varphi\right\rangle = \langle T_\iota, \varphi\rangle;$$

i.e., T is equal to T_ι in Ω_ι. ▮

Exercises

1. Let Ω be an open subset of $\mathbf{R}^n$, U an open subset of Ω, $T \in \mathfrak{D}'(\Omega)$ and T_U the restriction of T to U. Prove that $\text{Supp}\, T_U = U \cap \text{Supp}\, T$.

2. Show that if Ω is an open subset of $\mathbf{R}^n$ and U an open subset of Ω, then the map $T \mapsto T_U$ from $\mathfrak{D}'(\Omega)$ into $\mathfrak{D}'(U)$ is not necessarily surjective. (*Hint:* Consider $\Omega = \mathbf{R}$, $U =]0, \infty[$ and $T_f \in \mathfrak{D}'(U)$, where $f(x) = e^{1/x}$ for $x > 0$.)

3. Let $\mathfrak{F}$ be a filter on $\mathfrak{D}'(\Omega)$ and assume that for every compact subset K there exists $A \in \mathfrak{F}$ such that $\text{Supp}\, T \cap K = \emptyset$ for all $T \in A$. Prove that $\mathfrak{F}$ converges to zero.

§3. *Derivation*

Perhaps the most important property of distributions is that for each index j $(1 \leqq j \leqq n)$ one can define a linear map $\mathfrak{D}' \to \mathfrak{D}'$ which generalizes the usual partial derivation ∂_j. We shall denote this map by the same symbol ∂_j, since this abuse does not in general lead to confusion. To see how this map should be defined, let f be a once continuously differentiable function and let $T = T_f$ be the distribution we associated with it in Example 1.1. Then it is reasonable to require that the distribution $\partial_j T_f$ be associated with the function $\partial_j f$. If $\varphi \in \mathfrak{D}$, then, integrating by parts with respect to x_j and taking into account that φ vanishes outside a compact set, we have

$$\int_{\mathbf{R}^n} \partial_j f(x) \cdot \varphi(x)\, dx = -\int_{\mathbf{R}^n} f(x) \cdot \partial_j\varphi(x)\, dx.$$

If the above requirement is satisfied, we can rewrite this equation as

$$\langle \partial_j T, \varphi\rangle = -\langle T, \partial_j\varphi\rangle,$$

which we shall adopt as the general definition of $\partial_j T$. In other words, the map $\partial_j\colon \mathfrak{D}' \to \mathfrak{D}'$ will be the opposite of the transpose of the map $\partial_j\colon \mathfrak{D} \to \mathfrak{D}$. Before giving a formal definition, however, we must ascertain that the last map is continuous.

Proposition 1. *Let Ω be an open subset of $\mathbf{R}^n$. For each index j $(1 \leqq j \leqq n)$ the linear map $\varphi \mapsto \partial_j\varphi$ from $\mathfrak{D}(\Omega)$ into itself is continuous.*

Proof. It is sufficient to prove that for any compact subset K of Ω the map $\varphi \mapsto \partial_j\varphi$ from $\mathfrak{D}(K)$ into $\mathfrak{D}(\Omega)$ is continuous (Proposition 2.12.1). Let V be a neighborhood of 0 in $\mathfrak{D}(\Omega)$. There exist $\epsilon > 0$ and $k \in \mathbf{N}$ such that $V \cap \mathfrak{D}(K)$ contains the set

$$\{\varphi \mid |\partial^p\varphi(x)| \leqq \epsilon,\ |p| \leqq k\}.$$

The set

$$U = \{\varphi \mid |\partial^p\varphi(x)| \leqq \epsilon,\ |p| \leqq k + 1\}$$

is a neighborhood of 0 in $\mathfrak{D}(K)$ and $\varphi \in U$ implies $\partial_j\varphi \in V$. ▮

COROLLARY. *For every multi-index $p \in \mathbf{N}^n$ the map $\varphi \mapsto \partial^p\varphi$ is continuous from $\mathfrak{D}(\Omega)$ into itself.*

DEFINITION 1. *Let Ω be an open subset of $\mathbf{R}^n$ and $p \in \mathbf{N}^n$. We denote by ∂^p the linear map from $\mathfrak{D}'(\Omega)$ into $\mathfrak{D}'(\Omega)$ defined by*

$$\langle \partial^p T, \varphi \rangle = (-1)^{|p|}\langle T, \partial^p\varphi \rangle$$

for $T \in \mathfrak{D}'(\Omega)$, $\varphi \in \mathfrak{D}(\Omega)$. The image $\partial^p T = T^{(p)}$ of T is called the partial derivative of index p of T.

It follows from the corollary of Proposition 1 and the corollary of Proposition 3.12.3 that the map $\partial^p\colon \mathfrak{D}'(\Omega) \to \mathfrak{D}'(\Omega)$ is continuous. With the present definition every distribution has partial derivatives of all orders; this is one reason why distributions are useful. If $f \in \mathcal{E}^{(|p|)}(\Omega)$ and $g = \partial^p f$, then we see by integrating by parts $|p|$ times that $\partial^p T_f = T_g$. Now, $\mathcal{E}^{(|p|)}(\Omega)$ is dense in $\mathfrak{D}'(\Omega)$ by Proposition 1.3; hence ∂^p is the unique continuous map from $\mathfrak{D}'(\Omega)$ into itself which coincides with the usual partial derivation operator ∂^p on $\mathcal{E}^{(|p|)}(\Omega)$ (Proposition 2.9.5). We shall have a similar situation repeatedly: all operations introduced in this chapter can be defined either by transposition or by continuous extension. Let us also observe that $\partial_i\partial_j T = \partial_j\partial_i T$ for $T \in \mathfrak{D}'(\Omega)$, $1 \leqq i, j \leqq n$ since the analogous formula is valid for functions belonging to $\mathfrak{D}(\Omega)$. Clearly, $\mathrm{Supp}(\partial^p T) \subset \mathrm{Supp}\, T$ for any $p \in \mathbf{N}^n$.

EXAMPLE 1. Let Y be the Heaviside function on $\mathbf{R}$ defined by

$$Y(x) = \begin{cases} 1 & \text{if } x \geqq 0, \\ 0 & \text{if } x < 0. \end{cases}$$

The map

$$\varphi \mapsto \int_{-\infty}^{\infty} Y(x)\varphi(x)\, dx = \int_0^{\infty} \varphi(x)\, dx$$

from $\mathfrak{D}$ into $\mathbf{K}$ is continuous since

$$\left|\int_0^{\infty} \varphi(x)\, dx\right| \leqq a \cdot \max_x |\varphi(x)|$$

if the support of φ is contained in the interval $[-a, a]$. Thus Y defines a distribution which we shall also denote by Y, i.e.,

$$\langle Y, \varphi\rangle = \int_0^\infty \varphi(x)\,dx$$

for $\varphi \in \mathfrak{D}$. Observe that the definition of the distribution Y is not covered by Example 1.1 since the function Y is not continuous.

Now, for all $\varphi \in \mathfrak{D}$ we have by Definition 1

$$\begin{aligned}\langle \partial Y, \varphi\rangle &= -\langle Y, \partial\varphi\rangle = -\int_0^\infty \varphi'(x)\,dx \\ &= -\varphi(x)|_0^\infty = \varphi(0) = \langle \delta, \varphi\rangle;\end{aligned}$$

i.e., $Y' = \partial Y = \delta$. Note that the derivative of the function Y in the usual sense is the function whose value is 0 for $x \neq 0$ and is not defined at the origin.

EXAMPLE 2. If $p \in \mathbf{N}^n$, then the derivative $\partial^p \delta = \delta^{(p)}$ of the Dirac measure δ (Example 1.2) is defined by

$$\langle \delta^{(p)}, \varphi\rangle = (-1)^{|p|}\varphi^{(p)}(0),$$

where $\varphi \in \mathfrak{D}$ and $\varphi^{(p)} = \partial^p\varphi$.

EXAMPLE 3. More generally than in Example 1, let f be a function which is defined and continuous on $\mathbf{R}$ except for finitely many points $a_1 < a_2 < \cdots < a_s$. We also assume that at each point a_k $(1 \leqq k \leqq s)$ the one-sided limits

$$f(a_k + 0) = \lim_{\epsilon \downarrow 0} f(a_k + \epsilon)$$

and

$$f(a^k - 0) = \lim_{\epsilon \downarrow 0} f(a_k - \epsilon)$$

exist and are finite. Then f defines a distribution T_f if for $\varphi \in \mathfrak{D}$ we set

$$\langle T_f, \varphi\rangle = \int_{-\infty}^{\infty} f(x)\varphi(x)\,dx = \sum_{k=0}^{s} \int_{a_k}^{a_{k+1}} f(x)\varphi(x)\,dx$$

$(a_0 = -\infty, a_{s+1} = \infty)$, since

$$\left|\int_{-\infty}^{\infty} f(x)\varphi(x)\,dx\right| \leqq \max_x |\varphi(x)| \cdot \int_{-a}^{a} |f(x)|\,dx$$

if Supp φ is contained in the interval $[-a, a]$.

Let us now suppose that in each open interval $]a_k, a_{k+1}[$ $(0 \leqq k \leqq s)$ the function f has a continuous first derivative f' and at each point a_k $(1 \leqq k \leqq s)$ also the one-sided limits $f'(a_k + 0)$ and $f'(a_k - 0)$ exist and are finite. Then the distribution $T_{f'}$ is well defined and, denoting the jump

$f(a_k + 0) - f(a_k - 0)$ of f at a_k $(1 \leqq k \leqq s)$ by j_k, we have

$$\langle \partial T_f, \varphi \rangle = -\langle T_f, \partial\varphi \rangle = -\sum_{k=0}^{s} \int_{a_k}^{a_{k+1}} f(x)\varphi'(x)\,dx$$

$$= -\sum_{k=0}^{s} f(x)\varphi(x)\Big|_{a_k+0}^{a_{k+1}-0} + \sum_{k=0}^{s} \int_{a_k}^{a_{k+1}} f'(x)\varphi(x)\,dx$$

$$= \sum_{k=1}^{s} j_k\varphi(a_k) + \int_{-\infty}^{\infty} f'(x)\varphi(x)\,dx,$$

and therefore

$$\partial T_f = T_{f'} + \sum_{k=1}^{s} j_k \delta_{a_k}.$$

In particular, if f is continuous, we have $\partial T_f = T_{f'}$.

We shall now study in detail the maps $\partial_j\colon \mathfrak{D}' \to \mathfrak{D}'$, and for this we need to look more closely also at the maps $\partial_j\colon \mathfrak{D} \to \mathfrak{D}$. We start with the case $n = 1$.

The continuous linear map $\partial\colon \mathfrak{D}(\mathbf{R}) \to \mathfrak{D}(\mathbf{R})$ is injective since $\partial\varphi = 0$ implies that φ is a constant, but Supp φ is compact, and therefore $\varphi = 0$.

A function $\chi \in \mathfrak{D}(\mathbf{R})$ belongs to $H = \text{Im}(\partial)$ if and only if it satisfies the relation

$$\int_{-\infty}^{\infty} \chi(x)\,dx = 0. \tag{1}$$

Indeed, if $\chi = \partial\psi$ for some $\psi \in \mathfrak{D}(\mathbf{R})$, then

$$\int_{-\infty}^{\infty} \chi(x)\,dx = \int_{-\infty}^{\infty} \psi'(x)\,dx = \psi(x)\Big|_{-\infty}^{\infty} = 0.$$

Conversely, if χ satisfies (1), then the function ψ defined by

$$\psi(x) = \int_{-\infty}^{x} \chi(t)\,dt$$

belongs to $\mathfrak{D}(\mathbf{R})$ since ψ is a constant for large values of x, but because of (1) this constant must be zero. Clearly, $\chi = \partial\psi$.

By Example 1.1, the linear form $\chi \mapsto \int_{-\infty}^{\infty} \chi(x)\,dx$ is continuous on $\mathfrak{D}(\mathbf{R})$; therefore H is a closed hyperplane (Proposition 2.5.7) whose equation is (1).

Let φ_0 be a fixed element of $\mathfrak{D}(\mathbf{R})$ such that $\int_{-\infty}^{\infty} \varphi_0(x)\,dx = 1$. Every element $\varphi \in \mathfrak{D}(\mathbf{R})$ has a unique decomposition $\varphi = \lambda\varphi_0 + \chi$, where $\lambda \in \mathbf{K}$ and $\chi \in H$. Indeed, if we set $\lambda = \int_{-\infty}^{\infty} \varphi(x)\,dx$, then $\varphi - \lambda\varphi_0$ belongs to H, and if $\lambda_1\varphi_0 + \chi_1 = \lambda_2\varphi_0 + \chi_2$, then $(\lambda_1 - \lambda_2)\varphi_0 = \chi_2 - \chi_1 \in H$; hence $\lambda_1 = \lambda_2$ and $\chi_1 = \chi_2$.

Finally, the map $\chi = \partial\psi \mapsto \psi$ from H onto $\mathfrak{D}(\mathbf{R})$ is continuous; i.e., ∂ is a strict morphism. By Proposition 2.12.1 we have to prove that for any compact subset K of $\mathbf{R}$ the map $\chi \mapsto \psi$ from $H \cap \mathfrak{D}(K)$ into $\mathfrak{D}(\mathbf{R})$ is continuous. Let V be a neighborhood of 0 in $\mathfrak{D}(\mathbf{R})$ and let $L = [a, b]$ be a compact interval containing K. Then $V \cap \mathfrak{D}(L)$ contains a set of the form $\{\psi \mid |\partial^p\psi(x)| \leqq \epsilon, p \leqq m\}$, $\epsilon > 0$, $m \in \mathbf{N}$. Let U be the neighborhood of 0 in $H \cap \mathfrak{D}(K)$ defined by

$$\left\{\chi \mid |\chi(x)| \leqq \frac{\epsilon}{b-a}, |\partial^p\chi(x)| \leqq \epsilon, 0 \leqq p \leqq m-1\right\}$$

(if $m = 0$, the second condition should be omitted). Then $\chi \in U$ implies $\psi \in V$ since

$$|\psi(x)| \leqq \left|\int_a^x \chi(t)\,dt\right| \leqq \frac{\epsilon}{b-a}(b-a) = \epsilon$$

and

$$|\partial^p\psi(x)| = |\partial^{p-1}\chi(x)| \leqq \epsilon \qquad \text{for} \qquad 1 \leqq p \leqq m.$$

Collecting our results, we obtain

PROPOSITION 2. *The map* $\partial\colon \mathfrak{D}(\mathbf{R}) \to \mathfrak{D}(\mathbf{R})$ *is an injective strict morphism whose image* H *is a closed hyperplane. Given an element* $\varphi_0 \in \mathfrak{D}(\mathbf{R})$ *such that* $\int_{-\infty}^{\infty} \varphi_0(x)\,dx = 1$, *each element* $\varphi \in \mathfrak{D}(\mathbf{R})$ *has a unique decomposition*

$$\varphi = \lambda\varphi_0 + \chi, \tag{2}$$

where $\lambda = \int_{-\infty}^{\infty} \varphi(x)\,dx$ *and* $\chi \in H$.

Now we are able to prove the following result:

PROPOSITION 3. *The linear map* $\partial\colon \mathfrak{D}'(\mathbf{R}) \to \mathfrak{D}'(\mathbf{R})$ *is a surjective strict morphism whose kernel is the one-dimensional subspace of* $\mathfrak{D}'(\mathbf{R})$ *consisting of all distributions* T_f *associated with constant functions* f.

Proof. (a) Let $T \in \mathfrak{D}'(\mathbf{R})$ and $\partial T = S$. Using the decomposition (2), we have

$$\langle T, \varphi\rangle = \lambda\langle T, \varphi_0\rangle + \langle T, \chi\rangle = \lambda\langle T, \varphi_0\rangle - \langle S, \psi\rangle, \tag{3}$$

where $\chi = \partial\psi$. In particular, if $S = 0$, we have

$$\langle T, \varphi\rangle = \lambda\langle T, \varphi_0\rangle = \langle T, \varphi_0\rangle\int_{-\infty}^{\infty} \varphi(x)\,dx;$$

i.e., $T = T_f$, where $f(x) = \langle T, \varphi_0\rangle$.

(b) Let us prove that ∂ is surjective. Given $S \in \mathfrak{D}'(\mathbf{R})$, choose an arbitrary constant $\gamma \in \mathbf{K}$ and define T by

$$\langle T, \varphi\rangle = \gamma\lambda - \langle S, \psi\rangle,$$

where φ is arbitrary in $\mathfrak{D}(\mathbf{R})$ and λ, ψ have the same meaning as before.

Since the representation (2) is unique, T is a well-defined linear form on $\mathfrak{D}(\mathbf{R})$. Next, T is continuous since the maps $\varphi \mapsto \lambda$, $\lambda \mapsto \gamma\lambda$, $\varphi \mapsto \chi = \varphi - \lambda\varphi_0$, and $\psi \mapsto \langle S, \psi \rangle$ are known to be so and $\chi \mapsto \psi$ is continuous by Proposition 2. Finally,

$$\langle \partial T, \psi \rangle = -\langle T, \partial\psi \rangle = \langle S, \psi \rangle$$

for every $\psi \in \mathfrak{D}(\mathbf{R})$, i.e., $\partial T = S$. Observe that by (3) the distribution T is completely determined by the value $\gamma = \langle T, \varphi_0 \rangle$.

(c) With every $S \in \mathfrak{D}'(\mathbf{R})$ we associate that distribution $T \in \mathfrak{D}'(\mathbf{R})$ for which $\partial T = S$ and $\langle T, \varphi_0 \rangle = 0$, and we set $T = I(S)$. The map $I\colon \mathfrak{D}'(\mathbf{R}) \to \mathfrak{D}'(\mathbf{R})$ is clearly linear. By definition it satisfies

$$\partial I(S) = S \qquad \text{for all} \qquad S \in \mathfrak{D}'(\mathbf{R}) \tag{4}$$

and by (3) it satisfies

$$\langle I(S), \varphi \rangle = -\langle S, \psi \rangle \qquad \text{for all} \qquad S \in \mathfrak{D}'(\mathbf{R}) \quad \text{and} \quad \varphi \in \mathfrak{D}(\mathbf{R}). \tag{5}$$

Relation (5) means that $-I$ is the transpose of the continuous linear map $\varphi \mapsto \psi$. Hence by the corollary of Proposition 3.12.3 the map I is continuous. It follows from (4) therefore that ∂ is a strict morphism (Proposition 2.7.2). ▌

REMARK 1. $\mathfrak{D}'(\mathbf{R}) = N \oplus L$, where N consists of the constants and L of those distributions T for which $\langle T, \varphi_0 \rangle = 0$. If $\varphi = \lambda\varphi_0 + \chi \in \mathfrak{D}(\mathbf{R})$, $T = \gamma + U$, $U \in L$, then

$$\langle T, \varphi \rangle = \gamma\lambda + \langle U, \chi \rangle$$

(cf. Proposition 3.14.1).

REMARK 2. The preceding results hold if we replace $\mathbf{R}$ by a connected open subset of $\mathbf{R}$, i.e., by an open interval. They do not hold if the open set is not connected. Thus $\partial Y = 0$ in $\complement\{0\}$, but Y is not a constant there.

REMARK 3. In Volume 2 we shall establish similar results for elliptic differential operators on $\mathbf{R}^n$. For $n = 1$ the operator ∂ is elliptic.

COROLLARY. (a) *For every distribution $S \in \mathfrak{D}'(\mathbf{R})$ and every $p \in \mathbf{N}$ there exists a distribution $T \in \mathfrak{D}'(\mathbf{R})$ such that $\partial^p T = S$.*

(b) *If $T \in \mathfrak{D}'(\mathbf{R})$ is such that $\partial^p T = 0$, then $T = T_f$, where f is a polynomial of degree $\leqq p - 1$.*

Proof. (a) By Proposition 3 the assertion is true for $p = 1$. Assume that it is true for $p - 1$ $(p > 1)$. Then there exists $U \in \mathfrak{D}'(\mathbf{R})$ such that $\partial^{p-1} U = S$. By Proposition 3 there exists $T \in \mathfrak{D}'(\mathbf{R})$ such that $\partial T = U$. But then $\partial^p T = \partial^{p-1} U = S$.

(b) By Proposition 3 the assertion is true for $p = 1$. Assume that it is true for $p - 1$ $(p > 1)$ and let $T \in \mathfrak{D}'(\mathbf{R})$ be such that $\partial^p T = 0$. Set $S = \partial T$. Then $\partial^{p-1} S = 0$; hence by the induction hypothesis $S = T_g$, where g is a polynomial of degree $\leq p - 2$. Let f be a polynomial of degree $\leq p - 1$ such that $\partial f = g$. Then $\partial(T - T_f) = S - T_g = 0$; hence by Proposition 3 the distribution $T - T_f$ is associated with a constant. ∎

Before we pass to the case $n > 1$, we must introduce the translation of distributions. If f is a function defined on $\mathbf{R}^n$ and $h \in \mathbf{R}^n$, then we define the function $\tau_h f$ by $(\tau_h f)(x) = f(x - h)$ for all $x \in \mathbf{R}^n$. Let $f \in \mathfrak{C}$ and $g = \tau_h f$. Then we have for the associated distributions the relations

$$\begin{aligned}\langle T_g, \varphi\rangle &= \int_{\mathbf{R}^n} f(x - h)\varphi(x)\,dx \\ &= \int_{\mathbf{R}^n} f(x)\varphi(x + h)\,dx = \langle T_f, \tau_{-h}\varphi\rangle.\end{aligned}$$

This motivates

Definition 2. *Let $T \in \mathfrak{D}'$. The translate of T by the vector $h \in \mathbf{R}^n$ is the distribution $\tau_h T$ defined by*

$$\langle \tau_h T, \varphi\rangle = \langle T, \tau_{-h}\varphi\rangle$$

for all $\varphi \in \mathfrak{D}$.

The map $\tau_h: \mathfrak{D} \to \mathfrak{D}$ is an isomorphism since

$$\max_x |\partial^p \varphi(x)| = \max_x |\partial^p(\tau_h\varphi)(x)|$$

for all $p \in \mathbf{N}^n$. It follows that the map $\tau_h: \mathfrak{D}' \to \mathfrak{D}'$, which is the transpose of the map $\tau_{-h}: \mathfrak{D} \to \mathfrak{D}$, is an isomorphism from $\mathfrak{D}'$ onto itself. Clearly, $\tau_h^{-1} = \tau_{-h}$. Hence $\tau_h: \mathfrak{D}' \to \mathfrak{D}'$ is the transpose of the inverse (or equivalently the inverse of the transpose) of $\tau_h: \mathfrak{D} \to \mathfrak{D}$; i.e., we have

$$\langle \tau_h T, \tau_h \varphi\rangle = \langle T, \varphi\rangle$$

for all $T \in \mathfrak{D}'$ and $\varphi \in \mathfrak{D}$. In general, if E and F are two vector spaces and $u: E \to F$ an isomorphism, then the isomorphism

$$\check{u} = ({}^t u)^{-1} = {}^t(u^{-1}): E^* \to F^*,$$

which satisfies

$$\langle \check{u}(x'), u(x)\rangle = \langle x', x\rangle$$

for all $x' \in E'$ and $x \in E$, is said to be *contragredient* to u.

Example 4. If δ is the Dirac measure and $a \in \mathbf{R}^n$, then $\tau_a \delta = \delta_a$ (Example 1.2).

The operation of translation plays a role in the classical definition of the derivative of a function. In fact, the partial derivative $\partial_j\varphi$ at the point x is defined by

$$\lim_{t\to 0}\frac{\varphi(x+h)-\varphi(x)}{t}=\lim_{t\to 0}\frac{(\tau_{-h}\varphi)(x)-\varphi(x)}{t},$$

where $h=(h_1,\ldots,h_n)\in\mathbf{R}^n$ is given by $h_j=t$ and $h_k=0$ for $k\neq j$. We want to show that the situation is analogous if instead of pointwise convergence we consider convergence in the sense of our topologies in $\mathfrak{D}$ and $\mathfrak{D}'$.

PROPOSITION 4. *Let $h=(h_1,\ldots,h_n)\in\mathbf{R}^n$ be a vector parallel to the x_j-axis* (i.e., $h_k=0$ for $k\neq j$) *and write $h_j=t$. For each $\varphi\in\mathfrak{D}$ the function $t^{-1}(\tau_{-h}\varphi-\varphi)$ converges to $\partial_j\varphi$ in $\mathfrak{D}$ as $h\to 0$, and for each $T\in\mathfrak{D}'$ the distribution $t^{-1}(\tau_{-h}T-T)$ converges to ∂_jT in $\mathfrak{D}'$ as $h\to 0$.*

Proof. (a) Let us set $\psi_t=t^{-1}(\tau_{-h}\varphi-\varphi)$, $K=\operatorname{Supp}\varphi$, and let L be a compact neighborhood of K. For sufficiently small values of $|t|$ the support of ψ_t is contained in L.

For each $p\in\mathbf{N}^n$ the function $\partial^p\varphi$ is uniformly continuous on $\mathbf{R}^n$. Therefore given $\epsilon>0$, we obtain by using the mean-value theorem

$$|(\partial^p\psi_t)(x)-(\partial^p\partial_j\varphi)(x)|=|(\partial_j\partial^p\varphi)(x+\theta h)-(\partial_j\partial^p\varphi)(x)|\leqq\epsilon$$

$(0\leqq\theta\leqq 1)$, provided that $|t|$ is sufficiently small.

Let V be a neighborhood of 0 in $\mathfrak{D}$. Then $V\cap\mathfrak{D}(L)$ contains a set of the form $\{\chi\mid|\partial^p\chi(x)|\leqq\epsilon,\ |p|\leqq m\}$. Thus we see that $\psi_t-\partial_j\varphi\in V$ for sufficiently small $|t|$.

(b) For each $\varphi\in\mathfrak{D}$ we have

$$\begin{aligned}\langle t^{-1}(\tau_{-h}T-T),\varphi\rangle&=t^{-1}\langle\tau_{-h}T,\varphi\rangle-t^{-1}\langle T,\varphi\rangle\\&=t^{-1}\langle T,\tau_h\varphi\rangle-t^{-1}\langle T,\varphi\rangle=\langle T,t^{-1}(\tau_h\varphi-\varphi)\rangle.\end{aligned}$$

Since T is a continuous linear form on $\mathfrak{D}$, by part (a) the last expression tends to

$$\langle T,-\partial_j\varphi\rangle=\langle\partial_jT,\varphi\rangle$$

as $t\to 0$. The conclusion follows from Proposition 1.2. ∎

We shall also need the following result:

PROPOSITION 5. *Let $T\in\mathfrak{D}'$. The map $h\mapsto\tau_hT$ from $\mathbf{R}^n$ into $\mathfrak{D}'$ is continuous and differentiable; i.e., the limits*

$$\frac{\partial}{\partial h_j}(\tau_hT)=\lim_{t\to 0}t^{-1}(\tau_{h+\Delta h}T-\tau_hT)$$

exist and are continuous, where Δh denotes a vector parallel to the x_j-axis with component $(\Delta h)_j = t$ $(1 \leqq j \leqq n)$. Furthermore, we have

$$\frac{\partial}{\partial h_j}(\tau_h T) = -\partial_j(\tau_h T) = -\tau_h(\partial_j T).$$

Proof. Let A be a bounded set in $\mathfrak{D}$. For a fixed $h \in \mathbf{R}^n$ there exists a bounded set B_h in $\mathfrak{D}$ and $\epsilon > 0$ such that $\varphi \in A$, $k \in \mathbf{R}^n$ and $|k| < \epsilon$ imply

$$\tau_{-h-k}\varphi - \tau_{-h}\varphi \in |k|B_h.$$

We have therefore

$$|\langle \tau_{h+k}T - \tau_h T, \varphi\rangle| = |\langle T, \tau_{-h-k}\varphi - \tau_{-h}\varphi\rangle| \leqq 1,$$

i.e., $\tau_{h+k}T - \tau_h T \in A^{\circ}$ for $|k| < \epsilon$ provided that ϵ is sufficiently small. This proves the continuity of the map $h \mapsto \tau_h T$.

The existence of the partial derivatives and the relation

$$\frac{\partial}{\partial h_j}(\tau_h T) = -\partial_j(\tau_h T)$$

follow from Proposition 4 applied to $\tau_h T$. For $\varphi \in \mathfrak{D}$ we have

$$\begin{aligned}\langle \partial_j(\tau_h T), \varphi\rangle &= -\langle T, \tau_{-h}(\partial_j\varphi)\rangle \\ &= -\langle T, \partial_j(\tau_{-h}\varphi)\rangle = \langle \tau_h(\partial_j T), \varphi\rangle,\end{aligned}$$

and therefore $\partial_j(\tau_h T) = \tau_h(\partial_j T)$. The continuity of the map $h \mapsto \tau_h(\partial_j T)$ follows from the first part of the proof. ∎

Clearly, a function f is independent of the variable x_j if and only if $\tau_h f = f$ for every vector h parallel to the x_j-axis. Therefore it is reasonable to state

DEFINITION 3. *The distribution $T \in \mathfrak{D}'$ is independent of the variable x_j if $\tau_h T = T$ for all vectors $h \in \mathbf{R}^n$ parallel to the x_j-axis.*

PROPOSITION 6. *A distribution $T \in \mathfrak{D}'$ is independent of the variable x_j if and only if $\partial_j T = 0$.*

Proof. If $\tau_h T - T = 0$ for all h parallel to the x_j-axis, then $\partial_j T = 0$ by Proposition 4. Conversely, if $\partial_j T = 0$, then by Proposition 5 we have

$$\frac{\partial}{\partial h_j}(\tau_h T) = -\tau_h(\partial_j T) = 0,$$

and therefore the map $t \mapsto \tau_h T$, where $h = (h_1, \ldots, h_n)$, $h_j = t$, $h_k = 0$ for $k \neq j$, is constant. It follows that $\tau_h T = \tau_0 T = T$ for all such h. ∎

Now that we are in possession of the necessary notations and terminology, we can pass to the study of the map ∂_1. If $x = (x_1, \ldots, x_n)$ is a vector in $\mathbf{R}^n$, we shall also write x' or t for x_1 and x'' for the vector $(x_2, \ldots, x_n)$ of $\mathbf{R}^{n-1}$. Thus $x = (x_1, x'') = (x', x'') = (t, x'')$, and if f is a function defined on $\mathbf{R}^n$, we shall write $f(x', x'')$ or $f(t, x'')$ for $f(x)$.

A function $\chi \in \mathfrak{D} = \mathfrak{D}(\mathbf{R}^n)$ belongs to $M = \operatorname{Im}(\partial_1)$ if and only if it satisfies the relations

$$\int_{-\infty}^{\infty} \chi(t, x'')\, dt = 0 \tag{6}$$

for all $x'' \in \mathbf{R}^{n-1}$. Indeed, if $\chi = \partial_1\psi$ for some $\psi \in \mathfrak{D}$, then

$$\int_{-\infty}^{\infty} \chi(t, x'')\, dt = \int_{-\infty}^{\infty} \partial_1\psi(t, x'')\, dt = \psi(t, x'')\Big|_{t=-\infty}^{t=\infty} = 0.$$

Conversely, if the relations (6) are satisfied, then the function ψ defined by

$$\psi(x_1, x'') = \int_{-\infty}^{x_1} \chi(t, x'')\, dt$$

belongs to $\mathfrak{D}$ since if χ vanishes outside a cube of $\mathbf{R}^n$, then ψ also vanishes outside that cube. Clearly, $\partial_1\psi = \chi$.

For each $x'' \in \mathbf{R}^{n-1}$ the linear form $\chi \mapsto \int_{-\infty}^{\infty} \chi(t, x'')\, dt$ is continuous on $\mathfrak{D}$, and therefore M is the intersection of closed hyperplanes, i.e., a closed subspace of $\mathfrak{D}$.

If $\partial_1\varphi = 0$, then φ is a constant on every straight line parallel to the x_1-axis. But since φ has compact support, this constant is necessarily zero, i.e., $\varphi = 0$. Thus the map $\partial_1 \colon \mathfrak{D} \to \mathfrak{D}$ is injective.

Let φ_0 be a fixed element of $\mathfrak{D}(\mathbf{R})$ such that $\int_{-\infty}^{\infty} \varphi_0(t)\, dt = 1$. Every element $\varphi \in \mathfrak{D}$ has a unique decomposition $\varphi = \lambda\varphi_0 + \chi$, where $\lambda \in \mathfrak{D}(\mathbf{R}^{n-1})$ and $\chi \in M$. Indeed, if we set

$$\lambda(x'') = \int_{-\infty}^{\infty} \varphi(t, x'')\, dt,$$

then $\lambda \in \mathfrak{D}(\mathbf{R}^{n-1})$ and clearly $\varphi - \lambda\varphi_0 \in M$. Furthermore, if

$$\lambda_1\varphi_0 + \chi_1 = \lambda_2\varphi_0 + \chi_2,$$

then $(\lambda_1 - \lambda_2)\varphi_0 = \chi_2 - \chi_1 \in M$, hence $\lambda_1 = \lambda_2$ and $\chi_1 = \chi_2$.

Finally, the map $\chi = \partial_1\psi \mapsto \psi$ from M onto $\mathfrak{D}$ is continuous; i.e., ∂_1 is a strict morphism. By Proposition 2.12.1 we must prove that for any compact subset K of $\mathbf{R}^n$ the map $\chi \mapsto \psi$ from $M \cap \mathfrak{D}(K)$ into $\mathfrak{D}$ is continuous. Let V be a neighborhood of 0 in $\mathfrak{D}$ and let

$$L = \{x \mid |x_j| \leqq a,\ 1 \leqq j \leqq n\}$$

be a cube containing K. Then $V \cap \mathfrak{D}(L)$ contains a set of the form

$$\{\psi \mid |\partial^p\psi(x)| \leqq \epsilon,\ |p| \leqq m\}, \qquad \epsilon > 0, \quad m \in \mathbf{N}.$$

Let U be the neighborhood of 0 in $M \cap \mathfrak{D}(K)$ composed of all functions χ which satisfy the inequalities

$$|\partial^p \chi(x)| \leqq \frac{\epsilon}{2a} \quad \text{if} \quad |p| \leqq m \quad \text{and} \quad p_1 = 0$$

and the inequalities

$$|\partial^p \chi(x)| \leqq \epsilon \quad \text{if} \quad |p| \leqq m - 1.$$

Then $\chi \in U$ implies $\psi \in V$. Indeed, if $p_1 = 0$, we have

$$|\partial^p \psi(x)| = \left| \int_{-a}^{x_1} \partial^p \chi(t, x'')\, dt \right| \leqq \frac{\epsilon}{2a} 2a = \epsilon$$

since $\chi \in M$ implies that

$$\int_{-\infty}^{\infty} \partial^p \chi(t, x'')\, dt = 0$$

for all $x'' \in \mathbf{R}^{n-1}$ and all $p \in \mathbf{N}^n$ with $p_1 = 0$. On the other hand, if $p_1 > 0$, we can write $\partial^p = \partial^q \partial_1$ with $q \in \mathbf{N}^n$, $|q| = |p| - 1$, and therefore we have $|\partial^p \psi(x)| = |\partial^q \chi(x)| \leqq \epsilon$.

Collecting our results, we obtain

PROPOSITION 7. *The map $\partial_1 \colon \mathfrak{D} \to \mathfrak{D}$ is an injective strict morphism whose image M is a closed subspace of $\mathfrak{D}$. Given an element $\varphi_0 \in \mathfrak{D}(\mathbf{R})$ such that $\int_{-\infty}^{\infty} \varphi_0(t)\, dt = 1$, each element $\varphi \in \mathfrak{D}$ has a unique decomposition*

$$\varphi = \lambda \varphi_0 + \chi, \tag{7}$$

where $\lambda \in \mathfrak{D}(\mathbf{R}^{n-1})$ is defined by $\lambda(x'') = \int_{-\infty}^{\infty} \varphi(t, x'')\, dt$ and $\chi \in M$. In particular, $\chi \in M$ if and only if $\int_{-\infty}^{\infty} \chi(t, x'')\, dt = 0$ for all $x'' \in \mathbf{R}^{n-1}$.

Now we are able to prove the following result:

PROPOSITION 8. *The linear map $\partial_1 \colon \mathfrak{D}' \to \mathfrak{D}'$ is a surjective strict morphism whose kernel consists of all distributions which are independent of the variable x_1.*

Proof. (a) We know from Proposition 6 that $\mathrm{Ker}(\partial_1)$ consists of all distributions independent of x_1.

(b) If $T \in \mathfrak{D}'$ and $\partial_1 T = S$, then using the decomposition (7), we have

$$\langle T, \varphi \rangle = \langle T, \lambda \varphi_0 \rangle + \langle T, \chi \rangle = \langle T, \lambda \varphi_0 \rangle - \langle S, \psi \rangle, \tag{8}$$

where $\chi = \partial_1 \psi$. If $\partial_1 T = 0$, we can define a distribution $G \in \mathfrak{D}'(\mathbf{R}^{n-1})$ by setting

$$\langle G, \lambda \rangle = \langle T, \lambda \varphi_0 \rangle \tag{9}$$

for $\lambda \in \mathfrak{D}(\mathbf{R}^{n-1})$, and the map $T \mapsto G$ is clearly an isomorphism from $\mathrm{Ker}(\partial_1)$ onto $\mathfrak{D}'(\mathbf{R}^{n-1})$.

(c) Let us prove that ∂_1 is surjective. Given $S \in \mathfrak{D}'$, choose an arbitrary distribution $G \in \mathfrak{D}'(\mathbf{R}^{n-1})$ and define T by

$$\langle T, \varphi \rangle = \langle G, \lambda \rangle - \langle S, \psi \rangle,$$

where φ is arbitrary in $\mathfrak{D}$ and λ and ψ have the same meaning as before. Because of the uniqueness of the decomposition (7), the linear form T is well-defined on $\mathfrak{D}$. It is also continuous since the maps $\varphi \mapsto \lambda$, $\lambda \mapsto \langle G, \lambda \rangle$, $\varphi \mapsto \chi = \varphi - \lambda\varphi_0$, and $\psi \mapsto \langle S, \psi \rangle$ are clearly continuous, and $\chi \mapsto \psi$ is continuous by Proposition 7. Finally,

$$\langle \partial_1 T, \psi \rangle = -\langle T, \partial_1 \psi \rangle = \langle S, \psi \rangle$$

for every $\psi \in \mathfrak{D}$; i.e., $\partial_1 T = S$. Observe that by (8) and (9) the distribution T is completely determined by the distribution G.

(d) With every $S \in \mathfrak{D}'$ we associate that distribution $T \in \mathfrak{D}'$ for which $\partial_1 T = S$ and $G = 0$ and set $T = I_1(S)$. The map $I_1 \colon \mathfrak{D}' \to \mathfrak{D}'$ is clearly linear. We have by definition

$$\partial_1 I_1(S) = S \qquad \text{for all} \qquad S \in \mathfrak{D}' \tag{10}$$

and by (8) and (9)

$$\langle I_1(S), \varphi \rangle = -\langle S, \psi \rangle \qquad \text{for all} \qquad S \in \mathfrak{D}' \quad \text{and} \quad \varphi \in \mathfrak{D}. \tag{11}$$

Relation (11) shows that $-I_1$ is the transpose of the continuous linear map $\varphi \mapsto \psi$. Hence by the corollary of Proposition 3.12.3 the map I_1 is continuous. It follows therefore from (10) that ∂_1 is a strict morphism (Proposition 2.7.2). ∎

Remark 4. The last two propositions show that $\mathfrak{D}(\mathbf{R}^n)$ is the direct sum of two closed subspaces $M = \mathrm{Im}(\partial_1)$ and $P \simeq \mathfrak{D}(\mathbf{R}^{n-1})$ and that $\mathfrak{D}'(\mathbf{R}^n)$ is the direct sum of two closed subspaces $N = \mathrm{Ker}(\partial_1)$ and Q, where $M^\perp = N$ and $P^\perp = Q$.

Remark 5. If $n > 1$, then the image of the map $\partial_1 \colon \mathfrak{D} \to \mathfrak{D}$ is not a hyperplane (and has, in fact, infinite codimension), the kernel of the map $\partial_1 \colon \mathfrak{D}' \to \mathfrak{D}'$ is not finite dimensional, and the operator ∂_1 is not elliptic.

Proposition 9. *Let k be an integer, $1 \leqq k \leqq n$, and $S_1, S_2, \ldots, S_k$ distributions on $\mathbf{R}^n$. There exists a distribution $T \in \mathfrak{D}'$ for which*

$$\partial_j T = S_j \qquad (1 \leqq j \leqq k) \tag{12}$$

if and only if the distributions S_j satisfy the $k(k-1)/2$ relations

$$\partial_i S_j = \partial_j S_i \qquad (1 \leqq i, j \leqq k, i \neq j). \tag{13}$$

Proof. (a) If a distribution T satisfying the relations (12) exists, then we have

$$\partial_i S_j = \partial_i \partial_j T = \partial_j \partial_i T = \partial_j S_i$$

for $1 \leqq i, j \leqq k$.

(b) Next we show that if $S \in \mathfrak{D}'$ is independent of the variables $x_m, \ldots, x_n$ $(1 \leqq m \leqq n)$, then there exists $T \in \mathfrak{D}'$ which is independent of the variables $x_m, \ldots, x_n$ such that $\partial_1 T = S$. We know from the proof of Proposition 8 that if $\varphi = \lambda\varphi_0 + \chi$, where $\chi = \partial_1 \psi$, is the decomposition (7) of $\varphi \in \mathfrak{D}$, then the distribution T defined by $\langle T, \varphi \rangle = -\langle S, \psi \rangle$ satisfies the relations $\partial_1 T = S$. Now $\partial_j \chi = \partial_1 \partial_j \psi$, and

$$\partial_j \varphi = \partial_j \lambda \cdot \varphi_0 + \partial_j \chi \qquad (1 < j \leqq n)$$

is the decomposition (7) of $\partial_j \varphi$. We have therefore $\langle T, \partial_j \varphi \rangle = -\langle S, \partial_j \psi \rangle$; and if $\partial_j S = 0$ for $m \leqq j \leqq n$, then

$$\langle \partial_j T, \varphi \rangle = -\langle T, \partial_j \varphi \rangle = \langle S, \partial_j \psi \rangle = -\langle \partial_j S, \psi \rangle = 0,$$

i.e., $\partial_j T = 0$ for $m \leqq j \leqq n$.

(c) Now we prove by induction on k that the conditions (13) are also sufficient. For $k = 1$ this follows from Proposition 8. Assume that the assertion holds for $k - 1$ and let $S_1, \ldots, S_k$ be distributions satisfying (13). Then there exists a distribution T_1 such that $\partial_j T_1 = S_j$ for $1 \leqq j \leqq k - 1$. The distribution $\partial_k T_1 - S_k$ is independent of the variables $x_1, \ldots, x_{k-1}$ since

$$\partial_j(\partial_k T_1 - S_k) = \partial_k \partial_j T_1 - \partial_j S_k = \partial_k S_j - \partial_j S_k = 0$$

by (13). It follows from part (b) that there exists a distribution T_2, independent of the variables $x_1, \ldots, x_{k-1}$, such that $\partial_k T_2 = \partial_k T_1 - S_k$. Setting $T = T_1 - T_2$, we have $\partial_j T = \partial_j T_1 = S_j$ for $1 \leqq j \leqq k - 1$ and $\partial_k T = \partial_k T_1 - \partial_k T_2 = S_k$. ∎

Remark 6. For $k = n$ Proposition 9 can be stated more beautifully in terms of *currents*, i.e., differential forms with distribution coefficients: if

$$S = S_1\, dx_1 + S_2\, dx_2 + \cdots + S_n\, dx_n$$

is a current of degree 1, then there exists a distribution (= current of degree 0) T such that

$$S = dT = \partial_1 T \cdot dx_1 + \partial_2 T \cdot dx_2 + \cdots + \partial_n T \cdot dx_n$$

if and only if

$$dS = \sum_{i<j} (\partial_i S_j - \partial_j S_i)\, dx_i \wedge dx_j = 0.$$

More generally, let V be a differentiable manifold. Let Z^p be the vector space over $\mathbf{K}$ of all currents α of degree p on V such that $d\alpha = 0$, and let B^p be the subspace of Z^p formed by those currents α for which there exists a current β of degree $p - 1$ such that $\alpha = d\beta$. Then the quotient space Z^p/B^p is isomorphic to the p-th cohomology vector space $H^p(V; \mathbf{K})$ of V. This is one form of de Rham's theorem [87, 45, 36, 17].

We conclude this section with a very simple regularity theorem (cf. Exercise 10.8). We shall establish later much stronger results in this direction.

PROPOSITION 10 (du Bois Reymond's lemma). *Let f and g be two continuous functions on $\mathbf{R}^n$ and T_f and T_g the corresponding distributions. If $\partial_j T_f = T_g$, then the partial derivative $\partial_j f$ exists and $\partial_j f = g$.*

Proof. For simplicity of notations take $j = 1$ and let us define the function h by

$$h(x) = \int_0^{x_1} g(t, x'')\, dt,$$

where $x'' = (x_2, \ldots, x_n)$ as above. The function h is continuous, $\partial_1 h$ exists and $\partial_1 h = g$. Clearly, $\partial_1 T_h = T_g$, and therefore $\partial_1(T_f - T_h) = 0$. The function $u = f - h$ is continuous as the difference of two continuous functions. Since $\partial_1 T_u = 0$, by Proposition 6 we have $\tau_h T_u = 0$ and hence also $\tau_h u = 0$ for all vectors h parallel to the x_1-axis; i.e., the function u is independent of x_1. Thus $\partial_1 f = \partial_1 h = g$. ∎

EXERCISES

1. (a) Prove that $T \in \mathfrak{D}'$ satisfies the relations $\partial_j T = 0$ for $1 \leqq j \leqq k$ ($k \leqq n$) if and only if $\tau_h T = 0$ for all vectors $h = (h_1, \ldots, h_n) \in \mathbf{R}^n$ such that $h_{k+1} = \cdots = h_n = 0$; i.e., T is independent of the variables $x_1, \ldots, x_k$.

(b) For any vector $x = (x_1, \ldots, x_n) \in \mathbf{R}^n$ write

$$x' = (x_1, \ldots, x_k) \in \mathbf{R}^k, \qquad x'' = (x_{k+1}, \ldots, x_n) \in \mathbf{R}^{n-k},$$

and

$$x = (x', x'').$$

Prove that the map $\partial_1 \ldots \partial_k \colon \mathfrak{D} \to \mathfrak{D}$ is injective and

$$\chi \in M = \operatorname{Im}(\partial_1 \ldots \partial_k) = \bigcap_{j=1}^{k} \operatorname{Im}(\partial_j)$$

if and only if

$$\int_{\mathbf{R}^k} \chi(t, x'')\, dt = 0$$

for all $x'' \in \mathbf{R}^{n-k}$. Let $\varphi_0 \in \mathfrak{D}(\mathbf{R}^k)$ be such that $\int_{\mathbf{R}^k} \varphi_0(t)\, dt = 1$. Show that every $\varphi \in \mathfrak{D}$ has a unique decomposition given by $\varphi(x) = \lambda(x'')\varphi_0(x') + \chi(x)$, where $\lambda \in \mathfrak{D}(\mathbf{R}^{n-k})$ and $\chi \in M$. Prove that $\partial_1 \ldots \partial_k \colon \mathfrak{D} \to \mathfrak{D}$ is a strict morphism.

(c) Show that the linear map $\partial_1 \ldots \partial_k \colon \mathfrak{D}' \to \mathfrak{D}'$ is a surjective strict morphism whose kernel consists of all distributions which are independent of the variables $x_1, \ldots, x_k$.

(d) Show that if $\partial_j T = 0$ for $1 \leqq j \leqq n$, then $T = T_f$, where f is a constant.

(e) Show that if $\partial^p T = 0$ for all p such that $|p| \leqq m$, then $T = T_f$, where f is a polynomial of degree $\leqq m - 1$.

2. Prove that Propositions 4 and 5 hold if the spaces $\mathfrak{D}$ and $\mathfrak{D}'$ are replaced by the spaces $\mathcal{E}$ and $\mathcal{E}'$.

§4. *Distributions of finite order*

In this section we shall consider distributions which are continuous linear forms on the spaces $\mathfrak{D}^m(\Omega)$ introduced in Example 2.12.8. To establish that these are normal spaces of distributions we must consider the operation of convolution in a simple special case. We shall study convolutions more generally in §9.

Let f be a continuous function and g a continuous function with compact support, both defined on $\mathbf{R}^n$. For each vector $x \in \mathbf{R}^n$ the function $y \mapsto f(x - y)g(y)$ is continuous on $\mathbf{R}^n$, and its support is contained in that of g. Therefore the integral

$$h(x) = \int_{\mathbf{R}^n} f(x - y)g(y)\, dy$$

exists and defines the convolution $h = f * g$ of the two functions f and g. By a change of variables we see that h is also given by

$$h(x) = \int_{\mathbf{R}^n} f(y)g(x - y)\, dy.$$

We have

$$\operatorname{Supp} h \subset \operatorname{Supp} f + \operatorname{Supp} g. \tag{1}$$

Indeed, by Proposition 2.10.5 the set $A = \operatorname{Supp} f + \operatorname{Supp} g$ is closed. If $x \notin A$, then $x = x - y + y$, where either $x - y \notin \operatorname{Supp} f$ or $y \notin \operatorname{Supp} g$, for all $y \in \mathbf{R}^n$. It follows that $h(x) = 0$, i.e., $x \notin \operatorname{Supp} h$.

Let us now assume that f also has compact support. Then it follows from (1) that $\operatorname{Supp} h$ is compact. Furthermore, f is then uniformly continuous on $\mathbf{R}^n$, and it follows from

$$\begin{aligned} |h(x_1) - h(x_2)| &= \left| \int_{\mathbf{R}^n} \{f(x_1 - y) - f(x_2 - y)\} g(y)\, dy \right| \\ &\leqq \sup_y |f(x_1 - y) - f(x_2 - y)| \cdot \int_{\mathbf{R}^n} |g(y)|\, dy \end{aligned}$$

that h is continuous on $\mathbf{R}^n$. We see in a similar fashion that if $f \in \mathfrak{D}^m$ $(0 \leqq m \leqq \infty)$, then $h \in \mathfrak{D}^m$ and $\partial^p h = \partial^p f * g$ for $|p| \leqq m$.

Now we can prove

PROPOSITION 1. *Let Ω be an open subset of $\mathbf{R}^n$ and $m \in \mathbf{N}$. Then $\mathfrak{D}^m(\Omega)$ is a normal space of distributions* (Definition 2.3).

Proof. The map $\mathfrak{D}(\Omega) \hookrightarrow \mathfrak{D}^m(\Omega)$ is continuous by Example 2.12.8. The map $\varphi \mapsto T_\varphi$ from $\mathfrak{D}^m(\Omega)$ into $\mathfrak{D}'(\Omega)$ is composed of the maps

$$\mathfrak{D}^m(\Omega) \hookrightarrow \mathfrak{B}_0^m(\Omega) \hookrightarrow \mathfrak{B}^m(\Omega) \hookrightarrow \mathcal{E}^m(\Omega) \hookrightarrow \mathfrak{C}(\Omega),$$

which are continuous by Examples 2.12.8, 2.5.10, 2.5.5, 2.5.7, and of the map $f \mapsto T_f$ from $\mathfrak{C}(\Omega)$ into $\mathfrak{D}'(\Omega)$, which is continuous by Proposition 1.4.

It remains for us to prove that the image of $\mathfrak{D}(\Omega)$ is everywhere dense in $\mathfrak{D}^m(\Omega)$. Let $\rho \in \mathfrak{D}$ be a function having properties (a), (b), and (c) listed in Example 1.4, and let ρ_ϵ be the function defined there. Given $\varphi \in \mathfrak{D}^m(\Omega)$, let K be a compact neighborhood of $\operatorname{Supp} \varphi$ contained in Ω and let $\eta > 0$. Then $\rho_\epsilon * \varphi \in \mathfrak{D}(K)$ for sufficiently small $\epsilon > 0$, and for $|p| \leqq m$ we have

$$\partial^p(\rho_\epsilon * \varphi)(x) - (\partial^p\varphi)(x) = \int \rho_\epsilon(y)\{(\partial^p\varphi)(x - y) - (\partial^p\varphi)(x)\}\,dy;$$

hence

$$|\partial^p(\rho_\epsilon * \varphi)(x) - (\partial^p\varphi)(x)| \leqq \max_{|y| \leqq \epsilon} |(\partial^p\varphi)(x - y) - (\partial^p\varphi)(x)| \leqq \eta,$$

for $x \in K$ and sufficiently small ϵ. This means that $\rho_\epsilon * \varphi$ converges to φ in $\mathfrak{D}^m(K)$. Since $\mathfrak{D}^m(K) \hookrightarrow \mathfrak{D}^m(\Omega)$ is continuous, $\rho_\epsilon * \varphi$ tends to φ also in $\mathfrak{D}^m(\Omega)$. ∎

Let us observe for later use (Lemma 8.1) that the last conclusion of the proof also holds for $m = \infty$.

It follows from Proposition 1 that the dual space $\mathfrak{D}'^m(\Omega)$ of $\mathfrak{D}^m(\Omega)$ is a space of distributions. We have the sequence

$$\mathfrak{D}(\Omega) \hookrightarrow \cdots \hookrightarrow \mathfrak{D}^{m+1}(\Omega) \hookrightarrow \mathfrak{D}^m(\Omega) \hookrightarrow \cdots \hookrightarrow \mathfrak{K}(\Omega)$$

of continuous maps (Example 2.12.8) and the image of each space in the next one is everywhere dense. By transposition we obtain the sequence

$$\mathfrak{K}'(\Omega) \to \cdots \to \mathfrak{D}'^m(\Omega) \to \mathfrak{D}'^{m+1}(\Omega) \to \cdots \to \mathfrak{D}'(\Omega),$$

where each map is injective (Corollary 2 of Proposition 3.12.2). Hence we can and *we shall consider each of these spaces as a linear subspace of the spaces following it.* This gives a meaning to

DEFINITION 1. *A distribution defined in the open subset Ω of $\mathbf{R}^n$ is said to be of order m if it belongs to $\mathfrak{D}'^m(\Omega)$ but not to $\mathfrak{D}'^{m-1}(\Omega)$. A distribution of order 0 is also called a* (Radon) *measure on Ω.*

Thus $\mathfrak{D}'^m(\Omega)$ is the space of all distributions of order $\leqq m$. A distribution which belongs to some space $\mathfrak{D}'^m(\Omega)$ is said to be of *finite order*. There exist distributions which are not of finite order (Exercise 1). The space $\mathcal{K}'(\Omega) = \mathfrak{D}'^0(\Omega)$ of all measures will sometimes be denoted by $\mathfrak{M}(\Omega)$. If $\varphi \in \mathcal{K}(\Omega)$ and $\mu \in \mathfrak{M}(\Omega)$, then the value $\langle \mu, \varphi \rangle$ of the canonical bilinear form at (μ, φ) is traditionally denoted by

$$\int_\Omega \varphi(x)\, d\mu(x) \tag{2}$$

and called the *integral* of the function φ with respect to the measure μ.

A distribution $T \in \mathfrak{D}'(\Omega)$ is of order $\leqq m$ if it is continuous on $\mathfrak{D}(\Omega)$ for the (coarser) topology induced on it by $\mathfrak{D}^m(\Omega)$. By Proposition 2.5.2 and the definition of the semi-norms on $\mathfrak{D}^m(K)$ (Example 2.4.9) this is so if and only if for every compact subset K of Ω there exists a positive number M such that

$$|\langle T, \varphi \rangle| \leqq M \max_{|p| \leqq m} \max_{x} |\partial^p \varphi(x)|$$

for all $\varphi \in \mathfrak{D}(K)$. The same inequality holds then for all $\varphi \in \mathfrak{D}^m(K)$.

Example 1. If $f \in \mathcal{C}(\Omega)$, then the corresponding distribution T_f (Example 1.1) is a measure, and therefore we shall denote it also by μ_f or $d\mu_f$. If $\Omega = \mathbf{R}^n$ and $f(x) \equiv 1$, then the corresponding measure will be called the *Lebesgue measure*, which is sometimes denoted by dx in harmony with the notation (2).

Example 2. The distribution δ_a, introduced in Example 1.2, is a measure as we have already pointed out. If $|p| = m$, then $\partial^p \delta_a = \delta_a^{(p)}$ (Example 3.2) is a distribution of order m.

Example 3. The distribution v.p. $1/x$ (Example 1.3) is of order 1.

The space $\mathfrak{D}^m(\Omega)$ is complete (Example 2.12.8), barrelled (Example 3.6.4), and bornological (Example 3.7.2), but for finite m it is not reflexive. The space $\mathfrak{D}'^m(\Omega)$ equipped with the topology $\beta(\mathfrak{D}'^m(\Omega), \mathfrak{D}^m(\Omega))$ is complete (Proposition 3.7.6), barrelled, and bornological (Example 3.16.2).

Proposition 2. *If Ω is an open subset of $\mathbf{R}^n$, then $\mathfrak{D}^F(\Omega)$ (Example 2.12.8) is a normal space of distributions. Its dual $\mathfrak{D}'^F(\Omega)$ consists of all distributions of finite order.*

Proof. (a) We know (Example 2.12.8) that the identity map

$$\mathfrak{D}(\Omega) \to \mathfrak{D}^F(\Omega)$$

is continuous. On the other hand, the map $\varphi \mapsto T_\varphi$ from $\mathfrak{D}^F(\Omega)$ into $\mathfrak{D}'(\Omega)$ factors into the maps $\mathfrak{D}^F(\Omega) \hookrightarrow \mathfrak{D}^m(\Omega)$ ($m \in \mathbf{N}$) and $\mathfrak{D}^m(\Omega) \to \mathfrak{D}'(\Omega)$, of which the first is continuous by definition and the second by Proposition 1.

(b) The second assertion is an immediate consequence of the corollary of Proposition 3.14.5. ▮

PROPOSITION 3. *If Ω is an open subset of $\mathbf{R}^n$ and $m \in \mathbf{N}$, then $\mathcal{E}^m(\Omega)$ is a normal space of distributions, and its dual $\mathcal{E}'^m(\Omega)$ is the space of all distributions of order m with compact support; i.e., $\mathcal{E}'^m(\Omega) = \mathcal{E}'(\Omega) \cap \mathcal{D}'^m(\Omega)$.*

Proof. (a) The maps

$$\mathcal{D}(\Omega) \hookrightarrow \mathcal{E}(\Omega) \hookrightarrow \mathcal{E}^m(\Omega) \qquad \text{and} \qquad \mathcal{E}^m(\Omega) \hookrightarrow \mathcal{C}(\Omega) \to \mathcal{D}'(\Omega)$$

are continuous (Examples 2.5, 2.5.7, and Proposition 1.4). We see word for word as in Example 2.5 that $\mathcal{D}^m(\Omega)$ is dense in $\mathcal{E}^m(\Omega)$. Since $\mathcal{D}(\Omega)$ is dense in $\mathcal{D}^m(\Omega)$, it is *a fortiori* dense in $\mathcal{E}^m(\Omega)$.

(b) If $T \in \mathcal{E}'^m(\Omega)$, then the restrictions of T to $\mathcal{E}(\Omega)$ and $\mathcal{D}^m(\Omega)$ are both continuous; i.e., $T \in \mathcal{E}'(\Omega) \cap \mathcal{D}'^m(\Omega)$. Conversely, let T be a distribution of order m and with compact support K. Let $\chi \in \mathcal{D}(\Omega)$ be such that $\chi(x) = 1$ for all x in a neighborhood of K. For $\varphi \in \mathcal{E}^m(\Omega)$ we have $\chi\varphi \in \mathcal{D}^m(\Omega)$ and we define

$$\langle S, \varphi \rangle = \langle T, \chi\varphi \rangle.$$

We see exactly as in part (a) of the proof of Proposition 2.3 that S is a continuous linear form on $\mathcal{E}^m(\Omega)$ and coincides with T on $\mathcal{D}^m(\Omega)$. Thus T can be considered as a distribution belonging to $\mathcal{E}'^m(\Omega)$. ▮

COROLLARY. *Every distribution with compact support is of finite order.*

Proof. It follows from Example 2.11.4 and the corollary of Proposition 3.14.5 that if $T \in \mathcal{E}'(\Omega)$, then there exists a positive integer m such that $T \in \mathcal{E}'^m(\Omega)$. But then $T \in \mathcal{D}'^m(\Omega)$ by the proposition. ▮

The study of the derivation of distributions of finite order is based on

PROPOSITION 4. *Let Ω be an open subset of $\mathbf{R}^n$ and m a positive integer. For each index j ($1 \leqq j \leqq n$) the linear map $\varphi \mapsto \partial_j\varphi$ from $\mathcal{D}^{m+1}(\Omega)$ into $\mathcal{D}^m(\Omega)$ and from $\mathcal{E}^{m+1}(\Omega)$ into $\mathcal{E}^m(\Omega)$ is continuous.*

Proof. (a) (cf. Proposition 3.1) Let K be a compact subset of Ω and V a neighborhood of 0 in $\mathcal{D}^m(\Omega)$. There exists $\epsilon > 0$ such that $V \cap \mathcal{D}^m(K)$ contains the set

$$\{\varphi \mid |\partial^p\varphi(x)| \leqq \epsilon, |p| \leqq m\}.$$

The set

$$U = \{\varphi \mid |\partial^p\varphi(x)| \leqq \epsilon, |p| \leqq m + 1\}$$

is a neighborhood of 0 in $\mathcal{D}^{m+1}(K)$ and $\varphi \in U$ implies $\partial_j\varphi \in V$. Thus the map $\varphi \mapsto \partial_j\varphi$ is continuous from $\mathcal{D}^{m+1}(K)$ into $\mathcal{D}^m(\Omega)$, and thus by Proposition 2.12.1 also from $\mathcal{D}^{m+1}(\Omega)$ into $\mathcal{D}^m(\Omega)$.

(b) Let V be a neighborhood of 0 in $\mathcal{E}^m(\Omega)$ which we may assume to be of the form $\{\varphi \mid |\partial^p\varphi(x)| \leqq \epsilon, |p| \leqq m, x \in K\}$, where $\epsilon > 0$ and K is a compact subset of Ω. Then

$$U = \{\varphi \mid |\partial^p\varphi(x)| \leqq \epsilon, |p| \leqq m + 1, x \in K\}$$

is a neighborhood of 0 in $\mathcal{E}^{m+1}(\Omega)$, and $\varphi \in U$ implies $\partial_j\varphi \in U$. ∎

Observe that part (b) also proves that the map $\varphi \mapsto \partial_j\varphi$ is continuous from $\mathcal{E}(\Omega)$ into $\mathcal{E}(\Omega)$.

Corollary. *For every multi-index $p \in \mathbf{N}^n$ the map $\varphi \mapsto \partial^p\varphi$ is continuous from $\mathcal{D}^{m+|p|}(\Omega)$ into $\mathcal{D}^m(\Omega)$, from $\mathcal{E}^{m+|p|}(\Omega)$ into $\mathcal{E}^m(\Omega)$, and from $\mathcal{E}(\Omega)$ into $\mathcal{E}(\Omega)$.*

It follows from this corollary that the map $T \mapsto \partial^pT$ (Definition 3.1) is continuous from $\mathcal{D}'^m(\Omega)$ into $\mathcal{D}'^{m+|p|}(\Omega)$, from $\mathcal{E}'^m(\Omega)$ into $\mathcal{E}'^{m+|p|}(\Omega)$, and from $\mathcal{E}'(\Omega)$ into $\mathcal{E}'(\Omega)$ for either the topologies $\beta(E', E)$ (corollary of Proposition 3.12.3) or for the topologies $\kappa(E', E)$ (Proposition 3.12.7). In particular, if μ is a measure on Ω, then $\partial^p\mu$ is a distribution of order at most $|p|$. Conversely, we have the following result:

Theorem 1. *If $T \in \mathcal{D}'^m(\Omega)$, then there exists a finite family (μ_p) of measures ($p \in \mathbf{N}^n$, $|p| \leqq m$) such that $T = \sum_{|p|\leqq m} \partial^p\mu_p$.*

Proof. For each multi-index $p \in \mathbf{N}^n$ such that $|p| \leqq m$ let $\mathcal{K}_p(\Omega)$ be a space identical with $\mathcal{K}(\Omega)$. The map $u: \varphi \mapsto (\partial^p\varphi)_{|p|\leqq m}$ from $\mathcal{D}^m(\Omega)$ into the product space $\prod_{|p|\leqq m} \mathcal{K}_p(\Omega)$ is clearly linear and injective. It is continuous by the preceding corollary. Finally one sees that it is a strict morphism by observing that a fundamental system of neighborhoods of 0 in $\mathcal{D}^m(\Omega)$ is given by the sets V_h of functions φ satisfying $|\partial^p\varphi(x)| \leqq h(x)$ for $x \in \Omega$ and $|p| \leqq m$, where h is a continuous, strictly positive function on Ω (Exercise 2.12.9).

Setting

$$\langle L, (\partial^p\varphi)\rangle = \langle T, \varphi\rangle,$$

we define a continuous linear from L on $\operatorname{Im}(u)$. By the Hahn-Banach theorem (Proposition 3.1.1) there exists a continuous extension $\tilde{L}$ of L onto the whole space $\prod_{|p|\leqq m} \mathcal{K}_p(\Omega)$. By Proposition 3.14.1, for each p there exists a measure $(-1)^{|p|}\mu_p$ such that

$$\langle \tilde{L}, (\varphi_p)\rangle = \sum_{|p|\leqq m} (-1)^{|p|}\langle \mu_p, \varphi_p\rangle$$

for all $(\varphi_p) \in \prod_{|p|\leqq m} \mathcal{K}_p(\Omega)$. In particular, for $\varphi \in \mathcal{D}^m(\Omega)$ we have

$$\langle T, \varphi\rangle = \langle L, (\partial^p\varphi)\rangle = \sum_{|p|\leqq m} (-1)^{|p|}\langle \mu_p, \partial^p\varphi\rangle = \sum_{|p|\leqq m} \langle \partial^p\mu_p, \varphi\rangle;$$

i.e., $T = \sum_{|p|\leqq m} \partial^p\mu_p$. ∎

If a function φ vanishes in a neighborhood of the support of the distribution T, then $\langle T, \varphi \rangle = 0$. We can improve this result considerably (see also Proposition 7.7 and Exercise 9.4(c)).

THEOREM 2. *Let Ω be an open subset of $\mathbf{R}^n$ and $m \in \mathbf{N}$. If $T \in \mathcal{E}'^m(\Omega)$ and $\varphi \in \mathcal{E}^m(\Omega)$ is such that $\partial^p \varphi(x) = 0$ for all $x \in \operatorname{Supp} T$ and $|p| \leqq m$, then $\langle T, \varphi \rangle = 0$.*

Proof. (a) Set $K = \operatorname{Supp} T$ and denote by K_ϵ the set of all those points of $\mathbf{R}^n$ whose distance from K is $\leqq \epsilon$. For sufficiently small $\epsilon > 0$ the set K_ϵ is a compact neighborhood of K contained in Ω.

With the notations introduced in Chapter 2, §5 we can write Taylor's formula in the form

$$\varphi(x) = \sum_{|p| \leqq m} \frac{\partial^p \varphi(y)}{p!} (x - y)^p + R_m(x; y).$$

For every compact subset L of Ω and every $\eta > 0$ there exists $\theta > 0$ such that

$$\frac{|R_m(x; y)|}{|x - y|^m} < \eta$$

whenever $x, y \in L$ and $|x - y| < \theta$. Considering that $\partial^p \varphi(y) = 0$ for $y \in K$ and $|p| \leqq m$, we can choose $\epsilon > 0$ so small that

$$|\varphi(x)| \leqq \eta \epsilon^m \qquad \text{for all} \qquad x \in K_{4\epsilon}.$$

We can apply the same reasoning to the function $\partial^p \varphi$ for a fixed index $|p| \leqq m$. Observing that $\partial^q \partial^p \varphi(y) = 0$ for $y \in K$ and $|q| \leqq m - |p|$, and that only a finite number of choices have to be made, we can affirm the existence of a number $\epsilon > 0$ such that

$$|\partial^p \varphi(x)| \leqq \eta \epsilon^{m-|p|} \tag{3}$$

for $x \in K_{4\epsilon}$ and $|p| \leqq m$.

(b) By Proposition 2.12.5 there exists a function $\chi_\epsilon \in \mathfrak{D}(\Omega)$ such that $\chi_\epsilon(x) = 1$ for $x \in K_{2\epsilon}$, $\chi_\epsilon(x) = 0$ for $x \in \complement K_{3\epsilon}$, and $0 \leqq \chi_\epsilon(x) \leqq 1$ for all x. Let ρ_ϵ be the function employed in the proof of Proposition 1 and consider the convolution $\alpha_\epsilon = \chi_\epsilon * \rho_\epsilon$. We have $\alpha_\epsilon \in \mathfrak{D}(\Omega)$, $\operatorname{Supp} \alpha_\epsilon \subset K_{4\epsilon}$ by formula (1), and

$$\begin{aligned} \alpha_\epsilon(x) &= \int_{|y| \leqq \epsilon} \chi_\epsilon(x - y) \rho_\epsilon(y)\, dy \\ &= \int_{\mathbf{R}^n} \rho_\epsilon(y)\, dy = 1 \end{aligned}$$

for $x \in K_\epsilon$, since $x \in K_\epsilon$ and $|y| \leqq \epsilon$ imply $x - y \in K_{2\epsilon}$. We have furthermore

$$\begin{aligned} |\partial^p \alpha_\epsilon(x)| &= \left| \int_{\mathbf{R}^n} \chi_\epsilon(x - y)(\partial^p \rho_\epsilon)(y)\, dy \right| \\ &\leqq \int_{\mathbf{R}^n} |\partial^p \rho_\epsilon(y)|\, dy = \int_{\mathbf{R}^n} \frac{1}{\epsilon^n} \cdot \frac{1}{\epsilon^{|p|}} \left| \partial^p \rho\left(\frac{y}{\epsilon}\right) \right| dy \\ &= \epsilon^{-|p|} \int_{\mathbf{R}^n} |\partial^p \rho(x)|\, dx. \end{aligned}$$

Thus there exists a constant $c > 0$ such that

$$|\partial^p \alpha_\epsilon(x)| \leqq c\epsilon^{-|p|} \tag{4}$$

for all $x \in \Omega$ and $|p| \leqq m$.

(c) We have $(\alpha_\epsilon\varphi)(x) = \varphi(x)$ for $x \in K_\epsilon$ and $(\alpha_\epsilon\varphi)(x) = 0$ for $x \in \complement K_{4\epsilon}$. By the Leibniz formula (Proposition 2.5.3) $\partial^p(\alpha_\epsilon\varphi)$ is a linear combination of expressions of the form $\partial^q\alpha_\epsilon \cdot \partial^r\varphi$, where $q + r = p$. It follows therefore from (3) and (4) that there exists a constant $c' > 0$ such that

$$|\partial^p(\alpha_\epsilon\varphi)(x)| \leqq c'\epsilon^{m-|p|}\eta \leqq c'\eta \tag{5}$$

for all $x \in \Omega$ and $|p| \leqq m$ if ϵ is sufficiently small.

(d) Since $T \in \mathfrak{D}'^m(\Omega)$, there exists a constant $M > 0$ such that

$$|\langle T, \psi \rangle| \leqq M \max_{|p| \leqq m} \max_x |\partial^p \psi(x)| \tag{6}$$

for all $\psi \in \mathfrak{D}^m(K_{4\epsilon})$. Now $\operatorname{Supp}(1 - \alpha_\epsilon)\varphi \subset \complement K$, hence

$$\langle T, \varphi \rangle = \langle T, \alpha_\epsilon\varphi \rangle.$$

Since $\operatorname{Supp}(\alpha_\epsilon\varphi) \subset K_{4\epsilon}$, it follows from (5) and (6) that

$$|\langle T, \varphi \rangle| \leqq Mc'\eta.$$

Since $\eta > 0$ is arbitrary, we have $\langle T, \varphi \rangle = 0$. ∎

We know that $\operatorname{Supp} \partial^p \delta = \{0\}$ for all $p \in \mathbf{N}^n$. Conversely, we can deduce from the previous theorem the following important result.

PROPOSITION 5. *Every distribution whose support is $\{0\}$ is a finite linear combination of the Dirac measure and its derivatives.*

Proof. Let $\operatorname{Supp} T = \{0\}$. By the corollary to Proposition 3 we have $T \in \mathfrak{D}'^m$ for some $m \in \mathbf{N}$. For every $\varphi \in \mathfrak{D}$ Taylor's formula gives us the expression

$$\varphi(x) = \sum_{|p| \leqq m} c_p x^p + \psi(x),$$

where $c_p = \partial^p\varphi(0)/p!$, $\psi \in \mathcal{E}$, and $\partial^p\psi(0) = 0$ for $|p| \leqq m$. It follows from Theorem 2 that $\langle T, \psi\rangle = 0$ and therefore

$$\langle T, \varphi\rangle = \sum_{|p|\leqq m} \frac{1}{p!}\langle T, x^p\rangle\, \partial^p\varphi(0) = \sum_{|p|\leqq m} \frac{(-1)^{|p|}}{p!}\langle T, x^p\rangle\langle \partial^p\delta, \varphi\rangle.$$

Writing $\gamma_p = (-1)^{|p|}\langle T, x^p\rangle/p!$, we have $T = \sum_{|p|\leqq m} \gamma_p\, \partial^p\delta$. ∎

Exercises

1. Show that the linear form $T = \sum_{n\in\mathbf{N}} \partial^n\delta_n$, which for every $\varphi \in \mathcal{D}(\mathbf{R})$ is given by

$$\langle T, \varphi\rangle = \sum_{n\in\mathbf{N}} \varphi^{(n)}(n),$$

is a distribution on $\mathbf{R}$ but does not have finite order.

2. Show that if in Proposition 2.4 the distributions T_ι are of order $\leqq m$, then T is of order $\leqq m$.

3. As in Proposition 3.4, let $h = (h_1, \ldots, h_n) \in \mathbf{R}^n$ be a vector parallel to the x_j-axis and $h_j = t$.

(a) Show that if $\varphi \in \mathcal{D}^m$, $m \geqq 1$, then $t^{-1}(\tau_{-h}\varphi - \varphi)$ converges to $\partial_j\varphi$ in $\mathcal{D}^{m-1}$ as $h \to 0$.

(b) Show that if $T \in \mathcal{D}'^m$, $m \in \mathbf{N}$, then $t^{-1}(\tau_{-h}T - T)$ converges to ∂_jT in $\mathcal{D}'^{m+1}$ as $h \to 0$.

4. Show that if $\operatorname{Supp} T = \{0\}$, then the representation of T in the form $T = \sum_p \gamma_p\, \partial^p\delta$ (Proposition 5) is unique. (*Hint:* Consider $\langle T, x^p\rangle$.)

5. Show that if $T \in \mathcal{D}'^m(\Omega)$, $\varphi \in \mathcal{D}^m(\Omega)$ and $\operatorname{Supp} T \cap \operatorname{Supp} \varphi = \emptyset$, then $\langle T, \varphi\rangle = 0$.

§5. *Integrable distributions*

Denote by $\mathbf{1}$ the function which assumes the value 1 in every point of the open set Ω. In this section we shall consider distributions T for which the expression $\langle T, \mathbf{1}\rangle$ is defined. This expression will also be denoted by $\int_\Omega T$ and called the integral of T, which explains the title of the section. To motivate the terminology, let us observe that if $T = T_f$, then $\langle T, \mathbf{1}\rangle = \int_\Omega f(x)\, dx$ (cf. Remark 1).

PROPOSITION 1. *Let Ω be an open subset of $\mathbf{R}^n$. The Banach space* $\mathcal{C}_0(\Omega) = \mathcal{B}_0^0(\Omega)$ (Example 2.4.16 and Exercise 1.2.7) *is a normal space of distributions* (Definition 2.3).

Proof. The map $\mathcal{D}(\Omega) \hookrightarrow \mathcal{C}_0(\Omega)$ is composed of the maps

$$\mathcal{D}(\Omega) \hookrightarrow \mathcal{B}_0(\Omega) \hookrightarrow \mathcal{C}_0(\Omega),$$

which are continuous by Examples 2.12.8 and 2.5.7. The map $\varphi \mapsto T_\varphi$

from $\mathcal{C}_0(\Omega)$ into $\mathcal{D}'(\Omega)$ is composed of the maps $\mathcal{C}_0(\Omega) \hookrightarrow \mathcal{B}^0(\Omega) \hookrightarrow \mathcal{C}(\Omega)$, which are continuous by Examples 2.5.10, 2.5.5, and of the map $f \mapsto T_f$ from $\mathcal{C}(\Omega)$ into $\mathcal{D}'(\Omega)$, which is continuous by Proposition 1.4.

Let $\varphi \in \mathcal{C}_0(\Omega)$. Given $\epsilon > 0$, there exists a compact subset K of Ω such that $|\varphi(x)| \leqq \epsilon$ for $x \in \Omega \cap \complement K$. Let $\chi \in \mathcal{D}(\Omega)$ be such that $\chi(x) = 1$ for $x \in K$ and $0 \leqq \chi(x) \leqq 1$ for all $x \in \Omega$. Then $\chi\varphi \in \mathcal{K}(\Omega)$ and

$$|\varphi(x) - \chi(x)\varphi(x)| = |\{1 - \chi(x)\}\varphi(x)| \leqq \epsilon$$

for all $x \in \Omega$. This proves that the image of $\mathcal{K}(\Omega)$ is dense in $\mathcal{C}_0(\Omega)$. Since the image of $\mathcal{D}(\Omega)$ is dense in $\mathcal{K}(\Omega)$ (Proposition 4.1), and the map

$$\mathcal{K}(\Omega) \hookrightarrow \mathcal{C}_0(\Omega)$$

is continuous (Example 2.12.5), $\mathcal{D}(\Omega)$ is also dense in $\mathcal{C}_0(\Omega)$. ∎

It follows from Proposition 1 that the dual of $\mathcal{C}_0(\Omega)$ is a space of distributions. We shall denote this dual by $\mathcal{C}_0'(\Omega)$ or by $\mathcal{B}_0'^0(\Omega)$ or, for reasons which should be clear in a moment, also by $\mathfrak{M}^1(\Omega)$. Since the map

$$\mathcal{K}(\Omega) \hookrightarrow \mathcal{C}_0(\Omega)$$

is continuous, and the image of $\mathcal{K}(\Omega)$ is dense in $\mathcal{C}_0(\Omega)$, the map

$$\mathfrak{M}^1(\Omega) \to \mathfrak{M}(\Omega)$$

(Definition 4.1) obtained by transposition is injective (Corollary 2 of Proposition 3.12.2); i.e., every distribution belonging to $\mathfrak{M}^1(\Omega)$ can be considered as a measure. More precisely, a measure μ belongs to $\mathfrak{M}^1(\Omega)$ if and only if there exists a number $M \geqq 0$ such that

$$|\langle \mu, \varphi \rangle| \leqq M \max_x |\varphi(x)| \tag{1}$$

for all φ in $\mathcal{D}(\Omega)$ or in $\mathcal{K}(\Omega)$. If this is the case, then (1) holds for all $\varphi \in \mathcal{C}_0(\Omega)$ and the smallest value of M for which (1) holds is the norm $\|\mu\|$ of μ in the Banach space $\mathfrak{M}^1(\Omega)$.

EXAMPLE 1. The norm of the measure δ_a is 1.

REMARK 1. We shall see in Chapter 5 that the measures $\mu \in \mathfrak{M}^1(\Omega)$ are precisely those for which $\langle \mu, \mathbf{1} \rangle = \int_\Omega d\mu(x)$ is defined and finite ([10], Chapter IV, §4, No. 7, Proposition 12). Therefore the measures in $\mathfrak{M}^1(\Omega)$ are said to be *integrable* or of *finite total mass* (Bourbaki uses the unfortunate term "bounded").

PROPOSITION 2. *Let Ω be an open subset of $\mathbf{R}^n$, $m \in \mathbf{N}$, and j an integer such that $1 \leqq j \leqq n$. The linear map $\varphi \mapsto \partial_j\varphi$ is continuous from $\mathcal{B}_0^{m+1}(\Omega)$ into $\mathcal{B}_0^m(\Omega)$ and from $\mathcal{B}_0(\Omega)$ into $\mathcal{B}_0(\Omega)$* (Example 2.4.17).

Proof. Let V be a neighborhood of 0 in $\mathcal{B}_0^m(\Omega)$ which we may assume to be of the form

$$\{\varphi \mid |\partial^p\varphi(x)| \leqq \epsilon, x \in \Omega, |p| \leqq h\},$$

where $h \in \mathbf{N}$ and $h = m$ if m is finite. Then

$$U = \{\varphi \mid |\partial^p\varphi(x)| \leqq \epsilon, x \in \Omega, |p| \leqq h + 1\}$$

is a neighborhood of 0 in $\mathcal{B}_0^{m+1}(\Omega)$, and $\varphi \in U$ implies $\partial_j\varphi \in V$. ∎

COROLLARY. *For every multi-index $p \in \mathbf{N}^n$ the map $\varphi \mapsto \partial^p\varphi$ is continuous from $\mathcal{B}_0^{m+|p|}(\Omega)$ into $\mathcal{B}_0^m(\Omega)$ and from $\mathcal{B}_0(\Omega)$ into $\mathcal{B}_0(\Omega)$.*

PROPOSITION 3. *Let Ω be an open subset of $\mathbf{R}^n$ and $m \in \mathbf{N}$. If $(\mu_p)_{|p|\leqq m}$ is a family of measures belonging to $\mathfrak{M}^1(\Omega)$, then*

$$\varphi \mapsto \sum_{|p|\leqq m} (-1)^{|p|}\langle \mu_p, \partial^p\varphi\rangle \tag{2}$$

is a continuous linear form on $\mathcal{B}_0^m(\Omega)$, and conversely every continuous linear form on $\mathcal{B}_0^m(\Omega)$ is of the form (2), where the μ_p belong to $\mathfrak{M}^1(\Omega)$.

Proof. (a) Let $\epsilon > 0$ be a preassigned number, N the number of multi-indices $p \in \mathbf{N}^n$ such that $|p| \leqq m$, and $M = \max_{|p|\leqq m} \|\mu_p\|$. The set

$$V = \left\{\varphi \mid |\partial^p\varphi(x)| \leqq \frac{\epsilon}{NM}, x \in \Omega, |p| \leqq m\right\}$$

is a neighborhood of 0 in $\mathcal{B}_0^m(\Omega)$, and $\varphi \in V$ implies that

$$\left|\sum_{|p|\leqq m} (-1)^{|p|}\langle \mu_p, \partial^p\varphi\rangle\right| \leqq \sum_{|p|\leqq m} |\langle \mu_p, \partial^p\varphi\rangle| \leqq \sum_{|p|\leqq m} \|\mu_p\| \cdot \frac{\epsilon}{NM} \leqq \epsilon.$$

(b) Let L be a continuous linear form on $\mathcal{B}_0^m(\Omega)$. For each multi-index p such that $|p| \leqq m$, let $\mathcal{C}_0^{(p)}(\Omega)$ be a copy of the space $\mathcal{C}_0(\Omega)$. The map $u: \varphi \mapsto (\partial^p\varphi)_{|p|\leqq m}$ from $\mathcal{B}_0^m(\Omega)$ into $\prod_{|p|\leqq m} \mathcal{C}_0^{(p)}(\Omega)$ is clearly linear and injective. It is continuous by the preceding corollary, and considering the definition of the neighborhoods, it becomes obvious that it is a strict morphism. Setting

$$\langle F, (\partial^p\varphi)\rangle = \langle L, \varphi\rangle,$$

we define a continuous linear form F on $\mathrm{Im}(u)$. By the Hahn-Banach theorem (Proposition 3.1.1) there exists a continuous extension $\widetilde{F}$ of F onto the whole space $\prod_{|p|\leqq m} \mathcal{C}_0^{(p)}(\Omega)$. By Proposition 3.14.1 for each p there exists a measure $(-1)^{|p|}\mu_p \in \mathfrak{M}^1(\Omega)$ such that

$$\langle \widetilde{F}, (\varphi_p)\rangle = \sum_{|p|\leqq m} (-1)^{|p|}\langle \mu_p, \varphi_p\rangle$$

for all $(\varphi_p) \in \prod_{|p| \leq m} \mathcal{C}_0^{(p)}(\Omega)$. In particular, for $\varphi \in \mathcal{B}_0^m(\Omega)$ we have

$$\langle L, \varphi \rangle = \langle F, (\partial^p \varphi) \rangle = \sum_{|p| \leq m} (-1)^{|p|} \langle \mu_p, \partial^p \varphi \rangle. \blacksquare$$

COROLLARY. *The continuous linear forms on* $\mathcal{B}_0(\Omega)$ *are the maps of the form*

$$\varphi \mapsto \sum_{|p| \leq m} (-1)^{|p|} \langle \mu_p, \partial^p \varphi \rangle,$$

where m *is some positive integer and* $(\mu_p)_{|p| \leq m}$ *is a family of measures belonging to* $\mathfrak{M}^1(\Omega)$.

This follows from Example 2.11.4, the corollary of Proposition 3.14.5, and Proposition 3 (cf. the corollary of Proposition 4.3).

The distributions of the form $\sum_{|p| \leq m} \partial^p \mu_p$, where the μ_p belong to $\mathfrak{M}^1(\Omega)$ and $m \in \mathbf{N}$, will be called *integrable*. By the preceding corollary, to every continuous linear form on $\mathcal{B}_0(\Omega)$ there corresponds an integrable distribution, and every integrable distribution defines a continuous linear form on $\mathcal{B}_0(\Omega)$.

The image of the map $\mathcal{C}_0(\Omega) \hookrightarrow \mathcal{C}(\Omega)$ is everywhere dense since by Proposition 4.3 already $\mathfrak{D}(\Omega)$ is dense in $\mathcal{C}(\Omega)$. It follows that the map

$$\mathcal{C}'(\Omega) \to \mathfrak{M}^1(\Omega)$$

is injective; i.e., every measure with compact support is integrable.

EXERCISES

1. Show that if $T \in \mathcal{E}'^m(\Omega)$, then there exist measures μ_p $(|p| \leq m)$ whose support is contained in an arbitrary neighborhood of Supp T and such that

$$T = \sum_{|p| \leq m} \partial^p \mu_p.$$

(*Hint:* Consider an open neighborhood ω of Supp T and apply the method of the proof of Theorem 4.1 and Proposition 3 to $\mathcal{E}'^m(\omega)$.)

§6. *Multiplication*

Let $f \in \mathcal{C}(\Omega)$ and $\alpha \in \mathcal{E}(\Omega)$. Then the function αf belongs to $\mathcal{C}(\Omega)$, and we have (cf. Example 1.1)

$$\langle T_{\alpha f}, \varphi \rangle = \int_\Omega \alpha(x) f(x) \varphi(x)\, dx = \langle T_f, \alpha\varphi \rangle.$$

We want to define the product αT of a function $\alpha \in \mathcal{E}(\Omega)$ and a distribution $T \in \mathfrak{D}'(\Omega)$ in such a way that for $T = T_f$ we have $\alpha T_f = T_{\alpha f}$, and thus we are led to the following

DEFINITION 1. *Let Ω be an open subset of $\mathbf{R}^n$ and m either a positive integer or the symbol ∞. If $\alpha \in \mathcal{E}^m(\Omega)$ and $T \in \mathfrak{D}'^m(\Omega)$, then αT is defined by the relation*

$$\langle \alpha T, \varphi \rangle = \langle T, \alpha\varphi \rangle$$

for all $\varphi \in \mathfrak{D}^m(\Omega)$.

Since $\alpha\varphi \in \mathfrak{D}^m(\Omega)$, it is clear that αT is a well-defined linear form on $\mathfrak{D}^m(\Omega)$. That it is continuous, i.e., that $\alpha T \in \mathfrak{D}'^m(\Omega)$, follows from

PROPOSITION 1. *Let Ω be an open subset of $\mathbf{R}^n$, $0 \leqq m \leqq \infty$ and $\alpha \in \mathcal{E}^m(\Omega)$. The linear map $\varphi \mapsto \alpha\varphi$ from $\mathfrak{D}^m(\Omega)$ into $\mathfrak{D}^m(\Omega)$ is continuous.*

Proof. Let V be a neighborhood of 0 in $\mathfrak{D}^m(\Omega)$ and K a compact subset of Ω. There exists $\epsilon > 0$ and an integer $k \geqq 0$, where $k = m$ if m is finite, such that $V \cap \mathfrak{D}^m(K)$ contains the set $\{\varphi \mid |\partial^p\varphi(x)| \leqq \epsilon, |p| \leqq k\}$. By the Leibniz formula (Proposition 2.5.3) each $\partial^p(\alpha\varphi)$ is a linear combination of expressions of the form $\partial^q\alpha \cdot \partial^r\varphi$, where $q + r = p$. Since the functions $\partial^q\alpha$ ($|q| \leqq k$) are bounded on K, there exists $\eta > 0$ such that if φ belongs to the neighborhood $\{\varphi \mid |\partial^p\varphi(x)| \leqq \eta, |p| \leqq k\}$ of 0 in $\mathfrak{D}^m(K)$, then $\alpha\varphi \in V$. By Proposition 2.12.1 this proves the continuity of the map $\varphi \mapsto \alpha\varphi$. ▮

The map $\varphi \mapsto \langle T, \alpha\varphi \rangle$ is now indeed continuous since it is composed of the maps $\varphi \mapsto \alpha\varphi$ and $\psi \mapsto \langle T, \psi \rangle$.

By Definition 1 the map $T \mapsto \alpha T$ from $\mathfrak{D}'^m(\Omega)$ into $\mathfrak{D}'^m(\Omega)$ is the transpose of the map $\varphi \mapsto \alpha\varphi$ from $\mathfrak{D}^m(\Omega)$ into $\mathfrak{D}^m(\Omega)$, and therefore by the corollary of Proposition 3.12.3 it is continuous. If we imbed $\mathcal{C}(\Omega)$ into $\mathfrak{D}'(\Omega)$ with the help of the injection $f \mapsto T_f$, then the restriction of the map $T \mapsto \alpha T$ to $\mathcal{C}(\Omega)$ is the usual multiplication $f \mapsto \alpha f$. Since $\mathcal{C}(\Omega)$ is dense in $\mathfrak{D}'(\Omega)$ (Proposition 1.3), the map $T \mapsto \alpha T$ is the unique continuous extension of the map $f \mapsto \alpha f$ (Proposition 2.9.5).

EXAMPLE 1. $x\delta = 0$ since $\langle x\delta, \varphi \rangle = \langle \delta, x\varphi \rangle = (x\varphi)(0) = 0$.

EXAMPLE 2. If μ is a measure on Ω and α a continuous function on Ω, then the measure $\nu = \alpha\mu$, defined by

$$\int_\Omega \varphi(x)\, d\nu(x) = \int_\Omega \varphi(x)\alpha(x)\, d\mu(x)$$

for all $\varphi \in \mathcal{K}(\Omega)$, is also called the measure with *density* α with respect to μ.

PROPOSITION 2. *Let Ω be an open subset of $\mathbf{R}^n$ and $0 \leqq m \leqq \infty$. If $\alpha \in \mathcal{E}^m(\Omega)$ and $T \in \mathfrak{D}'^m(\Omega)$, then*

$$\operatorname{Supp}(\alpha T) \subset \operatorname{Supp} \alpha \cap \operatorname{Supp} T.$$

In particular, if either α or T has compact support, then αT has compact support.

Proof. We shall prove the equivalent relation

$$(\complement \operatorname{Supp} \alpha) \cup (\complement \operatorname{Supp} T) \subset \complement \operatorname{Supp}(\alpha T).$$

If $x \in \complement \operatorname{Supp} \alpha$, then there exists a neighborhood V of x such that $\alpha(y) = 0$ for all $y \in V$. If $\varphi \in \mathfrak{D}(\Omega)$ is such that $\operatorname{Supp} \varphi \subset V$, then $\alpha\varphi$ is identically 0, hence $\langle \alpha T, \varphi \rangle = \langle T, \alpha\varphi \rangle = 0$. Thus

$$x \in \complement \operatorname{Supp}(\alpha T).$$

If $x \in \complement \operatorname{Supp} T$, then there exists a neighborhood V of x such that $\operatorname{Supp} \varphi \subset V$ implies $\langle T, \varphi \rangle = 0$. But $\operatorname{Supp} \alpha\varphi \subset \operatorname{Supp} \varphi$; hence $\operatorname{Supp} \varphi \subset V$ implies $\langle \alpha T, \varphi \rangle = \langle T, \alpha\varphi \rangle = 0$ (Exercise 4.5), i.e., $x \in \complement \operatorname{Supp}(\alpha T)$. ∎

PROPOSITION 3. *Let Ω be an open subset of $\mathbf{R}^n$ and m a positive integer. If $\alpha \in \mathcal{E}(\Omega)$ and $T \in \mathfrak{D}'(\Omega)$ or if $\alpha \in \mathcal{E}^{m+1}(\Omega)$ and $T \in \mathfrak{D}'^m(\Omega)$, then*

$$\partial_j(\alpha T) = \partial_j\alpha \cdot T + \alpha \cdot \partial_j T$$

for $1 \leqq j \leqq n$.

Proof. For $\varphi \in \mathfrak{D}(\Omega)$ we have

$$\begin{aligned}
\langle \partial_j(\alpha T), \varphi \rangle &= -\langle \alpha T, \partial_j\varphi \rangle \\
&= -\langle T, \alpha \cdot \partial_j\varphi \rangle = -\langle T, \partial_j(\alpha\varphi) - \varphi \cdot \partial_j\alpha \rangle \\
&= \langle \partial_j T, \alpha\varphi \rangle + \langle \partial_j\alpha \cdot T, \varphi \rangle \\
&= \langle \alpha \cdot \partial_j T + \partial_j\alpha \cdot T, \varphi \rangle. \blacksquare
\end{aligned}$$

Exactly as in the proof of Proposition 2.5.3 we obtain the analogue of the Leibniz formula:

COROLLARY. *If $p \in \mathbf{N}^n$, $\alpha \in \mathcal{E}(\Omega)$ and $T \in \mathfrak{D}'(\Omega)$ or $\alpha \in \mathcal{E}^{m+|p|}(\Omega)$ and $T \in \mathfrak{D}'^m(\Omega)$, then*

$$\partial^p(\alpha T) = \sum_{q \leqq p} \binom{p}{q} \partial^q\alpha \cdot \partial^{p-q} T.$$

Next we prove the associativity and distributivity of the multiplication of distributions.

PROPOSITION 4. *Let Ω be an open subset of $\mathbf{R}^n$ and $0 \leqq m \leqq \infty$. If $\alpha \in \mathcal{E}^m(\Omega)$, $\beta \in \mathcal{E}^m(\Omega)$, $S \in \mathfrak{D}'^m(\Omega)$, and $T \in \mathfrak{D}'^m(\Omega)$, then*

$$(\alpha + \beta)T = \alpha T + \beta T, \qquad \alpha(S + T) = \alpha S + \alpha T,$$
$$(\alpha\beta)T = \alpha(\beta T).$$

Proof. If $\varphi \in \mathfrak{D}(\Omega)$, then

$$\begin{aligned}\langle(\alpha+\beta)T, \varphi\rangle &= \langle T, (\alpha+\beta)\varphi\rangle \\ &= \langle T, \alpha\varphi + \beta\varphi\rangle = \langle T, \alpha\varphi\rangle + \langle T, \beta\varphi\rangle \\ &= \langle \alpha T, \varphi\rangle + \langle \beta T, \varphi\rangle = \langle \alpha T + \beta T, \varphi\rangle,\end{aligned}$$

$$\begin{aligned}\langle \alpha(S+T), \varphi\rangle &= \langle S+T, \alpha\varphi\rangle \\ &= \langle S, \alpha\varphi\rangle + \langle T, \alpha\varphi\rangle = \langle \alpha S, \varphi\rangle + \langle \alpha T, \varphi\rangle \\ &= \langle \alpha S + \alpha T, \varphi\rangle,\end{aligned}$$

$$\begin{aligned}\langle(\alpha\beta)T, \varphi\rangle &= \langle T, (\alpha\beta)\varphi\rangle = \langle T, \beta(\alpha\varphi)\rangle = \langle \beta T, \alpha\varphi\rangle \\ &= \langle \alpha(\beta T), \varphi\rangle. \blacksquare\end{aligned}$$

REMARK 1. Let R be a ring with unit element 1 and M an additively written abelian group. We say that M is an *R-module* if there is given a map $(\alpha, T) \mapsto \alpha T$ from $R \times M$ into M such that $(\alpha+\beta)T = \alpha T + \beta T$, $\alpha(S+T) = \alpha S + \alpha T$, $\alpha(\beta T) = (\alpha\beta)T$, and $1 \cdot T = T$ for $\alpha, \beta \in R$ and $S, T \in M$. Thus $\mathfrak{D}'^m(\Omega)$ is an $\mathcal{E}^m(\Omega)$-module.

For $\alpha \in \mathcal{E}^m(\Omega)$, $T \in \mathfrak{D}'^m(\Omega)$ we define $T\alpha = \alpha T$. If $T_1, T_2, \ldots, T_k$ are k distributions of order $\leqq m$ such that at least $k-1$ of them belong to $\mathcal{E}^m(\Omega)$, then we can define inductively their product by

$$T_1 T_2 \cdots T_k = T_1(T_2 \cdots T_k).$$

This product is independent of the order of the factors, and we have for it the general rule of associativity

$$\left(\prod_{j=1}^{l} T_j\right) \cdot \left(\prod_{j=l+1}^{k} T_j\right) = \prod_{j=1}^{k} T_j$$

for any $1 \leqq l < k$. It is to be emphasized, however, that if T_1, T_2, and T_3 are distributions of order $\leqq m$ but T_1 and T_3 do not belong to $\mathcal{E}^m(\Omega)$, then $(T_1T_2)T_3$ and $T_1(T_2T_3)$ can be defined without being equal. For instance,

$$(\delta x) \text{ v.p.} \frac{1}{x} = 0 \cdot \text{v.p.} \frac{1}{x} = 0$$

by Example 1, but

$$\delta\left(x \cdot \text{v.p.} \frac{1}{x}\right) = \delta \mathbf{1} = \delta$$

since by Example 1.3 we have

$$\left\langle x \cdot \text{v.p.} \frac{1}{x}, \varphi \right\rangle = \lim_{\epsilon \to 0} \int_{|x|>\epsilon} \frac{x\varphi(x)}{x}\, dx = \int_{-\infty}^{\infty} \varphi(x)\, dx = \langle \mathbf{1}, \varphi\rangle.$$

We shall now study in detail the maps $T \mapsto x_j T$ from $\mathfrak{D}'$ into $\mathfrak{D}'$, and for this we need to look more closely also at the maps $\varphi \mapsto x_j\varphi$ from $\mathfrak{D}$ into $\mathfrak{D}$. We start with the case $n = 1$.

The continuous linear map u: $\varphi \mapsto x\varphi$ from $\mathfrak{D}(\mathbf{R})$ into $\mathfrak{D}(\mathbf{R})$ is injective since if $x\varphi(x) = 0$ for all $x \in \mathbf{R}$, then necessarily $\varphi(x) = 0$ for all $x \in \mathbf{R}$.

A function $\chi \in \mathfrak{D}(\mathbf{R})$ belongs to $H = \text{Im}(u)$ if and only if $\chi(0) = 0$. Indeed, if $\chi = x\psi$ for some $\psi \in \mathfrak{D}(\mathbf{R})$, then clearly $\chi(0) = 0$. Conversely, if $\chi(0) = 0$, then the function ψ, defined by

$$\psi(x) = \frac{\chi(x)}{x} = \frac{\chi(x) - \chi(0)}{x} \quad \text{if} \quad x \neq 0,$$

$$\psi(0) = \chi'(0),$$

belongs to $\mathfrak{D}(\mathbf{R})$ and $\chi = x\psi$.

By Example 1.2 the linear form $\chi \mapsto \chi(0)$ is continuous on $\mathfrak{D}(\mathbf{R})$. Hence H is a closed hyperplane (Proposition 2.5.7) whose equation is $\chi(0) = 0$.

Let φ_0 be a fixed element of $\mathfrak{D}(\mathbf{R})$ such that $\varphi_0(0) = 1$. Every element $\varphi \in \mathfrak{D}(\mathbf{R})$ has a unique decomposition $\varphi = \lambda\varphi_0 + \chi$, where $\lambda \in \mathbf{K}$ and $\chi \in H$. Indeed, if we set $\lambda = \varphi(0)$, then $\varphi - \lambda\varphi_0$ belongs to H, and if $\lambda_1\varphi_0 + \chi_1 = \lambda_2\varphi_0 + \chi_2$, then $(\lambda_1 - \lambda_2)\varphi_0 = \chi_2 - \chi_1 \in H$; hence

$$\lambda_1 = \lambda_2 \quad \text{and} \quad \chi_1 = \chi_2.$$

Finally, the map $\chi = x\psi \mapsto \psi$ from H onto $\mathfrak{D}(\mathbf{R})$ is continuous; i.e., u is a strict morphism. By Proposition 2.12.1 we have to prove that for any compact subset K of $\mathbf{R}$ the map $\chi \mapsto \psi$ from $H \cap \mathfrak{D}(K)$ into $\mathfrak{D}(\mathbf{R})$ is continuous. Let V be a neighborhood of 0 in $\mathfrak{D}(\mathbf{R})$. Then $V \cap \mathfrak{D}(K)$ contains a set of the form $\{\psi \mid |\partial^p\psi(x)| \leqq \epsilon, |p| \leqq m\}$, $\epsilon > 0$, $m \in \mathbf{N}$. Let U be the neighborhood of 0 in $H \cap \mathfrak{D}(K)$ defined by

$$\{\chi \mid |\partial^p\chi(x)| \leqq \epsilon, |p| \leqq m + 1\}.$$

Then $\chi \in U$ implies $\psi \in V$. Indeed, the formula

$$\chi(x) - \chi(0) = \int_0^x \chi'(u)\, du = \int_0^1 x\chi'(tx)\, dt$$

yields for $\chi(0) = 0$ and $\chi = x\psi$ the expression

$$\psi(x) = \int_0^1 \chi'(tx)\, dt.$$

Hence $|\partial^p\psi(x)| \leqq \max_{|u| \leqq |x|} |\partial^{p+1}\chi(u)| \leqq \epsilon$ for $p \leqq m$.

We collect our results:

PROPOSITION 5. *The map $\varphi \mapsto x\varphi$ from $\mathfrak{D}(\mathbf{R})$ into $\mathfrak{D}(\mathbf{R})$ is an injective strict morphism whose image H is a closed hyperplane. Given an element*

$\varphi_0 \in \mathfrak{D}(\mathbf{R})$ such that $\varphi_0(0) = 1$, each element $\varphi \in \mathfrak{D}(\mathbf{R})$ has a unique decomposition

$$\varphi = \lambda\varphi_0 + \chi, \tag{1}$$

where $\lambda = \varphi(0)$ and $\chi \in H$.

Now we are able to prove the following result:

PROPOSITION 6. *The linear map $v\colon T \mapsto xT$ is a surjective strict morphism from $\mathfrak{D}'(\mathbf{R})$ onto $\mathfrak{D}'(\mathbf{R})$ whose kernel is the one-dimensional subspace of $\mathfrak{D}'(\mathbf{R})$ consisting of all constant multiples of the Dirac measure.*

Proof. (a) Let $T \in \mathfrak{D}'(\mathbf{R})$ and $xT = S$. Using the decomposition (1), we have

$$\langle T, \varphi\rangle = \lambda\langle T, \varphi_0\rangle + \langle T, \chi\rangle = \lambda\langle T, \varphi_0\rangle + \langle S, \psi\rangle, \tag{2}$$

where $\chi = x\psi$. In particular, if $S = 0$, we have

$$\langle T, \varphi\rangle = \lambda\langle T, \varphi_0\rangle = \langle T, \varphi_0\rangle\varphi(0);$$

i.e., $T = \langle T, \varphi_0\rangle\delta$.

(b) Let us prove that v is surjective. Given $S \in \mathfrak{D}'(\mathbf{R})$, choose an arbitrary constant $\gamma \in \mathbf{K}$ and define T by

$$\langle T, \varphi\rangle = \gamma\lambda + \langle S, \psi\rangle,$$

where φ is arbitrary in $\mathfrak{D}(\mathbf{R})$ and λ, ψ have the same meaning as before. Since the representation (1) is unique, T is a well-defined linear form on $\mathfrak{D}(\mathbf{R})$. Next T is continuous since the maps $\varphi \mapsto \lambda$, $\lambda \mapsto \gamma\lambda$, $\varphi \mapsto \chi = \varphi - \lambda\varphi_0$, and $\psi \mapsto \langle S, \psi\rangle$ are known to be so, and $\chi \mapsto \psi$ is continuous by Proposition 5. Finally,

$$\langle xT, \psi\rangle = \langle T, x\psi\rangle = \langle S, \psi\rangle$$

for every $\psi \in \mathfrak{D}(\mathbf{R})$; i.e., $xT = S$. Observe that by (2) the distribution T is completely determined by the value $\gamma = \langle T, \varphi_0\rangle$.

(c) With every $S \in \mathfrak{D}'(\mathbf{R})$ we associate that distribution $T \in \mathfrak{D}'(\mathbf{R})$ for which $xT = S$ and $\langle T, \varphi_0\rangle = 0$, and we set $T = w(S)$. The map $w\colon \mathfrak{D}'(\mathbf{R}) \to \mathfrak{D}'(\mathbf{R})$ is clearly linear. By definition it satisfies

$$v(w(S)) = S \qquad \text{for all} \qquad S \in \mathfrak{D}'(\mathbf{R}), \tag{3}$$

and by (2) it satisfies

$$\langle w(S), \varphi\rangle = \langle S, \psi\rangle \qquad \text{for all} \qquad S \in \mathfrak{D}'(\mathbf{R}) \quad \text{and} \quad \varphi \in \mathfrak{D}(\mathbf{R}). \tag{4}$$

Relation (4) means that w is the transpose of the continuous linear map $\varphi \mapsto \psi$. Hence by the corollary of Proposition 3.12.3 the map w is continuous. It follows therefore from (3) that v is a strict morphism (Proposition 2.7.2). ∎

Let us now pass to the study of the maps $u_1: \varphi \mapsto x_1\varphi$ and $v_1: T \mapsto x_1T$ on $\mathbf{R}^n$. As in §2, if $x = (x_1, \ldots, x_n)$ is a vector in $\mathbf{R}^n$, we shall write x'' for the vector $(x_2, \ldots, x_n)$ of $\mathbf{R}^{n-1}$. Thus $x = (x_1, x'')$, and if f is a function defined on $\mathbf{R}^n$, we write $f(x) = f(x_1, x'')$.

A function $\chi \in \mathfrak{D} = \mathfrak{D}(\mathbf{R}^n)$ belongs to $M = \operatorname{Im}(u_1)$ if and only if

$$\chi(0, x'') = 0 \tag{5}$$

for all $x'' \in \mathbf{R}^{n-1}$. Indeed, if $\chi(x_1, x'') = x_1\psi(x_1, x'')$ for some $\psi \in \mathfrak{D}$, then $\chi(0, x'') = 0$. Conversely, if the relations (5) are satisfied, then the function ψ defined by

$$\psi(x_1, x'') = \frac{\chi(x_1, x'')}{x_1} \quad \text{if} \quad x_1 \neq 0,$$
$$\psi(0, x'') = \partial_1\chi(0, x'')$$

belongs to $\mathfrak{D}$ and $x_1\psi = \chi$.

For each $x'' \in \mathbf{R}^{n-1}$ the linear form $\chi \mapsto \chi(0, x'')$ is continuous on $\mathfrak{D}$, and therefore M is the intersection of closed hyperplanes, i.e., a closed subspace of $\mathfrak{D}$.

If $x_1\varphi = 0$, then $\varphi(x_1, x'') = 0$ for $x_1 \neq 0$, and since φ is continuous, $\varphi = 0$. Thus the map $u_1: \mathfrak{D} \to \mathfrak{D}$ is injective.

Let φ_0 be a fixed element of $\mathfrak{D}(\mathbf{R})$ such that $\varphi_0(0) = 1$. Every element $\varphi \in \mathfrak{D}$ has a unique decomposition $\varphi = \lambda\varphi_0 + \chi$, where $\lambda \in \mathfrak{D}(\mathbf{R}^{n-1})$ and $\chi \in M$. Indeed, if we set $\lambda(x'') = \varphi(0, x'')$, then $\lambda \in \mathfrak{D}(\mathbf{R}^{n-1})$, and clearly $\varphi - \lambda\varphi_0 \in M$. Furthermore, if $\lambda_1\varphi_0 + \chi_1 = \lambda_2\varphi_0 + \chi_2$, then

$$(\lambda_1 - \lambda_2)\varphi_0 = \chi_2 - \chi_1 \in M;$$

hence $\lambda_1 = \lambda_2$ and $\chi_1 = \chi_2$.

Finally, the map $\chi = x_1\psi \mapsto \psi$ from M onto $\mathfrak{D}$ is continuous; i.e., u_1 is a strict morphism. By Proposition 2.12.1 we have to prove that for any compact subset K of $\mathbf{R}^n$ the map $\chi \mapsto \psi$ from $M \cap \mathfrak{D}(K)$ into $\mathfrak{D}$ is continuous. Let V be a neighborhood of 0 in $\mathfrak{D}$. Then $V \cap \mathfrak{D}(K)$ contains a set of the form

$$\{\psi \mid |\partial^p\psi(x)| \leqq \epsilon, |p| \leqq m\}, \qquad \epsilon > 0, \qquad m \in \mathbf{N}.$$

Let U be the neighborhood of 0 in $M \cap \mathfrak{D}(K)$ defined by

$$\{\chi \mid |\partial^p\chi(x)| \leqq \epsilon, |p| \leqq m + 1\}.$$

Then $\chi \in U$ implies $\psi \in V$. Indeed, the formula

$$\chi(x_1, x'') - \chi(0, x'') = \int_0^{x_1} \partial_1\chi(u, x'')\, du = \int_0^1 x_1\partial_1\chi(tx_1, x'')\, dt$$

yields for $\chi(0, x'') = 0$ and $\chi = x_1\psi$ the expression

$$\psi(x_1, x'') = \int_0^1 \partial_1\chi(tx_1, x'')\, dt.$$

Hence $|\partial^p\psi(x)| \leqq \max_{|t|\leqq 1} |\partial^p\partial_1\chi(tx_1, x'')| \leqq \epsilon$ for $|p| \leqq m$.

Collecting our results, we obtain

PROPOSITION 7. *The map $\varphi \mapsto x_1\varphi$ is an injective strict morphism from $\mathfrak{D}$ into $\mathfrak{D}$ whose image M is a closed subspace of $\mathfrak{D}$. Given an element $\varphi_0 \in \mathfrak{D}(\mathbf{R})$ such that $\varphi_0(0) = 1$, each element has a unique decomposition*

$$\varphi = \lambda\varphi_0 + \chi, \tag{6}$$

where $\lambda \in \mathfrak{D}(\mathbf{R}^{n-1})$ is defined by $\lambda(x'') = \varphi(0, x'')$ and $\chi \in M$. In particular, $\chi \in M$ if and only if $\chi(0, x'') = 0$ for all $x'' \in \mathbf{R}^{n-1}$.

Now we are able to prove the following result:

PROPOSITION 8. *The linear map $v_1 \colon T \mapsto x_1T$ is a surjective strict morphism from $\mathfrak{D}'$ onto $\mathfrak{D}'$.*

Proof. (a) If $T \in \mathfrak{D}'$ and $x_1T = S$, then using the decomposition (6) we have

$$\langle T, \varphi\rangle = \langle T, \lambda\varphi_0\rangle + \langle T, \chi\rangle = \langle T, \lambda\varphi_0\rangle + \langle S, \psi\rangle, \tag{7}$$

where $\chi = x_1\psi$. If $x_1T = 0$, we can define a distribution $G \in \mathfrak{D}'(\mathbf{R}^{n-1})$ by setting

$$\langle G, \lambda\rangle = \langle T, \lambda\varphi_0\rangle \tag{8}$$

for $\lambda \in \mathfrak{D}(\mathbf{R}^{n-1})$, and the map $T \mapsto G$ is clearly a bijection from $\operatorname{Ker}(v_1)$ onto $\mathfrak{D}'(\mathbf{R}^{n-1})$.

(b) Let us prove that v_1 is surjective. Given $S \in \mathfrak{D}'$, choose an arbitrary distribution $G \in \mathfrak{D}'(\mathbf{R}^{n-1})$ and define T by

$$\langle T, \varphi\rangle = \langle G, \lambda\rangle + \langle S, \psi\rangle,$$

where φ is arbitrary in $\mathfrak{D}$ and λ and ψ have the same meaning as before. Because of the uniqueness of the decomposition (6), the linear form T is well-defined on $\mathfrak{D}$. It is also continuous since the maps $\varphi \mapsto \lambda$, $\lambda \mapsto \langle G, \lambda\rangle$, $\varphi \mapsto \chi = \varphi - \lambda\varphi_0$ and $\psi \mapsto \langle S, \psi\rangle$ are clearly so and $\chi \mapsto \psi$ is continuous by Proposition 7. Finally,

$$\langle x_1T, \psi\rangle = \langle T, x_1\psi\rangle = \langle S, \psi\rangle$$

for every $\psi \in \mathfrak{D}$; i.e., $x_1T = S$. Observe that by (7) and (8) the distribution T is completely determined by the distribution G.

(c) With every $S \in \mathfrak{D}'$ we associate that distribution $T \in \mathfrak{D}'$ for which $x_1T = S$ and $G = 0$ and set $T = w_1(S)$. The map $w_1 \colon \mathfrak{D}' \to \mathfrak{D}'$ is clearly linear. We have by definition

$$v_1(w_1(S)) = S \qquad \text{for all} \qquad S \in \mathfrak{D}' \tag{9}$$

and by (7) and (8)

$$\langle w_1(S), \varphi\rangle = \langle S, \psi\rangle \qquad \text{for all} \qquad S \in \mathfrak{D}' \ \text{ and } \ \varphi \in \mathfrak{D}. \tag{10}$$

Relation (10) shows that w_1 is the transpose of the continuous linear map $\varphi \mapsto \psi$. Hence by the corollary of Proposition 3.12.3 the map w_1 is continuous. It follows therefore from (9) that v_1 is a strict morphism (Proposition 2.7.2). ▮

REMARK 2. We shall determine later (Exercise 8.4) the kernel of the map $T \mapsto x_1 T$. The reader has no doubt observed the strong analogy between Propositions 2, 3, 7, 8 of §3 and Propositions 5, 6, 7, 8 of the present section. This analogy will be partly explained by the Fourier transformation in §11. In Chapter 7 we shall study the map $\varphi \mapsto P(x)\varphi$, where P is an arbitrary polynomial.

EXERCISES

1. Show that

$$x^l \cdot \left((-1)^{|p|}\frac{\partial^p \delta}{p!}\right) = \begin{cases} 0 & \text{if } \ l > p, \\ (-1)^{|p-l|}\dfrac{\partial^{|p-l|}\delta}{(p-l)!} & \text{if } \ l \leqq p, \end{cases}$$

where $l, p \in \mathbf{N}^n$.

2. (a) Let l be a positive integer. Show that $T \mapsto x^l T$ is a surjective strict morphism from $\mathfrak{D}'(\mathbf{R})$ onto $\mathfrak{D}'(\mathbf{R})$ whose kernel consists of all distributions of the form $\sum_{p \leqq l-1} \gamma_p \partial^p \delta$.

(b) Let $P(x)$ be a polynomial in one variable x. Show that $T \mapsto P(x)T$ is a surjective strict morphism from $\mathfrak{D}'(\mathbf{R})$ onto $\mathfrak{D}'(\mathbf{R})$ and determine its kernel.

§7. *Bilinear maps*

It follows from Proposition 6.4 that the map $(\alpha, T) \mapsto \alpha T$ from $\mathcal{E}^m(\Omega) \times \mathfrak{D}'^m(\Omega)$ into $\mathfrak{D}'^m(\Omega)$ is bilinear. In this section we want to examine the continuity properties of this map, but first we must discuss some general notions concerning bilinear maps.

Let us first recall the definition of a bilinear map. Let E, F, and G be three vector spaces over the same field $\mathbf{K}$. A map $(x, y) \mapsto b(x, y)$ from $E \times F$ into G is said to be bilinear if it satisfies the relations

$$\begin{aligned} b(\alpha x_1 + \beta x_2, y) &= \alpha b(x_1, y) + \beta b(x_2, y), \\ b(x, \alpha y_1 + \beta y_2) &= \alpha b(x, y_1) + \beta b(x, y_2) \end{aligned}$$

for all $\alpha, \beta \in \mathbf{K}$, $x, x_1, x_2 \in E$, and $y, y_1, y_2 \in F$. In other words, the map b is bilinear if for every $y \in F$ the map $b(\cdot, y)\colon x \mapsto b(x, y)$ from E into G is linear and for every $x \in E$ the map $b(x, \cdot)\colon y \mapsto b(x, y)$ from F into G is linear. Observe that in particular $b(x, 0) = b(0, y) = 0$ for all $x \in E$, $y \in F$. In the case $G = \mathbf{K}^1$ (Chapter 1, §5, p. 41) a bilinear map is simply a bilinear form (Chapter 1, §5, p. 43).

In analogy to Proposition 2.5.1 we have:

PROPOSITION 1. *Let E, F, and G be three topological vector spaces. A bilinear map $(x, y) \mapsto b(x, y)$ from $E \times F$ into G is continuous on the product space $E \times F$ if it is continuous at the point* $(0, 0)$.

Proof. Assume that b is continuous at $(0, 0)$. Let (x_0, y_0) be a point of $E \times F$ and W a neighborhood of 0 in G. We have to prove that there exists a neighborhood U of 0 in E and a neighborhood V of 0 in F such that $x \in U$ and $y \in V$ imply $b(x_0 + x, y_0 + y) - b(x_0, y_0) \in W$.

Let W_1 be a neighborhood of 0 in G such that $W_1 + W_1 + W_1 \subset W$. There exists a balanced neighborhood U_1 of 0 in E and a balanced neighborhood V_1 of 0 in F such that $x \in U_1$ and $y \in V_1$ imply $b(x, y) \in W_1$. Since U_1 is absorbing, there exists $\lambda > 1$ such that $x_0 \in \lambda U_1$. Similarly, there exists $\mu > 1$ such that $y_0 \in \mu V_1$. Set $U = (1/\mu)U_1$ and $V = (1/\lambda)V_1$. Then $U \subset U_1$ and $V \subset V_1$. Furthermore, if $y \in V$, then

$$b(x_0, y) = b\left(\frac{1}{\lambda}x_0, \lambda y\right) \in W_1.$$

Similarly, if $x \in U$, then

$$b(x, y_0) = b\left(\mu x, \frac{1}{\mu}y_0\right) \in W_1.$$

It follows from the identity

$$b(x_0 + x, y_0 + y) - b(x_0, y_0) = b(x_0, y) + b(x, y_0) + b(x, y)$$

that U and V satisfy our requirements. ∎

Many of the bilinear maps we shall encounter are not continuous. We must therefore consider some weaker forms of continuity.

DEFINITION 1. *Let E, F, and G be three topological vector spaces. A bilinear map $(x, y) \mapsto b(x, y)$ from $E \times F$ into G is said to be separately continuous if for every $y \in F$ the linear map $b(\cdot, y)\colon x \mapsto b(x, y)$ from E into G is continuous and for every $x \in E$ the linear map $b(x, \cdot)\colon y \mapsto b(x, y)$ from F into G is continuous.*

Every continuous bilinear map is separately continuous. Indeed, given a neighborhood W of 0 in G, there exists a neighborhood U of 0 in E and a neighborhood V of y in F such that $x \in U$ and $z \in V$ implies $b(x, z) \in W$ and in particular $x \in U$ implies $b(x, y) \in W$. By Proposition 2.5.1 this shows that $b(\cdot, y)$ is continuous. We see in the same way that $b(x, \cdot)$ is continuous.

We shall see that conversely not every separately continuous bilinear map is continuous. We have, however, the following result:

THEOREM 1. *Let E be a metrizable and barrelled locally convex space, F a metrizable topological vector space, and G a locally convex space. Every separately continuous bilinear map from $E \times F$ into G is continuous.*

Proof. Let $b: E \times F \to G$ be a separately continuous bilinear map. Since $E \times F$ is metrizable, by Proposition 1 it is sufficient to show that if (x_n, y_n) is a sequence of points in $E \times F$ which converges to $(0, 0)$, then $b(x_n, y_n)$ converges to 0 in G.

Let W be a balanced, closed, convex neighborhood of 0 in G and let T be the set of all those points $x \in E$ for which $b(x, y_n) \in W$ for every index n. We show that T is a barrel.

(a) T is absorbing. Indeed, let $a \in E$. Since the linear map $b(a, \cdot)$ is continuous, we have $b(a, y_n) \in W$ for $n \geqq N$. The set of points $b(a, y_n)$ with $n < N$ is finite, hence bounded in G, and therefore there exists $\lambda > 1$ such that $b(a, y_n) \in \lambda W$ for $n < N$. Thus

$$b\left(\frac{1}{\lambda} a, y_n\right) \in W \qquad \text{for all } n;$$

i.e., $a \in \lambda T$.

(b) T is balanced. Indeed, let $a \in T$ and $|\alpha| \leqq 1$. Then

$$b(\alpha a, y_n) = \alpha b(a, y_n) \in \alpha W \subset W$$

for all n; i.e., $\alpha a \in T$.

(c) T is convex. Indeed, let $a_1, a_2 \in T$, $\alpha \geqq 0$, $\beta \geqq 0$, $\alpha + \beta = 1$. Then

$$b(\alpha a_1 + \beta a_2, y_n) = \alpha b(a_1, y_n) + \beta b(a_2, y_n) \subset \alpha W + \beta W \subset W$$

for all n; i.e., $\alpha a_1 + \beta a_2 \in T$.

(d) T is closed. Indeed, let $a \in \overline{T}$. There exists a sequence (a_m) of points of T which converges to a (Exercise 2.8.1(b)). Since $b(\cdot, y_n)$ is continuous for every n, the sequence $b(a_m, y_n)$ tends to $b(a, y_n)$ as $m \to \infty$. Now $b(a_m, y_n) \in W$, and since W is closed, we have $b(a, y_n) \in W$ for every n; i.e., $a \in T$.

Since E is barrelled, T is a neighborhood of 0. Consequently, there exists $N_1 \in \mathbf{N}$ such that $x_n \in T$ for $n \geqq N_1$. Hence $b(x_n, y_n) \in W$ for $n \geqq N_1$. ∎

REMARK 1. It is essential to suppose that E is metrizable ([9], Chapter III, §4, Exercise 5). If, however, we suppose that E is not only barrelled but a Baire space, then it is not necessary to assume that it is metrizable (*loc. cit.*, Exercise 1).

We shall now introduce a kind of continuity which is intermediate between continuity proper and separate continuity.

DEFINITION 2. *Let E, F, and G be three topological vector spaces and $\mathfrak{S}$ a collection of bounded subsets of E. We say that a bilinear map*

$$b: E \times F \to G$$

is hypocontinuous with respect to $\mathfrak{S}$ (or $\mathfrak{S}$-hypocontinuous) *if it is separately continuous and satisfies the following condition:*

(HC) *For every neighborhood W of 0 in G and every set $A \in \mathfrak{S}$ there exists a neighborhood V of 0 in F such that $x \in A$ and $y \in V$ imply $b(x, y) \in W$.*

Clearly, if G is locally convex, then an $\mathfrak{S}$-hypocontinuous bilinear map is also hypocontinuous with respect to any collection (a) through (f) of subsets of F obtained from $\mathfrak{S}$ in Proposition 3.4.2.

If $\mathfrak{T}$ is a collection of bounded subsets of F, we define in an analogous way the bilinear maps which are hypocontinuous with respect to $\mathfrak{T}$ (or $\mathfrak{T}$-hypocontinuous). If b is hypocontinuous with respect to both $\mathfrak{S}$ and $\mathfrak{T}$, then we also say that b is $(\mathfrak{S}, \mathfrak{T})$-hypocontinuous. If $\mathfrak{S}$ is the collection of all bounded subsets of E and $\mathfrak{T}$ is the collection of all bounded sets in F, then the $(\mathfrak{S}, \mathfrak{T})$-hypocontinuous maps will simply be called hypocontinuous.

If the sets belonging to $\mathfrak{S}$ cover E and the sets belonging to $\mathfrak{T}$ cover F, then every bilinear map b which satisfies condition (HC) and the analogous condition for $\mathfrak{T}$ is automatically separately continuous. Indeed, if for instance $y \in F$ and $y \in B \in \mathfrak{T}$, then there exists for every neighborhood W of 0 in G a neighborhood U of 0 in E such that $x \in U$ implies $b(x, y) \in W$; i.e., $b(\cdot, y)$ is continuous.

On the other hand, every continuous bilinear map $(x, y) \mapsto b(x, y)$ is $(\mathfrak{S}, \mathfrak{T})$-hypocontinuous with respect to any pair $(\mathfrak{S}, \mathfrak{T})$ of collections of bounded sets. Indeed, for every neighborhood W of 0 in G there exists a neighborhood U of 0 in E and a neighborhood V of 0 in F such that $x \in U$ and $y \in V$ imply $b(x, y) \in W$. If $A \in \mathfrak{S}$, there exists $\lambda > 0$ such that $\lambda A \subset U$. Therefore $x \in A$ and $y \in \lambda V$ imply

$$b(x, y) = b\left(\lambda x, \frac{1}{\lambda} y\right) \in W.$$

Since λV is a neighborhood of 0 in F, this proves that b is $\mathfrak{S}$-hypocontinuous and we can see in a similar fashion that it is also $\mathfrak{T}$-hypocontinuous.

EXAMPLE 1. Let E be an infrabarrelled locally convex Hausdorff space and E' its dual equipped with the topology $\beta(E', E)$. The canonical bilinear form $(x, x') \mapsto \langle x, x' \rangle$ is then hypocontinuous on $E \times E'$. Indeed, let $\epsilon > 0$ and let A be a bounded subset of E. The polar A° of A is a neighborhood of 0 in E' for $\beta(E', E)$, and $x \in A$, $x' \in \epsilon A^\circ$ imply

$$|\langle x, x' \rangle| \leqq \epsilon.$$

If B is a $\beta(E', E)$-bounded subset of E', then it is equicontinuous (Proposition 3.6.6), and B° is a neighborhood of 0 in E. If $x \in \epsilon B^\circ$ and $x' \in B$, we have $|\langle x, x' \rangle| \leqq \epsilon$.

If E is a normed space, then we know that the canonical bilinear form is continuous on $E \times E'$ (Chapter 1, §7, p. 57). Conversely, *if the canonical bilinear form is continuous, then E is normable.* Indeed, if this is the case, then there exists a neighborhood U of 0 in E and a bounded subset A of E such that $x \in U$ and $x' \in A^\circ$ imply $|\langle x, x' \rangle| \leqq 1$. Thus $U \subset A^{\circ\circ}$, and since $A^{\circ\circ}$ is the balanced, convex, $\sigma(E, E')$-closed hull of A (Theorem 3.3.1), it is also bounded. Therefore U is bounded, and it follows from Proposition 2.6.1 that E is normable.

REMARK 2. It can be shown that not every separately continuous bilinear map is hypocontinuous ([9], Chapter III, §4, Exercise 4).

Let us prove some simple properties of hypocontinuous linear maps.

PROPOSITION 2. *Let E, F and G be three topological spaces, $\mathfrak{S}$ a collection of bounded subsets of E and $b: E \times F \to G$ an $\mathfrak{S}$-hypocontinuous bilinear map. Let $A \in \mathfrak{S}$ and B be a bounded subset of F. Then:*

(a) *The image of $A \times B$ under the map b is bounded in G.*
(b) *The map b is continuous on $A \times F$.*

Proof. Let W be an arbitrary neighborhood of 0 in G and determine the neighborhood V of 0 in F so that $x \in A$, $y \in V$ imply $b(x, y) \in W$.

(a) There exists $\lambda > 0$ such that $\lambda B \subset V$. If $x \in A$ and $y \in B$, then $\lambda b(x, y) = b(x, \lambda y) \in W$.

(b) Let $(x_0, y_0) \in A \times F$. For every point $(x, y) \in E \times F$ we have

$$b(x, y) - b(x_0, y_0) = b(x - x_0, y_0) + b(x, y - y_0).$$

Now $y - y_0 \in V$ implies $b(x, y - y_0) \in W$ for every $x \in A$. On the other hand, the linear map $b(\cdot, y_0): E \to G$ is continuous. Hence there exists a neighborhood U of 0 in E such that $x - x_0 \in U$ implies

$$b(x - x_0, y_0) \in W.$$

Therefore $(x, y) \in A \times F$, $x \in x_0 + U$, $y \in y_0 + V$ imply

$$b(x, y) - b(x_0, y_0) \in W + W. \blacksquare$$

PROPOSITION 3. *Let E, F and G be three topological vector spaces, $\mathfrak{S}$ a collection of bounded subsets of E, $\mathfrak{T}$ a collection of bounded subsets of F, and $b: E \times F \to G$ an $(\mathfrak{S}, \mathfrak{T})$-hypocontinuous bilinear map. If $A \in \mathfrak{S}$ and $B \in \mathfrak{T}$, then b is uniformly continuous on $A \times B$.*

Proof. Let W be an arbitrary neighborhood of 0 in G. Determine the neighborhood U of 0 in E such that $x \in U$, $y \in B$ imply $b(x, y) \in W$ and the neighborhood V of 0 in F such that $x \in A$ and $y \in V$ imply $b(x, y) \in W$. If $(x_1, y_1), (x_2, y_2) \in A \times B$, then $x_1 - x_2 \in U$ and $y_1 - y_2 \in V$ imply

$$b(x_1, y_1) - b(x_2, y_2) = b(x_1, y_1 - y_2) + b(x_1 - x_2, y_2) \in W + W. \blacksquare$$

THEOREM 2. *Let E be a topological vector space, F a barrelled locally convex space, G a locally convex space and $\mathfrak{S}$ any collection of bounded subsets of E. Every separately continuous bilinear map $b: E \times F \to G$ is $\mathfrak{S}$-hypocontinuous.*

Proof. Let W be a balanced, closed, convex neighborhood of 0 in G and $A \in \mathfrak{S}$. Denote by T the set of all points $y \in F$ for which $b(x, y) \in W$ for all $x \in A$. Let us show that T is a barrel.

(a) T is absorbing. Indeed, let $y \in F$. The linear map $b(\cdot, y)$ is continuous. Hence there exists a neighborhood U of 0 in E such that $x \in U$ implies $b(x, y) \in W$. There exists $\lambda > 0$ such that $\lambda A \subset U$. If $x \in A$, then we have $b(x, \lambda y) = b(\lambda x, y) \in W$; i.e., $\lambda y \in T$.

(b) T is balanced. Indeed, let $y \in T$ and $|\alpha| \leqq 1$. Then

$$b(x, \alpha y) = \alpha b(x, y) \in \alpha W \subset W$$

for all $x \in A$; i.e., $\alpha y \in T$.

(c) T is convex. Indeed, let $y_1, y_2 \in T$, $\alpha \geqq 0$, $\beta \geqq 0$, $\alpha + \beta = 1$. Then $b(x, \alpha y_1 + \beta y_2) = \alpha b(x, y_1) + \beta b(x, y_2) \in \alpha W + \beta W \subset W$ for all $x \in A$; i.e., $\alpha y_1 + \beta y_2 \in T$.

(d) T is closed. Indeed, let $y \in \overline{T}$. Let x be an arbitrary element of A and Z an arbitrary neighborhood of 0 in G. Since the linear map $b(x, \cdot)$ is continuous, there exists a neighborhood V of 0 in F such that $y_1 \in y + V$ implies $b(x, y_1) \in b(x, y) + Z$. By hypothesis $(y + V) \cap T \neq \emptyset$, and therefore $(b(x, y) + Z) \cap W \neq \emptyset$. Since Z was arbitrary, this means that $b(x, y) \in \overline{W} = W$. Since x was arbitrary in A, we have $y \in T$.

Since F is barrelled, T is a neighborhood of 0 in F; i.e., condition (HC) of Definition 2 is satisfied. $\blacksquare$

Now we return to the multiplication of distributions. First we must prove the following result:

PROPOSITION 4. *Let Ω be an open subset of $\mathbf{R}^n$ and $0 \leqq m \leqq \infty$. The bilinear map $(\alpha, \varphi) \mapsto \alpha\varphi$ from $\mathcal{E}^m(\Omega) \times \mathcal{D}^m(\Omega)$ into $\mathcal{D}^m(\Omega)$ is hypocontinuous.*

Proof. Since the spaces $\mathcal{E}^m(\Omega)$ and $\mathcal{D}^m(\Omega)$ are barrelled, by Theorem 2 it suffices to show that the map $(\alpha, \varphi) \mapsto \alpha\varphi$ is separately continuous. We know from Proposition 6.1 that for every $\alpha \in \mathcal{E}^m(\Omega)$ the map $\varphi \mapsto \alpha\varphi$ is continuous.

Let $\varphi \in \mathcal{D}^m(\Omega)$ and $K = \operatorname{Supp} \varphi$. Clearly, $\operatorname{Supp} \alpha\varphi \subset K$ for every $\alpha \in \mathcal{E}^m(\Omega)$. Let V be a neighborhood of 0 in $\mathcal{D}^m(\Omega)$. There exist an $\epsilon > 0$ and an integer $k \geqq 0$, where $k = m$ if m is finite, such that $V \cap \mathcal{D}^m(K)$ contains the set $\{\psi \mid |\partial^p\psi(x)| \leqq \epsilon, |p| \leqq k\}$. By the Leibniz formula (Proposition 2.5.3) $\partial^p(\alpha\varphi)$ is a linear combination of expressions of the form $\partial^q\alpha \cdot \partial^r\varphi$, where $q + r = p$. Since each $\partial^r\varphi$ is bounded, there exists an $\eta > 0$ such that if α belongs to the neighborhood

$$\{\alpha \mid |\partial^p\alpha(x)| \leqq \eta, x \in K, |p| \leqq k\}$$

of 0 in $\mathcal{E}^m(\Omega)$, then $\alpha\varphi \in V$. Thus the map $\alpha \mapsto \alpha\varphi$ is continuous. ∎

PROPOSITION 5. *Let Ω be an open subset of $\mathbf{R}^n$ and $0 \leqq m \leqq \infty$. Denote by $\mathcal{D}_\kappa'^m(\Omega)$ the space $\mathcal{D}'^m(\Omega)$ equipped with the topology $\kappa(\mathcal{D}'^m(\Omega), \mathcal{D}^m(\Omega))$ of uniform convergence on compact subsets of $\mathcal{D}^m(\Omega)$, and by $\mathcal{D}_\beta'^m(\Omega)$ the space $\mathcal{D}'^m(\Omega)$ equipped with the strong topology $\beta(\mathcal{D}'^m(\Omega), \mathcal{D}^m(\Omega))$.*

(a) *The bilinear map $(\alpha, T) \mapsto \alpha T$ from $\mathcal{E}^m(\Omega) \times \mathcal{D}_\kappa'^m(\Omega)$ into $\mathcal{D}_\kappa'^m(\Omega)$ is hypocontinuous with respect to the compact subsets of $\mathcal{E}^m(\Omega)$.*

(b) *The bilinear map $(\alpha, T) \mapsto \alpha T$ from $\mathcal{E}^m(\Omega) \times \mathcal{D}_\kappa'^m(\Omega)$ into $\mathcal{D}_\beta'^m(\Omega)$ is hypocontinuous with respect to the equicontinuous subsets of $\mathcal{D}'^m(\Omega)$.*

Proof. (a) Let W be a neighborhood of 0 in $\mathcal{D}_\kappa'^m(\Omega)$ and A a compact subset of $\mathcal{E}^m(\Omega)$. We can assume that $W = C^\circ$, where C is a compact subset of $\mathcal{D}^m(\Omega)$. It follows from Propositions 3 and 4 that the set $M = AC$ is compact in $\mathcal{D}^m(\Omega)$. Thus $V = M^\circ$ is a neighborhood of 0 in $\mathcal{D}_\kappa'^m(\Omega)$. Furthermore, $\alpha \in A$, $T \in V$, and $\varphi \in C$ imply

$$|\langle \alpha T, \varphi\rangle| = |\langle T, \alpha\varphi\rangle| \leqq 1;$$

i.e., $\alpha T \in C^\circ = W$.

(b) This time let $W = D^\circ$ be a neighborhood of 0 in $\mathcal{D}_\beta'^m(\Omega)$, where D is a bounded subset of $\mathcal{D}^m(\Omega)$, and let B be an equicontinuous subset of $\mathcal{D}'^m(\Omega)$. Then $N = B^\circ$ is a neighborhood of 0 in $\mathcal{D}^m(\Omega)$. It follows from Proposition 4 that there exists a neighborhood U of 0 in $\mathcal{E}^m(\Omega)$ such that $\alpha \in U$ and $\varphi \in D$ imply $\alpha\varphi \in N$. Hence if $\alpha \in U$, $T \in B$, and $\varphi \in D$, then

$$|\langle \alpha T, \varphi\rangle| = |\langle T, \alpha\varphi\rangle| \leqq 1;$$

i.e., $\alpha T \in D^\circ = W$. ∎

Since the map $\mathcal{D}_\beta'^m(\Omega) \hookrightarrow \mathcal{D}_\kappa'^m(\Omega)$ is continuous, it follows from part (b) that *a fortiori* the bilinear map $(\alpha, T) \mapsto \alpha T$ from $\mathcal{E}^m(\Omega) \times \mathcal{D}_\kappa'^m(\Omega)$ into $\mathcal{D}_\kappa'^m(\Omega)$ is hypocontinuous with respect to the equicontinuous subsets of $\mathcal{D}'^m(\Omega)$.

For $m = \infty$ the spaces $\mathcal{E}(\Omega)$ and $\mathcal{D}(\Omega)$ are Montel spaces, and the topology of uniform convergence on compact sets of $\mathcal{D}(\Omega)$ coincides with the strong topology of $\mathcal{D}'(\Omega)$. Hence the bilinear map $(\alpha, T) \mapsto \alpha T$ from $\mathcal{E}(\Omega) \times \mathcal{D}'(\Omega)$ into $\mathcal{D}'(\Omega)$ is hypocontinuous. For finite m we can prove a stronger result concerning the strong topology:

Proposition 6. *If $m \in \mathbf{N}$, then the bilinear map $(\alpha, T) \mapsto \alpha T$ from $\mathcal{E}^m(\Omega) \times \mathcal{D}_\beta'^m(\Omega)$ into $\mathcal{D}_\beta'^m(\Omega)$ is continuous.*

Proof. Let W be a neighborhood of 0 in $\mathcal{D}_\beta'^m(\Omega)$ which we may assume to be of the form $W = B^\circ$, where B is a bounded subset of $\mathcal{D}^m(\Omega)$. There exists a compact subset K of Ω and numbers $\mu_p > 0$ for $|p| \leqq m$ such that $\operatorname{Supp} \varphi \subset K$ and $|\partial^p \varphi(x)| \leqq \mu_p$ for $\varphi \in B$, $x \in \Omega$, and $|p| \leqq m$.

Let U be the neighborhood $\{\alpha \mid |\partial^p \alpha(x)| \leqq 1, x \in K, |p| \leqq m\}$ of 0 in $\mathcal{E}^m(\Omega)$. The set $A = UB$ is bounded in $\mathcal{D}^m(\Omega)$ since for $\alpha \in U$, $\varphi \in B$ we have $\operatorname{Supp}(\alpha\varphi) \subset \operatorname{Supp} \varphi \subset K$ and

$$|\partial^p(\alpha\varphi)(x)| \leqq \sum_{q \leqq p} \binom{p}{q} |\partial^q \alpha(x)| \cdot |\partial^{p-q}\varphi(x)| \leqq \nu_p$$

for $|p| \leqq m$, where we have set

$$\nu_p = \sum_{q \leqq p} \binom{p}{q} \mu_{p-q}.$$

Thus $V = A^\circ$ is a neighborhood of 0 in $\mathcal{D}_\beta'^m(\Omega)$, and furthermore $\alpha \in U$, $T \in V$, $\varphi \in B$ imply

$$|\langle \alpha T, \varphi \rangle| = |\langle T, \alpha\varphi \rangle| \leqq 1;$$

i.e., $\alpha T \in B^\circ = W$. ∎

Remark 3. The hypocontinuity of the map

$$(\alpha, T) \mapsto \alpha T$$

from $\mathcal{E}(\Omega) \times \mathcal{D}'(\Omega)$ into $\mathcal{D}'(\Omega)$ can be seen very easily as follows. Since $\mathcal{E}(\Omega)$ and $\mathcal{D}'(\Omega)$ are barrelled, by Theorem 2 it is sufficient to prove that the map is separately continuous. We know from Proposition 6.1 that the map $T \mapsto \alpha T$ is continuous. To prove that $\alpha \mapsto \alpha T$ is also continuous, it suffices—since $\mathcal{E}(\Omega)$ is metrizable—to prove that if (α_n) is a sequence in $\mathcal{E}(\Omega)$ which converges to 0, then $\alpha_n T$ converges to 0 in $\mathcal{D}'(\Omega)$. If $\varphi \in \mathcal{D}(\Omega)$, then $(\alpha_n \varphi)$ converges to 0 in $\mathcal{D}(\Omega)$; hence $\langle \alpha_n T, \varphi \rangle = \langle T, \alpha_n \varphi \rangle$ converges to 0 in $\mathbf{K}$. It follows from Proposition 1.2 that $(\alpha_n T)$ converges to 0 in $\mathcal{D}'(\Omega)$.

It follows from Proposition 6.2 that the bilinear map $(\alpha, T) \mapsto \alpha T$ maps both $\mathcal{E}^m(\Omega) \times \mathcal{E}'^m(\Omega)$ and $\mathcal{D}^m(\Omega) \times \mathcal{D}'^m(\Omega)$ into $\mathcal{E}'^m(\Omega)$. We leave it as an exercise for the reader to establish continuity properties similar to the above ones for these maps (Exercises 6, 7, 8).

We conclude this section with some consequences of Theorem 4.2.

PROPOSITION 7. *Let Ω be an open subset of $\mathbf{R}^n$ and $m \in \mathbf{N}$. If $T \in \mathfrak{D}'^m(\Omega)$, and $\varphi \in \mathfrak{D}^m(\Omega)$ is such that $\partial^p\varphi(x) = 0$ for all $x \in \operatorname{Supp} T$ and $|p| \leqq m$, then $\langle T, \varphi\rangle = 0$.*

Proof. Let $\alpha \in \mathfrak{D}(\Omega)$ be such that $\alpha(x) = 1$ for all $x \in \operatorname{Supp} \varphi$. Then $\alpha T \in \mathcal{E}'^m(\Omega)$ and $\langle \alpha T, \varphi\rangle = \langle T, \alpha\varphi\rangle = \langle T, \varphi\rangle$. Since $\operatorname{Supp}(\alpha T) \subset \operatorname{Supp} T$, we have $\partial^p\varphi(x) = 0$ for all $x \in \operatorname{Supp}(\alpha T)$ and $|p| \leqq m$. The conclusion follows from Theorem 4.2. ∎

PROPOSITION 8. *Let Ω be an open subset of $\mathbf{R}^n$ and $0 \leqq m \leqq \infty$. If $\alpha \in \mathcal{E}^m(\Omega)$, $T \in \mathfrak{D}'^m(\Omega)$, and $\partial^p\alpha(x) = 0$ for $x \in \operatorname{Supp} T$ and $|p| \leqq m$, then $\alpha T = 0$.*

Proof. Let (Ω_j) be a sequence of relatively compact open subsets covering Ω and let (β_j) be a locally finite infinitely differentiable partition of unity subordinated to (Ω_j). The support of $\beta_j T$ is contained in that of β_j (Proposition 6.2) and is in particular compact. It follows from the corollary of Proposition 4.3 that there exists an $m_j \in \mathbf{N}$ such that $\beta_j T \in \mathfrak{D}'^{m_j}(\Omega)$ (we choose $m_j = m$ if m is finite). For an arbitrary $\varphi \in \mathfrak{D}^{m_j}(\Omega)$ we have $\partial^p(\alpha\varphi)(x) = 0$ for all $x \in \operatorname{Supp}(\beta_j T)$ and $|p| \leqq m_j$, and therefore by Proposition 7 (or Theorem 4.2) we have $\langle \beta_j(\alpha T), \varphi\rangle = \langle \beta_j T, \alpha\varphi\rangle = 0$, i.e., $\beta_j(\alpha T) = 0$.

Now $\sum_{j=1}^k \beta_j$ converges to $\mathbf{1}$ in $\mathcal{E}^m(\Omega)$ as $k \to \infty$ and therefore $\sum_{j=1}^k \beta_j(\alpha T)$ converges to αT in $\mathfrak{D}'^m(\Omega)$ by Proposition 5(b). Hence $\alpha T = 0$. ∎

EXERCISES

1. Let E, F and G be three locally convex vector spaces, $(q_\iota)_{\iota \in I}$ a saturated family of semi-norms defining the topology of E, $(r_\lambda)_{\lambda \in L}$ a saturated family of semi-norms defining the topology of F, and $(s_\mu)_{\mu \in M}$ a family of semi-norms defining the topology of G. Show that a bilinear map $(x, y) \mapsto b(x, y)$ from $E \times F$ into G is continuous if and only if for every semi-norm s_μ there exist a semi-norm q_ι, a semi-norm r_λ, and a number $M > 0$ such that

$$s_\mu(b(x, y)) \leqq M q_\iota(x) r_\lambda(y)$$

for all $x \in E$, $y \in F$. (*Hint:* See the proof of Proposition 2.5.2.)

2. Let E, F, G be three topological vector spaces and $\mathfrak{B}$ a set of bilinear maps from $E \times F$ into G. Show that $\mathfrak{B}$ is equicontinuous if it is equicontinuous at the point $(0, 0)$.

3. Show that the map $(\alpha, \varphi) \mapsto \alpha\varphi$ from $\mathcal{C}(\Omega) \times \mathcal{K}(\Omega)$ into $\mathcal{K}(\Omega)$ is not continuous.

4. Denote by $\mathfrak{M}_\kappa(\Omega)$ the space of all measures on Ω equipped with the topology $\kappa(\mathfrak{M}(\Omega), \mathcal{K}(\Omega))$. Show that the map $(\alpha, \mu) \mapsto \alpha\mu$ from $\mathcal{C}(\Omega) \times \mathfrak{M}_\kappa(\Omega)$ into $\mathfrak{M}_\kappa(\Omega)$ is not continuous.

5. Show that the map $(\alpha, T) \mapsto \alpha T$ from $\mathcal{E}(\Omega) \times \mathcal{D}'(\Omega)$ into $\mathcal{D}'(\Omega)$ is not continuous.

6. Show that for $0 \leqq m \leqq \infty$ the map $(\alpha, \beta) \mapsto \alpha\beta$ from $\mathcal{E}^m(\Omega) \times \mathcal{E}^m(\Omega)$ into $\mathcal{E}^m(\Omega)$ is continuous.

7. Similarly as in Proposition 5, denote by $\mathcal{E}_\kappa'^m(\Omega)$ the space $\mathcal{E}'^m(\Omega)$ equipped with the topology $\kappa(\mathcal{E}'^m(\Omega), \mathcal{E}^m(\Omega))$ of uniform convergence on compact subsets of $\mathcal{E}^m(\Omega)$, and by $\mathcal{E}_\beta'^m(\Omega)$ the same space equipped with the strong topology $\beta(\mathcal{E}'^m(\Omega), \mathcal{E}^m(\Omega))$.

(a) Show that the bilinear map $(\alpha, T) \mapsto \alpha T$ from $\mathcal{E}^m(\Omega) \times \mathcal{E}_\kappa'^m(\Omega)$ into $\mathcal{E}_\kappa'^m(\Omega)$ is hypocontinuous with respect to the compact subsets of $\mathcal{E}^m(\Omega)$.

(b) Show that the bilinear map $(\alpha, T) \mapsto \alpha T$ from $\mathcal{E}^m(\Omega) \times \mathcal{E}_\kappa'^m(\Omega)$ into $\mathcal{E}_\beta'^m(\Omega)$ is hypocontinuous with respect to the equicontinuous subsets of $\mathcal{E}'^m(\Omega)$. (*Hint:* Use Exercise 6 and the proof of Proposition 5.)

8. (a) Show that the bilinear map $(\alpha, T) \mapsto \alpha T$ from $\mathcal{D}^m(\Omega) \times \mathcal{D}_\kappa'^m(\Omega)$ into $\mathcal{E}_\kappa'^m(\Omega)$ is hypocontinuous with respect to the compact subsets of $\mathcal{D}^m(\Omega)$.

(b) Show that the bilinear map $(\alpha, T) \mapsto \alpha T$ from $\mathcal{D}^m(\Omega) \times \mathcal{D}_\kappa'^m(\Omega)$ into $\mathcal{E}_\beta'^m(\Omega)$ is hypocontinuous with respect to the equicontinuous subsets of $\mathcal{D}'^m(\Omega)$.

9. Show that if $T \in \mathcal{D}'(\Omega)$, and $\varphi \in \mathcal{D}(\Omega)$ is such that $\partial^p\varphi(x) = 0$ for $x \in \operatorname{Supp} T$ and all $p \in \mathbf{N}^n$, then $\langle T, \varphi\rangle = 0$.

10. Let E and F be two distinguished Fréchet spaces and G a locally convex Hausdorff space. Equip their duals E', F', and G' with the strong topologies $\beta(E', E)$, $\beta(F', F)$, and $\beta(G', G)$. Prove that every separately continuous bilinear map $b: E' \times F' \to G'$ is continuous.

(*Hint:* By Proposition 3.16.1 and Theorem 2 the map b is hypocontinuous with respect to the bounded subsets of F'. Let C be a bounded subset of G. For each $x' \in E'$ and $z \in C$ the map $y' \mapsto \langle z, b(x', y')\rangle$ is a continuous linear form on F'. Hence there exists $v_z(x') \in F''$ such that $\langle z, b(x', y')\rangle = \langle v_z(x'), y'\rangle$ for all $y' \in F'$.

Next show that the set of linear maps $v_z: E' \to F''$ is equicontinuous if $z \in C$ and F'' is equipped with $\epsilon(F'', F')$. For this show that if V is a neighborhood of 0 in F, then there exists a bounded subset A of E such that $x' \in A^\circ$ and $y' \in V^\circ$ imply $b(x', y') \in C^\circ$, and thus $z \in C$ and $x' \in A^\circ$ imply $v_z(x') \in V^{\circ\circ}$.

Finally, if $(V_n)_{n\in\mathbf{N}}$ is a fundamental sequence of balanced, closed, convex neighborhoods of 0 in F'', for each $n \in \mathbf{N}$ choose a bounded set $A_n \subset E$ such that $x' \in A_n^\circ$ implies $v_z(x') \in V_n$ for $z \in C$. By Proposition 2.6.3 there exists a bounded set $A \subset E$ and a sequence $\lambda_n > 0$ such that $A_n \subset \lambda_n A$ for all $n \in \mathbf{N}$. Show that $x' \in A^\circ$ implies $v_z(x') \in \lambda_n V_n$ for all $z \in C$ and $n \in \mathbf{N}$, i.e., that $D = \bigcup_{z\in C} v_z(A^\circ)$ is bounded in F''. There exists a bounded set B in F such that $B^{\circ\circ} \supset D$. If $x' \in A^\circ$ and $y' \in B^\circ$, then $b(x', y') \in C^\circ$).

§8. *Tensor product*

We begin with the purely algebraic definition of the tensor product of two vector spaces E and F over the same field $\mathbf{K}$. We show that there exist a vector space P over $\mathbf{K}$ and a bilinear map $\beta: E \times F \to P$ having the following "universal property": for any vector space G over $\mathbf{K}$ and any bilinear map $b: E \times F \to G$ there exists a unique linear map $u: P \to G$ such that

$$u(\beta(x, y)) = b(x, y)$$

for every $x \in E$ and $y \in F$. We prove furthermore that two such spaces P are isomorphic.

For each pair $(x, y) \in E \times F$ let $\mathbf{K}_{(x,y)}$ denote a copy of $\mathbf{K}$ considered as a one-dimensional vector space over itself. The vectors of the external direct sum (cf. Chapter 2, §12)

$$\mathbf{K}^{(E \times F)} = \coprod_{(x,y) \in E \times F} \mathbf{K}_{(x,y)}$$

are the families $(\lambda_{(x,y)})_{(x,y) \in E \times F}$, where $\lambda_{(x,y)} \in \mathbf{K}$ and $\lambda_{(x,y)} = 0$ except for finitely many indices (x, y). We will find it convenient to denote also by (x, y) the element $(\epsilon_{(s,t)})$ of $\mathbf{K}^{(E \times F)}$ given by $\epsilon_{(x,y)} = 1$ and $\epsilon_{(s,t)} = 0$ if $(s, t) \neq (x, y)$. The elements (x, y) then form a basis of $\mathbf{K}^{(E \times F)}$, and $(\lambda_{(x,y)})$ can be written as a "formal linear combination" $\sum \lambda_{(x,y)}(x, y)$.

Let I be the subspace of $\mathbf{K}^{(E \times F)}$ generated by all elements of the form

$$(x_1 + x_2, y) - (x_1, y) - (x_2, y), \qquad (\alpha x, y) - \alpha(x, y), \tag{1}$$

$$(x, y_1 + y_2) - (x, y_1) - (x, y_2), \qquad (x, \alpha y) - \alpha(x, y), \tag{2}$$

where $x_1, x_2, x \in E$, $y_1, y_2, y \in F$ and $\alpha \in \mathbf{K}$. We set $P = \mathbf{K}^{(E \times F)}/I$, and $\beta(x, y)$ will be the image in P of the element $(x, y) \in \mathbf{K}^{(E \times F)}$ under the canonical surjection. Now suppose that we have a vector space G and a bilinear map $b: E \times F \to G$. We define a map $v: \mathbf{K}^{(E \times F)} \to G$ as follows:

$$v\left(\sum \lambda_{(x,y)}(x, y)\right) = \sum \lambda_{(x,y)} b(x, y).$$

It is immediately verified that v is a linear map and that it vanishes on all elements of the form (1) and (2). Hence v vanishes on I and defines a linear map $u: P \to G$ for which

$$u\left(\sum \lambda_{(x,y)} \beta(x, y)\right) = \sum \lambda_{(x,y)} b(x, y),$$

and in particular $u(\beta(x, y)) = b(x, y)$ for every $x \in E$, $y \in F$. Also u is completely determined by this last condition since the elements $\beta(x, y)$ generate P.

Suppose that we have a second space Q and a bilinear map $\gamma: E \times F \to Q$ with the same property. Then there exists a linear map $f: P \to Q$ such that $f(\beta(x, y)) = \gamma(x, y)$ for every $x \in E$, $y \in F$ and a linear map $g: Q \to P$ such that $g(\gamma(x, y)) = \beta(x, y)$ for every $x \in E$, $y \in F$. We have

$$g(f(\beta(x, y))) = \beta(x, y) \qquad \text{and} \qquad f(g(\gamma(x, y))) = \gamma(x, y)$$

for all $x \in E$, $y \in F$. Now the condition that for every bilinear map $b: E \times F \to G$ there exists a *unique* linear map $u: Q \to G$ such that

$$u(\gamma(x, y)) = b(x, y)$$

for all $x \in E$, $y \in F$ implies in particular that the elements $\gamma(x, y)$ generate Q. Since the elements $\beta(x, y)$ generate P, we have $g \circ f = 1_P$, $f \circ g = 1_Q$, and f is an isomorphism.

The vector space P constructed in the preceding proof is called the tensor product of the vector spaces E and F and will be denoted by $E \otimes_{\mathbf{K}} F$ or simply $E \otimes F$ if there is no doubt about the field $\mathbf{K}$. The bilinear map

$$\beta: E \times F \to E \otimes F$$

is called the canonical map, and the image $\beta(x, y)$ of an element $(x, y) \in E \times F$ will be denoted by $x \otimes y$. Every element of $E \otimes F$ is a finite linear combination of elements of the form $x \otimes y$, and we have

$$(x_1 + x_2) \otimes y = x_1 \otimes y + x_2 \otimes y, \qquad (\alpha x) \otimes y = \alpha(x \otimes y),$$
$$x \otimes (y_1 + y_2) = x \otimes y_1 + x \otimes y_2, \qquad x \otimes (\alpha y) = \alpha(x \otimes y)$$

for all $x_1, x_2, x \in E$, $y_1, y_2, y \in F$, $\alpha \in \mathbf{K}$. In particular, the representation of an element of $E \otimes F$ as the linear combination of elements of the form $x \otimes y$ is not unique.

Let $\mathbf{K}^1$ be the field $\mathbf{K}$ considered as a one-dimensional vector space over itself. The bilinear map $(\alpha, x) \mapsto \alpha x$ from $\mathbf{K}^1 \times E$ into E defines a linear map $u: \mathbf{K}^1 \otimes E \to E$ which for elements of the form $\alpha \otimes x$ is given by

$$u(\alpha \otimes x) = \alpha x.$$

Let us show that u is an isomorphism. In the first place, u is surjective since given $x \in E$, we have $u(1 \otimes x) = x$. On the other hand, u is injective. Indeed, let $z \in \mathbf{K}^1 \otimes E$ be such that $u(z) = 0$. We can write z in the form

$$z = \sum_{k=1}^{n} \alpha_k \otimes x_k,$$

where the x_k are linearly independent. Then $u(z) = \sum_{k=1}^{n} \alpha_k x_k = 0$ implies $\alpha_k = 0$ for $1 \leqq k \leqq n$; i.e., $z = 0$.

Next let $(f_\mu)_{\mu \in M}$ be an algebraic basis of the vector space F. We want to prove that every element of $E \otimes F$ can be written uniquely in the form $\sum_{\mu \in M} x_\mu \otimes f_\mu$, where $x_\mu \in E$ and $x_\mu = 0$ except for finitely many indices μ. To show that every element of $E \otimes F$ admits such a representation, it is sufficient to consider elements of the form $x \otimes y$. If $y = \sum \eta_\mu f_\mu$, then $x \otimes y = \sum \eta_\mu x \otimes f_\mu$, where $\eta_\mu x \in E$. To prove uniqueness, consider the bilinear map $b: E \times F \to E^{(M)}$ defined as follows: for any element $(x, y) \in E \times F$ write $y = \sum \eta_\mu f_\mu$ and set $b(x, y) = (\eta_\mu x)_{\mu \in M}$. The map b is well-defined because (f_μ) is an algebraic basis of F. Let $u: E \otimes F \to E^{(M)}$ be the linear map associated with b, i.e., such that

$u(x \otimes y) = b(x, y)$ for all $x \in E$, $y \in F$. If $z = \sum_{\mu \in M} x_\mu \otimes f_\mu$, we have $u(z) = (x_\mu)_{\mu \in M}$. It is sufficient to verify this for a single term $x \otimes f_\nu$, and for such a term we have $u(x \otimes f_\nu) = b(x, f_\nu) = (x_\mu)_{\mu \in M}$, where $x_\nu = x$ and $x_\mu = 0$ if $\mu \neq \nu$. Now, to prove the uniqueness in question, let $z = \sum x_\mu \otimes f_\mu = 0$. Then $u(z) = (x_\mu) = 0$, i.e., $x_\mu = 0$ for every $\mu \in M$.

Let $(e_\lambda)_{\lambda \in L}$ be an algebraic basis of E and $(f_\mu)_{\mu \in M}$ an algebraic basis of F. We prove that $(e_\lambda \otimes f_\mu)_{(\lambda,\mu) \in L \times M}$ is an algebraic basis of $E \otimes F$. In the first place, every element of $E \otimes F$ is a linear combination of the $e_\lambda \otimes f_\mu$ since if $x = \sum_{\lambda \in L} \xi_\lambda e_\lambda$ and $y = \sum_{\mu \in M} \eta_\mu f_\mu$, then

$$x \otimes y = \sum_{(\lambda,\mu) \in L \times M} \xi_\lambda \eta_\mu (e_\lambda \otimes f_\mu).$$

To prove linear independence, let

$$\sum_{(\lambda,\mu) \in L \times M} \alpha_{\lambda\mu}(e_\lambda \otimes f_\mu) = \sum_{\mu \in M} \Big(\sum_{\lambda \in L} \alpha_{\lambda\mu} e_\lambda\Big) \otimes f_\mu = 0.$$

Then by the preceding result we have

$$\sum_{\lambda \in L} \alpha_{\lambda\mu} e_\lambda = 0$$

for each $\mu \in M$; and since (e_λ) is free, it follows that $\alpha_{\lambda\mu} = 0$ for all $\lambda \in L$ and $\mu \in M$.

In particular, if E and F are finite-dimensional we have

$$\dim E \otimes F = \dim E \cdot \dim F.$$

Now let E, F, G, H be four vector spaces and $u: E \to G$, $v: F \to H$ two linear maps. The map $(x, y) \mapsto u(x) \otimes v(y)$ from $E \times F$ into $G \otimes H$ is clearly bilinear. There exists therefore a unique linear map $w: E \otimes F \to G \otimes H$ which satisfies $w(x \otimes y) = u(x) \otimes v(y)$ for $x \in E$, $y \in F$. We call w the tensor product of the linear maps u and v and denote it by $u \otimes v$. Thus

$$(u \otimes v)(x \otimes y) = u(x) \otimes v(y) \qquad \text{for all} \qquad x \in E, y \in F.$$

If u and v are linear forms, then $u \otimes v$ is a linear map $E \otimes F \to \mathbf{K} \otimes \mathbf{K}$. But we know that $\alpha \otimes \beta \mapsto \alpha\beta$ is an isomorphism from $\mathbf{K} \otimes \mathbf{K}$ onto $\mathbf{K}$ and therefore we shall regard $u \otimes v$ as the linear form on $E \otimes F$ given by

$$(u \otimes v)(x \otimes y) = u(x)v(y) \qquad \text{for} \qquad x \in E, y \in F.$$

Let us now introduce the notations we shall use in most of the remainder of this section. We shall consider a euclidean space $\mathbf{R}^k$ of dimension k and a euclidean space $\mathbf{R}^l$ of dimension l. The vectors of the first space will be denoted by x and those of the second by y (with eventual subscripts or primes). Ξ will denote an open subset of $\mathbf{R}^k$ and H an open subset of $\mathbf{R}^l$, and we will consider $\Xi \times \mathrm{H}$ as an open subset of $\mathbf{R}^n$ $(n = k + l)$.

If p and q are multi-indices, $p = (p_1, \ldots, p_k) \in \mathbf{N}^k$, $q = (q_1, \ldots, q_l) \in \mathbf{N}^l$, then ∂_x^p, ∂_y^q will denote the derivations

$$\frac{\partial^{|p|}}{\partial x_1^{p_1} \cdots \partial x_k^{p_k}}, \qquad \frac{\partial^{|q|}}{\partial y_1^{q_1} \cdots \partial y_l^{q_l}}$$

respectively. Finally s and t will denote either positive integers or the symbol ∞. If s or t or both $= \infty$, then $s + t = \infty$.

We define a bilinear map from the space $\mathfrak{D}(\Xi) \times \mathfrak{D}(\mathrm{H})$ into $\mathfrak{D}(\Xi \times \mathrm{H})$ by associating with every pair (φ, ψ) of functions the function

$$(x, y) \mapsto \varphi(x)\psi(y).$$

To this bilinear map there corresponds a unique linear map

$$j: \mathfrak{D}(\Xi) \otimes \mathfrak{D}(\mathrm{H}) \to \mathfrak{D}(\Xi \times \mathrm{H})$$

which associates with each element

$$\sum \alpha_{\lambda\mu}\varphi_\lambda \otimes \psi_\mu \qquad (\varphi_\lambda \in \mathfrak{D}(\Xi), \quad \psi_\mu \in \mathfrak{D}(\mathrm{H}))$$

the function

$$(x, y) \mapsto \sum \alpha_{\lambda\mu}\varphi_\lambda(x)\psi_\mu(y).$$

The map j is injective. Indeed, let $\zeta \in \mathfrak{D}(\Xi) \otimes \mathfrak{D}(\mathrm{H})$ be such that $j(\zeta) = 0$. We can write ζ in the form $\zeta = \sum \alpha_{\lambda\mu}\varphi_\lambda \otimes \psi_\mu$, where the family $(\varphi_\lambda)_{\lambda \in L}$ is free in $\mathfrak{D}(\Xi)$ and the family $(\psi_\mu)_{\mu \in M}$ is free in $\mathfrak{D}(\mathrm{H})$. Now $j(\zeta) = 0$ means that

$$\sum \alpha_{\lambda\mu}\varphi_\lambda(x)\psi_\mu(y) = 0$$

for all $x \in \Xi$ and $y \in \mathrm{H}$. Hence we have $\sum \alpha_{\lambda\mu}\varphi_\lambda(x) = 0$ for all $\mu \in M$ and $x \in \Xi$, and therefore $\alpha_{\lambda\mu} = 0$ for all $\lambda \in L$, $\mu \in M$; i.e., $\zeta = 0$.

We shall call j the canonical injection, and we shall usually identify $\mathfrak{D}(\Xi) \otimes \mathfrak{D}(\mathrm{H})$ with its image in $\mathfrak{D}(\Xi \times \mathrm{H})$. Our next aim is to show that $\mathfrak{D}(\Xi) \otimes \mathfrak{D}(\mathrm{H})$ is dense in $\mathfrak{D}^m(\Xi \times \mathrm{H})$ for $0 \leqq m \leqq \infty$. This will be accomplished with the help of

LEMMA 1 (Henri Cartan). *Let Ω be an open subset of $\mathbf{R}^n$ and $0 \leqq m \leqq \infty$. Let ρ be a function in $\mathfrak{D}$ having properties* (a), (b), (c) *listed in Example* 1.4 *and let ρ_ϵ be the function defined there. Denote by $\mathfrak{F}(\Omega)$ the set of all functions $\tau_h\rho_\epsilon$* (cf. Definition 3.2), *where $h \in \mathbf{R}^n$ and $\epsilon > 0$ vary so that*

$$\mathrm{Supp}(\tau_h\rho_\epsilon) \subset \Omega.$$

Then $\mathfrak{F}(\Omega)$ is total in $\mathfrak{D}^m(\Omega)$.

Proof. In the proof of Proposition 4.1 we saw that the convolutions $\rho_\epsilon * \varphi$, where $\varphi \in \mathfrak{D}^m(\Omega)$ and $\epsilon > 0$ is small, are dense in $\mathfrak{D}^m(\Omega)$. Hence it is sufficient to prove that every such convolution can be approximated in $\mathfrak{D}^m(\Omega)$ by finite linear combinations of the functions $\tau_h\rho_\epsilon$. We may clearly assume that φ is not identically 0.

Let K be a compact neighborhood of Supp φ contained in Ω and let $\epsilon > 0$ be so small that any ball with radius ϵ that meets Supp φ is contained in K. Then $\rho_\epsilon * \varphi \in \mathfrak{D}(K)$. Let V be a neighborhood of 0 in $\mathfrak{D}^m(\Omega)$. The intersection $V \cap \mathfrak{D}^m(K)$ contains a set of the form

$$\{\alpha \mid |\partial^p\alpha(x)| \leqq \eta, |p| \leqq h\},$$

where $\eta > 0$, $h \in \mathbf{N}$, and $h = m$ if m is finite. We can cover $\mathbf{R}^n$ by cubes, no two of which have an interior point in common and such that

$$|\partial^p\rho_\epsilon\,(x - y') - \partial^p\rho_\epsilon\,(x - y'')| < \eta\cdot\left\{\int_\Omega |\varphi(x)|\,dx\right\}^{-1}$$

for $x \in \mathbf{R}^n$ and $|p| \leqq h$ if y' and y'' belong to the same cube. Only a finite number of these cubes, say $C_1, C_2, \ldots, C_l$, meet Supp φ. In each set $C_i \cap$ Supp φ pick a point y_i and define the function $\chi \in \mathfrak{D}(K)$ by

$$\chi(x) = \sum_{i=1}^{l} \rho_\epsilon(x - y_i)\int_{C_i} \varphi(y)\,dy.$$

Then

$$|\partial^p\chi(x) - \partial^p(\rho_\epsilon * \varphi)(x)| = \left|\sum_{i=1}^{l}\int_{C_i}\{\partial^p\rho_\epsilon(x - y_i) - \partial^p\rho_\epsilon(x - y)\}\varphi(y)\,dy\right|$$

$$\leqq \frac{\eta}{\int_\Omega |\varphi(x)|\,dx}\sum_{i=1}^{l}\int_{C_i}|\varphi(y)|\,dy = \eta$$

for $x \in \mathbf{R}^n$ and $|p| \leqq h$; i.e., $\chi \in \rho_\epsilon * \varphi + V$. ∎

Now we can prove

PROPOSITION 1. *For* $0 \leqq m \leqq \infty$ *the space* $\mathfrak{D}(\Xi) \otimes \mathfrak{D}(\mathrm{H})$ *is dense in* $\mathfrak{D}^m(\Xi \times \mathrm{H})$.

Proof. We know that there exist functions $\rho \in \mathfrak{D}(\mathbf{R}^k)$ and $\sigma \in \mathfrak{D}(\mathbf{R}^l)$ such that ρ satisfies conditions (a), (b), (c) of Example 1.4 on $\mathbf{R}^k$ and σ satisfies these conditions on $\mathbf{R}^l$. The function $(x, y) \mapsto \sqrt{2}^n\rho(\sqrt{2}\,x)\sigma(\sqrt{2}\,y)$ satisfies conditions (a), (b), (c) of Example 1.4 on $\mathbf{R}^n = \mathbf{R}^k \times \mathbf{R}^l$. Hence by Lemma 1 the finite linear combinations of the functions

$$(x, y) \mapsto \rho(\sqrt{2}\,(x - x_0))\sigma(\sqrt{2}\,(y - y_0))$$

are dense in $\mathfrak{D}^m(\Xi \times \mathrm{H})$. ∎

Let $S \in \mathfrak{D}'^s(\Xi)$ and $T \in \mathfrak{D}'^t(\mathrm{H})$. We have identified $S \otimes T$ with the linear form on $\mathfrak{D}^s(\Xi) \otimes \mathfrak{D}^t(\mathrm{H})$, which for $\varphi \in \mathfrak{D}^s(\Xi)$ and $\psi \in \mathfrak{D}^t(\mathrm{H})$ is given by

$$\langle S \otimes T, \varphi \otimes \psi\rangle = \langle S, \varphi\rangle\langle T, \psi\rangle.$$

We want to show that $S \otimes T$ is a continuous linear form on $\mathfrak{D}(\Xi) \otimes \mathfrak{D}(\mathrm{H})$ for the topology induced by $\mathfrak{D}^{s+t}(\Xi \times \mathrm{H})$. First, however, let us introduce some more notations which will be very useful in the sequel. If

$$\chi \in \mathfrak{D}^m(\Xi \times \mathrm{H}),$$

then for each $y \in \mathrm{H}$ we denote by $\chi(\cdot, y)$ the function $x \mapsto \chi(x, y)$ defined for $x \in \Xi$ (or $x \in \mathbf{R}^k$), and similarly $\chi(x, \cdot)$ denotes the function

$$y \mapsto \chi(x, y).$$

Let K be a compact subset of Ξ and L a compact subset of H. If

$$\operatorname{Supp} \chi \subset K \times L,$$

then $\operatorname{Supp} \chi(\cdot, y) \subset K$ for all y, and in particular $\chi(\cdot, y) \in \mathfrak{D}^m(\Xi)$. Similarly, $\operatorname{Supp} \chi(x, \cdot) \subset L$ for all x and $\chi(x, \cdot) \in \mathfrak{D}^m(\mathrm{H})$.

If $S \in \mathfrak{D}'^s(\Xi)$ and $\varphi \in \mathfrak{D}^s(\mathrm{H})$, we shall also write $\int_\Xi S(x)\varphi(x)\,dx$ instead of $\langle S, \varphi\rangle$. This notation, used by Schwartz ([88], I, p. 71), has the advantage of showing the variable x. Of course, one has to bear in mind that S is not a function and in particular that the symbol $S(x)$ has no meaning. If now $\chi \in \mathfrak{D}^m(\Xi \times \mathrm{H})$, then the function

$$y \mapsto \langle S, \chi(\cdot, y)\rangle = \int_\Xi S(x)\chi(x, y)\,dx \tag{3}$$

will naturally be denoted by $\int_\Xi S(x)\chi(x, \cdot)\,dx$.

Lemma 2. *If* $\chi \in \mathfrak{D}^{s+t}(K \times L)$ *and* $S \in \mathfrak{D}'^s(\Xi)$, *then* (3) *belongs to* $\mathfrak{D}^t(L)$.

Proof. If $y \notin L$, then $\chi(x, y) = 0$ for every x, and therefore

$$\int_\Xi S(x)\chi(x, y)\,dx = 0.$$

Thus the support of (3) is contained in L.

Next the function (3) is continuous. Indeed, the map $y \mapsto \chi(\cdot, y)$ from H into $\mathfrak{D}^s(K)$ is continuous since this means that the functions

$$y \mapsto \partial_x^p \chi(x, y),$$

where $p \in \mathbf{N}^k$, $|p| \leqq h$, $x \in \Xi$, $h \geqq 0$ is an integer, $h = s$ if s is finite, are uniformly equicontinuous; this is a consequence of the uniform continuity of the functions $(x, y) \mapsto \partial_x^p \chi(x, y)$. It follows that the map

$y \mapsto \chi(\cdot, y)$ from H into $\mathfrak{D}^s(\Xi)$ is continuous, and since S is continuous on $\mathfrak{D}^s(\Xi)$, we obtain the continuity of (3).

To show that (3) has continuous partial derivatives of orders $\leqq t$ it is enough to show that for $|q| \leqq t$ we have

$$\partial_y^q \int_\Xi S(x)\chi(x, \cdot)\, dx = \int_\Xi S(x) \cdot \partial_y^q \chi(x, \cdot)\, dx, \tag{4}$$

for if so, then by what we have just seen the function

$$y \mapsto \partial_y^q \int_\Xi S(x)\chi(x, y)\, dx$$

will be continuous.

To prove (4) we can restrict ourselves by recursion to a derivation ∂_j of order 1 with respect to one variable y_j. Let h be a vector in $\mathbf{R}^l$ parallel to the y_j-axis and $y_0 \in$ H. The function

$$x \mapsto h_j^{-1}\{\chi(x, y_0 + h) - \chi(x, y_0)\}$$

tends to the function

$$x \mapsto (\partial_j\chi)(x, y_0)$$

in $\mathfrak{D}^s(\Xi)$ as $|h| \to 0$ because the expressions

$$\begin{aligned} h_j^{-1}\{\partial_x^p\chi(x, y_0 + h) - \partial_x^p\chi(x, y_0)\} &- \partial_x^p\, \partial_j\chi(x, y_0) \\ &= \partial_x^p\, \partial_j\chi(x, y_0 + \theta h) - \partial_x^p\, \partial_j\chi(x, y_0) \end{aligned}$$

$(0 < \theta < 1)$ can be made arbitrarily small, uniformly in x, by taking $|h|$ sufficiently small, since the function $(x, y) \mapsto \partial_x^p\, \partial_j\chi(x, y)$ is uniformly continuous. We have therefore

$$\begin{aligned} \partial_j \int_\Xi & S(x)\chi(x, y_0)\, dx \\ &= \lim \frac{1}{h_j}\left\{\int_\Xi S(x)\chi(x, y_0 + h)\, dx - \int_\Xi S(x)\chi(x, y_0)\, dx\right\} \\ &= \lim \int_\Xi S(x)\, \frac{\chi(x, y_0 + h) - \chi(x, y_0)}{h_j}\, dx \\ &= \int_\Xi S(x) \cdot \lim \frac{1}{h_j}\, \{\chi(x, y_0 + h) - \chi(x, y_0)\}\, dx \\ &= \int_\Xi S(x)\, \partial_j\chi(x, y_0)\, dx. \blacksquare \end{aligned}$$

LEMMA 3. *If $S \in \mathfrak{D}'^s(\Xi)$ and $T \in \mathfrak{D}'^t(\mathrm{H})$, then the linear form*

$$\chi \mapsto \int_{\mathrm{H}} T(y) \left(\int_\Xi S(x)\chi(x, y)\, dx\right) dy$$

is continuous on $\mathfrak{D}^{s+t}(\Xi \times \mathrm{H})$.

Proof. For each compact subset K of Ξ there exists a constant $M_K > 0$ and an integer $a \leqq s$ such that

$$|\langle S, \varphi\rangle| \leqq M_K \max_{|p| \leqq a} \max_{x} |\partial^p \varphi(x)|$$

for all $\varphi \in \mathfrak{D}^s(\Xi)$. Similarly, for each compact subset L of H there exists a constant $N_L > 0$ and an integer $b \leqq t$ such that

$$|\langle T, \psi\rangle| \leqq N_L \max_{|q| \leqq b} \max_{y} |\partial^q \psi(y)|$$

for all $\psi \in \mathfrak{D}^t(\mathrm{H})$. If $\chi \in \mathfrak{D}^{s+t}(K \times L)$, then we have

$$\left|\int_\Xi S(x)\, \partial_y^q \chi(x, y)\, dx\right| \leqq M_K \max_{|p| \leqq a} \max_{x} |\partial_x^p\, \partial_y^q \chi(x, y)|$$

for every $y \in \mathrm{H}$, and $|q| \leqq t$, $q \in \mathbf{N}^l$. Hence by (4)

$$\begin{aligned}\left|\int_\mathrm{H} T(y) \left(\int_\Xi S(x)\chi(x, y)\, dx\right) dy\right| &\leqq N_L \max_{|q| \leqq b} \max_{y} \left|\partial_y^q \int_\Xi S(x)\chi(x, y)\, dx\right| \\ &= N_L \max_{|q| \leqq b} \max_{y} \left|\int_\Xi S(x)\, \partial_y^q \chi(x, y)\, dx\right| \\ &\leqq M_K N_L \max_{|p| \leqq a} \max_{|q| \leqq b} \max_{x,y} |\partial_x^p\, \partial_y^q \chi(x, y)| \\ &\leqq M_K N_L \max_{|p+q| \leqq a+b} \max_{x,y} |\partial_x^p\, \partial_y^q \chi(x, y)|. \ \blacksquare\end{aligned}$$

If χ is of the form $\chi(x, y) = \varphi(x)\psi(y)$ with $\varphi \in \mathfrak{D}(\Xi)$ and $\psi \in \mathfrak{D}(\mathrm{H})$, i.e., $\chi = j(\varphi \otimes \psi)$, then

$$\int_\mathrm{H} T(y) \left(\int_\Xi S(x)\chi(x, y)\, dx\right) dy = \langle S, \varphi\rangle\langle T, \psi\rangle,$$

and it follows from Lemma 3 that $S \otimes T$ is continuous on $\mathfrak{D}(\Xi) \otimes \mathfrak{D}(\mathrm{H})$ for the topology induced by $\mathfrak{D}^{s+t}(\Xi \times \mathrm{H})$. By Propositions 1 and 2.9.5 the linear form $S \otimes T$ has a unique continuous extension to $\mathfrak{D}^{s+t}(\Xi \times \mathrm{H})$. Thus the following definition becomes meaningful.

DEFINITION 1. *Given two distributions $S \in \mathfrak{D}'^s(\Xi)$ and $T \in \mathfrak{D}'^t(\mathrm{H})$, the unique distribution*

$$R \in \mathfrak{D}'^{s+t}(\Xi \times \mathrm{H})$$

which satisfies

$$\langle R, \varphi \otimes \psi\rangle = \langle S, \varphi\rangle\langle T, \psi\rangle$$

for all $\varphi \in \mathfrak{D}^s(\Xi)$ and $\psi \in \mathfrak{D}^t(\mathrm{H})$ will be called the tensor product of the distributions S and T and denoted by $S \otimes T$.

Our discussions leading to Definition 1 prove the following result.

PROPOSITION 2. *If* $S \in \mathfrak{D}'^s(\Xi)$ *and* $T \in \mathfrak{D}'^t(\mathrm{H})$, *then for every* $\chi \in \mathfrak{D}^{s+t}(\Xi \times \mathrm{H})$ *we have*

$$\begin{aligned}\langle S \otimes T, \chi\rangle &= \int_{\Xi} S(x)\left(\int_{\mathrm{H}} T(y)\chi(x, y)\, dy\right) dx \\ &= \int_{\mathrm{H}} T(y)\left(\int_{\Xi} S(x)\chi(x, y)\, dx\right) dy.\end{aligned}$$

EXAMPLE 1. Let $f \in \mathcal{C}(\Xi)$, $g \in \mathcal{C}(\mathrm{H})$ and let T_f, T_g be the corresponding distributions (Example 1.1). Then

$$\begin{aligned}\langle T_f \otimes T_g, \varphi \otimes \psi\rangle &= \langle T_f, \varphi\rangle\langle T_g, \psi\rangle \\ &= \int f(x)\varphi(x)\, dx \int g(y)\psi(y)\, dy \\ &= \int f(x)g(y)\varphi(x)\psi(y)\, dx\, dy.\end{aligned}$$

If, in accordance with the identification made before Lemma 1, we denote by $f \otimes g$ the function $(x, y) \mapsto f(x)g(y)$ defined on $\Xi \times \mathrm{H}$, then we have

$$T_f \otimes T_g = T_{f \otimes g}.$$

EXAMPLE 2. Let δ_x be the Dirac measure (Example 1.2) on $\mathbf{R}^k$ and δ_y the Dirac measure on $\mathbf{R}^l$. Then

$$\langle \delta_x \otimes \delta_y, \varphi \otimes \psi\rangle = \langle \delta_x, \varphi\rangle\langle \delta_y, \psi\rangle = \varphi(0)\psi(0);$$

i.e., $\delta_x \otimes \delta_y$ is the Dirac measure δ on $\mathbf{R}^n = \mathbf{R}^k \times \mathbf{R}^l$.

Let us prove a few simple properties of the tensor product of distributions.

PROPOSITION 3. *If* $S \in \mathfrak{D}'^s(\Xi)$ *and* $T \in \mathfrak{D}'^t(\mathrm{H})$, *then*

$$\operatorname{Supp} S \otimes T = \operatorname{Supp} S \times \operatorname{Supp} T.$$

Proof. If $(x, y) \in \operatorname{Supp} S \times \operatorname{Supp} T$, then for every neighborhood U of x there exists a function $\varphi \in \mathfrak{D}(\Xi)$ with support contained in U such that $\langle S, \varphi\rangle \neq 0$, and for every neighborhood V of y there exists a function $\psi \in \mathfrak{D}(\mathrm{H})$ with support contained in V such that $\langle T, \psi\rangle \neq 0$. But then $\varphi \otimes \psi$ has its support contained in the arbitrarily small neighborhood $U \times V$ of (x, y), and $\langle S \otimes T, \varphi \otimes \psi\rangle = \langle S, \varphi\rangle\langle T, \psi\rangle \neq 0$. Thus

$$(x, y) \in \operatorname{Supp} S \otimes T.$$

Conversely, suppose that $(x, y) \notin \operatorname{Supp} S \times \operatorname{Supp} T$. Then either $x \notin \operatorname{Supp} S$ or $y \notin \operatorname{Supp} T$. To fix our ideas, let us assume that the first one is the case. There exists an open neighborhood U of x which does not meet $\operatorname{Supp} S$. If $\chi \in \mathfrak{D}(\Xi \times \mathrm{H})$ has its support contained in $U \times \mathrm{H}$, then for each $y \in \mathrm{H}$ the function $\chi(\cdot, y)$ has its support contained in U, and therefore by Proposition 2 we have

$$\langle S \otimes T, \chi \rangle = \int_{\mathrm{H}} T(y) \left(\int_{\Xi} S(x)\chi(x, y)\, dx \right) dy = 0.$$

Since $U \times \mathrm{H}$ is a neighborhood of (x, y), we have $(x, y) \notin \operatorname{Supp} S \otimes T$. ▮

PROPOSITION 4. *If $S \in \mathfrak{D}'^s(\Xi)$, $T \in \mathfrak{D}'^t(\mathrm{H})$, $p \in \mathbf{N}^k$ and $q \in \mathbf{N}^l$, then*

$$\partial^{(p,q)}(S \otimes T) = \partial_x^p S \otimes \partial_y^q T,$$

where

$$\partial^{(p,q)} = \frac{\partial^{|p+q|}}{\partial x_1^{p_1} \cdots \partial x_k^{p_k} \partial y_1^{q_1} \cdots \partial y_l^{q_l}}.$$

Proof. For $\varphi \in \mathfrak{D}(\Xi)$ and $\psi \in \mathfrak{D}(\mathrm{H})$ we have

$$\begin{aligned}
\langle \partial^{(p,q)}(S \otimes T), \varphi \otimes \psi \rangle &= (-1)^{|p+q|} \langle S \otimes T, \partial^{(p,q)}(\varphi \otimes \psi) \rangle \\
&= (-1)^{|p+q|} \langle S \otimes T, \partial_x^p \varphi \otimes \partial_y^q \psi \rangle \\
&= (-1)^{|p|} \langle S, \partial_x^p \varphi \rangle \cdot (-1)^{|q|} \langle T, \partial_y^q \psi \rangle \\
&= \langle \partial_x^p S, \varphi \rangle \langle \partial_y^q T, \psi \rangle = \langle \partial_x^p S \otimes \partial_y^q T, \varphi \otimes \psi \rangle.
\end{aligned}$$

The assertion follows from Proposition 1. ▮

EXAMPLE 3. With the notations of Example 2 and Proposition 4 we have

$$\partial^{(p,q)} \delta = \partial_x^p \delta_x \otimes \partial_y^q \delta_y.$$

Observe that $\partial_x^p \delta_x$ is of order $|p|$, $\partial_y^q \delta_y$ of order $|q|$, and $\partial^{(p,q)} \delta$ of order $|p| + |q|$.

PROPOSITION 5. *If $\alpha \in \mathcal{E}^{s+t}(\Xi)$, $\beta \in \mathcal{E}^{s+t}(\mathrm{H})$, $S \in \mathfrak{D}'^s(\Xi)$, and $T \in \mathfrak{D}'^t(\mathrm{H})$, then*

$$(\alpha \otimes \beta)(S \otimes T) = (\alpha S) \otimes (\beta T). \tag{5}$$

Proof. Clearly, $\alpha \otimes \beta \in \mathcal{E}^{s+t}(\Xi \times \mathrm{H})$. For $\varphi \in \mathfrak{D}^s(\Xi)$ and $\psi \in \mathfrak{D}^t(\mathrm{H})$ we have

$$\begin{aligned}
\langle (\alpha \otimes \beta)(S \otimes T), \varphi \otimes \psi \rangle &= \langle S \otimes T, (\alpha \otimes \beta)(\varphi \otimes \psi) \rangle \\
&= \langle S \otimes T, (\alpha\varphi) \otimes (\beta\psi) \rangle = \langle S, \alpha\varphi \rangle \langle T, \beta\psi \rangle \\
&= \langle \alpha S, \varphi \rangle \langle \beta T, \psi \rangle = \langle (\alpha S) \otimes (\beta T), \varphi \otimes \psi \rangle.
\end{aligned}$$

The assertion follows again from Proposition 1. ▮

REMARK 1. If only $\alpha \in \mathcal{E}^s(\Xi)$ and $\beta \in \mathcal{E}^t(\mathrm{H})$, then formula (5) *defines* $(\alpha \otimes \beta)(S \otimes T)$ as a distribution belonging to $\mathfrak{D}'^{s+t}(\Xi \times \mathrm{H})$.

Let $E_1, \ldots, E_h$ be h vector spaces over the same field $\mathbf{K}$. Similarly as in the case of two factors, we can define the tensor product

$$E_1 \otimes \cdots \otimes E_h = \bigotimes_{j=1}^{h} E_j$$

of the spaces E_j $(1 \leqq j \leqq h)$. For this we recall that a map $v\colon \prod_{j=1}^{h} E_j \to G$ into some vector space G over $\mathbf{K}$ is said to be h-linear if for each j and any choice of the vectors $x_1, \ldots, x_{j-1}, x_{j+1}, \ldots, x_h$, the map $x_j \mapsto v(x_1, \ldots, x_j, \ldots, x_h)$ is linear. Then

$$\bigotimes_{j=1}^{h} E_j$$

is defined as the vector space (unique up to an isomorphism) which has the following "universal property": there exists an h-linear map

$$\mu\colon \prod_{j=1}^{h} E_j \to \bigotimes_{j=1}^{h} E_j$$

such that given a vector space G over $\mathbf{K}$ and an h-linear map $v\colon \prod_{j=1}^{h} E_j \to G$, there exists a unique linear map

$$u\colon \bigotimes_{j=1}^{h} E_j \to G$$

which satisfies the relation $v = u \circ \mu$. The elements of $\bigotimes_{j=1}^{h} E_j$ are linear combinations of the elements

$$\bigotimes_{j=1}^{h} x_j = x_1 \otimes \cdots \otimes x_h = \mu(x_1, \ldots, x_h)$$

$(x_j \in E_j,\ 1 \leqq j \leqq h)$, and u is defined by

$$u(x_1 \otimes \cdots \otimes x_h) = v(x_1, \ldots, x_h).$$

For any integer r such that $1 \leqq r < h$, the space

$$\left(\bigotimes_{j=1}^{r} E_j\right) \otimes \left(\bigotimes_{j=r+1}^{h} E_j\right)$$

is canonically isomorphic to

$$\bigotimes_{j=1}^{h} E_j$$

("associativity" of the tensor product). If all E_j are finite dimensional, then

$$\dim \bigotimes_{j=1}^{h} E_j = \prod_{j=1}^{h} \dim E_j.$$

Finally, if E_j and F_j are vector spaces and $u_j\colon E_j \to F_j$ linear maps $(1 \leqq j \leqq h)$, then the linear map

$$\bigotimes_{j=1}^{h} u_j = u_1 \otimes \cdots \otimes u_h\colon \bigotimes_{j=1}^{h} E_j \to \bigotimes_{j=1}^{h} F_j$$

is defined by

$$\left(\bigotimes_{j=1}^{h} u_j\right)\left(\bigotimes_{j=1}^{h} x_j\right) = \bigotimes_{j=1}^{h} u_j(x_j).$$

Consider now h euclidean spaces $\mathbf{R}^{n_j}$ $(1 \leqq j \leqq h)$, and in each space $\mathbf{R}^{n_j}$ an open set Ω_j. If $\varphi_j \in \mathfrak{D}(\Omega_j)$, then the function

$$(x_1, \ldots, x_h) \mapsto \varphi_1(x_1) \cdots \varphi_h(x_h)$$

$(x_j \in \Omega_j,\ 1 \leqq j \leqq h)$ belonging to $\mathfrak{D}(\prod_{j=1}^{h} \Omega_j)$ will be identified with the element

$$\bigotimes_{j=1}^{h} \varphi_j = \varphi_1 \otimes \cdots \otimes \varphi_h$$

of $\bigotimes_{j=1}^{h} \mathfrak{D}(\Omega_j)$. With this identification

$$\bigotimes_{j=1}^{h} \mathfrak{D}(\Omega_j)$$

becomes a dense subspace of $\mathfrak{D}^m(\prod_{j=1}^{h} \Omega_j)$ for $0 \leqq m \leqq \infty$ (cf. Proposition 1). If for each index j $(1 \leqq j \leqq h)$ we are given a distribution $S_j \in \mathfrak{D}'^{m_j}(\Omega_j)$, then there exists a unique distribution

$$\bigotimes_{j=1}^{h} S_j = S_1 \otimes \cdots \otimes S_h$$

belonging to $\mathfrak{D}'^{\Sigma m_j}(\prod_{j=1}^{h} \Omega_j)$ which satisfies the condition

$$\left\langle \bigotimes_{j=1}^{h} S_j, \bigotimes_{j=1}^{h} \varphi_j \right\rangle = \prod_{j=1}^{h} \langle S_j, \varphi_j \rangle$$

for all $\varphi_j \in \mathfrak{D}^{m_j}(\Omega_j)$, $1 \leqq j \leqq h$. In particular, we have

$$\left(\bigotimes_{j=1}^{r} S_j\right) \otimes \left(\bigotimes_{j=r+1}^{h} S_j\right) = \bigotimes_{j=1}^{h} S_j$$

for any integer r such that $1 \leqq r < h$. The simple verification of these facts is left to the reader who should also state and prove the analogues of Propositions 2 through 5 for more than two factors.

Example 4. Denote by Y_j the Heaviside function (Example 3.1) with respect to the variable x_j. Then $Y_1 \otimes \cdots \otimes Y_n$ is the function on $\mathbf{R}^n$

which takes the value 1 if $x_j \geqq 0$ for all j $(1 \leqq j \leqq n)$ and the value 0 otherwise. We have

$$\partial_1 \cdots \partial_n(Y_1 \otimes \cdots \otimes Y_n) = (\partial_1 Y_1) \otimes \cdots \otimes (\partial_n Y_n) = \delta.$$

To conclude this section we shall investigate the continuity of the map $(S, T) \mapsto S \otimes T$, which is obviously bilinear and which by Propositions 3 and 4.3 maps $\mathcal{E}'^s(\Xi) \times \mathcal{E}'^t(\mathrm{H})$ into $\mathcal{E}'^{s+t}(\Xi \times \mathrm{H})$. We can see similarly as in Lemma 2 that if $S \in \mathcal{E}'^s(\Xi)$ and $\chi \in \mathcal{E}^{s+t}(\Xi \times \mathrm{H})$, then the function $y \mapsto \int_\Xi S(x)\chi(x, y)\,dx$ belongs to $\mathcal{E}^t(\mathrm{H})$ and formula (4) holds for $|q| \leqq t$. If furthermore $T \in \mathcal{E}'^t(\mathrm{H})$, then the analogue of Proposition 2 holds.

PROPOSITION 6. *The bilinear map* $(S, T) \mapsto S \otimes T$ *from* $\mathcal{E}'^s(\Xi) \times \mathcal{E}'^t(\mathrm{H})$ *into* $\mathcal{E}'^{s+t}(\Xi \times \mathrm{H})$ *is continuous if we equip* $\mathcal{E}'^s(\Xi)$, $\mathcal{E}'^t(\mathrm{H})$, *and* $\mathcal{E}'^{s+t}(\Xi \times \mathrm{H})$ *with their strong topologies.*

Proof. Let C be a bounded subset of $\mathcal{E}^{s+t}(\Xi \times \mathrm{H})$. We have to find a bounded subset A of $\mathcal{E}^s(\Xi)$ such that the set B of all functions

$$y \mapsto \int_\Xi S(x)\chi(x, y)\,dx,$$

where S varies in A° and χ in C, is bounded in $\mathcal{E}^t(\mathrm{H})$. Then $S \in A^\circ$ and $T \in B^\circ$ will imply $S \otimes T \in C^\circ$.

If K is an arbitrary compact subset of Ξ, L a compact subset of H, and $|p + q| \leqq s + t$, then there exists $\nu(p, q, K, L) > 0$ such that $|\partial_x^p \partial_y^q \chi(x, y)| \leqq \nu(p, q, K, L)$ for $\chi \in C$, $x \in K$ and $y \in L$.

Let (L_j) be an increasing sequence of compact subsets of H whose union is H (cf. Example 2.6.3). For each $j \in \mathbf{N}$ denote by A_j the set of all functions $x \mapsto \partial_y^q \chi(x, y)$, where $\chi \in C$, $y \in L_j$ and $|q| \leqq \min(t, j)$. The set A_j is bounded in $\mathcal{E}^s(\Xi)$ since if K is a compact subset of Ξ and $|p| \leqq s$, then

$$|\partial_x^p \partial_y^q \chi(x, y)| \leqq \max_{|q| \leqq j} \nu(p, q, K, L_j)$$

for all $x \in K$. By Proposition 2.6.3 there exists a bounded subset A of $\mathcal{E}^s(\Xi)$ and a sequence (λ_j) of positive numbers such that $A_j \subset \lambda_j A$ for all $j \in \mathbf{N}$.

Now let $S \in A^\circ$ and $\chi \in C$. For $q \in \mathbf{N}^l$, $|q| \leqq t$, and $y \in L_j$ we have

$$\left|\partial_y^q \int_\Xi S(x)\chi(x, y)\,dx\right| = \left|\int_\Xi S(x)\,\partial_y^q \chi(x, y)\,dx\right| \leqq \lambda_j,$$

provided that $j > |q|$. Thus B is indeed bounded in $\mathcal{E}^t(\mathrm{H})$. ∎

PROPOSITION 7. *The bilinear map* $(S, T) \mapsto S \otimes T$ *from* $\mathcal{D}'^s(\Xi) \times \mathcal{D}'^t(\mathrm{H})$ *into* $\mathcal{D}'^{s+t}(\Xi \times \mathrm{H})$ *is continuous if we equip* $\mathcal{D}'^s(\Xi)$, $\mathcal{D}'^t(\mathrm{H})$, *and*

$$\mathcal{D}'^{s+t}(\Xi \times \mathrm{H})$$

with their strong topologies.

Proof. Let C be a bounded set in $\mathfrak{D}^{s+t}(\Xi \times \mathrm{H})$. For each $q \in \mathbf{N}^l$, $|q| \leqq t$ the functions

$$\partial_y^q \chi(\cdot, y): x \mapsto \partial_y^q \chi(x, y),$$

where χ varies in C and y in H, form a bounded set A_q in $\mathfrak{D}^s(\Xi)$. Indeed, there exists a compact subset K of Ξ and a compact subset L of H such that $\operatorname{Supp} \chi \subset K \times L$, and in particular $\operatorname{Supp} \chi(\cdot, y) \subset K$ for all $\chi \in C$. Furthermore, for each $p \in \mathbf{N}^k$, $|p| \leqq s$ the expression $|\partial_x^p \partial_y^q \chi(x, y)|$ is bounded as (x, y) varies in $\Xi \times \mathrm{H}$.

We also see that $A_q \subset \mathfrak{D}^s(K)$. Hence by Proposition 2.6.3 there exist a bounded subset A of $\mathfrak{D}^s(K)$ and a family (λ_q) of positive numbers such that $A_q \subset \lambda_q A$ for $q \in \mathbf{N}^l$, $|q| \leqq t$. The set A is also bounded in $\mathfrak{D}^s(\Xi)$, and $U = A^{\circ}$ is a neighborhood of 0 in $\mathfrak{D}'^s(\Xi)$. If $S \in U$, then

$$\left| \int_{\Xi} S(x)\, \partial_y^q \chi(x, y)\, dx \right| \leqq \lambda_q \tag{6}$$

for all $\chi \in C$.

By Lemma 2 for every $S \in U$ and $\chi \in C$ the function $\int_\Xi S(x)\chi(x, \cdot)\, dx$ belongs to $\mathfrak{D}^t(\mathrm{H})$ and has its support contained in L; and by (4) and (6) we have

$$\max_y \left| \partial_y^q \int_{\Xi} S(x)\chi(x, y)\, dx \right| \leqq \lambda_q$$

for $|q| \leqq t$. In other words, the set

$$B = \left\{ \int_{\Xi} S(x)\chi(x, \cdot)\, dx \mid S \in U, \chi \in C \right\}$$

is bounded in $\mathfrak{D}^t(\mathrm{H})$. Thus $V = B^{\circ}$ is a neighborhood of 0 in $\mathfrak{D}'^t(\mathrm{H})$, and $S \in U$, $T \in V$ imply $|\langle S \otimes T, \chi\rangle| \leqq 1$ for all $\chi \in C$; i.e., $S \otimes T \in C^{\circ}$. ∎

Next we investigate the continuity of $(S, T) \mapsto S \otimes T$ for the topology of uniform convergence on compact sets (cf. also Exercise 8).

PROPOSITION 8 (Dieudonné). *The bilinear map* $(S, T) \mapsto S \otimes T$ *from* $\mathfrak{D}'^s(\Xi) \times \mathfrak{D}'^t(\mathrm{H})$ *into* $\mathfrak{D}'^{s+t}(\Xi \times \mathrm{H})$ *is continuous if we equip these spaces with the topologies* $\kappa(\mathfrak{D}'^s(\Xi), \mathfrak{D}^s(\Xi))$, $\kappa(\mathfrak{D}'^t(\mathrm{H}), \mathfrak{D}^t(\mathrm{H}))$, *and*

$$\kappa(\mathfrak{D}'^{s+t}(\Xi \times \mathrm{H}), \mathfrak{D}^{s+t}(\Xi \times \mathrm{H})).$$

Proof. We may assume that both s and t are finite. Let C be a relatively compact subset of $\mathfrak{D}^{s+t}(\Xi \times \mathrm{H})$. We must find a relatively compact subset A of $\mathfrak{D}^s(\Xi)$ and a relatively compact subset B of $\mathfrak{D}^t(\mathrm{H})$ such that $S \in A^{\circ}$ and $T \in B^{\circ}$ imply $S \otimes T \in C^{\circ}$. By Proposition 2 this amounts to finding B such that as χ varies in C and T in B° the set of functions $\int_{\mathrm{H}} T(y)\chi(\cdot, y)\, dy$ is a relatively compact subset A of $\mathfrak{D}^s(\Xi)$.

By Proposition 3.9.12 the functions $\chi \in C$ have their supports contained in a set of the form $K \times L$, where K is a compact subset of Ξ and L a compact subset of H. Furthermore, for every $p \in \mathbf{N}^k$ and $q \in \mathbf{N}^l$ such that $|p + q| \leqq s + t$ the set of functions $\partial_x^p \partial_y^q \chi$, where χ varies in C, is uniformly equicontinuous and uniformly bounded. For every integer $h > 0$ there exists therefore $\epsilon(h) > 0$ such that

$$|\partial_x^p \, \partial_y^q \chi(x', y) - \partial_x^p \, \partial_y^q \chi(x'', y)| \leqq \frac{1}{h^2}$$

if $|x' - x''| \leqq \epsilon(h)$.

Let B be the subset of $\mathfrak{D}^t(\mathrm{H})$ consisting of all functions

$$\varphi_{p,x}\colon y \mapsto \partial_x^p \chi(x, y),$$

where $\chi \in C$, $|p| \leqq s$, $x \in \Xi$, and of all functions

$$\varphi_{h,p,x',x''}\colon y \mapsto h\{\partial_x^p \chi(x', y) - \partial_x^p \chi(x'', y)\},$$

where $\chi \in C$, $h > 0$, $|p| \leqq s$, $x', x'' \in \Xi$, $|x' - x''| \leqq \epsilon(h)$. By Proposition 3.9.12 the set B is relatively compact:

(1) The support of every function belonging to B is contained in L.

(2) Let $|q| \leqq t$. The set of all functions $\partial_y^q \varphi_{p,x}$ is clearly equicontinuous. On the other hand,

$$|\partial_y^q \varphi_{h,p,x',x''}(y)| \leqq \frac{1}{h} \tag{7}$$

for every $\varphi_{h,p,x',x''}$ and $y \in \mathrm{H}$. Given an integer $r > 0$ we have

$$|\partial_y^q \varphi_{h,p,x',x''}(y') - \partial_y^q \varphi_{h,p,x',x''}(y'')| \leqq \frac{2}{h} \leqq \frac{1}{r},$$

provided that $h \geqq 2r$. There exists $\eta = \eta(r)$ such that

$$|\partial_x^p \, \partial_y^q \chi(x, y') - \partial_x^p \, \partial_y^q \chi(x, y'')| \leqq \frac{1}{4r^2}$$

for all $x \in \Xi$, provided $|y' - y''| \leqq \eta$. It follows that

$$\begin{aligned}
&|\partial_y^q \varphi_{h,p,x',x''}(y') - \partial_y^q \varphi_{h,p,x',x''}(y'')| \\
&\quad = h|\partial_x^p \, \partial_y^q \chi(x', y') - \partial_x^p \, \partial_y^q \chi(x'', y') - \partial_x^p \, \partial_y^q \chi(x', y'') + \partial_x^p \, \partial_y^q \chi(x'', y'')| \\
&\quad \leqq h|\partial_x^p \, \partial_y^q \chi(x', y') - \partial_x^p \, \partial_y^q \chi(x', y'')| + h|\partial_x^p \, \partial_y^q \chi(x'', y') - \partial_x^p \, \partial_y^q (x'', y'')| \\
&\quad \leqq h\left(\frac{1}{4r^2} + \frac{1}{4r^2}\right) = \frac{h}{2r^2} < \frac{1}{r},
\end{aligned}$$

provided that $h < 2r$ and $|y' - y''| \leqq \eta$.

(3) It follows from (7) and the hypotheses on C that B is bounded.

Finally, again using Proposition 3.9.12, we prove that the set A is relatively compact:

(1) The support of every function belonging to A is contained in K.

(2) Let $|p| \leqq s$. Then we have

$$\left|\partial_x^p \int_{\mathrm{H}} T(y)\chi(x', y)\,dy - \partial_x^p \int_{\mathrm{H}} T(y)\chi(x'', y)\,dy\right| = \left|\int_{\mathrm{H}} T(y) \cdot \partial_x^p\{\chi(x', y) - \chi(x'', y)\}\,dy\right| \leqq \frac{1}{h}$$

for $T \in B^{\circ}$ and $\chi \in C$, provided that $|x' - x''| \leqq \epsilon(h)$.

(3) If $|p| \leqq s$, then

$$\left|\partial_x^p \int_{\mathrm{H}} T(y)\chi(x, y)\,dy\right| = \left|\int_{\mathrm{H}} T(y)\,\partial_x^p \chi(x, y)\,dy\right| \leqq 1$$

if $T \in B^{\circ}$, $\chi \in C$, and $x \in \Xi$. ∎

Exercises

1. Show that if the linear maps $u: E \to G$ and $v: F \to H$ are injective, then the map $u \otimes v: E \otimes F \to G \otimes H$ is injective.

2. Prove the existence and uniqueness of the tensor product

$$\bigotimes_{j=1}^{h} E_j$$

and verify the assertions made about it in the text.

3. Prove that the image of $\mathfrak{D}'^s(\Xi) \times \mathfrak{D}'^t(\mathrm{H})$ under the map $(S, T) \mapsto S \otimes T$ is total in $\mathfrak{D}'^{s+t}(\Xi \times \mathrm{H})$.

4. (a) Using the notations of Proposition 6.8, show that $x_1 T = 0$ if and only if $T = \delta_{x_1} \otimes G$, where $G \in \mathfrak{D}'(\mathbf{R}^{n-1})$.

(b) Show that the kernel of the map $T \mapsto x_1^l T$ from $\mathfrak{D}'(\mathbf{R}^n)$ into $\mathfrak{D}'(\mathbf{R}^n)$ is the set of distributions of the form

$$\Big(\sum_{p \leqq l-1} \gamma_p \partial_1^p \delta_{x_1}\Big) \otimes G,$$

where $G \in \mathfrak{D}'(\mathbf{R}^{n-1})$.

5. Show that the distribution $T \in \mathfrak{D}'$ is independent of the variables $x_1, \ldots, x_k$ (Exercise 3.1(a)) if and only if it is of the form $T = \mathbf{1}_{x'} \otimes G_{x''}$, where $x' = (x_1, \ldots, x_k) \in \mathbf{R}^k$, $x'' = (x_{k+1}, \ldots, x_n)$, $\mathbf{1}_{x'}$ is the function which takes the constant value 1 on $\mathbf{R}^k$, and $G \in \mathfrak{D}'(\mathbf{R}^{n-k})$.

6. Show that the bilinear map $(\varphi, \psi) \mapsto \varphi \otimes \psi$ from $\mathcal{E}(\Xi) \times \mathcal{E}(\mathrm{H})$ into $\mathcal{E}(\Xi \times \mathrm{H})$ is continuous.

7. Show that the bilinear map $(\varphi, \psi) \mapsto \varphi \otimes \psi$ from $\mathfrak{D}(\Xi) \times \mathfrak{D}(\mathrm{H})$ into $\mathfrak{D}(\Xi \times \mathrm{H})$ is hypocontinuous.

8. Prove that the bilinear map $(S, T) \mapsto S \otimes T$ from $\mathcal{E}'^s(\Xi) \times \mathcal{E}'^t(\mathrm{H})$ into $\mathcal{E}'^{s+t}(\Xi \times \mathrm{H})$ is continuous if we equip these spaces with the topology of uniform convergence on compact sets.

§9. *Convolution*

At the beginning of §4 we introduced the convolution $h = f * g$ of two functions $f \in \mathcal{C}$ and $g \in \mathcal{K}$ with the help of the formula

$$h(x) = \int_{\mathbf{R}^n} f(x - y)g(y)\, dy, \qquad x \in \mathbf{R}^n.$$

We have seen that if both f and g have compact support, then h is continuous and also has compact support. In particular, h defines a distribution T_h which for $\varphi \in \mathfrak{D}$ is given by

$$\langle T_h, \varphi\rangle = \int_{\mathbf{R}^n} \varphi(x) \int_{\mathbf{R}^n} f(x - y)g(y)\, dy\, dx.$$

If we change variables in the iterated integral and consider that the function $(x, y) \mapsto f(x)g(y)$ has compact support in $\mathbf{R}^{2n}$, then we obtain

$$\langle T_h, \varphi\rangle = \int_{\mathbf{R}^n} \int_{\mathbf{R}^n} f(x)g(y)\varphi(x + y)\, dx\, dy. \tag{1}$$

Since the function $(x, y) \mapsto f(x)g(y)$ was identified with the tensor product $f \otimes g$ (Example 8.1), we are led to

DEFINITION 1. *Let $S \in \mathcal{E}'(\mathbf{R}^n)$ and $T \in \mathcal{E}'(\mathbf{R}^n)$ be two distributions with compact support on $\mathbf{R}^n$. For $\varphi \in \mathfrak{D}(\mathbf{R}^n)$ denote by φ^Δ the function belonging to $\mathcal{E}(\mathbf{R}^{2n})$ defined by $(x, y) \mapsto \varphi(x + y)$. The convolution $S * T$ of the distributions S and T is defined by*

$$\langle S * T, \varphi\rangle = \langle S \otimes T, \varphi^\Delta\rangle.$$

The fact that $S * T$ is a distribution will follow from

PROPOSITION 1. *The linear map $\varphi \mapsto \varphi^\Delta$ from $\mathfrak{D}(\mathbf{R}^n)$ into $\mathcal{E}(\mathbf{R}^{2n})$ is continuous.*

Proof. By Proposition 2.12.1 we must prove that for every compact subset K of $\mathbf{R}^n$ the map $\varphi \mapsto \varphi^\Delta$ from $\mathfrak{D}(K)$ into $\mathcal{E}(\mathbf{R}^{2n})$ is continuous. Let V be a neighborhood of 0 in $\mathcal{E}(\mathbf{R}^{2n})$. Then V contains a set of the form

$$\{\psi \mid |\partial^r\psi(z)| \leqq \epsilon,\ r \in \mathbf{N}^{2n},\ |r| \leqq m,\ z \in M\},$$

where $m \in \mathbf{N}$ and M is a compact subset of $\mathbf{R}^{2n}$. The set

$$U = \{\varphi \mid |\partial^p\varphi(x)| \leqq \epsilon,\ |p| \leqq 2m\}$$

is a neighborhood of 0 in $\mathfrak{D}(K)$, and $\varphi \in U$ implies $\varphi^{\triangle} \in V$ since

$$\partial_x^p \partial_y^q \varphi^{\triangle}(x, y) = (\partial^{p+q}\varphi)(x + y),$$

where, similarly as in the preceding section,

$$\partial_x^p = \frac{\partial^{|p|}}{\partial x_1^{p_1} \cdots \partial x_n^{p_n}}, \qquad \partial_y^q = \frac{\partial^{|q|}}{\partial y_1^{q_1} \cdots \partial y_n^{q_n}},$$

$p, q \in \mathbf{N}^n$. ∎

Since $S \otimes T \in \mathcal{E}'(\mathbf{R}^{2n})$ the map $\varphi \mapsto \langle S \otimes T, \varphi^{\triangle}\rangle$ is a continuous linear form on $\mathfrak{D}(\mathbf{R}^n)$; i.e., $S * T$ is a distribution on $\mathbf{R}^n$. With the notations introduced in §8 we can write

$$\begin{aligned}\langle S * T, \varphi\rangle &= \int_{\mathbf{R}^n} S(x) \left(\int_{\mathbf{R}^n} T(y)\varphi(x + y)\, dy\right) dx \\ &= \int_{\mathbf{R}^n} T(y) \left(\int_{\mathbf{R}^n} S(x)\varphi(x + y)\, dx\right) dy.\end{aligned} \tag{2}$$

We shall see in Proposition 2 that $\operatorname{Supp} S * T \subset \operatorname{Supp} S + \operatorname{Supp} T$; hence $S * T$ has compact support.

The difficulty in defining the convolution for two arbitrary distributions resides in the fact that for $\varphi \in \mathfrak{D}$ the function $\varphi^{\triangle}$ has compact support only if φ is identically zero. Indeed, $\operatorname{Supp} \varphi^{\triangle}$ is the union of the affine hyperplanes of $\mathbf{R}^{2n}$ whose equations are $x + y = \xi$, where ξ varies in $\operatorname{Supp} \varphi$. In §6 we defined the product of two distributions, but the less regular one factor became, the more regular we had to assume the other factor. In particular, if one factor was a completely arbitrary distribution, then we had to assume that the other factor was an infinitely differentiable function. In the present situation two distributions whose convolution we want to define cannot both "grow arbitrarily fast at infinity." Thus if one of them is completely arbitrary, we must assume that the other has compact support. We shall now define the convolution $S * T$ of two distributions S and T such that for every $\varphi \in \mathfrak{D}(\mathbf{R}^n)$ the support of $S \otimes T$ intersects $\operatorname{Supp} \varphi^{\triangle}$ in a compact set. Observe that if $S = f$ and $T = g$ belong to $\mathcal{C}(\mathbf{R}^n)$ and this condition is satisfied, then the integration in (1) is extended over a compact subset of $\mathbf{R}^n \times \mathbf{R}^n$, and therefore $h = f * g$ is well-defined. In §11 and §2 of Chapter 6 we shall see some other cases in which the convolution of two distributions is defined.

The basis of our definition is the important remark that *we can give a meaning to the scalar product $\langle T, \varphi\rangle$ even if $T \in \mathfrak{D}'(\mathbf{R}^n)$ and $\varphi \in \mathcal{E}(\mathbf{R}^n)$, provided that the set $C = \operatorname{Supp} T \cap \operatorname{Supp} \varphi$ is compact.* Let $\chi \in \mathfrak{D}(\mathbf{R}^n)$ be such that $\chi(x) = 1$ for all x belonging to a neighborhood of C. We define

$$\langle T, \varphi\rangle = \langle T, \chi\varphi\rangle. \tag{3}$$

The expression on the right-hand side has a meaning since $\chi\varphi \in \mathfrak{D}(\mathbf{R}^n)$.

Next let us show that this definition of $\langle T, \varphi \rangle$ is independent of the choice of χ. Indeed, let $\psi \in \mathfrak{D}(\mathbf{R}^n)$ be another function which is equal to 1 on a neighborhood of C. Then

$$\operatorname{Supp}(\chi - \psi)\varphi \subset \operatorname{Supp}(\chi - \psi) \cap \operatorname{Supp} \varphi \subset \complement C \cap \operatorname{Supp} \varphi \subset \complement \operatorname{Supp} T,$$

and therefore $\langle T, (\chi - \psi)\varphi \rangle = 0$; i.e., $\langle T, \chi\varphi \rangle = \langle T, \psi\varphi \rangle$.

If either T or φ has compact support, then $\langle T, \varphi \rangle$ has the same meaning as before, since in this case we can choose χ to be equal to 1 on a neighborhood of Supp T (cf. the proof of Proposition 2.3) or of Supp φ.

If $C = \operatorname{Supp} T \cap \operatorname{Supp} \varphi$ is compact, then the formula

$$\langle \partial_j T, \varphi \rangle = -\langle T, \partial_j \varphi \rangle$$

is still valid. Indeed, let $\chi \in \mathfrak{D}(\mathbf{R}^n)$ be equal to 1 in an open set U containing C. Since $\operatorname{Supp} \partial_j T \subset \operatorname{Supp} T$ and $\operatorname{Supp} \partial_j \varphi \subset \operatorname{Supp} \varphi$, we have

$$\begin{aligned} \langle \partial_j T, \varphi \rangle = \langle \partial_j T, \chi\varphi \rangle &= -\langle T, \partial_j(\chi\varphi) \rangle \\ &= -\langle T, \partial_j\chi \cdot \varphi \rangle - \langle T, \chi \cdot \partial_j \varphi \rangle \\ &= -\langle T, \chi \cdot \partial_j\varphi \rangle = -\langle T, \partial_j \varphi \rangle \end{aligned}$$

because $\partial_j \chi$ vanishes in U. We can prove similarly that

$$\langle \alpha T, \varphi \rangle = \langle T, \alpha\varphi \rangle$$

for every $\alpha \in \mathcal{E}(\mathbf{R}^n)$.

Now we can return to the definition of the convolution of two distributions. Let A and B be two closed subsets of $\mathbf{R}^n$ and for any compact subset K of $\mathbf{R}^n$ denote by K^Δ the subset of $\mathbf{R}^n \times \mathbf{R}^n$ formed by all points (x, y) such that $x + y \in K$. The following two conditions are equivalent:

(Σ) *For every compact subset K of $\mathbf{R}^n$ the set $(A \times B) \cap K^\Delta$ is compact in $\mathbf{R}^n \times \mathbf{R}^n$.*

(Σ') *For every compact subset K of $\mathbf{R}^n$ the set $A \cap (K - B)$ is compact in $\mathbf{R}^n$.*

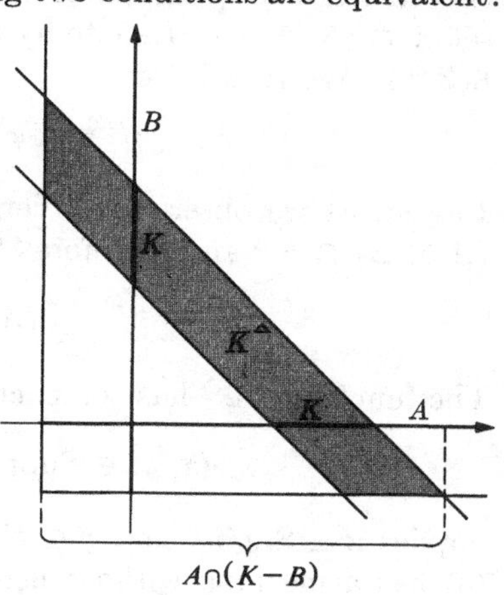

Indeed, if $(A \times B) \cap K^\Delta$ is compact, then it is contained in some set $M \times N$, where M and N are compact subsets of $\mathbf{R}^n$. If $x \in A \cap (K - B)$, then there exists some $y \in B$ such that $x + y \in K$; i.e., $(x, y) \in K^\Delta$. Since also $(x, y) \in A \times B$, we have $(x, y) \in M \times N$ and in particular $x \in M$. Hence

$A \cap (K - B) \subset M$. Conversely, assume that

$$M = A \cap (K - B)$$

is compact. If $(x, y) \in (A \times B) \cap K^{\Delta}$, then $x \in A$ and $x + y \in K$. Hence $x \in M$ and $y \in K - x \subset K - M$. Now $N = K - M$ is compact, and $(A \times B) \cap K^{\Delta} \subset M \times N$, which proves that (Σ') implies (Σ). We also see from the proof that $A \cap (K - B)$ is the first projection and $B \cap (K - A)$ the second projection of $(A \times B) \cap K^{\Delta}$.

DEFINITION 2. *Let S and T be two distributions on $\mathbf{R}^n$ such that*

$$A = \operatorname{Supp} S \quad \textit{and} \quad B = \operatorname{Supp} T$$

satisfy condition (Σ). The convolution $S * T$ is defined by

$$\langle S * T, \varphi \rangle = \langle S \otimes T, \varphi^{\Delta} \rangle \tag{4}$$

for every $\varphi \in \mathfrak{D}(\mathbf{R}^n)$.

We must prove again that $S * T$ is indeed a distribution, i.e., a continuous linear form on $\mathfrak{D}(\mathbf{R}^n)$. If K is a compact subset of $\mathbf{R}^n$, by condition (Σ) the set $K_0 = (A \times B) \cap K^{\Delta}$ is compact in $\mathbf{R}^{2n}$. Let $\alpha \in \mathfrak{D}(\mathbf{R}^{2n})$ be such that $\alpha(x, y) = 1$ on a neighborhood of K_0. By virtue of Proposition 2.12.1 and the definition (3) of the right-hand side of (4) it is sufficient to prove that $\varphi \mapsto \langle S \otimes T, \alpha\varphi^{\Delta} \rangle$ is a continuous linear form on $\mathfrak{D}(K)$. The map $\varphi \mapsto \varphi^{\Delta}$ from $\mathfrak{D}(K)$ into $\mathcal{E}(\mathbf{R}^{2n})$ is continuous by Proposition 1, and the map $\varphi^{\Delta} \mapsto \alpha\varphi^{\Delta}$ from $\mathcal{E}(\mathbf{R}^{2n})$ into $\mathfrak{D}(\mathbf{R}^{2n})$ by Proposition 7.4. Since $S \otimes T \in \mathfrak{D}'(\mathbf{R}^{2n})$, the assertion follows.

REMARK 1. Let S and T be two distributions whose supports A and B satisfy conditions (Σ) and (Σ'). Let $\varphi \in \mathfrak{D}(\mathbf{R}^n)$ have support K and choose $\chi \in \mathfrak{D}(\mathbf{R}^n)$ such that $\chi(x) = 1$ on a neighborhood of the compact set $A \cap (K - B)$. Denote by χ^* the function $(x, y) \mapsto \chi(x)$ belonging to $\mathcal{E}(\mathbf{R}^{2n})$. We then have

$$\langle S * T, \varphi \rangle = \langle S \otimes T, \chi^* \varphi^{\Delta} \rangle. \tag{5}$$

Indeed, as we observed earlier, $A \cap (K - B)$ is the first projection of $(A \times B) \cap K^{\Delta}$ and therefore $\chi^*(x, y) = 1$ on a neighborhood of

$$(A \times B) \cap K^{\Delta}.$$

The function $\chi^*\varphi^{\Delta}$ has compact support since

$$(x, y) \in \operatorname{Supp}(\chi^*\varphi^{\Delta}) \subset \operatorname{Supp} \chi^* \cap K^{\Delta}$$

implies $x \in \operatorname{Supp} \chi$ and $y \in K - \operatorname{Supp} \chi$. Thus the right-hand side of (5) has a meaning without using definition (3), in spite of the fact that the support of χ^* is not compact. By Proposition 8.2 we have the relations

$$\begin{aligned}\langle S * T, \varphi\rangle &= \int_{\mathbf{R}^n} S(x)\chi(x)\left(\int_{\mathbf{R}^n} T(y)\varphi(x+y)\,dy\right)dx \\ &= \int_{\mathbf{R}^n} T(y)\left(\int_{\mathbf{R}^n} S(x)\chi(x)\varphi(x+y)\,dx\right)dy.\end{aligned}$$

REMARK 2. It is clear that if $A = \operatorname{Supp} S$ is compact, then A and $B = \operatorname{Supp} T$ satisfy condition (Σ') for any distribution T. In this case we can choose $\chi \in \mathfrak{D}(\mathbf{R}^n)$ independently of φ such that $\chi(x) = 1$ on a neighborhood of A. Furthermore, relation (2) holds in this case for every $\varphi \in \mathfrak{D}(\mathbf{R}^n)$: the function $x \mapsto \int T(y)\varphi(x+y)\,dy$ does not necessarily have a compact support, but this is compensated by the fact that S does. Similarly, the fact that T does not have a compact support is compensated by the fact that the function $y \mapsto \int S(x)\varphi(x+y)\,dx$ has the compact support $\operatorname{Supp} \varphi - A$ (cf. Exercise 1).

REMARK 3. Another important case where the supports of S and T manifestly satisfy condition (Σ) is when both are contained in a translate of the "positive cone" of $\mathbf{R}^n$, i.e., in a set defined by inequalities of the form $x_1 \geqq \alpha_1, \ldots, x_n \geqq \alpha_n$.

REMARK 4. According to the definition given in (3), the scalar product $\langle S * T, \varphi\rangle$ has a meaning whenever $\operatorname{Supp} \varphi^{\Delta}$ meets $\operatorname{Supp} S \times \operatorname{Supp} T$ in a compact set.

Let us now prove some properties of convolutions.

PROPOSITION 2. *Let S and T be two distributions such that $A = \operatorname{Supp} S$ and $B = \operatorname{Supp} T$ satisfy condition (Σ). Then $\operatorname{Supp} S * T \subset A + B$.*

LEMMA 1. *If A and B are two closed subsets of $\mathbf{R}^n$ which satisfy condition (Σ), then $A + B$ is closed.*

Proof. Let $z \in \overline{A + B}$. There exists a sequence of points

$$(x_n, y_n) \in A \times B$$

such that $x_n + y_n$ tends to z. The set K whose points are $x_n + y_n$ $(n \in \mathbf{N})$ and z is compact, and therefore by our hypothesis the set

$$L = A \cap (K - B)$$

is compact. Since $x_n \in L$, there exists a subsequence (x_{n_k}) which converges to some point $x_0 \in A$. Consequently,

$$y_{n_k} = x_{n_k} + y_{n_k} - x_{n_k}$$

converges to some point $y_0 \in B$ and $z = x_0 + y_0 \in A + B$. ▮

Proof of Proposition 2. Let $z \notin A + B$. By Lemma 1 there exists a neighborhood V of z which does not meet $A + B$. Let $\varphi \in \mathfrak{D}(\mathbf{R}^n)$ have its support contained in V. Then Supp φ^Δ does not meet

$$A \times B = \operatorname{Supp} S \otimes T$$

(Proposition 8.3). Hence $\langle S * T, \varphi \rangle = \langle S \otimes T, \varphi^\Delta \rangle = 0$; i.e., z does not belong to Supp $S * T$. ∎

COROLLARY. *If both S and T have compact support, then $S * T$ has compact support.*

PROPOSITION 3. *Let $S \in \mathfrak{D}'$ and $A = \operatorname{Supp} S$. Suppose that the support of $\varphi \in \mathfrak{D}$ is contained in the open subset Ω of $\mathbf{R}^n$. If T_1 and T_2 are two distributions which are equal in $\Omega - A$, then $\langle S * T_1, \varphi \rangle = \langle S * T_2, \varphi \rangle$, provided that these scalar products have a meaning.*

REMARK 5. The set $\Omega - A$ is open as the union of the open sets $\Omega - a$, $a \in A$.

REMARK 6. The proposition can be stated in the following form: the value of $S * T$ in Ω depends only on the value of T in $\Omega - A$. In particular, if A is contained in a "small" neighborhood of 0, then the value of $S * T$ in Ω depends only on the value of T in a "small" neighborhood of $\overline{\Omega}$.

Proof of Proposition 3. Set $K = \operatorname{Supp} \varphi$. Clearly, all we have to prove is that $(A \times \operatorname{Supp}(T_1 - T_2)) \cap K^\Delta = \emptyset$. Now $(x, y) \in K^\Delta$ implies $x + y \in K \subset \Omega$. On the other hand, $y \in \operatorname{Supp}(T_1 - T_2)$ implies

$$y \in \complement(\Omega - A),$$

and therefore $x + y \notin \Omega$ for all $x \in A$. ∎

PROPOSITION 4. *Let s and t denote positive integers or the symbol ∞. If $S \in \mathfrak{D}'^s$, $T \in \mathfrak{D}'^t$ and the sets $A = \operatorname{Supp} S$, $B = \operatorname{Supp} T$ satisfy condition $(\sum)$, then $S * T$ belongs to $\mathfrak{D}'^{s+t}$.*

Proof. We must prove that for each compact subset K of $\mathbf{R}^n$ the map $\varphi \mapsto \langle S * T, \varphi \rangle$ is a continuous linear form on $\mathfrak{D}^{s+t}(K)$. By assumption the set $K_0 = (A \times B) \cap K^\Delta$ is compact in $\mathbf{R}^{2n}$. If $\alpha \in \mathfrak{D}(\mathbf{R}^{2n})$ is such that $\alpha(x, y) = 1$ on a neighborhood of K_0, then by definition

$$\langle S * T, \varphi \rangle = \langle S \otimes T, \alpha\varphi^\Delta \rangle.$$

Now we see exactly as in the proof of Proposition 1 that $\varphi \mapsto \varphi^\Delta$ is a continuous linear map from $\mathfrak{D}^{s+t}(K)$ into $\mathcal{E}^{s+t}(\mathbf{R}^{2n})$. By Proposition 7.4 the map $\varphi^\Delta \mapsto \alpha\varphi^\Delta$ from $\mathcal{E}^{s+t}(\mathbf{R}^{2n})$ into $\mathfrak{D}^{s+t}(\mathbf{R}^{2n})$ is continuous. Finally, $S \otimes T$ is a continuous linear form on $\mathfrak{D}^{s+t}(\mathbf{R}^{2n})$ (Definition 8.1). ∎

Proposition 5. *The bilinear map* $(S, T) \mapsto S * T$ *from* $\mathcal{E}'^s \times \mathcal{E}'^t$ *into* $\mathcal{E}'^{s+t}$ *is continuous if we equip* $\mathcal{E}'^s$, $\mathcal{E}'^t$, *and* $\mathcal{E}'^{s+t}$ *with their strong topologies.*

Proof. It is clear that the map is bilinear, and it follows from Proposition 4.3, the corollary of Proposition 2, and Proposition 4 that it maps $\mathcal{E}'^s \times \mathcal{E}'^t$ into $\mathcal{E}'^{s+t}$.

Next let us prove that the map $\varphi \mapsto \varphi^\triangle$ from $\mathcal{E}^{s+t}(\mathbf{R}^n)$ into $\mathcal{E}^{s+t}(\mathbf{R}^{2n})$ is continuous. Indeed, let V be a neighborhood of 0 in $\mathcal{E}^{s+t}(\mathbf{R}^{2n})$. There exists a compact subset M of $\mathbf{R}^{2n}$, an $\epsilon > 0$, and an integer h ($h = s + t$ if $s + t$ is finite) such that V contains the set

$$\{\psi \mid |\partial^r \psi(z)| \leqq \epsilon, r \in \mathbf{N}^{2n}, |r| \leqq h, z \in M\}.$$

Let K be the image of M in $\mathbf{R}^n$ under the continuous map $(x, y) \mapsto x + y$. Then K is compact, $U = \{\varphi \mid |\partial^p \varphi(x)| \leqq \epsilon, |p| \leqq h, x \in K\}$ is a neighborhood of 0 in $\mathcal{E}^{s+t}(\mathbf{R}^n)$, and $\varphi \in U$ implies $\varphi^\triangle \in V$.

Now let C be a bounded subset of $\mathcal{E}^{s+t}(\mathbf{R}^n)$. The image $C^\triangle$ of C under the continuous map $\varphi \mapsto \varphi^\triangle$ is a bounded subset of $\mathcal{E}^{s+t}(\mathbf{R}^{2n})$. By Proposition 8.6 there exists a neighborhood U of 0 in $\mathcal{E}'^s$ and a neighborhood V of 0 in $\mathcal{E}'^t$ such that $S \in U$ and $T \in V$ imply $S \otimes T \in (C^\triangle)^\circ$. Since $\langle S * T, \varphi\rangle = \langle S \otimes T, \varphi^\triangle\rangle$ by Definition 1, $S \in U$ and $T \in V$ imply $S * T \in C^\circ$. ∎

Proposition 6. *Let* K *be a compact subset of* $\mathbf{R}^n$ *and let* $\mathcal{E}'^s(K)$ *be the subspace of* $\mathcal{E}'^s$ *formed by the distributions whose support is contained in* K, *equipped with the topology induced by the strong topology of* $\mathcal{E}'^s$. *The bilinear map* $(S, T) \mapsto S * T$ *from* $\mathcal{E}'^s(K) \times \mathcal{D}'^t$ *into* $\mathcal{D}'^{s+t}$ *is continuous.*

Proof. It follows from Remark 2 that $S * T$ is well-defined and from Propositions 4.3 and 4 that $S * T \in \mathcal{D}'^{s+t}$.

Choose $\chi \in \mathcal{D}$ so that $\chi(x) = 1$ on a neighborhood of K. According to Remark 2 the distribution $S * T$ is defined by

$$\langle S * T, \varphi\rangle = \langle S \otimes T, \chi^* \varphi^\triangle\rangle,$$

where $\varphi \in \mathcal{D}^{s+t}$. Let C be a bounded subset of $\mathcal{D}^{s+t}$. The image C_1 under the continuous map $\varphi \mapsto \chi^*\varphi^\triangle$ is a bounded subset of $\mathcal{E}^{s+t}(\mathbf{R}^{2n})$. By Proposition 8.7 there exists a neighborhood U of 0 in $\mathcal{D}'^s$ and a neighborhood V of 0 in $\mathcal{D}'^t$ such that $S \in U$ and $T \in V$ imply $S \otimes T \in C_1^\circ$. Now the injection $\mathcal{D}^s \hookrightarrow \mathcal{E}^s$ is continuous (Examples 2.12.8, 2.5.3, 2.5.10, and 2.5.5), and thus the transpose $\mathcal{E}'^s \to \mathcal{D}'^s$ is also continuous, and in particular $U \cap \mathcal{E}'^s(K)$ is a neighborhood of 0 for the topology induced by $\mathcal{E}'^s$. Since $S \in U \cap \mathcal{E}'^s(K)$ and $T \in V$ imply $S * T \in C^\circ$, the conclusion follows. ∎

PROPOSITION 7. *The bilinear map* $(S, T) \mapsto S * T$ *from* $\mathcal{E}'^s \times \mathcal{D}'^t$ *into* $\mathcal{D}'^{s+t}$ *is hypocontinuous if we equip* $\mathcal{E}'^s$, $\mathcal{D}'^t$, *and* $\mathcal{D}'^{s+t}$ *with their strong topologies.*

Proof. Again by Remark 2 the distribution $S * T$ is well-defined; by Propositions 4.3 and 4 it belongs to $\mathcal{D}'^{s+t}$.

(a) Let A be a bounded subset of $\mathcal{E}'^s$ and W a neighborhood of 0 in $\mathcal{D}'^{s+t}$. Since $\mathcal{E}^s$ is barrelled, the set A is equicontinuous; i.e., there exists a neighborhood Y of 0 in $\mathcal{E}^s$ such that $\varphi \in Y$ and $S \in A$ imply

$$|\langle S, \varphi\rangle| \leqq 1.$$

We may assume that Y is of the form $\{\varphi \mid |\partial^p\varphi(x)| \leqq \epsilon, x \in K, |p| \leqq h\}$, where K is a compact subset of $\mathbf{R}^n$. It follows that $A \subset \mathcal{E}'^s(K)$ since if $\varphi \in \mathcal{E}^s$ is such that $\operatorname{Supp} \varphi \cap K = \emptyset$, then $\lambda\varphi \in Y$ for all $\lambda \in \mathbf{K}$; hence for any $S \in A$ we have $|\langle S, \lambda\varphi\rangle| \leqq 1$ for all $\lambda \in \mathbf{K}$, i.e., $\langle S, \varphi\rangle = 0$. Clearly, A is bounded in $\mathcal{E}'^s(K)$. According to Proposition 6 there exists a neighborhood U of 0 in $\mathcal{E}'^s(K)$ and a neighborhood V_1 of 0 in $\mathcal{D}'^t$ such that $S \in U$ and $T \in V_1$ imply $S * T \in W$. On the other hand, there exists $\lambda > 0$ such that $A \subset \lambda U$. If we set $V = (1/\lambda)V_1$, then $S \in A$ and $T \in V$ imply $(1/\lambda)S \in U$ and $\lambda T \in V_1$; hence

$$S * T = \left(\frac{1}{\lambda} S\right) * (\lambda T) \in W.$$

(b) Let B be a bounded subset of $\mathcal{D}'^t$ and C a bounded subset of $\mathcal{D}^{s+t}$. If K is a compact subset of $\mathbf{R}^n$, then for each $p \in \mathbf{N}^n$ with $|p| \leqq s$ the set C_p of all functions

$$\partial_x^p \varphi^{\triangle}(x, \cdot) : y \mapsto (\partial^p\varphi)(x + y),$$

where φ varies in C and x in K, is bounded in $\mathcal{D}^t$. It follows that for each $p \in \mathbf{N}^n$ with $|p| \leqq s$ the set of functions

$$x \mapsto \partial_x^p \int T(y)\varphi(x + y)\, dy$$

is bounded as x varies in K, T in B, and φ in C. This means that the set A of functions

$$x \mapsto \int T(y)\varphi(x + y)\, dy$$

is bounded in $\mathcal{E}^s$ as T varies in B and φ in C. If $S \in A^{\circ}$ and $T \in B$, we have

$$|\langle S * T, \varphi\rangle| = \left|\int S(x) \left(\int T(y)\varphi(x + y)\, dy\right) dx\right| \leqq 1$$

for all $\varphi \in C$; i.e., $S * T \in C^{\circ}$. ∎

Consider h distributions T_j on $\mathbf{R}^n$, and set $A_j = \operatorname{Supp} T_j$ $(1 \leqq j \leqq h)$. For a compact subset K of $\mathbf{R}^n$ denote now by $K^\triangle$ the subset of $\mathbf{R}^{nh}$ formed by all points $(x_1, \ldots, x_h)$ such that $x_j \in \mathbf{R}^n$ and $x_1 + \cdots + x_h \in K$. Suppose that the A_j satisfy the following generalization of condition $(\sum)$:

$(\sum_h)$ *For every compact subset K of $\mathbf{R}^n$ the set $(\prod_{j=1}^h A_j) \cap K^\triangle$ is compact in $\mathbf{R}^{nh}$.*

Then we define the convolution

$$\mathop{\star}_{j=1}^{h} T_j = T_1 * \cdots * T_h$$

of the distributions T_j by

$$\left\langle \mathop{\star}_{j=1}^{h} T_j, \varphi \right\rangle = \left\langle \bigotimes_{j=1}^{h} T_j, \varphi^\triangle \right\rangle$$

for all $\varphi \in \mathfrak{D}(\mathbf{R}^n)$, where $\varphi^\triangle \in \mathfrak{D}(\mathbf{R}^{nh})$ is defined by

$$\varphi^\triangle(x_1, \ldots, x_h) = \varphi(x_1 + \cdots + x_h).$$

Condition $(\sum_h)$ is clearly satisfied if all the sets A_j, except possibly one, are compact. More generally:

LEMMA 2. (a) *If the sets $A_1, \ldots, A_h$ satisfy condition $(\sum_h)$ and the sets $B_1, \ldots, B_k$ are compact, then the sets $A_1, \ldots, A_h, B_1, \ldots, B_k$ satisfy condition $(\sum_{h+k})$.*

(b) *If the sets $A_1, \ldots, A_h$ satisfy condition $(\sum_h)$, then any subfamily having l elements $(2 \leqq l \leqq h)$ satisfies condition $(\sum_l)$.*

Proof. (a) If $(x_1, \ldots, x_h, y_1, \ldots, y_k)$ belongs to the subset

$$C = \left\{\left(\prod_{i=1}^{h} A_i\right) \times \left(\prod_{j=1}^{k} B_j\right)\right\} \cap K^\triangle$$

of $\mathbf{R}^{nh} \times \mathbf{R}^{nk}$, then $x_1 + \cdots + x_h \in K - B_1 - \cdots - B_k$. Hence by hypothesis the first projection of C is a compact subset of $\mathbf{R}^{nh}$. But the second projection of C is also compact since it is contained in $\prod_{j=1}^k B_j$.

(b) For the simplicity of notation let $A_1, \ldots, A_l$ $(l < h)$ be the subfamily in question. Let K be a compact subset of $\mathbf{R}^n$, pick a point a_j in each of the sets A_j with $l < j \leqq h$, and set $L = K + a_{l+1} + \cdots + a_h$. Then L is compact and by hypothesis $D = (\prod_{j=1}^h A_j) \cap L^\triangle$ is compact in $\mathbf{R}^{nh} = \mathbf{R}^{nl} \times \mathbf{R}^{n(h-l)}$. Now it can be seen, similarly as in the proof of the equivalence of the conditions $(\sum)$ and $(\sum')$, that the projection of D onto the first factor $\mathbf{R}^{nl}$ is the set

$$\left(\prod_{j=1}^{l} A_j\right) \cap (L - A_{l+1} - \cdots - A_h)^\triangle.$$

Hence this set is compact; on the other hand, it contains the set

$$\left(\prod_{j=1}^{l} A_j\right) \cap K^{\Delta}. \blacksquare$$

PROPOSITION 8. (a) *Let S and T be two distributions on $\mathbf{R}^n$ such that $A = \operatorname{Supp} S$ and $B = \operatorname{Supp} T$ satisfy condition $(\sum)$. Then $S * T = T * S$.*

(b) *Let R, S, and T be three distributions on $\mathbf{R}^n$ whose supports satisfy condition $(\sum_3)$. Then*

$$(R * S) * T = R * (S * T) = R * S * T.$$

Proof. (a) If $\varphi \in \mathfrak{D}$ and $K = \operatorname{Supp} \varphi$, choose $\alpha \in \mathfrak{D}(\mathbf{R}^{2n})$ such that $\alpha(x, y) = 1$ on a neighborhood of $(A \times B) \cap K^{\Delta}$. Then we have by Definition 2 and Proposition 8.2

$$\begin{aligned}\langle S * T, \varphi\rangle &= \langle S \otimes T, \alpha\varphi^{\Delta}\rangle \\ &= \int S(x) \left(\int T(y)\alpha(x, y)\varphi(x + y)\, dy\right) dx \\ &= \int T(y) \left(\int S(x)\alpha(x, y)\varphi(x + y)\, dx\right) dy \\ &= \langle T \otimes S, \alpha\varphi^{\Delta}\rangle = \langle T * S, \varphi\rangle.\end{aligned}$$

(b) If $A = \operatorname{Supp} R$, $B = \operatorname{Supp} S$, $C = \operatorname{Supp} T$, then for any compact set $K \subset \mathbf{R}^n$ the first projection

$$M = A \cap (K - B - C) \subset \mathbf{R}^n$$

and the second projection

$$N = B \cap (K - A - C) \subset \mathbf{R}^n$$

of the set $D = (A \times B \times C) \cap K^{\Delta} \subset \mathbf{R}^n \times \mathbf{R}^n \times \mathbf{R}^n$ are compact.

Let $\varphi \in \mathfrak{D}$ with $K = \operatorname{Supp} \varphi$ and choose the functions $\chi \in \mathfrak{D}$ and $\psi \in \mathfrak{D}$ so that $\chi(x) = 1$ on a neighborhood of M and $\psi(y) = 1$ on a neighborhood of N. Denote by γ^* the function defined on $\mathbf{R}^{3n}$ by

$$(x, y, z) \mapsto \chi(x)\psi(y).$$

The following properties can be verified immediately:

(i) $\gamma^*(x, y, z) = 1$ on a neighborhood of D.

(ii) $\gamma^*\varphi^{\Delta}$ has compact support in $\mathbf{R}^{3n}$.

(iii) For each $x \in A$ we have $\psi(y) = 1$ on a neighborhood of

$$B \cap (\operatorname{Supp}(\tau_{-x}\varphi) - C).$$

(iv) $\chi(x) = 1$ on a neighborhood of $A \cap (K - \operatorname{Supp} S * T)$.

It follows from Remark 1 and the analogous fact concerning three factors that

$$
\begin{aligned}
\langle R * S * T, \varphi\rangle &= \langle R \otimes S \otimes T, \gamma^* \varphi^{\Delta}\rangle \\
&= \int R(x)\chi(x) \left(\int S(y)\psi(y) \left(\int T(z)\varphi(x+y+z)\, dz\right) dy\right) dx \\
&= \int R(x)\chi(x) \left(\int (S * T)(y)\varphi(x+y)\, dy\right) dx \\
&= \langle R * (S * T), \varphi\rangle;
\end{aligned}
$$

i.e., $R * S * T = R * (S * T)$.

To prove that $R * S * T = (R * S) * T$ we choose $\eta,\ \theta \in \mathfrak{D}$ so that $\eta(y) = 1$ on a neighborhood of $B \cap (K - A - C)$ and $\theta(z) = 1$ on a neighborhood of $C \cap (K - A - B)$ and use the relation

$$
\begin{aligned}
\langle R * S * T, \varphi\rangle &= \int R(x) \left(\int S(y)\eta(y) \left(\int T(z)\theta(z)\varphi(x+y+z)\, dz\right) dy\right) dx \\
&= \int (R * S)(x) \left(\int T(z)\theta(z)\varphi(x+z)\, dz\right) dx \\
&= \langle (R * S) * T, \varphi\rangle. \blacksquare
\end{aligned}
$$

In view of Lemma 2(b) we obtain by induction from Proposition 8(b) the general associativity rule

$$\left(\mathop{\star}_{j=1}^{l} T_j\right) * \left(\mathop{\star}_{j=l+1}^{h} T_j\right) = \mathop{\star}_{j=1}^{h} T_j$$

for any l such that $1 \leqq l < h$, provided that the T_j are distributions whose supports satisfy condition (Σ_h).

Remark 7. If R, S and T are three distributions whose supports do not satisfy condition (Σ_3), then $(R * S) * T$ and $R * (S * T)$ can both be defined without being equal. Thus in the case $n = 1$ we have

$$(\mathbf{1} * \partial\delta) * Y = \mathbf{0} * Y = 0$$

since $\mathbf{1} * \partial\delta = \partial\mathbf{1} = 0$ (cf. Proposition 10), but

$$\mathbf{1} * (\partial\delta * Y) = \mathbf{1} * \delta = \mathbf{1}$$

(cf. Propositions 9 and 10).

Proposition 9. *For any distribution* $T \in \mathfrak{D}'$ *we have* $\delta * T = T * \delta = T$.

Proof. For $\varphi \in \mathfrak{D}$ we have

$$
\begin{aligned}
\langle T * \delta, \varphi\rangle &= \int T(x) \left(\int \delta(y)\varphi(x+y)\, dy\right) dx \\
&= \int T(x)\varphi(x)\, dx = \langle T, \varphi\rangle. \blacksquare
\end{aligned}
$$

It follows from Proposition 8 that $\mathcal{E}'$ is a commutative ring if convolution is taken for multiplication, and by Proposition 9 the Dirac measure δ is the unit element of $\mathcal{E}'$. It also follows from Proposition 8 that $\mathcal{D}'$ is an $\mathcal{E}'$-module (Remark 6.1) if the multiplication between elements of $\mathcal{E}'$ and $\mathcal{D}'$ is again the convolution.

PROPOSITION 10. *Let S and T be two distributions such that $A = \operatorname{Supp} S$ and $B = \operatorname{Supp} T$ satisfy condition $(\sum)$. Then for every partial derivation ∂_j $(1 \leqq j \leqq n)$ we have*

$$\partial_j(S * T) = \partial_j S * T = S * \partial_j T.$$

Proof. Since $\operatorname{Supp}(\partial_j S) \subset A$, the last two terms are well-defined. For $\varphi \in \mathcal{D}$ let $\chi \in \mathcal{D}$ be such that $\chi(x) = 1$ on a neighborhood of

$$A \cap (\operatorname{Supp} \varphi - B).$$

We have

$$\begin{aligned} \langle \partial_j(S * T), \varphi \rangle &= -\langle S * T, \partial_j \varphi \rangle \\ &= -\int S(x)\chi(x) \left(\int T(y)(\partial_j \varphi)(x + y)\, dy \right) dx \\ &= \int S(x)\chi(x) \left(\int (\partial_j T)(y)\varphi(x + y)\, dy \right) dx \\ &= \langle S * \partial_j T, \varphi \rangle; \end{aligned}$$

i.e., $\partial_j(S * T) = S * \partial_j T$. The other relation follows now from Proposition 8(a). ∎

With the notations introduced in Chapter 2, §5, a polynomial of degree $\leqq m$ in the indeterminates $\xi_1, \ldots, \xi_n$ is an expression of the form

$$P(\xi) = \sum_p \alpha_p \xi^p,$$

where p varies in $\mathbf{N}^n$, $\alpha_p \in \mathbf{K}$, $\alpha_p = 0$ for $|p| > m$, and $\xi = (\xi_1, \ldots, \xi_n)$. It is exactly of degree m if there exists at least one p with $|p| = m$ such that $\alpha_p \neq 0$. We shall usually identify the polynomial $P(\xi)$ with the polynomial function $\xi \mapsto P(\xi)$ defined on $\mathbf{K}^n$. For any $q = (q_1, \ldots, q_n) \in \mathbf{N}^n$ we denote by $P^{(q)}(\xi)$ the derivative

$$\frac{\partial^{|q|}}{\partial \xi_1^{q_1} \cdots \partial \xi_n^{q_n}} P(\xi).$$

Let us introduce the notation

$$D_j = \frac{1}{2\pi i} \frac{\partial}{\partial x_j} = \frac{1}{2\pi i} \partial_j$$

for $j = 1, \ldots, n$, the significance of which will only appear in §11. If

$p \in \mathbf{N}^n$, then D^p will obviously stand for

$$D^p = D_1^{p_1} \cdots D_n^{p_n} = \frac{1}{(2\pi i)^{|p|}} \partial^p = \frac{1}{(2\pi i)^{|p|}} \partial_1^{p_1} \cdots \partial_n^{p_n}$$
$$= \frac{1}{(2\pi i)^{|p|}} \frac{\partial^{|p|}}{\partial x_1^{p_1} \cdots \partial x_n^{p_n}}.$$

If $P(\xi)$ is a polynomial of degree m and we replace each ξ_j by D_j, then we obtain the *partial differential operator* $P(D)$ *with constant coefficients* and of *order* m. If for $x \in \mathbf{R}^n$ and $\xi \in \mathbf{R}^n$ we write

$$\langle x, \xi \rangle = x_1 \xi_1 + \cdots + x_n \xi_n,$$

then we have

$$P(D) e^{2\pi i \langle x, \xi \rangle} = P(\xi) \cdot e^{2\pi i \langle x, \xi \rangle}. \tag{6}$$

By virtue of Propositions 9 and 10 we have

$$P(D)T = P(D)\delta * T.$$

We shall often identify the differential operator $P(D)$ with the distribution $P(D)\delta$, whereby differentiation becomes a convolution. The relation between the function $\xi \mapsto P(\xi)$ and the distribution $P(D)\delta$ will be made more explicit in §11. The support of $P(D)\delta$ is the origin, and we know from Proposition 4.5 that every distribution whose support is the origin is of the form $P(D)\delta$. Let us observe that if the distributions $P(D)\delta$ and $Q(D)\delta$ are identified with the differential operators $P(D)$ and $Q(D)$, then the convolution $P(D)\delta * Q(D)\delta$ is identified with the composed operator $P(D) \circ Q(D)$ which corresponds to the product $P(\xi)Q(\xi)$ of the polynomials $P(\xi)$ and $Q(\xi)$. Thus the commutative ring of all distributions of the form $P(D)\delta$, with the convolution as multiplication, is isomorphic to the ring of all polynomials $P(\xi)$ with the ordinary product as multiplication (cf. Theorem 11.3).

A distribution E such that

$$P(D)E = \delta$$

is called a *fundamental* (or elementary) *solution* of the partial differential operator $P(D)$. If we know E, we can solve the partial differential equation $P(D)S = T$, where S is an unknown distribution and T a given distribution with compact support. Indeed, $S = E * T$ will be a solution since by Propositions 9 and 10 we have

$$P(D)(E * T) = (P(D)E) * T = \delta * T = T.$$

Every other solution is of the form $S + H$, where H is an arbitrary solution of the homogeneous equation $P(D)H = 0$.

EXAMPLE 1. Consider the Laplace operator

$$\Delta = \frac{\partial^2}{\partial x_1^2} + \cdots + \frac{\partial^2}{\partial x_n^2}.$$

If $n \geqq 3$ a fundamental solution of Δ is given by the function

$$E = -\frac{\Gamma[(n-2)/2]}{4\pi^{n/2}} \frac{1}{r^{n-2}},$$

where $r = (x_1^2 + \cdots + x_n^2)^{1/2}$. Observe that this function is not continuous at the origin, but by Proposition 1.2 it nevertheless defines a distribution since it is improperly integrable.

For $\varphi \in \mathfrak{D}$ we have

$$\langle \Delta E, \varphi \rangle = \langle E, \Delta\varphi \rangle = -\frac{\Gamma[(n-2)/2]}{4\pi^{n/2}} \lim_{\epsilon \to 0} \int_{r \geqq \epsilon} \frac{(\Delta\varphi)(x)}{r^{n-2}}\, dx.$$

We apply Green's identity

$$\int_\Omega (u\, \Delta v - v\, \Delta u)\, dx = \int_{\partial\Omega} \left(u \frac{\partial v}{\partial n} - v \frac{\partial u}{\partial n} \right) d\omega,$$

where $\partial/\partial n$ indicates derivation in the direction of the outward normal of the boundary $\partial\Omega$ of Ω. For Ω we take the domain between the spheres $r = \epsilon$ and $r = R$, where R is so large that Supp φ is contained in the ball $r < R$. We obtain

$$\int_{r \geqq \epsilon} \frac{\Delta\varphi}{r^{n-2}}\, dx = \int_{r \geqq \epsilon} \varphi\Delta\left(\frac{1}{r^{n-2}}\right) dx + \int_{r=\epsilon} \varphi \frac{\partial}{\partial r}\left(\frac{1}{r^{n-2}}\right) \epsilon^{n-1}\, d\sigma$$
$$- \int_{r=\epsilon} \frac{1}{r^{n-2}} \frac{\partial\varphi}{\partial r} \epsilon^{n-1}\, d\sigma,$$

where $d\sigma$ denotes the surface element on the unit sphere $\mathbf{S}_{n-1}$. The first integral on the right-hand side is zero since $1/r^{n-2}$ is harmonic, i.e., $\Delta(1/r^{n-2}) = 0$, outside the origin. Since

$$\frac{\partial}{\partial r} \frac{1}{r^{n-2}} = -(n-2)\epsilon^{1-n}$$

on $r = \epsilon$, the second integral equals

$$-(n-2) \int_{r=\epsilon} \varphi(x)\, d\sigma = -(n-2)|\mathbf{S}_{n-1}| \cdot \frac{1}{|\mathbf{S}_{n-1}|} \int_{r=\epsilon} \varphi(r\sigma)\, d\sigma,$$

where $|\mathbf{S}_{n-1}| = 2\pi^{n/2}/\Gamma(n/2)$ is the surface area of $\mathbf{S}_{n-1}$. As $\epsilon \to 0$ this expression tends to

$$-(n-2)|\mathbf{S}_{n-1}| \cdot \varphi(0) = -\frac{4\pi^{n/2}}{\Gamma[(n-2)/2]} \langle \delta, \varphi \rangle.$$

Finally, the third integral is dominated in absolute value by an expression of the form const $\cdot\, \epsilon \int d\sigma$, hence tends to zero as $\epsilon \to 0$. It follows that

$$\langle \Delta E, \varphi \rangle = \langle \delta, \varphi \rangle;$$

i.e., $\Delta E = \delta$, as asserted.

If T is a distribution with compact support, then a solution S of the Poisson equation $\Delta S = T$ is given by the Newtonian potential

$$S = U^T = E * T$$

of T. If T is a sufficiently regular function f, then its *Newtonian potential* U^f is given by

$$U^f(x) = -\frac{\Gamma[(n-2)/2]}{4\pi^{n/2}} \int_{\mathbf{R}^n} \frac{f(y)}{|x-y|^{n-2}}\, dy.$$

We shall return to these questions in Chapter 6.

It is one of the great triumphs of the theory of distributions that every partial differential operator with constant coefficients possesses a fundamental solution. We shall prove in Chapter 7 this result obtained in 1953 independently by Ehrenpreis and Malgrange ([46], Section 3.1).

Let us prove the following useful generalization of Leibniz's formula (Corollary of Proposition 6.3):

PROPOSITION 11. *Let $P(D)$ be a partial differential operator with constant coefficients of order m. For $\alpha \in \mathcal{E}(\Omega)$ and $T \in \mathcal{D}'(\Omega)$ or for $\alpha \in \mathcal{E}^{k+m}(\Omega)$ and $T \in \mathcal{D}'^k(\Omega)$ we have*

$$P(D)(\alpha T) = \sum_{q \in \mathbf{N}^n} \frac{D^q \alpha}{q!} P^{(q)}(D) T. \tag{7}$$

REMARK 8. If $P(D) = D^p$, i.e., $P(\xi) = \xi^p$, then

$$P^{(q)}(\xi) = \frac{p!}{(p-q)!} \xi^{p-q},$$

i.e.,

$$P^{(q)}(D) = \frac{p!}{(p-q)!} D^{p-q},$$

and since

$$\frac{p!}{q!(p-q)!} = \binom{p}{q},$$

we obtain the formula of the corollary of Proposition 6.3 [multiplied by the constant $(2\pi i)^{-|p|}$]. Let us also observe that since $P^{(q)}(D) = 0$ for $|q| > m$, the sum on the right-hand side of (7) is finite.

Proof of Proposition 11. It follows from Proposition 6.3 that

$$P(D)(\alpha T) = \sum_q (D^q\alpha)(Q_q(D)T), \tag{8}$$

where the $Q_q(D)$ are differential operators independent of α and T. To determine these operators $Q_q(D)$ we choose $\xi,\ \eta \in \mathbf{R}^n$, $\alpha(x) = e^{2\pi i\langle x,\xi\rangle}$, and $T = T_f$, where $f(x) = e^{2\pi i\langle x,\eta\rangle}$. Substituting in (8), using (6), and dividing both sides by $e^{2\pi i\langle x,\, \xi+\eta\rangle}$, we obtain

$$P(\xi + \eta) = \sum_q \xi^q Q_q(\eta). \tag{9}$$

On the other hand, we have by Taylor's formula

$$P(\xi + \eta) = \sum_q \frac{\xi^q}{q!} P^{(q)}(\eta). \tag{10}$$

Comparing (9) and (10), we see that

$$Q_q(\eta) = \frac{1}{q!} P^{(q)}(\eta). \ \blacksquare$$

PROPOSITION 12. *Let S and T be two distributions whose supports satisfy condition* ($\sum$) *and let h be a vector of* $\mathbf{R}^n$. *Then we have* (cf. Definition 3.2)

$$\tau_h(S * T) = \tau_h S * T = S * \tau_h T.$$

Proof. (a) Let us first assume that S has compact support. For $\varphi \in \mathfrak{D}$ we have then

$$\begin{aligned}\langle \tau_h(S * T), \varphi\rangle &= \langle S * T, \tau_{-h}\varphi\rangle \\ &= \int S(x) \left(\int T(y)\varphi(x + y + h)\, dy\right) dx \\ &= \int S(x) \left(\int (\tau_h T)(y)\varphi(x + y)\, dy\right) dx \\ &= \langle S * \tau_h T, \varphi\rangle;\end{aligned}$$

i.e., $\tau_h(S * T) = S * \tau_h T$. We see similarly that $\tau_h(S * T) = \tau_h S * T$.

(b) In the general case, since the support of δ is compact, we have by Proposition 9 and part (a) of the proof

$$\tau_h(S * T) = \tau_h\big(\delta * (S * T)\big) = \tau_h\delta * (S * T).$$

But by Lemma 2 and Proposition 8(b) we further have

$$\tau_h\delta * (S * T) = (\tau_h\delta * S) * T = \tau_h(\delta * S) * T = \tau_h S * T.$$

The other relation follows from Proposition 8(a). $\blacksquare$

Let $S \in \mathfrak{D}'$ be a fixed distribution. By virtue of Proposition 7 the linear map $u: T \mapsto S * T$ from $\mathcal{E}'$ into $\mathfrak{D}'$ is continuous. Furthermore, by Proposition 10 it commutes with partial derivations; i.e.,

$$\partial_j u(T) = u(\partial_j T)$$

for $1 \leqq j \leqq n$. We want to prove the converse result:

THEOREM 1. *Let u be a continuous linear map from $\mathcal{E}'$ into $\mathfrak{D}'$ which commutes with partial derivations. There exists a distribution $S \in \mathfrak{D}'$ such that $u(T) = S * T$ for every $T \in \mathcal{E}'$.*

The proof proper of this theorem will be preceded by two auxiliary results, the first of which is interesting in itself.

PROPOSITION 13. *The translates $\tau_h \delta = \delta_h$ of the Dirac measure δ form a total subset of $\mathfrak{D}'$ and of $\mathcal{E}'$.*

Proof. Since $\mathfrak{D}'$ is reflexive, every continuous linear form on $\mathfrak{D}'$ is given by $T \mapsto \langle T, \varphi \rangle$ for some $\varphi \in \mathfrak{D}$. If $\langle \delta_h, \varphi \rangle = \varphi(h) = 0$ for every $h \in \mathbf{R}^n$, then $\varphi = 0$, which proves that the set $\{\delta_h \mid h \in \mathbf{R}^n\}$ is total in $\mathfrak{D}'$ (Proposition 1.6.2 and Exercise 3.1.4). For $\mathcal{E}'$ the proof is similar. ∎

LEMMA 3. *Let u be a continuous linear map from $\mathcal{E}'$ into $\mathfrak{D}'$. Then u commutes with partial derivations if and only if it commutes with translations.*

Proof. (a) If u commutes with translations, i.e.,

$$\tau_h u(T) = u(\tau_h T) \tag{11}$$

for all $h \in \mathbf{R}^n$ and $T \in \mathcal{E}'$, then for any vector h parallel to the x_j-axis we have

$$t^{-1}(\tau_h u(T) - u(T)) = u(t^{-1}\{\tau_h T - T\}),$$

where $t = h_j$. Since u is continuous, we have by Proposition 3.4 and Exercise 3.2

$$\partial_j u(T) = u(\partial_j T).$$

(b) Suppose that u commutes with partial derivations, let $T \in \mathcal{E}'$, $\varphi \in \mathfrak{D}$, and define

$$g(h) = \langle u(\tau_h T), \tau_h \varphi \rangle,$$

where h varies in $\mathbf{R}^n$. We have

$$\frac{\partial}{\partial h_j} g(h) = \left\langle \frac{\partial}{\partial h_j} u(\tau_h T), \tau_h \varphi \right\rangle + \left\langle u(\tau_h T), \frac{\partial}{\partial h_j} \tau_h \varphi \right\rangle.$$

Since u is continuous, we have

$$\frac{\partial}{\partial h_j} u(\tau_h T) = u\left(\frac{\partial}{\partial h_j} \tau_h T\right). \tag{12}$$

It follows from Proposition 3.5 and Exercise 3.2 that

$$u\left(\frac{\partial}{\partial h_j} \tau_h T\right) = u\left(-\frac{\partial}{\partial x_j} \tau_h T\right); \tag{13}$$

and since u commutes with partial derivations, we have

$$u\left(-\frac{\partial}{\partial x_j} \tau_h T\right) = -\frac{\partial}{\partial x_j} u(\tau_h T). \tag{14}$$

From (12), (13), and (14) we obtain

$$\begin{aligned} \frac{\partial}{\partial h_j} g(h) &= \left\langle -\frac{\partial}{\partial x_j} u(\tau_h T), \tau_h \varphi \right\rangle + \left\langle u(\tau_h T), \frac{\partial}{\partial h_j} \tau_h \varphi \right\rangle \\ &= \left\langle u(\tau_h T), \left(\frac{\partial}{\partial x_j} + \frac{\partial}{\partial h_j}\right) \tau_h \varphi \right\rangle = 0 \end{aligned}$$

since obviously

$$\left(\frac{\partial}{\partial x_j} + \frac{\partial}{\partial h_j}\right) \tau_h \varphi = 0.$$

It follows that g is a constant; i.e.,

$$\langle u(\tau_h T), \tau_h \varphi \rangle = \langle u(T), \varphi \rangle$$

for every $h \in \mathbf{R}^n$. For $\psi \in \mathfrak{D}$ set $\varphi = \tau_{-h}\psi$, i.e., $\psi = \tau_h \varphi$. Then

$$\langle u(\tau_h T), \psi \rangle = \langle u(T), \tau_{-h}\psi \rangle = \langle \tau_h u(T), \psi \rangle,$$

which proves that (11) is satisfied for every $h \in \mathbf{R}^n$. ▮

Proof of Theorem 1. For a fixed distribution $T \in \mathcal{E}'$ the map

$$R \mapsto u(R * T) - R * u(T)$$

is continuous from $\mathcal{E}'$ into $\mathfrak{D}'$ and therefore its kernel N is closed. On the other hand, it follows from the hypothesis and from Lemma 3 that

$$u(\delta_h * T) = u(\tau_h T) = \tau_h u(T) = \delta_h * u(T);$$

i.e., all distributions δ_h belong to N. By Proposition 13 we have $N = \mathcal{E}'$, i.e., $u(R * T) = R * u(T)$ for all $R \in \mathcal{E}'$ and $T \in \mathcal{E}'$. Set $S = u(\delta)$. Then $u(T) = u(\delta * T) = u(\delta) * T = S * T$ for every $T \in \mathcal{E}'$. ▮

Corollary. *Let u be a continuous linear map from $\mathfrak{D}'$ into $\mathfrak{D}'$ which commutes with partial derivations. There exists a distribution with compact support $S \in \mathcal{E}'$ such that $u(T) = S * T$ for every $T \in \mathfrak{D}'$.*

Proof. Since $\mathcal{E}$ is a normal space of distributions (Example 2.5) the map $\mathcal{E}' \hookrightarrow \mathfrak{D}'$ is continuous, and therefore by Theorem 1 there exists $S \in \mathfrak{D}'$ such that $u(T) = S * T$ for every $T \in \mathcal{E}'$.

Suppose that Supp S is not compact. Then there exists a sequence of points $x_n \in \operatorname{Supp} S$ such that $|x_n| \to \infty$ as $n \to \infty$. The sequence of distributions $c_n \delta_{-x_n}$, where the c_n are constants, tends to 0 in $\mathfrak{D}'$, since if B is a bounded subset of $\mathfrak{D}$, then there exists a compact subset K of $\mathbf{R}^n$ such that Supp $\varphi \subset K$ for $\varphi \in B$ (Example 2.12.6), and therefore

$$\langle c_n \delta_{-x_n}, \varphi \rangle = c_n \varphi(-x_n) = 0$$

for all $\varphi \in B$ if n is sufficiently large. On the other hand, the sequence

$$u(c_n \delta_{-x_n}) = S * c_n \delta_{-x_n} = c_n \tau_{-x_n} S$$

does not tend to zero if we choose the constants c_n suitably. Indeed, for every n there exists a function $\psi_n \in \mathfrak{D}$ having its support in the ball $B_1(x_n) = \{x \mid |x - x_n| \leqq 1\}$ and such that $\langle S, \psi_n \rangle = 1$. We can write $\psi_n = c_n \tau_{x_n} \varphi_n$, where $\varphi_n \in \mathfrak{D}$; the support of φ_n is contained in the ball $B_1 = \{x \mid |x| \leqq 1\}$, and c_n is chosen so that $|\partial^p \varphi_n(x)| \leqq 1$ for $|p| \leqq n$. Then the set $\{\varphi_n \mid n \in \mathbf{N}\}$ is bounded in $\mathfrak{D}$ but

$$\langle c_n \tau_{-x_n} S, \varphi_n \rangle = \langle S, c_n \tau_{x_n} \varphi_n \rangle = \langle S, \psi_n \rangle = 1$$

for all n. We have therefore arrived at a contradiction with the continuity of u, and so Supp S must be compact.

Finally, we have $u(T) = S * T$ for every $T \in \mathfrak{D}'$ since the map

$$T \mapsto u(T) - S * T$$

from $\mathfrak{D}'$ into $\mathfrak{D}'$ is continuous (Proposition 7), and its kernel contains $\mathcal{E}'$ which is dense in $\mathfrak{D}'$. ∎

We conclude this section with a relation between tensor product and convolution.

Proposition 14. *Let Q and R be two distributions belonging to $\mathfrak{D}'(\mathbf{R}^k)$ whose supports satisfy condition (Σ). Similarly, let S and T be two distributions belonging to $\mathfrak{D}'(\mathbf{R}^l)$ whose supports satisfy condition (Σ). Then the supports of the distributions $Q \otimes S$ and $R \otimes T$ belonging to $\mathfrak{D}'(\mathbf{R}^{k+l})$ satisfy condition (Σ), and we have*

$$(Q * R) \otimes (S * T) = (Q \otimes S) * (R \otimes T). \tag{15}$$

Proof. (a) Write $A = \operatorname{Supp} Q \otimes S$ and $B = \operatorname{Supp} R \otimes T$. Let K be a compact subset of $\mathbf{R}^{k+l} = \mathbf{R}^k \times \mathbf{R}^l$. Its two projections $K_1 \subset \mathbf{R}^k$ and $K_2 \subset \mathbf{R}^l$ are compact, and $K \subset K_1 \times K_2$. The isomorphism

$$(x_1, y_1; x_2, y_2) \mapsto (x_1, x_2; y_1, y_2)$$

from $(\mathbf{R}^k \times \mathbf{R}^l) \times (\mathbf{R}^k \times \mathbf{R}^l)$ onto $(\mathbf{R}^k \times \mathbf{R}^k) \times (\mathbf{R}^l \times \mathbf{R}^l)$ maps the subset $(A \times B) \cap K^\Delta$ of $\mathbf{R}^{k+l} \times \mathbf{R}^{k+l}$ into the set

$$\{(\operatorname{Supp} Q \times \operatorname{Supp} R) \cap K_1^\Delta\} \times \{(\operatorname{Supp} S \times \operatorname{Supp} T) \cap K_2^\Delta\}. \tag{16}$$

Indeed,

$$(x_1, y_1) \in A = \operatorname{Supp} Q \times \operatorname{Supp} S, \qquad (x_2, y_2) \in B = \operatorname{Supp} R \times \operatorname{Supp} T,$$
$$(x_1 + x_2, y_1 + y_2) \in K$$

imply

$$(x_1, x_2) \in \operatorname{Supp} Q \times \operatorname{Supp} R, \qquad x_1 + x_2 \in K_1,$$
$$(y_1, y_2) \in \operatorname{Supp} S \times \operatorname{Supp} T, \qquad y_1 + y_2 \in K_2.$$

Since by assumption the two factors in (16) are compact, the sets A and B satisfy condition $(\sum)$.

(b) Consider now $\varphi \in \mathfrak{D}(\mathbf{R}^k)$ and $\psi \in \mathfrak{D}(\mathbf{R}^l)$. Let $\chi \in \mathfrak{D}(\mathbf{R}^k)$ be such that $\chi(x) = 1$ on a neighborhood U of $\operatorname{Supp} Q \cap (\operatorname{Supp} \varphi - \operatorname{Supp} R)$ and $\zeta \in \mathfrak{D}(\mathbf{R}^l)$ such that $\zeta(y) = 1$ on a neighborhood V of

$$\operatorname{Supp} S \cap (\operatorname{Supp} \psi - \operatorname{Supp} T).$$

Then

$$\chi \otimes \zeta \in \mathfrak{D}(\mathbf{R}^{k+l}) \qquad \text{and} \qquad \chi(x)\zeta(y) = 1$$

on the neighborhood $U \times V$ of the set

$$\{\operatorname{Supp} Q \cap (\operatorname{Supp} \varphi - \operatorname{Supp} R)\} \times \{\operatorname{Supp} S \cap (\operatorname{Supp} \psi - \operatorname{Supp} T)\},$$

which contains $A \cap (\operatorname{Supp} \varphi \otimes \psi - B)$ as can be checked immediately. We therefore have

$$\begin{aligned}
&\langle (Q * R) \otimes (S * T), \varphi \otimes \psi \rangle \\
&= \langle Q * R, \varphi \rangle \langle S * T, \psi \rangle \\
&= \left(\int Q(x)\chi(x) \left(\int R(\xi)\varphi(x + \xi)\, d\xi \right) dx \right) \left(\int S(y)\zeta(y) \left(\int T(\eta)\psi(y + \eta)\, d\eta \right) dy \right) \\
&= \int (Q \otimes S)(x, y)\chi(x)\zeta(y) \left(\int (R \otimes T)(\xi, \eta)(\varphi \otimes \psi)(x + \xi, y + \eta)\, d\xi\, d\eta \right) dx\, dy \\
&= \langle (Q \otimes S) * (R \otimes T), \varphi \otimes \psi \rangle,
\end{aligned}$$

from where relation (15) follows. ∎

Exercises

1. Let $S \in \mathfrak{D}'$ and $\varphi \in \mathfrak{D}$. Show that the support of the function

$$y \mapsto \int S(x)\varphi(x+y)\,dx$$

is the set $\operatorname{Supp} S - \operatorname{Supp} \varphi$.

2. Let X and Y be two locally compact spaces. A continuous map $f: X \to Y$ is said to be *proper* if for every compact subset K of Y the subset $f^{-1}(K)$ is compact in X.

(a) Let $X' = X \cup \{\omega_1\}$ and $Y' = Y \cup \{\omega_2\}$ be the Alexandrov compactifications (Exercise 2.10.7) of the spaces X and Y. Given a map $f: X \to Y$, let $f': X' \to Y'$ be defined by $f'(x) = f(x)$ for $x \in X$ and $f(\omega_1) = \omega_2$. Show that the continuous map $f: X \to Y$ is proper if and only if the map $f': X' \to Y'$ is continuous.

(b) Let A and B be two closed subsets of $\mathbf{R}^n$. Show that A and B satisfy condition $(\sum)$ if and only if the $(x, y) \mapsto x + y$ from $A \times B$ into $\mathbf{R}^n$ is proper.

3. Prove that the bilinear map $(S, T) \mapsto S * T$ from $\mathcal{E}'^s \times \mathcal{E}'^t$ into $\mathcal{E}'^{s+t}$ is continuous if we equip $\mathcal{E}'^s$, $\mathcal{E}'^t$, and $\mathcal{E}'^{s+t}$ with the topology of uniform convergence on compact sets. (*Hint:* See Exercise 8.8 and the proof of Proposition 5.)

4. Let $T \in \mathfrak{D}'(\mathbf{R}^n)$ and $\varphi \in \mathcal{E}(\mathbf{R}^n)$ be such that $\operatorname{Supp} T \cap \operatorname{Supp} \varphi$ is compact.

(a) Show that $\langle T, \varphi \rangle = \langle \varphi T, 1 \rangle$ could have served as the definition of $\langle T, \varphi \rangle$.

(b) Show that if $\operatorname{Supp} T \cap \operatorname{Supp} \varphi = \emptyset$, then $\langle T, \varphi \rangle = 0$.

(c) Show that if $\partial^p \varphi(x) = 0$ for all $x \in \operatorname{Supp} T$ and $p \in \mathbf{N}^n$, then $\langle T, \varphi \rangle = 0$.

5. Let λ be the function $x \mapsto a_1x_1 + \cdots + a_nx_n$ and ϵ the function $x \mapsto e^{\lambda(x)}$. Prove the formulas

$$\epsilon(S * T) = (\epsilon S) * (\epsilon T)$$

and

$$\lambda(S * T) = (\lambda S) * T + S * (\lambda T).$$

§10. *Regularization*

One of the main applications of convolution is the regularization of a distribution T, i.e., the approximation of T by expressions of the form $\rho_\epsilon * T$, where $\rho_\epsilon \in \mathfrak{D}$ is the function we have considered several times (Example 1.4, Proposition 4.1). Since ρ_ϵ tends to δ as $\epsilon \to 0$ (Example 1.4), by Propositions 9.7 and 9.9 the distribution $\rho_\epsilon * T$ tends to T. Now it will turn out that $\rho_\epsilon * T$ is (a distribution identified with) an infinitely differentiable function. Hence if we let ϵ run through the values $1/k$ $(k \geqq 1)$, we will have an effective method to approximate a distribution by a *sequence* of infinitely differentiable functions. We know of course that any distribution is the limit of a filter of infinitely differentiable functions (Proposition 1.3).

Let us introduce a notation which will be useful in the sequel. If f is a function defined on $\mathbf{R}^n$, we denote by $\check{f}$ the function $x \mapsto f(-x)$. Clearly

$\check{\check{f}} = f$ and the map $\varphi \mapsto \check{\varphi}$ is an automorphism of any space $\mathfrak{D}^m$ or $\mathcal{E}^m$ $(0 \leqq m \leqq \infty)$. Also

$$(\partial^p \varphi)^{\vee} = (-1)^{|p|}\, \partial^p \check{\varphi}, \qquad \tau_h \check{\varphi} = (\tau_{-h}\varphi)^{\vee}.$$

If f is continuous, we have for any $\varphi \in \mathfrak{D}$

$$\begin{aligned}\langle \check{f}, \varphi\rangle &= \int_{\mathbf{R}^n} f(-x)\varphi(x)\, dx \\ &= \int_{\mathbf{R}^n} f(x)\varphi(-x)\, dx = \langle f, \check{\varphi}\rangle.\end{aligned}$$

This leads us to define for $T \in \mathfrak{D}'(\mathbf{R}^n)$ the distribution $\check{T}$ by

$$\langle \check{T}, \varphi\rangle = \langle T, \check{\varphi}\rangle,$$

where $\varphi \in \mathfrak{D}(\mathbf{R}^n)$. Clearly, $\check{\check{T}} = T$ and the map $T \mapsto \check{T}$ is an automorphism of any space $\mathfrak{D}'^m$ or $\mathcal{E}'^m$ $(0 \leqq m \leqq \infty)$; in fact, $T \mapsto \check{T}$ is the transpose of the map $\varphi \mapsto \check{\varphi}$. We have

$$(\partial^p T)^{\vee} = (-1)^{|p|}\, \partial^p \check{T}, \qquad \tau_h \check{T} = (\tau_{-h} T)^{\vee}.$$

One verifies immediately that for $\alpha \in \mathcal{E}^m$ and $T \in \mathfrak{D}'^m$ we have

$$(\alpha T)^{\vee} = \check{\alpha}\check{T}$$

and that if S and T are two distributions whose supports satisfy condition (Σ), then $(S * T)^{\vee} = \check{S} * \check{T}$.

As in §8, the letters s and t will denote either a positive integer or the symbol ∞. If $s = \infty$, then $s - t$ will be equal to ∞ for any t.

PROPOSITION 1. *Let $\alpha \in \mathcal{E}^s$, $T \in \mathfrak{D}'^t$, where $s \geqq t$, and suppose that the supports of α and T satisfy condition (Σ) of §9. Then $\alpha * T$ is a function belonging to $\mathcal{E}^{s-t}$ which for any $x \in \mathbf{R}^n$ is given by*

$$(\alpha * T)(x) = \langle T, \tau_x \check{\alpha}\rangle. \tag{1}$$

Proof. The expression $\langle T, \tau_x\check{\alpha}\rangle$ is well-defined by formula (3) of §9. Indeed, if we write $A = \operatorname{Supp} \alpha$ and $B = \operatorname{Supp} T$, then

$$\operatorname{Supp}(\tau_x \check{\alpha}) = x - A,$$

and condition (Σ) implies that $B \cap (x - A)$ is compact.

Next let us prove that the map $x \mapsto \tau_x\check{\alpha}$ from $\mathbf{R}^n$ into $\mathcal{E}^s$ is continuous. Let V be a neighborhood of 0 in $\mathcal{E}^s$ which we may assume to be of the form

$$V = \{\varphi \mid |\partial^p \varphi(x)| \leqq \epsilon,\ |p| \leqq h,\ x \in K\},$$

where h is a positive integer, $h = s$ if s is finite, and K is a compact subset of $\mathbf{R}^n$. Let U be a compact neighborhood of the point $a \in \mathbf{R}^n$. For any $x \in \mathbf{R}^n$ we have

$$\partial_y^p\{\tau_x\check{\alpha}(y) - \tau_a\check{\alpha}(y)\} = (-1)^{|p|}\{(\partial^p\alpha)(x - y) - (\partial^p\alpha)(a - y)\}.$$

If x varies in U and y in K, then $x - y$ varies in the compact set $U - K$. Now $\partial^p\alpha$ is uniformly continuous on $U - K$. Hence if we choose $|x - a|$ sufficiently small, we will have

$$|(\partial^p\alpha)(x - y) - (\partial^p\alpha)(a - y)| \leqq \epsilon$$

for all $y \in K$ and $|p| \leqq h$; i.e., $\tau_x\check{\alpha} \in \tau_a\check{\alpha} + V$.

Let C be a compact subset of $\mathbf{R}^n$ and $\chi \in \mathfrak{D}$ such that $\chi(x) = 1$ on a neighborhood of the compact set $B \cap (C - A)$. For $x \in C$ we have $\operatorname{Supp}(\tau_x\check{\alpha}) \subset C - A$ and therefore

$$\langle T, \tau_x\check{\alpha}\rangle = \langle T, \chi \cdot \tau_x\check{\alpha}\rangle. \tag{2}$$

From Proposition 7.4 and what we have just seen it follows that the map $x \mapsto \chi \cdot \tau_x\check{\alpha}$ from C into $\mathfrak{D}^s$ is continuous. But $T \in \mathfrak{D}'^t \subset \mathfrak{D}'^s$; thus the function $x \mapsto \langle T, \chi \cdot \tau_x\check{\alpha}\rangle$ is continuous on C, and hence the function $x \mapsto \langle T, \tau_x\check{\alpha}\rangle$ is continuous on $\mathbf{R}^n$.

Keeping the same notations, let us prove relation (1). If $\varphi \in \mathfrak{D}(C)$, then by Remark 9.1 we have

$$\begin{aligned}\langle \alpha * T, \varphi\rangle &= \langle \alpha \otimes T, \chi^*\varphi^\Delta\rangle \\ &= \int T(y)\chi(y)\left(\int \alpha(x)\varphi(x + y)\, dx\right) dy \\ &= \int T(y)\chi(y)\left(\int \varphi(x)\alpha(x - y)\, dx\right) dy \\ &= \int \varphi(x)\left(\int T(y)\chi(y)\alpha(x - y)\, dy\right) dx \\ &= \int \langle T, \tau_x\check{\alpha}\rangle\varphi(x)\, dx,\end{aligned}$$

which proves (1).

Finally, we prove that $x \mapsto \langle T, \tau_x\check{\alpha}\rangle$ is $s - t$ times continuously differentiable. It is sufficient to prove that the relation

$$\partial^p\langle T, \tau_x\check{\alpha}\rangle = (-1)^{|p|}\langle T, \tau_x(\partial^p\alpha)\check{}\,\rangle \tag{3}$$

holds for $|p| \leqq s - t$ since $\partial^p\alpha \in \mathcal{E}^{s-|p|}$, $T \in \mathfrak{D}'^{s-|p|}$, and therefore the function $x \mapsto \langle T, \tau_x(\partial^p\alpha)\check{}\,\rangle$ is continuous as we know already. By recursion it is sufficient to prove (3) for $|p| = 1$. If j is an index such that $1 \leqq j \leqq n$, denote by $h = (h_1, \ldots, h_n)$ a vector such that $h_j = t$,

$h_k = 0$ for $k \neq j$, and take t so small that $x - h \in C$. Then we have by (2)

$$\partial_j\langle T, \tau_x\check{\alpha}\rangle = \lim_{t\to 0} t^{-1}\{\langle T, \chi \cdot \tau_{x-h}\check{\alpha}\rangle - \langle T, \chi \cdot \tau_x\check{\alpha}\rangle\}$$
$$= \langle T, \chi \cdot \lim_{t\to 0} t^{-1}\{\tau_{x-h}\check{\alpha} - \tau_x\check{\alpha}\}\rangle.$$

Now $\tau_{x-h}\check{\alpha} = \tau_{-h}(\tau_x\check{\alpha})$. On the other hand, it can be seen as in the proof of Proposition 3.4 (cf. also Exercise 3.2) that for $\varphi \in \mathcal{E}^m$ the function $t^{-1}(\tau_{-h}\varphi - \varphi)$ converges to $\partial_j\varphi$ in $\mathcal{E}^{m-1}$. Thus we obtain

$$\partial_j\langle T, \tau_x\check{\alpha}\rangle = \langle T, \partial_j(\tau_x\check{\alpha})\rangle.$$

Since $\partial_j(\tau_x\check{\alpha}) = \tau_x(\partial_j\check{\alpha}) = -\tau_x(\partial_j\alpha)\check{}$, we have proved formula (3) for $|p| = 1$. ∎

Remark 1. Proposition 1 can be used to give a definition of convolution which does not use the tensor product ([46], Definitions 1.6.1 and 1.6.2). First, one defines $T * \alpha$ for $\alpha \in \mathfrak{D}$ and $T \in \mathfrak{D}'$ by formula (1). Next one proves that if $u: \mathfrak{D} \to \mathcal{E}$ is a continuous linear map which commutes with translations, then there exists a unique distribution $R \in \mathfrak{D}'$ such that $u(\varphi) = R * \varphi$ for every $\varphi \in \mathfrak{D}$ (Exercise 2). Finally, if $S \in \mathfrak{D}'$ and $T \in \mathcal{E}'$, then $\varphi \mapsto S * (T * \varphi)$ is a continuous linear map from $\mathfrak{D}$ into $\mathcal{E}$ which commutes with translations. Hence there exists a unique $R \in \mathfrak{D}'$ such that $S * (T * \varphi) = R * \varphi$ for all φ, and one sets $R = S * T$.

Proposition 2. *Let ρ be a continuous function defined on $\mathbf{R}^n$ which has the following properties*

(a) Supp $\rho \subset B_1$,
(b) $\rho(x) \geqq 0$ *for all* $x \in \mathbf{R}^n$,
(c) $\int_{\mathbf{R}^n} \rho(x)\, dx = 1$,

where $B_1 = \{x \mid |x| \leqq 1\}$ is the unit ball of $\mathbf{R}^n$. For $\epsilon > 0$ define the function ρ_ϵ by

$$\rho_\epsilon(x) = \frac{1}{\epsilon^n}\rho\left(\frac{x}{\epsilon}\right).$$

Then ρ_ϵ converges to the Dirac measure δ in the space $\mathcal{E}'^m$ $(0 \leqq m \leqq \infty)$ equipped with the topology $\kappa(\mathcal{E}'^m, \mathcal{E}^m)$ of uniform convergence on compact subsets of $\mathcal{E}^m$, as $\epsilon \to 0$.

Proof. Denote by M the subset of $\mathcal{E}'^m$ formed by the ρ_ϵ with $0 < \epsilon \leqq 1$ and the Dirac measure δ. For any $\epsilon \leqq 1$ and $\varphi \in \mathcal{E}^m$ we have

$$|\langle\rho_\epsilon, \varphi\rangle| = \frac{1}{\epsilon^n}\left|\int \rho\left(\frac{x}{\epsilon}\right)\varphi(x)\, dx\right|$$
$$= \left|\int \rho(x)\varphi(\epsilon x)\, dx\right| \leqq \max_{|x|\leqq 1} |\varphi(x)|;$$

hence M is weakly bounded. Since $\mathcal{E}^m$ is barrelled (Example 3.6.3), the set M is equicontinuous (Proposition 3.6.2). It follows from Proposition 3.9.8 that in order to establish the proposition we have to prove that for any $\varphi \in \mathcal{E}^m$ we have

$$\lim_{\epsilon \to 0} \int \rho_\epsilon(x)\varphi(x)\, dx = \varphi(0).$$

This can be seen exactly as in Example 1.4. ∎

Remark 2. Unless $m = \infty$ the functions ρ_ϵ do not tend to δ in $\mathcal{E}'^m$ for the strong topology $\beta(\mathcal{E}'^m, \mathcal{E}^m)$ (cf. Exercise 3).

Proposition 3. *For every $T \in \mathfrak{D}'^t$ the map $S \mapsto S * T$ from $\mathcal{E}'^s$ into $\mathfrak{D}'^{s+t}$ is continuous if we equip the spaces $\mathcal{E}'^s$ and $\mathfrak{D}'^{s+t}$ with the topologies $\kappa(\mathcal{E}'^s, \mathcal{E}^s)$ and $\kappa(\mathfrak{D}'^{s+t}, \mathfrak{D}^{s+t})$ of uniform convergence on compact sets.*

Proof. We may assume that s is finite, since otherwise the proposition follows from Proposition 9.7. Let W be a neighborhood of 0 in $\mathfrak{D}'^{s+t}$, which we may assume to be of the form $W = C^\circ$, where C is a compact subset of $\mathfrak{D}^{s+t}$. Denote by A the set of all functions

$$x \mapsto \int T(y)\varphi(x + y)\, dy = \langle T, \tau_{-x}\varphi\rangle,$$

where φ varies in C. It follows from Proposition 1 that $A \subset \mathcal{E}^s$.

Let us prove that A is a compact subset of $\mathcal{E}^s$. In the first place, for every $x \in \mathbf{R}^n$ and $|p| \leqq s$ the set $\tau_{-x}\partial^p C$ is compact in $\mathfrak{D}^{s+t-|p|}$. Since T is continuous on $\mathfrak{D}^{s+t-|p|}$, the set $T(\tau_{-x}\partial^p C)$ is compact in $\mathbf{K}$; i.e., $\partial^p\langle T, \tau_{-x}\varphi\rangle$ is bounded as φ varies in C.

Next, for every $\epsilon > 0$ there exists a neighborhood V of 0 in $\mathfrak{D}^t$ such that $|\langle T, \psi\rangle| \leqq \epsilon$ for $\psi \in V$. We may assume that V is defined by a set of inequalities

$$|\partial^q\psi(x)| \leqq \epsilon_k \quad \text{for} \quad |x| \geqq k - 1 \quad \text{and} \quad |q| \leqq m_k,$$

where $\epsilon_k \geqq \epsilon_{k+1}$ and $m_k \leqq m_{k+1} \leqq t$ $(k \geqq 1)$. Let $p \in \mathbf{N}^n$ be such that $|p| \leqq s$ and let x be a point of $\mathbf{R}^n$. Since for $|p + q| \leqq s + t$ the set $\{\partial^{p+q}\varphi \mid \varphi \in C\}$ is equicontinuous and the supports of the $\varphi \in C$ are contained in a compact subset of $\mathbf{R}^n$ (Proposition 3.9.12), there exists $\eta > 0$ such that

$$|\partial_y^{p+q}\varphi(x' + y) - \partial_y^{p+q}\varphi(x + y)| < \epsilon_k$$

provided that $\varphi \in C$, $|y| \geqq k - 1$, $|q| \leqq m_k$, and $|x' - x| \leqq \eta$ $(k \geqq 1)$. In other words, the functions

$$y \mapsto \partial^p\varphi(x' + y) - \partial^p\varphi(x + y)$$

belong to V as φ varies in C and $|x' - x| \leqq \eta$. Hence

$$\left|\partial^p \int T(y)\varphi(x' + y)\, dy - \partial^p \int T(y)\varphi(x + y)\, dy\right| \leqq \epsilon$$

provided that $|x' - x| \leqq \eta$ and $\varphi \in C$. It follows now from Proposition 3.9.11 that A is indeed compact.

If $S \in A^{\circ}$, then

$$|\langle S * T, \varphi\rangle| = \left|\int S(x) \left(\int T(y)\varphi(x + y)\, dy\right) dx\right| \leqq 1$$

for all $\varphi \in C$; i.e., $S * T \in C^{\circ} = W$. ∎

It follows from Proposition 1 that if $\rho \in \mathfrak{D}$, then for any $T \in \mathfrak{D}'$ the convolution $\rho_\epsilon * T$ is an infinitely differentiable function which will be called a *regularization* of T. If $T \in \mathfrak{D}'^t$, then by Propositions 2 and 3 the regularizations converge to T in $\mathfrak{D}'^t$ as $\epsilon \to 0$. For $t = \infty$ this was pointed out at the beginning of this section.

We want to develop an analogous procedure for the approximation by infinitely differentiable functions of distributions defined in an open subset of $\mathbf{R}^n$. First, however, we must introduce one more notation. If f is a function defined on $\mathbf{R}^n$, in particular if $f \in \mathfrak{C}$, we set

$$\operatorname{Tr} f = f(0).$$

If $\alpha \in \mathcal{E}^m$ and $T \in \mathfrak{D}'^m$ are such that $\operatorname{Supp} \check{\alpha}$ and $\operatorname{Supp} T$ satisfy condition (Σ), then it follows from Proposition 1 that

$$\langle T, \alpha\rangle = \operatorname{Tr}(\check{\alpha} * T).$$

Observe that the left-hand side in this formula is well defined since condition (Σ) implies in particular that $\operatorname{Supp} T \cap \operatorname{Supp} \alpha$ is compact. If $S \in \mathfrak{D}'^s$ and $T \in \mathfrak{D}'^t$ are such that their supports satisfy condition (Σ), then for any $\varphi \in \mathfrak{D}^{s+t}$ we have

$$\langle S * T, \varphi\rangle = \langle T, \check{S} * \varphi\rangle. \tag{4}$$

Indeed, $\operatorname{Supp} T$ and $\operatorname{Supp}(\check{S} * \varphi)^{\check{}} = \operatorname{Supp} S - \operatorname{Supp} \varphi$ clearly satisfy condition (Σ) and $\check{S} * \varphi \in \mathcal{E}^t$ by Proposition 1. Furthermore,

$$\langle S * T, \varphi\rangle = \operatorname{Tr}(S * T * \check{\varphi}) = \operatorname{Tr}\big(T * (\check{S} * \varphi)^{\check{}}\big) = \langle T, \check{S} * \varphi\rangle.$$

Now we are in the position to prove

PROPOSITION 4. *Let Ω be an open subset of $\mathbf{R}^n$ and $T \in \mathfrak{D}'^m(\Omega)$. There exists a sequence of functions belonging to $\mathcal{E}(\Omega)$ which converges to T in the space $\mathfrak{D}'^m(\Omega)$ equipped with the topology $\kappa\big(\mathfrak{D}'^m(\Omega), \mathfrak{D}^m(\Omega)\big)$.*

Proof. Let $(K_k)_{k \geqq 1}$ be a sequence of compact subsets of Ω such that $K_k \subset K_{k+1}$ and that every compact subset of Ω is contained in some K_k (cf. Example 2.6.3). For each $k \geqq 1$ let α_k be a function belonging to $\mathfrak{D}(\Omega)$ such that $\alpha_k(x) = 1$ in a neighborhood of K_k. The distribution $\alpha_k T$ has its support contained in a compact subset of Ω and can therefore be considered as a distribution defined on $\mathbf{R}^n$. If the function ρ which figures in Proposition 2 belongs to $\mathfrak{D}$, then by Proposition 1 the functions

$$\rho_{1/k} * (\alpha_k T) \tag{5}$$

($k \geqq 1$) are in $\mathcal{E}$.

For $\varphi \in \mathfrak{D}^m(\Omega)$ we have by (4)

$$\langle \rho_{1/k} * (\alpha_k T), \varphi \rangle = \langle T, \alpha_k(\check{\rho}_{1/k} * \varphi) \rangle. \tag{6}$$

Now $\check{\rho}_{1/k} * \varphi$ converges to φ in $\mathfrak{D}^m(\Omega)$ as $k \to \infty$ (cf. the proof of Proposition 4.1), and α_k converges to $\mathbf{1}$ in $\mathcal{E}^m(\Omega)$ as $k \to \infty$. It follows from Proposition 7.4 that $\alpha_k(\check{\rho}_{1/k} * \varphi)$ tends to φ in $\mathfrak{D}^m(\Omega)$ as $k \to \infty$. Hence (6) converges to $\langle T, \varphi \rangle$ as $k \to \infty$; i.e., the sequence (5) converges to T weakly in $\mathfrak{D}'^m(\Omega)$. It follows from Proposition 3.9.8 that (5) converges to T also in the topology of uniform convergence on compact sets. ∎

Exercises

1. (a) Show that if $\alpha \in \mathfrak{D}^s$ and $T \in \mathcal{E}'^t$, where $s \geqq t$, then $\alpha * T \in \mathfrak{D}^{s-t}$.

(b) Let $T \in \mathcal{E}'$. Show that $\varphi \mapsto T * \varphi$ is a continuous map from $\mathfrak{D}$ into $\mathfrak{D}$ whose transpose is the map $S \mapsto \check{T} * S$. Apply this result to the particular cases $T = \tau_h \delta$ (cf. Definition 3.2) and $T = P(D)\delta$, where $P(D)$ is a partial differential operator with constant coefficients (§9).

2. Show that if $u: \mathfrak{D} \to \mathcal{E}$ is a continuous linear map such that $\tau_h u(\varphi) = u(\tau_h \varphi)$ for every $h \in \mathbf{R}^n$ and $\varphi \in \mathfrak{D}$, then there exists a unique distribution $R \in \mathfrak{D}'$ such that $u(\varphi) = R * \varphi$ for every $\varphi \in \mathfrak{D}$. (*Hint:* The map $\check{\varphi} \mapsto \operatorname{Tr} u(\varphi)$ is a distribution.)

3. Show that the functions ρ_ϵ of Proposition 2 do not tend to δ in $\mathfrak{C}'$ for the strong topology $\beta(\mathfrak{C}', \mathfrak{C})$. (*Hint:* In $\mathfrak{C}$ there exists a fundamental system of bounded sets B such that each B contains a function φ with $\int \rho_\epsilon(x)\varphi(x)\,dx \geqq 0$ for all $\epsilon > 0$ but $\varphi(0) = -1$.)

4. Show that the bilinear map $(\alpha, T) \mapsto \alpha * T$ from $\mathfrak{D}^s \times \mathcal{E}'^t$ into $\mathfrak{D}^{s-t}$ (cf. Exercise 1(a)) is hypocontinuous if we equip $\mathcal{E}'^t$ with the topology $\beta(\mathcal{E}'^t, \mathcal{E}^t)$.

5. Show that the bilinear map $(\alpha, T) \mapsto \alpha * T$ from $\mathfrak{D}^s \times \mathfrak{D}'^t$ into $\mathcal{E}^{s-t}$ is hypocontinuous if we equip $\mathfrak{D}'^t$ with the topology $\beta(\mathfrak{D}'^t, \mathfrak{D}^t)$.

6. Show that the bilinear map $(\alpha, T) \mapsto \alpha * T$ from $\mathcal{E}^s \times \mathcal{E}'^t$ into $\mathcal{E}^{s-t}$ is hypocontinuous if we equip $\mathcal{E}'^t$ with the topology $\beta(\mathcal{E}'^t, \mathcal{E}^t)$.

7. Show that the bilinear map $(S, T) \mapsto S * T$ from $\mathcal{E}'^s \times \mathfrak{D}'^t$ into $\mathfrak{D}'^{s+t}$ is hypocontinuous with respect to the equicontinuous subsets of $\mathcal{E}'^s$ and the com-

pact subsets of $\mathfrak{D}'^t$ if these spaces are equipped with the topologies $\kappa(\mathcal{E}'^s, \mathcal{E}^s)$, $\kappa(\mathfrak{D}'^t, \mathfrak{D}^t)$, and $\kappa(\mathfrak{D}'^{s+t}, \mathfrak{D}^{s+t})$. (*Hint:* Apply Proposition 8.8 and use the proof of Proposition 3.)

8. Show that if f and g belong to $\mathcal{C}(\Omega)$ and $\partial_j T_f = T_g$ for some j $(1 \leqq j \leqq n)$, then $\partial_j f$ exists and is equal to g. (*Hint:* First reduce to the case when f has compact support. Then

$$\partial_j(\rho_\epsilon * f) = \partial_j(\rho_\epsilon * T_f) = \rho_\epsilon * T_g = \rho_\epsilon * g.$$

Finally, $\rho_\epsilon * f$ tends to f and $\rho_\epsilon * g$ to g uniformly as $\epsilon \to 0$.)

§11. *Fourier transform*

We conclude this chapter with a concept which is of basic importance in the applications of distributions to the theory of partial differential equations, in the theory of probability and many other branches of analysis.

For our purposes it will be useful to introduce two copies of the n-dimensional Euclidean space $\mathbf{R}^n$ and to denote them by X and Ξ. Vectors of X will be denoted by italic letters $x = (x_1, \ldots, x_n), y, \ldots$, vectors of Ξ by Greek ones $\xi = (\xi_1, \ldots, \xi_n), \eta, \ldots$ The spaces X and Ξ will be considered as forming a dual system with respect to the bilinear form

$$(x, \xi) \mapsto \langle x, \xi\rangle = x_1\xi_1 + \cdots + x_n\xi_n.$$

DEFINITION 1. *The Fourier transform $\mathfrak{F}\varphi$ or $\hat{\varphi}$ of a function $\varphi \in \mathcal{S}(X)$ is the function defined by*

$$\hat{\varphi}(\xi) = \int_X \varphi(x)e^{-2\pi i\langle x,\xi\rangle}\,dx.$$

We shall prove that $\hat{\varphi}$ exists, that it belongs to $\mathcal{S}(\Xi)$, and that the map $\mathfrak{F}\colon \varphi \mapsto \hat{\varphi}$ is an isomorphism from $\mathcal{S}(X)$ onto $\mathcal{S}(\Xi)$ whose inverse $\overline{\mathfrak{F}}\colon \hat{\varphi} \mapsto \varphi$ is given by

$$\varphi(x) = \int_\Xi \hat{\varphi}(\xi)e^{2\pi i\langle x,\xi\rangle}\,d\xi.$$

In the first place, repeated derivation under the integral sign yields for any $p \in \mathbf{N}^n$ the relation

$$D^p\hat{\varphi}(\xi) = (-1)^{|p|}\int_X x^p\varphi(x)e^{-2\pi i\langle x,\xi\rangle}\,dx. \tag{1}$$

Since for $|x_j| \geqq 1$ we have

$$\begin{aligned} |x^p\varphi(x)| &= \left|\frac{x^p}{(1+|x|^2)^{|p|+n}}(1+|x|^2)^{|p|+n}\varphi(x)\right| \\ &\leqq \frac{|x|^{|p|}}{(1+|x|^2)^{|p|+n}}\left|\max_{x\in X}(1+|x|^2)^{|p|+n}\varphi(x)\right| \end{aligned} \tag{2}$$

the integrand in (1) is absolutely integrable, and therefore $\hat{\varphi}$ exists and is infinitely differentiable ([2], Theorem 14–24, p. 443). On the other hand, φ and all its derivatives tend to zero as $|x| \to \infty$, and therefore repeated integration by parts yields for any $q \in \mathbf{N}^n$ the relation

$$\xi^q \hat{\varphi}(\xi) = \int_X D^q \varphi(x) e^{-2\pi i \langle x, \xi \rangle}\, dx. \tag{3}$$

Combining (1) and (3), we have

$$\xi^q D^p \hat{\varphi}(\xi) = (-1)^{|p|} \int_X D^q (x^p \varphi(x)) e^{-2\pi i \langle x, \xi \rangle}\, dx \tag{4}$$

for every p and q in $\mathbf{N}^n$. Since by the Leibniz formula $D^q(x^p\varphi)$ is a linear combination of terms $x^r \cdot D^s\varphi$, it follows from (2), (4), and Proposition 2.5.2 that $\hat{\varphi} \in \mathcal{S}(\Xi)$ and that the map $\varphi \mapsto \hat{\varphi}$ from $\mathcal{S}(X)$ into $\mathcal{S}(\Xi)$ is continuous (cf. Example 2.5.2).

Proposition 1. *For $m \in \mathbf{N}$ and $k \in \mathbf{Z}$ the space $\mathcal{S}_k^m$* (Example 2.4.11) *is a normal space of distributions* (Definition 2.3).

Proof. The injection $\mathfrak{D} \hookrightarrow \mathcal{S}_k^m$ is clearly continuous (cf. Example 2.12.8). On the other hand, the injection $\mathcal{S}_k^m \hookrightarrow \mathcal{E}^m$ is continuous since if K is a compact subset of $\mathbf{R}^n$ and $p \in \mathbf{N}^n$, $|p| \leqq m$, then for every $\varphi \in \mathcal{S}_k^m$ we have

$$\max_{x \in K} |\partial^p \varphi(x)| \leqq \max_{x \in K} (1 + |x|^2)^{-k} \cdot \max_{x \in \mathbf{R}^n} (1 + |x|^2)^k |\partial^p \varphi(x)|.$$

Since the injection $\mathcal{E}^m \hookrightarrow \mathfrak{D}'$ is also continuous (Proposition 4.3), we see that $\mathcal{S}_k^m \hookrightarrow \mathfrak{D}'$ is continuous.

Let $\zeta \in \mathfrak{D}$ be such that $\zeta(x) = 1$ for $|x| \leqq 1$, $\zeta(x) = 0$ for $|x| \geqq 2$, and $0 \leqq \zeta(x) \leqq 1$ for all $x \in \mathbf{R}^n$. Denote by ζ_j the function $x \mapsto \zeta(x/j)$ and furthermore write

$$C = \max_{|p| \leqq m} \max_x |\partial^p \zeta(x)| + 1.$$

If $\varphi \in \mathcal{S}_k^m$, then $\zeta_j \varphi \in \mathfrak{D}^m$ and

$$\begin{aligned} &\max_x (1 + |x|^2)^k |\partial^p \{\varphi(x) - \zeta_j(x)\varphi(x)\}| \\ &\qquad = \max_x (1 + |x|^2)^k \left| \sum_q \binom{p}{q} \partial^{p-q}(1 - \zeta(x/j)) \cdot \partial^q \varphi(x) \right| \\ &\qquad \leqq C \sum_q \binom{p}{q} \max_{|x| \geqq j} (1 + |x|^2)^k |\partial^q \varphi(x)|. \end{aligned}$$

By the definition of the space $\mathcal{S}_k^m$ this last expression is arbitrarily small if we choose j sufficiently large, and thus $\mathfrak{D}^m$ is dense in $\mathcal{S}_k^m$. Since $\mathfrak{D}$ is dense in $\mathfrak{D}^m$ (Proposition 4.1) and the map $\mathfrak{D}^m \hookrightarrow \mathcal{S}_k^m$ is clearly continuous, $\mathfrak{D}$ is also dense in $\mathcal{S}_k^m$. ∎

PROPOSITION 2. *Let $m \in \mathbf{N}$ and $k \in \mathbf{Z}$. The dual $\mathcal{S}_k'^m$ of $\mathcal{S}_k^m$ can be identified with the space of all distributions of the form*

$$T = (1 + |x|^2)^k \sum_{|p| \leq m} \partial^p \mu_p,$$

where each μ_p is a measure which belongs to $\mathfrak{M}^1$ (§5).

Proof. If L is a continuous linear form on $\mathcal{S}_k^m$, then its restriction T to $\mathfrak{D}$ is a distribution since $\mathfrak{D} \hookrightarrow \mathcal{S}_k^m$ is continuous. Furthermore, the map $L \mapsto T$ is injective since $\mathfrak{D}$ is dense in $\mathcal{S}_k^m$.

Denote by u the isomorphism $\varphi \mapsto (1 + |x|^2)^k \varphi$ from $\mathcal{S}_k^m$ onto $\mathcal{B}_0^m$ (Example 2.5.8), and by $\check{u}\colon \mathcal{S}_k'^m \to \mathcal{B}_0'^m$ the contragredient isomorphism. For any $\varphi \in \mathcal{S}_k^m$ we have

$$\langle L, \varphi \rangle = \langle \check{u}(L), u(\varphi) \rangle = \langle \check{u}(L), (1 + |x|^2)^k \varphi \rangle.$$

By Proposition 5.3 there exists a family $(\mu_p)_{|p| \leq m}$ of measures belonging to $\mathfrak{M}^1$ such that $\check{u}(L) = \sum_{|p| \leq m} \partial^p \mu_p$. For $\varphi \in \mathfrak{D}$ we have

$$\langle L, \varphi \rangle = \left\langle (1 + |x|^2)^k \sum_{|p| \leq m} \partial^p \mu_p, \varphi \right\rangle;$$

i.e.,

$$T = (1 + |x|^2)^k \sum_{|p| \leq m} \partial^p \mu_p.$$

Conversely, if $(\mu_p)_{|p| \leq m}$ is a family of measures belonging to $\mathfrak{M}^1$, then

$$\varphi \mapsto \left\langle \sum_{|p| \leq m} \partial^p \mu_p, (1 + |x|^2)^k \varphi \right\rangle$$

is a continuous linear form on $\mathcal{S}_k^m$ whose restriction to $\mathfrak{D}$ is the distribution $(1 + |x|^2)^k \sum_{|p| \leq m} \partial^p \mu_p$. ∎

PROPOSITION 3. *The space $\mathcal{S}$ is a normal space of distributions whose dual $\mathcal{S}'$ can be identified with the space of all distributions of the form*

$$(1 + |x|^2)^k \sum \partial^p \mu_p,$$

where $k \in \mathbf{N}$ and (μ_p) is a finite family of measures belonging to $\mathfrak{M}^1$.

Proof. If we take into account Example 2.11.6, then the continuity of the maps $\mathfrak{D} \hookrightarrow \mathcal{S} \hookrightarrow \mathfrak{D}'$ follows from the continuity of the maps

$$\mathfrak{D} \hookrightarrow \mathcal{S}_k^m \hookrightarrow \mathfrak{D}'$$

for all $m \in \mathbf{N}$ and $k \in \mathbf{Z}$. Let $\varphi \in \mathcal{S}$ and V a neighborhood of 0 in $\mathcal{S}$. We may assume that V is the inverse image of a neighborhood U of 0 in some space $\mathcal{S}_k^m$. By Proposition 1 there exists $\psi \in \mathfrak{D}$ such that $\psi \in \varphi + U$, and therefore $\psi \in \varphi + V$; i.e., $\mathfrak{D}$ is dense in $\mathcal{S}$.

The assertion concerning $\mathcal{S}'$ follows from Example 2.11.6, the Corollary of Proposition 3.14.5, and Proposition 2. ∎

The elements of $\mathcal{S}'$ are called *temperate* (or slowly increasing) distributions. If f is an absolutely integrable continuous function on $\mathbf{R}^n$, then $T_f \in \mathfrak{M}^1$ since

$$|\langle T_f, \varphi\rangle| = \left|\int f(x)\varphi(x)\,dx\right| \leqq \max_x |\varphi(x)| \int |f(x)|\,dx.$$

By Proposition 3 the measure T_f belongs to $\mathcal{S}'$, and in particular $\varphi \mapsto T_\varphi$ is a map from $\mathcal{S}$ into $\mathcal{S}'$. We can prove exactly as in Example 1.1 that this map is injective, and therefore we shall usually consider $\mathcal{S}$ as a subspace of $\mathcal{S}'$. Since the space $\mathcal{S}$ is reflexive (Example 3.9.5 or 3.15.4), the proof of Proposition 1.3 shows that $\mathcal{S}$ is dense in $\mathcal{S}'$.

We want to define the Fourier transform as a map from $\mathcal{S}'(X)$ into $\mathcal{S}'(\Xi)$. To motivate our definition, consider a function f belonging to $\mathcal{S}(X)$ and a function φ belonging to $\mathcal{S}(\Xi)$. Then we have

$$\langle \mathfrak{F}f, \varphi\rangle = \langle f, \mathfrak{F}\varphi\rangle \tag{5}$$

since both sides are equal to

$$\int_X \int_\Xi f(x)\varphi(x)e^{-2\pi i\langle x,\xi\rangle}\,dx\,d\xi.$$

This leads us to

DEFINITION 2. *The Fourier transform $\mathfrak{F}T$ or $\hat{T}$ of a distribution $T \in \mathcal{S}'(X)$ is the image of T under the transpose of the map $\mathfrak{F}\colon \mathcal{S}(\Xi) \to \mathcal{S}(X)$; i.e.,*

$$\langle \mathfrak{F}T, \varphi\rangle = \langle T, \mathfrak{F}\varphi\rangle$$

for all $\varphi \in \mathcal{S}(\Xi)$.

It is of course an abuse of notation to denote the transpose ${}^t\mathfrak{F}$ of $\mathfrak{F}$ by the same letter $\mathfrak{F}$. It is, however, analogous to the abuse committed with the symbol ∂^p, and it is very convenient, since by (5) the operation ${}^t\mathfrak{F}$ defined on $\mathcal{S}'(X)$ coincides with $\mathfrak{F}$ on $\mathcal{S}(X)$. It follows from the continuity of $\mathfrak{F}\colon \mathcal{S}(\Xi) \to \mathcal{S}(X)$ and the corollary of Proposition 3.12.3 that the map $\mathfrak{F}\colon \mathcal{S}'(X) \to \mathcal{S}'(\Xi)$ is continuous. Since $\mathcal{S}$ is dense in $\mathcal{S}'$, the map $\mathfrak{F}\colon \mathcal{S}'(X) \to \mathcal{S}'(\Xi)$ is the unique continuous extension of $\mathfrak{F}\colon \mathcal{S}(X) \to \mathcal{S}(\Xi)$.

EXAMPLE 1. $\mathfrak{F}(\delta) = \mathbf{1}$ since for $\varphi \in \mathcal{S}(\Xi)$ we have

$$\langle \mathfrak{F}(\delta), \varphi\rangle = \langle \delta, \mathfrak{F}\varphi\rangle = \hat{\varphi}(0) = \int_\Xi \varphi(\xi)\,d\xi = \langle \mathbf{1}, \varphi\rangle.$$

EXAMPLE 2. It follows from the identity

$$\max_{x\in\mathbf{R}^n} |x^r D^q(D^p\varphi)(x)| = \max_{x\in\mathbf{R}^n} |x^r D^{p+q}\varphi(x)|$$

that the linear map $\varphi \mapsto D^p\varphi$ is continuous from $\mathcal{S}$ into $\mathcal{S}$, and therefore the map $T \mapsto D^pT$ is continuous from $\mathcal{S}'$ into $\mathcal{S}'$. It follows similarly from Leibniz's formula that $\varphi \mapsto x^p\varphi$ is a continuous map from $\mathcal{S}$ into $\mathcal{S}$ and that $T \mapsto x^pT$ is a continuous map from $\mathcal{S}'$ into $\mathcal{S}'$.

For $T \in \mathcal{S}'(X)$ and $p \in \mathbf{N}^n$ we have

$$\mathfrak{F}(D^pT) = \xi^p\mathfrak{F}(T). \tag{6}$$

Indeed, for $\varphi \in \mathcal{S}(\Xi)$ we have

$$\langle \mathfrak{F}(D^pT), \varphi\rangle = \langle D^pT, \mathfrak{F}\varphi\rangle = (-1)^{|p|}\langle T, D^p\mathfrak{F}\varphi\rangle.$$

Now (1) can be rewritten as

$$D^p\mathfrak{F}\varphi = (-1)^{|p|}\mathfrak{F}(\xi^p\varphi);$$

hence

$$\langle \mathfrak{F}(D^pT), \varphi\rangle = \langle T, \mathfrak{F}(\xi^p\varphi)\rangle = \langle \xi^p\mathfrak{F}(T), \varphi\rangle,$$

from which the desired conclusion follows.

If $T = P(D)\delta = \sum \alpha_p D^p\delta$ is a partial differential operator with constant coefficients (§9), then by (6) we have $\mathfrak{F}(T)(\xi) = \sum \alpha_p \xi^p = P(\xi)$. This is the relation between the operator $P(D)$ and the polynomial function $\xi \mapsto P(\xi)$ alluded to in §9 after formula (6).

EXAMPLE 3. For $T \in \mathcal{S}'(X)$ and $p \in \mathbf{N}^n$ we have

$$\mathfrak{F}\big((-x)^pT\big) = D^p\mathfrak{F}(T).$$

Indeed, for $\varphi \in \mathcal{S}(\Xi)$ we have

$$\langle \mathfrak{F}\big((-x)^pT\big), \varphi\rangle = \langle T, (-x)^p\mathfrak{F}\varphi\rangle.$$

Now (3) can be rewritten as

$$(-x)^p\mathfrak{F}\varphi = (-1)^{|p|}\mathfrak{F}(D^p\varphi).$$

Hence

$$\langle \mathfrak{F}\big((-x)^pT\big), \varphi\rangle = (-1)^{|p|}\langle T, \mathfrak{F}(D^p\varphi)\rangle = \langle D^p\mathfrak{F}(T), \varphi\rangle.$$

EXAMPLE 4. For $T \in \mathcal{S}'(X)$ and $h \in \mathbf{R}^n$ we have

$$\mathfrak{F}(\tau_hT) = e^{-2\pi i\langle h,\xi\rangle}\mathfrak{F}(T)$$

and

$$\mathfrak{F}(e^{2\pi i\langle h,x\rangle}T) = \tau_h\mathfrak{F}(T).$$

For $\varphi \in \mathcal{S}(\Xi)$ we have

$$\begin{aligned}\mathfrak{F}(\tau_h\varphi)(x) &= \int_\Xi \varphi(\xi - h)e^{-2\pi i\langle x,\xi\rangle}\,d\xi \\ &= \int_\Xi \varphi(\xi)e^{-2\pi i\langle x,\xi+h\rangle}\,d\xi = e^{-2\pi i\langle x,h\rangle}(\mathfrak{F}\varphi)(x)\end{aligned}$$

and
$$\{\tau_{-h}\mathfrak{F}(\varphi)\}(x) = \int_{\Xi} \varphi(\xi)e^{-2\pi i\langle x+h,\xi\rangle}\,d\xi = \mathfrak{F}\big(e^{-2\pi i\langle h,\xi\rangle}\varphi(\xi)\big)(x).$$

Hence
$$\langle \mathfrak{F}(\tau_h T), \varphi\rangle = \langle T, \tau_{-h}\mathfrak{F}(\varphi)\rangle = \langle T, \mathfrak{F}(e^{-2\pi i\langle h,\xi\rangle}\varphi)\rangle = \langle e^{-2\pi i\langle h,\xi\rangle}\mathfrak{F}(T), \varphi\rangle,$$

which proves the first formula, and
$$\langle \mathfrak{F}(e^{2\pi i\langle h,x\rangle}T), \varphi\rangle = \langle T, e^{2\pi i\langle h,x\rangle}\mathfrak{F}(\varphi)\rangle = \langle T, \mathfrak{F}(\tau_{-h}\varphi)\rangle = \langle \tau_h\mathfrak{F}(T), \varphi\rangle,$$

which proves the second formula.

In the next proposition we shall consider h Euclidean spaces $\mathbf{R}^{n_j}$ $(1 \leqq j \leqq h)$, and of each $\mathbf{R}^{n_j}$ we shall consider two copies X_j and Ξ_j which will form a dual system as indicated at the beginning of this section. Then $\prod_{j=1}^h X_j$ and $\prod_{j=1}^h \Xi_j$ are both isomorphic to $\mathbf{R}^n$ $(n = n_1 + \cdots + n_h)$, and they form a dual system with respect to the bilinear form
$$(x, \xi) \mapsto \langle x, \xi\rangle = \sum_{j=1}^h \langle x_j, \xi_j\rangle$$
$(x_j \in X_j,\ \xi_j \in \Xi_j)$.

PROPOSITION 4. *If* $T_j \in \mathcal{S}'(X_j)$ *for* $1 \leqq j \leqq h$, *then*
$$\bigotimes_{j=1}^h T_j \in \mathcal{S}'\Big(\prod_{j=1}^h X_j\Big)$$
and
$$\mathfrak{F}\Big(\bigotimes_{j=1}^h T_j\Big) = \bigotimes_{j=1}^h \mathfrak{F}(T_j). \tag{7}$$

Proof. Consider a family $(\varphi_j)_{1\leqq j\leqq h}$ of functions such that $\varphi_j \in \mathcal{S}(X_j)$. If x_j denotes a variable vector in X_j and ∂^{p_j} a partial derivation with respect to the components of x_j, then we have
$$\partial^p(\varphi_1 \otimes \cdots \otimes \varphi_h) = \partial^{p_1}\varphi_1 \otimes \cdots \otimes \partial^{p_h}\varphi_h$$
and
$$(1 + |x|^2)^k \left|\partial^p \bigotimes_{j=1}^h \varphi_j(x)\right| \leqq \prod_{j=1}^h (1 + |x_j|^2)^k|\partial^p\, \varphi_j(x)|.$$

Hence
$$\bigotimes_{j=1}^h \varphi_j \in \mathcal{S}\Big(\prod_{j=1}^h X_j\Big); \qquad \text{i.e.,} \qquad \bigotimes_{j=1}^h \mathcal{S}(X_j) \subset \mathcal{S}\Big(\prod_{j=1}^h X_j\Big).$$

Also $\bigotimes_{j=1}^{h} \mathcal{S}(X_j)$ is dense in $\mathcal{S}(\prod_{j=1}^h X_j)$ since $\bigotimes_{j=1}^{h} \mathcal{D}(X_j)$ is dense in $\mathcal{D}(\prod_{j=1}^h X_j)$ (Proposition 8.1), and $\mathcal{D}(\prod_{j=1}^h X_j)$ is dense in $\mathcal{S}(\prod_{j=1}^h X_j)$ (Proposition 3).

For $\varphi_j \in \mathcal{S}(X_j)$ we have

$$\mathcal{F}\left(\bigotimes_{j=1}^{h} \varphi_j\right)(\xi) = \int_{\Pi X_j} \left(\bigotimes_{j=1}^{h} \varphi_j\right)(x) e^{-2\pi i\langle x, \xi\rangle}\, dx$$

$$= \prod_{j=1}^{h} \int_{X_j} \varphi_j(x_j) e^{-2\pi i \langle x_j, \xi_j\rangle}\, dx_j = \left(\bigotimes_{j=1}^{h} \hat{\varphi}_j\right)(\xi).$$

The identity

$$\left\langle \bigotimes_{j=1}^{h} T_j, \bigotimes_{j=1}^{h} \varphi_j \right\rangle = \prod_{j=1}^{h} \langle T_j, \varphi_j\rangle$$

shows that

$$\bigotimes_{j=1}^{h} T_j \in \mathcal{S}'\left(\prod_{j=1}^{h} X_j\right).$$

Finally, we have

$$\left\langle \mathcal{F}\left(\bigotimes_{j=1}^{h} T_j\right), \bigotimes_{j=1}^{h} \varphi_j \right\rangle = \left\langle \bigotimes_{j=1}^{h} T_j, \mathcal{F}\left(\bigotimes_{j=1}^{h} \varphi_j\right)\right\rangle$$

$$= \left\langle \bigotimes_{j=1}^{h} T_j, \bigotimes_{j=1}^{h} \mathcal{F}(\varphi_j)\right\rangle = \prod_{j=1}^{h} \langle T_j, \mathcal{F}(\varphi_j)\rangle$$

$$= \prod_{j=1}^{h} \langle \mathcal{F}(T_j), \varphi_j\rangle = \left\langle \bigotimes_{j=1}^{h} \mathcal{F}(T_j), \bigotimes_{j=1}^{h} \varphi_j \right\rangle,$$

from which (7) follows. ∎

EXAMPLE 5. We have

$$\mathcal{F}(e^{-\pi|x|^2}) = e^{-\pi|\xi|^2}; \tag{8}$$

i.e., the function $x \mapsto e^{-\pi|x|^2}$ is its own Fourier transform (cf. Exercise 2). Since for $x = (x_1, \ldots, x_n) \in X = \mathbf{R}^n$ we have

$$e^{-\pi|x|^2} = \prod_{j=1}^{n} e^{-\pi x_j^2},$$

by virtue of Proposition 4 it is sufficient to prove (8) in the case $n = 1$.

Writing $\varphi(x) = e^{-\pi|x|^2}$, we have

$$\hat{\varphi}(\xi) = \int_{-\infty}^{\infty} e^{-\pi x^2 - 2\pi i x\xi}\, dx = e^{-\pi\xi^2} \int_{-\infty}^{\infty} e^{-\pi(x+i\xi)^2}\, dx,$$

and it is sufficient to show that the last integral is equal to 1. For $\xi = 0$ this integral is the so-called Gauss integral, and the following calculation shows that its value is indeed 1:

$$\left(\int_{-\infty}^{\infty} e^{-\pi x^2}\,dx\right)^2 = \left(\int_{-\infty}^{\infty} e^{-\pi x^2}\,dx\right)\left(\int_{-\infty}^{\infty} e^{-\pi y^2}\,dy\right)$$
$$= \int_{-\infty}^{\infty}\int_{-\infty}^{\infty} e^{-\pi(x^2+y^2)}\,dx\,dy = \int_0^{\infty}\int_0^{2\pi} e^{-\pi r^2} r\,dr\,d\theta$$
$$= 2\pi\left[-\frac{1}{2\pi}e^{-\pi r^2}\right]_0^{\infty} = 1,$$

where we have set $r^2 = x^2 + y^2$.

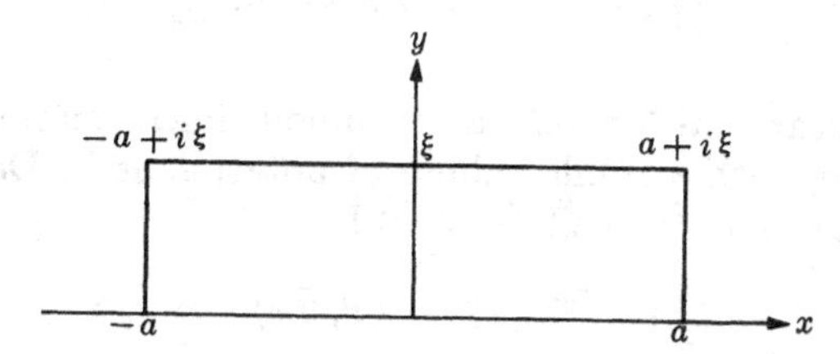

Now $z \mapsto e^{-\pi z^2}$ is an entire function of the complex variable $z = x + iy$, and therefore its integral around the rectangle with vertices $-a$, a, $a + i\xi$, $-a + i\xi$ is zero ($a > 0$). Now the integrals on the two vertical sides tend to zero as $a \to \infty$ since for instance

$$\left|\int_a^{a+i\xi} e^{-\pi z^2}\,dz\right| = \left|i\int_0^{\xi} e^{-\pi(a^2-y^2)}e^{-2\pi i a y}\,dy\right|$$
$$\leqq |\xi|e^{-\pi(a^2-\xi^2)}.$$

We have therefore

$$\int_{-\infty}^{\infty} e^{-\pi(x+i\xi)^2}\,dx = \lim_{a\to\infty}\int_{-a+i\xi}^{a+i\xi} e^{-\pi z^2}\,dz$$
$$= \lim_{a\to\infty}\int_{-a}^{a} e^{-\pi z^2}\,dz = \int_{-\infty}^{\infty} e^{-\pi x^2}\,dx = 1.$$

EXAMPLE 6. For $h \in \mathbf{R}^n$ we have

$$\mathcal{F}(e^{2\pi i\langle h,x\rangle}) = \tau_h\delta = \delta_h.$$

In the first place, by virtue of Example 4 it is sufficient to consider $h = 0$, i.e., to prove

$$\mathcal{F}(\mathbf{1}) = \delta.$$

Next $\mathbf{1} = \mathbf{1}_{x_1} \otimes \cdots \otimes \mathbf{1}_{x_n}$ and $\delta = \delta_{x_1} \otimes \cdots \otimes \delta_{x_n}$ (cf. Example 8.2), and therefore by Proposition 4 it is sufficient to consider $n = 1$.

Set $\mathfrak{F}(\mathbf{1}) = T$. By Example 2 we have $\xi T = \mathfrak{F}(D\mathbf{1}) = 0$, and therefore by Proposition 6.6 the distribution T is a constant multiple of the Dirac measure; i.e., $T = \lambda\delta$. To determine the factor λ observe that we have

$$\langle \mathfrak{F}(\mathbf{1}), \varphi \rangle = \lambda \langle \delta, \varphi \rangle = \langle \mathbf{1}, \mathfrak{F}\varphi \rangle$$

for every $\varphi \in \mathcal{S}(\mathbf{R})$. Choosing $\varphi = e^{-\pi\xi^2}$ we obtain by Example 5

$$\lambda = \int_{-\infty}^{\infty} e^{-\pi x^2}\, dx = 1.$$

Together with the Fourier transform $\mathfrak{F}$ we also must consider the *conjugate Fourier transform* $\overline{\mathfrak{F}}$, which for $\varphi \in \mathcal{S}(X)$ is defined by

$$(\overline{\mathfrak{F}}\varphi)(\xi) = \int_X \varphi(x) e^{2\pi i \langle x, \xi \rangle}\, dx. \tag{9}$$

It is perfectly clear that $\overline{\mathfrak{F}}$ is also a continuous linear map from $\mathcal{S}(X)$ into $\mathcal{S}(\Xi)$. We define, with a similar abuse of notation as in Definition 2, the continuous linear map $\overline{\mathfrak{F}}\colon \mathcal{S}'(X) \to \mathcal{S}'(\Xi)$ by

$$\langle \overline{\mathfrak{F}}T, \varphi \rangle = \langle T, \overline{\mathfrak{F}}\varphi \rangle$$

for all $T \in \mathcal{S}'(X)$ and $\varphi \in \mathcal{S}(\Xi)$. This map $\overline{\mathfrak{F}}$ coincides on $\mathcal{S}(X)$ with the map defined by (9) and is its unique continuous extension to $\mathcal{S}'(X)$. The formulas

$$\begin{aligned} \overline{\mathfrak{F}}(\delta) &= \mathbf{1}, & \overline{\mathfrak{F}}(\mathbf{1}) &= \delta, \\ \overline{\mathfrak{F}}(D^p T) &= (-\xi)^p \overline{\mathfrak{F}}(T), & \overline{\mathfrak{F}}(x^p T) &= D^p \overline{\mathfrak{F}}(T), \\ \overline{\mathfrak{F}}(\tau_h T) &= e^{2\pi i \langle h, \xi \rangle} \overline{\mathfrak{F}}(T), & \overline{\mathfrak{F}}(e^{-2\pi i \langle h, x \rangle} T) &= \tau_h \overline{\mathfrak{F}}(T), \\ \overline{\mathfrak{F}}(e^{-\pi |x|^2}) &= e^{-\pi |\xi|^2}, & \overline{\mathfrak{F}}(e^{-2\pi i \langle h, x \rangle}) &= \delta_h \end{aligned}$$

can be established exactly like the corresponding formulas for the transform $\mathfrak{F}$.

We now arrive at one of the main results of the theory.

THEOREM 1. *The Fourier transform $\mathfrak{F}$ is an isomorphism from $\mathcal{S}(X)$ onto $\mathcal{S}(\Xi)$ whose inverse is given by $\overline{\mathfrak{F}}$. Similarly, $\mathfrak{F}$ is an isomorphism from $\mathcal{S}'(X)$ onto $\mathcal{S}'(\Xi)$ whose inverse is given by $\overline{\mathfrak{F}}$.*

Proof. We know that $\mathfrak{F}\colon \mathcal{S}(X) \to \mathcal{S}(\Xi)$ is a continuous linear map. If $\varphi \in \mathcal{S}(X)$, then for any $h \in X$ we have by Example 6

$$\begin{aligned} \varphi(h) &= \langle \delta_h, \varphi \rangle = \langle \mathfrak{F}(e^{2\pi i \langle h, \xi \rangle}), \varphi \rangle \\ &= \langle e^{2\pi i \langle h, \xi \rangle}, \mathfrak{F}\varphi \rangle \end{aligned}$$

or, changing notations,

$$\varphi(x) = \int_\Xi \hat{\varphi}(\xi) e^{2\pi i \langle x, \xi \rangle}\, d\xi;$$

i.e., $\bar{\mathcal{F}}\mathcal{F}\varphi = \varphi$. We see exactly in the same way that for every $\psi \in \mathcal{S}(\Xi)$ we have $\mathcal{F}\bar{\mathcal{F}}\psi = \psi$. It follows that $\mathcal{F}$ is injective (since $\mathcal{F}\varphi = 0$ implies $\varphi = \bar{\mathcal{F}}\mathcal{F}\varphi = 0$), surjective (since given $\psi \in \mathcal{S}(\Xi)$, we have $\mathcal{F}(\bar{\mathcal{F}}\psi) = \psi$), and that $\bar{\mathcal{F}}$ is the inverse of $\mathcal{F}$. Since $\bar{\mathcal{F}}: \mathcal{S}(\Xi) \to \mathcal{S}(X)$ is a continuous linear map, $\mathcal{F}$ is an isomorphism.

For $T \in \mathcal{S}'(X)$ and $S \in \mathcal{S}'(\Xi)$ we have $\bar{\mathcal{F}}\mathcal{F}T = T$ and $\mathcal{F}\bar{\mathcal{F}}S = S$ since

$$\langle \bar{\mathcal{F}}\mathcal{F}T, \varphi \rangle = \langle \mathcal{F}T, \bar{\mathcal{F}}\varphi \rangle = \langle T, \mathcal{F}\bar{\mathcal{F}}\varphi \rangle = \langle T, \varphi \rangle$$

and

$$\langle \mathcal{F}\bar{\mathcal{F}}S, \psi \rangle = \langle S, \bar{\mathcal{F}}\mathcal{F}\psi \rangle = \langle S, \psi \rangle$$

for any $\varphi \in \mathcal{S}(X)$ and $\psi \in \mathcal{S}(\Xi)$. It follows, similarly as above, that the continuous linear map $\mathcal{F}: \mathcal{S}'(X) \to \mathcal{S}'(\Xi)$ is bijective, its inverse is the continuous linear map $\bar{\mathcal{F}}: \mathcal{S}'(\Xi) \to \mathcal{S}'(X)$, and $\mathcal{F}$ is an isomorphism. ∎

Next let us investigate the multiplication of a temperate distribution by an infinitely differentiable function. We first prove

PROPOSITION 5. *If* $\alpha \in \mathcal{O}_M$ (Example 2.4.15) *and* $\varphi \in \mathcal{S}$, *then* $\alpha\varphi \in \mathcal{S}$. *Conversely, if* $\alpha \in \mathcal{E}$ *is such that* $\alpha\varphi \in \mathcal{S}$ *for all* $\varphi \in \mathcal{S}$, *then* $\alpha \in \mathcal{O}_M$. *We have* $\mathcal{O}_M = \bigcap_{m \in \mathbf{N}} \mathcal{O}_C^m$ (Example 2.12.9). *Finally, for each* $\alpha \in \mathcal{O}_M$ *the map* $\varphi \mapsto \alpha\varphi$ *from* $\mathcal{S}$ *into* $\mathcal{S}$ *is continuous.*

Proof. (a) If $\varphi \in \mathcal{S}$, then $(1 + |x|^2)^k \, \partial^p \varphi \in \mathcal{S}$ for every $k \in \mathbf{N}$ and $p \in \mathbf{N}^n$. Now $(1 + |x|^2)^{k+1} \, \partial^p(\alpha\varphi)$ is a linear combination of functions of the form $(1 + |x|^2)^{k+1} \, \partial^q\varphi \cdot \partial^r\alpha$. If $\alpha \in \mathcal{O}_M$, then by the definition of $\mathcal{O}_M$ each such function is bounded. Thus

$$(1 + |x|^2)^k |\partial^p(\alpha\varphi)(x)| \leqq \frac{C}{1 + |x|^2},$$

which shows that $\alpha\varphi \in \mathcal{S}$.

(b) If $\alpha\varphi \in \mathcal{S}$ for every $\varphi \in \mathcal{S}$, then it follows from the formula

$$\partial_j\alpha \cdot \varphi = \partial_j(\alpha\varphi) - \alpha \cdot \partial_j\varphi$$

that $\partial_j\alpha \cdot \varphi \in \mathcal{S}$ for every $\varphi \in \mathcal{S}$. We see by induction that $\partial^p\alpha \cdot \varphi \in \mathcal{S}$ for every $\varphi \in \mathcal{S}$ and in particular $\alpha \in \mathcal{O}_M$.

(c) A function α belongs to $\bigcap_{m \in \mathbf{N}} \mathcal{O}_C^m$ if and only if it is infinitely differentiable, and for every $m \in \mathbf{N}$ there exists an integer $k(m)$ such that for $|p| \leqq m$ the function

$$x \mapsto (1 + |x|^2)^{k(m)} |\partial^p\alpha(x)|$$

tends to zero as $|x| \to \infty$. We may obviously assume that

$$0 \geqq k(m) \geqq k(m+1) \to -\infty.$$

Suppose that $\alpha \in \bigcap_{m \in \mathbf{N}} \mathcal{O}_C^m$ and let $\varphi \in \mathcal{S}$, $|p| = m$. Then

$$\varphi(x)\, \partial^p \alpha(x) = (1 + |x|^2)^{-k(m)} \varphi(x) \cdot (1 + |x|^2)^{k(m)}\, \partial^p \alpha(x),$$

which shows that $\alpha \in \mathcal{O}_M$.

(d) Consider a function $\alpha \in \mathcal{O}_M$ and suppose that it does not belong to $\bigcap_{m \in \mathbf{N}} \mathcal{O}_C^m$. Then for some $p \in \mathbf{N}^n$ there exists a sequence of points $x_j \in \mathbf{R}^n$ with $|x_j| + 2 \leqq |x_{j+1}|$ such that

$$|(1 + |x_j|^2)^{-j}\, \partial^p \alpha(x_j)| \geqq 1$$

for every $j \in \mathbf{N}$. Let $\rho \in \mathfrak{D}$ have the properties (i) through (iv) listed in Chapter 2, §12 and consider the function φ defined by

$$\varphi(x) = \sum_{j \in \mathbf{N}} \frac{\rho(x - x_j)}{(1 + |x_j|^2)^j}.$$

Since the supports of the terms are disjoint, the function is infinitely differentiable. Furthermore, if $k \in \mathbf{N}$ and $p \in \mathbf{N}^n$, then for

$$|x_j| - 1 \leqq |x| \leqq |x_j| + 1$$

we have

$$(1 + |x|^2)^k\, \partial^p \varphi(x) = \frac{(1 + |x|^2)^k}{(1 + |x_j|^2)^k} \frac{\partial^p \rho(x - x_j)}{(1 + |x_j|^2)^{j-k}}.$$

Since

$$\frac{1 + |x|^2}{1 + |x_j|^2} \leqq 6 \qquad \text{and} \qquad \max |\partial^p \rho(x - x_j)| = \max |\partial^p \rho(x)|,$$

we see by taking $j > k$ sufficiently large that $\varphi \in \mathcal{S}$. On the other hand,

$$\varphi(x_j) |\partial^p \alpha(x_j)| = \rho(0) \cdot (1 + |x_j|^2)^{-j} |\partial^p \alpha(x_j)| \geqq \rho(0) > 0$$

for all $j \in \mathbf{N}$, which contradicts the assumption $\alpha \in \mathcal{O}_M$.

(e) Let V be a neighborhood of 0 in $\mathcal{S}$ which we may assume to be defined by the inequalities

$$(1 + |x|^2)^k |\partial^p \psi(x)| \leqq \epsilon \qquad \text{for} \qquad |p| \leqq m,$$

where k, $m \in \mathbf{N}$ and $\epsilon > 0$. If $\alpha \in \mathcal{O}_M$, then by the Leibniz formula we can determine $\eta > 0$ so that, with the notation of part (c), the inequalities

$$(1 + |x|^2)^{k - k(m)} |\partial^p \varphi(x)| \leqq \eta$$

for $|p| \leqq m$ imply $\alpha\varphi \in V$. ▮

THEOREM 2. *If $\alpha \in \mathcal{O}_M$, then $\alpha T \in \mathcal{S}'$ for every $T \in \mathcal{S}'$, and conversely if $\alpha \in \mathcal{E}$ is such that $\alpha T \in \mathcal{S}'$ for every $T \in \mathcal{S}'$, then $\alpha \in \mathcal{O}_M$. For each $\alpha \in \mathcal{O}_M$ the map $T \mapsto \alpha T$ from $\mathcal{S}'$ into $\mathcal{S}'$ is continuous.*

Proof. If $\alpha \in \mathcal{O}_M$ and $T \in \mathcal{S}'$, then by Definition 6.1 we have

$$\langle \alpha T, \varphi \rangle = \langle T, \alpha\varphi \rangle \tag{10}$$

for every $\varphi \in \mathfrak{D}$. By Proposition 5 the map $\varphi \mapsto \alpha\varphi$ from $\mathcal{S}$ into $\mathcal{S}$ is continuous, and since $T \in \mathcal{S}'$, the map $\varphi \mapsto \langle T, \alpha\varphi \rangle$ is also continuous; i.e., αT is a continuous linear form on $\mathfrak{D}$ for the topology induced by $\mathcal{S}$. Thus αT can be extended by continuity to all of $\mathcal{S}$, and the linear form so extended belongs to $\mathcal{S}'$. Relation (10) then holds for all $\varphi \in \mathcal{S}$; i.e., the map $T \mapsto \alpha T$ is the transpose of the map $\varphi \mapsto \alpha\varphi$. It follows from the corollary of Proposition 3.12.3 that $T \mapsto \alpha T$ is continuous.

Let $\alpha \in \mathcal{E}$ be such that $\alpha T \in \mathcal{S}'$ for every $T \in \mathcal{S}'$. Then for any $\varphi \in \mathcal{S}$ the map $T \mapsto \langle \alpha T, \varphi \rangle$ is a continuous linear form on $\mathcal{S}'$; i.e., there exists $\psi \in \mathcal{S}$ such that $\langle \alpha T, \varphi \rangle = \langle T, \psi \rangle$ for all $T \in \mathcal{S}'$ and in particular for all $T \in \mathfrak{D}$. Hence $\alpha\varphi = \psi \in \mathcal{S}$, and thus $\alpha \in \mathcal{O}_M$ by Proposition 5. ∎

To conclude this section and also this chapter, we consider the convolution $S * T$ of a distribution S and of a temperate distribution T. If $S \in \mathcal{E}'$, then $S * T$ is of course well-defined, and it is easy to see that $S * T \in \mathcal{S}'$. We shall see, however, that $S * T$ is meaningful and belongs to $\mathcal{S}'$ even if S belongs to the larger space $\mathcal{O}'_C$, which we must investigate first.

One proves exactly like Proposition 3 the following result:

PROPOSITION 6. *For any $k \in \mathbf{Z}$ the space $\mathcal{S}_k$ (Example 2.4.12) is a normal space of distributions whose dual $\mathcal{S}'_k$ can be identified with the space of all distributions of the form $(1 + |x|^2)^k \sum \partial^p \mu_p$, where (μ_p) is a finite family of measures belonging to $\mathfrak{M}^1$.*

In particular, $\mathcal{S}_0 = \mathcal{B}_0$ is a normal space of distributions. Its dual $\mathcal{B}'_0$ (denoted by $\mathfrak{D}'_{L^1}$ in [82]) consists of all integrable distributions (§5).

COROLLARY. *The space $\mathcal{O}_C$ (Example 2.12.9) is a normal space of distributions whose dual $\mathcal{O}'_C$ can be identified with the space of all distributions T such that for every $k \in \mathbf{Z}$ the distribution $(1 + |x|^2)^k T$ is integrable.*

Proof. The continuity of $\mathfrak{D} \hookrightarrow \mathcal{O}_C$ follows from the continuity of $\mathfrak{D} \hookrightarrow \mathcal{S}_k$ (Proposition 6) and $\mathcal{S}_k \hookrightarrow \mathcal{O}_C$. The continuity of $\mathcal{O}_C \hookrightarrow \mathfrak{D}'$ follows from the continuity of $\mathcal{S}_k \hookrightarrow \mathfrak{D}'$ (Proposition 6) and Proposition 2.12.1.

Let $\varphi \in \mathcal{O}_C$ and V a neighborhood of 0 in $\mathcal{O}_C$. Then $\varphi \in \mathcal{S}_k$ for some k and $V \cap \mathcal{S}_k$ is a neighborhood of 0 in $\mathcal{S}_k$. By Proposition 6 there exists $\psi \in \mathfrak{D}$ such that $\psi \in \varphi + V \cap \mathcal{S}_k$ and *a fortiori* $\psi \in \varphi + V$. Thus $\mathfrak{D}$ is dense in $\mathcal{O}_C$.

Finally, a linear form on $\mathcal{O}_C$ is continuous if and only if its restriction to each $\mathcal{S}_k$ is continuous (Proposition 2.12.1), i.e., if $T = (1 + |x|^2)^k T_k$ with $T_k \in \mathcal{B}_0'$. ∎

The elements of $\mathcal{O}_C'$ are called *rapidly decreasing* distributions.

REMARK 1. A function α belongs to $\mathcal{O}_C$ if and only if it is infinitely differentiable and there exists an integer k such that for every $p \in \mathbf{N}^n$ the function

$$x \mapsto (1 + |x|^2)^k |\partial^p \alpha(x)|$$

tends to zero as $|x| \to \infty$. Thus $\mathcal{O}_C \subset \mathcal{O}_M$ (cf. Proposition 5), and it is easy to see that $\mathcal{O}_C \neq \mathcal{O}_M$ (Exercise 6). The map $\mathcal{O}_C \hookrightarrow \mathcal{O}_M$ is continuous. Indeed, by Proposition 2.12.1 we must prove that for each $k \in \mathbf{Z}$ the map $\mathcal{S}_k \hookrightarrow \mathcal{O}_M$ is continuous. If $\varphi \in \mathcal{S}$, then there exists $\mu > 0$ such that $(1 + |x|^2)^{-k} |\varphi(x)| \leqq \mu$; hence

$$\max_x |\varphi(x)\, \partial^p \alpha(x)| \leqq \mu \cdot \max_x |(1 + |x|^2)^k\, \partial^p \alpha(x)|$$

for $p \in \mathbf{N}^n$ and $\alpha \in \mathcal{S}_k$.

Since by Example 2.5.6 the map $\mathcal{O}_M \hookrightarrow \mathcal{E}$ is continuous, it follows that $\mathcal{O}_C \hookrightarrow \mathcal{E}$ is continuous. On the other hand, the image of $\mathcal{O}_C$ is dense in $\mathcal{E}$ since already the image of $\mathcal{D}$ is dense in $\mathcal{E}$ (Example 2.5). Hence by Corollary 2 of Proposition 3.12.2 the map $\mathcal{E}' \to \mathcal{O}_C'$ is injective; i.e., every distribution with compact support is rapidly decreasing.

PROPOSITION 7. *If we define the convolution $T * \varphi$ of $T \in \mathcal{S}'$ and $\varphi \in \mathcal{S}$ by*

$$(T * \varphi)(x) = \langle T, \tau_x \check{\varphi} \rangle,$$

then the function $T * \varphi$ belongs to $\mathcal{O}_C$. Furthermore, for each $T \in \mathcal{S}'$ the map $\varphi \mapsto T * \varphi$ from $\mathcal{S}$ into $\mathcal{O}_C$ is continuous.

REMARK 2. It follows from Proposition 10.1 that if Supp T and Supp φ satisfy condition (Σ) of §9, then $T * \varphi$ coincides with the convolution in the sense of Definition 9.2.

LEMMA 1. *For every $x \in \mathbf{R}^n$ and $y \in \mathbf{R}^n$ we have*

$$1 + |y|^2 \leqq 2(1 + |x|^2)(1 + |x + y|^2).$$

Proof. For every real α and β we have $(\alpha + \beta)^2 \leqq 2\alpha^2 + 2\beta^2$ since $2\alpha^2 + 2\beta^2 - (\alpha + \beta)^2 = \alpha^2 + \beta^2 - 2\alpha\beta = (\alpha - \beta)^2 \geqq 0$. Hence

$$|y|^2 = |x + y - x|^2 \leqq 2|x + y|^2 + 2|x|^2$$

and

$$\begin{aligned} 1 + |y|^2 &\leqq 1 + 2|x|^2 + 2|x + y|^2 \\ &\leqq 2(1 + |x|^2)(1 + |x + y|^2). \end{aligned}$$ ∎

LEMMA 2 (cf. Proposition 3.4). *Let $h = (h_1, \ldots, h_n) \in \mathbf{R}^n$ be such that $h_j = t$ and $h_k = 0$ for $k \neq j$. If $\varphi \in \mathcal{S}$, then $\psi_t = t^{-1}(\tau_{-h}\varphi - \varphi)$ converges to $\partial_j\varphi$ in $\mathcal{S}$ as $t \to 0$.*

Proof. Let V be the neighborhood of 0 in $\mathcal{S}$ defined by the inequalities $(1 + |x|^2)^k|\partial^p\chi(x)| \leqq \epsilon$ for $|p| \leqq m$, where $k \in \mathbf{Z}$, $\epsilon > 0$, and $m \in \mathbf{N}$. There exists $\rho > 0$ such that $(1 + |x|^2)^k|\partial^p\varphi(x)| \leqq \frac{1}{2}\epsilon$ for $|x| > \rho$ and $|p| \leqq m + 1$. Since $\partial^p\psi_t(x) = (\partial_j\, \partial^p\varphi)(x + \theta h)$ with $0 \leqq \theta \leqq 1$, if we take $|h| \leqq 1$, we will have

$$(1 + |x|^2)^k|\partial^p\psi_t(x)| \leqq \tfrac{1}{2}\epsilon \tag{11}$$

and in particular

$$(1 + |x|^2)^k|\partial^p\, \partial_j\varphi(x)| \leqq \tfrac{1}{2}\epsilon \tag{12}$$

for $|x| > \rho + 1$ and $|p| \leqq m$. On the other hand, we have

$$\begin{aligned}(1 + |x|^2)^k|\partial^p\{\psi_t(x) - (\partial_j\varphi)(x)\}| \\ = (1 + |x|^2)^k|\partial_j\, \partial^p\varphi(x + \theta h) - \partial_j\, \partial^p\varphi(x)| \leqq \epsilon\end{aligned} \tag{13}$$

for all $|x| \leqq \rho + 1$, provided that $|h|$ is sufficiently small. It follows from (11), (12), and (13) that

$$(1 + |x|^2)^k|\partial^p\{\psi_t(x) - \partial_j\varphi(x)\}| \leqq \epsilon$$

for $x \in \mathbf{R}^n$ and $|p| \leqq m$, i.e., that $\psi_t \in \partial_j\varphi + V$ for sufficiently small $|t|$. ∎

Proof of Proposition 7. (a) By Lemma 1 we have

$$(1 + |y|^2)^k|\partial^p(\tau_x\check{\varphi})(y)| \leqq 2^k(1 + |x|^2)^k(1 + |x - y|^2)^k|\partial^p\varphi(x - y)|$$

for $y \in \mathbf{R}^n$, $k \in \mathbf{Z}$, $p \in \mathbf{N}^n$. Hence for every $x \in \mathbf{R}^n$ the function $\tau_x\check{\varphi}$ belongs to $\mathcal{S}$, and thus $\langle T, \tau_x\check{\varphi}\rangle$ is well-defined.

Let us prove that the map $x \mapsto \tau_x\check{\varphi}$ from $\mathbf{R}^n$ into $\mathcal{S}$ is continuous. Let

$$V = \{\psi \mid (1 + |y|^2)^k|\partial^p\psi(y)| \leqq \epsilon, y \in \mathbf{R}^n, |p| \leqq m\}$$

be a neighborhood of 0 in $\mathcal{S}$, where $k \in \mathbf{Z}$, $\epsilon > 0$, and $m \in \mathbf{N}$. Let a be a point of $\mathbf{R}^n$ and let us consider points $x \in \mathbf{R}^n$ such that $|x - a| \leqq 1$. If we choose $\rho > 0$ so large that

$$(1 + |y|^2)^k|\partial^p\varphi(y)| \leqq \tfrac{1}{2}8^{-k}(1 + |a|^2)^{-k}\epsilon$$

for $|y| > \rho$ and $|p| \leqq m$, then we have by Lemma 1

$$\begin{aligned}(1 + |y|^2)^k|\partial^p\varphi(x - y)| &\leqq 2^k(1 + |x|^2)^k(1 + |x - y|^2)^k|\partial^p\varphi(x - y)| \\ &\leqq 8^k(1 + |a|^2)^k(1 + |x - y|^2)^k|\partial^p\varphi(x - y)| \leqq \tfrac{1}{2}\epsilon\end{aligned} \tag{14}$$

provided $|y| > \rho + |a| + 1$ and $|p| \leqq m$. In the compact set

$$|y| \leqq \rho + 2|a| + 2$$

the function $\partial^p \varphi$ is uniformly continuous. Hence there exist $0 < \delta < 1$ such that

$$(1 + |y|^2)^k |(\partial^p \varphi)(x - y) - (\partial^p \varphi)(a - y)| \leqq \epsilon \tag{15}$$

for $|y| \leqq \rho + |a| + 1$, $|p| \leqq m$, provided that $|x - a| \leqq \delta$. It follows from (14) and (15) that

$$\begin{aligned} &|(1 + |y|^2)^k \, \partial^p \{\tau_x \check{\varphi}(y) - \tau_a \check{\varphi}(y)\}| \\ &\qquad = |(1 + |y|^2)^k \{(\partial^p \varphi)(x - y) - (\partial^p \varphi)(a - y)\}| \leqq \epsilon \end{aligned}$$

for $y \in \mathbf{R}^n$, $|p| \leqq m$, and $|x - a| \leqq \delta$; i.e.,

$$\tau_x \check{\varphi} \in \tau_a \check{\varphi} + V$$

for $|x - a| \leqq \delta$.

Since T is a continuous linear form on $\mathcal{S}$, we see that

$$T * \varphi : x \mapsto \langle T, \tau_x \check{\varphi} \rangle$$

is continuous on $\mathbf{R}^n$.

Next we shall prove the formula

$$\partial^p \langle T, \tau_x \check{\varphi} \rangle = (-1)^{|p|} \langle T, \tau_x (\partial^p \varphi)\check{\ } \rangle \tag{16}$$

for $p \in \mathbf{N}^n$. Since $\partial^p \varphi \in \mathcal{S}$ for all $p \in \mathbf{N}^n$, it will then follow by what we have just proved that $T * \varphi$ has continuous partial derivatives of all orders. By recursion it is sufficient to prove (16) for $|p| = 1$. Let

$$h = (h_1, \ldots, h_n) \in \mathbf{R}^n$$

be such that $h_j = t$ and $h_k = 0$ for $k \neq j$. Then

$$\begin{aligned} \partial_j \langle T, \tau_x \check{\varphi} \rangle &= \lim_{t \to 0} t^{-1} \{ \langle T, \tau_{x-h} \check{\varphi} \rangle - \langle T, \tau_x \check{\varphi} \rangle \} \\ &= \langle T, \lim_{t \to 0} t^{-1} \{ \tau_{-h}(\tau_x \check{\varphi}) - \tau_x \check{\varphi} \} \rangle. \end{aligned}$$

Now by Lemma 2 the function $t^{-1}\{\tau_{-h}(\tau_x \check{\varphi}) - \tau_x \check{\varphi}\}$ converges to

$$\partial_j (\tau_x \check{\varphi}) = \tau_x (\partial_j \check{\varphi}) = -\tau_x (\partial_j \varphi)\check{\ }$$

in $\mathcal{S}$ as $t \to 0$. Thus we obtain

$$\partial_j \langle T, \tau_x \check{\varphi} \rangle = -\langle T, \tau_x (\partial_j \varphi)\check{\ } \rangle,$$

which is (16) when $|p| = 1$.

Finally, we must show that $T * \varphi \in \mathcal{O}_C$. By Proposition 3 we have $T = (1 + |y|^2)^k T_0$ for some $k \in \mathbf{Z}$ and $T_0 \in \mathcal{B}_0'$, and

$$(1 + |x|^2)^{-k} \partial^p \langle T, \tau_x \check{\varphi} \rangle = \int T_0(y)(1 + |x|^2)^{-k}(1 + |y|^2)^k(\partial^p \varphi)(x - y)\, dy. \tag{17}$$

By Lemma 1 we have

$$\begin{aligned} \max_y\, (1 + |x|^2)^{-k}(1 + |y|^2)^k|\partial^p\varphi(x - y)| &\leqq 2^k \max_y\, (1 + |x - y|^2)^k|\partial^p\varphi(x - y)| \\ &= 2^k \max_y\, (1 + |y|^2)^k|\partial^p\varphi(y)|. \end{aligned} \tag{18}$$

It follows from this estimate and Leibniz's formula that for each $p \in \mathbf{N}^n$ the set of functions

$$y \mapsto (1 + |x|^2)^{-k}(1 + |y|^2)^k\, \partial^p\varphi(x - y) \tag{19}$$

is bounded in $\mathcal{B}_0$ as x varies in $\mathbf{R}^n$. Thus by (17) for every $p \in \mathbf{N}^n$ the function

$$x \mapsto (1 + |x|^2)^{-k}\, \partial^p \langle T, \tau_x \check{\varphi} \rangle$$

is bounded on $\mathbf{R}^n$; i.e., $T * \varphi \in \mathcal{S}_{-k-1} \subset \mathcal{O}_C$.

(b) If W is a neighborhood of 0 in $\mathcal{O}_C$, then $W \cap \mathcal{S}_{-k-1}$ contains a set of the form

$$\{\chi \mid (1 + |x|^2)^{-k-1}|\partial^p\chi(x)| \leqq \eta,\ |p| \leqq l\}.$$

On the other hand, there exists a neighborhood U of 0 in $\mathcal{B}_0$ such that $\psi \in U$ implies $|\langle T_0, \psi \rangle| \leqq \eta$. It follows from (18) and Leibniz's formula that there exists a neighborhood Y of 0 in $\mathcal{S}$, such that if $\varphi \in Y$, then for $|p| \leqq l$ and $x \in \mathbf{R}^n$ the functions (19) belong to U. But then

$$\begin{aligned} &(1 + |x|^2)^{-k-1}|\partial^p(T * \varphi)(x)| \\ &\quad \leqq (1 + |x|^2)^{-1} \left| \int T_0(y)(1 + |x|^2)^{-k}(1 + |y|^2)^k(\partial^p\varphi)(x - y)\, dy \right| \leqq \eta \end{aligned}$$

for $|p| \leqq l$; i.e., $\varphi \in Y$ implies $T * \varphi \in W$. ∎

Definition 3. *The convolution $S * T$ of a distribution $S \in \mathcal{O}_C'$ and a distribution $T \in \mathcal{S}'$ is given by*

$$\langle S * T, \varphi \rangle = \langle S, \check{T} * \varphi \rangle$$

for all $\varphi \in \mathcal{S}$.

Since by Proposition 7 the convolution $\check{T} * \varphi$ belongs to $\mathcal{O}_C$, the right-hand side of the preceding formula is well-defined. Furthermore, since

$$\varphi \mapsto \check{T} * \varphi$$

is continuous by Proposition 7 and S is a continuous linear form on $\mathcal{O}_C$, the map $\varphi \mapsto \langle S, \check{T} * \varphi \rangle$ is continuous on $\mathcal{S}$; i.e., $S * T \in \mathcal{S}'$. It follows from formula (4) of §10 that if Supp S and Supp T satisfy condition (Σ), then the new and the old definitions of $S * T$ coincide.

Finally, we state the following fundamental result, whose proof is postponed to Volume 2 (Chapter 6, §1).

THEOREM 3. *The Fourier transform $\mathcal{F}$ maps the space $\mathcal{O}'_C(X)$ isomorphically onto the space $\mathcal{O}_M(\Xi)$. Furthermore, for $S \in \mathcal{O}'_C(X)$ and $T \in \mathcal{S}'(X)$ we have*

$$\mathcal{F}(S * T) = \mathcal{F}(S) \cdot \mathcal{F}(T).$$

The inverse of the isomorphism $\mathcal{F}: \mathcal{O}'_C(X) \to \mathcal{O}_M(\Xi)$ is, of course, the isomorphism $\overline{\mathcal{F}}: \mathcal{O}_M(\Xi) \to \mathcal{O}'_C(X)$. Also $\mathcal{F}$ establishes an isomorphism between $\mathcal{O}_M(X)$ and $\mathcal{O}'_C(\Xi)$. If $\alpha \in \mathcal{O}_M(X)$ and $T \in \mathcal{S}'(X)$, then

$$\mathcal{F}(\alpha T) = \mathcal{F}(\alpha) * \mathcal{F}(T).$$

EXERCISES

1. Prove that the Fourier transform of the function $x \mapsto e^{-k|x|^2}$ defined on X is the function

$$\xi \to \sqrt{\pi/k}^{\,n} e^{-(\pi^2/k)|\xi|^2}$$

defined on Ξ. Letting $k \to \infty$, deduce another proof of the formula $\mathcal{F}(1) = \delta$.

2. For every $k \in \mathbf{N}$ define the Hermite function $\mathcal{H}_k$ on $\mathbf{R}$ by

$$(-1)^k \sqrt{k!}\, \pi^{k/2} \mathcal{H}_k(x) = e^{\pi x^2} \frac{d^k}{dx^k} e^{-2\pi x^2}.$$

Show that $\mathcal{F}(\mathcal{H}_k) = (-i)^k \mathcal{H}_k$.

3. Prove that for every $\varphi \in \mathcal{S}$ the map $\alpha \mapsto \alpha\varphi$ from $\mathcal{O}_M$ into $\mathcal{S}$ is continuous.

4. Prove that the bilinear map $(\alpha, \beta) \mapsto \alpha\beta$ from $\mathcal{O}_M \times \mathcal{O}_M$ into $\mathcal{O}_M$ is continuous.

5. Show that the function $x \mapsto e^{i\pi x^2}$ defined on $\mathbf{R}$ belongs to $\mathcal{O}_M$ but not to $\mathcal{O}_C$.

6. Assuming Theorem 3, prove the following assertions:

(a) For every $S \in \mathcal{O}'_C$ the linear map $T \mapsto S * T$ from $\mathcal{S}'$ into $\mathcal{S}'$ is continuous.
(b) If $R \in \mathcal{O}'_C$ and $S \in \mathcal{O}'_C$, then $R * S \in \mathcal{O}'_C$.
(c) If $R \in \mathcal{O}'_C$, $S \in \mathcal{O}'_C$, and $T \in \mathcal{S}'$, then $(R * S) * T = R * (S * T)$.
(d) If $S \in \mathcal{O}'_C$ and $T \in \mathcal{S}'$, then for every j $(1 \leq j \leq n)$ we have

$$\partial_j(S * T) = \partial_j S * T = S * \partial_j T$$

and for every $h \in \mathbf{R}^n$ we have

$$\tau_h(S * T) = \tau_h S * T = S * \tau_h T.$$

(e) If $Q \in \mathcal{O}'_C(\mathbf{R}^k)$, $R \in \mathcal{S}'(\mathbf{R}^k)$, $S \in \mathcal{O}'_C(\mathbf{R}^l)$, and $T \in \mathcal{S}'(\mathbf{R}^l)$, then

$$Q \otimes S \in \mathcal{O}'_C(\mathbf{R}^{k+l})$$

and

$$(Q * R) \otimes (S * T) = (Q \otimes S) * (R \otimes T).$$

7. Show that the strong topology on $\mathcal{O}'_C$ is the coarsest topology for which the maps $T \mapsto (1 + |x|^2)^k T$ from $\mathcal{O}'_C$ into $\mathcal{B}'_0$ are continuous, where $\mathcal{B}'_0$ is equipped with its strong topology and $k \in \mathbf{Z}$.

8. Show that for every $\varphi \in \mathcal{S}$ the map $T \mapsto T * \varphi$ from $\mathcal{S}'$ into $\mathcal{O}_C$ is continuous.

9. Show that $(T, \varphi) \mapsto T * \varphi$ is a separately continuous linear map from $\mathcal{O}'_C \times \mathcal{S}$ into $\mathcal{S}$ and from $\mathcal{O}'_C \times \mathcal{O}_C$ into $\mathcal{O}_C$.

Bibliography

This short bibliography contains, besides the works quoted in the text, mainly references which might be useful to the reader for further study. Additional bibliography can be found in [15, 19, 52, 63]. A few papers of great historical interest and some recent publications (with the reference to their review in Mathematical Reviews) are also included.

1. AHLFORS, LARS V., *Complex Analysis. An Introduction to the Theory of Analytic Functions of One Complex Variable.* McGraw-Hill, New York-Toronto-London, 1953.
2. APOSTOL, TOM M., *Mathematical Analysis. A Modern Approach to Advanced Calculus.* Addison-Wesley, Reading, Mass., 1957.
3. ARSOVE, M. G.; EDWARDS, R. E., "Generalized bases in topological linear spaces." *Studia Math.*, **19** (1960), 95–113. MR **22** #5871.
4. BAOUENDI, MOHAMED SALAH, "Division des distributions dans $\mathcal{O}_M$." *C. R. Acad. Sci., Paris,* **258** (1964), 1978–1980. MR **28** #4351.
5. BAUER, HEINZ, *Konvexität in topologischen Vektorräumen, insbesondere: Integraldarstellung in konvexen, kompakten Mengen,* Vorlesung im Wintersemester 1963/64 an der Universität Hamburg, ausgearbeitet von Ulrich Krause.
6. BOAS, RALPH P., JR., *Entire functions.* Academic Press, New York, 1954.
7. BOURBAKI, N., *Eléments de mathématiques.* Livre II. *Algèbre,* chapitre 2, *Algèbre linéaire.* 3rd ed. Actualités Scientifiques et Industrielles 1236; Hermann, Paris, 1962.
8. BOURBAKI, N., *Eléments de mathématiques.* Livre III. *Topologie générale.* Actualités Scientifiques et Industrielles 1142, 1143, 1235, 1045, 1084, 1196; Hermann, Paris, 1953–1961.
9. BOURBAKI, N., *Eléments de mathématiques.* Livre V. *Espaces vectoriels topologiques.* Actualités Scientifiques et Industrielles 1189, 1229, 1230; Hermann, Paris, 1953, 1955.
10. BOURBAKI, N., *Eléments de mathématiques.* Livre VI. *Intégration.* Actualités Scientifiques et Industrielles 1175, 1244, 1281, 1306; Hermann, Paris, 1952, 1956, 1959, 1963.
11. BOURBAKI, N., "Sur certains espaces vectoriels topologiques." *Ann. Inst. Fourier* (Grenoble), **2** (1950), 5–16.
12. BRACE, JOHN W., "Compactness in the weak topology." *Math. Mag.*, **28** (1955), 125–134.

13. CAMPOS FERREIRA, JAIME, "Les espaces de Schwartz et les espaces d'applications linéaires continues." *Portugal. Math.*, **18** (1959), 1–32. MR **22** #2874.

14. COLLINS, HERON S., "Completeness, full completeness and *k* spaces." *Proc. Amer. Math. Soc.*, **6** (1955), 832–835.

15. DAY, MAHLON M., *Normed Linear Spaces.* Ergebnisse der Mathematik und ihrer Grenzgebiete, Neue Folge, Heft 21. Springer-Verlag, Berlin-Göttingen-Heidelberg, 1958.

16. DAY, MAHLON M., "On the basis problem in normed spaces." *Proc. Amer. Math. Soc.* **13** (1962), 655–658. MR **25** #1435.

17. DE RHAM, GEORGES, *Variétés différentiables. Formes, courants, formes harmoniques.* Actualités Scientifiques et Industrielles 1222; Hermann, Paris, 1955.

18. DIEUDONNÉ, JEAN, "La dualité dans les espaces vectoriels topologiques." *Ann. Sci. Ecole Norm. Sup.*, **59** (1942), 107–139.

19. DIEUDONNÉ, JEAN, "Recent developments in the theory of locally convex vector spaces." *Bull. Amer. Math. Soc.*, **59** (1953), 495–512.

20. DIEUDONNÉ, JEAN; SCHWARTZ, LAURENT, "La dualité dans les espaces ($\mathfrak{F}$) et ($\mathfrak{L}\mathfrak{F}$)." *Ann. Inst. Fourier* (Grenoble), **1** (1949), 61–101.

21. DIXMIER, JACQUES, *Les algèbres d'opérateurs dans l'espace hilbertien* (*algèbres de von Neumann*). Cahiers Scientifiques, fasc. 25, Gauthier-Villars, Paris, 1957.

22. DOUADY, ADRIEN, "Parties compactes d'un espace de fonctions continues à support compact." *C. R. Acad. Sci. Paris*, **257** (1963), 2788–2791. MR **27** #6121.

23. EDMUNDS, D. E., "$\sum$-symmetric locally convex spaces." *Proc. Amer. Math. Soc.*, **14** (1963), 697–700. MR **27** #4055.

24. EDWARDS, R. E., "Integral bases in inductive limit spaces." *Pacific J. Math.*, **10** (1960), 797–812. MR **22** #5868.

25. EDWARDS, R. E., *Functional Analysis: Theory and Applications.* Holt, Rinehart and Winston, New York, 1965.

26. FAN, KY, "On the Krein-Milman theorem." *Proc. Sympos. Pure Math.*, Vol. VII, pp. 211–219. Amer. Math. Soc., Providence, R. I., 1963. MR **27** #4056.

27. FRIEDMAN, AVNER, *Generalized Functions and Partial Differential Equations.* Prentice-Hall, Englewood Cliffs, N. J., 1963. MR **29** #2672.

28. FULLERTON, ROBERT E., "Geometrical characterizations of certain function spaces." *Proc. Internat. Sympos. Linear Spaces* (Jerusalem, 1960), pp. 227–236. Jerusalem Academic Press, Jerusalem; Pergamon, Oxford; 1961.

29. GÁL, I. S., "On sequences of operations in complete vector spaces." *Amer. Math. Monthly*, **60** (1953), 527–538.

30. GEL'FAND, I. M.; ŠILOV, G. E., *Obobščennye funkcii* [Generalized functions]. I. *Obobščennye funkcii i deĭstvija nad nimi* [Generalized functions and operations on them]. Gosudarstvennoe izdatel'stvo fiziko-matematičeskoĭ literatury, Moscow, 1958. MR **20** #4182.
English translation: GEL'FAND, I. M.; SHILOV, G. E., *Generalized functions.* Vol. I: *Properties and operations.* Academic Press, New York-London, 1964. MR **29** #3869.

German translation: GELFAND, I. M.; SCHILOW, G. E., *Verallgemeinerte Funktionen (Distributionen). I. Verallgemeinerte Funktionen und das Rechnen mit ihnen.* VEB Deutscher Verlag der Wissenschaften, Berlin, 1960.
French translation: GUELFAND, I. M.; CHILOV, G. E., *Les distributions.* Collection Universitaire de Mathématiques, 8. Dunod, Paris, 1962. MR **24** #A2235.

31. GEL'FAND, I. M.; ŠILOV, G. E., *Obobščennye funkcii* [Generalized functions]. II. *Prostranstva osnovnyh i obobščennyih funkcii* [Spaces of fundamental and generalized functions]. Gosudarstvennoe izdatel'stvo fiziko-matematičeskoĭ literatury, Moscow, 1958. MR **21** #5142 a.
German translation: GELFAND, I. M.; SCHILOW, G. E., *Verallgemeinerte Funktionen (Distributionen). II. Lineare topologische Räume, Räume von Grundfunktionen und verallgemeinerten Funktionen.* VEB Deutscher Verlag der Wissenschaften, Berlin, 1962. MR **26** #6765.
French translation: GUELFAND, I. M.; CHILOV, G. E., *Les distributions.* Tome 2, *Espaces fondamentaux.* Collection Universitaire de Mathématiques, 15. Dunod, Paris, 1964. MR **28** #4354.
32. GEL'FAND, I. M.; ŠILOV, G. E., *Obobščennye funkcii* [Generalized functions]. III. *Nekotorye voprosy teorii differencialnyh uravnenii* [Some questions in the theory of differential equations]. Gosudarstvennoe izdatel'stvo fiziko-matematičeskoĭ literatury, Moscow, 1958. MR **21** #5142 b.
German translation: GELFAND, I. M.; SCHILOW, G. E., *Verallgemeinerte Funktionen (Distributionen). III. Einige Fragen zur Theorie der Differentialgleichungen.* VEB Deutscher Verlag der Wissenschaften, Berlin, 1964.
33. GEL'FAND, I. M.; VILENKIN, N. JA., *Obobščennye funkcii* [Generalized functions]. IV. *Nekotorye primenenija garmoničeskogo analiza, osnaščennye gil'bertovy prostranstva* [Some applications to harmonic analysis, equipped Hilbert spaces]. Gosudarstvennoe izdatel'stvo fiziko-matematičeskoĭ literatury, Moscow, 1961. MR **26** #4173.
English translation: GEL'FAND, I. M.; VILENKIN, N. YA., *Applications to Harmonic Analysis.* Academic Press, New York-London, 1964.
German translation: GELFAND, I. M.; WILENKIN, N. J., *Verallgemeinerte Funktionen (Distributionen). IV. Einige Anwendungen der harmonischen Analyse, Gelfandsche Raumtripel.* VEB Deutscher Verlag der Wissenschaften, Berlin, 1964.
34. GEL'FAND, I. M.; GRAEV, M. I.; VILENKIN, N. JA., *Obobščennye funkcii* [Generalized functions]. V. *Integral'naja geometrija i svjazannye s neĭ voprosy teorii predstavlenii* [Integral geometry and related questions in the theory of representations]. Gosudarstvennoe izdatel'stvo fiziko-matematičeskoĭ literatury, Moscow, 1962. MR **28** #3324.
35. GIL DE LAMADRID, JESÚS, "Complementation of manifolds in topological vector spaces." *Arch. Math.*, **14** (1963), 59–61. MR **28** #3306.
36. GODEMENT, ROGER, *Topologie algébrique et théorie des faisceaux.* Actualités Scientifiques et Industrielles 1252; Hermann, Paris, 1958.
37. GOFFMANN, CASPER; PEDRICK, GEORGE, *A First Course in Functional Analysis.* Prentice-Hall, Englewood Cliffs, N. J., 1965.
38. GOLDBERG, SEYMOUR, "On Dixmier's theorem concerning conjugate spaces." *Math. Ann.*, **147** (1962), 244–247.

39. GOULD, G. G., "Locally unbounded topological fields and box topologies of products of vector spaces." *J. London Math. Soc.*, **36** (1961), 273–281. MR **24** #A413.

40. GROTHENDIECK, ALEXANDRE, "Critères de compacité dans les espaces fonctionnels généraux." *Amer. J. Math.*, **74** (1952), 168–186.

41. GROTHENDIECK, ALEXANDRE, "Sur les espaces ($\mathfrak{F}$) et ($\mathfrak{D}\mathfrak{F}$)." *Summa Brasil. Math.*, **3** (1954), 57–123.

42. GROTHENDIECK, ALEXANDRE, "Produits tensoriels topologiques et espaces nucléaires." *Mem. Amer. Math. Soc.*, **16** (1955).

43. GROTHENDIECK, ALEXANDRE, *Espaces vectoriels topologiques*, 2nd ed. Sociedade de Matemática de São Paulo, São Paulo, 1958.

44. HALMOS, PAUL R., *Introduction to Hilbert Space and the Theory of Spectral Multiplicity*. Chelsea, New York, 1951.

45. HIRZEBRUCH, F., *Topological Methods in Algebraic Geometry*. Die Grundlehren der Mathematischen Wissenschaften, Bd. 131. Springer-Verlag, New York, 1966.

46. HÖRMANDER, LARS, *Linear Partial Differential Operators*. Die Grundlehren der mathematischen Wissenschaften, Bd. 116. Springer-Verlag, Berlin-Göttingen-Heidelberg, 1963.

47. HUSAIN, TAQDIR, *The Open-mapping and Closed-graph Theorems in Topological Vector Spaces*. Clarendon Press, Oxford, 1965. MR **31** #2589.

48. KANTOROVIČ, L. V.; AKILOV, G. P., *Funkcional'nyĭ analiz v normirovannyh prostransvah* [Functional analysis in normed spaces]. Gosudarstvennoe izdatel'stvo fiziko-matematičeskoĭ literatury, Moscow, 1959. MR **22** #9837. English translation: KANTOROVICH, L. V.; AKHILOV, G. P., *Functional Analysis in Normed Spaces*. Pergamon, Oxford, 1964.

49. KELLEY, JOHN L.; NAMIOKA, ISAAC, *Linear Topological Spaces*. With the collaboration of W. F. Donoghue, Jr., Kenneth R. Lucas, B. J. Pettis, Ebbe Thue Poulsen, G. Bailey Price, Wendy Robertson, W. R. Scott, Kennan T. Smith. D. van Nostrand, New York-London-Toronto, 1963. MR **29** #3851.

50. KŌMURA, YUKIO, "On linear topological spaces." *Kumamoto J. Sci.*, Ser. A, **5**, 148–157 (1962). MR **27** #1800.

51. KŌMURA, YUKIO, "Some examples on linear topological spaces." *Math. Ann.*, **153** (1964), 150–162.

52. KÖTHE, GOTTFRIED, *Topologische lineare Räume*. I. Die Grundlehren der mathematischen Wissenschaften, Bd. 107. Springer-Verlag, Berlin-Göttingen-Heidelberg, 1960. MR **24** #A411.

53. LEVIN, V. L., "Non-degenerate spectra of locally convex spaces." *Dokl. Akad. Nauk SSSR*, **135** (1960), 12–15 (Russian); translated as *Soviet Math. Dokl.* **1**, 1227–1230. MR **22** #4932.

54. LEVIN, V. L., "On a theorem of A. I. Plessner." *Uspehi Mat. Nauk*, **16** (1961), no. 5 (101), 177–179 (Russian). MR **24** #A2820.

55. LEVIN, V. L., "*B*-completeness conditions for ultrabarrelled and barrelled spaces." *Dokl. Akad. Nauk SSSR*, **145** (1962), 273–275 (Russian). MR **25** #3347.

56. LEVIN, V. L., "On a class of locally convex spaces." *Dokl. Akad. Nauk SSSR*, **145** (1962), 35–37 (Russian). MR **25** #5375.

57. LINDENSTRAUSS, JORAM, "On a problem of Nachbin concerning extension of operators." *Israel J. Math.*, **1** (1963), 75–84. MR **28** #478.
58. LIONS, J. L., *Equations différentielles opérationnelles et problèmes aux limites.* Die Grundlehren der mathematischen Wissenschaften, Bd. 111. Springer-Verlag, Berlin-Göttingen-Heidelberg, 1961.
59. LUXEMBURG, W. A. J., "On closed linear subspaces and dense linear subspaces of locally convex topological linear spaces." *Proc. Internat. Sympos. Linear Spaces* (Jerusalem, 1960), pp. 307–318. Jerusalem Academic Press, Jerusalem; Pergamon, Oxford; 1961. MR **25** #2404.
60. MAHOWALD, MARK; GOULD, GERALD, "Quasi-barrelled locally convex spaces." *Proc. Amer. Math. Soc.*, **11** (1960), 811–816. MR **23** #A499.
61. MAHOWALD, M., "Barrelled spaces and the closed graph theorem." *J. London Math. Soc.*, **36** (1961), 108–110. MR **23** #A2728.
62. MAKAROV, B. M., "Some pathological properties of inductive limits of *B*-spaces." *Uspehi Mat. Nauk*, **18** (1963), no. 3 (111), 171–178 (Russian). MR **27** #2839.
63. MARINESCU, G., *Espaces vectoriels pseudotopologiques et théorie des distributions.* Hochschulbücher für Mathematik, Bd. 59. VEB Deutscher Verlag der Wissenschaften, Berlin, 1963. MR **29** #3878.
64. MARTINEAU, ANDRÉ, "Sur une propriété caractéristique d'un produit de droites." *Arch. Math.*, **11** (1960), 423–426. MR **23** #A3981.
65. MERGELYAN, S. N., "Weighted approximation by polynomials." *Amer. Math. Soc. Transl.*, (2), **10** (1958), 59–106.
66. MOCHIZUKI, NOZOMU, "On fully complete spaces." *Tōhoku Math. J.*, (2), **13** (1961), 485–490. MR **26** #5397.
67. NACHBIN, LEOPOLDO, "Topological vector spaces of continuous functions." *Proc. Nat. Acad. Sci. U.S.A.*, **40** (1954), 471–474.
68. NACHBIN, LEOPOLDO, "Some problems in extending and lifting continuous linear transformations." *Proc. Internat. Sympos. Linear Spaces* (Jerusalem, 1960), pp. 340–350. Jerusalem Academic Press, Jerusalem; Pergamon, Oxford; 1961. MR **24** #2826.
69. NACHBIN, LEOPOLDO, "Sur l'approximation polynomiale ponderée des fonctions réelles continues." *Atti della 2a riunione del Groupement de Mathématiciens d'expression latine* (Firenze, 26–30 settembre 1961 – Bologna, 1–3 ottobre 1961), pp. 42–58. Cremonese, Roma, 1963.
70. NACHBIN, LEOPOLDO, "Résultats récents et problèmes de nature algébrique en théorie de l'approximation." *Proceedings of the International Congress of Mathematicians*, pp. 379–384, Stockholm, 1962.
71. NACHBIN, LEOPOLDO, "Weighted approximation over topological spaces and the Bernstein problem over finite dimensional vector spaces." *Topology*, **3** (1964), suppl. 1, 125–130. MR **28** #4286.
72. NACHBIN, LEOPOLDO, "Weighted approximation for algebras and modules of continuous functions: Real and self-adjoint complex cases." *Ann. of Math.*, (2) **81** (1965), 289–302. MR **31** #628.
73. NACHBIN, LEOPOLDO, *Elements of Approximation Theory.* Department of Mathematics, University of Rochester, Rochester, N. Y., 1964. Notas de Matemática no. 33, Instituto de Matemática Pura e Aplicada, Rio de Janeiro, 1965.

74. Peetre, Jaak, *Elliptic Partial Differential Equations of Higher Order.* The Institute for Fluid Dynamics and Applied Mathematics, Lecture series, no. 40, University of Maryland, College Park, Md., 1962.
75. Pták, Vlastimil, "Completeness and the open mapping theorem." *Bull. Soc. Math. France,* **86** (1958), 41–74.
76. Pták, Vlastimil, "On the closed graph theorem." *Czechoslovak Math. J.,* **9** (84) (1959), 523–527. MR **22** #5864.
77. Robertson, Alex. P.; Robertson, Wendy, "On the closed graph theorem." *Proc. Glasgow Math. Assoc.,* **3** (1956), 9–12.
78. Robertson, A. P.; Robertson, W. J., *Topological Vector Spaces.* Cambridge Tracts in Mathematics and Mathematical Physics, no. 53. Cambridge University Press, Cambridge-New York, 1964. MR **28** #5318.
79. Schaefer, Helmut, "Zur komplexen Erweiterung linearer Räume." *Arch. Math.,* **10** (1959), 363–365. MR **22** #171.
80. Schwartz, Laurent, *Étude des sommes d'exponentielles réelles et imaginaires.* Actualités Scientifiques et Industrielles 959; Hermann, Paris, 1959.
81. Schwartz, Laurent, *Théorie des distributions.* I. 2nd ed. Actualités Scientifiques et Industrielles 1091; Hermann, Paris, 1957.
82. Schwartz, Laurent, *Théorie des distributions.* II. Actualités Scientifiques et Industrielles 1122; Hermann, Paris, 1951.
83. Schwartz, Laurent, *Produits tensoriels topologiques et espaces nucléaires.* Séminaire, Institut Henri Poincaré, Paris, 1953–54.
84. Schwartz, Laurent, "Espaces de fonctions différentiables à valeurs vectorielles." *J. Analyse Math.,* **4** (1954/55), 88–148.
85. Schwartz, Laurent, *Lectures on Complex Analytic Manifolds.* Tata Institute of Fundamental Research, Bombay, 1955.
86. Schwartz, Laurent, *Ecuaciones diferenciales parciales elípticas.* Universidad Nacional de Colombia, Departamento de Matemáticas y Estadística, Bogotá, 1956.
87. Schwartz, Laurent, *Variedades analíticas complejas.* Universidad Nacional de Colombia, Departamento de Matemáticas y Estadística, Bogotá, 1956.
88. Schwartz, Laurent, "Théorie des distributions à valeurs vectorielles. I.–II." *Ann. Inst. Fourier* (Grenoble), **7** (1957), 1–141; **8** (1959), 1–209. MR **21** #6534; **22** #8322.
89. Schwartz, Laurent, *Méthodes mathématiques pour les sciences physiques.* Avec le concours de Denise Huet. Hermann, Paris, 1961.
90. Schwartz, Laurent, *Convergence de distributions dont les dérivées convergent.* Studies in mathematical analysis and related topics, pp. 364–372. Stanford University Press, Stanford, Calif., 1962. MR **28** #4355.
91. Schwartz, Laurent, *Some Applications of the Theory of Distributions.* Lectures on Modern Mathematics. I., pp. 23–58. John Wiley & Sons, New York, 1963. MR **31** #2611.
92. Schwartz, Laurent, "Sous-espaces hilbertiens d'espaces vectoriels topologiques et noyaux associés. (Noyaux reproduisants)." *J. Analyse Math.,* **13** (1964), 115–256. MR **31** #3835.
93. Schwartz, Laurent, *Functional Analysis.* New York University, Courant Institute of Mathematical Sciences. New York, 1964.

94. Shibata, Toshio, "Adjoint space and dual space." *Proc. Japan Acad.*, **36** (1960), 261–266. MR **22** #9835.
95. Shirota, Taira, "On locally convex vector spaces of continuous functions." *Proc. Japan Acad.*, **30** (1954), 292–298.
96. Simons, S., "The bornological space associated with $\mathbf{R}^I$." *J. London Math. Soc.*, **36** (1961), 461–473. MR **24** #A2822.
97. Słowikowski, W., "On continuity of inverse operators." *Bull. Amer. Math. Soc.*, **67** (1961), 467–470. MR **24** #A1590.
98. Słowikowski, W., "Quotient spaces and the open map theorem." *Bull. Amer. Math. Soc.*, **67** (1961), 498–500. MR **24** #A1591.
99. Stampacchia, Guido, "Formes bilinéaires coercitives sur les ensembles convexes." *C. R. Acad. Sci. Paris*, **258** (1964), 4413–4416. MR **29** #3864.
100. Treves, J. F., *Lectures on Linear Partial Differential Equations with Constant Coefficients.* Notas de Matemática no. 27, Instituto de Matemática Pura e Aplicada, Rio de Janeiro, 1961. MR **27** #5020.
101. Waelbroeck, Lucien, "Le complété et le dual d'un espace localement convexe." *Bull. Soc. Math. Belg.*, **16** (1964), 393–406. MR **31** #3842.
102. Welland, R. R., "Metrizable Köthe spaces." *Proc. Amer. Math. Soc.*, **11** (1960), 580–587. MR **22** #12369.
103. Widder, David Vernon, *The Laplace Transform.* Princeton Mathematical Series, vol. 6, Princeton University Press, Princeton, N. J., 1941.
104. Wilansky, Albert, *Functional Analysis.* Blaisdell, New York, 1964.
105. Yosida, Kosaku, *Functional Analysis.* Die Grundlehren der mathematischen Wissenschaften, Bd. 123. Springer-Verlag, Berlin-Göttingen-Heidelberg, 1965.
106. Zaanen, A. C., "Banach function spaces." *Proc. Internat. Sympos. Linear Spaces* (Jerusalem, 1960), pp. 448–452. Jerusalem Academic Press, Jerusalem; Pergamon, Oxford; 1961. MR **25** #426.
107. Goldberg, Seymour, *Unbounded Linear Operators. Theory and Applications.* McGraw-Hill, New York, 1966.
108. Phelps, Robert R., *Lectures on Choquet's Theorem.* Van Nostrand Mathematical Studies #7; D. van Nostrand, Princeton, 1966.
109. Schaefer, Helmut H., *Topological Vector Spaces.* Macmillan, New York, 1966.
110. Whitley, Robert, "Projecting m onto c_0." *Amer. Math. Monthly*, **73** (1966), 285–286.

INDEX OF NOTATIONS

TABLES

INDEX

INDEX OF NOTATIONS

The notations introduced on pages 1 through 3 are not listed.

Topologies

(F and G are two paired vector spaces)

Spaces

TABLES

1. Inclusion relations among some function spaces introduced in the text. Every injection map is continuous; $m \leq m'$.

$$
\begin{array}{ccccccc}
\mathcal{K}(\Omega) & \subset & \mathcal{C}_0(\Omega) & \subset & \mathcal{B}^0(\Omega) & \subset & \mathcal{C}(\Omega) \\
\cup & & \cup & & \cup & & \cup \\
\mathcal{D}^m(\Omega) & \subset & \mathcal{B}_0^m(\Omega) & \subset & \mathcal{B}^m(\Omega) & \subset & \mathcal{E}^m(\Omega) \\
\cup & & \cup & & \cup & & \cup \\
\mathcal{D}^{m'}(\Omega) & \subset & \mathcal{B}_0^{m'}(\Omega) & \subset & \mathcal{B}^{m'}(\Omega) & \subset & \mathcal{E}^{m'}(\Omega) \\
\cup & & \cup & & \cup & & \cup \\
\mathcal{D}(\Omega) & \subset & \mathcal{B}_0(\Omega) & \subset & \mathcal{B}(\Omega) & \subset & \mathcal{E}(\Omega)
\end{array}
$$

$$
\begin{array}{ccccccccccccccccc}
 & & & & & & k = 0 & & & & & & & & & & \\
 & & & & k > 0 & & \Big\downarrow & & k' < 0 & & & & & & & & \\
 & & & & \downarrow & & \downarrow & & \downarrow & & & & & & & & \\
\mathcal{K} & \subset & \mathcal{S}^0 & \subset & \mathcal{S}_k^0 & \subset & \mathcal{C}_0 & \subset & \mathcal{S}_{k'}^0 & & \subset & & & & \mathcal{O}_C^0 & \subset & \mathcal{C} \\
\cup & & \cup & & \cup & & \cup & & \cup & & & & & & \cup & & \cup \\
\mathcal{D}^m & \subset & \mathcal{S}^m & \subset & \mathcal{S}_k^m & \subset & \mathcal{B}_0^m & \subset & \mathcal{S}_{k'}^m & & \subset & & & & \mathcal{O}_C^m & \subset & \mathcal{E}^m \\
\cup & & \cup & & \cup & & \cup & & \cup & & & & & & \cup & & \cup \\
\mathcal{D}^{m'} & \subset & \mathcal{S}^{m'} & \subset & \mathcal{S}_k^{m'} & \subset & \mathcal{B}_0^{m'} & \subset & \mathcal{S}_{k'}^{m'} & & \subset & & & & \mathcal{O}_C^{m'} & \subset & \mathcal{E}^{m'} \\
\cup & & \cup & & \cup & & \cup & & \cup & & & & & & \cup & & \cup \\
\mathcal{D} & \subset & \mathcal{S} & \subset & \mathcal{S}_k & \subset & \mathcal{B}_0 & \subset & \mathcal{S}_{k'} & \subset & \mathcal{O}_C & \subset & & & \mathcal{O}_M & \subset & \mathcal{E}
\end{array}
$$

2. Properties of some function spaces ($m \in \mathbf{N}$); the numbers refer to pages.

	semi-norms	normable	metriz-able	complete	barrelled	borno-logical	Montel	Schwartz
$\mathcal{B}^m(\Omega)$	92	110		136	214	222	—	—
$\mathcal{B}(\Omega)$	92	—	116	136	214	222	—	—
$\mathcal{B}_0^m(\Omega)$	91	110		137	214	222	—	—
$\mathcal{B}_0(\Omega)$	91	—	116	137	214	222	—	—
$\mathcal{C}(\Omega)$	89	110		136	214	222	—	—
$\mathcal{D}^m(K)$	90	110		137	214	222	—	—
$\mathcal{D}(K)$	90	—	116	137	214	222	240	282
$\mathcal{D}^m(\Omega)$		—	—	172	215	222	—	—
$\mathcal{D}(\Omega)$	171	—	—	165	215	222	241	282
$\mathcal{E}^m(\Omega)$	89	—	116	136	214	222	—	—
$\mathcal{E}(\Omega)$	89	—	116	136	214	222	239	281
$H(\Omega)$	238	—	238	238	238	238	239	285
$\mathcal{K}(K)$	90	110		137	214	222	—	—
$\mathcal{K}(\Omega)$		—	—	164	215	222	—	—
$\mathcal{O}_C^m$		—			215	222	—	—
$\mathcal{O}_C$		—	—	*	215	222	*	*
$\mathcal{O}_M$	91	—	—	137	*	*	*	*
$\mathcal{S}_k^m$	90	110		137	214	222	—	—
$\mathcal{S}_k$	90	—	116	154	214	222	—	—
$\mathcal{S}^m$	90	—	116	154	214	222	—	—
$\mathcal{S}$	91	—	116	154	214	222	240	283

3. Relations among various types of locally convex spaces.

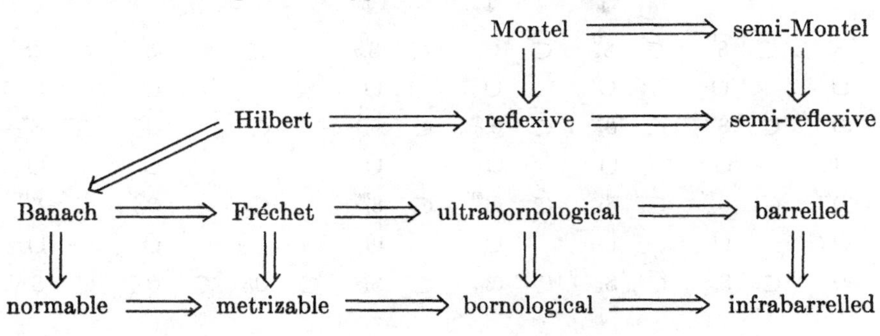

* See [42], Chapter II, p. 131

Index

The terms introduced on pages 1 through 3 are not listed.

ABCDE69876